Proteases: Structure and Function

Klaudia Brix • Walter Stöcker
Editors

Proteases: Structure and Function

Springer

Editors
Klaudia Brix
School of Engineering and Science
Jacobs University
Bremen
Germany

Walter Stöcker
Institute of Zoology
Johannes Gutenberg University
Mainz
Germany

ISBN 978-3-7091-0884-0 ISBN 978-3-7091-0885-7 (eBook)
DOI 10.1007/978-3-7091-0885-7
Springer Wien Heidelberg New York Dordrecht London

Library of Congress Control Number: 2013956226

Printed on acid-free paper

Springer is part of Springer Science+Business Media (www.springer.com)

Foreword

The first chapter "**Protease Families, Evolution, and Mechanism of Action**" provides us with a comprehensive and refreshing view over protease and peptidase family relationships. It is written by Neil Rawlings, THE expert, when it comes to proteolytic enzyme classification. He has pioneered and initiated with Alan Barrett MEROPS, an invaluable database for protease researchers, that has been maintained and updated for several years now. The first chapter of this book therefore introduces and explains the logics of this exceptional information resource and compendium on proteases. The conclusion of Neil Rawlings' chapter considers classification based on the comparison of sequences and structural motifs of proteases that evolved over millions of years with their (substrate specificities and) types of catalysis during proteolysis.

The second chapter "**Kinetics of the Interaction of Peptidases with Substrates and Modifiers**" focuses on the tools provided by enzyme kinetics for characterizing peptidase-mediated turnover of substrates. Enzymatic reactions eventually take place in the context of interactions of the specific enzymes with their modifiers, which may function as activators and inhibitors. Thus, the modifiers of specific enzymes—here, proteases—are of high relevance when discussing enzyme kinetics conceptually. Poised by their educational and didactical talents, the experts Antonio Baici, Marko Novinec, and Brigita Lenarčič teamed up in writing a chapter that will serve as an excellent guide for researchers investigating the biology of proteolysis and the biochemistry of proteolytic enzymes encountering substrates in the cellular environment. This chapter is an invaluable source to the quantitative kinetic analysis of proteases for both the well-versed researcher and beginner in the field of peptidases, their substrates, and modifiers.

In the third chapter "**Compartmentalization of Proteolysis**," the cell biologists Klaudia Brix, Christopher J. Scott, and Margarete M. Heck discuss the challenges that proteolytic enzymes face when acting in various compartments of eukaryotic cells. These enzymes need to be precisely targeted and adapted to their individual work space. In this context, the authors explain the blueprint of demands and necessities of cells in coping with proteolysis for maintenance of protein homeostasis. They outline the biochemical conditions for the cleaving environment(s) and the spatial concept of cellular compartmentalization required for intracellular and extracellular juxta-membrane proteolytic processing. They address the important issue of radical decision making, which is essential at times of busy trafficking and turnover of substrates. Finally, the control of proteolysis by endogenous inhibitors and new developments including intra-membrane proteolysis are presented.

The fourth chapter "**Cathepsins: Getting in Shape for Lysosomal Proteolysis**" is dedicated to describe the molecular and cellular requirements of correct targeting, sorting, and trafficking of lysosomal cathepsins. Disorders affecting the biosynthesis and transport of cathepsins manifest with severe symptoms ranging from lysosomal storage disorders of several kinds to osteoporosis and Papillon–Lefèvre syndrome. The authors Ann H. Erickson, Ciro Isidoro, Lukas Mach, and John S. Mort combine biochemical, cell biological, and structural expertise to outline most instructively how different cell types have found solutions for loading the compartments of the endocytic pathway with aspartic and cysteine cathepsins. The authors also explain comprehensively the very individual and specific roles of these proteolytic enzymes beyond general protein degradation and turnover.

The fifth chapter "**Limited and Degradative Proteolysis in the Context of Posttranslational Regulatory Networks: Current Technical and Conceptional Advances**" explains technologies that allow assessing protease functions in degradomics approaches which aim at characterization of protease activities in the complexity of their network partners—substrates and inhibitors—and with regard to physiological and pathophysiological backgrounds. Thus, *in vivo* validation techniques of substrates identified by *in vitro* degradomics are outlined, for instance by employing activity-based probes or knockout mouse models. In addition, the interrelatedness of proteolytic cleavage as a crucial posttranslational modification is discussed with respect to diverse other posttranslational modifications like glycosylation, phosphorylation, or ubiquitinylation. This chapter, written by the next-generation experts Stefan Tholen, Maria M. Koczorowska, Zon Weng Lai, Joern Dengjel, and Oliver Schilling, also revisits important traditional concepts ranging from cellular protein turnover to protein fates and protein homeostasis in general.

The sixth chapter "**Exploring Systemic Functions of Lysosomal Proteases: The Perspective of Genetically Modified Mouse Models**" deals with a specific proteolytic network by highlighting experimental approaches in which cysteine cathepsins and/or their endogenous inhibitors have been tested by targeted gene knockout and phenotypic rescue experiments by knockin studies. The seniors in the list of authors Martina Gansz, Ursula Kern, Christoph Peters, and Thomas Reinheckel belong to the pioneers in generating genetic mouse models in which cysteine cathepsins are lacking or over-expressed systemically or in a cell type- and tissue-specific manner. Specific examples depicted in this chapter concern the biological significance of cathepsins for epidermal keratinocyte homeostasis in the skin and for the cardiovascular system including pathological features of both atherosclerosis and heart development. Of note, effects of cathepsin deficiencies, overexpression, compensatory functional redundancy, and misbalance with anti-proteolytic factors such as the endogenous cystatins/stefins illustrate strikingly how important it is to investigate proteases and their inhibitors in their cell type-specific context of a given tissue. Even more important, the genetically modified mouse models uncover how systems biology is urgently needed to provide us with insights into the importance of proteolysis for specific processes, regardless of whether looking at developmental, physiological, pathophysiological, or regenerative roles of proteases and their inhibitors.

In the seventh chapter "**Astacins: Proteases in Development and Tissue Differentiation**" structural biology meets function of a versatile group of proteases, namely the astacins. These proteases act from the very beginning of life right starting with sperm–egg fusion, early embryonic development, organogenesis, and hatching processes. Moreover, astacins are crucial in the adult for peri-cellular signaling, tissue homeostasis, and regeneration, implicating that malfunctions may cause neurodegenerative diseases, fibrosis, or cancer. This astonishing group of proteases comprises small, single domain enzymes up to giant proteases like meprin multimers reaching MDa size. The versatility of astacins has been studied in a variety of biological systems throughout the animal kingdom. This chapter interprets this fascinating group of enzymes combining molecular biology and zoology, as provided by Walter Stöcker, with the beauty and explanatory power of structural biology, as contributed by the protein crystallographer F. Xavier Gomis-Rüth.

The eighth chapter "**Proteases in Death Pathways**" takes us from the very beginning of life of a multicellular organism through the crucial phases of development and tissue homeostasis, where programmed cell death is an essential prerequisite. The responsible proteases, caspases, are an extremely well-studied protease family, but they bear mysteries and secrets that need to be uncovered since their dysfunction may initiate a most dangerous landslide of severe pathologies ranging from inflammation and cancer to neurodegenerative diseases. Again, structure meets function in this chapter, and the young researcher Andreas Flütsch has teamed up with the most experienced researcher Markus Grütter in an approach that combines cell and structural biology and which links underpinning topics related to caspases with application and therapeutic potential of, e.g., DARPins that overcome the problems of specific targeting of caspases. Programmed cell death and its several modalities come along lively in this chapter.

The ninth chapter "**ADAM Proteases in Physiology and Pathophysiology: Cleave to Function in Health or to Cause Disease**" covers the sheddases ADAM17 and ADAM10. The authors Joachim Grötzinger and Stefan Rose-John provide us with deep insights into the structure–function relationships of these multi-substrate proteases. ADAMs cleave a huge variety of soluble protein substrates like cytokines, growth factors, or adhesion molecules besides being seemingly made for shedding of transmembrane receptor proteins. These proteolytic enzymes are multi-domain transmembrane proteins themselves, and it is believed that conformational changes, i.e., bending of the active side back towards plasma membrane-near regions, facilitate their broad cleavage repertoire that includes even the processing of transmembrane domains. This chapter highlights once more the importance of creative animal model design in cases where knockout approaches are not applicable.

The authors introduce an approach called EXITS (exon induced translational stop) to generate animals with strongly reduced ADAM17 activity throughout all tissues. Such hypomorphic animal models bear the promise to clarify how ADAMs realize their functions in development as sheddases of numerous receptors initiating vital signaling pathways like for instance those operating via the EGF receptor. It goes without saying that the ADAM proteases fulfill uncountable functions in regulating development, tissue homeostasis, and regeneration. Likewise it is still more than difficult to therapeutically target ADAMS for treatment of diseases in which they are involved, be it during the initial or progressive phases of inflammation or other immune system related disorders, and in cancer.

The tenth chapter "**Proteases in the Nervous System**" covers comprehensively the proteolytic enzymes with a major but not exclusive function in the nervous system. The expert authors Holger Cynis, Stefan Lichtenthaler, Leona Wagner, and Hans-Ulrich Demuth begin with a survey of APP processing enzymes and discuss evidences and contradictory findings when claiming specific proteolytic enzymes as alpha-, beta-, and gamma-secretases. While there is general agreement within the scientific community about which enzymes are alpha-secretases, namely ADAM10, ADAM17, and other metallopeptidases, the issue of beta-secretases is a lot more controversial. Gamma-secretase is an unusual aspartyl protease acting as heterotetrameric complex in catalyzing regulated intra-membrane proteolysis of TMPs such as APP and another dozens of substrates, many of them belonging to the receptors that initiate signaling during developmental processes. However, in a more provocative view considering the enormous turnover of APP (i.e., 8 % per hour), it is not the generation of Aβ but its degradation which will become crucial for understanding AD and other neurodegenerative disorders. Thus, the authors deal with a variety of enzymes involved in Aβ catabolism. An in-depth description of the structural and functional features, tissue distribution, regulation, and therapeutic potentials of targeting proline-specific peptidases reveals thorough insights into neuropeptide processing enzymes as well as their involvement in general peptide processing tasks. A number of enzyme families are highlighted which are often not covered due to their uniqueness such as prolylcarboxypeptidase, X-prolylaminopeptidases, prolidase, and prolinase.

The eleventh chapter "**Proteases in the Mammalian Digestive System**" takes us to another aspect of protein and peptide substrate processing, namely in the mammalian digestive system. The long-standing expertise of the authors S. Gaylen Bradley, Toni M. Antalis, and Judith S. Bond introduces and highlights the proteolytic enzymes acting along the alimentary tract in this very well-digestible and thus "tasty" chapter. Differential expression patterns and peculiar activities characterize the gastrointestinal proteases which comprise a variety of enzymes ranging from enteropeptidase to meprins and including matriptase as one of the crucial markers of colon cancer. However, a number of other proteases like DPs or the serine protease TMPRSS4 are also discussed. Without doubt, it is the importance of the epithelial lining of the GI tract as a barrier between the outside world and the body's interior which implies that the tasks and teamwork of proteolytic enzymes in the mammalian digestive tract are up for more than simple degradation. It is the beauty of cleaving with precision and maintaining the body healthy and well protected from intruders that is enabled by a fine-tuned differential expression and activity of proteolytic enzymes harbored in the digestive system.

The twelfth chapter "**Calpains in Health and Disease**" introduces a remarkable family of cytosolic enzymes that are activated upon increased calcium concentration, i.e., typically following traumatic injury or upon prolonged chemical and mechanical stress. The activity of one of the family members, skeletal muscle specific CAPN3/p94, even depends on both calcium and sodium ions. Calpains have recently experienced a nomenclature reform, which was integrated into this chapter while it was in press. Calpains are found throughout the animals and plant kingdoms. Calpains and their endogenous inhibitors, the calpastatins, function in such diverse processes as cell motility and cell death, vesicular transport, and even pH-sensing in yeast or body size determination of nematodes. Calpains are considered parts of signal transduction cascades and cleave a brought range of cytosolic substrates, thereby releasing smaller fragments or peptides that can, in turn, serve as biomarkers of the site of the insult. Thus, while calcium hyperactivated calpains contribute to massive degradation of signaling receptors and enzymes, transcription factors, as well as cytoskeletal and other structural proteins in, e.g., brain injury/stroke, they bear an enormous potential as theranostics. In stroke, calpain inhibitors may even prove neuro-protective agents. Calpains also exhibit vital tasks in skeletal muscle and the so-called calpainopathies are covered comprehensively. Moreover, calpains bear tasks in the GI tract where they protect the stomach from stress-induced ulcer formation, and they are considered contributing to glucose control in diabetes and obesity. The underlying theme of this chapter is to shed light on a protease family which is both protective and destructive. Therefore, activators and inhibitors of calpains must both be considered in physiological and pathological situations. The expert authors John Anagli, Kevin K.W. Wang, Yasuko Ono, and Hiroyuki Sorimachi, who have acquired a wealth of knowledge about this intriguing family of proteolytic enzymes, suggest novel therapeutic approaches by not only discussing novel biomarkers for diagnostics but also considering calpain inhibition for treatment of ischemic or traumatic cell and tissue injury.

The thirteenth chapter "**Metalloproteinases in Cartilage Matrix Breakdown: The Roles in Rheumatoid Arthritis and Osteoarthritis**" by Hideaki Nagase and Gillian Murphy summarizes and critically discusses the impressive knowledge of the ECM-degrading enzyme families comprising MMPs, ADAMs, and ADAMTSs in the context of rheumatoid arthritis and osteoarthritis. Thus, destructive ECM degradation as a response to inflammation destroys cartilage upon stress, injury, or excessive load but also during aging. Numerous proteases have been attributed to function in this severe pathophysiology that is a major burden not only for aging patient populations. The metalloproteinases discussed in this chapter are considered key players in degradation of a variety of substrates ranging from triple-helical collagen to proteoglycans like aggrecans. The ADAM proteases are viewed here from a different angle compared to previous chapters of this book, in that their destructive rather than regulatory nature of proteolytic action becomes an obvious feature in RA and OA. The cell surface vicinity required for degradation of cartilage is being much better understood now by studying receptors like LRP1 that keeps ADAMTS5 and endogenous inhibitors like TIMP3 in the pericellular environment, and by uncovering additional molecular control mechanisms. Finally, the authors provide at least a glimpse of light at the horizon for treatment options by pointing to the problems of using metalloproteinase inhibitors and by highlighting some recent developments that promise better potential for success, i.e., chondroprotective agents like pentosan polysulfate acting as an exosite inhibitor of aggrecanases.

The fourteenth chapter "**MMP-Mediated Collagen Remodeling and Vessel Functions**" covers functions of MMPs that are usually not considered in the first place when it comes to this versatile ECM remodeling group of proteolytic enzymes. The expert team of Agnès Noel and Nor Eddine Sounni sets a focus on the vascular and lymphatic systems in a refreshing manner. They highlight the cells, which are required to maintain homeostasis, and focus on the importance of the dynamics and integrity of intercellular junctions. In detail, the main roles of, e.g., MMP12, a factor in atherosclerotic plaque rupture, are described besides a number of equally important pathological roles of this and other MMPs. A prominent part of this chapter is attributed to MT1-MMP, a key angio-modulating enzyme controlling vessel maturation and/or regression in healthy and pathological conditions. In their outlook the authors emphasize the importance of understanding the basic mechanistic functions of these relevant proteases, which are essential to design specific therapeutics targeting the lymphatic system and the vasculature *in vivo*.

The fifteenth chapter "**Proteases in Cancer: Significance for Invasion and Metastasis**" is key and a highlight of the book. The authors Bonnie F. Sloane, Karin List, Barbara Fingleton, and Lynn Matrisian come from different protease worlds, but are closely connected by their common interest in tumor biology. The huge wealth of information in this field of research has been put together in an excellently clear and understandable fashion. The red line guiding through this chapter is a schematic figure visualizing the stages of tumorigenesis to which the various groups of proteases are correlated. The authors very systematically discuss each relevant group of proteases from MMPs, ADAMs, ADAMTSs, over-secreted and membrane-anchored serine proteases, and cysteine cathepsins to threonine proteases (i.e., the proteasome) in the context of tumor initiation, tumor promotion, invasion and metastasis, angiogenesis, inflammation, and immunity.

Bremen, Germany Klaudia Brix
Mainz, Germany Walter Stöcker

Contents

Chapter 1
Protease Families, Evolution and Mechanism of Action

Neil D. Rawlings

1.1 Proteolytic Enzymes

A proteolytic enzyme releases an amino acid or a peptide from a protein or larger peptide. Proteolysis is one of the final stages of post-translational processing for many proteins, and is unidirectional. It can occur immediately after biosynthesis or some time later when the protein has been directed to the place where it is required to function. Very few proteins are resistant to proteolysis, and ultimately nearly all proteins will be broken down to their component amino acids for re-use, either by the organism itself or by a predator or detritivore. Proteolysis is required to remove the initiating methionine from newly synthesized, cytoplasmic proteins; to remove signal peptides from proteins targeted to the cell's secretory pathway; to remove targeting signals from proteins targeted to specific organelles such as the mitochondrion or chloroplast; to remove propeptides from enzymes, hormones and receptors that are synthesized as precursors, so that these are activated; to release individual proteins and peptides from polyproteins; to release bioactive peptides from protein precursors; to release proteins from the cell surface ("shedding"); to switch off the signals that peptides and proteins initiate by degrading either them or the proteins they bind to; to recycle amino acids by degrading the proteins; to destroy potentially lethal or toxic proteins from parasites and pathogens; to release antigenic peptides from parasites and pathogens; and to obtain amino acids from food proteins. Pathogens and parasites also use proteolytic enzymes to invade their hosts, and to inactivate any host protein that could harm them or interfere with their reproduction. There are also many pathological situations where the degradation of a protein leads to disease, such as tumour invasion, rheumatoid arthritis or Alzheimer's disease.

N.D. Rawlings (✉)
The Wellcome Trust Sanger Institute, Wellcome Trust Genome Campus, Hinxton, Cambridgeshire CB10 1SA, UK
e-mail: ndr@sanger.ac.uk

K. Brix and W. Stöcker (eds.), *Proteases: Structure and Function*,
DOI 10.1007/978-3-7091-0885-7_1, © Springer-Verlag Wien 2013

Table 1.1 Counts of proteolytic enzymes in model organisms

Species	Number of proteolytic enzymes
Homo sapiens	600
Mus musculus	660
Drosophila melanogaster	477
Caenorhabditis elegans	359
Arabidopsis thaliana	678
Saccharomyces cerevisiae	117
Escherichia coli	405
Bacillus subtilis	190
Thermoplasma volcanium	43

Counts exclude homologues that are known or thought not to be proteolytic enzymes because essential residues have been replaced, pseudogenes and fragments

Almost all organisms require proteolytic enzymes to function. There are some viruses, which do not encode proteolytic enzymes in their genomes, but even these hi-jack host enzymes to do the processing for them. Table 1.1 shows total number of proteolytic enzymes for selected organisms. The number of proteolytic enzymes in a bacterium depends on the sequences derived from bacteriophages and plasmids. Because bacteriophages and plasmids encode their own unique proteolytic enzymes, the more strains of a bacterium that are sequenced the greater the total number of proteolytic enzymes there will be. Similarly, in a eukaryote, the number of proteolytic enzymes is affected by how retrotransposons are treated, because each retrotransposon may encode a proteolytic enzyme.

The terms "proteolytic enzyme", "protease", "proteinase" and "peptidase" are often thought to be synonymous, but this is not the case. The term "proteolytic" was first used in 1877 (Foster 1877) and applied to pepsin. "Proteolysis" first occurred in 1880 (Roberts 1880), again in conjunction with digestion in the stomach. The term "protease" was invented by Vines in 1903 (Vines 1903); the term "proteinase" came a few years later (Fischer 1907), when it was used as a synonym for "protease". The term "peptidase" was first used in 1918 (Petersen and Short 1918), and in 1923 a peptidase was distinguished from a true protease: a protease degraded proteins, whereas a peptidase hydrolysed only peptides or "peptones". Such a distinction would not be acceptable now, because it is now known that a proteinase can also cleave peptides.

1.1.1 Peptidases

A peptidase is defined by the Nomenclature Committee of the International Union of Biochemistry and Molecular Biology as a hydrolase acting on a peptide bond. In Enzyme Nomenclature, hydrolases are class 3 and peptidases are subclass 3.4. Nearly all proteolytic enzymes are hydrolases. However, there are other catalytic mechanisms by which a peptide bond can be broken.

The simplest peptidase molecule has two subdomains separated by the active site [often a "cleft", but sometimes termed a "canyon" if the cleft is long or deep, as it is in thimet oligopeptidase (Ray et al. 2004)]. A cartoon of a typical peptidase structure is shown in Fig. 1.1. Within the active site are the residues responsible for catalysis, which can be a catalytic dyad or triad. Sometimes an additional residue is required to stabilize the "oxyanion hole" in the intermediate that forms between the enzyme and substrate, so up to four residues may be essential. Lining the active site are substrate-binding sites into which one amino acid of the substrate can fit. Which amino acids are able to fit into which binding site determines the specificity of the peptidase. Specificity can be directed to residues either side of the cleavage site of a substrate, but it is unusual for there to be more than four binding sites. However, specificity can be directed towards an amino acid three or four residues away from the cleavage site, in either direction. However, a binding pocket should not be thought of a rigid; there is now evidence of co-operation between binding sites, so that occupation of one binding site by a particular amino acid allows binding of another amino acid in a neighbouring binding site (Schilling et al. 2011). Each binding pocket in the peptidase is named; those on the N-terminal side of the scissile bond are numbered S1, S2, S3, etc, with S1 being closest to the scissile bond and S3 being furthest. On the C-terminal side, pockets are number S1′, S2′, S3′ etc. Residues in the substrate are named P1, P2, P3, etc. from the scissile bond towards the N-terminus and P1′, P2′, P3′, etc from the scissile bond towards the C-terminus [(Schechter and Berger 1968), see Fig. 1.2].

When specificity is quite simple, such as being directed towards arginyl or lysyl bonds, then to make the peptidase more selective in terms of the substrates it cleaves, extra binding sites may exist, that may even be on additional domains or subunits. All of the peptidases in the blood coagulation cascade have specificity for arginyl or lysyl bonds, and all are members of the same peptidase family (S1). Other domains, such as kringles, are required to provide extra contacts so that one peptidase activates only the next component in the cascade. The binding sites outside the active site cleft are known as "exosites".

1.1.1.1 Classification by Specificity

Peptidases have been divided into two broad groups, those that depend upon and act upon a free N- or C-terminus, known as "exopeptidases", and those that act on internal peptide bonds, irrespective of whether one or both termini are blocked or otherwise modified, known as "endopeptidases". This distinction is still maintained in *Enzyme Nomenclature*.

Exopeptidases can be further divided according to whether they act at the N- or C-terminus, and by the number of amino acids released during one cleavage. An exopeptidase that releases one N-terminal amino acid is an aminopeptidase (EC sub-subclass 3.4.11). If it releases a dipeptide from the N-terminus it is a dipeptidyl-peptidase (EC sub-subclass 3.4.14), and if a tripeptide is released a tripeptidyl-peptidase (also included in EC sub-subclass 3.4.14). An exopeptidase that releases a

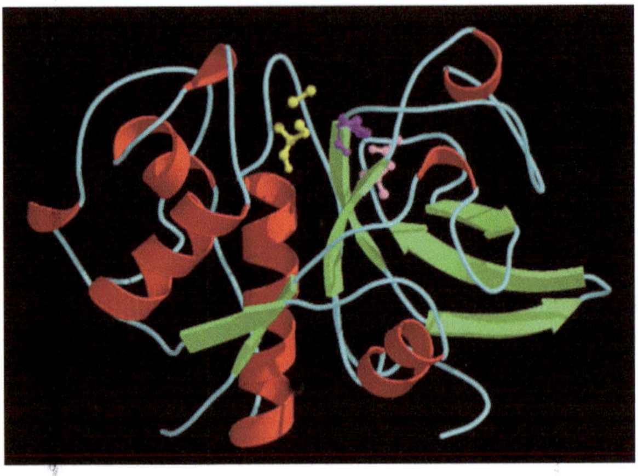

Fig. 1.1 Structure of chymopapain. A Richardson diagram (Richardson 1985) of the structure of chymopapain (subfamily C1A) is shown. The image is derived from Protein Data Bank entry 1YAL. The diagram has been made using the MOLSCRIPT (Kraulis 1991) and RENDER (Bacon and Anderson 1988) software. Helices are shown as *red coils*, beta strands as *green arrows* and *turns* and unstructured regions as a *cyan wire*. The active site residues are shown in *ball-and-stick* representation: Cys159 (thio-methylated) in *yellow*, His293 in *purple* and Asn313 in *pink*

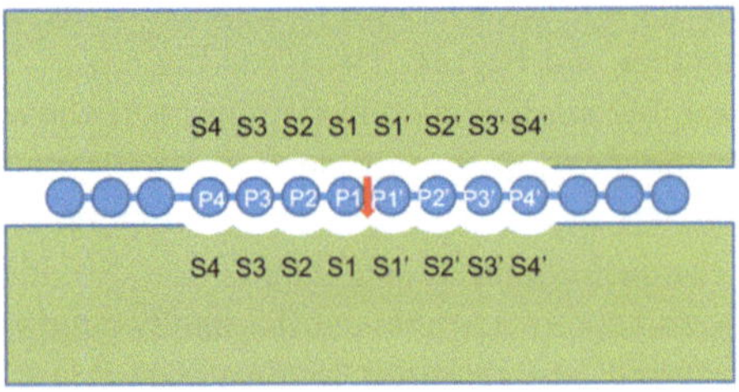

Fig. 1.2 Peptidase binding site and substrate residue nomenclature. The nomenclature according to Schechter and Berger (1968) is shown. The enzyme is represented as a *green rectangle*, with the substrate threaded through the central active site. Residues in the substrate as shown as *beads on a string*. The *red arrow* indicates the scissile bond. Eight substrate binding pockets in the peptidase are shown and are named S4–S1 and S1′–S4′. Residues in the substrate that occupy these binding pockets are named P4–P1 and P1′–P4′

single amino acid from the C-terminus is a carboxypeptidase (EC sub-subclasses 3.4.16, 3.4.17 and 3.4.18); and if a dipeptide is released it is a peptidyl-dipeptidase (EC sub-subclass 3.4.15). An exopeptidase that only cleaves a dipeptide is known as a dipeptidase (EC sub-subclass 3.4.13). These activities are shown in Fig. 1.3. The term "tripeptidase" has been used in the literature to describe a peptidase that cleaves tripeptides; however, this could be an activity ascribed to an aminopeptidase [for example aminopeptidase T (Strauch and Miller 1983)] or a carboxypeptidase (depending on where cleavage occurs), hence the term "tripeptidase" is not recommended (Fig. 1.3).

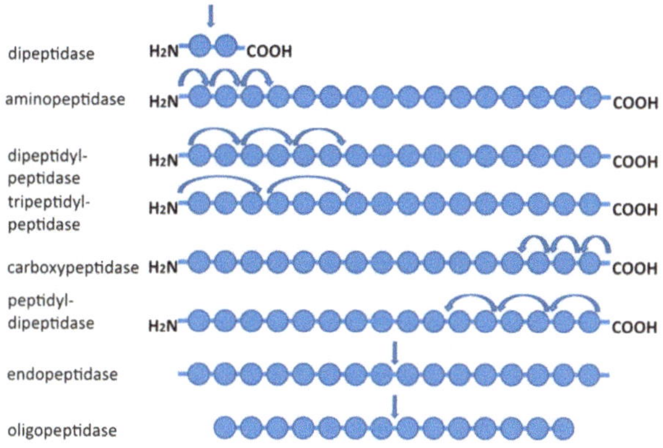

Fig. 1.3 Classification by specificity. The figure shows the catalytic activity of different peptidases according to their *Enzyme Nomenclature* sub-subclasses. Substrates are shown as *beads on a string* with a bead representing an amino acid. *Arrows* indicate cleavage positions

Endopeptidases are not easily grouped in the same way as exopeptidases. The only major distinction is between those that act on peptides and proteins of any size and those that act on peptides that are short; the latter are termed "oligopeptidases".

There is another group of peptidases known as "omega-peptidases". These do not act on standard peptide bonds, but on peptide bonds where one of the amino acids is modified or on peptide bonds that are not made by the alpha carbon. Figure 1.4 shows a variety of such omega-peptidases.

Isopeptide bonds are formed between a carboxy terminus of one protein and the amino group on the side chain of an amino acid (usually lysine) on another protein. Several proteins are targeted for degradation or translocation by the addition of small proteins such as ubiquitin, sumo and NEDD8. There are peptidases that remove these targeting proteins so that they can be re-used, and these peptidases are cleaving isopeptide bonds (Komander 2010). Usually, these peptidases are also able to remove targeting proteins attached to the N-terminal amino group, which is a normal peptide bond. Hence, there is no EC category specifically for isopeptidases.

1.1.1.2 Classification by Catalytic Type

An important characteristic of a peptidase is its "catalytic type". The catalytic type can often help in determining the pH at which the peptidase will operate and for the inhibitors that will inactivate it. All peptidases, exopeptidases and endopeptidases, can be classified according to their catalytic type.

Hydrolysis by a peptidase has often been described as an acid-base reaction. In some peptidases, nucleophilic attack on the peptide bond of the substrate leads to the formation of a temporary complex known as the acyl intermediate. This rapidly breaks down and a proton is transferred to a residue in the peptidase known as the general base, allowing a water molecule to hydrolyse the peptide bond. The nature of the nucleophilic attack differs between the different peptidase catalytic types.

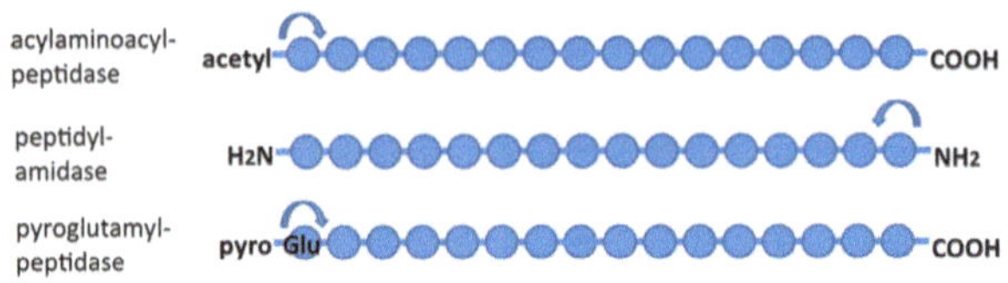

Fig. 1.4 Omega-peptidases. The figure shows some examples of the catalytic activity of omega peptidases. Substrates are shown as *beads on a string* with a bead representing an amino acid. *Arrows* indicate cleavage positions

Peptidases can be classified by catalytic type, and, at present, there are six different types. These can be grouped into those in which the nucleophile is a side-chain of an amino acid ("protein nucleophiles") and those in which the nucleophile is an activated water molecule ("water nucleophiles"). Protein nucleophiles can be the hydroxyl of a serine or threonine residue, or the thiol of a cysteine, and the catalytic types are known as serine-, threonine- or cysteine-type. A water nucleophile can either be activated by the side chains of amino acids (aspartates or glutamates), or by a metal ion bound by amino acid side-chains. Thus the catalytic types are known as aspartyl-, glutamyl- or metallo-peptidases. There is a seventh category, those of unknown catalytic type. Proteolytic enzymes that are not peptidases can employ asparagine as a nucleophile (see below).

Typically, threonine- and metallopeptidases have a neutral pH optimum. Aspartic, glutamic and cysteine peptidases have an acidic pH optimum, and serine peptidases have a neutral to basic pH optimum. However, there are always exceptions: meprin B is a metallopeptidase with an acidic pH optimum; renin and human immunodeficiency virus retropepsin are aspartic peptidases with a neutral pH optimum; calpain and caspases are cysteine peptidases with a neutral pH optimum.

Aspartic peptidases are typically inhibited by pepstatin, diazoacetylnorleucine methyl ester (DAN) and 1,2-epoxy-3-(p-nitrophenoxy) propane (EPNP). Cysteine peptidases are inhibited by compounds that chelate thiols, such as iodoacetic acid, iodoacetamide and *N*-ethylmaleimide. Serine peptidases are inhibited by diisopropyl fluorophosphate (DFP), phenylmethane sulfonylfluoride (PMSF) and 3,4-dichloroisocoumarin (DCI). Metallopeptidases are inhibited by metal chelators such as ethylenediaminetetraacetic acid (EDTA) and 1-10 phenanthroline. However, because some intracellular peptidases are thiol-dependent, inhibition by thiol chelators does not necessarily imply that they are cysteine peptidases. An example of a thiol-dependent metallopeptidase is thimet oligopeptidase (Tisljar and Barrett 1990). Similarly, metal ions can be structurally important and not be involved in catalysis, and still be susceptible to metal chelators so that the enzyme is inhibited. An example is the calcium-dependent, serine peptidase proteinase K, which is inhibited by EDTA (Bajorath et al. 1988). Because the glutamate peptidase from *Scytalidium lignicolum* is inhibited by EPNP (Tsuru 1987), for many years it was thought to be an unusual aspartic peptidase.

The main problem with either of these two classification systems, classification by reaction catalysed or classification by catalytic type, is that neither system matches the evolutionary origins of peptidases. Peptidases that are of different catalytic types may perform similar reactions, and peptidases of the same catalytic type may not be homologous. This will be discussed further below.

1.1.1.3 Classification by Homology

MEROPS Families

Proteins are best classified into domain families, and the Pfam database has led the way (Punta et al. 2012). Most proteins consist of a series of domains that can be likened to beads on a string, with each bead representing a different domain—or multiple copies of the same domain. The simplest peptidase in its mature form consists of a single protein domain. However, there are numerous examples of peptidases that are multidomain proteins. Fortunately, very few peptidases carry more than one domain essential for peptide bond hydrolysis.

The *MEROPS* classification of peptidases was introduced in 1993 (Rawlings and Barrett 1993). When comparing sequences to group them into peptidase families, it is important to consider only the domain bearing the active site residues (the "peptidase unit") and to discard any other domains. Sequences with homologous peptidase units are grouped into a peptidase family. The first character in the family name denotes the catalytic type ("A" for aspartic, "C" for cysteine, "G" for glutamic, "M" for metallo, "S" for serine, "T" for threonine and "U" for unknown). So far, no peptidase family exists containing peptidases of different catalytic types.

A family is built around a peptidase that has been well characterized biochemically, which is termed the "type example". This is often a peptidase for which the sequence, tertiary structure and specificity is known; e.g. the type example for peptidase family S1 is chymotrypsin. The sequence of the peptidase unit of each type example is submitted to a BlastP search of either the NCBI protein sequence database (Sayers et al. 2011) or UniProt (Anonymous 2010). The sequences of peptidase units are determined to be homologous if the expect value in a BlastP (Altschul et al. 1990) or FastA search (Lipman and Pearson 1985) is less than 0.001.

Not all members of a family will necessarily be returned as homologues in a single BlastP search because transitive relationships may exist. If protein A (the type example) returns protein B in a BlastP search but not protein C, and if a BlastP search with protein B returns proteins A and C, then the relationship between proteins A and C is transitive. Sometimes a transitive relationship is found between groups of peptidases that had been considered to be separate families; these are united in a single family (retaining the family name with the lowest number) but the separate groups become subfamilies. Other methods such as Psi-Blast searches (Altschul et al. 1997) are used to find these distant relationships. A subfamily name is the family name with a sequential letter added at the end. For example, family M10 contains subfamilies M10A (the matrixins) and M10B (the serralysins).

MEROPS Clans

A single BlastP search is unlikely to find all the homologues of a single protein sequence, not only because of transitive relationships but because some sequences are so distantly related that the similarity can only be found by comparing tertiary structures. If two peptidases from different families have similar tertiary structures (in the peptidase unit) then the two families are grouped into a clan. Tertiary structures are compared using the DALI search engine (Holm and Sander 1995), and structures are assumed to be related if the Z score is greater than 6.00. A clan is therefore a higher taxon than a family and one clan can contain many families. A clan has a two-character name in which the initial letter indicates the catalytic type, just as it does for a family, with the letter "P" for a clan that contain peptidases of different catalytic types. The second character is a sequential letter.

There are three clans (PA, PB and PC) each containing peptidases of different catalytic types. Clan PA includes 20 different families, 6 of them of cysteine peptidases from viruses, such as picornains 2A and 3C from poliovirus (family C3), and 14 families of serine peptidases, including the chymotrypsin family (S1) and other viral peptidases such as togavirin (S3) and flavivirin (S7). The relationship between viral peptidases and chymotrypsin was first postulated by Bazan and Fletterick (1988). Clan PB contains five families of cysteine peptidases, a single serine peptidase family and four families of threonine peptidases, including proteasome components (family T1). All the enzymes in clan PB are Ntn-hydrolases (Ntn = N-terminal nucleophile) where residue 1 in the mature protein bears the nucleophile (the thiol of a cysteine or the hydroxyl of a serine or threonine). Most of the mature enzymes are not themselves peptidases, but are synthesized as precursors which are autolytically processed to remove an N-terminal propeptide and to leave the active site residue at the N-terminus. It is the autolytic activity that makes the precursor a peptidase.

When no crystal structure has been solved for any member of a family, it may still be possible to assign that family to a clan according to the order of active site residues in the protein sequence. There are also some clans which contain a single peptidase family because the structure of a member is known, but where that structure shows no similarity to that of any other peptidase. For example, the structure of omptin (family A26), a membrane proteinase from *Escherichia coli*, is unique (Vandeputte-Rutten et al. 2001) and family A26 is the only member of clan AF.

Some clans are divided into subclans if there are characteristics that link some of the families together. For example, all the families of serine peptidases in clan PA are grouped in subclan PA(S) and all the families of cysteine peptidases in subclan PA(C). In clan MA, some of the peptidases in the families bind the single metal ion (usually zinc) by two histidines in a His-Glu-Xaa-Xaa-Glu motif (where Xaa is any amino acid) with the third ligand being a glutamate C-terminal to this motif; these are known as Glu-zincins and are grouped together in subclan MA(E). Peptidases in some of the other families have an extended zinc-binding motif, usually His-Glu-Xaa-Xaa-His-Xaa-Xaa-Gly-Xaa-Xaa-His, in which all three histidines are metal ligands, and also have an essential methionine conserved; these are known as Met-zincins and are grouped in subclan MA(M).

Is there a level of homology higher than that of clan? There is evidence that peptidases in clan CE, which includes adenain the adenovirus endopeptidase (family C5), have a structure that is a permutation of the structure of papain (family C1, clan CA) (Ding et al. 1996). These peptidases are in different clans because active site residues are not in the same order. In papain, four residues are important with two in each subdomain; the glutamine that helps stabilize the oxyanion hole index and the nucleophilic Cys in the N-terminal subdomain, and the histidine (general base) and the asparagine which orientates the histidine ring correctly in the second domain. So the order of essential residues is Gln, Cys, His and Asn. In adenain, the first and second subdomains have swapped positions, so the order is His, Asp or Glu (replacing the Asn of papain), Gln and Cys. Comparisons of the structures of metallocarboxypeptidases (family M14, clan MC) and leucine aminopeptidases (family M17, clan MF) suggest a common evolutionary origin (Artymiuk et al. 1992), despite the fact that the carboxypeptidase binds a single zinc ion (with three ligands) and the aminopeptidase binds two manganese ions (with five ligands).

Clans and families of proteolytic enzymes are shown in Table 1.2.

MEROPS Identifiers

An individual peptidase is also given a unique identifier consisting of the family name (padded with a zero if it is less than three characters), a dot, and a sequential three-digit number. Thus, chymotrypsin A has the *MEROPS* identifier S01.001, and is a member of subfamily S1A, family S1, subclan PA(S) and clan PA. The *MEROPS* classification system is thus hierarchical: identifier, family and clan, with clan at the top of the hierarchy. The clan containing most families is MA with 38; the next biggest are clans CA (27) and PA (20). The families containing most identifiers (and thus most biochemically characterized peptidases that are known to be different) are S1 (456 identifiers), M12 (178), C1 (144) and S8 (143).

There are special *MEROPS* identifiers for (a) non-peptidase homologues, that are proteins known or suspected not be peptidases because active site residues are not conserved (in the form fam.9*, where "fam" is the three-character family name); (b) for pseudogenes (in the form fam.P*); and (c) for peptidase complexes where there is more than one peptidase unit (in the form Xfam-*). Examples of non-peptidase homologues are S01.971 for azurocidin and M12.950 for ADAM2. An example of a pseudogene is A01.P02 for the human chymosin pseudogene. Examples of peptidase complexes are XM12-001 for meprin A that consists of a heterodimers of two subunits in family M12; XS26-001 for the signal peptidase complex which includes two subunits in family S26; and XT01-001 for the 20S proteasome complex (containing 14 different subunits in family T1). The tricorn peptidase complex consists of subunits from different peptidase families (M1, S33 and S45) and so has the special indentifier XP01-001.

Table 1.2 Clans and families of proteolytic enzymes

Clan	Subclan	Family	Subfamily	MEROPS identifier	Type example	Accession	Organism	Active site residues	Metal ligands
AA		A1	A1A	A01.001	Pepsin A	P00790	*Homo sapiens*	D/H, Y/F/S, D	
			A1B	A01.040	Nepenthesin	Q766C3	*Nepenthes gracilis*	D, Y, D	
		A2	A2A	A02.001	HIV-1 retropepsin	P04586	Human immunodeficiency virus 1	D	
			A2B	A02.022	Ty3 transposon peptidase	Q99315	*Saccharomyces cerevisiae*	D	
			A2C	A02.021	Gypsy transposon peptidase	P10401	*Drosophila melanogaster*	D	
			A2D	A02.054	Osvaldo retrotransposon peptidase		*Drosophila buzzatii*	D	
			A2E	A02.051	Retrotransposon peptidase		*Schizosaccharomyces pombe*	D	
			A2G	A02.052	Retrotransposon 17.6 peptidase	P20825	*Drosophila melanogaster*	D	
			A2H	A02.063	Walleye dermal sarcoma virus retropepsin		Walleye dermal sarcoma virus	D	
		A3	A3A	A03.001	Caulimovirus peptidase	P03554	Cauliflower mosaic virus	D	
			A3B	A03.002	Bacilliform virus peptidase	P27502	Rice tungro bacilliform virus	D	
		A9		A09.001	Spumapepsin	P14350	Human spumaretrovirus	D	
		A11	A11A	A11.001	Copia transposon peptidase	P04146	*Drosophila melanogaster*	D	
			A11B	A11.003	Ty1 transposon peptidase	Q12501	*Saccharomyces cerevisiae*	D	
		A28		A28.001	DNA-damage inducible protein 1	P40087	*Saccharomyces cerevisiae*		
		A32		A32.001	PerP peptidase		*Caulobacter crescentus*	D	
		A33		A33.001	skin SASPase		*Mus musculus*	D	
AC		A8		A8.001	signal peptidase II	P00804	*Escherichia coli*	D, D	
AD		A22	A22A	A22.001	presenilin 1	P49768	*Homo sapiens*	D, D	
			A22B	A22.003	Impas 1 peptidase	Q8TCT9	*Homo sapiens*	D, D	
		A24	A24A	A24.001	Type 4 prepilin peptidase 1	P22610	*Pseudomonas aeruginosa*	D, D	

		A24B	A24.016	Preflagellin peptidase		*Methanococcus maripaludis*	D, D
AE	A25		A25.001	gpr peptidase	P22321	*Bacillus megaterium*	D, D, K/H
	A31		A31.001	HybD peptidase	P37182	*Escherichia coli*	E/D, D, H
	A26		A26.001	Omptin	P09169	*Escherichia coli*	D, D, D, H
A-	A5		A05.001	Thermopsin	P17118	*Sulfolobus acidocaldarius*	
	A36		A36.001	Sporulation factor SpoIIGA	P13801	*Bacillus subtilis*	D
CA	C1	C1A	C01.001	Papain	P00784	*Carica papaya*	Q, C, H, N/Q
		C1B	C01.085	Bleomycin hydrolase	Q01532	*Saccharomyces cerevisiae*	Q, C, H, N
	C2	C2A	C02.002	Calpain-2	P17655	*Homo sapiens*	Q, C, H, N
		C2B	C02.022	Tpr peptidase	P25806	*Porphyromonas gingivalis*	Q, C, H, N
	C6		C06.001	Helper component peptidase	P18247	Potato virus Y	C, H
	C10		C10.001	Streptopain	Q5X9P3	*Streptococcus pyogenes*	C, H
	C12		C12.001	Ubiquitinyl hydrolase-L1	P09936	*Homo sapiens*	Q, C, H, D
	C16	C16A	C16.001	Papain-like peptidase 1	P19751	Murine hepatitis virus	C, H, D
		C16B	C16.006	Papain-like peptidase 2	P19751	Murine hepatitis virus	C, H, D
	C19		C19.015	Ubiquitin-specific peptidase 14	P54578	*Homo sapiens*	N, C, H, D/N
	C28		C28.001	L-peptidase		Foot-and-mouth disease virus	N, C, H, D
	C31		C31.001	Cysteine peptidase alpha	Q83017	Lactate-dehydrogenase-elevating virus	C, H
	C32		C32.001	Arterivirus cysteine peptidase	Q04561	Porcine reproductive and respiratory syndrome virus	C, H
	C39		C39.001	Bacteriocin-processing peptidase	P36497	*Pediococcus acidilactici*	Q, C, H, D/E
	C47		C47.001	Staphopain A	Q8NVS9	*Staphylococcus aureus*	Q, C, H, N
	C51		C51.001	D-alanyl-glycyl peptidase	P24556	*Staphylococcus aureus*	C, H
	C54		C54.003	Autophagin-1	Q9Y4P1	*Homo sapiens*	Y, C, D, H

(continued)

Table 1.2 (continued)

Clan	Subclan	Family	Subfamily	MEROPS identifier	Type example	Accession	Organism	Active site residues	Metal ligands
		C58	C58A	C58.001	YopT peptidase	O68703	*Yersinia pestis*	C, H, D	
			C58B	C58.003	HopN1 peptidase		*Pseudomonas syringae*	C, H, D	
		C64		C64.001	Cezanne deubiquitinylating peptidase	Q6GQQ9	*Homo sapiens*	C, H	
		C65		C65.001	Otubain-1	Q96FW1	*Homo sapiens*	D, C, H	
		C66		C66.001	IdeS peptidase		*Streptococcus pyogenes*	C, H, D/E, D/N	
		C67		C67.001	CylD protein	Q9NQC7	*Homo sapiens*	Q, C, H, D	
		C71		C71.001	Pseudomurein endoisopeptidase Pei		Methanobacterium phage psiM2	C, H, D	
		C76		C76.001	UL36 deubiquitinylating peptidase	P10220	Human herpesvirus 1	C, H	
		C78		C78.001	UfSP1 peptidase	Q9CZP0	*Mus musculus*	Y, C, D, H	
		C83		C83.001	Gamma-glutamylcysteine dipeptidyltranspeptidase		*Nostoc sp.* PCC 7120	Q, C, H, D	
		C85		C85.001	DUBA deubiquitinylating enzyme	Q96G74	*Homo sapiens*	D, C, H	
		C86		C86.001	Ataxin-3	P54252	*Homo sapiens*	C, H	
		C87		C87.001	Nairovirus deubiquitinylating peptidase		Crimean-Congo hemorrhagic fever virus	C, H, D	
		C88		C88.001	OTU1 peptidase	P43558	*Saccharomyces cerevisiae*	D, C, H	
CD		C11		C11.001	Clostripain	P09870	*Clostridium histolyticum*	H, C	
		C13		C13.001	Legumain	P49046	*Canavalia ensiformis*	H, C	
		C14	C14A	C14.001	Caspase-1	P43527	*Rattus norvegicus*	H, C	
			C14B	C14.035	Metacaspase Yca1	Q08601	*Saccharomyces cerevisiae*	H, C	
		C25		C25.001	Gingipain R	P28784	*Porphyromonas gingivalis*	H, C	
		C50		C50.001	Separase	Q03018	*Saccharomyces cerevisiae*	H, C	
		C80		C80.001	RTX self-cleaving toxin		*Vibrio cholerae*	H, C	
CE		C5		C05.001	Adenain	P03252	Human adenovirus type 2	H, D/E, Q, C	

		C48		C48.001	Ulp1 peptidase	Q02724	*Saccharomyces cerevisiae*	H, D, Q, C	
		C55		C55.001	YopJ protein	P31498	*Yersinia pseudotuberculosis*	H, E, Q, C	
		C57		C57.001	I7L processing peptidase	P12926	Vaccinia virus	H, D/N, Q, C	
		C63		C63.001	Processing peptidase	Q65228	African swine fever virus	H, N, Q, C	
		C79		C79.001	ElaD peptidase	Q47013	*Escherichia coli*	H, C	
CF		C15		C15.001	Pyroglutamyl-peptidase I (prokaryote)	P46107	*Bacillus amyloliquefaciens*	E, C, H	
CL		C60	C60A	C60.001	Sortase A		*Staphylococcus aureus*	H, C	
			C60B	C60.002	Sortase B		*Staphylococcus aureus*	H, C	
		C82		C82.001	L,D-transpeptidase		*Enterococcus faecium*	H, C	
CM		C18		C18.001	Peptidase 2	P26664	Hepatitis C virus	H, E, C	
CN		C9		C09.001	nsP2 peptidase	P03317	Sindbis virus	C, H	
CO		C40		C40.001	Dipeptidyl-peptidase VI (bacteria)	P39043	*Bacillus sphaericus*	C, H, H/N	
C-		C7		C07.001	p29 peptidase	P10941	*Cryphonectria* hypovirus	C, H	
		C8		C08.001	p48 peptidase	Q9YTU2	*Cryphonectria* hypovirus 1	C, H	
		C21		C21.001	Tymovirus peptidase	P28477	Turnip yellow mosaic virus	C, H	
		C23		C23.001	Carlavirus peptidase		Apple stem pitting virus	C, H	
		C27		C27.001	Rubella virus peptidase	Q99IE5	Rubella virus	C, H	
		C33		C33.001	Nsp2 cysteine peptidase	P19811	Equine arteritis virus	C, H	
		C36		C36.001	Papain-like peptidase		Beet necrotic yellow vein virus	C, H	
		C42		C42.001	Papain-like peptidase	Q08534	Beet yellows virus	C, H	
		C53		C53.001	Pestivirus Npro peptidase	P19712	Classical swine fever virus	E, H, C	
		C70		C70.001	AvrRpt2 peptidase	Q6LAD6	*Pseudomonas syringae*	C, H, D	
		C74		C74.001	Pestivirus NS2 peptidase	P19711	Bovine viral diarrhea virus 1	H, E, C	
		C75		C75.001	AgrB peptidase	P0C1P7	*Staphylococcus aureus*	H, C	
		C84		C84.001	prtH peptidase		*Bacteroides forsythus*		
GA		G1		G01.001	Scytalidoglutamic peptidase	P15369	*Scytalidium lignicolum*	Q, E	
GB		G2		G02.001	Pre-neck appendage protein	P20345	Bacteriophage phi-29	D, E	
MA	MA(E)	M1		M01.001	Aminopeptidase N	P15144	*Homo sapiens*	E, Y	H, H, E

(continued)

Table 1.2 (continued)

Clan	Subclan	Family	Subfamily	MEROPS identifier	Type example	Accession	Organism	Active site residues	Metal ligands
MA	MA(E)	M2		M02.001	Angiotensin-converting enzyme peptidase unit 1	P12821	*Homo sapiens*	E	H, H, E
		M3	M3A	M03.001	Thimet oligopeptidase	P24155	*Rattus norvegicus*	E	H, H, E
			M3B	M03.007	Oligopeptidase F	Q9CEV7	*Lactococcus lactis*	E	H, H, E
		M4		M04.001	Thermolysin	P00800	*Bacillus thermoproteolyticus*	E, H	H, H, E
		M5		M05.001	Mycolysin	P20910	*Streptomyces cacaoi*	E	H, H, E
		M9	M9A	M09.001	Bacterial collagenase V	P43154	*Vibrio alginolyticus*	E	H, H, E
			M9B	M09.003	Bacterial collagenase H		*Clostridium histolyticum*	E	H, H, E
		M13		M13.001	Neprilysin	P08473	*Homo sapiens*	E, D	H, H, E
		M26		M26.001	IgA1-specific metallopeptidase	Q59986	*Streptococcus sanguinis*	E	H, H
		M27		M27.001	Tentoxilysin	P04958	*Clostridium tetani*	E, R, Y	H, H, E
		M30		M30.001	Hyicolysin	Q08002	*Staphylococcus hyicus*	E	H, H, E
		M32		M32.001	Carboxypeptidase Taq	P42663	*Thermus aquaticus*	E	H, H, E
		M34		M34.001	Anthrax lethal factor	P15917	*Bacillus anthracis*	E, Y	H, H, E
		M36		M36.001	Fungalysin	P46075	*Aspergillus fumigatus*	E	H, H
		M41		M41.001	FtsH peptidase		*Escherichia coli*	E	H, H, D
		M48	M48A	M48.001	Ste24 peptidase	P47154	*Saccharomyces cerevisiae*	E	H, H, E
			M48B	M48.002	HtpX peptidase	P23894	*Escherichia coli*	E	H, H, E
		M56		M56.001	BlaR1 peptidase	P18357	*Staphylococcus aureus*	E	H, H, E
		M60		M60.001	Enhancin		*Lymantria dispar* nucleopolyhedrovirus	E	H, H
		M61		M61.001	Glycyl aminopeptidase		*Sphingomonas capsulata*	E	H, H, E
		M78		M78.001	ImmA peptidase		*Bacillus subtilis*	E	H, H, E
		M85		M85.001	NleC peptidase		*Escherichia coli*	E	H, H
MA	MA(M)	M6		M06.001	Immune inhibitor A	P23382	*Bacillus thuringiensis*	E	H, H, D

	M7		M07.001	Snapalysin	P0A3Z8	*Streptomyces lividans*	E	H, H, D
	M8		M08.001	Leishmanolysin	P08148	*Leishmania major*	E, M	H, H, H
	M10	M10A	M10.001	Matrix metallopeptidase-1	P03956	*Homo sapiens*	E	H, H, H
		M10B	M10.051	Serralysin	P23694	*Serratia marcescens*	E, Y	H, H, H
		M10C	M10.020	Fragilysin	P54355	*Bacteroides fragilis*	E	H, H, H
	M11		M11.001	Gametolysin	P31178	*Chlamydomonas reinhardtii*	E	H/Q, H, H
	M12	M12A	M12.001	Astacin	P07584	*Astacus astacus*	E	H, H, H
		M12B	M12.141	Adamalysin	P34179	*Crotalus adamanteus*	E	H, H, H
	M35		M35.002	Deuterolysin	P46073	*Aspergillus flavus*	E	H, H, D
	M43	M43A	M43.001	Cytophagalysin		*Cytophaga sp.*	E	H, H, H
		M43B	M43.004	Pappalysin-1	Q13219	*Homo sapiens*	E, M	H, H, H
	M54		M54.001	Archaelysin	Q57729	*Methanocaldococcus jannaschii*	E	H, H, H
	M57		M57.001	prtB proetin		*Myxococcus xanthus*	E	H, H, H
	M64		M64.001	IgA peptidase		*Clostridium ramosum*	E	H, H, D
	M66		M66.001	StcE peptidase	O82882	*Escherichia coli*	E	H, H, H
	M72		M72.001	Peptidyl-Asp metallopeptidase		*Pseudomonas aeruginosa*	E	H, H, H
	M80		M80.001	Wss1 peptidase	P38838	*Saccharomyces cerevisiae*	E	H, H, H
	M84		M84.001	mpriBi peptidase		*Bacillus intermedius*	E	H, H, H
MC	M14	M14A	M14.001	Carboxypeptidase A1	P15085	*Homo sapiens*	R, E	H, E, H
		M14B	M14.005	Carboxypeptidase E	P04836	*Bos taurus*	R, E	H, E, H
		M14C	M14.008	Gamma-D-glutamyl-(L)-meso-diaminopimelate peptidase I		*Bacillus sphaericus*	E	H, E, H
		M14D	M14.027	Cytosolic carboxypeptidase 6		*Homo sapiens*	E	H, E, H
MD	M15	M15A	M15.001	Zinc D-Ala-D-Ala carboxy-peptidase	P00733	*Streptomyces albus*	H	H, D, H
		M15B	M15.010	VanY D-Ala-D-Ala carboxy-peptidase	P37711	*Enterococcus faecium*	E	H, D, H
		M15C	M15.020	Ply118 D-Ala-D-Glu peptidase	Q37976	Bacteriophage A118	D	H, D, H
		M15D	M15.011	VanX D-Ala-D-Ala dipeptidase	Q06241	*Enterococcus faecium*	E	H, D, H
	M74		M74.001	Murein endopeptidase	P0C0T5	*Escherichia coli*	H	H, D, H
ME	M16	M16A	M16.001	Pitrilysin	P05458	*Escherichia coli*	E, E	H, H, E

(continued)

Table 1.2 (continued)

Clan	Subclan	Family	Subfamily	MEROPS identifier	Type example	Accession	Organism	Active site residues	Metal ligands
			M16B	M16.003	Mitochondrial processing peptidase beta-subunit	P10507	*Saccharomyces cerevisiae*	E, E	H, H, E
			M16C	M16.009	Eupitrilysin	Q5JRX3	*Homo sapiens*	E, E	H, H, E
		M44		M44.001	Pox virus metallopeptidase	P68493	Vaccinia virus	E	H, H, E
MF		M17		M17.001	Leucyl aminopeptidase (animal)	P00727	*Bos taurus*	K, R	K, D, D, D, E
MG		M24	M24A	M24.001	Methionyl aminopeptidase 1	P0AE18	*Escherichia coli*	H	D, D, H, E, E
			M24B	M24.004	Aminopeptidase P (bacteria)	P15034	*Escherichia coli*	H, H, H	D, D, H, E, E
MH		M18		M18.001	Aminopeptidase I	P14904	*Saccharomyces cerevisiae*	D, E	H, D, E, D, H
		M20	M20A	M20.001	Glutamate carboxypeptidase	P06621	*Pseudomonas sp.*	D, E	H, D, E, E/D, H
			M20B	M20.003	Peptidase T	P29745	*Escherichia coli*	D, E	H, D, E, D, H
			M20C	M20.007	Xaa-His dipeptidase	P15288	*Escherichia coli*	D, E	H, D, E, D, H
			M20D	M20.008	Carboxypeptidase Ss1	P58156	*Sulfolobus solfataricus*	E	D/E/C, D/H, E/H, H
		M28	M28A	M28.003	Aminopeptidase S	P80561	*Streptomyces griseus*	D, E	H, D, E, E/D, H
			M28B	M28.010	Glutamate carboxypeptidase II	Q04609	*Homo sapiens*	D, E	H, D, E, E/D, H
			M28C	M28.005	IAP aminopeptidase	P10423	*Escherichia coli*	D, E	H, D, E, E/D, H

		M28E	M28.002	Aminopeptidase Ap1	Q01693	*Vibrio proteolyticus*	D, E	H, D, E, D, H
	M42		M42.001	Glutamyl aminopeptidase (bacterium)	Q48677	*Lactococcus lactis*	D, E	H, D, E, D/E, H
MJ	M19		M19.001	Membrane dipeptidase	P16444	*Homo sapiens*		H, D, E, H, H
	M38		M38.001	Isoaspartyl dipeptidase (metallo-type)	P39377	*Escherichia coli*	D	H, H, K, H, H
MK	M22		M22.001	O-sialoglycoprotein peptidase	P36175	*Mannheimia haemolytica*		H, H
MM	M50	M50A	M50.001	S2P peptidase	O43462	*Homo sapiens*	E	H, H, D
		M50B	M50.002	Sporulation factor SpoIVFB	P26937	*Bacillus subtilis*	E	H, H, D
MN	M55		M55.001	D-aminopeptidase DppA	P26902	*Bacillus subtilis*	H	D, E, H, H, E
MO	M23	M23A	M23.001	Beta-lytic metallopeptidase	P27458	*Achromobacter lyticus*	H	H, D, H
		M23B	M23.004	Lysostaphin	P10547	*Staphylococcus simulans*	H	H, D, H
MP	M67	M67A	M67.001	Poh1 peptidase	P43588	*Saccharomyces cerevisiae*	E	H, H, D
		M67B	M67.010	JAMM-like protein		*Archaeoglobus fulgidus*	E	H, H, D
		M67C	M67.006	AMSH deubiquitinating peptidase	O95630	*Homo sapiens*	E	H, H, D
MQ	M29		M29.001	Aminopeptidase T	P23341	*Thermus aquaticus*	Y	E, E, H, H, D
M-	M49		M49.001	Dipeptidyl-peptidase III	O55096	*Rattus norvegicus*	E	H, H, E
	M73		M73.001	Camelysin		*Bacillus cereus*		
	M75		M75.001	Imelysin		*Pseudomonas aeruginosa*		
	M76		M76.001	Atp23 peptidase		*Homo sapiens*	E	H, H
	M77		M77.001	Tryptophanyl aminopeptidase 7-DMATS-type peptidase		*Aspergillus fumigatus*		
	M79		M79.001	Prenyl peptidase 2	Q03530	*Saccharomyces cerevisiae*		
	M81		M81.001	MlrC peptidase		*Sphingomonas sp.* ACM-3962		
	M82		M82.001	PrsW peptidase	P50738	*Bacillus subtilis*		

(continued)

Table 1.2 (continued)

Clan	Subclan	Family	Subfamily	MEROPS identifier	Type example	Accession	Organism	Active site residues	Metal ligands
NA		N1		N01.001	Nodavirus peptidase	P12870	Flock house virus	D, N	
		N2		N02.001	Tetravirus coat protein		*Nudaurelia capensis* omega virus	E, N	
		N8		N08.001	Poliovirus capsid VP0-type self-cleaving protein	P03300	Human poliovirus 1	N/Q	
NB		N6		N06.001	YscU protein	P69987	*Yersinia pseudotuberculosis*	N	
NC		N7		N07.001	Reovirus type 1 coat protein	P11077	Mammalian orthoreovirus 1	N	
ND		N4		N04.001	Tsh-associated self-cleaving domain	Q47692	*Escherichia coli*	N, Y, E, R	
NE		N5		N05.001	Picobirnavirus self-cleaving protein	Q50LE5	Human picobirnavirus	N	
PA	PA(C)	C3	C3A	C03.001	Enterovirus picornain 3C	P03299	Human poliovirus 1	H, E/D, C	
			C3B	C03.020	Enterovirus picornain 2A	P03301	Human poliovirus 1	H, E/D, C	
			C3C	C03.008	Aphthovirus picornain 3C		Foot-and-mouth disease virus	H, E/D, C	
			C3D	C03.003	Comovirus picornain 3C	P03600	Cowpea mosaic virus	H, E/D, C	
			C3E	C03.005	Hepatovirus picornain 3C	P14553	Hepatitis A virus	H, D, C	
			C3F	C03.023	Parechovirus picornain 3C	Q66578	Human parechovirus 1	H, C	
			C3G	C03.024	Waikavirus peptidase	Q91PP5	Rice tungro spherical virus	H, E/D, C	
		C4		C04.001	Nuclear-inclusion-a peptidase	P13529	Plum pox virus	H, D, C	
		C24		C24.001	Calicivirus 3C-like peptidase	Q86117	Rabbit hemorrhagic disease virus	H, D/E, C	
		C30		C30.004	Main peptidase	Q9IW06	Transmissible gastroenteritis virus	H, C	
		C37		C37.001	Calicivirin	Q04544	Southampton virus	H, C	
		C62		C62.001	3C-like peptidase		Gill-associated virus	H, C	
PA	PA(S)	S1	S1A	S01.001	Chymotrypsin A	P00766	*Bos taurus*	H, D, S	
			S1B	S01.269	Glutamyl peptidase I	P0C1U8	*Staphylococcus aureus*	H, D, S	

			S1C	S01.273	DegP peptidase	P0C0V0	*Escherichia coli*	H, D, S
			S1D	S01.280	Lysyl peptidase (bacteria)	P15636	*Achromobacter lyticus*	H, D, S
			S1E	S01.261	Streptogrisin A	P00776	*Streptomyces griseus*	H, D, S
			S1F	S01.109	Astrovirus serine peptidase	Q9IFX3	Human astrovirus	H, D, S
		S3		S03.001	Togavirin	P03316	Sindbis virus	H, D, S
		S6		S06.001	IgA1-specific serine peptidase		*Neisseria gonorrhoeae*	H, D, S
		S7		S07.001	Flavivirin	Q9YRV3	Yellow fever virus	H, D, S
		S29		S29.001	Hepacivirin	P26664	Hepatitis C virus	H, D, S
		S30		S30.001	Potyvirus P1 peptidase	P13529	Plum pox virus	H, D, S
		S31		S31.001	Pestivirus NS3 polyprotein peptidase	Q01499	Bovine viral diarrhea virus 1	H, D, S
		S32		S32.001	Arterivirus serine peptidase	P19811	Equine arteritis virus	H, D, S
		S39	S39A	S39.001	Sobemovirus peptidase		Cocksfoot mottle virus	H, D, S
			S39B	S39.002	Luteovirus peptidase	P11622	Potato leaf roll luteovirus	H, D, S
		S46		S46.001	Dipeptidyl-peptidase 7		*Porphyromonas gingivalis*	H, D, S
		S55		S55.001	SpoIVB peptidase		*Bacillus subtilis*	H, D, S
		S64		S64.001	Ssy5 peptidase	P47002	*Saccharomyces cerevisiae*	H, D, S
		S65		S65.001	Torovirus picornain-like cysteine peptidase	P0C6F4	Breda virus	H, E, S
		S75		S75.001	Bafinivirus serine peptidase		White bream virus	H, D, S
PB	PB(C)	C44		C44.001	Amidophosphoribosyltransferase precursor	Q06203	*Homo sapiens*	C
		C45		C45.001	Acyl-coenzyme A:6-aminopenicillanic acid acyl-transferase precursor	P15802	*Penicillium chrysogenum*	C
		C59		C59.001	Penicillin V acylase precursor	P12256	*Bacillus sphaericus*	C
		C69		C69.001	Dipeptidase A	Q48558	*Lactobacillus helveticus*	C
		C89		C89.001	Acid ceramidase precursor		*Homo sapiens*	C, R, D
PB	PB(S)	S45		S45.001	Penicillin G acylase precursor	P06875	*Escherichia coli*	S
		T1	T1A	T1.002	Archaean proteasome, beta component	P28061	*Thermoplasma acidophilum*	T
			T1B	T1.006	HslV component of HslUV peptidase		*Escherichia coli*	T/S
		T2		T2.001	Glycosylasparaginase precursor	P20933	*Homo sapiens*	T

(continued)

Table 1.2 (continued)

Clan	Subclan	Family	Subfamily	MEROPS identifier	Type example	Accession	Organism	Active site residues	Metal ligands
		T3		T3.001	Gamma-glutamyltransferase 1 (bacterial-type)	P18956	*Escherichia coli*	T/S	
		T6		T6.001	Polycystin-1	P98161	*Homo sapiens*	T	
PC	PC(C)	C26		C26.001	Gamma-glutamyl hydrolase	Q62867	*Rattus norvegicus*	C, H	
		C56		C56.001	PfpI peptidase	Q51732	*Pyrococcus furiosus*	E, C, H	
PC	PC(S)	S51		S51.001	Dipeptidase E	P0A7C6	*Escherichia coli*	S, H, E	
PD	PD(C)	C46		C46.001	Hedgehog protein	Q02936	*Drosophila melanogaster*	C, S/T, H	
PD	PD(N)	N9		N09.001	Intein-containing V-type proton ATPase catalytic subunit A	P17255	*Saccharomyces cerevisiae*	C/S/T, N, C/S/T	
		N10		N10.002	Intein-containing replicative DNA helicase precursor	Q55418	*Synechocystis sp.* PCC 6803	C, N, C/S/T	
		N11		N11.001	Intein-containing chloroplast ATP-dependent endopeptidase	P42379	*Chlamydomonas eugametos*	C, N	
SB		S8	S8A	S08.001	Subtilisin Carlsberg	P00780	*Bacillus licheniformis*	D, H, N, S	
			S8B	S08.070	Kexin	P13134	*Saccharomyces cerevisiae*	D, H, N/D, S	
		S53	S53	S53.001	Sedolisin	P42790	*Pseudomonas sp.* 101	E, D, D, S	
SC		S9	S9A	S09.001	Prolyl oligopeptidase	P23687	*Sus scrofa*	S, D, H	
			S9B	S09.003	Dipeptidyl-peptidase IV (eukaryote)	P27487	*Homo sapiens*	S, D, H	
			S9C	S09.004	Acylaminoacyl-peptidase	P13798	*Homo sapiens*	S, D, H	
			S9D	S09.021	Glutamyl peptidase (plant)		*Arabidopsis thaliana*	S, D, H	
		S10		S10.001	Carboxypeptidase Y	P00729	*Saccharomyces cerevisiae*	S, D, H	
		S15		S15.001	Xaa-Pro dipeptidyl-peptidase	Q02W78	*Lactococcus lactis*	S, D, H	
		S28		S28.001	Lysosomal Pro-Xaa carboxy-peptidase	P42785	*Homo sapiens*	S, D, H	
		S33		S33.001	Prolyl aminopeptidase	P42786	*Neisseria gonorrhoeae*	S, D, H	
		S37		S37.001	PS-10 peptidase		*Streptomyces lividans*	S, D, H	

SE	S11		S11.001	D-Ala-D-Ala carboxypeptidase A	Q05523	*Geobacillus stearothermophilus*	S, K, S
	S12		S12.001	D-Ala-D-Ala carboxypeptidase B	P15555	*Streptomyces lividans*	S, K, Y
	S13		S13.001	D-Ala-D-Ala peptidase C	P24228	*Escherichia coli*	S, K, S
SF	S24		S24.001	Repressor LexA	P0A7C2	*Escherichia coli*	S, K
	S26	S26A	S26.001	Signal peptidase I	P00803	*Escherichia coli*	S, K
		S26B	S26.010	Signalase (animal) 21 kDa component	P15367	*Saccharomyces cerevisiae*	S, H
		S26C	S26.014	TraF peptidase	Q03450	*Escherichia coli*	S, K
SH	S21		S21.002	Cytomegalovirus assemblin	P16753	Human herpesvirus 5	H, S, H
SJ	S16		S16.001	Lon-A peptidase	P0A9M0	*Escherichia coli*	S, K/R
	S50		S50.001	Birnavirus Vp4 peptidase	Q703G9	Infectious pancreatic necrosis virus	S, K
	S69		S69.001	VP4 peptidase		Tellina virus 1	S, K
SK	S14		S14.001	Peptidase Clp (type 1)	Q1RF98	*Escherichia coli*	S, H, D
	S41	S41A	S41.001	C-terminal processing peptidase-1	P23865	*Escherichia coli*	S, K
		S41B	S41.005	Tricorn core peptidase (archaea)	P96086	*Thermoplasma acidophilum*	S, H, S, E
	S49	S49A	S49.001	Signal peptide peptidase A	P08395	*Escherichia coli*	K, S, S
		S49B	S49.003	Protein C	P03711	Bacteriophage lambda	S, S, K
SP	S59		S59.001	Nucleoporin 145		*Homo sapiens*	H, S
SQ	S58		S58.001	DmpA aminopeptidase		*Ochrobactrum anthropi*	S
SR	S60		S60.001	Lactoferrin	P02788	*Homo sapiens*	K, S
SS	S66		S66.001	Murein tetrapeptidase LD-carboxypeptidase		*Pseudomonas aeruginosa*	S, E, H
ST	S54		S54.001	Rhomboid-1 (Diptera)	P20350	*Drosophila melanogaster*	S, H
S-	S48		S48.001	HetR peptidase	Q9F5Y3	*Anabaena variabilis*	S
	S62		S62.001	Influenza A PA peptidase	P21427	Influenza A virus	S
	S63		S63.001	EGF-like module containing mucin-like hormone receptor-like 2	Q9UHX3	*Homo sapiens*	S/T
	S68		S68.001	PIDD auto-processing protein unit 1	Q9HB75	*Homo sapiens*	H, S

(continued)

Table 1.2 (continued)

Clan	Subclan	Family	Subfamily	MEROPS identifier	Type example	Accession	Organism	Active site residues	Metal ligands
		S71		S71.001	MUC1 self-cleaving mucin	P15941	*Homo sapiens*	S	
		S72		S72.001	Dystroglycan		*Homo sapiens*	S	
		S73		S73.001	gpO peptidase	P25478	Enterobacteria phage P2	D, H, S	
		S74		S74.001	Endosialidase CIMCD self-cleaving protein		Enterobacteria phage K1F		
T-		T5		T5.001	Ornithine acetyltransferase precursor	Q04728	*Saccharomyces cerevisiae*	T	
U-		U9		U09.001	Prohead peptidase	P06807	Enterobacteria phage T4		
		U32		U32.001	Collagenase	P33437	*Porphyromonas gingivalis*		
		U35		U35.001	Prohead peptidase	P49860	Bacteriophage HK97		
		U40		U40.001	Protein P5 murein endopeptidase	P07582	Bacteriophage phi-6		
		U49		U49.001	Lit peptidase	P11072	*Escherichia coli*		
		U56		U56.001	Homomultimeric peptidase		*Thermotoga maritima*		
		U57		U57.001	yabG protein	P37548	*Bacillus subtilis*		
		U62		U62.001	Microcin-processing peptidase 1	P0AFK0	*Escherichia coli*		
		U69		U69.001	AIDA-I self-cleaving autotransporter protein	Q03155	*Escherichia coli*		

The *MEROPS* classification at the clan, subclan (where applicable), family, and subfamily (where applicable) levels are shown. For each family or subfamily, the type example is shown, showing the *MEROPS* identifier, source organism, source protein sequence, range of the peptidase unit, active site residues and metal ligands (for metallopeptidases). Where they exist, names of subclans and subfamilies are shown. None of the families listed under A-, C-, M-, S-, T- and U- can be assigned to a clan

An identifier is only created for peptidases that are unique and biochemically characterized. Initially, the identifier is applied to the sequence of a peptidase from a single species, known as the "holotype", but is then extended to encompass what we believe to be the same peptidase in other species. his assignment is done by creating an alignment for each family or subfamily from all the protein sequences of the peptidase units using the MUSCLE software (Edgar 2004), and a phylogenetic tree is generated from this alignment using the UPGMA algorithm implemented in the QUICKTREE software (Howe et al. 2002). Fragments, non-peptidase homologues and pseudogenes are excluded from the alignment and tree. In each tree the holotypes are found, then the *MEROPS* identifier for each holotype is extended to all sequences derived from a common node in the tree provided that no other holotype is included in that set and that the mean percentage difference is not less than 50 %. An identifier is not applied to a sequence if the difference in sequence length compared to that of the holotype is 50 or more residues, so that sequences given the same identifier will possess the same domain architecture. This means that it is rare for a *MEROPS* identifier to be applied to peptidases from more than one superkingdom of organisms.

An exception to the rule that a *MEROPS* identifier is applied only to a biochemically characterized peptidase occurs when identifiers are applied to all the peptidases from the completely sequenced proteome of a very well known, model organism. At present, identifiers have been created for all the peptidases present in human, mouse, *Drosophila melanogaster*, *Caenorhabditis elegans*, *Arabidopsis thaliana*, *Saccharomyces cerevisiae* and *Escherichia coli*. These identifiers take the form fam.A* (or fam.B* in the case of family S1 where more than a hundred such identifiers have been created) so that they can be easily distinguished from the usual identifiers representing characterized peptidases. When the peptidase is eventually characterized, the special *MEROPS* identifier will be replaced with one in the normal form. These identifiers have been created to help us cope with the flood of data derived from complete genome sequences. When the *MEROPS* classification began, almost all homologues could be assigned to an identifier. In the intervening 18 years, more than 2,000 organisms have had their complete genomes sequenced and their complete proteomes predicted, and there are now nearly 149,000 sequences in the *MEROPS* collection representing active peptidases. Nearly 81,000 of these can not be assigned to a *MEROPS* identifier because they are too dissimilar from the sequences of existing holotypes. The special identifiers from the model organisms have been extended to over 7,000 sequences.

1.1.2 Catalytic Mechanisms

The position of the active site residues (catalytic residues and/or metal ligands) is usually conserved between homologous structures. In the case of methionyl aminopeptidase (family M24), which is derived from an ancient gene duplication and subsequent fusion, one half of the molecule is structurally similar to the other half, even though the sequence similarity is undetectable, and the positions of the metal ligands in one half of the molecule are equivalent to the positions in the other half (Roderick and Matthews 1988). Because every cellular organism possesses methionyl aminopeptidase (which removes the initiating methionine from some cytoplasmic proteins) and these are not only all homologous but all possess the same quasi-twofold symmetry, the gene duplication/fusion event must be very ancient indeed, probably dating back to the universal ancestor.

1.1.2.1 Serine Peptidases

There may be similarities between the catalytic mechanisms of peptidases from one clan and those from another. In most serine peptidases there is a catalytic triad consisting of a serine, which bears the nucleophilic hydroxyl group, a histidine, which acts as the general base, and an aspartate which is believed to orientate the imidazolium ring of the histidine so that the Ser and His are close and the hydroxyl group is activated. This mechanism is seen in peptidases as diverse as trypsin [family S1, subclan PA(S)], subtilisin (family S8, clan SB) and carboxypeptidase Y (family S10, clan SC). Even though the order of residues in the sequence differs between one family and another, the active site geometries are so similar that they can be superimposed (Liao et al. 1992). In subclan PA(S) the order is His, Asp and Ser; in clan SB it is Asp, His and Ser; and in clan SC it is Ser, Asp and His. In all, there are 30 families of serine peptidases with a Ser/His/Asp catalytic triad. In ten families, lysine can replace histidine as the general base, in which case a catalytic dyad is all that may be required. Examples of serine peptidases with an active site lysine include the D-Ala-D-Ala carboxypeptidases (clan SE), signal peptidase 1 (family S26, clan SF), Lon-A peptidase (family S16, clan SJ) and signal peptide peptidase A (family S49, clan SK). Members of clan SE employ a catalytic triad in which the third residue is either another serine or a tyrosine.

1.1.2.2 Threonine Peptidases

Threonine peptidases are all presumed to have an N-terminal threonine bearing both the nucleophile (the hydroxyl on the side chain) and the general base (the amino group). There are also N-terminal nucleophile (Ntn) hydrolases in which the N-terminus is a serine or cysteine (when the thiol group is the nucleophile), including families C44 (glutamine PRPP amidotransferase), C45 (acyl-coenzyme A:6-aminopenicillanic-acid-acyltransferase precursor), C59 (penicillin V acylase), C69 (dipeptidase A from *Lactobacillus*), C89 (acid ceramidase), S45 (penicillin acylase) and S58 (DmpA aminopeptidase); all of these families except S58 are structurally related and are members of clan PB, which also includes the proteasome (family T1). The structure for family S58 has been resolved, and shown to be unique amongst peptidases (Cheng and Grishin 2005). Many of the Ntn-hydrolases are enzymes other than peptidases, and peptidase activity is restricted to a single processing event to release the propeptide from the precursor. Releasing the propeptide exposes the new nucleophilic N-terminal residue, and it is assumed that it is this same residue that attacks the preceding peptide bond in the precursor. Processing has been prevented by site-directed mutation of the assumed catalytic residue in DmpA aminopeptidase (family S58) (Fanuel et al. 1999) and gamma-glutamyltransferase (family T3) (Inoue et al. 2000). Some homologues, such as asparagine synthase from *Escherchia coli* (UniProt Q84DG0), do not possess a propeptide and activation merely requires removal of the N-terminal initiating methionine (by a methionyl aminopeptidase); these homologues are therefore not peptidases, even though the catalytic residue is conserved.

1.1.2.3 Cysteine Peptidases

Seventy-two families of cysteine peptidases have an essential Cys/His dyad in which the cysteine bears the nucleophilic thiol and the histidine acts as a general base. Peptidases from clan CD have only the catalytic dyad. A third residue may be involved which again is thought to be important for positioning the histidine ring. This third residue may be asparagine (as it is in papain and calpain), aspartate (as it is in adenain and some de-ubiquitinating enzymes) or glutamate (as it is in pyroglutamyl-peptidase I). A fourth residue may also be important for stabilizing the acyl intermediate by helping to form the oxyanion hole. This fourth residue can be a glutamine (as it is in papain and calpain), an asparagine (as it is in some deubiquitinating enzymes and the L-peptidase from foot-and-mouth disease virus), or a tyrosine (autophagins and UfSP1 peptidase). In members of clans CA, CF, CN and CO the order of catalytic dyad is Cys, His, whereas in clans CD, CE, CL and CM it is His, Cys.

1.1.2.4 Aspartic Peptidases

Aspartic peptidases are all endopeptidases. Water is bound by two aspartates in most members of the known families, although in some families the active peptidase is a homodimer with each monomer providing one of the aspartates. Clan AA includes the families of pepsin (A1) and the HIV-1 retropepsin (A2). The structure of pepsin clearly shows an ancient gene duplication/fusion event (Tang et al. 1978), as has been observed in chymopapain and methionyl aminopeptidase, with each subdomain bearing one of the active site aspartates. The structure of retropepsin, however, represents a single pepsin subdomain, and so the peptidase is only active as a homodimer (Navia et al. 1989). There has been considerable debate about which architecture represents the ancestral state: the retropepsin-like single domain or the pepsin like two-domain. Until recently, bacterial homologues in family A1 were unknown, but this is no longer the case (Rawlings and Bateman 2009); also several families of peptidases from eukaryotes and bacteria have been discovered that are similar to the retropepsin single-domain architecture, notably families A28

(DNA-damage inducible protein DDI1) (Sirkis et al. 2006) and A32 (Chen et al. 2006). Although this has not resolved the argument, it has pushed back the time at which divergence occurred to before the universal ancestor of eukaryotes and bacteria. Retropepsin-like peptidases are not the only aspartic peptidases that are active only as dimers. The histoaspartic peptidase from *Plasmodium falciparum* is exceptional in that one of the pair of aspartics is not conserved and replaced by a histidine. Unusually, this histidine is not essential for catalytic activity (Parr et al. 2008), in the apoenzyme binds a zinc ion and to be active the peptidase must dimerize (Bhaumik et al. 2009).

Aspartic peptidases have a predominantly acidic pH optimum, but there are exceptions, such as renin in family A1 and retropepsins in family A2 and membrane-bound peptidases such as the presenilins (family A22 subfamily A), signal peptide peptidases (family A22 subfamily B) and omptin (family A26).

1.1.2.5 Glutamic Peptidases

There are only two families of glutamic peptidases. Peptidases in family G1 have a catalytic dyad consisting of the nucleophilic glutamate and a glutamine, which provides electrophilic assistance and oxyanion stabilisation (Fujinaga et al. 2004). Peptidases in family G2 are precursors that act autolytically, which in the case of bacteriophage Phi29 leads to the release of the tailspike protein. The catalytic dyad consists of an aspartate and a glutamate, in which the glutamate activates a water molecule (Schulz and Ficner 2011).

1.1.2.6 Metallopeptidases

The metal ion in a metallopeptidase is usually zinc, but methionyl aminopeptidase 1 (family M24) binds cobalt (Roderick and Matthews 1993), aminopeptidase P (family M24) binds manganese (Cottrell et al. 2000), and the VMP3 peptidase from *Volvox* apparently binds copper (Heitzer and Hallmann 2002). Some metallopeptidases bind two metal ions, both of which are important for catalysis ("co-catalytic"). Peptidases with two metal ions are restricted to clans MF, MG, MH, MN and MQ. Each metal ion is tetrahedrally co-ordinated by three amino acid side chain ligands and a water molecule. In peptidases with co-catalytic metal ions, one amino acid binds both metals, so there are five amino acid ligands. Ligands are usually histidines, glutamates or aspartates, but a lysine is a metal ligand in leucyl aminopeptidase (Burley et al. 1992). The HmrA peptidase from *Staphylococcus aureus* (family M20) has been shown to use cysteine as a zinc ligand (Botelho et al. 2011), which previously had only been known as a ligand for structural metal ions.

The sequence motif His-Glu-Xaa-Xaa-His (HEEXH) includes two of the three metal ligands and is found in members of clans MA and MM. The motif is also found, but rarely conserved, in other proteins, so presence of the motif alone does not imply the protein is a metallopeptidase. In clan MA the third metal ligand is a glutamate as it is in thermolysin, or a histidine or aspartate in the extended motif HEXXHXXGXX(H,D), which occurs in matrix metallopeptidases, serralysins, astacins and reprolysins. Family M50 is the only family in clan MM and includes integral membrane proteins such as the S2P protease. The tertiary structure for a homologue from *Methanocaldococcus jannaschii* has been solved (Feng et al. 2007), showing that the structure is different to that of members of clan MA. Variations of this HEXXH motif occur in members of clan ME (which includes pitrilysin), where the motif is HXXEH (Finch et al. 1986), and clan MK (which includes sialoglycoprotease), where the motif is HXEXH (Abdullah et al. 1991), though in this clan the motif is poorly conserved and may not represent the metal binding site.

1.1.3 Comparisons of Classifications

Comparing the different classification systems shows that the major distinction between an exopeptidase and an endopeptidase is not relevant in terms of peptidase chemistry or evolution. Although most aminopeptidases are of metallo-type, for example membrane alanyl aminopeptidase (family M1), leucyl aminopeptidase (M17) and methionyl aminopeptidase (M24), there are also cysteine (e.g. aminopeptidase C, family C1) or serine type (e.g. aminopeptidase DmpB, family S12; prolyl aminopeptidase, family S33; DmpA aminopeptidase, family S58). Similarly, carboxypeptidases are principally of metallo-type (e.g. carboxypeptidase A, family M14; D-Ala-D-Ala carboxypeptidases, family M15; glutamate carboxypeptidase, family M20; carboxypeptidase Taq, family M32), but those of serine-type exist (e.g. carboxypeptidase Y, family S10; D-Ala-D-Ala carboxypeptidases, families S11, S12 and S13; lysosomal Pro-Xaa carboxypeptidase, family S28). Dipeptidases are mainly of metallo type (e.g. vanX D-Ala-D-Ala dipeptidase, family M15; membrane dipeptidase, family M19; carnosine dipeptidase I, family M20; Xaa-Pro dipeptidase, family M24; isoaspartyl dipeptidase, family M38) but can be cysteine (dipeptidase A, family C69), serine (dipeptidase E, family S51) or threonine-type (isoaspartyl dipeptidase, family T2). Dipeptidyl-peptidases can be of cysteine (e.g. dipeptidyl-peptidase I, family C1), metallo (e.g. dipeptidyl-peptidase III, family M49), or serine type (e.g. dipeptidyl-peptidase IV, family S9; Xaa-Pro dipeptidyl-peptidase, family S15; dipeptidyl-peptidase II, family S28; dipeptidyl-peptidase 7, family S46). Peptidyl-dipeptidases, however, are only known to be metalloenzymes (e.g. angiotensin-converting enzyme, family M2; peptidyl-dipeptidase Dcp, family M3).

The distinction between endo- and exopeptidases also does not apply at the family level. For example, the papain family (C1) includes mainly endopeptidases, but there are aminopeptidases, such as bleomycin hydrolase and aminopeptidase C, and a dipeptidyl-peptidase. Some peptidases in the family have both endo- and exopeptidases activities: cathepsin B can act as a peptidyl-dipeptidase (Musil et al. 1991) and cathepsin H can act as an aminopeptidase (Kirschke et al. 1977). Family S12 includes peptidases involved in processing the crosslinker peptide of bacterial cell walls or its precursor, which can be either carboxypeptidases or aminopeptidases. Family M20 includes carboxypeptidases (e.g. glutamate carboxypeptidase), dipeptidases (e.g. carnosine dipeptidase I) as well as an endopeptidase (HMRA peptidase [11]). Similarly, family M28 contains a mixture of carboxypeptidases (e.g. glutamate carboxypeptidase II) and aminopeptidases (aminopeptidase Y). Family M24 includes aminopeptidase (such as methionyl aminopeptidase), dipeptidases (such as Xaa-pro dipeptidase) and aminopeptidase P.

Endopeptidases can be of any catalytic type. The following examples show that functional simiarities exist between peptidases from different families, clans and catalytic types.

1.1.4 Integral Membrane Peptidases

Many peptidases are associated with cell membranes. The attachment can be due to an N-terminal membrane-spanning region, a C-terminal membrane-spanning region, or a looser association with the membrane via a C-terminal glycosylphosphatidinylinositol moiety. In addition, there are a number of peptidases that have multiple transmembrane regions that are integral membrane proteins, with the active site either within the membrane or close to the surface of the membrane. An active site within a membrane is unusual, because hydrolysis occurs in an aqueous environment and membranes are hydrophobic. From solved tertiary structures, integral membrane peptidases create a pocket within the membrane that is water-filled and where hydrolysis can occur. There are several peptidases with multiple transmembrane regions, however, very few of these have the active site within the membrane. A list of peptidases with multiple transmembrane domains is shown in Table 1.3. Obtaining crystals of these proteins is difficult because of their location in the membrane, but structures have been solved for omptin (Vandeputte-Rutten et al. 2001), DapE peptidase (Badger et al. 2005), FtsH peptidase (Suno et al. 2006), HtpX peptidase (PDB:3CQB, unpublished), the MJ0392 protein from *Methanocaldococcus jannaschii* (an M50B homologue) (Feng et al. 2007), leader peptidase (Paetzel et al. 1998), signal peptide peptidase A (Kim et al. 2008), and GlpG peptidase (an S54 homologue) (Wang et al. 2006). Only the peptidase domain has been solved for ComA protein from *Streptococcus mutans* (a C39 homologue) (Ishii et al. 2010). Crystal structures are known for homologues of families S8 and M24, but not for membrane-associated proteins. It can be seen from Table 1.3 that integral membrane proteins can be of almost any catalytic type.

Table 1.3 Peptidases with multiple-pass transmembrane regions

Peptidase	Family	Location of active site	UniProt
Signal peptidase II	A8	Periplasmic	P00804
Presenilin	A22A	Transmembrane	P49768
Signal peptide peptidase	A22B	Transmembrane	Q8TCT7
Type 4 prepilin peptidase 1	A24A	Periplasmic	P25960
Preflagellin peptidase	A24B	Extracellular	Q58312
Omptin	A26	Extracellular	P09169
Sporulation factor SpoIIGA	A36	Cytoplasmic	P13801
Bacteriocin-processing peptidase	C39	Extracellular	P36497
AgrB peptidase	C75	Transmembrane?	P0C1P7
PH0974 dipeptidase	M24B	Cytoplasmic	P65810
FtsH peptidase	M41	Cytoplasmic	P0AAI3
Ste24	M48A	Cytoplasmic	P47154
HtpX peptidases	M48B	Cytoplasmic	P23894
S2P protease	M50A	Transmembrane	043462
Sporulation factor SpoIVFB	M50B	Transmembrane	P26937
RseP peptidase	M50B	Transmembrane	051145
BlaR1 peptidase	M56	Cytoplasmic	P12287
CAAX prenyl protease 2	M79	Cytoplasmic	Q03530
TagC peptidase (*Dictyostelium discoideum*)	S8A	Extracellular	Q23868
Leader peptidase	S26A	Periplasmic	P00803
Signal peptidase SipW (*Bacillus*)	S26B	Periplasmic	P54506
Signal peptide peptidase A	S49A	Periplasmic	P08395
Rhomboids	S54	Transmembrane	P20350
Lit peptidase (*Escherichia coli*)	U49	Unknown	P11072

The name of the peptidase, the *MEROPS* family to which is belongs and the cellular location of the active site are shown. The location of the active site has been derived from the known or predicted active site residues and the cellular topology as shown in the relevant UniProt entry

1.1.5 Self-Cleaving Proteins

Most proteins undergo some form of processing to generate the mature form. This may be as simple as removing the initiating methionine or a signal or other targeting peptide. Many proteins require activation by proteolysis, to remove an N- or C-terminal propeptide, or to convert a single chain protein into multiple chains. Although most of these processing events are performed by enzymes, some proteins are capable of processing themselves.

Viral polyproteins have to be processed to release the individual proteins, and very often the peptidases responsible for this are part of the polyprotein. Perhaps the best known of these is the HIV retropepsin, which releases itself from the polyprotein and then processes it at at least nine sites (Dunn and Rao 2004). Picornain 2A (family C3) from entero- and rhinoviruses, part of the *pol* polyprotein, is able to release itself and then inactivate host proteins by proteolysis, such as components of the eukaryotic translation initiation factor 4 gamma (Gradi et al. 2003, 2004). However, there are some viral peptidases that do no more than release themselves, and have no further proteolytic activity. An example is togavirin (family S3) from the Sindbis virus, which becomes a component of the mature virus particle once it has released itself. The single cleavage happens intramolecularly in *cis*, whereas usually proteolysis occurs in *trans*—intermolecularly where an enzyme molecule cleaves a substrate molecule. Further proteolytic activity is impossible because the C-terminal Trp remains in the P1 subsite of the enzyme. The tertiary structure shows that togavirin is structurally related to trypsin (Tong et al. 1993).

Both bacterial and eukaryote proteins are able to process themselves, and after processing cease to be peptidases. Processing can happen in *cis* or in *trans*. Peptidase families containing proteins that are known only to process their own precursors are C46 (hedgehog protein precursor), C56 (protein DJ-1), C59 (penicillin V acylase precursor), C80 (*Clostridium* toxins), G2 (bacteriophage Phi29 tailspike protein), S24 (repressor LexA), S63 (EGF-like module containing mucin-like hormone receptor-like 2), S68 (PIDD auto-processing protein), S71 (mucin 1), T2 (glycosylasparaginase precursor); T3 (gamma-glutamyltransferase 1), T5 (ornithine acetyltransferase precursor) and T6 (polycystin-1 precursor). Proteins that only cleave themselves are of several catalytic types, but mostly those with protein nucleophiles. The bacteriophage Phi29 tailspike protein precursor is the only known self-cleaving protein to use activated water as the nucleophile; there are no aspartic or metallopeptidases that are only self-cleaving.

1.1.6 Relationships to Other Hydrolases

It must also be borne in mind that some peptidases have the same tertiary structure as other hydrolytic enzymes. Peptidase clan SC contains six families, including that of prolyl oligopeptidase (S9), carboxypeptidase Y (S10) and prolyl aminopeptidase (S33), and members of the clan possess a fold known as the alpha/beta hydrolase fold. There are a large number of enzymes besides peptidases that share this fold, and it is clear that an architecture that provides the environment for an active site can be utilised for many different catalytic purposes (Holmquist 2000). A bizarre example of this multiple use of an active site applies to the isoaspartyl dipeptidase from *Escherichia coli* (family M38, clan MJ). There are only two peptidase in clan MJ which includes enzymes such as dihydro-orotases, and dihydropyrimindases (Jozic et al. 2003).

1.2 Other Proteolytic Enzymes

Mechanisms other than hydrolysis exist by which a peptide bond can be broken. The enzymes responsible are lyases and transferases.

There are a number of families of self-cleaving proteins that utilize a mechanism peculiar to asparagine. Under the right conditions, asparagine can be induced to form a succinimide ring. If this occurs to an asparagine within a peptide, then the peptide bond is broken when the succinimide ring forms, releasing a C-terminal peptide. The asparagine is nucleophilic, representing a seventh catalytic type. Ten families of these asparagine peptide lyases have been discovered so far, named N1–N11 (see Table 1.2).

Gram-negative bacteria have special mechanisms for the secretion of proteins through the outer cell membrane. The tsh protein is synthesized with a large C-terminal domain known as an autotransporter that forms a pore in the membrane. The N-terminal domain is threaded through the pore. When an acidic residue and an asparagine are in close contact, the asparagine forms a succinimide, the peptide bond is broken, and the N-terminal domain is released to the surrounding medium (Tajima et al. 2010). The autotransporter domain of the tsh protein is the type example for family N4.

Several viral proteins also undergo processing at an asparagine residue. The purpose of this cleavage is not known for the capsid proteins from *Nodavirus* (family N1), *Tetravirus* (N2), *Picobirnavirus* (N5) and *Poliovirus* (N8), but in *Reovirus* (N7) the released peptide is myristoylated at the N-terminus and migrates to the host erythrocyte cell membrane and contributes to pore formation and ultimately penetration by viral particles (Zhang et al. 2009).

A third group of proteins that undergo processing at asparagine residues are proteins containing inteins. An intein is derived from a piece of parasitic DNA that encodes for an endonuclease. The endonuclease cleaves the host DNA at a very specific site: the site where the intein DNA will be inserted. This occurs within a protein coding gene, within the protein coding region so that the "host" protein is split into two pieces, each known as an extein. When the protein is synthesized, the intein can extract itself and then splice together the exteins to form a normal, functional protein. The intein is then free to act as an endonuclease. Two cleavage events are required to release the intein, neither of which is a hydrolytic event. The first cleavage occurs at the end of the first extein domain and the N-terminus of the intein. The first residue of the intein and the first residue of the second extein is either a serine, cysteine or threonine. The peptide bond between the first extein and the intein is rearranged to form a linear ester or thioester (depending on the nature of the first residue of the intein), then a transesterification reaction transfers the first extein to the first residue of the second extein, forming a branched intermediate with two N-termini and a single C-terminus. This intermediate is unstable. The asparagine at the end of the intein cyclises, excising the intein. The ester or thioester bond linking the extein domains undergoes an acyl rearrangement to form a normal peptide bond, thus splicing the extein domains together (Raghavan and Minnick 2009). The first cleavage is the action of a transaminase (*Enzyme Nomenclature* subclass EC 2.6), the second the action of a lyase. Three families of intein-containing proteins have so far been recognized (see Table 1.2).

Another unusual proteolytic enzyme is hedgehog precursor (family C46), which is also autocatalytic. The N-terminal effector domain when released is important for normal, segmental development in the *Drosophila* embryo (mutation causes the hedgehog phenotype). The structure of this domain has been solved and shows surprising similarity to metallo-type D-Ala-D-Ala carboxypeptidases (family M15), including the zinc-binding motif, but it is not known to be a peptidase. The structure of the C-terminal processing domain has also been solved, and this show similarity to the intein domains mentioned above (Hall et al. 1997). The N-terminal residue of the autoprocessing protein is a cysteine, which is essential for autoprocessing. The released hedgehog domain has an essential cholesterol moiety attached to the C-terminus which happens during autolysis (Mann and Beachy 2000). Cholesterol acts as a second nucleophile attacking the thioester bond formed between the C-terminal glycine of the hedgehog domain and the N-terminal cysteine of the autoprocessing domain during the first stage of the processing process (Hall et al. 1997). Thus, processing in the hedgehog protein precursor is also not a hydrolytic reaction.

In the biosynthesis of the hormone α-melanotropin, peptidylglycine monooxygenase (EC 1.14.17.3) converts the C-terminal glycine to glyoxylate, which is then released by peptidylamidoglycolate lyase (EC 4.3.2.5), leaving a peptide amide. The final reaction is the action of a lyase, acting on a carbon-nitrogen bond eliminating ammonia. Both activities are performed by the same protein, known as peptidyl-glycine alpha-amidating monooxygenase. This protein has two domains, the N-terminal one being the monooxygenase and the C-terminal domain being the lyase (Katopodis et al. 1991). It is only the combined operation of both domains that gives the appearance of a proteolytic reaction, consequently this enzyme is not included in the *MEROPS* classification.

1.3 Conclusions

Although the majority of known proteolytic enzymes are peptidases, there are a number of autolytic cleavages which do not require hydrolysis, and are therefore not peptidases. There are at least seven different catalytic types of proteolytic enzymes, where the nucleophile is either a water hydroxyl group attached to an aspartate or glutamate residue or to one or two metal ions; is the thiol or hydroxyl group of a cysteine, serine or threonine side chain; or cleavage is brought about by cyclization of an asparagine residue to form a succinimide. Proteolytic enzymes with an asparagine nucleophile are not peptidases. Peptidases can be functionally grouped into exopeptidases and endopeptidases, and exopeptidases can be further subdivided depending upon which terminus of the protein or peptide is being attacked and the number of residues released during a single cleavage event. However, because enzymes of the same catalytic type can be unrelated to one another and closely related enzymes can have very different specificities, classification by catalytic type or specificity bears little if any relationship to the evolution of the enzymes. Classification by sequence and structural homology better reflects the evolutionary relationships and origins of proteolytic enzymes. The classification and nomenclature of proteolytic enzymes is maintained in the *MEROPS* database (http://merops.sanger.ac.uk) (Rawlings et al. 2010).

Acknowledgements This work was supported by the Wellcome Trust [grant number WT077044/Z/05/Z].

References

Abdullah KM, Lo RY, Mellors A (1991) Cloning, nucleotide sequence, and expression of the Pasteurella haemolytica A1 glycoprotease gene. J Bacteriol 173:5597–5603

Altschul SF, Gish W, Miller W, Myers EW, Lipman DJ (1990) Basic local alignment search tool. J Mol Biol 215:403–410

Altschul SF, Madden TL, Schaffer AA, Zhang J, Zhang Z, Miller W, Lipman DJ (1997) Gapped BLAST and PSI-BLAST: a new generation of protein database search programs. Nucleic Acids Res 25:3389–3402

Anonymous (2010) The universal protein resource (UniProt) in 2010. Nucleic Acids Res 38: D142–D148

Artymiuk PJ, Grindley HM, Park JE, Rice DW, Willett P (1992) Three-dimensional structural resemblance between leucine aminopeptidase and carboxypeptidase A revealed by graph-theoretical techniques. FEBS Lett 303:48–52

Bacon D, Anderson WF (1988) A fast algorithm for rendering space-filling molecule pictures. J Mol Graph 6:219–220

Badger J, Sauder JM, Adams JM, Antonysamy S, Bain K, Bergseid MG, Buchanan SG, Buchanan MD, Batiyenko Y, Christopher JA, Emtage S, Eroshkina A, Feil I, Furlong EB, Gajiwala KS, Gao X, He D, Hendle J, Huber A, Hoda K, Kearins P, Kissinger C, Laubert B, Lewis HA, Lin J, Loomis K, Lorimer D, Louie G, Maletic M, Marsh CD, Miller I, Molinari J, Muller-Dieckmann HJ, Newman JM, Noland BW, Pagarigan B, Park F, Peat TS, Post KW, Radojicic S, Ramos A, Romero R, Rutter ME, Sanderson WE, Schwinn KD, Tresser J, Winhoven J, Wright TA, Wu L, Xu J, Harris TJ (2005) Structural analysis of a set of proteins resulting from a bacterial genomics project. Proteins 60:787–796

Bajorath J, Hinrichs W, Saenger W (1988) The enzymatic activity of proteinase K is controlled by calcium. Eur J Biochem 176:441–447

Bazan JF, Fletterick RJ (1988) Viral cysteine proteases are homologous to the trypsin-like family of serine proteases: structural and functional implications. Proc Natl Acad Sci U S A 85:7872–7876

Bhaumik P, Xiao H, Parr CL, Kiso Y, Gustchina A, Yada RY, Wlodawer A (2009) Crystal structures of the histo-aspartic protease (HAP) from Plasmodium falciparum. J Mol Biol 388:520–540

Botelho TO, Guevara T, Marrero A, Arede P, Fluxa VS, Reymond JL, Oliveira DC, Gomis-Ruth FX (2011) Structural and functional analyses reveal that Staphylococcus aureus antibiotic resistance factor HmrA is a zinc-dependent endopeptidase. J Biol Chem 286:25697–25709

Burley SK, David PR, Sweet RM, Taylor A, Lipscomb WN (1992) Structure determination and refinement of bovine lens leucine aminopeptidase and its complex with bestatin. J Mol Biol 224:113–140

Chen JC, Hottes AK, McAdams HH, McGrath PT, Viollier PH, Shapiro L (2006) Cytokinesis signals truncation of the PodJ polarity factor by a cell cycle-regulated protease. EMBO J 25:377–386

Cheng H, Grishin NV (2005) DOM-fold: a structure with crossing loops found in DmpA, ornithine acetyltransferase, and molybdenum cofactor-binding domain. Protein Sci 14:1902–1910

Cottrell GS, Hooper NM, Turner AJ (2000) Cloning, expression, and characterization of human cytosolic aminopeptidase P: a single manganese(II)-dependent enzyme. Biochemistry 39:15121–15128

Ding J, McGrath WJ, Sweet RM, Mangel WF (1996) Crystal structure of the human adenovirus proteinase with its 11 amino acid cofactor. EMBO J 15:1778–1783

Dunn BM, Rao M (eds) (2004) Human immunodeficiency virus 1 retropepsin. Elsevier, London

Edgar RC (2004) MUSCLE: a multiple sequence alignment method with reduced time and space complexity. BMC Bioinformatics 5:113

Fanuel L, Goffin C, Cheggour A, Devreese B, Van Driessche G, Joris B, Van Beeumen J, Frere JM (1999) The DmpA aminopeptidase from Ochrobactrum anthropi LMG7991 is the prototype of a new terminal nucleophile hydrolase family. Biochem J 341(Pt 1):147–155

Feng L, Yan H, Wu Z, Yan N, Wang Z, Jeffrey PD, Shi Y (2007) Structure of a site-2 protease family intramembrane metalloprotease. Science 318:1608–1612

Finch PW, Wilson RE, Brown K, Hickson ID, Emmerson PT (1986) Complete nucleotide sequence of the Escherichia coli ptr gene encoding protease III. Nucleic Acids Res 14:7695–7703

Fischer MH (1907) Physiology of alimentation. John Wiley, London

Foster M (1877) Text Bk Physiol ii iv, 276 (Anonymous)

Fujinaga M, Cherney MM, Oyama H, Oda K, James MN (2004) The molecular structure and catalytic mechanism of a novel carboxyl peptidase from Scytalidium lignicolum. Proc Natl Acad Sci U S A 101:3364–3369

Gradi A, Svitkin YV, Sommergruber W, Imataka H, Morino S, Skern T, Sonenberg N (2003) Human rhinovirus 2A proteinase cleavage sites in eukaryotic initiation factors (eIF) 4GI and eIF4GII are different. J Virol 77:5026–5029

Gradi A, Foeger N, Strong R, Svitkin YV, Sonenberg N, Skern T, Belsham GJ (2004) Cleavage of eukaryotic translation initiation factor 4GII within foot-and-mouth disease virus-infected cells: identification of the L-protease cleavage site in vitro. J Virol 78:3271–3278

Hall TM, Porter JA, Young KE, Koonin EV, Beachy PA, Leahy DJ (1997) Crystal structure of a Hedgehog autoprocessing domain: homology between Hedgehog and self-splicing proteins. Cell 91:85–97

Heitzer M, Hallmann A (2002) An extracellular matrix-localized metalloproteinase with an exceptional QEXXH metal binding site prefers copper for catalytic activity. J Biol Chem 277:28280–28286

Holm L, Sander C (1995) Dali: a network tool for protein structure comparison. Trends Biochem Sci 20:478–480

Holmquist M (2000) Alpha/Beta-hydrolase fold enzymes: structures, functions and mechanisms. Curr Protein Pept Sci 1:209–235

Howe K, Bateman A, Durbin R (2002) QuickTree: building huge Neighbour-Joining trees of protein sequences. Bioinformatics 18:1546–1547

Inoue M, Hiratake J, Suzuki H, Kumagai H, Sakata K (2000) Identification of catalytic nucleophile of Escherichia coli gamma-glutamyltranspeptidase by gamma-monofluorophosphono derivative of glutamic acid: N-terminal thr-391 in small subunit is the nucleophile. Biochemistry 39:7764–7771

Ishii S, Yano T, Ebihara A, Okamoto A, Manzoku M, Hayashi H (2010) Crystal structure of the peptidase domain of Streptococcus ComA, a bifunctional ATP-binding cassette transporter involved in the quorum-sensing pathway. J Biol Chem 285:10777–10785

Jozic D, Kaiser JT, Huber R, Bode W, Maskos K (2003) X-ray structure of isoaspartyl dipeptidase from E.coli: a dinuclear zinc peptidase evolved from amidohydrolases. J Mol Biol 332:243–256

Katopodis AG, Ping DS, Smith CE, May SW (1991) Functional and structural characterization of peptidylamidoglycolate lyase, the enzyme catalyzing the second step in peptide amidation. Biochemistry 30:6189–6194

Kim AC, Oliver DC, Paetzel M (2008) Crystal structure of a bacterial signal Peptide peptidase. J Mol Biol 376:352–366
Kirschke H, Langner J, Wiederanders B, Ansorge S, Bohley P, Hanson H (1977) Cathepsin H: an endoaminopeptidase from rat liver lysosomes. Acta Biol Med Ger 36:185–199
Komander D (2010) Mechanism, specificity and structure of the deubiquitinases. Subcell Biochem 54:69–87
Kraulis PJ (1991) MOLSCRIPT: a program to produce both detailed and schematic plots of protein structures. J Appl Crystallogr 24:946–950
Liao DI, Breddam K, Sweet RM, Bullock T, Remington SJ (1992) Refined atomic model of wheat serine carboxypeptidase II at 2.2-A resolution. Biochemistry 31:9796–9812
Lipman DJ, Pearson WR (1985) Rapid and sensitive protein similarity searches. Science 227:1435–1441
Mann RK, Beachy PA (2000) Cholesterol modification of proteins. Biochim Biophys Acta 1529:188–202
Musil D, Zucic D, Turk D, Engh RA, Mayr I, Huber R, Popovic T, Turk V, Towatari T, Katunuma N et al (1991) The refined 2.15 A X-ray crystal structure of human liver cathepsin B: the structural basis for its specificity. EMBO J 10:2321–2330
Navia MA, Fitzgerald PM, McKeever BM, Leu CT, Heimbach JC, Herber WK, Sigal IS, Darke PL, Springer JP (1989) Three-dimensional structure of aspartyl protease from human immunodeficiency virus HIV-1. Nature 337:615–620
Paetzel M, Dalbey RE, Strynadka NC (1998) Crystal structure of a bacterial signal peptidase in complex with a beta-lactam inhibitor. Nature 396:186–190
Parr CL, Tanaka T, Xiao H, Yada RY (2008) The catalytic significance of the proposed active site residues in Plasmodium falciparum histoaspartic protease. FEBS J 275:1698–1707
Petersen W, Short CA (1918) On the relation of the serum ereptase (peptidase) titer to the clinical course in pneumonia. J Infect Dis 191:147–153
Punta M, Coggill PC, Eberhardt RY, Mistry J, Tate J, Boursnell C, Pang N, Forslund K, Ceric G, Clements J, Heger A, Holm L, Sonnhammer EL, Eddy SR, Bateman A, Finn RD (2012) The Pfam protein families database. Nucleic Acids Res 40:D290–D301
Raghavan R, Minnick MF (2009) Group I introns and inteins: disparate origins but convergent parasitic strategies. J Bacteriol 191:6193–6202
Rawlings ND, Barrett AJ (1993) Evolutionary families of peptidases. Biochem J 290:205–218
Rawlings ND, Bateman A (2009) Pepsin homologues in bacteria. BMC Genomics 10:437
Rawlings ND, Barrett AJ, Bateman A (2010) MEROPS: the peptidase database. Nucleic Acids Res 38:D227–D233
Ray K, Hines CS, Coll-Rodriguez J, Rodgers DW (2004) Crystal structure of human thimet oligopeptidase provides insight into substrate recognition, regulation, and localization. J Biol Chem 279:20480–20489
Richardson JS (1985) Schematic drawings of protein structures. Methods Enzymol 115:359–380
Roberts W (1880) The Lumleian lectures on the digestive ferments, and the preparation and use of artificially digested food. Br Med J 1880:575–577
Roderick SL, Matthews BW (1988) Crystallization of methionine aminopeptidase from Escherichia coli. J Biol Chem 263:16531
Roderick SL, Matthews BW (1993) Structure of the cobalt-dependent methionine aminopeptidase from Escherichia coli: a new type of proteolytic enzyme. Biochemistry 32:3907–3912
Sayers EW, Barrett T, Benson DA, Bolton E, Bryant SH, Canese K, Chetvernin V, Church DM, DiCuccio M, Federhen S, Feolo M, Fingerman IM, Geer LY, Helmberg W, Kapustin Y, Landsman D, Lipman DJ, Lu Z, Madden TL, Madej T, Maglott DR, Marchler-Bauer A, Miller V, Mizrachi I, Ostell J, Panchenko A, Phan L, Pruitt KD, Schuler GD, Sequeira E, Sherry ST, Shumway M, Sirotkin K, Slotta D, Souvorov A, Starchenko G, Tatusova TA, Wagner L, Wang Y, Wilbur WJ, Yaschenko E, Ye J (2011) Database resources of the National Center for Biotechnology Information. Nucleic Acids Res 39:D38–D51

Schechter I, Berger A (1968) On the active site of proteases. 3. Mapping the active site of papain; specific peptide inhibitors of papain. Biochem Biophys Res Commun 32:898–902

Schilling O, Huesgen PF, Barre O, Auf dem Keller U, Overall CM (2011) Characterization of the prime and non-prime active site specificities of proteases by proteome-derived peptide libraries and tandem mass spectrometry. Nat Protoc 6:111–120

Schulz EC, Ficner R (2011) Knitting and snipping: chaperones in beta-helix folding. Curr Opin Struct Biol 21:232–239

Sirkis R, Gerst JE, Fass D (2006) Ddi1, a eukaryotic protein with the retroviral protease fold. J Mol Biol 364:376–387

Strauch KL, Miller CG (1983) Isolation and characterization Salmonella typhimurium mutants lacking a tripeptidase (peptidase T). J Bacteriol 154:763–771

Suno R, Niwa H, Tsuchiya D, Zhang X, Yoshida M, Morikawa K (2006) Structure of the whole cytosolic region of ATP-dependent protease FtsH. Mol Cell 22:575–585

Tajima N, Kawai F, Park SY, Tame JR (2010) A novel intein-like autoproteolytic mechanism in autotransporter proteins. J Mol Biol 402:645–656

Tang J, James MN, Hsu IN, Jenkins JA, Blundell TL (1978) Structural evidence for gene duplication in the evolution of the acid proteases. Nature 271:618–621

Tisljar U, Barrett AJ (1990) A distinct thimet peptidase from rat liver mitochondria. FEBS Lett 264:84–86

Tong L, Wengler G, Rossmann MG (1993) Refined structure of Sindbis virus core protein and comparison with other chymotrypsin-like serine proteinase structures. J Mol Biol 230:228–247

Tsuru D (1987) An epoxide-reactive residue in Scytalidium lignicolum acid protease B: possible involvement of glutamic acid residue in the catalytic activity. Adv Biosci 65:257–264

Vandeputte-Rutten L, Kramer RA, Kroon J, Dekker N, Egmond MR, Gros P (2001) Crystal structure of the outer membrane protease OmpT from Escherichia coli suggests a novel catalytic site. EMBO J 20:5033–5039

Vines SH (1903) Proteolytic enzymes in plants. Ann Bot (London) 17:237–264

Wang Y, Zhang Y, Ha Y (2006) Crystal structure of a rhomboid family intramembrane protease. Nature 444:179–180

Zhang L, Agosto MA, Ivanovic T, King DS, Nibert ML, Harrison SC (2009) Requirements for the formation of membrane pores by the reovirus myristoylated micro1N peptide. J Virol 83:7004–7014

Chapter 2
Kinetics of the Interaction of Peptidases with Substrates and Modifiers

Antonio Baici, Marko Novinec, and Brigita Lenarčič

2.1 Introduction

A significant proportion of past and present research on peptidases was/is dedicated to the interactions between enzymes, substrates and modifiers, many of which have a direct bearing to human health. Any study of the efficacy of modifiers aimed at modulating the activity of peptidases begins *in vitro*, for practical reasons using synthetic substrates, which after hydrolysis of a susceptible peptide bond produce a measurable signal proportional to the concentration of hydrolyzed substrate. Enzyme kinetics provides the tools to accomplish this task, which aims at elucidating the underlying kinetic mechanisms of action. This information is necessary when formulating hypotheses on the mechanisms of action of the peptidases with naturally occurring substrates and modifiers, as well with synthetic or semisynthetic modifiers intended to be used as drugs.

The kinetic tools for characterizing substrate turnover by peptidase and interactions with inhibitors and activators are dispersed between numerous specialized publications. Often, important kinetic methods are part of studies whose emphasis is placed on the biological properties of the enzymes and the 'technical' part is overlooked. In other instances, kinetic theories are published in journals with predominantly theoretical character and are overlooked as well. Particular methods that are not treated in specialized books can be found in the specific literature. Yet, finding these methods and putting them to work is a responsibility left to the end user. In this chapter, enzyme kinetic concepts relevant to peptidases will be discussed, while general theories can be found in excellent books dedicated to this topic (Cornish-Bowden 2004; Fersht 1977; Segel 1975).

A. Baici (✉)
Department of Biochemistry, University of Zurich, 8057 Zurich, Switzerland
e-mail: abaici@bioc.uzh.ch

M. Novinec • B. Lenarčič
Department of Chemistry and Biochemistry, University of Ljubljana, 1000 Ljubljana, Slovenia
e-mail: marko.novinec@fkkt.uni-lj.si; Brigita.Lenarcic@fkkt.uni-lj.si

K. Brix and W. Stöcker (eds.), *Proteases: Structure and Function*,
DOI 10.1007/978-3-7091-0885-7_2,

2.2 Symbols, Nomenclature, Conventions and Software Used

In the interest of unambiguous communication in research and teaching, especially in reporting results for publication, a consistent use of nomenclature and symbols in enzyme kinetics is highly recommended. We follow here the recommendations of the Nomenclature Committee of the International Union of Biochemistry (International Union of Biochemistry 1979, 1982). Recommended and other symbols are summarized in Table 2.1.

GraphPad Prism version 5.04 for Windows, GraphPad Software (San Diego, California, USA) was used for regression analysis and graphical representations. Numerical simulations of time-dependent processes were performed with Matlab® and Simulink® (MathWorks, Natick, Massachusetts, USA). Model fitting by a combination of numerical simulation and non-linear regression was performed with KinTek Explorer 2.5 software (KinTek Corporation, Austin, Texas, USA). The examples of kinetic experiments shown in this chapter are either unpublished originals from the laboratories of the authors or were provided by colleagues as acknowledged in the figure legends.

2.3 Kinetics of Enzyme-Catalyzed Peptide Bond Hydrolysis

The peptide bond hydrolases or *peptidases* (EC 3.4.x.y) are subdivided into seven catalytic types according to their catalytic mechanism in aspartic, cysteine, glutamic, metallo, asparagine, serine and threonine peptidases (Rawlings et al. 2010). These enzymes catalyze the hydrolytic cleavage of peptide bonds in proteins and peptides of various sizes down to dipeptides using diverse chemical approaches for performing this task at the molecular level. Though, all peptidases share a common catalytic strategy which consists of a proton carrier for delivering the proton from an attacking nucleophile to the leaving group of the peptide being cleaved. In this process, shown in Scheme 2.1, the peptide substrate (S) reacts with water (W) to generate two products, P_N and P_C, which denote the N-terminal and the C-terminal peptide products with respect to the cleaved bond. Only the residues R_1 and R_2 pertaining to the scissile peptide bond are shown in Scheme 2.1. These represent the P_1 and P_1' amino acid side chains of the substrate that bind to the S_1 and S_1' pockets of the enzyme, respectively (Schechter and Berger 1967).

Therefore, the reaction mechanism of peptide bond hydrolysis involves two substrates and two products. The sequence of kinetic events in serine, threonine and cysteine peptidases is shown in Scheme 2.2a in Cleland's notation (Cleland 1963). Without entering the details of the mechanism at the molecular and atomic level, the enzyme binds first the (poly)peptide, a covalently modified enzyme is formed and one of the two products, i.e. the C-terminal part of the original substrate, is released (Polgár 2004a, b). Next, water reacts as the second substrate with the

Table 2.1 Symbols for enzyme kinetics

Symbol	Meaning
E	Enzyme
EI, EX	Adsorptive enzyme-inhibitor, enzyme-modifier complex
E·I, E·X	Reversible enzyme-inhibitor, enzyme-modifier X complex
E-I	Inactivated enzyme (irreversible inhibition by covalent interaction)
ES	Enzyme-substrate adsorptive complex (Michaelis complex)
I	Reversible inhibitor or inactivator (=irreversible inhibitor)
k_{cat}	Catalytic constant. Unit: s^{-1}
K_m	Michaelis constant. Unit: M
k_{cat}/K_m	Specificity constant. Units: $M^{-1}\ s^{-1}$
K_s	Dissociation constant of the enzyme-substrate complex. Unit: M
K_i	Dissociation constant of the enzyme-inhibitor complex. Unit: M
k_n, k_{-n}	Rate constants for the nth step of enzyme-catalyzed reactions, positive in the forward and negative in the reverse reaction. Units: $M^{-1}\ s^{-1}$ for second-order, s^{-1} for first-order
P	Product, symbolically also for more than one product
S	Substrate
v	Generic reaction velocity (reaction rate)
v_i	Reaction velocity in presence of inhibitors or inactivators
v_x	Reaction velocity in presence of a modifier X
v_s	Reaction velocity at steady-state
v_z	Reaction velocity at the beginning of a reaction (z = zero time)
v_0	Reaction velocity in the absence of modifiers. The index is 'zero', not the letter 'o'
v_∞	Reaction velocity following the exponential phase in temporary inactivation
V	Limiting rate, recommended symbol for 'maximum velocity' V_{max}. Units: $M\ s^{-1}$
X	Generic modifier
λ	First-order rate constant of an exponential process. Unit: s^{-1}
σ	$[S]/K_m$, dimensionless
t	Time, whose unit will always be the second (s)

The concentration of any species, indicated by a letter enclosed in square brackets has dimensions of mol dm^{-3} = M. Without additional specification it indicates the concentration of the free species, i.e. not bound to other species. For instance, [S] = free substrate concentration, $[E]_t$ = total enzyme concentration

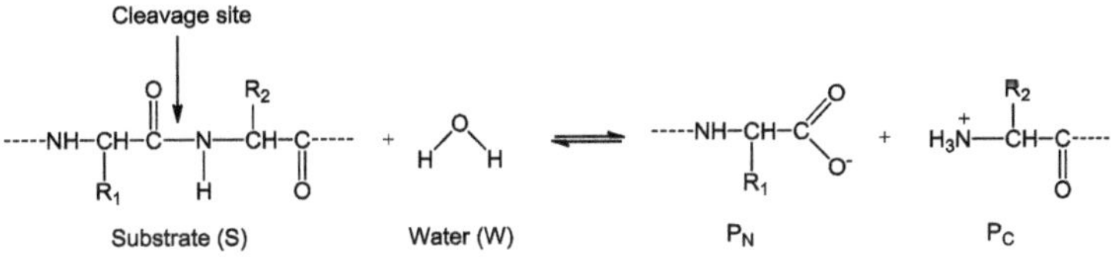

Scheme 2.1 Peptide bond hydrolysis

modified enzyme whereby the second product, the N-terminal part of the substrate, is released and the enzyme is restored to its original state. This mechanism is known as substituted-enzyme, double-displacement or Ping Pong Bi Bi mechanism. In aspartic peptidases the nucleophile is a water molecule coordinated by the carboxyl groups of two aspartate residues (James 2004). Following substrate binding

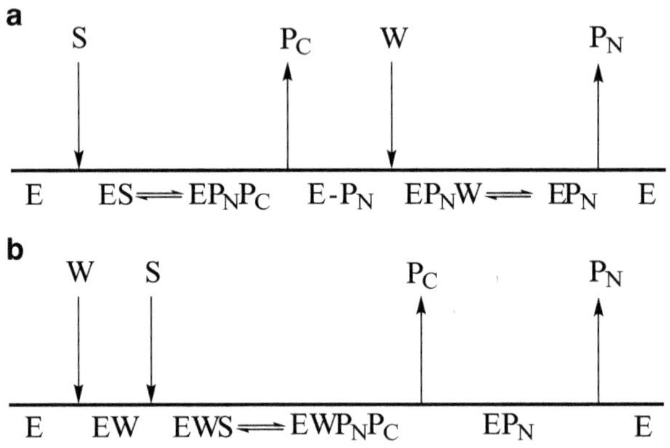

Scheme 2.2 Kinetic mechanisms of peptidases. (**a**) Serine, threonine and cysteine peptidases; (**b**) aspartic and metallo peptidases

(Scheme 2.2b), a tetrahedral intermediate is formed, followed by release of the two cleavage products in the same order of Scheme 2.2a. Also in the metallopeptidases the nucleophile is water, which is bound to a zinc ion, and the mechanism can be sketched (only kinetically) as shown in Scheme 2.2b despite of chemical events differing from those of the aspartic peptidases (Auld 2004; Tallant et al. 2010). For example, in the carboxypeptidase A mechanism the C-terminal product is cleaved off the substrate but retains a salt bridge to a glutamic acid residue until the N-terminal product has left the active center of the enzyme and another water molecule has been bound by the zinc ion. The kinetic mechanism of aspartic and metallopeptidases is therefore of an ordered type.

Since water is in great excess, the mechanisms in Scheme 2.2 can be reduced to that in Scheme 2.3 as an ordered Uni Bi mechanism in Cleland's nomenclature, in which water is only formally omitted (Cleland 1963).

In the notation of enzyme kinetics, which includes kinetic constants, Scheme 2.3 is written as shown in Scheme 2.4. Here a further simplification was tacitly introduced by making the reverse reactions of the second and third steps irreversible. Without ignoring the principle of microscopic reversibility, in hydrolytic enzymes this is justified after considering the exergonic character of peptide bond hydrolysis, the high energy barrier of the reverse reaction and the presence of excess water. Again for practical reasons, the mechanism in Scheme 2.4 is often written in the oversimplified form shown in Scheme 2.5, where the second and third steps of Scheme 2.4 are taken together in an apparently single step with the kinetic constant k_{cat}. This corresponds, only formally, to a Briggs-Haldane mechanism, for which the rate is given by the Michaelis-Menten equation with v as steady-state velocity and $V = k_{cat}[E]_t$ as limiting rate (Eq. 2.1).

$$v = \frac{V[S]}{K_m + [S]}. \tag{2.1}$$

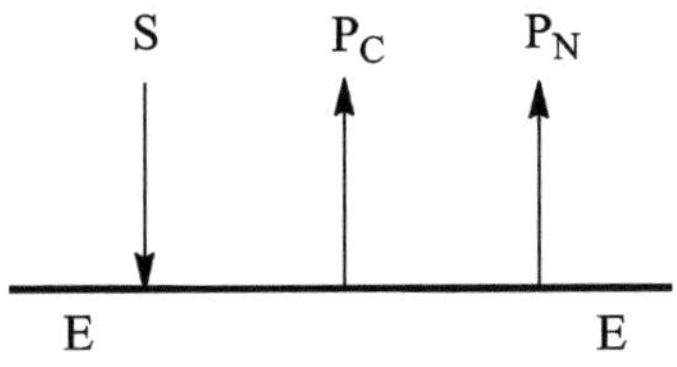

Scheme 2.3 Simplification the peptidase mechanisms to an ordered Uni Bi sequence

$$\mathrm{E+S} \underset{k_{-1}}{\overset{k_1}{\rightleftharpoons}} \mathrm{ES} \xrightarrow{k_2} \underset{\substack{+\\ \mathrm{P_C}}}{\mathrm{EP_N}} \xrightarrow{k_3} \mathrm{E+P_N}$$

Scheme 2.4 Kinetic notation of the simplified ordered Uni By mechanism

$$\mathrm{E+S} \underset{k_{-1}}{\overset{k_1}{\rightleftharpoons}} \mathrm{ES} \xrightarrow{k_{\mathrm{cat}}} \mathrm{E+P}$$

Scheme 2.5 Short form of the peptidolytic reaction as Uni Uni mechanism

The catalytic and the Michaelis constants for the mechanism in Scheme 2.4 and Eq. (2.1) are given by

$$k_{\mathrm{cat}} = \frac{k_2 k_3}{k_2 + k_3}; \quad K_{\mathrm{m}} = \frac{k_3(k_{-1} + k_2)}{k_1(k_2 + k_3)}. \tag{2.2}$$

Keeping in mind the significance of the last simplification, we will continue the treatment of the kinetics of peptidase action using this formalism including also modifiers such as inhibitors and activators.

2.4 Basic Tools for Kinetic Data Analysis

In this section, after introducing an indispensable method to assess enzyme stability during kinetic assays, the basic notions of graphical and mathematical analysis of kinetic data are introduced.

2.4.1 Checking Enzyme Stability During Assays

Steady-state measurements of substrate turnover are limited to relatively short times, during which the substrate concentration is assumed to change insignificantly from its initial value. Letting reactions to proceed to completeness allows exploiting the full information contained in progress curves because the tangent in any point of a curve represents a velocity, which is then evaluated in a broad range of substrate concentrations, i.e. from its initial value towards zero or to a value dictated by thermodynamic equilibrium. Besides the conceptual difference between steady-state and progress curve methods in dealing with changing substrate and modifier concentrations, another significant factor is the time scale of the measurements. Steady-state experiments can be set up to record only a short, 'initial' part of the whole reaction, in which the measured signal changes linearly with time. Recording an entire progress curve can however take a considerable time, during which the reactants, but more importantly the enzymes, may be subjected to changes that affect their concentration as active species. The instability of several peptidases after dilution in assay buffers for kinetic measurements is well known. Therefore, any study based on progress curves must first ascertain the stability of the enzyme over the whole measurement time. For this purpose, Selwyn developed a straightforward method, which consists in measuring the time course of the reaction under consideration keeping all conditions identical but at two or more enzyme concentrations (Selwyn 1965). The underlying principle is based on the following general form of the rate for any enzyme-catalyzed reaction

$$\frac{\mathrm{d}[\mathrm{P}]}{\mathrm{d}t} = [\mathrm{E}] \cdot f\{[\mathrm{S}], [\mathrm{X}], [\mathrm{P}]\}, \tag{2.3}$$

with integral given by

$$[\mathrm{E}] \cdot t = f\{[\mathrm{P}]\}. \tag{2.4}$$

The analytical form of this integrated equation depends on the particular system but, independently of its complexity, the product concentration [P] only depends on the enzyme concentration multiplied by time. This property is understood intuitively by considering that doubling the concentration of enzyme, i.e. the catalyst, the reaction rate doubles. Thus, for instance, the amount of product generated by an enzyme concentration [E] = 10 nM at $t = 120$ s will be the same as with [E] = 20 nM and $t = 60$ s. Accordingly, a series of progress curves measured at several different values of the enzyme concentration, while keeping all other variables constant, produces a single trace in a plot of [P] versus [E]·t if the enzyme concentration remains constant during the measuring time. If, conversely, the enzyme is denatured or its concentration changes for any other cause, [E] becomes a function of time and plotting [P] versus [E]·t generates as many different curves as the different initial enzyme concentrations used for collecting data. The Selwyn method is illustrated in Fig. 2.1 for testing the suitability of assays lasting a relatively long time in the case of enzyme instability during the assay time (elastase-2) and in the case of enzyme stability (HIV-1 retropepsin). The example with elastase-2 clearly shows that the conditions are inadequate for performing assays over the time indicated, while the assay for HIV-1 retropepsin is appropriate.

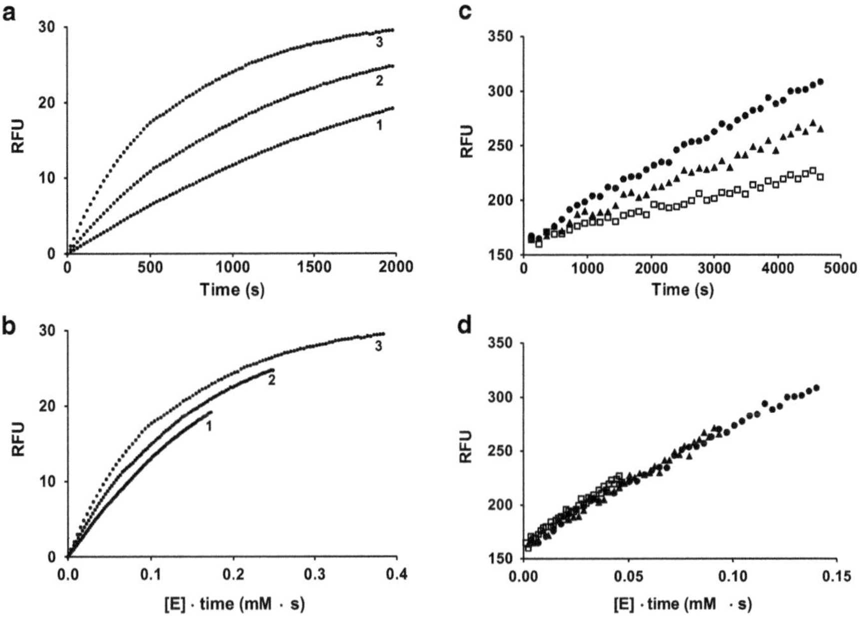

Fig. 2.1 The Selwyn test as a diagnostic tool for time-dependent loss of enzyme activity in continuous assays. (**a**) and (**b**): human elastase-2 (leukocyte elastase, EC 3.4.21.37) with 520 μM MeOSuc-Ala-Ala-Pro-Val-7-(4-methyl)coumarylamide as substrate in 50 mM Tris/HCl buffer, 150 mM NaCl, pH = 7.50, 25 °C; the enzyme concentrations in the assays were 88, 126 and 194 nM in traces 1, 2, and 3, respectively. (**c**) and (**d**): Q7K mutant of HIV-1 retropepsin (HIV-1 protease, EC 3.4 23.16) with 10 μM DABCYL-g-Abu-Ser-Gln-Asn-Tyr-Pro-Ile-Val-Gln-EDANS as substrate. The buffer was 50 mM sodium acetate, 1 mM EDTA, 2.5 mM DTT, 1 M NaCl, 2.5 % glycerol (v/v), 10 % DMSO (v/v), 0.1 % nonidet P-40 (v/v), pH = 4.7, 25 °C. Enzyme concentrations (nM): *squares* 10, *triangles* 20, *circles* 30. RFU = relative fluorescence units, directly proportional to product concentration. Data kindly provided by Dr. H. Roschitzki-Voser and A. Flütsch, Department of Biochemistry, University of Zurich

2.4.2 Graphical Analysis

Graphical methods were the first to be utilized to extract kinetic information from experiments, while statistical procedures were developed in more recent times and gained popularity after the appearance of commercial software which integrates mathematical methods and drawing facilities. Instead of being obsolete, graphical analysis of kinetic data is very useful for the preliminary screening of experimental results with the aim of identifying a mechanism, or a limited number of mechanisms, and of obtaining approximate values of kinetic parameters, which can be further used for final refinement by mathematical methods. Since curved plots cannot be used for 'manual' calculations, the original rate equations must first be linearized using a variety of algebraic manipulations. The slopes of straight lines can thus be calculated and useful intercepts with the Cartesian coordinates can be guessed directly or by extrapolation. The subjectivity in drawing straight lines through experimental points, which are affected by inevitable errors, can in principle be overcome by linear regression procedures. However, their applicability is subject to mathematical rules that cannot be ignored, particularly concerning error structure and distribution (see below). If these rules are ignored, the calculated kinetic parameters make little sense because their statistical significance cannot be guessed. This poses limitations in reporting reliable results for publication as well as for the purposes of good manufacturing practice and good laboratory practice, notably in pharmaceutical sciences.

2.4.3 Regression Analysis

Linear and non-linear regression methods are widely used in enzyme kinetics. A thorough review of the basics of regression analysis, which belongs to the group of parametric statistical methods, has been published by Johansen and Lumry (1961). The authors put emphasis on kinetic applications and pointed out the importance of knowing the structure and distribution of experimental errors as a prerequisite for the applicability of regression analysis. Before utilizing any regression procedure the following criteria must be satisfied (pp. 355–356 in Cornish-Bowden 2004):

- The errors are normally distributed, i.e. distributed according to a bell-shaped Gauss-curve.
- The variance of the independent variable is zero, i.e. there are only errors in the dependent variable. For instance, in Michaelis-Menten kinetics [S] is assumed to be error-free while errors are associated with the rate v.
- The weight of the errors is known.
- The value and sign of a given error is not related to the values and signs of other errors. In mathematical language: the errors are stochastically independent from each other.[1]
- Systematic errors can be neglected, i.e. the distribution curve of each error has the mean value zero. In practice, we have no doubts that we are using the correct equation.

Unfortunately and recurrently, these mathematical requirements are disregarded. Unlike linear regression, which goes straight to calculating the best fit of slope and intercept in a set of data that describe a linear dependence of the variables using the least-squares approach, non-linear regression requires initial estimates of the parameters. These are used in an iterative procedure until a minimum is reached for the sum of the squared deviations.

---

[1] Stochastics, 'the art of guessing' (Greek stóchos = guess), comprises the mathematical theories of probability and statistics. If the happening of an event is not influenced by the happening of another event the two events occur at random and are said to be stochastically independent.

The precision of the best-fit values obtained by non-linear regression analysis is typically reported as the standard errors derived during the regression procedure and are based on the covariance matrix. The values of the standard errors are used to compute confidence intervals, which represent a better criterion to judge the goodness of fit. Thus, if calculations are performed with software that computes confidence intervals, these should be included in reporting results since standard errors may unacceptably underestimate the real errors if the parameters are not well constrained by data. In any case, analysis of the mutual dependency of all estimated parameters is essential, even if small values of the standard errors would suggest an excellent fit. With equations containing two or more parameters it can happen that their values calculated by non-linear regression are not unique, meaning that another set of parameters would 'nicely' fit data as well. Also, one or more parameters may be redundant, suggesting that a simpler model would better describe the data. Software packages, such as for instance GraphPad Prism, parameter dependency is calculated as part of goodness of fit assessment. See also the comments on FitSpace for the Global Kinetic Explorer in Sect. 2.4.4.

The adherence to or deviation of fitted curves from a model can be assessed by methods that are part of commercial software, such as GraphPad Prism. These include residual analysis and the runs test. Residuals are the vertical distances between experimental points and the fitted curve and may be positive or negative. Any non-random, i.e. systematic distribution of the residuals, should be carefully examined and the fit procedure repeated with alternative models. A 'run' is defined as a sequence of points, which are located either above or below the best fit curve calculated by regression analysis. If the model produces a poor fit to data, clusters of points either above or below the curve are observed and the total number of expected runs, which can be calculated, is smaller than that expected from randomly distributed errors.

A comment to a frequent flaw in describing the application of regression analysis, which is not merely a semantic issue, is required at this stage. In the literature, sentences like the following are typically found: '... data were fitted to equation X', or '... parameters shown in Table Y were obtained by fitting experimental results to equation Z'. These expressions suggest that the available data were manipulated until they reached the desired fit to a model. The correct sentences should read instead: '... equation X was fitted to data', and '... equation Z was fitted to experimental results to calculate the parameters shown in Table Y'.

2.4.4 *Numerical Integration and Global Fit of Progress Curves*

Non-linear regression is based on the fulfillment of the criteria discussed in the preceding section. Moreover, integrated rate equations and their utilization in non-linear regression procedures are often based on restrictive assumptions that can or cannot be satisfied experimentally. Also, in many instances analytical integrals of rate equations do not exist at all. In such cases computational limitations can be overcome by performing numerical instead of analytical integration of rate equations. This procedure supplies a value of the area under a curve with an approximation that depends on the algorithm chosen and supplies an almost exact solution. One or more differential equations can be integrated numerically at the same time to produce simulated concentration vs. time profiles. Such curves, generated with initial guesses of the parameters of the considered equations, can be compared with experimental data, the squared differences between them can be calculated, and this procedure can be iterated until reaching a minimum of the squared deviations. The algorithms to perform the necessary calculations by numerical methods (solvers) have been implemented in several software packages, of which just two are commented here. Simulink® operates within the numerical computing software MATLAB®[2] and offers block libraries that are used to represent symbolically time-dependent processes. Eight solvers with fixed or variable step size are available to cover a broad band of applications in any field of physics, chemistry and engineering that requires numerical integration of differential equations of nonstiff and stiff problems (for a stiff differential equation some numerical methods for its solution are unstable unless the step size used for integration is very small). Simulink is a valuable tool for educational purposes in enzyme kinetics because it offers a friendly working interface and compels the user to put hands on by planning and programming the whole numerical integration procedure, from writing differential equations based on a kinetic model to making connections between kinetic paths in even very complex mechanisms. Simulink can also be used to perform parameter optimization by combining numerical integration and non-linear regression. Unfortunately, two major drawbacks are slow performance using conventional desktop computers and the lack of statistical information of the best-fit parameters, such as standard errors and confidence intervals.

A software package fully dedicated to enzyme kinetics is the KinTek Global Kinetic Explorer[3] (Johnson 2009; Johnson et al. 2009a, b). KinTek does not require particular efforts from the part of the user and kinetic models are entered with letters connected by the = sign. The necessary differential equations are set up and all rate constants for forward and reverse reaction directions are displayed automatically. A graphical user interface allows scrolling the values of the parameters to be calculated and to directly visualize the results on screen in real time. This is particularly useful for guessing the initial values that are necessary for the fitting procedure. After entering experimental data, which can either be single curves or sets of curves and even different sets of experiments, simulation is performed by numerical integration. The curve resulting from numerical integration is compared with the experimental curve and the sum of the squared deviations is calculated. The procedure is iterated until a minimum is reached. The strongest feature of

[2] http://www.mathworks.com

[3] http://www.kintek-corp.com/KGExplorer

KinTek is the FitSpace Explorer, which 'calculates the dependence of the sum square error on each pair of parameters while allowing all remaining parameters to be adjusted in seeking the best fit'. Results are displayed graphically as three-dimensional plots, which reveal all relationships between parameters and show whether the set of fitted parameters is unique and well constrained by the data.

2.5 Calculation of k_{cat} and K_m

Low molecular mass peptide substrates carrying fluorogenic or chromogenic leaving groups, or containing internally quenched fluorescent moieties, are typically used to evaluate the ability of peptidases to catalyze the cleavage of peptide bonds. Among the purposes of this investigation are the determination of substrate specificity and the identification of adequate substrates for assessing the action of modifiers *in vitro*. To accomplish these tasks it is necessary to determine the kinetic parameters k_{cat} and K_m either individually, from which their ratio can then be calculated, or k_{cat}/K_m can be measured directly. It is superfluous to mention that such measurements should be carried out as precisely as possible using the most appropriate methods. The ratio k_{cat}/K_m, referred to as the 'specificity constant' (Fersht 1977), is the parameter of choice for assessing the competence of an enzyme in performing catalysis preferentially on a given substrate in the presence of others (pp. 36–39 in Cornish-Bowden 2004). The ratio k_{cat}/K_m, which has also been called 'catalytic efficiency', 'catalytic potential' and 'performance constant', is a misleading parameter if used to compare the catalytic effectiveness of two enzymes on the same substrate (Eisenthal et al. 2007; Koshland 2002).

2.5.1 Graphical Analysis

The direct linear plot created by Eisenthal and Cornish-Bowden (Cornish-Bowden and Eisenthal 1974; Eisenthal and Cornish-Bowden 1974) in undoubtedly the most robust and trustworthy among other graphical procedures for calculating the kinetic parameters of substrate turnover by enzymes (Sect. 2.6 in Cornish-Bowden 2004). In this method, the Michaelis-Menten equation (2.1) is rearranged by considering K_m and V as variables and the measured v values and known substrate concentrations [S] as constants

$$V = v + \frac{v}{[\mathrm{S}]} K_m. \tag{2.5}$$

The straight lines described by Eq. (2.5) have v as ordinate intercept and $v/[\mathrm{s}]$ as slope in the parameter space defined by V and K_m. In the absence of errors, the direct linear plot consists of a sheaf of straight lines intersecting at a common point,

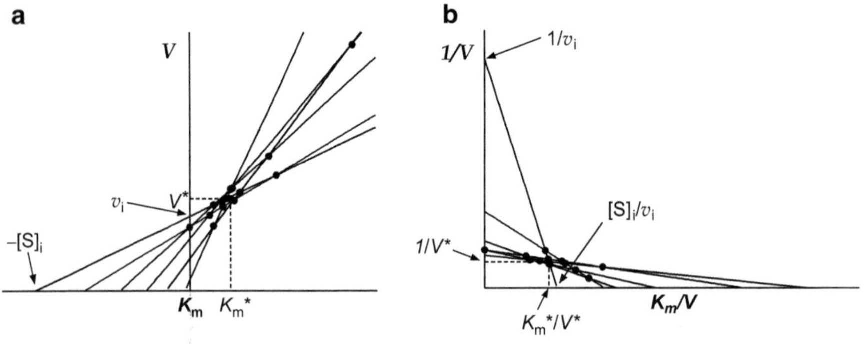

Fig. 2.2 The direct linear plot. The same observations were plotted in the original form (**a**) and in the reciprocal variant (**b**). In both cases the maximum number of intercepts for n observations is given by $n(n - 1)/2$. In panel (**a**) the best estimates of the parameters V and K_m, denoted by an *asterisk*, are obtained by reading out the coordinates of the median of the intersections on the axes of the diagram. Similarly in panel (**b**) the best estimates of $1/V$ and K_m/V are denoted by *asterisks* and determined from the coordinates of the median of all intersections (*dots*), from which the parameters are calculated

while with experimental data multiple intersection points result from inevitable errors. The maximum number of intersections is $n(n - 1)/2$, where n is the number of observations. The direct linear plot belongs to the group of distribution-free or non-parametric statistical methods. With respect to parametric methods, for whose correct application normally distributed errors are a prerequisite, non-parametric approaches do not depend on error distribution. For this property, distribution-free methods are less sensitive to the presence of outliers. While parametric statistics considers the sample mean as the best-fit, distribution-free methods make use of the median as the best estimator. This value can be visually identified by sorting all numbers under consideration in increasing order and taking the figure which lies in the middle of the series. For an odd number of data the median is placed 'in the middle' and for an even number of data the median is calculated as the mean of the two 'central' values.

In the ideal case, intersections occur in the first quadrant as in the representation of Fig. 2.2a. Negative estimates of V and/or K_m may be observed when any intersection of the lines occurs in the second or third quadrant. Second quadrant intercepts are treated as they would occur at face value (i.e. the abscissa coordinate is multiplied by -1) and third quadrant intercepts are considered as being both large and positive. This problem can be fixed considering a practical variant of the direct linear plot, which consists in drawing straight lines from intercepts $[S]_i/v_i$ and $1/v_i$ instead of $-[S]_i$ and v_i (Fig. 2.2b) (Cornish-Bowden and Eisenthal 1978).With this alternative method the intersections can be identified easier than in the original plot and any intersection in the second or third quadrant does not need a particular treatment.

Since both variants of this plot become crowded with increasing number of data, it is possible to skip the graphical representation and to calculate instead the intersections analytically. This is easily accomplished with the aid of a spreadsheet calculation program, which can be programmed to solve n systems of pairs of linear equations from n measurements and to show the medians of all intersections for both axes. Confidence intervals of the median can be estimated from the ranked intersections to show the statistical significance of the measurements (Cornish-Bowden et al. 1978).

2.5.2 *Non-linear Regression Analysis*

The Michaelis-Menten equation is a friendly one, since initial estimates are easily calculated from data, e.g. by taking the largest value of the measured velocities as an estimate or V and the median of the substrate concentrations to estimate K_m. The example in Fig. 2.3 shows the fit of Eq. (2.1) to a set of measurements for evaluating V and K_m for a substrate of bovine α-chymotrypsin. The figure shows the best fit curve and the 95 % confidence band considering equal weights for all data points. The fit by non-linear regression was also performed using the 'automatic outlier elimination' option of GraphPad Prism with results listed for comparison in Table 2.2. The same data were also evaluated with the direct linear plot (Sect. 2.5.1), with the appropriate 95 % confidence intervals (Cornish-Bowden et al. 1978). Inspection of the best fit values obtained by the three methods shows that the apparent 'outlier', i.e. the point outside the 95 % confidence band in Fig. 2.3, has a larger impact on K_m than on V, with the larger discrepancy between ordinary regression and outlier elimination. Automatic outlier elimination should be used with caution, especially because such eliminations may be rather arbitrary.

Another variant is robust non-linear regression, which is based on a Lorentzian instead on Gaussian distribution of errors and makes the fit less sensitive to outliers. At least as implemented by GraphPad Prism, this is however a qualitative method to assess the effect of possible outliers without generating standard errors and confidence intervals. This method applied to the data in Fig. 2.3 gave in any case $V = 0.137$ (μM s^{-1}) and $K_m = 30.8$ μM, which are very close to the values obtained with the direct linear plot.

When the k_{cat}/K_m ratio needs to be calculated from the individual values of the two parameters together with their standard errors (SE) or standard deviations (SD), the SE or SD associated with k_{cat}/K_m can be calculated with the Fenner formula (Fenner 1931)

$$\frac{\overline{x}_1 \pm s_{\overline{x}_1}}{\overline{x}_2 \pm s_{\overline{x}_2}} = \frac{\overline{x}_1}{\overline{x}_2} \pm \frac{1}{\overline{x}_2^2}\sqrt{\overline{x}_1^2 s_{\overline{x}_2}^2 + \overline{x}_2^2 s_{\overline{x}_1}^2}, \tag{2.6}$$

where $\overline{x}_i \pm s_{\overline{x}_i}$ indicate the means with their associated 'errors'. In the same paper Fenner also described how to calculate the errors of sums, subtractions

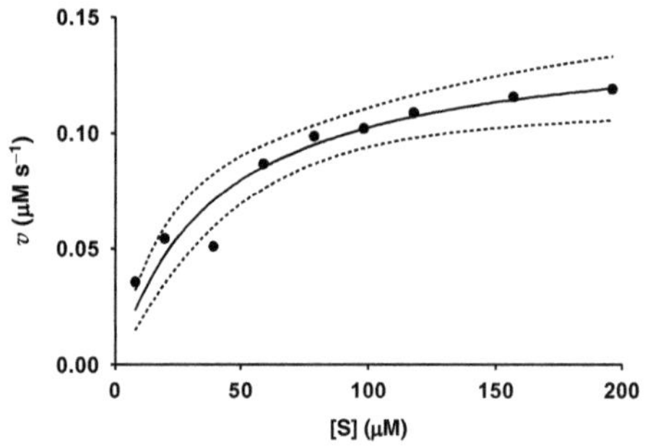

Fig. 2.3 Fit of the Michaelis-Menten equation (2.1) to data by non-linear regression. The *solid line* represents the best fit and the *dashed lines* show the 95 % confidence band. Measurements performed with bovine α-chymotrypsin and N-succinyl-Ala-Ala-Pro-Phe-p-nitroanilide as substrate in 0.1 M Tris/HCl buffer, pH = 7.80, 25 °C. The α-chymotrypsin active site concentration was determined by burst kinetics and the enzyme concentration in the assay was 2.8 nM

Table 2.2 Comparison of kinetic parameters for the data in Fig. 2.3 calculated by non-linear regression fit of Eq. (2.1) to data and with the direct linear plot

	V (µM s^{-1})	K_m (µM)
NLR	0.144 ± 0.013 (0.114–0.174)	40.3 ± 11.2 (13.9–66.7)
NLR, outlier elim.	0.134 ± 0.004 (0.125–0.143)	27.6 ± 3.0 (20.2–34.9)
Direct linear plot	0.138 (0.127–0.149)	31.0 (20.9–44.8)

The ± values represent standard errors from regression analysis and those in parentheses correspond to 95 % confidence intervals

NLR non-linear regression, *outlier elim.* automatic outlier elimination

and multiplications of mean values with associated errors. For example, from $k_{cat} = 14.7 \pm 0.7$ s^{-1} and $K_m = (95 \pm 4) \times 10^{-6}$ M the ratio of the two parameters is calculated as $k_{cat}/K_m = 154\ 737 \pm 9\ 836$ M^{-1} s^{-1}. For publication purposes, this last value would be conveniently rounded up e.g. as 154 700 ± 9 800 M^{-1} s^{-1}.

2.5.3 *First-Order Kinetics*

In some instances the substrates of peptidases, such as proteins with a large molecular mass, are not amenable to large excursions in their concentrations for calculating kinetic parameters by any of the methods illustrated in this section. Equation (2.1) reduces to a first-order equation if [S] $\ll K_m$, with the first-order rate constant V/K_m, and to a zero-order equation for [S] $\gg K_m$, with $v = V$. First-order kinetics can thus be exploited for the direct calculation of V/K_m from experiments performed at low substrate concentrations. An illustrative example is the degradation of the four human IgG subclasses, named IgG1 to IgG4, by elastase-2 (human leukocyte elastase, EC 3.4.21.37) into discrete fragments (Baici et al. 1980).

Following incubation of monoclonal IgGs with elastase-2 for various times, the fragments were separated by denaturing polyacrylamide gel electrophoresis, the bands on the gel were identified by immunochemistry and their intensities, proportional to concentrations, were determined by gel scanner densitometry. The first enzymatic cleavage of IgG occurred in the hinge region and this represented the primary hydrolytic degradation of the protein, which was followed by other proteolytic events that occurred at slower rates. Thus, following the disappearance of IgG in the first part of reaction progress offers a mean for characterizing the kinetic features of the most susceptible peptide bond. The band intensities for the disappearing IgG-band at various incubation times were plotted and a single exponential was fitted to data to obtain first-order rate constants corresponding to V/K_m. These values, divided by the titrated enzyme active site concentration, gave second-order constants k_{cat}/K_m of 12.3, 4.3, 63.3 and 1.2 $M^{-1}\ s^{-1}$ for IgG1, IgG2, IgG3 and IgG4, respectively. Considering the limits of the method, these numbers provide a practical semiquantitative criterion for comparing the relative susceptibilities of the IgG subclasses to proteolysis. The same method can be used to characterize the kinetics of limited proteolysis, i.e. a single cut in precursor proteins. Precision can be enhanced by measuring the variation of protein concentration with time e.g. by HPLC.

2.5.4 *The Integrated Michaelis-Menten Equation*

Equation (2.1) can be integrated to give

$$\frac{[P]}{t} = V + \frac{K_m}{t}\ln\left\{1 - \frac{[P]}{[S]_t}\right\}, \tag{2.7}$$

which can be written in different equivalent forms but remains in any case an implicit equation (Orsi and Tipton 1979). Although V and K_m can be calculated as intercept and slope from the straight line obtained in a plot of $[P]/t$ vs. $\ln(1-[P]/[S]_t)/t$, the procedure cannot give statistically reliable values of the parameters because the errors associated with [P] appear in both the dependent and the independent variable (Johansen and Lumry 1961). Several alternative graphical methods are known that use integrated equations for calculating kinetic parameters including modifiers as well as substrate and product inhibition (Orsi and Tipton 1979), but are subjected to the same statistical limitation. Due to this restriction, the use of the integrated Michaelis-Menten equation instead of another approach for calculating statistically valid kinetic parameters remains a matter of taste. In general however, progress curves contain much more information than initial velocity measurements and exploiting the entire course of enzymatic reactions depends on how the progress curves are treated for quantitative purposes, e.g. using numerical integration methods and disregarding analytical integration as described in the next subsection.

2.5.5 *Numerical Integration of the Differential Equation*

As an alternative to analytical integration of the Michaelis-Menten equation, numerical integration can be performed. We illustrate this procedure with an example of human cathepsin K and a synthetic dipeptide substrate. Data were collected as described in the legend of Fig. 2.4 and the Michaelis-Menten equation was fitted by numerical integration combined with non-linear regression using the KinTek Global Kinetic Explorer program. With reference to Scheme 2.5, k_1 was kept fixed at the value of 100 $\mu M^{-1} s^{-1}$, while k_{cat} and k_{-1} were globally fitted using all traces. During the 500 s of the experiment in panel (b) the assays passed the Selwyn test (Sect. 2.4.1). The global fit to data collected for either 20 or 500 s was not perfect, as shown by the best-fit traces systematically deviating from data for some of the curves. This is a characteristic of global fitting and represents the rule rather than the exception even with high quality data collected as precisely as possible. The issue is that global fitting is not permissive to even small deviations from ideal progress curves, i.e. data should not contain errors and should perfectly adhere to the model. Nevertheless, the best-fit values of the kinetic parameters calculated by globally fitting data in panels (a) and (b) of Fig. 2.4 were similar. The slopes of the curves in panel (a) were used to calculate the kinetic parameters with the direct linear plot (Sect. 2.5.1) and by non-linear regression (Sect. 2.5.2), giving reasonably comparable results as shown in Table 2.3.

A limitation of the application of numerical integration to data as those in Fig. 2.4 is that error estimates cannot be obtained for k_{cat} and K_m because the individual rate constants cannot in general be calculated with certainty and they are affected by large standard errors. However, this does not preclude the calculation of good estimates of for k_{cat} and K_m because the same values will be obtained by any combination of the parameter sets estimated by numerical integration (Johnson 2009).

2.6 Classical and Tight-Binding Enzyme Modification

Enzyme inhibitors can be divided into two broad types depending on whether the enzyme-inhibitor complex can dissociate back to free enzyme and inhibitor (reversible) or not (irreversible). Irreversible inhibition, in the following called 'inactivation', will be treated in Sect. 2.8. Reversible inhibitors can either bind very quickly to enzymes, e.g. as a diffusion controlled process, or it may take a relatively long time to complete the formation of the enzyme-inhibitor complex. Morrison (Morrison 1982) proposed a classification of reversible enzyme inhibitors based on the rate of formation of the E·I complex and on the relative concentrations of enzyme and inhibitor as shown in Table 2.4.

Classical reversible inhibitors rapidly associate and dissociate for $[I]_t > [E]_t$, so that the condition $[I] \approx [I]_t$ could be taken as granted. However, the relative

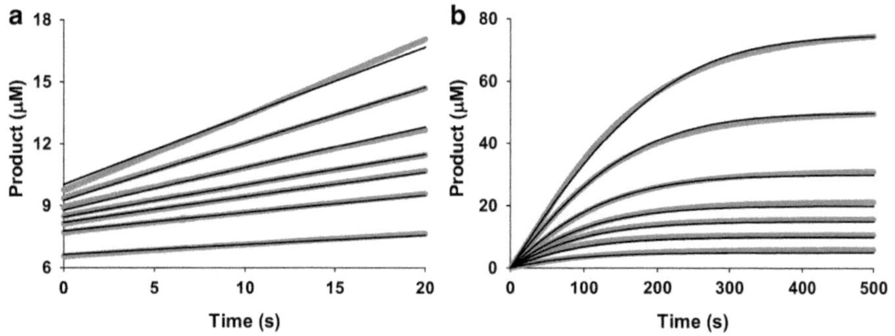

Fig. 2.4 Global fit of reaction progress curves. Human Cathepsin K was rapidly mixed in a stopped-flow apparatus with Z-Phe-Arg-7-(4-methyl)coumarylamide. Two sets of measurements were performed using the same solutions of enzyme and substrate kept in separate syringes with measuring times of 20 s (panel **a**) and 500 s (panel **b**). The concentrations after mixing were: enzyme = 20 nM of titrated active sites, substrate 5, 10, 15, 20, 30, 50 and 75 μM (from *bottom* to *top traces*). 100 mM sodium phosphate buffer, 5 mM dithiothreitol, pH = 6.00, temperature 25 ± 1 °C. *Thick traces* represent data and *thin continuous lines* best fit curves

Table 2.3 Comparison of values of the kinetic parameters from the data in Fig. 2.4 calculated by numerical integration of the Michaelis-Menten equation, conventional non-linear regression or with the direct linear plot

	V (μM s^{-1})	K_m (μM)
NI (Fig. 2.4a)	0.57	50.8
NI (Fig. 2.4b)	0.62	48.6
NLR	0.53 ± 0.02 (0.48–0.58)	50.7 ± 4.0 (40.9–60.5)
DLP	0.49 (0.33–0.68)	42.2 (24.6–65.1)

The ± values represent standard errors from regression analysis and those in parentheses correspond to 95 % confidence intervals

NI numerical integration with KinTek software, *NLR* non-linear regression, *DLP* direct linear plot

magnitudes of $[E]_t$, $[I]_t$ and K_i are not considered in this classification and a classical inhibitor becomes a tight-binder if the enzyme concentration is raised to the order of magnitude of K_i. For high-affinity modifiers inhibition occurs with $[I]_t \approx [E]_t$ at the low enzyme concentrations typically used *in vitro* with $[I]_t$ in the order of magnitude of its K_i value, so that $[I]_t \approx [E]_t \approx K_i$. Thus, Morrison's relationships between $[E]_t$ and $[I]_t$ for the tight-binding and the slow, tight-binding classes of inhibitors in Table 2.4, originally formulated as $[I]_t \approx [E]_t$, are more coherent by changing them to $[I]_t \approx [E]_t$ and K_i. In this section we will deal with the fast acting classes of inhibitors while the slow processes will be discussed in Sects. 2.7 and 2.8.

Enzyme activation can be divided into essential and non-essential. As intuitively indicated by these adjectives, essential activation denotes the compelling presence of an activating partner in order for an enzyme to exert catalysis, while non-essential activation is an elective property of some substances that bind enzymes and enhance thereby their activity over the threshold observed in their absence.

Table 2.4 Classification of reversible enzyme inhibitors [modified after Morrison (1982)]

Type of inhibition	Relationship between $[E]_t$ and $[I]_t$	Rate of formation of the inhibited complex
Classical	$[I]_t \gg [E]_t$	Fast
Tight-binding	$[I]_t \approx [E]_t$ and K_i	Fast
Slow-binding	$[I]_t \gg [E]_t$	Slow
Slow, tight-binding	$[I]_t \approx [E]_t$ and K_i	Slow

For the tight-binding and the slow, tight-binding cases the original definition $[I]_t \approx [E]_t$ has been changed into $[I]_t \approx [E]_t$ and K_i

2.6.1 The General Modifier Mechanism

When classical inhibitors bind their target enzymes and there is no further product release for $[I] \to \infty$, enzyme activity is driven to zero and the inhibition is called linear, complete or full. In some instances an ESI-complex exists and is catalytically active, in which case the inhibition is called hyperbolic or partial. The terms 'linear' and 'hyperbolic' refer to the form of plots of $1/v$ versus [I]. Since the name 'inhibition' without further specification is generic, the adjectives 'linear' or 'hyperbolic' as well and the inhibition type (competitive, uncompetitive or mixed) are added.

Non-essential activation is kinetically similar to hyperbolic inhibition, while an essential activator can be seen as a co-substrate. Therefore, classical inhibition and activation can be conveniently described by the general modifier mechanism proposed by Botts and Morales (1953) and treated in a uniform way with a common equation. The general modifier mechanism for unireactant enzymes, and consequently also for peptidases under the conventions discussed in Sect. 2.3, is shown in Scheme 2.6. The ternary ES·X complex between enzyme, substrate and modifier X can be formed through the paths E → ES → ES·X or E → E·X → ES·X. This is logical because the free energy change of the two paths is the same and thus their overall equilibrium dissociation constant is also the same, i.e. $K_s \alpha K_x = K_x \alpha K_s$ and the scheme represents a thermodynamic box. When Scheme 2.6 is written for an inhibitor, X and K_x may be substituted by I and K_i, respectively, while for an activator these symbols may be changed into A and K_a.

The steady-state rate equation for the general modifier mechanism contains terms in $[S]^2$, $[X]^2$, $[S]^2[X]$ and $[S][X]^2$, which poses limits to its practical application for extracting individual rate constants from experiments. However, assuming that binding of X to E and ES is at quasi-equilibrium, while for the fluxes around ES and ES·X (the catalytic steps) the steady-state assumption is valid, equilibrium dissociation constants can be used in place of individual rate constants as shown in Scheme 2.6 to derive the rate equation, which is given by

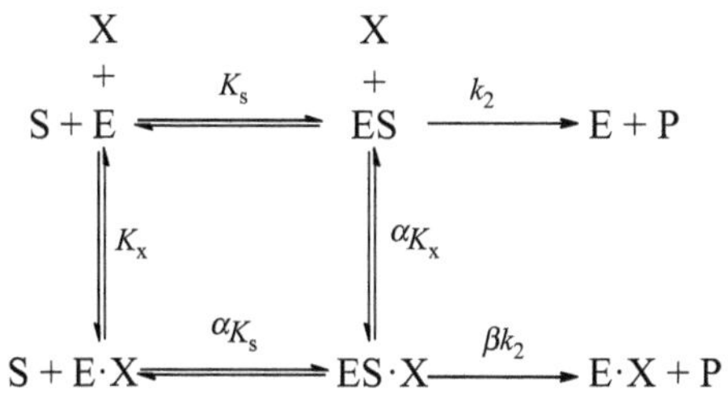

Scheme 2.6 The general modifier mechanism

Table 2.5 Reversible modifier mechanisms as particular cases of the general modifier mechanism

α	β	Modifier mechanism
$\alpha = \infty$	$\beta = 0$	Linear competitive inhibition
$0 < \alpha < \infty$	$\beta = 1$	Hyperbolic competitive inhibition
$\alpha \to 0, K_x \to \infty$	$\beta = 0$	Linear uncompetitive inhibition
$0 < \alpha < 1$	$0 < \beta < 1$ and $\alpha = \beta$	Hyperbolic uncompetitive inhibition
$1 \leq \alpha < \infty$	$\beta = 0$	Linear mixed inhibition
$1 \leq \alpha < \infty$	$0 < \beta < 1$	Hyperbolic mixed inhibition
$1 \leq \alpha < \infty$	$\beta > 1$	Non-essential activation
$0 < \alpha < 1$	$0 < \beta$ and $\alpha < \beta$	Hyperbolic mixed inhibition or non-essential activation depending on [S]

$$\frac{v}{V} = \frac{\sigma\left(1 + \beta\frac{[X]}{\alpha K_x}\right)}{1 + \frac{[X]}{K_x} + \sigma\left(1 + \frac{[X]}{\alpha K_x}\right)} \tag{2.8}$$

The symbol σ for $[S]/K_m$ is used here and in the following for practical reasons. Equation (2.8) applies to classical inhibition types, either linear or hyperbolic, and to activators. The coefficient α specifies the position of the equilibria and the character of the modifier, which for inhibitors can be competitive, uncompetitive or mixed. With $\beta = 0$ inhibitors are linear, with $0 < \beta \leq 1$ inhibitors are hyperbolic, while $\beta > 1$ indicates non-essential activation. Various combinations of the α and β coefficients characterize a large number of diverse reversible modifier mechanisms (Fontes et al. 2000) as shown in Table 2.5. Many such combinations have never been observed for modifiers of peptidases but they may occur for other enzyme classes. For instance, the last row of Table 2.5 contains, among conditions under which only activation is observed, also those that generate either inhibition or activation depending on substrate concentration. Linear uncompetitive inhibition has only purely theoretical character for peptidases but it is considered because, blended with competitive inhibition, it is necessary to describe all gradations of the interesting category of mixed inhibitors. To maintain a consistent nomenclature, for linear uncompetitive inhibition in Table 2.5 we can figure out that $K_i \to \infty$ and $\alpha \to 0$ at the same time, so that αK_i has a finite value, but obviously individual values of α and K_i cannot be measured.

2.6.2 *The Specific Velocity Plot*

A graphical method for analyzing the kinetics of enzyme modification according to the general modifier mechanism is the specific velocity plot (Baici 1981). The plot is based on the equation

$$\frac{v_0}{v_X} = \frac{[X]\left(\dfrac{1}{\alpha K_X} - \dfrac{1}{K_X}\right)}{1 + \beta\dfrac{[X]}{\alpha K_X}} \frac{\sigma}{1+\sigma} + \frac{1 + \dfrac{[X]}{K_X}}{1 + \beta\dfrac{[X]}{\alpha K_X}}, \tag{2.9}$$

in which the rates in the absence (v_0) and presence of the modifier (v_X) are normalized as a dimensionless ratio, which is a function of the dimensionless ratio $\sigma/(1 + \sigma)$, known as specific velocity. Equation (2.9) is a handy tool that describes any particular case of the general modifier mechanism (Scheme 2.6) as straight lines for inhibitors and activators, provided the quasi-equilibrium assumption is valid. Any consistent deviation from linearity should be considered individually and substrate inhibition, violation of the quasi-equilibrium assumption or others causes evaluated.

The function represented by Eq. (2.9) is defined in the interval $0 < \sigma/(1 + \sigma) < 1$. For $[S] \to 0$, $\sigma/(1 + \sigma) \to 0$ and for $[S] \to \infty$, $\sigma/(1 + \sigma) \to 1$. When plotting v_0/v_i versus $\sigma/(1 + \sigma)$ it is helpful to draw two ordinates: the first for $\sigma/(1 + \sigma) = 0$ and the second for $\sigma/(1 + \sigma) = 1$. For $[S] = K_m$, $\sigma/(1 + \sigma) = 0.5$ (Fig. 2.5a). Under the assumptions made above, linear or hyperbolic mechanisms of enzyme modification yield straight lines which have a common intersection with ordinate equal to 1 and abscissa corresponding to $(\alpha - \beta)/(\alpha - 1)$. The case $\alpha = 1$ with slope $= 0$ is the only exception to this rule. Otherwise the slope of the lines depends on the relationship between α and β as illustrated by the simulated examples in Fig. 2.6, where the modifier type can be recognized at a glance from the slope of the lines and the position of the abscissa intercept. For instance, inhibition types which are linear competitive or have a predominantly competitive component ($K_i < \alpha K_i$), the common intersection point lies on the right side of the plot. For calculating α, β and K_i the extrapolated intersections of the straight lines with the two ordinates are used. Defining 'a' the intersection with the left ordinate $[\sigma/(1 + \sigma) = 0]$ and 'b' the intersection with the right ordinate $[\sigma/(1 + \sigma) = 1]$, it follows from Eq. (2.9) that

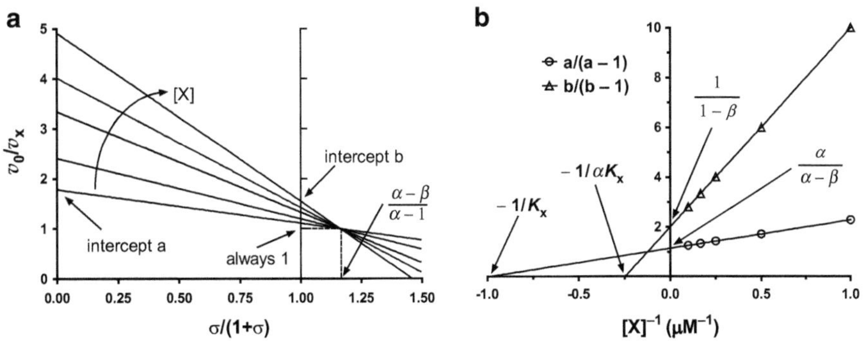

Fig. 2.5 Properties of the specific velocity plot illustrated with a simulated example of hyperbolic mixed inhibition. The primary and secondary plots are shown in panels (**a**) and (**b**), respectively. Independently of the mechanism (linear, mixed, inhibition or activation), with $\alpha \neq 1$ the family of *straight lines* in the primary plots intersect at a common point, with ordinate = 1 and abscissa = $(\alpha - \beta)/(\alpha - 1)$. As shown here, the primary and the secondary plot allow the calculation of the modifier equilibrium constant as well as of the α and β coefficients

$$\mathrm{a} = \frac{1 + \dfrac{[\mathrm{X}]}{K_\mathrm{X}}}{1 + \dfrac{\beta[\mathrm{X}]}{\alpha K_\mathrm{X}}}; \quad \mathrm{b} = \frac{1 + \dfrac{[\mathrm{X}]}{\alpha K_\mathrm{X}}}{1 + \dfrac{\beta[\mathrm{X}]}{\alpha K_\mathrm{X}}} \tag{2.10}$$

and straightforward rearrangement of these relationships yields

$$\frac{\mathrm{a}}{\mathrm{a}-1} = \frac{\alpha K_\mathrm{X}}{\alpha - \beta}\frac{1}{[\mathrm{X}]} + \frac{\alpha}{\alpha - \beta} \tag{2.11}$$

$$\frac{\mathrm{b}}{\mathrm{b}-1} = \frac{\alpha K_\mathrm{X}}{1 - \beta}\frac{1}{[\mathrm{X}]} + \frac{1}{1 - \beta}. \tag{2.12}$$

The plots of a/(a−1) or b/(b−1) versus 1/[X] are straight lines, from which α, β and K_i can be calculated (Fig. 2.5).

The simulated examples in Fig. 2.6 show the specific velocity plots and replots for linear competitive, linear mixed and hyperbolic mixed inhibition as well for non-essential activation.

The specific velocity plot is a plain graphical method for diagnostic purposes and for semiquantitative analysis of classical, reversible enzyme modification. The plot is superior to the double reciprocal plot in revealing subtle differences between competitive and mixed inhibitors and is therefore well suited for analyzing the action of allosteric effectors. We wish however to emphasize that this method is unsuitable for regression analysis. Namely, the ratio $\sigma/(1 + \sigma)$ is nothing else than v_0/V, which means that v_0 is part of the dependent as well as of the independent variable. This violates one of the fundamental principles of statistical methods for data analysis, i.e. that only the dependent variable is affected by errors (Sect. 2.4.3). Nevertheless, the specific velocity plot provides good estimates of inhibition or activation constants as well as the α and β

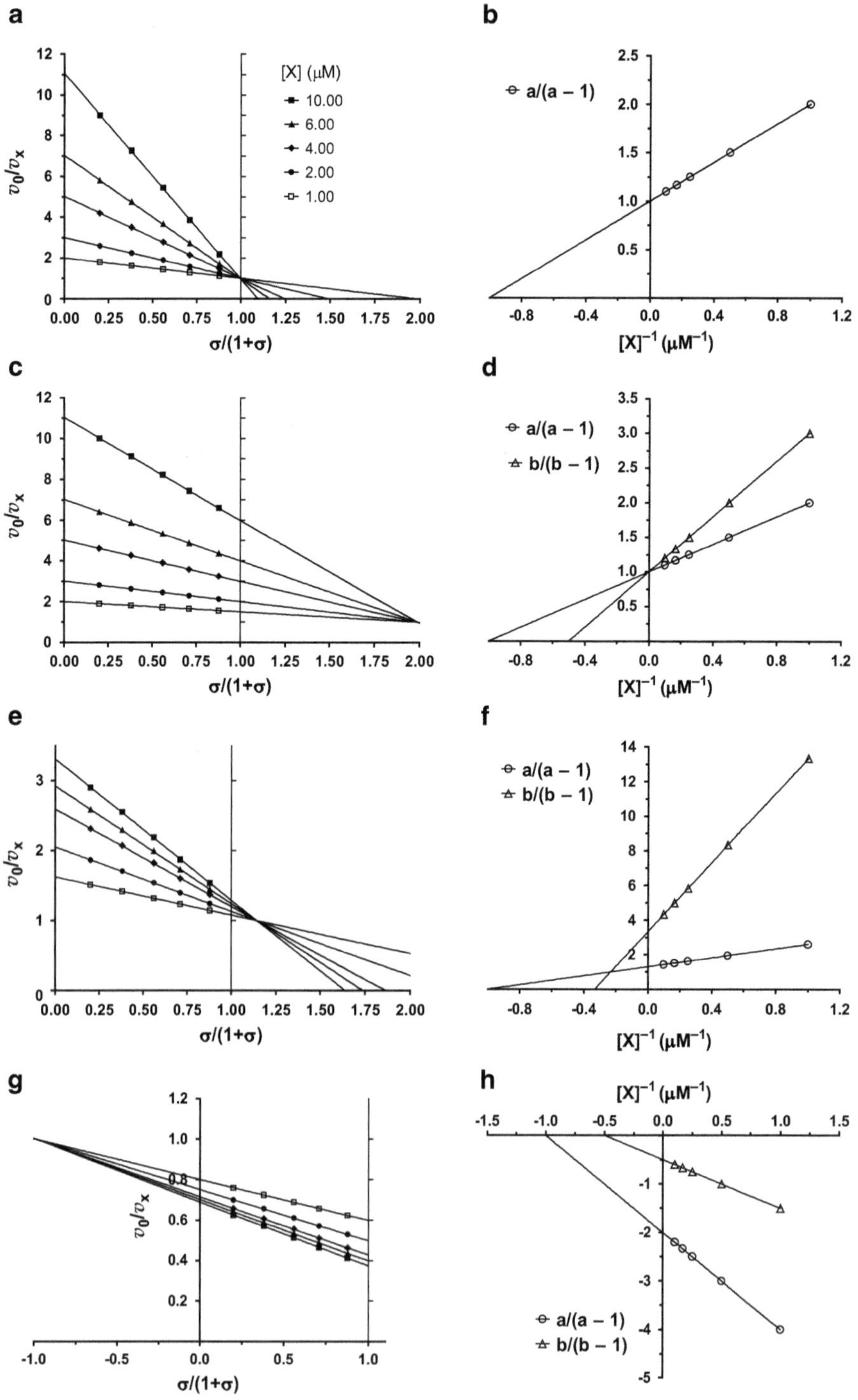

Fig. 2.6 Simulated examples of the specific velocity plot (**a**, **c**, **e**, **g**) and replots (**b**, **d**, **f**, **h**). The four examples were simulated with the following common parameters: $k_{cat} = 20\ s^{-1}$, $[E]_t = 0.025$

coefficients of the general modifier mechanism, which can be further used for refining calculations using Eq. (2.8) considering the modifier concentration as the independent variable.

2.6.3 A General Equation for Classical and Tight-Binding Systems

The treatment of the general modifier mechanism and of the specific velocity plot tacitly assumed that the concentration of free and total modifier was about the same ($[X] \approx [X]_t$) in every instance. Here we consider the case of tight inhibitor binding and examine first the relationships between $[I]_t$, $[E]_t$ and K_i. If comparable concentrations of enzyme and inhibitor give rise to appreciable inhibition and if this process is diffusion controlled, i.e. it occurs very rapidly, we are faced with tight-binding inhibition. This concept must be stated more precisely because it can for instance happen that, with 1 mM enzyme and inhibitor, full inhibition occurs because $K_i = 1$ μM. However, with inhibitor and enzyme at a concentration of 0.01 μM the degree of inhibition will be lower. The condition for tight-binding inhibition is therefore precisely defined with $[I]_t \approx [E]_t$ and K_i (Table 2.4).

A rate equation for tight-binding inhibition, valid for both linear and hyperbolic inhibition mechanisms, has been derived by Baici (1987). A regrettable error in this paper, which prevented the universal use of this equation for any mechanism, has been amended by Szedlacsek et al. (1988), who used symbols different from those of the original. To keep consistency with the symbols used in this chapter, the equation is written here using the same notation of the general modifier mechanism:

$$v_i = \frac{v_0}{2}\left[\frac{\alpha+\sigma-\beta(1+\sigma)}{\alpha+\sigma}\right]\left\{\begin{array}{l}\sqrt{\left[\left(\frac{1+\sigma}{\alpha+\sigma}\frac{\alpha K_i}{[E]_t}+\frac{[I]_t}{[E]_t}-1\right)^2+4\frac{1+\sigma}{\alpha+\sigma}\frac{\alpha K_i}{[E]_t}\right]} \\ +\frac{\alpha+\sigma+\beta(1+\sigma)}{\alpha+\sigma-\beta(1+\sigma)}-\frac{1+\sigma}{\alpha+\sigma}\frac{\alpha K_i}{[E]_t}-\frac{[I]_t}{[E]_t}\end{array}\right\}. \quad (2.13)$$

Equation (2.13), which contains the total concentrations of inhibitor and enzyme, can be used for calculating the inhibition constant of both classical and

Fig. 2.6 (continued) μM, from which $V = 0.5$ μM s^{-1}, $K_m = 20$ μM, $K_x = 1$ μM. Additionally: (**a**) and (**b**) (linear competitive inhibition) $\alpha = \infty$ and $\beta = 0$; (**c**) and (**d**) (linear mixed inhibition) $\alpha = 2$ and $\beta = 0$; (**e**) and (**f**) (hyperbolic mixed inhibition) $\alpha = 3$ and $\beta = 0.7$; (**g**) and (**h**) (non-essential activation) $\alpha = 2$ and $\beta = 3$. The same five inhibitor concentrations (μM) shown in panel (**a**) were used for all examples

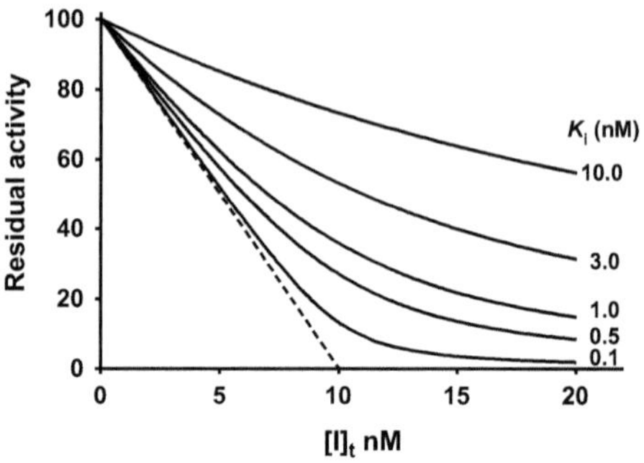

Fig. 2.7 Tight-binding inhibition. Simulated curves of residual activity versus total inhibitor concentration calculated with Eq. (2.13) and the following parameters: $v_0 = 100$, $\alpha = \infty$ (in practice a large value such as 10^9), $\beta = 0$, $\sigma = 1$, $[E]_t = 10$ nM, K_i five values (nM) as shown in the figure, $[I]_t$ continuously varied between 0 and 20 nM. The *dashed line* represents the titration curve, which can be obtained with an irreversible inhibitor

tight-binding inhibitors that may be linear or hyperbolic. It is indispensable whenever the concentrations of enzyme and inhibitor are in the same order of magnitude and the value of K_i prevents the validity of the assumption $[I] \approx [I]_t$. Precise calculations require of course the knowledge of the active sites concentration of the enzyme.

The relationship between $[I]_t$, $[E]_t$ and K_i is illustrated in Fig. 2.7, which shows residual activity profiles as a function of the total concentration of a linear competitive inhibitor at fixed enzyme concentration and variable K_i. Depending on the magnitude of the inhibition constant, the residual activity profile comes close to the titration curve, the dashed straight line in Fig. 2.7, which can be obtained with an irreversible inhibitor.

The properties of Fig. 2.7 can be exploited for determining the active concentration of a protein inhibitor of peptidases using an enzyme whose active sites concentration has been previously measured by titration with an irreversible inhibitor. This procedure is explained in Fig. 2.8 for a preparation of the human thyroglobulin type-1 domain of testican 3 (TST3) as a linear competitive inhibitor of human cathepsin B with a known $K_i = 13.6$ nM. The enzyme was previously titrated with the inactivator E-64 and used at a known final concentration in a further experiment with variable TST3 amounts, whose concentration as protein was known. The residual activity in the presence of a fluorogenic substrate and with a titrated cathepsin B concentration in the assays of 10.0 nM is plotted in Fig. 2.8 versus the total inhibitor concentration as protein, measured photometrically. It is immediately seen that any attempt at extrapolating the unknown TST3 concentration from the curve is useless (the 'true' titration line would correspond to the straight dashed line). Hence, Eq. (2.13) was fitted to data by treating the enzyme concentration as the sole parameter to be optimized, i.e. all parameters in the equation were set as known (see the legend of Fig. 2.8), while only $[E]_t$ was allowed to float during non-linear regression. The best fit value for $[E]_t$ was 13.0 ± 2.9 nM.

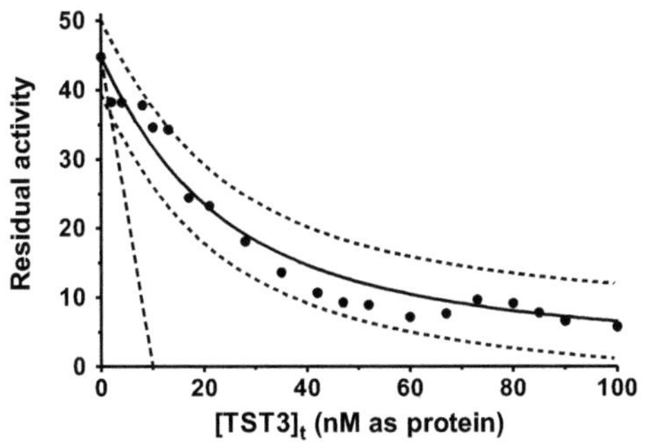

Fig. 2.8 Titration of human cathepsin B with known active site concentration with TST3 as reversible inhibitor. Residual activity (fluorescence reading units) plotted versus TST3 total concentration as protein. Equation (2.13) was fitted to data (*black dots*) with the following fixed parameters: $v_0 = 44.8, \alpha = 10^9$ and $\beta = 0$ (linear competitive inhibitor), $\sigma = 0.12, K_i = 13.6$ nM. The enzyme active site concentration was known (10.0 nM) but was considered as the sole parameter to be fitted with the purpose of measuring the unknown active site concentration of the inhibitor and the best fit (*solid line*) gave $[E]_t = 13.0 \pm 2.9$ nM. The *dashed straight line* indicates the titration curve that would be obtained with an irreversible inhibitor, and the *dashed bent curves* show the 95 % confidence band. The runs test suggested no significant deviation from the model

Since the true enzyme concentration was 10.0 nM, the protein concentration of TST3 was multiplied by the factor 10/13 = 0.77 to obtain its active site concentration, i.e. the inhibitor preparation was 77 % active.

An example of kinetic analysis, in which the specific velocity plot for diagnostic purposes and of Eq. (2.13) for quantitative analysis were combined is the inhibition of caspase-2 by a designed ankyrin repeat. This modifier behaves as an allosteric effector of caspase-2 and acts as a hyperbolic mixed inhibitor. The kinetic mechanism fits neatly with the crystal structure of the complex (Schweizer et al. 2007).

2.6.4 Non-productive Binding and Substrate Inhibition

The substrates of peptidases are oligomers or large polymers. Apart the physiologically relevant cases of limited proteolysis, in which just one peptide bond is cleaved within a polypeptide, large protein substrates may be cleaved at multiple sites, i.e. wherever 'specific' susceptible bonds are recognized by the enzymes. For practical reasons, *in vitro* measurements are performed with synthetic oligopeptides of low molecular mass, which in virtue of their less bulky structure may be expected to bind the enzyme not only in the productive way leading to peptide bond cleavage, but possibly in one or more additional unproductive ways that do not lead to proteolytic breakdown. This situation can be described kinetically as shown in Scheme 2.7, which is analogous to linear competitive inhibition. If more than one non-specific binding site exists Scheme 2.7 symbolizes the average of all of them.

Scheme 2.7 Non-productive substrate binding

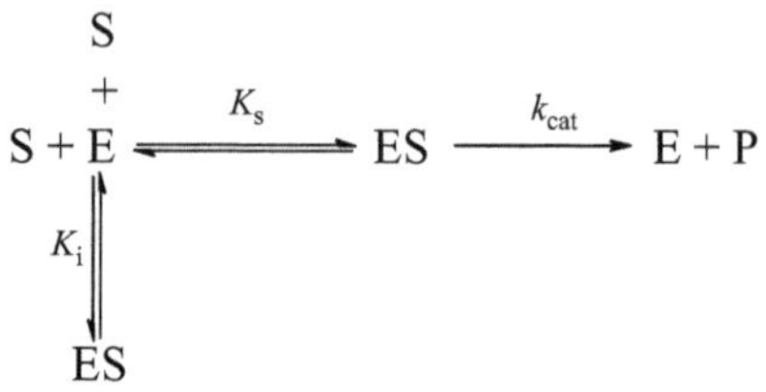

This inhibition is of the linear competitive type and its rate equation can be obtained from the general modifier mechanism, Eq. (2.8), by setting $\alpha = \infty$ and $\beta = 0$. The explicit expression $[\mathrm{S}]/K_\mathrm{m}$ for σ is used to clearly describe the properties of this system:

$$v = \frac{V[\mathrm{S}]}{K_\mathrm{m}\left(1 + \frac{[\mathrm{S}]}{K_\mathrm{i}}\right) + [\mathrm{S}]}. \tag{2.14}$$

The 'substrate' is thus at the same time substrate and inhibitor and there is no way to appreciate that inhibition exist from velocity measurements at various substrate concentrations because a rectangular hyperbola is obtained with or without such an inhibitory effect. Equation (2.14) can namely be rearranged to

$$v = \frac{\dfrac{V}{1+\dfrac{K_\mathrm{m}}{K_\mathrm{i}}}[\mathrm{S}]}{\dfrac{K_\mathrm{m}}{1+\dfrac{K_\mathrm{m}}{K_\mathrm{i}}} + [\mathrm{S}]}. \tag{2.15}$$

According to Cornish-Bowden (p. 137 in Cornish-Bowden 2004), calling $\widetilde{V}$ and $\widetilde{K}_\mathrm{m}$ the parameters that would be observed in the absence of non-productive binding, Eq. (2.15) is equivalent to the Michaelis-Menten equation

$$v = \frac{V[\mathrm{S}]}{K_\mathrm{m} + [\mathrm{S}]},$$

in which the limiting rate and the Michaelis constant are divided by the same factor

$$V = \frac{\widetilde{V}}{1 + \dfrac{\widetilde{K}_\mathrm{m}}{K_\mathrm{i}}}, \quad K_\mathrm{m} = \frac{\widetilde{K}_\mathrm{m}}{1 + \dfrac{\widetilde{K}_\mathrm{m}}{K_\mathrm{i}}}. \tag{2.16}$$

The relevant aspect is that $V/K_\mathrm{m} = \widetilde{V}/\widetilde{K}_\mathrm{m}$ and thus the competitive inhibitory nature of substrates that bind non-productively will remain unobserved (see also Fig. 2.9). Therefore, the measured values of the kinetic parameters will be biased by an unknown factor with respect to parameters expected from the binding mode that

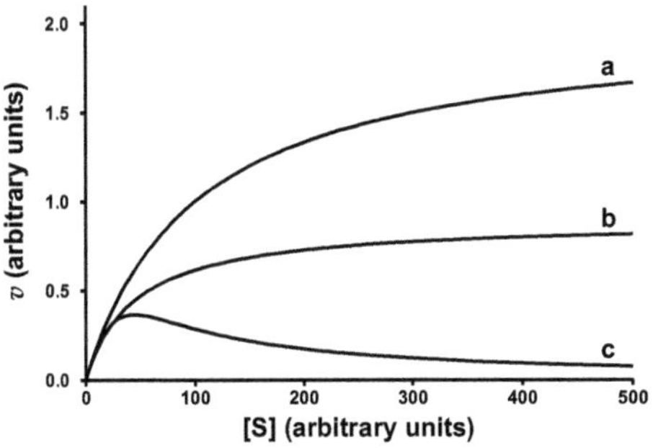

Fig. 2.9 Non-productive binding and substrate inhibition. The three curves were simulated with the common parameters $V = 2$ and $K_m = 100$. a—Michaelis-Menten equation (2.1); b—non-productive binding, Eq. (2.14), with $K_i = 80$; c—substrate inhibition, Eq. (2.17), with $K_{si} = 20$ (parameters and concentrations in arbitrary units)

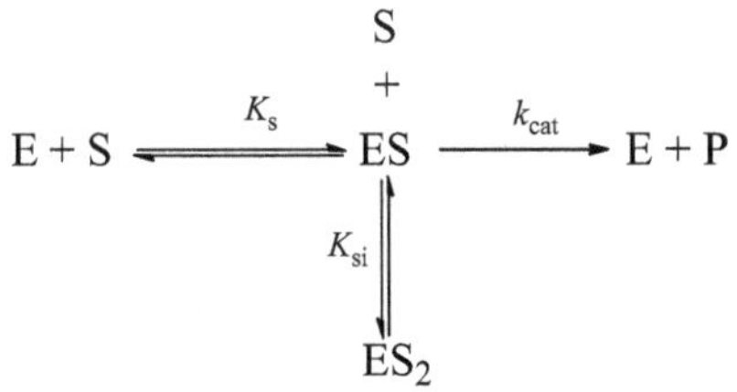

Scheme 2.8 Uncompetitive substrate inhibition

results in peptide bond cleavage. This means that comparing k_{cat}, K_m and k_{cat}/K_m for different peptide substrates may not always reflect the authentic 'specificity' of the considered peptidase.

The term 'substrate inhibition' is reserved to the case in which a substrate molecule binds to an already formed ES complex according to the mechanism in Scheme 2.8.

Oligopeptide synthetic substrates of peptidases are very often involved in substrate inhibition of this type when substrate is in excess. The rate equation for this mechanism is deduced as a particular case of the general modifier mechanism, Eq. (2.8), with $\beta = 0$, [S] replacing [X], and αK_i called now K_{si} to indicate the inhibitory action of the substrate

$$v = \frac{V'[S]}{K'_m + [S] + \frac{[S]^2}{K_{si}}}. \tag{2.17}$$

The primed symbols on limiting rate and Michaelis constant emphasize the fact that they cannot be considered Michaelis parameters because the rate has a limit of zero for $[S] \to \infty$, not V', and K'_m does not correspond to the substrate concentration for which $v{=}V'/2$. In fact, the squared substrate concentration in the denominator of Eq. (2.17) forces the rate approaching zero at infinite [S]. Non-productive binding and substrate inhibition are illustrated graphically in Fig. 2.9. In curve b it is shown that non-productive binding is characterized by a rectangular hyperbola, for which any inhibitory substrate effect cannot be suspected in the absence of additional information. Curve c describes substrate inhibition and curve a shows the 'regular' profile that would be obtained in the absence of any inhibitory effects.

2.7 Slow-Binding Inhibition

The nomenclature of the 'slow' mechanisms adopted here is that of Table 2.4. Chemical reasons for the slow inhibiting behavior include: (1) the formation of an intermediate with structural analogy to the transition state during catalysis; (2) fast binding of the inhibitor to the enzyme but true inhibition is only accomplished after slow rearrangement of the enzyme conformation; (3) a molecule exists in equilibrium between two or more chemical forms of which only one is the true inhibitor, which may have a small concentration so that the second-order association reaction $E + I \rightarrow EI$ may proceed very slowly; (4) the enzyme exists in two interconverting conformational states, of which only one binds the inhibitor. The kinetic information that can be extracted from slow-binding and slow, tight-binding inhibition experiments is larger than that obtained from conventional steady-state measurements of fast-acting inhibitors. In fact, besides overall inhibition constants, also second-order rate constants for the formation of the inhibited enzyme and rate constants for the dissociation of the E·I complex can be calculated. Two approaches are possible: to use the integrated rate equation for diagnosing the mechanism and calculating the parameters, or to simulate the process by numerical integration using all individual kinetic constants starting from initial guesses and then fitting iteratively the integrated progress curves to data by non-linear regression (Sect. 2.4.4). For slow-binding systems both methods have the own advantages and limitations. Integrated rate equations can only be derived under restrictive assumptions or cannot be derived at all, but provide valuable information about the underlying mechanism if the assumptions can be kept under control experimentally. Furthermore, the dependencies on [I] of the parameters calculated by regression analysis represent a powerful diagnostic tool for model discrimination. Numerical integration does not depend on restrictive assumptions and can combine the information from different sets of experiments fitted globally to exploit the full statistical power of the method. Model discrimination can only be performed by running individually the relevant models with the same data and comparing the statistical outputs to find the best matching set. The major drawback of numerical integration is in many instances the large number of individual kinetic constants that must be fitted for some mechanisms and the impossibility of finding a unique array of constants that matches the model. In other words, despite superb mathematics, it is often impossible to decide which mechanism best describes the data and to evaluate all of the constants for the system because these are not well

Scheme 2.9 The general case of linear competitive, slow-binding inhibition

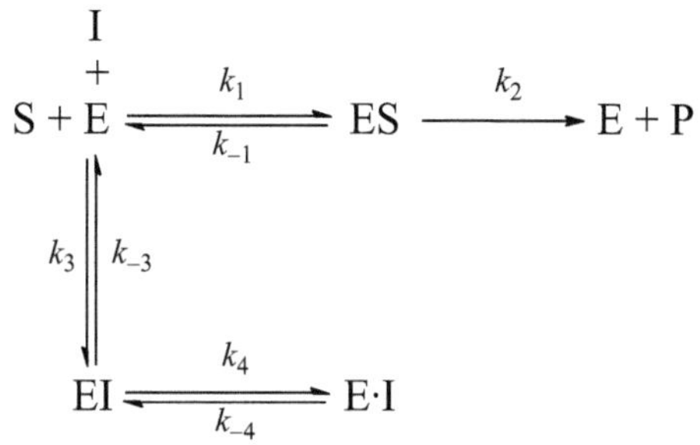

constrained by data. The two-step mechanism of linear competitive, slow-binding inhibition shown in Scheme 2.9 illustrates this problem.

Whenever a linear competitive inhibitor manifests slow-binding behavior and neither for the inhibitor nor for the enzyme, equilibria between different forms are known from chemical or other evidence, the mechanism in Scheme 2.9 should be considered first without making any assumption about the relative rates of equilibration of the two steps, the existence of only one step or any other simplifying aspect. The attainment of the steady-state for this mechanism is preceded by two exponential phases. Thus, the first logical approach is to fit the generic Eq. (2.18) (single exponential followed by a linear increase of the function) and Eq. (2.19) (double exponential followed by linear increase) to data and to determine which one fits the data better than the other. The A-factors in these equations are the amplitudes of the exponential phases and d is a displacement on the ordinate. It is anticipated that one of the two phases may be relatively rapid and/or may have a small amplitude. Thus, conventional methods for data acquisition may be inadequate because the time 'lost' for mixing solutions and activating the recording of reaction progress may take a relatively long time with loss of precious data. In such cases the use of a stopped-flow apparatus is highly recommended.

$$Y = A_1\left(1 - e^{-\lambda_1 t}\right) + kt + d \tag{2.18}$$

$$Y = A_1\left(1 - e^{-\lambda_1 t}\right) + A_2\left(1 - e^{-\lambda_2 t}\right) + kt + d \tag{2.19}$$

If it can be established that Eq. (2.18) fits the primary data (progress curves) better than Eq. (2.19), it can be assumed that one of the two steps of the mechanism in Scheme 2.9 is much more rapid than the other or that the mechanism degenerates to a single-step reaction between enzyme and inhibitor. The same reasoning holds for other mechanisms characterized by just one slow step and, correspondingly, the integrated equation for all these systems contains only one exponential phase. For these mechanisms, described in Sect. 2.7.1, an integrated rate equation can be obtained. For the general case shown in Scheme 2.9 an integrated rate equation has been derived under restrictive conditions, but due to its complexity it can hardly be used in practice (Kuzmič 2008). However, this and any other slow-binding mechanism can be analyzed by numerical integration as described with an example in Sect. 2.7.3.

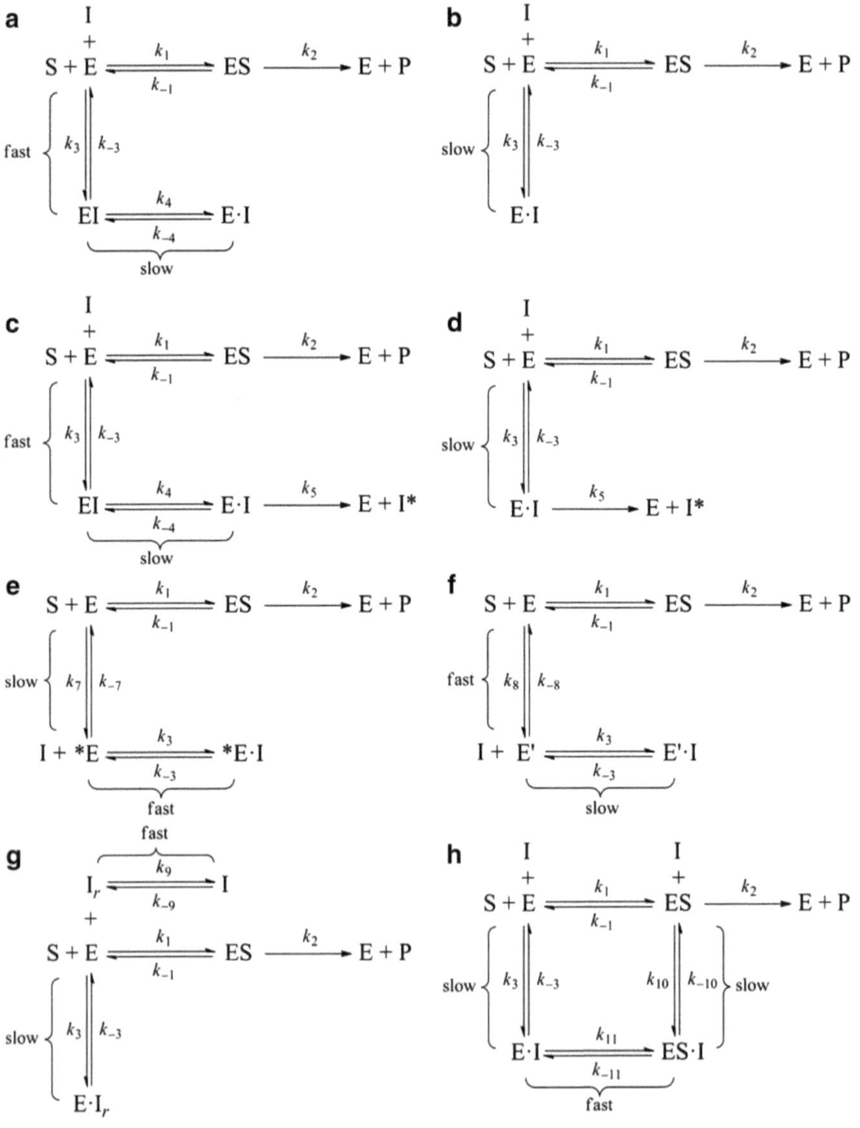

Scheme 2.10 Mechanisms for slow-binding inhibition of peptidases. The numbering of kinetic constants is consistently kept throughout the mechanisms to identify the same or similar paths

2.7.1 Integrated Rate Equations: Their Usefulness and Limits

A common trend in the literature has been to consider the two mechanisms shown in panels (a) and (b) of Scheme 2.10 as the major representatives of slow-binding inhibition systems. While these might really be the most frequently encountered mechanisms, in most instances analysis was conducted on the assumption of an underlying linear competitive mechanism without demonstrating this fact by adequate experiments. In principle, any other of the mechanisms shown in Scheme 2.10 may be responsible for the sluggishness of the inhibition process, a matter of fact that must be considered.

For all mechanisms in Scheme 2.10 (but see the comments below on the temporary inhibition mechanisms c and d) the integrated rate equation is given by

$$[\mathrm{P}] = v_\mathrm{s}t - \frac{(v_\mathrm{s} - v_\mathrm{z})}{\lambda}\left(1 - e^{-\lambda t}\right) + d, \tag{2.20}$$

where v_s and v_z are the velocities at steady-state and at time zero, respectively, λ is an apparent first-order constant that describes the exponential phase, and d a displacement on the ordinate of the progress curve that accounts for any non-zero value of the signal proportional to [P] at the beginning of the reaction, e.g. absorbance or fluorescence of the substrate before hydrolysis of the susceptible peptide bond. This equation has originally been published by Frieden in the context of enzyme hysteresis (Frieden 1970). The derivation of Eq. (2.20) is subordinated to the following restrictive assumptions: (1) steady-state conditions are set up very rapidly for the flux around E and ES; (2) non-steady state conditions exist for the steps marked 'slow' and rapid equilibrium conditions exist for the steps marked 'fast' in Scheme 2.10; (3) $[\mathrm{S}]_\mathrm{t} \gg [\mathrm{E}]_\mathrm{t}$ so that there is no depletion of free substrate through binding to the enzyme; (4) measurements are performed for a time that does not involve substantial turnover of substrate and accumulation of product; (5) $[\mathrm{I}]_\mathrm{t}$ is at least 10 times greater than $[\mathrm{E}]_\mathrm{t}$, meaning that the condition $[\mathrm{I}] \approx [\mathrm{I}]_\mathrm{t}$ is valid throughout and tight-binding is not present.

With reference to Scheme 2.10, mechanism b is simply a degenerated form of mechanism a because the concentration of the EI complex is kinetically insignificant. Mechanisms c and d represent temporary inhibition, in which the E·I complex breaks down to free enzyme that is put back to bind substrate and inhibitor, and to an inactive form of the inhibitor I*, in general a proteolytically cleaved form. We wish to emphasize that the integrated rate equation (2.20) for mechanisms c and d is valid only under the additional restrictive assumption that the decay of E·I into E + I* is a slow process and that [I*] remains much smaller than $[\mathrm{I}]_\mathrm{t}$ during the observation time of the experiment. This means that the inhibitor is present at a sufficiently high concentration, which acts as a sink to satisfy also the conditions $[\mathrm{I}] \approx [\mathrm{I}]_\mathrm{t}$ and $[\mathrm{I}]_\mathrm{t} \gg [\mathrm{E}]_\mathrm{t}$. A safe experimental judgment for this condition is that the steady-state line represented by the straight line in trace b of Fig. 2.10 does not yet start to bend up (see below). Mechanism d has been demonstrated for instance in the reaction of cathepsin L with the thyroglobulin type-1 domain of human testican (Meh et al. 2005). Mechanisms e and f, in which the inhibitor binds to a rare form of the enzyme that fluctuates between two conformational states, are very similar and differ only for the position of the slow step. Mechanism e has been put forward in a non-proteolytic context (Duggleby et al. 1982) and mechanism f has been discussed in the frame of thermolysin inhibition by phosphonates (Bartlett and Marlowe

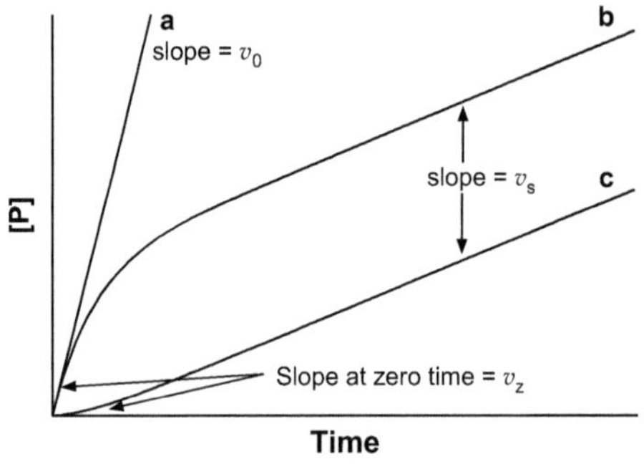

Fig. 2.10 Progress curves for slow-binding inhibition according to Eq. (2.20). a—enzyme incubated with substrate alone; b—reaction started by adding enzyme to a mixture of substrate and inhibitor; c—reaction started by adding substrate to enzyme preincubated with inhibitor for a sufficiently long time to allow formation of the inhibited complex

1987). Following the observation that peptidases may be present in different conformational states with distinct kinetic properties, e.g. cathepsin K at pH = 7.40 (Novinec et al. 2010), these two mechanisms should be included among the candidates in slow inhibition processes of peptidases.

In mechanism g the enzyme reacts with a rare form of the inhibitor (I_r), which exists in equilibrium with other species (Bartlett and Marlowe 1987). An example for peptidases is the inhibition of cathepsin B by leupeptin, which equilibrates between the free aldehyde, the aldehyde hydrate and a cyclic carbinolamine. Only the aldehyde form, which makes up merely 2 % of the total leupeptin concentration, is inhibitory. Thus, the slow-binding behavior of leupeptin is due to the low concentration of the inhibitor, which lowers the rate of E·I formation, and to the stability of an intermediate tetrahedral hemithioacetal (Schultz et al. 1989). Finally, mechanism h represents a linear, slow-binding mixed inhibition type, which has been demonstrated for adenosine deaminase (Cha 1976). This mechanism should be considered methodically when studying slow inhibition of peptidases, in particular by allosteric effectors.

Progress curves, as product concentration versus time conforming to the integrated equation (2.20), are shown in Fig. 2.10. In trace (b), in which reaction is started by adding enzyme to a mixture of substrate and inhibitor, slow inhibition is characterized by an exponential phase that is followed by a linear steady-state with slope v_S. This is the standard type of progress curves normally seen in the literature. Trace c in Fig. 2.10 results when enzyme and inhibitor are preincubated at high concentration to allow complex formation and reaction is started by diluting this mixture into a substrate-containing solution. The slopes of traces b and c are the same. The type of experiment as in trace c gives direct information on the dissociation constant of the inhibited complex and is precious for model discrimination. The slopes of traces b and c will not remain linear during the entire course of the reaction but will result in a horizontal plateau when all substrate has been used up. For the temporary inhibition mechanisms c and d in Scheme 2.10 the slope of

Table 2.6 Expressions of the parameters in Eq. (2.20) for mechanisms a and b in Scheme 2.10

Mechanism a		Mechanism b	
$\lambda = k_{-4} + \dfrac{k_4[\mathrm{I}]}{K_\mathrm{i}\left(1 + \dfrac{[\mathrm{S}]}{K_\mathrm{m}}\right) + [\mathrm{I}]}$ $K_\mathrm{i} = \dfrac{k_{-3}}{k_3}$	(2.21)	$\lambda = k_{-3} + \dfrac{k_3[\mathrm{I}]}{1 + \dfrac{[\mathrm{S}]}{K_\mathrm{m}}}$	(2.22)
$v_\mathrm{z} = \dfrac{V[\mathrm{S}]}{K_\mathrm{m}\left(1 + \dfrac{[\mathrm{I}]}{K_\mathrm{i}}\right) + [\mathrm{S}]}$	(2.23)	$v_\mathrm{z} = v_0 = \dfrac{V[\mathrm{S}]}{K_\mathrm{m} + [\mathrm{S}]}$	(2.24)
$v_\mathrm{s} = \dfrac{V[\mathrm{S}]}{K_\mathrm{m}\left(1 + \dfrac{[\mathrm{I}]}{K_\mathrm{i}^*}\right) + [\mathrm{S}]}$ $K_\mathrm{i}^* = K_\mathrm{i}\left(\dfrac{k_{-4}}{k_4 + k_{-4}}\right)$	(2.25)	$v_\mathrm{s} = \dfrac{V[\mathrm{S}]}{K_\mathrm{m}\left(1 + \dfrac{[\mathrm{I}]}{K_\mathrm{i}}\right) + [\mathrm{S}]}$	(2.26)

The expressions of v_z and v_0 apply to assays in which the reaction is started by adding enzyme to a solution containing substrate and inhibitor

trace b first slowly increases to a level equal to v_0 (trace a) and then declines to zero following substrate depletion. The form of trace c depends on the time the enzyme had been preincubated with inhibitor before adding substrate to start recording the proteolytic reaction. Of course, the proportion of degraded inhibitor I* will increase with preincubation time and the form of the progress curves will vary. This experiment with temporary reversible inhibitors has thus no quantitative value but is highly diagnostic to detect temporary inhibition since in the extreme case of a very long preincubation, adding substrate will result in no inhibition with a slope corresponding to trace a, provided the enzyme has not lost activity for some reason during the incubation time.

While the various mechanisms have the same common integrated rate equation (2.20), what distinguishes them from one another are the expressions of v_s, v_z, and λ. These expressions, distributed in a number of specialized papers and reviews, are summarized below for practical consultation: Tables 2.6 and 2.7 list λ, v_z and v_s for mechanisms a–d of Scheme 2.10, while Table 2.8 lists only the values of λ for mechanisms e–h. If the assumptions made above for deriving Eq. (2.20) can be guaranteed experimentally and data are collected with precision, discrimination of the mechanisms in Scheme 2.10 can be accomplished by analyzing the dependency of v_s, v_z and λ on [I]. The relevant equations for these variables are then used to extract kinetic constants. The distinction between mechanisms a and b is straightforward because λ depends hyperbolically on [I] in mechanism a and linearly in mechanism b, and v_z depends on [I] in mechanism a, while it is independent of [I] in mechanism b. The discrimination of mechanisms a and b from their corresponding temporary inhibition mechanisms c and d cannot be made on the basis of the

Table 2.7 Expressions of the parameters in Eq. (2.20) for the temporary inhibition mechanisms c and d in Scheme 2.10

Mechanism c	Mechanism d
$\lambda = k_{-4} + k_5 + \dfrac{k_4[\mathrm{I}]}{K_\mathrm{i}\left(1+\dfrac{[\mathrm{S}]}{K_\mathrm{m}}\right)+[\mathrm{I}]}$ (2.27)	$\lambda = k_{-3} + k_5 + \dfrac{k_3 K_\mathrm{m}}{K_\mathrm{m}+[\mathrm{S}]}[\mathrm{I}]$ (2.28)
$K_\mathrm{i} = \dfrac{k_{-3}}{k_3}$	
v_z, the same as Eq. (2.23)	v_z, the same as Eq. (2.24)
$v_\mathrm{s} = \dfrac{V[\mathrm{S}]}{K_\mathrm{m}\left\{1+\dfrac{[\mathrm{I}]}{K^*_{\mathrm{i,\ temp}}}\right\}+[\mathrm{S}]}$ (2.29)	$v_\mathrm{s} = \dfrac{V[\mathrm{S}]}{K_\mathrm{m}\left(1+\dfrac{[\mathrm{I}]}{K_{\mathrm{i,\ temp}}}\right)+[\mathrm{S}]}$ (2.30)
$K^*_{\mathrm{i,temp}} = K_\mathrm{i}\dfrac{k_{-4}+k_5}{k_4+k_{-4}+k_5}$	$K_{\mathrm{i,temp}} = \dfrac{k_{-3}+k_5}{k_3}$

The expressions of v_z and v_0 apply to assays in which the reaction is started by adding enzyme to a solution containing substrate and inhibitor. These equations are only valid under the assumption $[\mathrm{I}^*] \ll [\mathrm{I}]_\mathrm{t}$

Table 2.8 Expressions for λ in Eq. (2.20) for mechanisms e, f, g, and h in Scheme 2.10

Mechanism e	Mechanism f
$\lambda = \dfrac{k_7}{1+\dfrac{[\mathrm{S}]}{K_\mathrm{m}}} + \dfrac{k_{-7}}{1+\dfrac{[\mathrm{I}]}{K_\mathrm{i}}}$ (2.31)	$\lambda = k_{-3} + k_3\dfrac{K_\mathrm{e}[\mathrm{I}]}{1+K_\mathrm{e}+\dfrac{[\mathrm{S}]}{K_\mathrm{m}}}$ (2.32)
	$K_\mathrm{e} = \dfrac{[\mathrm{E}']}{[\mathrm{E}]}$
Mechanism g	Mechanism h
$\lambda = k_{-3} + k_3\dfrac{K_r[\mathrm{I}]}{1+\dfrac{[\mathrm{S}]}{K_\mathrm{m}}}$ (2.33)	$\lambda = \dfrac{k_{-3}+\dfrac{k_{-10}k_{11}}{k_{-11}}[\mathrm{S}]}{1+\dfrac{k_{11}}{k_{-11}}[\mathrm{S}]} + \dfrac{k_3+k_{10}\dfrac{[\mathrm{S}]}{K_\mathrm{m}}}{1+\dfrac{[\mathrm{S}]}{K_\mathrm{m}}}[\mathrm{I}]$ (2.34)
$K_r = \dfrac{[\mathrm{I}_r]}{[\mathrm{I}]} \ll 1$	

dependencies of v_s, v_z and λ upon [I] because the shape of the functions are the same, but they can readily be distinguished from one another by preincubating enzyme and inhibitor for increasing times and starting reactions by adding substrate (trace c in Fig. 2.10). The slope of the steady-state portion of the curve will be independent of the preincubation time for mechanisms a and b but will increase with incubation time for mechanisms c and d as a consequence of inhibitor degradation to I*. Mechanism e poses no diagnostic problems because it is the only one for which λ depends hyperbolically on [I] but in the reverse direction, i.e. decreasing for increasing [I]. Furthermore, for mechanism e, the asymptote of λ for $[\mathrm{I}] \to \infty$ equals $k_7/(1 + [\mathrm{S}]/K_\mathrm{m})$ and thus depends on substrate concentration.

In the other mechanisms for which λ has a hyperbolic dependence on [I] (mechanisms a and c), λ increases for increasing [I] and the asymptote is independent on [S]. More puzzling is the diagnosis of mechanisms f, g and h on the basis of the dependence of λ upon [I], which is linear as it is for mechanisms b and d. However, the slopes of the steady-state lines of progress curves will reveal the mixed-type nature of mechanism h, which can then be diagnosed and the constants determined. As a general comment, the vast majority of publications on slow-binding inhibition report progress curves measured at a fixed enzyme and substrate concentration and variable [I]. However, it would be very useful to perform the same experiments also keeping [I] constant for variable [S] in order to determine if the inhibitor has a competitive or a mixed character. Finally, mechanisms f and g cannot be diagnosed on the basis of kinetics alone since they would be mostly ascribed to mechanism b. This shows that any information about enzyme properties, in particular the existence of interconverting enzyme conformations, should be considered. An example of how to recognize enzyme conformers is the already mentioned work on cathepsin K (Novinec et al. 2010).

For space reasons we cannot show here graphical examples of the mechanisms of slow-binding inhibition discussed in this section. The reader might find useful a review of diagnostic methods for slow reversible and irreversible inhibition (Baici et al. 2009).

2.7.2 Slow, Tight-Binding Inhibition

When the condition $[I] \approx [I]_t$ cannot be met Eq. (2.20) is not usable and the inhibition is of the tight-binding type. Integrating the rate equation for such systems is not always possible. Out of the mechanisms in Scheme 2.10, integrated rate equations for the tight-binding condition have been published for case b (Cha 1980; Williams et al. 1979) and for case h (Cha 1976), while for mechanism a and for its more general counterpart shown in Scheme 2.9 analytical integration is not possible. In any case, the integrated rate equations for slow, tight-binding are subjected to restrictive assumptions. These equations are not reproduced here because the method described below in Sect. 2.7.3 is free of such limitations and supersedes analytical integration.

2.7.3 Numerical Integration Coupled to Non-linear Regression

When an integrated rate equation can either not be obtained or the assumptions/ conditions listed in Sect. 2.7.1 cannot be realized experimentally, or if tight-binding further complicates the situation, numerical integration is the method of choice for

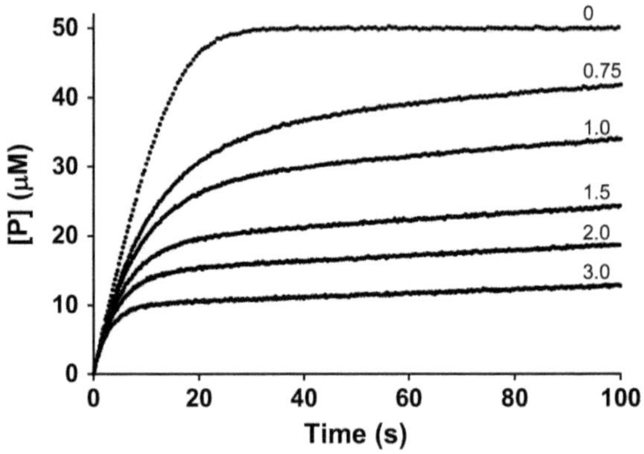

Fig. 2.11 Simulated data for slow, tight-binding inhibition. The simulation was performed with Simulink software using the following kinetic constants and total concentrations: $k_1 = 100\ \mu M^{-1}\ s^{-1}$, $k_{-1} = 1{,}990\ s^{-1}$, $k_2 = 10\ s^{-1}$ (these constants give $K_m = 20\ \mu M$), $k_3 = 0.4\ \mu M^{-1}\ s^{-1}$, $k_{-3} = 0.003\ s^{-1}$, $[E]_t = 0.5\ \mu M$, $[S]_t = 50\ \mu M$. Numbers next to the curves represent $[I]_t$ in μM units

data analysis. The best way to illustrate this method is to work out an example. We deal here with mechanism b in Scheme 2.10 for which a set of 'data' was simulated with Simulink as shown in Fig. 2.11. Some noise was also added to the simulated curves to mimic a real experiment. In this example we assume that the enzyme is not particularly stable in diluted solution over the period of 10–30 min typically necessary to measure slow-binding inhibition reactions. To overcome this problem experiments can be performed using a high enzyme concentration in order to shorten the reaction time and the progress curves are conveniently measured with a stopped-flow apparatus (less than 2 min in Fig. 2.11). The details of the kinetic constants and initial reactant concentrations are described in the figure legend. Since $[E]_t = 0.5\ \mu M$, with the kinetic constants of the system we have also tight-binding, meaning that a considerable portion of the added inhibitor binds to the enzyme and the condition $[I] \approx [I]_t$ is invalid. Figure 2.11 shows that during reaction, depending on inhibitor concentration, 20–80 % of substrate is converted to product. Thus, with both inhibitor and substrate depletion, the integrated rate equation cannot be used neither in the form of Eq. (2.20) nor in the specially adapted form for slow, tight-binding inhibition (Cha 1980; Williams et al. 1979). Despite this fact, we compare now the results that can be obtained with the 'illegal' use of Eq. (2.20) with those from numerical integration. We start by fitting Eq. (2.20) to the data of Fig. 2.11 as shown in Fig. 2.12. Judging from the good superimposition of the fitted curves and data we could conclude that the fitting procedure is successful, but let us examine the values of the kinetic constants that we can extract from this procedure. The fit gives v_s, v_z and λ, from which the kinetic constants can be obtained from a plot of λ versus the inhibitor concentration (Fig. 2.13, see also Eq. (2.22) for λ in Table 2.6, mechanism b). The values of the kinetic constants are shown in Fig. 2.13 and summarized in Table 2.9 for comparison with the error-free data and with the method of numerical integration.

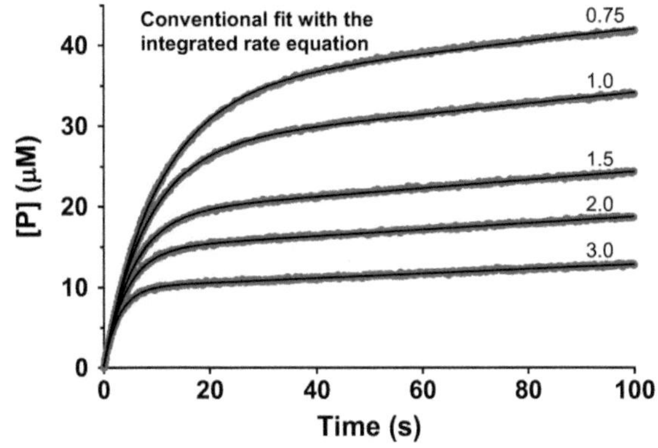

Fig. 2.12 Fit of Eq. (2.20) to the data in Fig. 2.11 using the five curves in the presence of inhibitor. The values of λ obtained for each inhibitor concentration are used to extract kinetic constants (Fig. 2.13). Numbers next to the curves represent $[I]_t$ in μM units. *Thick, noisy traces* represent data and *thin continuous lines* best fit curves

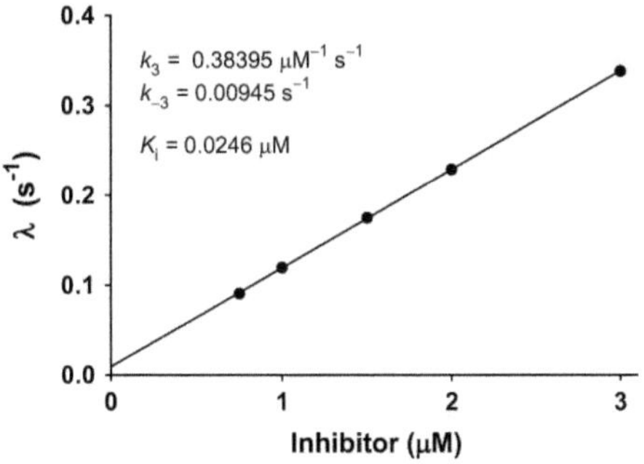

Fig. 2.13 Calculation of kinetic constants using the values of λ obtained from the best fit in Fig. 2.12. According to the equations for mechanism b in Table 2.6, k_3 and k_{-3} can be calculated from the slope and intercept of the *straight line*, respectively. $K_i = k_{-3}/k_3$

We evaluate now the same rate constants by numerical integration using KinTek software. The fitted curves, which include the control curve in the absence of inhibitor, are shown in Fig. 2.14. The quality of the fit was ascertained by calculating the confidence contours from globally fitting all progress curves. A limited range of mutual dependence of k_3 and k_{-3} indicated that these constants were determined precisely and were well constrained by the data. Kinetic constants with associated errors are summarized in Table 2.9.

We see from this example the basic difference between conventional fitting by non-linear regression using integrated rate equations (Table 2.9 method B) and fitting by numerical integration (Table 2.9 method C). Comparing the rate constants calculated by non-linear regression or by numerical integration with the reference values (Table 2.9 method A) we see that the rate constant k_3 is estimated fairly well

Table 2.9 Kinetic constants and error analysis for the example of slow-, tight-binding inhibition (primary data in Fig. 2.11)

Methods and properties	k_3 (μM^{-1} s^{-1})	k_{-3} (s^{-1})	$K_i = k_{-3}/k_3$ (nM)
(A) Error-free values simulated with Simulink	0.4	0.003	7.5
(B) Integrated rate equation			
Best fit	0.38395	0.00945	24.6
Standard error	0.00165	0.00086	2.2
95 % confidence intervals	0.3787–0.3892	0.0067–0.0122	
(C) Numerical integration			
Best fit	0.3995	0.002973	7.44
Standard error	4.27×10^{-4}	7.69×10^{-5}	0.19
95 % confidence intervals	0.397–0.402	0.00285–0.00312	

All decimals from calculations are shown for the parameters to compare the precision of the methods

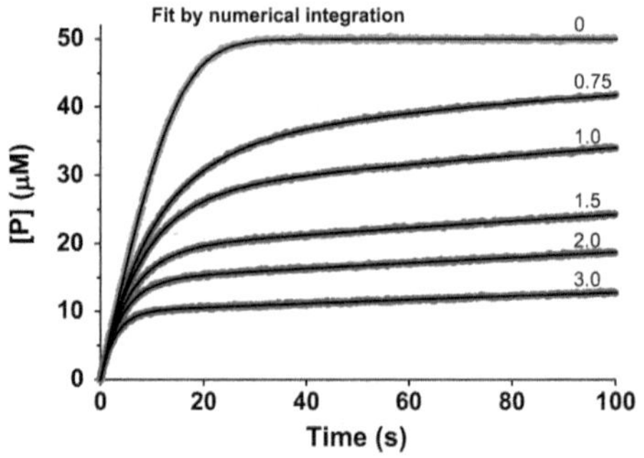

Fig. 2.14 Global fit of all curves in Fig. 2.11, including the curve with substrate alone, by numerical integration using KinTek software. Numbers next to the curves represent $[I]_t$ in μM units. *Thick, noisy traces* represent data and *thin continuous lines* best fit curves

by non-linear fitting and very good by numerical integration. On the contrary, the estimate of the rate constant k_{-3} by non-linear regression is poor, whereas we obtain a value very close to the true one by numerical integration. This discrepancy is clearly seen in the value of K_i, which is overestimated by the 'illegal' method B (24.6 nM), while numerical integration gives a satisfactory estimate of 7.44 nM. The small deviations of method C with respect to error free values (A) are due to the noise introduced in the curves.

2.7.4 The Serpin Inhibition Mechanism

Serpins are a superfamily of peptidase inhibitors found in a variety of organisms ranging from human to plants and viruses and share a unique mechanism of action. They inhibit mostly serine peptidases, from which the name 'serpin' derives

Scheme 2.11 The serpin inhibition mechanism

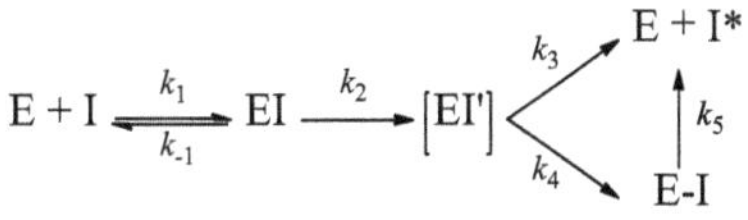

(SERine Peptidase INhibitors), while some also exhibit cross-class inhibition, e.g. the serpin CrmA inhibits granzyme B as well as caspases (Komiyama et al. 1994). The serpin molecule can exist in multiple conformations of which only one has inhibitory activity. In this canonical conformation, the 'reactive center loop', which is the main determinant of inhibitory activity and specificity, is exposed on the top of the serpin molecule and can interact with the active site of its target.

Serpins behave as mechanism-based (suicide) inhibitors that act by trapping the target enzyme in a covalent, essentially irreversible, enzyme-inhibitor complex. The major role in target recognition is performed by the reactive center loop which hence determines the specificity of each serpin. The basic kinetic mechanism of serpin inhibition is shown in Scheme 2.11. First, an adsorptive non-covalent EI complex is formed by insertion of the reactive center loop into the peptidase active site. From here the reaction proceeds analogous to a substrate cleavage reaction through a tetrahedral intermediate to the formation of an acyl-enzyme complex EI′ with an overall rate constant k_2. At this point the reaction can either proceed with rate constant k_3 to yield cleaved serpin I* and recycled enzyme, or the EI′ complex can rearrange with rate constant k_4 to a kinetically trapped, covalently bound E-I complex. E-I can further break down into free enzyme and cleaved serpin in a slow process with rate constant k_5.

Considering that k_5 is much smaller than k_3, the ratio of serpin molecules needed to inhibit one enzyme molecule is defined as the stoichiometry of inhibition (SI)

$$\mathrm{SI} = \frac{k_3 + k_4}{k_4}. \tag{2.35}$$

The value of SI depends on the ratio between k_3 and k_4 and can theoretically take any value larger than or equal to 1. Experimentally, SI values are determined from the ratio between cleaved serpin and stable, covalent E-I complex, though caution is advised in these experiments due to the high sensitivity of the SI to experimental conditions *in vitro*, such as ionic strength, temperature, pH, etc. The rate of complex formation can be determined from kinetic assays using low molecular mass synthetic substrates. However, one must take into account that the overall reaction rate of covalent complex formation (enzyme inactivation) includes both pathways in the mechanism. Hence, we can only measure an apparent overall association rate constant k_{app}, which is defined as

$$k_{\mathrm{app}} = \frac{k_2}{K_{\mathrm{m}}} \frac{1}{\mathrm{SI}}, \tag{2.36}$$

with k_2 as overall rate constant for the formation of the acyl-enzyme complex, which is the rate-limiting step in the mechanism, and K_{m} is the Michaelis constant

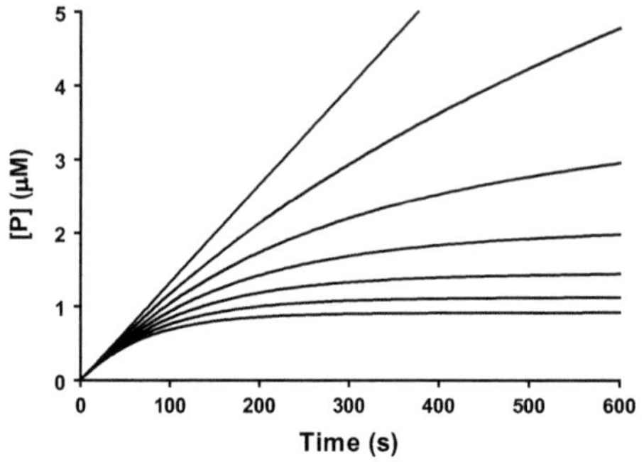

Fig. 2.15 Simulation of progress curves for the inhibition of human elastase-2 (EC 3.4.21.37) by wild-type α1-peptidase inhibitor. The simulation was performed with KinTek software with 1.0 nM enzyme and 500 µM MeOSuc-AAPV-p-nitroanilide as substrate. Published values of kinetic parameters were: $k_{\mathrm{app}} = 1.2 \times 10^7\ \mathrm{M}^{-1}\ \mathrm{s}^{-1}$ (Hopkins et al. 1993), $k_{\mathrm{cat}} = 17\ \mathrm{s}^{-1}$ and $K_{\mathrm{m}} = 0.14$ mM (Nakajima et al. 1979). Inhibitor concentrations were 0, 1, 2, 3, 4, 5 and 6 nM from *top* to *bottom*

for the reaction with the substrate used to monitor reaction. A simulated example of progress curves typically obtained in such experiments is shown in Fig. 2.15.

2.8 Enzyme Inactivation

Enzyme inactivation results from the (mostly) slow reaction of irreversible modifiers with catalytically essential elements of enzymes. The covalent compound formed in the reaction is called here E-I to allow identification towards a reversible E·I complex at a glance. Examples are the inactivation of serine peptidases by diisopropyl fluorophosphate and the inactivation of cysteine peptidases by the *trans*-epoxysuccinic acid derivative E-64. The kinetic treatment of enzyme inactivation is the same as that of slow-binding inhibition with the difference that E-I does not dissociate back to its components. Representative examples of inactivation mechanisms relevant to peptidases are shown in Scheme 2.12, where the numbering of paths, where applicable, was kept in line with the corresponding reversible mechanisms in Scheme 2.10. Mechanisms a–d in Scheme 2.12 match the reversible counterparts with the same identification labels. The lack of reversibility in the reaction between E and I and the absence of a steady-state phase simplifies the integrated rate equation for mechanisms a and b to

$$[\mathrm{P}] = \frac{v_{\mathrm{z}}}{\lambda}\left(1 - e^{-\lambda t}\right) + d, \tag{2.37}$$

where v_{z}, λ and d have the same meaning as in Eq. (2.20). For the temporary inactivation mechanisms c and d the integrated rate equation must consider enzyme recycling, which results in continuously re-feeding the flux around E and ES. The integrated rate equation for mechanisms c and d is given by

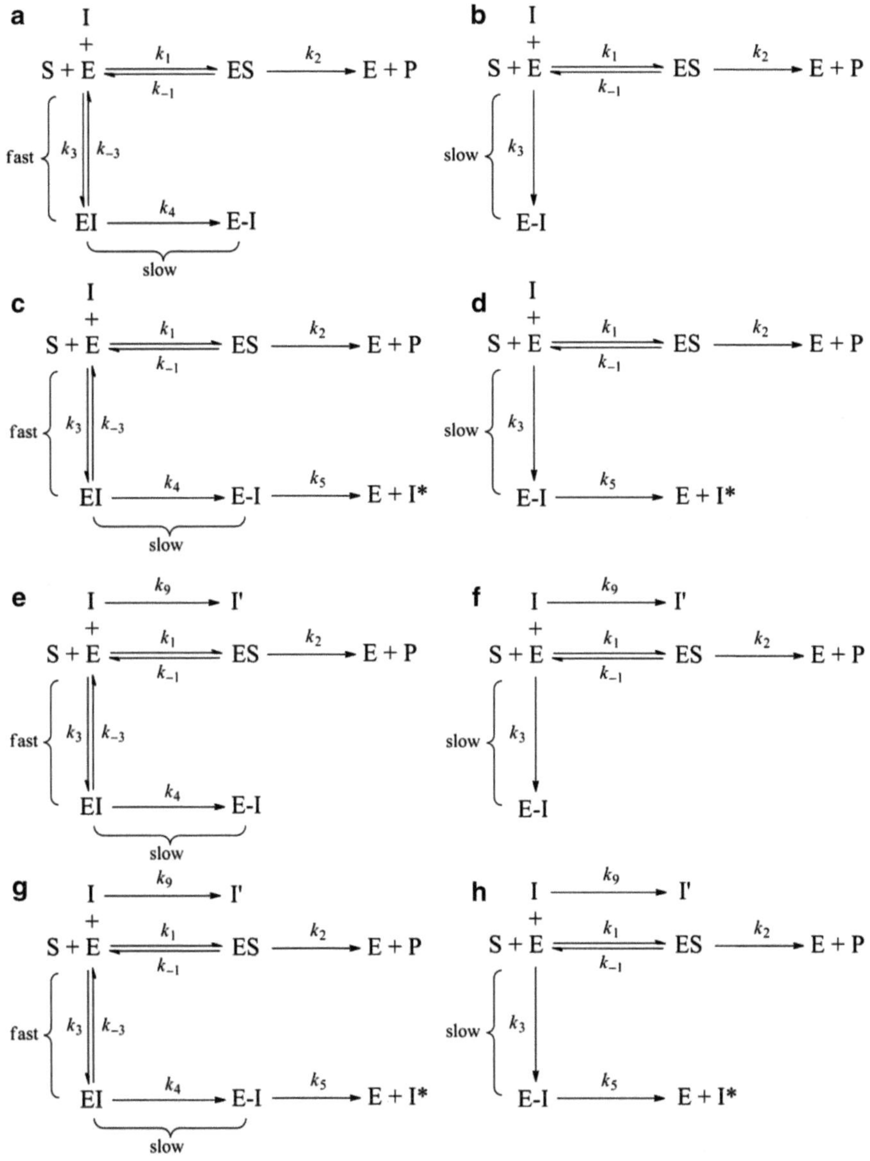

Scheme 2.12 Mechanisms for enzyme inactivation relevant to peptidases. The numbering of kinetic constants is consistently kept throughout the mechanisms to identify the same or similar paths and to match numbering in Scheme 2.10

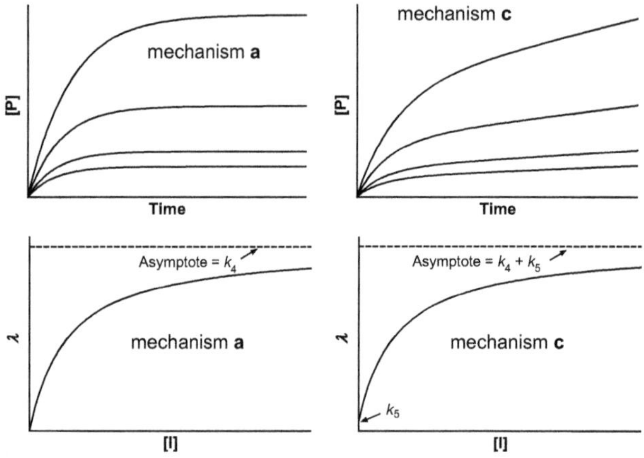

Fig. 2.16 Progress curves and the dependency of λ on inactivator concentration for mechanisms a and c in Scheme 2.12, as indicated

Table 2.10 Expressions of the parameters in Eq. (2.37) for mechanisms a and b in Scheme 2.12

Mechanism a		Mechanism b	
$\lambda = \dfrac{k_4[\mathrm{I}]}{K_\mathrm{i}\left(1+\dfrac{[\mathrm{S}]}{K_\mathrm{m}}\right)+[\mathrm{I}]}$ $K_\mathrm{i} = \dfrac{k_{-3}}{k_3}; \quad k_i = \dfrac{k_4}{K_\mathrm{i}}$	(2.39)	$\lambda = \dfrac{k_3[\mathrm{I}]}{1+\dfrac{[\mathrm{S}]}{K_\mathrm{m}}}$	(2.40)
v_z, the same as Eq. (2.23)		v_z, the same as Eq. (2.24)	

It is implicitly intended that reactions are started by adding enzyme to a solution containing substrate and inhibitor. k_3 and $k_\mathrm{i} = k_4/K_\mathrm{i}$, with units $\mathrm{M}^{-1}\,\mathrm{s}^{-1}$, are second-order constant of enzyme inactivation in mechanisms a and b, respectively, used to report enzyme inactivation results

$$[\mathrm{P}] = v_\infty t + \frac{v_\mathrm{z} - v_\infty}{\lambda}\left(1 - e^{-\lambda t}\right) + d, \qquad (2.38)$$

in which v_∞ substitutes v_s of the reversible counterpart (Baici et al. 2009). v_∞ has the meaning of a velocity at the end of the exponential phase, which is obtained mathematically by setting $t = \infty$ in $e^{-\lambda t}$. Typical progress curves for mechanism a and its temporary counterpart c of Scheme 2.12, together with the dependencies of λ on [I] are shown in Fig. 2.16. The expressions of v_∞, v_z and λ for mechanisms a, b and c, d are shown in Tables 2.10 and 2.11, respectively.

While the diagnosis of mechanism a is straightforward for the hyperbolic dependence of λ on [I] and the curve starting at the origin of the coordinates, which makes a distinction from the other mechanisms, temporary inactivation mechanism c is characterized by progress curves that cannot be distinguished from reversible slow-binding inhibition. A practical way for distinguishing

Table 2.11 Expressions of the parameters in Eq. (2.38) for mechanisms c and d in Scheme 2.12

Mechanism c		Mechanism d	
$\lambda = k_5 + \dfrac{k_4[\mathrm{I}]}{K_i\left(1+\dfrac{[\mathrm{S}]}{K_m}\right)+[\mathrm{I}]}$	(2.42)	$\lambda = k_5 + \dfrac{k_3[\mathrm{I}]}{1+\dfrac{[\mathrm{S}]}{K_m}}$	(2.43)
$K_i = \dfrac{k_{-3}}{k_3}$			
v_z, the same as Eq. (2.23)		v_z, the same as Eq. (2.24)	
$v_\infty = \dfrac{V[\mathrm{S}]}{K_m\left\{1+\dfrac{[\mathrm{I}]}{K_i}\left(1+\dfrac{k_4}{k_5}\right)\right\}+[\mathrm{S}]}$	(2.44)	$v_\infty = \dfrac{V[\mathrm{S}]}{K_m\left(1+\dfrac{k_3}{k_5}[\mathrm{I}]\right)+[\mathrm{S}]}$	(2.45)

Reactions started by adding enzyme to a solution containing substrate and inhibitor

mechanism c in Scheme 2.12 from slow-binding inhibition mechanism a in Scheme 2.10 is to preincubate enzyme and inactivator and to start reaction by adding substrate. The steady-state slope is independent of preincubation time for the reversible mechanism but does depend on time in temporary inactivation.

The one-step inactivation mechanism b is characterized by a linear dependence of λ on [I] with the line passing through the origin of the coordinates, as suggested by Eq. (2.40). The distinction of the temporary inactivation mechanism d from slow-binding inhibition in one step (mechanism b in Scheme 2.10) is again possible by preincubating enzyme with inactivator and starting reaction with substrate.

Low molecular mass compounds designed as irreversible inhibitors of peptidases may occasionally undergo non-enzymatic degradation in the assay solution, e.g. by hydrolysis to an inactive molecule. This fact, described by mechanisms e and f in Scheme 2.12, must be taken into account in the calculation of kinetic constants. To further complicate the picture, it may happen that an inactivator molecule undergoes spontaneous decomposition and is at the same time temporary as shown by mechanisms g and h. While molecules exhibiting the properties of these two last mechanisms can hardly have a practical value as inactivators of peptidases, and the labor involved in determining all constants may result in a mere academic exercise, the experimenter must be prepared to recognize this possibility.

For mechanisms e–h in Scheme 2.12 integrated equations can either not be derived or are exceedingly complex and hence the method of choice for elucidating their kinetic behavior is to go straight to numerical integration with the method illustrated in Sect. 2.7.3. Yet, *Topham* developed a method for mechanisms e and f, which may be felt knotty by less experienced end users, but is nevertheless very useful (Topham 1990). For instance, the following integrated rate equation obtained by Maclaurin series expansion, has been successfully applied to acetylcholinesterase inactivators acting with mechanism f in Scheme 2.12 (Baici et al. 2009):

$$[\mathrm{P}] = \frac{v_z}{k_9} e^{-\left\{\frac{k_3 K_\mathrm{m}[\mathrm{I}]_\mathrm{t}}{k_9(K_\mathrm{m}+[\mathrm{S}])}\right\}} \left\{ k_9 t + \sum_{i=1}^{\infty} \left\{ \frac{k_3 K_\mathrm{m}[\mathrm{I}]_\mathrm{t}}{k_9(K_\mathrm{m}+[\mathrm{S}])} \right\}^i \frac{\left[1-\left(e^{-k_9 t}\right)^i\right]}{i \cdot i!} \right\}. \tag{2.41}$$

In Eq. (2.41) $v_z = v_0$ and the second-order inactivation constant k_3, as well the first-order decay constant k_9 of the unstable inactivator, can be evaluated by non-linear regression. The number of terms in the Maclaurin series expansion depends on the value of k_9: with k_9 as small as 0.001 s^{-1} expansion to the tenth term is required, while the third term is sufficient for $k_9 = 0.005$ s^{-1}.

Details on enzyme inactivation, with analysis and diagnostics of mechanisms accompanied by real kinetic measurements for most of the mechanisms in Scheme 2.12 have been published for a series of inactivators of acetylcholinesterase (Baici et al. 2009). As a warning for the reader, in this paper the indices of some kinetic constants differ from those in this chapter.

2.9 Further Concepts Relevant to Peptidases

2.9.1 Measuring 'Invisible' Kinetic Parameters

To investigate the interaction between peptidases and macromolecular substrates devoid of measurable signals for following reaction development, a progress curve method can be of help (Baici 1990). The macromolecular substrates can either be soluble proteins or even insoluble components of the extracellular matrix. The principle is based on incubating the enzyme with a synthetic fluorogenic substrate, the *reporter substrate*, in the presence of the macromolecular soluble substrate or finely powdered insoluble substrate such as elastin. With insoluble substrates the progress curves resemble those in the presence of slow-binding inhibitors and the information that can be extracted from the measurements are the mechanism and the rates of adsorption and desorption of the enzyme to/from the insoluble substrate, as studied e.g. with elastase-2 (Baici 1990) and the cathepsins K, L and S (Novinec et al. 2007). With soluble proteins substrates, the macromolecules can be formally treated as being competitive inhibitors but the 'inhibition constant' measured corresponds to K_m of the most susceptible peptide bond (Baici 1990). With this method, kinetic constants for the adsorption of cathepsins K, L and S to elastins from three different sources have been calculated with models similar to those used in interfacial enzyme catalysis, with the surface area of the insoluble substrate replacing concentrations (Novinec et al. 2007). This approach is also very useful to assess the action of peptidase inhibitors in presence of naturally occurring protein substrates and the physiological significance of these interactions, which pose serious problems to the pharmacological control of extracellular matrix-degrading peptidases (Baici 1998).

2.9.2 Double Enzyme-Modifier Interactions

The characterization of enzyme modifiers *in vitro*, either inhibitors or activators, starts with the analysis of the behavior of one modifier in presence of the target enzyme and a suitable substrate. However, *in vivo* it is conceivable that exogenous modifiers, e.g. therapeutically active inhibitors, may compete with endogenous inhibitors or other molecules capable of interacting with the enzyme. We developed a rigorous mathematical model to describe the behavior of inhibitors and activators in multiple-interaction systems (Schenker and Baici 2009). A general kinetic equation can describe apparently different phenomena ranging from inhibition to activation in any of 126 combinations of two enzyme modifiers. Paradoxical or otherwise unpredictable effects resulting from the action of two modifiers on the same enzyme can be analyzed and modeled with this general method.

2.9.3 Reporting Kinetic Results as IC_{50}

In reporting results of enzyme inactivation the useful parameter is the second-order inactivation constant (units M^{-1} s^{-1}), which corresponds to k_3 in the one-step mechanisms and to $k_i = k_4/K_i$ in the two-step mechanisms (Sect. 2.8). IC_{50}, the concentration of a *reversible* inhibitor at which v_0 is reduced by 50 %, depends on the inhibition type (Chou 1974; Naqui 1983). If applied *correctly*, IC_{50} can be used for reporting results (Cortés et al. 2001). However, IC_{50} to characterize *irreversible* inhibition does not make sense. Unfortunately, the 'potency' of irreversible inhibitors is often reported in the literature as IC_{50}. Enzyme inactivation is a time-dependent phenomenon: allowing *sufficient time* for the formation of E-I, enzyme activity is driven to zero if $[I]_t \geq [E]_t$. The reaction velocity corresponds to $\frac{1}{2}v_0$ only when $[I]_t = \frac{1}{2}[E]_t$. Thus, for enzyme inactivators, IC_{50} corresponds to one half the enzyme concentration used in the assay and has no further physical meaning.

References

Auld DS (2004) Catalytic mechanisms for metallopeptidases. In: Barrett AJ, Rawlings ND, Woessner JF Jr (eds) Aspartic and metallo peptidases, vol 1, Handbook of proteolytic enzymes. Elsevier, London, pp 268–289

Baici A (1981) The specific velocity plot. A graphical method for determining inhibition parameters for both linear and hyperbolic enzyme inhibitors. Eur J Biochem 119:9–14

Baici A (1987) Graphical and statistical analysis of hyperbolic, tight-binding inhibition. Biochem J 244:793–796

Baici A (1990) Interaction of human leukocyte elastase with soluble and insoluble protein substrates. A practical kinetic approach. Biochim Biophys Acta 1040:355–364

Baici A (1998) Inhibition of extracellular matrix-degrading endopeptidases: problems, comments, and hypotheses. Biol Chem 379:1007–1018

Baici A, Knöpfel M, Fehr K, Skvaril F, Böni A (1980) Kinetics of the different susceptibility of the four human immunoglobulin G subclasses to proteolysis by human lysosomal elastase. Scand J Immunol 12:41–50

Baici A, Schenker P, Wächter M, Rüedi P (2009) 3-Fluoro-2,4-dioxa-3-phosphadecalins as inhibitors of acetylcholinesterase. A reappraisal of kinetic mechanisms and diagnostic methods. Chem Biodivers 6:261–282

Bartlett PA, Marlowe CK (1987) Possible role for water dissociation in the slow binding of phosphorus-containing transition-state-analogue inhibitors of thermolysin. Biochemistry 26:8553–8561

Botts J, Morales M (1953) Analytical description of the effects of modifiers and of enzyme multivalency upon the steady state catalyzed reaction rate. Trans Faraday Soc 49:696–707

Cha S (1976) Tight-binding inhibitors—III. A new approach for the determination of competition between tight-binding inhibitors and substrates. Inhibition of adenosine deaminase by coformycin. Biochem Pharmacol 25:2695–2702

Cha S (1980) Tight-binding inhibitors—VII. Extended interpretation of the rate equation. Experimental designs and statistical methods. Biochem Pharmacol 29:1779–1789

Chou T (1974) Relationships between inhibition constants and fractional inhibition in enzyme-catalyzed reactions with different numbers of reactants, different reaction mechanisms, and different types and mechanisms of inhibition. Mol Pharmacol 10:235–247

Cleland WW (1963) The kinetics of enzyme-catalyzed reactions with two or more substrates or products. I. Nomenclature and rate equations. Biochim Biophys Acta 67:104–136

Cornish-Bowden A (2004) Fundamentals of enzyme kinetics. Portland Press, London

Cornish-Bowden A, Eisenthal R (1974) Statistical considerations in the estimation of enzyme kinetic parameters by the direct linear plot and other methods. Biochem J 139:721–730

Cornish-Bowden A, Eisenthal R (1978) Estimation of Michaelis constants and maximum velocity from the direct linear plot. Biochim Biophys Acta 523:268–272

Cornish-Bowden A, Porter WR, Trager WF (1978) Evaluation of distribution-free confidence limits for enzyme kinetic parameters. J Theor Biol 74:163–175

Cortés A, Cascante M, Cárdenas ML, Cornish-Bowden A (2001) Relationships between inhibition constants, inhibitor concentrations for 50% inhibition and types of inhibition: new ways of analysing data. Biochem J 357:263–268

Duggleby RG, Attwood PV, Wallace JC, Keech DB (1982) Avidin is a slow-binding inhibitor of pyruvate carboxylase. Biochemistry 21:3364–3370

Eisenthal R, Cornish-Bowden A (1974) The direct linear plot. A new graphical procedure for estimating enzyme kinetic parameters. Biochem J 139:715–720

Eisenthal R, Danson MJ, Hough DW (2007) Catalytic efficiency and k_{cat}/K_M: a useful comparator? Trends Biotechnol 25:247–249

Fenner G (1931) Das Genauigkeitsmass von Summen, Differenzen, Produkten und Quotienten der Beobachtungsreihen. Naturwissenschaften 19:310

Fersht A (1977) Enzyme structure and mechanism. Freeman, New York

Fontes R, Ribeiro JM, Sillero A (2000) Inhibition and activation of enzymes. The effect of a modifier on the reaction rate and on kinetic parameters. Acta Biochim Pol 47:233–257

Frieden C (1970) Kinetic aspects of regulation of metabolic processes. The hysteretic enzyme concept. J Biol Chem 245:5788–5799

Hopkins PCR, Carrell RW, Stone SR (1993) Effects of mutations in the hinge region of serpins. Biochemistry 32:7650–7657

International Union of Biochemistry (1979) Units of enzyme activity. Eur J Biochem 97:319–320

International Union of Biochemistry (1982) Symbolism and terminology in enzyme kinetics. Recommendations 1981. Eur J Biochem 128:281–291

James MNG (2004) Catalytic pathway of aspartic peptidases. In: Barrett AJ, Rawlings ND, Woessner JF Jr (eds) Aspartic and metallo peptidases, vol 1, Handbook of proteolytic enzymes. Elsevier, London, pp 12–19

Johansen G, Lumry R (1961) Statistical analysis of enzymic steady-state rate data. C R Trav Lab Carlsberg 32:185–214

Johnson KA (2009) Fitting enzyme kinetic data with Kintek Global Kinetic Explorer. Methods Enzymol 467:601–626

Johnson KA, Simpson ZB, Blom T (2009a) FitSpace Explorer: an algorithm to evaluate multidimensional parameter space in fitting kinetic data. Anal Biochem 387:30–41

Johnson KA, Simpson ZB, Blom T (2009b) Global Kinetic Explorer: a new computer program for dynamic simulation and fitting of kinetic data. Anal Biochem 387:20–29

Komiyama T, Ray CA, Pickup DJ, Howard AD, Thornberry NA, Peterson EP, Salvesen G (1994) Inhibition of interleukin-1-β converting-enzyme by the cowpox virus serpin Crma. An example of cross-class inhibition. J Biol Chem 269:19331–19337

Koshland DE (2002) The application and usefulness of the ratio k_{cat}/K_m. Bioorg Chem 30:211–213

Kuzmič P (2008) A steady state mathematical model for stepwise "slow-binding" reversible enzyme inhibition. Anal Biochem 380:5–12

Meh P, Pavšič M, Turk V, Baici A, Lenarčič B (2005) Dual concentration-dependent activity of thyroglobulin type-1 domain of testican: specific inhibitor and substrate of cathepsin L. Biol Chem 386:75–83

Morrison JF (1982) The slow-binding and slow, tight-binding inhibition of enzyme-catalysed reactions. Trends Biochem Sci 7:102–105

Nakajima K, Powers JC, Ashe BM, Zimmerman M (1979) Mapping the extended substrate binding site of cathepsin G and human leukocyte elastase. Studies with peptide substrates related to the α1-protease inhibitor reactive site. J Biol Chem 254:4027–4032

Naqui A (1983) What does I50 mean? Biochem J 215:429–430

Novinec M, Grass RN, Stark WJ, Turk V, Baici A, Lenarčič B (2007) Interaction between human cathepsins K, L and S and elastins: mechanism of elastinolysis and inhibition by macromolecular inhibitors. J Biol Chem 282:7893–7902

Novinec M, Kovačič L, Lenarčič B, Baici A (2010) Conformational flexibility and allosteric regulation of cathepsin K. Biochem J 429:379–389

Orsi BA, Tipton KF (1979) Kinetic analysis of progress curves. Methods Enzymol 63:159–183

Polgár L (2004a) Catalytic mechanisms of cysteine peptidases. In: Barrett AJ, Rawlings ND, Woessner JF Jr (eds) Cysteine, serine and threonine peptidases, vol 2, Handbook of proteolytic enzymes. Elsevier, London, pp 1072–1079

Polgár L (2004b) Catalytic mechanisms of serine and threonine peptidases. In: Barrett AJ, Rawlings ND, Woessner JF Jr (eds) Cysteine, serine and threonine peptidases, vol 2, Handbook of proteolytic enzymes. Elsevier, London, pp 1440–1448

Rawlings ND, Barrett AJ, Bateman A (2010) MEROPS: the peptidase database. Nucleic Acids Res 38:D227–D233

Schechter I, Berger A (1967) On the size of the active sites in proteases. I. Papain. Biochem Biophys Res Commun 27:157–162

Schenker P, Baici A (2009) Simultaneous interaction of enzymes with two modifiers: reappraisal of kinetic models and new paradigms. J Theor Biol 261:318–329

Schultz RM, Varma-Nelson P, Ortiz R, Kozlowski KA, Orawski AT, Pagast P, Frankfater A (1989) Active and inactive forms of the transition-state analog protease inhibitor leupeptin: explanation of the observed slow binding of leupeptin to cathepsin B and papain. J Biol Chem 264:1497–1507

Schweizer A, Roschitzki-Voser H, Amstutz P, Briand C, Gulotti-Georgieva M, Prenosil E, Binz HK, Capitani G, Baici A, Plückthun A, Grütter MG (2007) Inhibition of caspase-2 by a designed ankyrin repeat protein: specificity, structure, and inhibition mechanism. Structure 15:625–636

Segel IH (1975) Enzyme kinetics. Behavior and analysis of rapid equilibrium and steady-state enzyme systems. Wiley, New York

Selwyn MJ (1965) A simple test for inactivation of an enzyme during assay. Biochim Biophys Acta 105:193–195

Szedlacsek SE, Ostafe V, Serban M, Vlad MO (1988) A re-evaluation of the kinetic equations for hyperbolic tight-binding inhibition. Biochem J 254:311–312

Tallant C, Marrero A, Gomis-Ruth FX (2010) Matrix metalloproteinases: fold and function of their catalytic domains. Biochim Biophys Acta 1803:20–28

Topham CM (1990) A generalized theoretical treatment of the kinetics of an enzyme-catalysed reaction in the presence of an unstable irreversible modifier. J Theor Biol 145:547–572

Williams JW, Morrison JF, Duggleby RG (1979) Methotrexate, a high-affinity pseudosubstrate of dihydrofolate reductase. Biochemistry 18:2567–2573

Chapter 3
Compartmentalization of Proteolysis

Klaudia Brix, Christopher J. Scott, and Margarete M.S. Heck

3.1 Introduction

Compartmentalization of proteolysis is essential for homeostasis in all tissues and organs, not only at the level of communication of cells with their environment, but also at the sub-cellular level (Fig. 3.1). Within the context of intracellular compartmentalization, it is important to consider the definition of organelles as sub-cellular regions limited by biological membranes that form distinct compartments featuring specific biochemical environments (Fig. 3.2).

Compartmentalization, in a cellular context, distinguishes hydrophilic sub-cellular regions of defined properties that are separated from one another by hydrophobic membranes (Palade 1964, 1966; Clegg 1991; Lipowsky 1995). Proteolysis, the catalytic hydrolysis (and thus cleavage) of peptide bonds, was traditionally thought to require hydrophilic environments (Gerlt and Babbitt 2001). Lipid-enriched, hydrophobic conditions were not considered conducive to proteolytic catalysis where a molecule of water plays a central role (Chaps. 1 and 2; Brown et al. 2000). However, we have learned during the past decades that such (strict) proposals are too narrow and no longer fully valid (Weihofen and Martoglio 2003; Wolfe and Kopan 2004; Ha 2007). Undisputedly, evolution has found a huge variety of solutions to the demands of our body in modifying proteins—be they

K. Brix (✉)
School of Engineering and Science, Research Center MOLIFE – Molecular Life Science, Jacobs University Bremen, Campus Ring 6, Research II, 28759 Bremen, Germany
e-mail: k.brix@jacobs-university.de

C.J. Scott
School of Pharmacy, Queen's University Belfast, 97 Lisburn Road, Belfast BT9 7BL, UK
e-mail: c.scott@qub.ac.uk

M.M.S. Heck
Queen's Medical Research Institute, Centre for Cardiovascular Science, University of Edinburgh, 47 Little France Crescent, Edinburgh EH16 4TJ, UK
e-mail: margarete.heck@ed.ac.uk

K. Brix and W. Stöcker (eds.), *Proteases: Structure and Function*,
DOI 10.1007/978-3-7091-0885-7_3, © Springer-Verlag Wien 2013

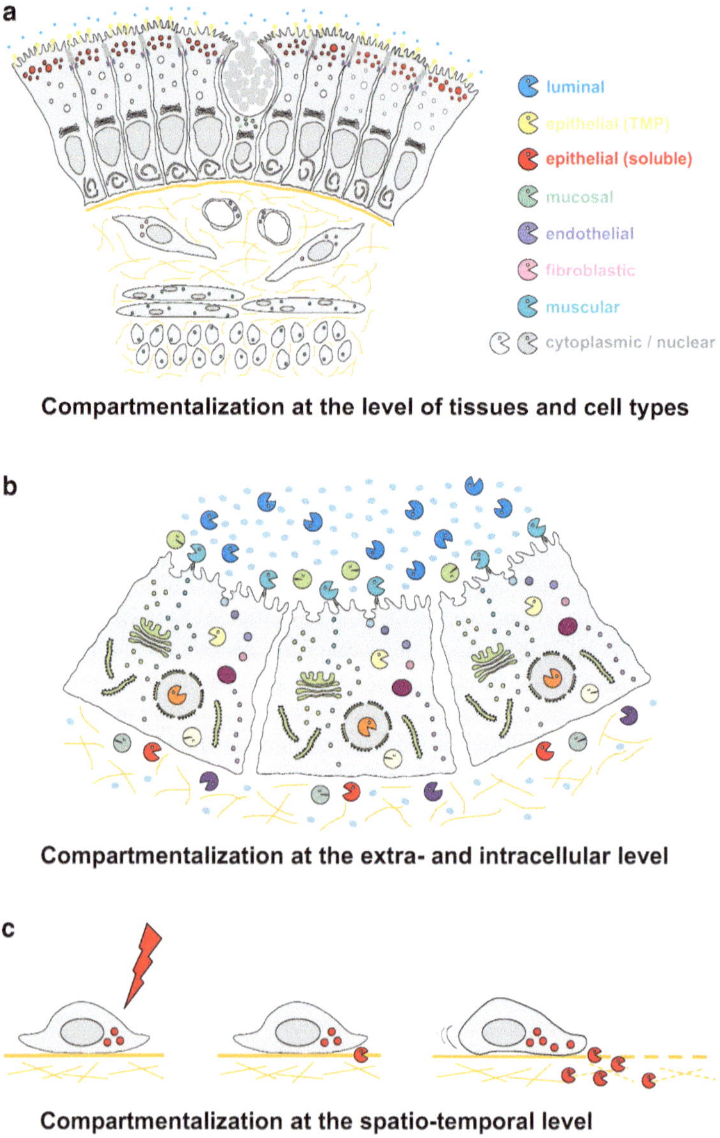

Fig. 3.1 Schematic representation of compartmentalized proteolysis. Proteases perform their tasks at different locations within cells or tissues. Functions of distinct proteolytic enzymes differ depending on the position they fill in the complex network of proteases and where those meet their natural substrates. Compartmentalization occurs at the level of tissues consisting of different cell types (**a**) and at the sub-cellular level (**b**). Regulation of proteolysis in space and time is possible by means of different distributions of proteases, their activators, their co-factors, their inhibitors and their substrates (**c**). Thus, compartmentalization of proteolysis is influenced by signalling events and thereby also depends on the cross-talk of cells with each other and their environment. Bolt signifies a signal from the environment

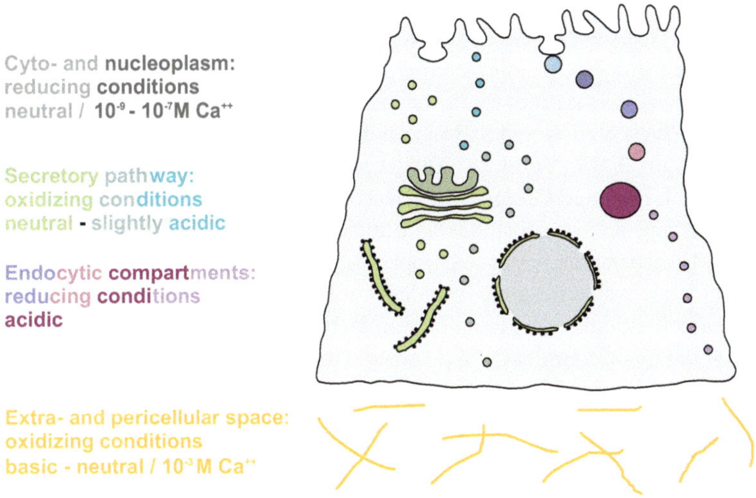

Fig. 3.2 Schematic representation of compartmentalization based on biochemical features of cellular organelles. Peptidases are faced with strikingly different biochemical conditions for substrate interaction and proteolytic cleavage depending on whether extra- or intracellular proteolysis takes place. The latter processes depend on where within the cellular compartments proteases engage in substrate cleavage; a number of examples of different biochemical conditions of proteolytic processing are depicted including oxidizing versus reducing, and acidic versus basic or neutral conditions. The ionic strength of certain compartments, e.g. the secretory granules, or the presence of metal ions are further well known factors contributing to the regulation of the activities of proteolytic enzymes

folded or unfolded, monomeric or multimeric—by protease-mediated hydrolysis that takes place in all possible environments including even the lipid bilayer itself (Chaps. 4 and 7–15; Fritz et al. 1973; Steiner et al. 1980; Fricker 1988; Travis and Fritz; 1991; Stubbs and Bode 1994; Bode et al. 1997; Chapman et al. 1997; Khan and James 1998; Galivan et al. 2000; Davies 2001; Chambers and Laurent 2002; Hooper 2002; Seiki 2002; Potempa et al. 2003; Schechter 2005; Arolas et al. 2007; Barrett and Rawlings 2007; Brix et al. 2008; Cobbe et al. 2009; Ovaere et al. 2009; Schweitzer and Naumann 2010).

Proteases were first defined by their presumably predominant functions as protein degrading enzymes. However, it has been conclusively shown that many proteases undertake a range of functions—and not simply as mere destructive enzymes. Although some proteases are essential in protein turnover, ensuring removal of unwanted or damaged proteins and producing pools of amino acids for *de novo* protein biosynthesis (Bonifacino and Weissman 1998; Reed 2003; Ciechanover and Iwai 2004), discrete cleavage or post-translational processing of target proteins is just as important. These processing events can promote protein folding, maturation, or achieve initiation, termination or modulation of protein functionality (Dobson 2003; Ciechanover 2005; Collins and Tansey 2006). They also engage in peptide processing, for example in antigen presentation, in the cellular defence of intruders and pathogens (Fineschi and Miller 1997; Nakagawa et al. 1998; Villadangos et al. 1999; Riese and Chapman 2000; Trombetta and Mellman 2005; Herget and Tampe 2007; Watts 2012). Additionally, it is worthy to note that some proteases can also have *non-proteolytic* roles, a topic that requires additional attention as we attempt to grasp the complexity of the world of proteolytic enzymes beyond their active sites (Chaps. 2, 5, 6, and 8; Reudelhuber et al. 1998; Blasi and Carmeliet 2002; Friedl and Wolf 2003; Yu et al. 2006; Lamkanfi et al. 2007; Uddin et al. 2008; Strongin 2010; Bhat and Greer 2011; Kwak et al. 2011).

Proteases can act on substrates in a number of ways: in a simple one-to-one fashion, or to cut specific bonds in complex substrates consisting of several proteins, or to sever high molecular mass proteins containing multiple domains. More than one protease may cleave a specific peptide bond in any one substrate at a time, further adding to the complexity of proteolytic control. It may be of further relevance to note that proteases can act as single molecules or as oligomers (e.g. the 26S proteasome). Increasingly, we are finding evidence for the need for proteases to complex with activators or co-factors (Fig. 3.3). Although these co-factors can be other proteins, they may also comprise lipids (invadolysin, secretases, SPPases) (McHugh et al. 2004; Cobbe et al. 2009; Di Cara et al. 2013), carbohydrates (cathepsin K, ADAMs, ADAM-TSs) or even nucleic acids (truncated cathepsin V, separases) (Chaps. 1 and 2; Bode and Huber 1992, 2000; Uhlmann 2003; Ong et al. 2007; Sun et al. 2009). Proteases may position themselves in a particular cellular context by interacting with scaffolding factors, for example in apoptosis (Chap. 8; Seiki 2002; Schweitzer and Naumann 2010; Kersse et al. 2011), or they may traffic along traditional and novel, sometimes unexpected, transport pathways to reach the location required for proteolytic activity (e.g. endo-lysosomal enzymes, "alternative" secretion of some convertases) (Chap. 4; Linke et al. 2002a; Buth et al. 2004, 2007; Brix and Jordans 2005; Sloane et al. 2006; Brix et al. 2008; Sameni et al. 2009) (Figs. 3.2, 3.4, and 3.5). Proteolytic activity may be controlled by physical compartmentalization of the protease as well as the presence of specific endogenous inhibitor(s).

In this chapter, we highlight some of the facets of compartmentalization of proteolysis—without claiming to be comprehensive in discussing the unlimited variations on the theme of proteolytic activity. We will start by discussing the seemingly most obvious aspect of compartmentalization of proteolysis, that is, the action of proteases in different cellular compartments.

3.2 Sweet or Savoury—Salty or Tart: Biochemical Conditions of the Cleaving Environment

Evolution has given us proteases that are able to hydrolyze peptide bonds in a plethora of diverse biochemical conditions (Figs. 3.1 and 3.2). For example, calpain activity depends on free calcium concentrations ranging from 10^{-9} M in resting cells to 10^{-7} M in activated cells (Chap. 12; Vanlangenakker et al. 2008; Sorimachi et al. 2010), while other proteases like matrix metalloproteinases (MMPs) exhibit

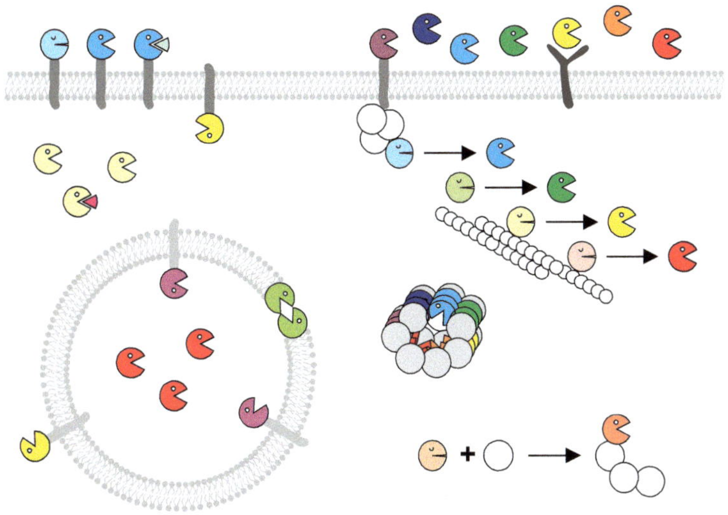

Fig. 3.3 Schematic representation of compartmentalization by lipid membranes, scaffolding factors, and complex formation. Proteases act as single molecules, in complexes, or in sequential modes. Proteolytic enzymes may be soluble, equipped with transmembrane domains, or interact with receptors and other scaffolding factors eventually initiating, terminating, or otherwise affecting the extent of substrate cleavage. Some proteases form multimeric complexes with other macromolecules such as other proteins, glycoproteins or proteoglycans in the extra- and pericellular space. In contrast, the formation of macromolecular assemblies within the cytosol results in substrate processing in e.g. the proteasome which provides a cleavage chamber with specific conditions, thus, separating the protease-substrate interaction from the surrounding cytosol without using the principle of compartmentalization by lipid membranes. Most proteolytic processes require hydrophilic conditions whereas others are enabled to take place near or at the level of lipid membranes for cleaving signal peptides or to carry out proteolysis within transmembrane domains

maximal activity in an extracellular environment characterized by 10^{-3} M free calcium. Furthermore, pH-values ranging from basic to highly acidic, and redox conditions from oxidizing (in the compartments of the secretory pathway and extracellular space) to reducing conditions (within endo-lysosomal compartments) are crucial determinants of protein folding and proteolytic activity (Chap. 2; Barrett and Kirschke 1981; Kirschke et al. 1995; McGrath 1999; Reinheckel et al. 2000; Pillay et al. 2002; Jordans et al. 2009; Zhou et al. 2010; Scott and Gruenberg 2011). Therefore, compartmentalization and complex formation not only generate efficient environments for proteolytic cleavage, but may modulate the local biochemical environment around an active protease.

One exquisite exemplification of this is in the proposed mechanism of Regulated Intramembrane Proteolysis (RIP) where proteases, like signal peptidases, sheddases, or the Alzheimer's Precursor Protein (APP)-cleaving secretases, acquire the ability to cleave near or within amphipathic, α-helical transmembrane domains (Chaps. 9 and 10; Annaert and De Strooper 1999; De Strooper and Annaert 2000; Steiner and Haass 2000; Urban et al. 2001; Urban and Freeman 2002; Ehrmann and Clausen 2004;

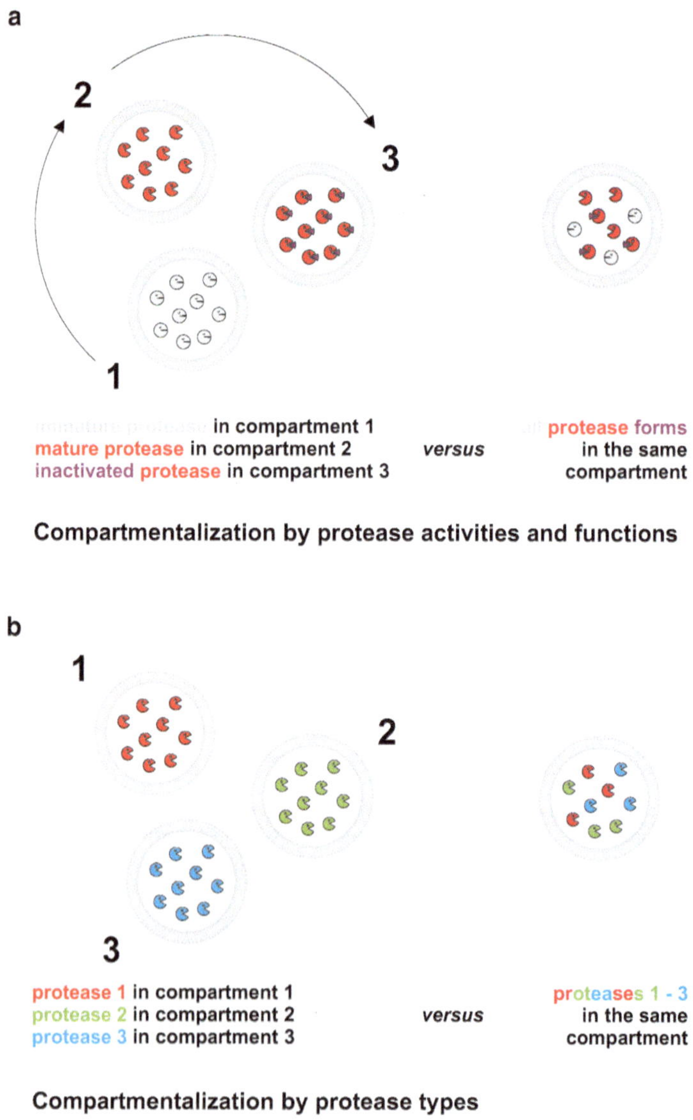

Fig. 3.4 Schematic representation of different principles of compartmentalization. (**a**) Proteolytic enzymes of mammalian cells are often synthesized as inactive precursors, the zymogen forms (1), that are matured by proteolytic processing or other means at the sub-cellular location where the active protease is required to function (2). In contrast, compartments equipped with inactivating factors such as endogenous protease inhibitors provide conditions that restrict proteolytic cleavages (3). Shuttling between the different compartments is realized by directed protein transport processes ensuring protease trafficking to the desired location for meeting with their substrates at a given time (*left panels*). In some cases, individual compartments house different forms of the same proteolytic enzymes which are regulated in their activity by interaction with activating or inhibiting factors to initiate or terminate proteolytic substrate cleavage (*right panel*). (**b**) Proteases belong to distinct protein families (for classification see, http://merops.sanger.ac.uk) which are sorted into different (*left*) or the same sub-cellular compartments of mammalian cells (*right*)

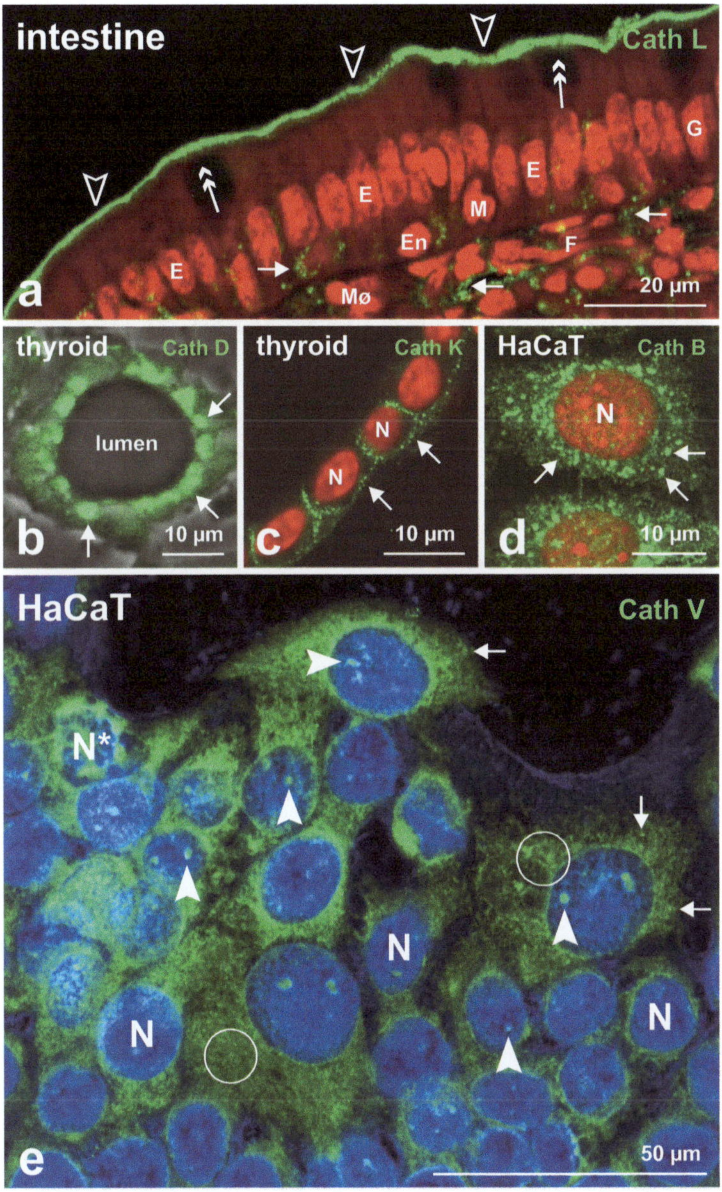

Fig. 3.5 Compartmentalization of cathepsins in mammalian cells. Confocal fluorescence micrographs of cryosections prepared from mouse tissue (**a–c**) and of cultured human cells (**d** and **e**) after immunolabelling with antibodies specific for the aspartic protease cathepsin D or the cysteine peptidases cathepsins B, K, L, or V, and fluorophore-conjugated secondary antibodies (*green*). Nuclei were counter-stained with Draq 5 (*red* in **a**, **c**, and **d**; *blue* in **e**) and tissue structures viewed in phase contrast (**b**); overlays of the single channels are displayed in false colours. Although all cathepsins depicted in this figure belong to the so-called "endo-lysosomal proteases", their distribution patterns differ dramatically between the tissues and individual cell types. (**a**) Cathepsin L is detectable within vesicles gathering in the peri-nuclear regions of fibroblasts (F) and macrophages (Mø) of the lamina propria as well as of enteroendocrine cells (En) of the

Parkin et al. 2004; Selkoe and Wolfe 2007; Freeman 2008; Murphy 2009; Lichtenthaler et al. 2011). The proteases involved in RIP, represented by soluble and peripheral membrane proteins, are proposed to form channel-like cleaving environments by arranging into multimeric complexes with other proteins. In contrast, turnover of transmembrane proteins in general, which includes extensive cleavage of hydrophobic transmembrane domains in addition to more straightforward degradation of the hydrophilic luminal and cytosolic protein domains, is enabled by their extraction from within lipid-rich environments (membranes) through recruitment by the ESCRT (Endosomal Sorting Complex Required for Transport) machinery (Gu et al. 1997; Gruenberg 2001; Hicke and Dunn 2003; Gruenberg and Stenmark 2004; Piper and Katzmann 2007; van Meer and de Kroon 2011). A variety of hydrolytic enzymes present in the endocytic compartments, then degrade the transmembrane proteins in addition to their other substrates, such as membrane lipids (Chap. 4; Kirschke et al. 1995; McGrath 1999; Brix 2005).

Cleavage of peptide bonds in the extracellular milieu (Fig. 3.1) requires enzymes that thrive in carbohydrate- or glycosaminoglycan-rich environments; conditions of

---

Fig. 3.5 (continued) mucosal epithelium in mouse intestine (*arrows*). While this staining pattern is expected from the classical view on this member of the papain-like peptidase family, cathepsin L is almost absent from M-cells (M) and even secreted from the enterocytes (E) and goblet cells (G) of the mucosa (*double headed arrows*), from where it reaches the intestinal lumen and re-associates with the apical plasma membrane of the intestine epithelial cells (*open arrowheads*). These distribution patterns point to different functions that cathepsin L fulfils in the various cell types found in the intestinal mucosa. (**b** and **c**) Cathepsins D and K belong to different families of enzymes since they act as aspartic or cysteine peptidases, respectively. The epithelial cells of the thyroid gland sort both proteases into vesicles (*arrows*) that differ in size and are destined to either the apical (**b**) or the basolateral poles of thyroid epithelial cells depending on their physiological states (**c**). Both of these cathepsins together with a number of other related peptidases cleave the prohormone thyroglobulin by endo- and exopeptidatic modes, thereby liberating the thyroid hormones 3′,3,5-triiodothyronine T3 and thyroxine T4. The precise cleavage patterns of the substrate depend on the position and activity of its processing enzyme in the various compartments of the endocytic pathway of thyrocytes and within the extracellular lumen of thyroid follicles. Thus, transport of proteases and substrate cleavage appear compartmentalized and tightly regulated in the polarized epithelial cells of the thyroid gland. (**d** and **e**) Human keratinocytes exhibit vesicular staining of the cysteine cathepsin B (*arrows* in **d**) whereas the cysteine cathepsin V is additionally and predominantly detected in an ER-like reticular pattern throughout the cytosol of HaCaT cells (*circles* in **e**). These localizations point to different maturation stages of the two cysteine cathepsins resulting from distinct transport routes under steady state conditions. Sub-compartmentalization is even more complex because molecular variants of cathepsin B and, even more prominent, cathepsin V variants lacking the signal peptides are detectable in the nuclei of proliferating keratinocytes (*arrowheads* in **e**). N denotes nuclei, and N* indicates an apoptotic cell in which the subcellular compartments appear condensed and structurally altered in comparison to the normal appearance of organelles in this keratinocyte cell line. For further details see references (Brix and Herzog 1994; Brix et al. 1996, 1997; Lemansky et al. 1998; Tepel et al. 2000; Linke et al. 2002a, b; Friedrichs et al. 2003; Buth et al. 2004, 2007; Mayer et al. 2006, 2008, 2009; Jordans et al. 2009; Vreemann et al. 2009; Tedelind et al. 2010, 2011; Arampatzidou et al. 2011a, 2012; Dauth et al. 2011b, 2012; Haugen et al. 2013; reviewed in Brix et al. 2001, 2008, 2011; Brix 2005; Brix and Jordans 2005; Arampatzidou et al. 2011b; Dauth et al. 2011a) elevated negative charge and, hence, salt-'enriching' conditions (Jacques 1979; Bergers and Coussens 2000; Blobel 2000a; Rosenblum and Kozarich 2003; Stamenkovic 2003; Strongin 2010; Sato and Takino 2010; Zogg and Brandstetter 2011). Sequential cleavages mediated by cascades of proteases acting on one or several protein substrates (perhaps including the proteases themselves) in a tightly regulated manner seem to explain how compartmentalization is achieved even in the peri-cellular environment (Chaps. 13–15; Chapman et al. 1994; Owen and Campbell 1995; Holmbeck et al. 1999; Sternlicht and Werb 2001; Seiki 2002; Ellis 2003; Mott and Werb 2004; Sounni and Noel 2005; Munshi and Stack 2006), where lipid membranes are not the limiting structure of the reaction compartment, but rather serve as an organizing structural support. The solution of cells to such problems of focalized proteolytic events in the peri-cellular surrounding is often achieved by arranging for protease delivery to the cleavage environment in a sequential and spatially regulated manner (Basbaum and Werb 1996; Murphy and Gavrilovic 1999; Blobel 2000a; Hooper et al. 2001; Itoh 2006; Owen 2008; Vreemann et al. 2009; Friedl and Wolf 2009; Murphy and Nagase 2011; Pagano and Reboud-Ravaux 2011). Upon initiation via appropriate signals, fine-tuning the sequence of proteolysis occurs via interaction with partners comprising not only substrates but also other proteases, membrane receptors, and protease inhibitors (Chaps. 2 and 15; Brix et al. 2001; Schenk and Quaranta 2003; Dano et al. 2005; Jordans et al. 2009; Sun et al. 2009; Das et al. 2010) (Fig. 3.3).

Moreover, the chromatin environment of the nucleus also presents demanding conditions, characterized by extraordinarily long charged polymers such as DNA and RNA (negatively charged), and histones (positively charged). In addition, nucleic acids can provide scaffolding platforms for proteolysis (Ong et al. 2007) where the proteolytic enzymes feature positively-charged molecular surfaces to facilitate interaction with DNA as exemplified by separases to cleave the cohesin "glue" holding sister chromatids together (Heck 1997; Lamond and Earnshaw 1998; Nasmyth et al. 2000; Chapman 2004; Goulet et al. 2004; Goulet and Nepveu 2004; Ruchaud et al. 2007; Hudson et al. 2009; He et al. 2009; Sun et al. 2009; Yanagida 2009; Cauwe and Opdenakker 2010).

These examples clearly illustrate that proteases are dependent on the properties of the cleavage environment and that proteases may react or be modulated by rapid and transient alterations of the biochemical conditions during signalling events (Squier 2006; Salvesen and Dixit 1997). However, proteases themselves may be regulated by proteolytic cleavage and by other potentially reversible post-translational modifications including acetylation, methylation, glycosylation, hydrocarbonylation (farnesylation, geranylation, myristoylation), oxidation, phosphorylation, sulfation, SUMOylation, or ubiquitylation (Chap. 5; Doucet et al. 2008). These findings have profound implications on our understanding of the role of proteolysis as, above all, proteases are the mediators of possibly the most important post-translational modification of proteins—cleavage! This is because proteolytic action is essentially non-reversible and therefore frequently represents 'points of no return' as exemplified by caspase activation during apoptosis (Chap. 8).

Thus, eukaryotic cells have devised an almost unlimited repertoire of proteases; including soluble enzymes, transmembrane and peripheral-membrane proteins, macromolecular assemblies in the cytosol (proteasomes) or in the extracellular environment (meprins, MMPs, plasmin/uPA/uPAR). These proteases have the ability to interact with a likewise versatile group of substrates ranging from dipeptides to supramolecular protein assemblies that can be modified by endo- or exo-peptidic cleavages (Chaps. 1, 2 and 5). The diversity of proteolytic enzymes is represented by some hundreds of molecules classified on the basis of the catalytic mechanism of action as aspartic, cysteine, glutamic, metallo, serine, or threonine proteases (Chap. 1; Barrett et al. 2001, 2003; Barrett 2004; Rawlings et al. 2004; Rawlings 2010). In addition, a number of peptidases have not yet been classified into any of the known enzyme families.

3.3 The Architectural Art of Compartmentalization of Proteolysis: Specified Rooms for Proteolytic Processing

Like a room plan that is defined when constructing a new house, the eukaryotic cell hosts a variety of compartments with specific, well-defined conditions for specific [proteolytic] activities.

The nucleus, like an office or a creativity centre, can be considered a compartment of planning where proteolytic processing of proteins is a rare event needed for reading and distribution of the blueprint dictating when proliferating cells progress through the cell cycle, or pause in G_0-phase in order to differentiate (Nasmyth et al. 2000; Pellman and Christman 2001; Goulet and Nepveu 2004). Therefore, histones, transcription factors, and structural proteins connecting sister chromatids are the main substrates known for nuclear proteases (Heck 1997; Goulet et al. 2004, 2007, 2008; Ong et al. 2007; Tedelind et al. 2010; Haugen et al. 2013). While more substrates will undoubtedly be identified in the future, it appears that protease activity in the nucleus of interphase cells is occasional and highly regulated. During mitosis, the nuclear envelope is disassembled (as a result of reversible phosphorylation cascades), and proteolysis follows the principles of the cytoplasm. Upon initiation of apoptosis by specific intrinsic or extrinsic factors, however, the nuclear envelope is also rearranged (with cleavage of nuclear lamins this time), eventually leading to programmed cell death (Wyllie et al. 1980).

The kitchen where meals are created can be compared with the centres of *de novo* protein biosynthesis such as the rough endoplasmic reticulum (rER), and the semi-autonomous organelles, mitochondria and plastids. Proteases of these compartments—perhaps with additional exceptions in the cytoplasm (see below)—are selective in their choice of substrates, i.e. they cut off short targeting sequences to ensure directionality of delivery of newly synthesized proteins (ER, mitochondria, plastids) and, together with chaperones, they sample and inspect protein domains for proper folding (cytosol, ER, mitochondria, plastids).

The proteases of the compartments of *de novo* protein biosynthesis are involved in maturation of newly synthesized proteins, and therefore enable limited proteolysis that is domain-driven, sequence-targeted and restrictive—rarely involving "arbitrary" cleavage along the length of the polypeptide chain. Only in conditions of overload or disease, resulting in massive misfolding of newly synthesized proteins, are the proteases of these compartments 'up-regulated' for enhanced protein processing. However, the cell will first take all measures possible to attempt enhanced folding or refolding assisted by chaperones that become up-regulated in the so-called Unfolded Protein Response (UPR). The misfolded proteins, however, like the kitchen garbage, are delivered to other compartments for storage and disposal in aggresomes and inclusion bodies (Kopito 2000; Singhvi and Garriga 2009). The improperly folded proteins will eventually be destined for degradation by proteasomes. This process, when involving the ER, is also known as ER-associated degradation (ERAD) during which newly synthesized but misfolded proteins of the ER lumen are retro-translocated for degradation in cytoplasmic and ER-associated proteasomes (Wiertz et al. 1996a, b; Meusser et al. 2005; Romisch 2005).

The post-ER compartments of the secretory pathway including the intermediate compartment, Golgi apparatus, *trans*-Golgi network (TGN) and secretory vesicles are crucially important for modification, further maturation, sorting and packaging of proteins (Farquhar and Palade 1981, 1998; Varki 1993; Rothman and Wieland 1996; Kaiser et al. 2002). Like in a living room, molecules involved in post-translational modifications of pro-proteins (such as glycosylation, phosphorylation, sulfation, or proteolytic processing) gather to perform the critical cellular actions in preparation for extracellular secretion of proteins that frequently takes place in response to communication of cells with their extracellular environment (Docherty and Steiner 1982; Zhou et al. 1999). Well-studied proteases known for their significance as proprotein-processing enzymes are furin and other proprotein-convertases (PCs) (Seidah et al. 1991; Garten et al. 1994; Schaller and Ryan 1996; Nakayama 1997; Steiner 1998; Seidah and Prat 2002; Thomas 2002; Rockwell and Thorner 2004; Henrich et al. 2005; Scamuffa et al. 2006; Creemers and Khatib 2008; Seidah 2011). The PCs interact in a transient and selective manner with a huge variety of soluble proteins thereby processing the pro-forms of secretory proteins that include ECM constituents as well as peptide or protein hormones (Steiner 1969; Steiner et al. 1980; Seidah and Chretien 1999; Fu et al. 2008).

The major sorting steps take place at the level of the TGN, a compartment that resembles endo-lysosomes with respect to acidity, and where cargo loading and packaging into secretory vesicles or secretory granules and other Golgi-derived vesicles occurs. Thus, proteins en-route to endo-lysosomes including the zymogen [inactive] forms of proteases such as the asparaginyl endopeptidase (AEP/legumain) or the cathepsins are also transported along the secretory route to the TGN. From here, they reach their next sub-cellular destinations in a non-processed and proteolytically inactive proform (Chap. 4; Mort and Buttle 1997; Schaschke et al. 1998; Khan and James 1998; Linke et al. 2002a, b; Mach 2002; Wiederanders et al. 2003; Burden et al. 2007; Buttle 2007; Brix et al. 2008). Some endo-lysosomal proteases, like cathepsins B and L, are believed to become sorted into and activated in secretory vesicles of pancreatic cells and in neurosecretory granules of pituitary cells (Tooze et al. 1991; Halangk et al. 2000; Hook et al. 2004, 2008; Meister et al. 2010; Funkelstein et al. 2010); these examples, however, are specific to certain cell types. Secretory granules and secretory vesicles of most eukaryotic cells may be considered as compartments where proteases that are almost ready to function (like chymase, tryptase, perforins) are stored together with a number of other secretory macromolecules in a highly concentrated and compacted fashion. Only very-limited proteolysis is taking place in these granules and vesicles, if at all (Gomez-Lazaro et al. 2010), and they may thus be considered analogous to hallways and corridors of houses that are used by both inhabitants and visitors simply for by-passing and through-traffic and possibly where coats are removed.

The dining room, designed for consumption of meals, serves as a reference point for the compartments of the endocytic pathway, i.e. endosomes and lysosomes (Gruenberg and Howell 1989; Schmid 1993; Gruenberg and Maxfield 1995; Mukherjee et al. 1997; Gu and Gruenberg 1999; Gagescu et al. 2000; Nichols and Lippincott-Schwartz 2001; Conner and Schmid 2003). A number of entry points exist for proteins reaching endo-lysosomes (Rechsteiner 1990; Seglen et al. 1996; Felberbaum-Corti et al. 2003; Lin et al. 2004; Klionsky 2007; Mizushima et al. 2008; Arias and Cuervo 2011; Chen and Klionsky 2011). Endo-lysosomes are multi-faceted and interchangeable sub-cellular compartments that perform numerous functions above their main catabolic tasks in the break-down of proteins and other internalized molecules (Brix 2005; Brix et al. 2008). Restricted and limited proteolysis in early endocytic compartments is extremely important in antigen processing for subsequent MHC class II dependent presentation (Riese and Chapman 2000). The early endosome is considered a ligand-unloading and a sorting station (not only for polarized epithelial cells), and it has receiving and distribution functions resembling those of the TGN in many respects (Howell et al. 1989; Sachse et al. 2002). Sorting may result in subsequent trafficking of proteins like receptors back to the plasma membrane for recycling, or further transport to the later endocytic compartments for transient storage or degradation (and thus, down-regulation) (Katzmann et al. 2002; Lin et al. 2004). The late endosome, however, is clearly the dining table where incoming proteins are taken apart and often broken down to the level of single amino acids. The menu (sequence) and the selection of substrates available on the dinner table will need one or many proteases to proteolytically process any given protein substrate (Tedelind et al. 2011). Thus, early, or possibly even all compartments along the endocytic pathway may be utilised—and, like the arrangement of our dinner tables, the number and types of cutlery will differ depending on the food served. The remnants of proteolytic processing will be stored in poorly accessible residual bodies often accumulating in lysosomes and in vacuoles [plant], which we may therefore compare with the trash bags of our house.

Peroxisomes of animal cells, glyoxysomes of plant cells, and the smooth ER of all eukaryotic cells are the major compartments for detoxification (Platta and Erdmann 2007) in which, in part, extremely oxidative environments enable beta-oxidation which facilitates protein degradation. The composition of peroxisomes and glyoxysomes does not appear to contain proteases. Yet, occasionally, there have been reports of proteolytic enzymes like insulin-degrading enzyme (IDE) that are involved in the degradation of oxidized proteins within peroxisomes (Authier et al. 1996; Morita et al. 2000). However, these compartments are certainly not considered protein-processing compartments of major impact on cellular functions, so we may compare them to the trash compactor or garbage disposal of a house that dispose of contaminating and potentially toxic components.

3.4 Proteases Facing the Extracellular Space

The outside deck/terrace of eukaryotic cells is represented by the plasma membrane with a variety of cellular appendages like cilia, flagella, or microvilli reaching into the extracellular surroundings (Kenny and Maroux 1982). Like the plasma membrane itself, these appendages can carry a multitude of proteolytic enzymes anchored as transmembrane proteins or, if soluble, by binding to cell surface receptors (Chaps. 7, 9 and 11; Bode et al. 1996; Rosenblum and Kozarich 2003; Mentlein 2004; List et al. 2006; Bugge et al. 2007; Turner and Nalivaeva 2007; Lopez-Otin and Bond 2008; Szabo and Bugge 2008; Sohail et al. 2008; Sterchi et al. 2008; Choi et al. 2009; Ramsay et al. 2009; Yao 2010; Hendrickx et al. 2011). Such co-ordination of proteolytic activity in the peri-cellular environment is well known for important contributions in protein breakdown and peptide processing, in particular, at the apical plasma membrane domains of enterocytes (Chap. 11; Kenny and Maroux 1982; Bank et al. 2008; Matteucci and Giampietro 2009; Yu et al. 2010; Arampatzidou et al. 2011a, 2012). A classic example of this is in the most vital event of fertilization, where proteases are secreted from within the sperm head acrosome for degradation and modification of the protective vitellin layer surrounding the oocyte (Blobel 2000b). The oocyte is eventually reached by the sperm for subsequent fusion and zygote formation through explosive, actin polymerization-driven extrusion of microvillus-like extensions.

In addition to outward-looking appendages of cells, invaginations of the plasma membrane also play a role in the dynamics of proteolysis. Some invaginations form during internalization of molecules destined to reach the compartments of the endocytic pathway. Others, like caveolae (little flask-like caves), seem to be less well connected to the cellular interior, but provide a specific biochemical environment in which protein processing is facilitated (Simionescu et al. 1972; Hajjar and Acharya 2000; Predescu et al. 2001; Kim and Hajjar 2002; Pelkmans and Helenius 2002; Navarro et al. 2004). Thus, plasma membrane indentations—caveolae and other non-caveolin coated microdomains—can provide areas of concentrated cell surface receptors for restricting and focalizing proteases to specified sub-domains of the cell surface (Owen and Campbell 1995; Mai et al. 2000; Cavallo-Medved and Sloane 2003; Gumy et al. 2010).

Invadopodia may be analogous to caveolae in their protease concentrating roles but are constructed so as to extend finger-like into the extracellular matrix (Chen 1996; Murphy and Gavrilovic 1999; Linder 2007; O'Brien and O'Connor 2008; Frittoli et al. 2011). Invadopodia have become prominently known for their distinctive composition and functional duality for being involved in both cell adhesion and extracellular matrix degradation through associated proteases. Hence, delicate and rapidly interchangeable cellular extensions that protrude and communicate with the surrounding environment like an easily remodelled patio may also contribute in multiple ways to peri- and extra-cellular proteolytic processes (Woodward et al. 2007; Brix et al. 2008; Korkmaz et al. 2008; Pham 2008; Stetler-Stevenson 2008; Sato and Takino 2010). The proteases found in such exposed positions are usually characterized by extended extracellular domains which are often heavily glycosylated to promote stability against proteolytic attacks, and like their soluble relatives, provide many platforms and binding sites for interaction with other macromolecules or regulatory factors (Gahmberg and Tolvanen 1996; Manon-Jensen et al. 2010; Cawston and Young 2010).

The extracellular space itself, the garden, also houses a myriad of proteases that are derived from the many secretions of different cells in a tissue (Chaps. 6, 13–15; Andrews 2000; Overall and Blobel 2007). Proteolytic enzymes in the extracellular space may either be secreted in active form or remain latent until activation triggers other proteolytic enzymes to convert them into active proteases that often interact with each other to cleave protein substrates by sequential modes (Brix and Herzog 1994; Brix et al. 1996, 2001; Tepel et al. 2000; Linke et al. 2002a; Mayer et al. 2009; Vreemann et al. 2009; Tedelind et al. 2011). The extracellular space can therefore be considered a rich reservoir of proteolytic enzymes that are not only crucial for tissue remodelling but, in part, also contribute to organization of extracellular matrix components, a function that is crucial during morphogenesis in embryonic development (using ADAMs [A Disintegrin And Metalloprotease], ADAM-TSs) (Chap. 9; White 2003; Noel et al. 2004; Apte 2004; Edwards et al. 2008; Kveiborg et al. 2008; Reiss and Saftig 2009; van Goor et al. 2009; Dikic and Schmidt 2010; Urban 2010; Saftig and Reiss 2011).

3.5 Democracy as an Answer for Radical Decision-Making Processes

Radical decision-making during peptide bond cleavage may be seen as the hallmark of proteolytic activitiy, yet proteases can hardly be thought of as dictators. Often proteases act in proteolytic networks in which it is not the individual enzyme that counts (Chaps. 2 and 5). Collaboration, including finely-tuned and highly-regulated actions amongst several proteases, is instrumental in paving the way to success. Clearly, cleaving substrates at the right time and place, and with the desired pace and the required specificity is crucial (Liu et al. 1999; Kidd et al. 2001; Rao 2003;

Baruch et al. 2004; Joyce and Hanahan 2004; Blum et al. 2005; Brix and Jordans 2005; Carlson and Cravatt 2007; Gocheva and Joyce 2007; Victor and Sloane 2007; Blum 2008; Brix et al. 2008; Jedeszko et al. 2008; Paulick and Bogyo 2008). We shall discuss these aspects of sub-compartmentalization by looking in more detail at the specialists. These include the endo-lysosomal proteases, which come in a most astounding array and act, not exclusively but optimally in endosomes and lysosomes of mammalian cells (Chap. 4; Kirschke et al. 1995; Chapman et al. 1997; McGrath 1999; Turk et al. 2001; Brix 2005; Mohamed and Sloane 2006; Brix et al. 2008; Reiser et al. 2010) (Fig. 3.5).

The compartments of the endocytic pathway contains a wide range of protein processing and degrading enzymes including the cysteine peptidase AEP/legumain that engages in proteolytic maturation of the proforms of other endo-lysosomal proteases (Barrett and Rawlings 2001; Ishidoh and Kominami 2002; Watts et al. 2005; Haugen et al. 2013), aspartic cathepsins D and E (Barrett 1979; Yamamoto 1995; Dunn et al. 1998; Tatnell et al. 1998; Rochefort et al. 2000; Tsukuba et al. 2000; Dash et al. 2003; Nakanishi 2003; Liaudet-Coopman et al. 2006; Zaidi et al. 2008; Hook et al. 2008; Nicotra et al. 2010), the serine cathepsins A and G (Travis 1988; Hiraiwa 1999; Pham 2006; Caughey 2007; Korkmaz et al. 2008; Meyer-Hoffert 2009; Kessenbrock et al. 2011), and the most versatile group of the cysteine cathepsins B, C, F, H, K, L, O, S, V, W, and X/Z in man (Chap. 6; Kirschke et al. 1995; McGrath 1999; Turk et al. 2001; Greenbaum et al. 2002; Brix 2005; Sloane et al. 2005; Mohamed and Sloane 2006; Gocheva and Joyce 2007; Brix et al. 2008; Reiser et al. 2010). Rodents express even more (placental) cysteine cathepsins (Sol-Church et al. 2002) and protozoa like *Tetrahymena* are clearly very extreme examples with dozens of these proteolytic enzymes. The cathepsins that can be found in endo-lysosomes of every eukaryotic cell are the cathepsins B, D, G, H, and L (Turk et al. 2000; Reinheckel et al. 2001; Vasiljeva et al. 2007; Brix et al. 2008; Reiser et al. 2010), while other cathepsins may be present only in certain cell types that are facing very specific challenges, e.g. osteoclasts express high levels of cathepsin K to facilitate bone matrix turnover (Bromme and Okamoto 1995; Saftig et al. 1998; Tepel et al. 2000; Lecaille et al. 2003; Desmarais et al. 2009; Podgorski 2009; Rachner et al. 2011). Nonetheless, all of these proteolytic enzymes can process their substrates by limited or by full proteolysis (Brix and Herzog 1994; Brix et al. 1996, 2001; Friedrichs et al. 2003; Dauth et al. 2011a, b).

Substrates reach the endocytic compartments not only by internalization from the extracellular space, but also by direct entry from the cytosol via chaperone-mediated or "classic/conventional" autophagy (Dice 2007). The outcome of substrate proteolysis by endo-lysosomal proteases depends on whether individual or multiple proteases act on the substrate with endo- and/or exo-peptidase modes of cleavage (Chaps. 1, 2, 6, 15; Barrett and Kirschke 1981; Kirschke et al. 1995; McGrath 1999; Tepel et al. 2000; Jordans et al. 2009). Moreover, proteolytic processing in endocytic compartments is aided by increasingly reductive conditions that support protein unfolding (Pillay et al. 2002; Jordans et al. 2009; Scott and Gruenberg 2011). In general, the hydrolytic enzymes of endo-lysosomes will team up so that almost all protein substrates with their many other, non-proteolytic post-translational modifications can be handled very efficiently. Interestingly enough, endo-lysosomal proteolytic enzymes are sometimes selective in their substrate choice, whereas others are less selective and cleave almost any protein substrate. The pH optimum of substrate cleavage by endo-lysosomal proteolytic enzymes spans an astonishingly wide range from neutral pH-values for cathepsin S down to the most acidic pH-values for cathepsin D (Takahashi and Tang 1981; Kirschke and Wiederanders 1994). Hence, protein processing and degradation by endo-lysosomal proteases is already initiated in the peri-cellular vicinity of specialized cells that secrete these enzymes in a regulated or non-regulated manner, and, in general, the activity of endo-lysosomal proteases is available in all compartments of the endocytic pathway (Brix and Herzog 1994; Brix et al. 2008). An example of this kind can be found in the thyroid gland, where a huge protein substrate (thyroglobulin) and its covalently cross-linked supramolecular assemblies are handled by the cathepsins B, K, L, and S acting in a temporal- and spatially-regulated, sequential manner for proteolytic liberation of iodinated thyronines from within the polypeptide chain (Brix et al. 1996, 2001; Friedrichs et al. 2003; Jordans et al. 2009; Dauth et al. 2011a).

The cocktail of proteases that is present in one or another sub-compartment of the endocytic pathway will determine the extent of protein substrate cleavages. Numerous questions therefore arise: what is an endosome and what is a lysosome? What governs or triggers transport of endo-lysosomal proteases into the interchangeable compartments of the endocytic pathway? And what are the targeting sequences and which mechanisms explain why some compartments retain fewer, while others contain more, of these proteolytic enzymes? The answers to these questions are partially derived from the study of mannose 6-phosphorylation of the pro-forms of endo-lysosomal proteases (Kornfeld 1992; Peters and von Figura 1994; Bresciani and von Figura 1996). This post-translational modification is maintained in the compartments of the secretory pathway, but partially also persists in the extracellular space. The well known cation-dependent and -independent mannose 6-phosphate receptors can deal with sorting at the level of the TGN, or with re-internalization should pro-forms become secreted (Peters et al. 1990; Hille et al. 1992; Koster et al. 1993; McIntyre et al. 1994; Pohlmann et al. 1995). However, this elegant molecular mechanism of endo-lysosomal enzyme trafficking is not realized in all cell types (Ludwig et al. 1994; Pohlmann et al. 1995; Tanaka et al. 2000). Therefore, the answers to these questions will not be universal, but vary from cell type to cell type (Linke et al. 2002a, b; Brix 2005; Brix and Jordans 2005; Brix et al. 2008; Tedelind et al. 2011).

3.6 Debating Clubs: Proteases and Their Inhibitors

Proteolysis is a result of balancing proteolytic and anti-proteolytic factors (Chaps. 6 and 15; Turk and Bode 1991; Basbaum and Werb 1996; Turk et al. 1997; Matrisian 1999; Coussens et al. 2002; Abrahamson et al. 2003; Turk et al. 2003; Kaiserman et al. 2006; Mohamed and Sloane 2006; Lopez-Otin and Matrisian 2007; Vasiljeva et al. 2007; Scott and Taggart 2010; Reiser et al. 2010). While this notion may have been formulated early on in the examination of proteolysis, it still remains an apt proposal. However, the temporal and spatial regulation of proteolysis must be considered as equally important to the balance with anti-proteolytic factors, if not more. Hence, a well orchestrated debating event between substrates, proteases, and their inhibitors is required for a given cellular process to dictate the final outcome of proteolytic cleavage. So far, many more investigations exist on the proposal of this house: "how and when and where to *start* proteolysis", rather than on: "how and when and where to *halt* the event". However, signalling may intervene (Mackie et al. 2002; Ichihara et al. 2006; Mockaitis and Estelle 2008; Murphy et al. 2009; Smith and Marshall 2010) and continued proteolysis may be required in some instances. Control is essential as if proteolysis is exceeded, it may lead to severe disease and eventual death. Therefore, inhibitors of proteases, that typically act intracellularly, are highly concentrated in extracellular fluids for prevention of premature proteolysis around a cell or tissue should the proteases happen to escape from cells in unplanned events (Travis and Salvesen 1983; Turk and Bode 1991; Nagase et al. 1996; Bode et al. 1999; Deveraux et al. 1999; Silverman et al. 2001; Salvesen and Duckett 2002; Abrahamson et al. 2003; Whisstock and Bottomley 2008; Drag and Salvesen 2010). Thus, termination of proteolytic cleavage is as important as its initiation and the Master of Ceremony that determines the sequence and types of proteolytic events taking place must consider protease inhibitors as essential factors in the compartmentalization of proteolysis.

Again, endo-lysosomal enzymes represent good examples for illustrating such control mechanisms by their endogenous inhibitors. Endo-lysosomal enzymes can become extremely dangerous for other cellular compartments or even for the entire cell if released into the extracellular space in an uncontrolled fashion. Therefore, endo-lysosomal proteolytic enzymes must be transported in a defined manner and premature activation is further shielded by potent 60–100 residue inhibitory N-terminal pro-peptide domains as in the case of cysteine cathepsin proteases (Chap. 4; Mach et al. 1994; Mach 2002; Wiederanders et al. 2003). These domains have two functions; firstly, they adopt their own clearly defined secondary structure which acts as a folding chaperone to facilitate the larger catalytic domain to fold. Secondly, the pro-peptide binds tightly across the active site of the catalytic domain in reverse orientation to normal substrates, thus acting as competitive inhibitors. This tight binding is pH dependent and as the inactive zymogen species is transported to the lysosomes, the propeptide will dissociate from the catalytic domain in the acidified lysosomal lumen, leaving the active site exposed and active and results in the irreversible proteolytic removal of the propeptide domain leaving the mature active protease domain (Pungercar et al. 2009; Linke et al. 2002a, b). However, despite this understanding of the role of the pro-peptide, the recent studies highlighting the presence of cysteine cathepsins in the nucleus, a consequence of downstream translational initiation (Chapman 2004; Goulet and Nepveu 2004; Goulet et al. 2004; Tedelind et al. 2010), highlights that these control mechanisms for the pro-peptide require further supplementation by the presence of inhibitors such as Stefin B in the nucleus (Ong et al. 2007; Ceru et al. 2010).

Cystatins and stefins are the most powerful counter-players of some of the endo-lysosomal proteases (Turk and Bode 1991; Turk et al. 1997, 2003; Abrahamson et al. 2003). Some families of these protease inhibitors reside within the cytosol whereas others are secreted into the extracellular space for abundant presence in extracellular fluids of tissues and body organs. Hence, the cysteine peptidase inhibitors are separated as safe-guarding molecules from the endo-lysosomal cysteine proteases by a membrane, the endo-lysosomal membrane and/or the plasma membrane. Very similar strategies are realized for serpins that inhibit both serine- and cysteine peptidases and which are abundantly present in the cyto- and nucleoplasm of eukaryotic cells (Silverman et al. 2001; Whisstock and Bottomley 2006, 2008). The aspartic peptidases of endo-lysosomes are kept in control by another group of cross-class inhibitors, the thyropins, which also interfere with cysteine peptidase activities (Chap. 2; Lenarcic and Bevec 1998; Novinec et al. 2006). Another regulatory mechanism of keeping the team of endo-lysosomal proteases well within their borders is reflected by the relatively strict pH-requirements for both proteolytic activity and stability of these enzymes, thereby explaining why many members of this large group of proteases can be kept in check, simply by compartmentalization to specific cellular regions of defined biochemical conditions (Mort et al. 1984; Mort and Buttle 1997; Jordans et al. 2009).

MMPs, meprins/BMPs, ADAMs and ADAM-TSs are best known for their proteolytic activities exerted in the extra- and peri-cellular space (Chaps. 7, 9, 11, 13–15; Brinckerhoff and Matrisian 2002; Coussens et al. 2002; Norman et al. 2003; Edwards et al. 2008; Reiss and Saftig 2009; Rosenberg 2009; Tallant et al. 2010). These enzymes are either soluble or transmembrane proteins. However, many act in sequence thereby establishing proteolytic cascades in which one metalloprotease activates the next (Chakraborti et al. 2003). These cascades of proteolytic activation often take place at the cell surface with the plasma membrane and its receptors serving as scaffolding support. Another excellent example of a similar kind is provided by the urokinase-type plasminogen activator (uPA) that interacts with the plasminogen activator receptor (uPAR), a transmembrane protein of the plasma membrane (Magdolen et al. 2000; Blasi and Carmeliet 2002; Behrendt 2004; Mondino and Blasi 2004; Shi and Stack 2007). Moreover, in a number of proteolytic processes that involve MMPs and PAs (Plasminogen Activators), the endogenous protease inhibitors TIMPs (tissue inhibitors of MMPs) and PAIs (PA inhibitors) become part of the proteolytic cascade and therefore fine-tune and regulate the peri-cellular activities of these proteolytic enzymes (Bode and Renatus 1997; Bode and Maskos 2001, 2003; Stetler-Stevenson 2008; Brew and Nagase 2010). Final termination of such proteolytic events that can involve multiple members of the same and also of related and unrelated protease families is considered to occur by down-regulation (Jiang et al. 2001), i.e. degradation by endo-lysosomal pathways.

3.7 Speed-Dating of Proteolytic Enzymes and Their Substrates

So far we have not described one of the most intriguing examples of sub-compartmentalization of proteolysis and of processivity of proteolytic cleavage—the processing of linear polypeptides by the very well organized protease-machines of the cytosol, the proteasomes (Groll and Clausen 2003).

The cytoplasm, as the largest sub-compartment of eukaryotic cells, bears a variety of complex protein-processing assemblies (De Mot et al. 1999; Brandstetter et al. 2001; Rosenblum and Kozarich 2003). The best understood of such cytoplasmic protease assemblies is the 26S proteasome that can function in cleaving a variety of different peptide-bonds, and which can even be considered a protein processing machine when acting in conjunction with 19S cap-structures that comprise energy-dependent unfolding factors (Wolf and Hilt 2004; Vierstra 2009; Stadtmueller and Hill 2011). The structure of the yeast proteasome has been determined in atomic detail to a resolution of 2.4 Å by X-ray crystallography which also discovered that the proteasome belongs to a new class of proteases, the threonine peptidases (Groll et al. 1997). These astonishingly versatile, and, in cellular immune response reactions, highly adaptable (Groettrup et al. 1996, 2001a, b, 2010; van den Eynde and Morel 2001; Kloetzel and Ossendorp 2004; Rivett and Hearn 2004; Driscoll and Dechowdhury 2010) protease complexes provide a hydrolytic environment with unique properties in that four heptameric rings assemble to form a cleavage chamber (Groll and Clausen 2003; Stadtmueller and Hill 2011) that concentrates, guides and shields protease substrates from surrounding influences without being enclosed by a lipid membrane. The fully functional proteasome is already the 20S assembly which is complemented by two 19S regulatory caps, one at each end to prevent inappropriate degradation in the larger 26S proteasome.

The precise localization of proteasomes is still a matter of debate (Rivett 1998; Brooks et al. 2000a, b; Wojcik and DeMartino 2003). While their cytoplasmic location is undisputed, it is less clear under which circumstances—and for which reasons, proteasomes may be detected in the nucleus. Proteasomes are often associated with the cytosolic leaflets of the ER membrane or the nuclear envelope where they engage in ERAD, the ER-associated degradation of misfolded protein intermediates that are destined for destruction in order to protect the cell from an overload and accumulation of unfolded proteins (Wiertz et al. 1996a, b). Likewise, proteasomes are detectable as constituents of the so-called aggresomes, non-membrane enclosed regions of the cytoplasm of cells that harbour inclusions of misfolded proteins resulting from protein over-expression (Kopito 2000). In all these latter cases, protein substrates may or may not require ubiquitinylation as a proteasome-targeting signal (Pickart 1997; Schwartz and Hochstrasser 2003; Ciechanover 2003; Tai and Schuman 2008).

The proteasome is thus a beautiful example where evolution has devised a complex protein assembly employing a variety of proteolytic activities to cleave substrates in a more or less specific fashion. The result may be release of peptides of defined length, suitable to enter the ER lumen via TAP-transporter for MHC class I dependent antigen presentation of virus-derived peptides (Abele and Tampe 2006, 2009; Hansen and Bouvier 2009), or tailor-made for further destruction by other proteolytic enzymes of the cytosol that feature amino- or carboxypeptidase activities (Gomis-Ruth 2008; Gomis-Ruth et al. 2012) in order to replenish the cellular pool of free amino acids.

3.8 How to Organise Proteolytic Actions in Busy Times Like Rapid Cell Cycle Progression, Remodelling of Cellular Components, or Cell Death

The cytoplasm is already considered as a compartment of *de novo* protein biosynthesis. However, it is also the cellular compartment of protein maintenance and turnover in response to signalling or mechanical damage of cytoskeletal proteins (Chap. 12; Goldberg and Dice 1974; Dice and Walker 1979; Dice 1987, 1990; Olson and Dice 1989; Rechsteiner and Hill 2005). It is also well understood that cell cycle progression depends on cyclic degradation and reformation of cyclins. Moreover, the cytoplasm contains many scaffolding factors, proteins and ribonucleoproteins, that have the ability to build and organize larger assemblies of macromolecules, such as the apoptosome that initiates a sequence of dramatic proteolytic events eventually leading to cell death in a tightly regulated and programmed manner (Chap. 8; Tschopp et al. 2003; Vanlangenakker et al. 2008; Declercq et al. 2009; Pop and Salvesen 2009; Vandenabeele et al. 2010; Krysko et al. 2011). Hence, while the cytoplasm may be the largest compartment of eukaryotic cells, it has established unique strategies to sub-compartmentalize proteolytic actions to certain required areas. As calcium waves emerging from the sperm entry point specify polarity of the zygote, and thus the body axis at the earliest stage of development, the cytoplasm of eukaryotic cells has found intriguing and elegant solutions to the primary question of this chapter, compartmentalization of proteolysis.

Besides proteasomes and a number of other proteolytic enzymes that are important contributors to guarantee the pool of free amino acids, other proteases are found abundantly in the cytosol: the procaspases (Earnshaw et al. 1999; Riedl and Shi 2004; He et al. 2009; Pop and Salvesen 2009; Drag and Salvesen 2010) and the calpains (Sorimachi et al. 1997, 2010; Reverter et al. 2001; Storr et al. 2011). Procaspases are cysteine peptidases cleaving after aspartic acid upon their proteolytic activation in the initiation of apoptotic pathways leading to programmed cell death (Alnemri et al. 1996). The assembly of caspases in apoptosomes is triggered by a variety of intrinsic and extrinsic signalling pathways (Wyllie et al. 1980; Budihardjo et al. 1999; Song and Steller 1999; Salvesen 2002; Riedl and Salvesen 2007; Kersse et al. 2011). Again, a sub-compartmentalization of their proteolytic activities can be conceived from the formation of restricted, caspase-enriched cytoplasmic areas where scaffolding factors serve as the platforms for apoptosome formation, and, as in the case of proteasomes, without the help of lipid membranes. Calpains (also cysteine peptidases) represent yet another solution to restricting proteolytic activities, as they are dependent on cytosolic free calcium in the range of nanomolar concentrations that exist only transiently in response to activation of cells via e.g. protein kinase C-mediated signalling (Sato and Kawashima 2001). Calpains are known to be involved in cytoskeletal protein remodelling in skeletal muscle cells under conditions of excessive contractile activity that may cause micro-damage to the muscle fibers (Chap. 12; Koohmaraie 1992).

Besides the above described mechanisms of restricting proteolytic activity of cytoplasmic proteases to certain areas of the cytoplasm, a number of 'safe-guarding' protease inhibitors can be found in certain areas of the cytoplasm as well (Turk et al. 2003; Kaiserman et al. 2006) (see also above). First, and foremost, the serpins which act as cross-class inhibitors of both serine and cysteine peptidases are found (Blasi 1993; Smirnova et al. 1994; Brunner and Preissner 1994; Pappot et al. 1995; Bailey et al. 2006; Kaiserman et al. 2006; Izuhara et al. 2008). Secondly, the cystatins are localised not only in all body fluids and in endocytic compartments, but other members of this inhibitor family are also present in the cytoplasm (Travis and Salvesen 1983; Turk and Bode 1991; Bode et al. 1999; Deveraux et al. 1999; Silverman et al. 2001; Abrahamson et al. 2003; Whisstock and Bottomley 2008). Cystatins are believed to protect cells from sudden death by inhibiting cysteine peptidases which not only reside and act normally in the cytoplasm, but which may become enriched in the cytosol if endo-lysosomes suffer membrane leakage or rupture (Chap. 8; Nixon and Cataldo 1993; McNeil and Steinhardt 1997; Gerasimenko et al. 2001; Degli Esposti 2008; Ivanova et al. 2008; Turk and Turk 2009). Thus, eukaryotic cells can further transiently modulate proteolytic activity through the expression and localisation of cognate inhibitors.

3.9 Themes Emerging from the Discovery of the New Kids on the Block

The role for selected proteolysis in ubiquitin biology is an area that is fast developing. Ubiquitin (and indeed related peptides such as SUMO, ISG15) is post-translationally added to proteins by E3 ubiquitin ligase complexes (Pickart and

Eddins 2004). This modification of a specific protein can have a broad range of effects, from simple targeting of a protein for degradation, to modification of its activity in a manner not dissimilar from the role of phosphorylation throughout biological networks. Ubiquitination forms a key signaling role in processes such as cell cycle regulation (Song and Rape 2008), enzyme activation/inactivation (Adhikari et al. 2007) and DNA repair (Kennedy and D'Andrea 2005).

Ubiquitin is a 76 residue peptide, the free C-terminus of which is conjugated onto lysine side chain amino groups on target proteins forming a pseudo-peptide bond (Pickart 2001). However, the form of this ubiquitination can vary substantially from a simple mono-ubiquitination to addition of a number of ubiquitins (poly-ubiquitination). This post-translational modification is further complicated in that ubiquitin itself has a number of lysine residues and a branched poly-ubiquitin can be created through these points—each of which has its own distinctive conformation and role (K48 branched polymers labels proteins for degeneration and K63 polymers activate target proteins) (Pickart and Fushman 2004).

Key to the role of ubiquitination signaling is the fact that it is reversible and removal of ubiquitin is coordinated by proteases called deubiquitinating enzymes (DUBs). To date, almost 100 DUB proteases have been identified, belonging to cysteine and metalloprotease families (Reyes-Turcu et al. 2009). However, it is interesting to note that no endogenous inhibitors of these proteases have been identified to date and control of their activity appears to use other mechanisms including transcriptional control, substrate recognition and physical compartmentalization (Zhao et al. 2008).

In addition to these mechanisms of control, within the cytoplasm where the majority of these proteases are located, a major source of control is the presence of binding partners and adaptor proteins. Indeed, many of the DUB proteases contain other domains in addition to their catalytic core to facilitate binding to other proteins, and in particular scaffold proteins where it is thought that the DUB forms part of a multi-protein complex (often with the cognate E3 ubiquitin ligase complex), coordinating its activity and substrate specificity (Marfany and Denuc 2008). In this regard, it is perhaps particularly striking that invadolysin, a metalloprotease coordinating mitotic progression, has recently been found to interact genetically with non-stop, a DUB (MM Heck, personal communication).

3.10 And Now, We Retire to the Study

We conclude that we have only glimpsed the tip of the ice-berg. Many more intriguing features of how proteolysis is compartmentalized in the realm of the cell are to be discovered, understood and set into context before we will fully grasp the complexity of proteolytic enzymes and their compartmentalization principles. Faulty localization, delayed or too early, too fast or too slow, extensive or non-productive interactions of proteases, substrates, inhibitors and other factors regulating proteolysis will result in dramatic effects to the cell—and likely in diseases detrimental to the functioning of an organism. Understanding the regulation of intra- and extra-cellular proteolysis will be an important step to assess the physiology of eu- and prokaryotic cells. In turn, this understanding will be crucial to clarifying the routes leading to disease, and ultimately, shed light on potential therapeutic pathways.

References

Abele R, Tampe R (2006) Modulation of the antigen transport machinery TAP by friends and enemies. FEBS Lett 580(4):1156–1163

Abele R, Tampe R (2009) Peptide trafficking and translocation across membranes in cellular signaling and self-defense strategies. Curr Opin Cell Biol 21(4):508–515

Abrahamson M, Alvarez-Fernandez M, Nathanson CM (2003) Cystatins. Biochem Soc Symp 70:179–199

Adhikari A, Xu M, Chen ZJ (2007) Ubiquitin-mediated activation of TAK1 and IKK. Oncogene 26(22):3214–3226

Alnemri ES, Livingston DJ, Nicholson DW, Salvesen G, Thornberry NA, Wong WW, Yuan J (1996) Human ICE/CED-3 protease nomenclature. Cell 87(2):171

Andrews NW (2000) Regulated secretion of conventional lysosomes. Trends Cell Biol 10 (8):316–321

Annaert W, De Strooper B (1999) Presenilins: molecular switches between proteolysis and signal transduction. Trends Neurosci 22(10):439–443

Apte SS (2004) A disintegrin-like and metalloprotease (reprolysin type) with thrombospondin type 1 motifs: the ADAMTS family. Int J Biochem Cell Biol 36(6):981–985

Arampatzidou M, Mayer K, Iolyeva ME, Gebre Asrat S, Ravichandran M, Günther R, Schüle T, Reinheckel T, Brix K (2011a) Studies of intestinal morphology and cathepsin B expressing in a transgenic mouse aiming at intestine-specific expression of CathB-EGFP. Biol Chem 392:983–993

Arampatzidou M, Rehders M, Dauth S, Yu DMT, Tedelind S, Brix K (2011b) Imaging of protease functions – current guide to spotting cysteine cathepsins in classical and novel scenes of action in mammalian epithelial cells and tissues. Ital J Anat Embryol 116(1):1–19

Arampatzidou M, Schütte A, Hansson GC, Saftig P, Brix K (2012) Effects of cathepsin K deficiency on intercellular junction proteins, luminal mucus layers, and extracellular matrix constituents in the mouse colon. Biol Chem 393:1391–1403

Arias E, Cuervo AM (2011) Chaperone-mediated autophagy in protein quality control. Curr Opin Cell Biol 23(2):184–189

Arolas JL, Vendrell J, Aviles FX, Fricker LD (2007) Metallocarboxypeptidases: emerging drug targets in biomedicine. Curr Pharm Des 13(4):349–366

Authier F, Posner BI, Bergeron JJ (1996) Insulin-degrading enzyme. Clin Invest Med 19 (3):149–160

Bailey CM, Khalkhali-Ellis Z, Seftor EA, Hendrix MJ (2006) Biological functions of maspin. J Cell Physiol 209(3):617–624

Bank U, Bohr UR, Reinhold D, Lendeckel U, Ansorge S, Malfertheiner P, Tager M (2008) Inflammatory bowel diseases: multiple benefits from therapy with dipeptidyl- and alanyl-aminopeptidase inhibitors. Front Biosci 13:3699–3713

Barrett AJ (1979) Cathepsin D: the lysosomal aspartic proteinase. Ciba Found Symp 75:37–50

Barrett AJ (2004) Bioinformatics of proteases in the MEROPS database. Curr Opin Drug Discov Devel 7(3):334–341

Barrett AJ, Kirschke H (1981) Cathepsin B, Cathepsin H, and cathepsin L. Methods Enzymol 80 (Pt C):535–561

Barrett AJ, Rawlings ND (2001) Evolutionary lines of cysteine peptidases. Biol Chem 382 (5):727–733

Barrett AJ, Rawlings ND (2007) 'Species' of peptidases. Biol Chem 388(11):1151–1157

Barrett AJ, Rawlings ND, O'Brien EA (2001) The MEROPS database as a protease information system. J Struct Biol 134(2–3):95–102

Barrett AJ, Tolle DP, Rawlings ND (2003) Managing peptidases in the genomic era. Biol Chem 384(6):873–882

Baruch A, Jeffery DA, Bogyo M (2004) Enzyme activity–it's all about image. Trends Cell Biol 14 (1):29–35

Basbaum CB, Werb Z (1996) Focalized proteolysis: spatial and temporal regulation of extracellular matrix degradation at the cell surface. Curr Opin Cell Biol 8(5):731–738

Behrendt N (2004) The urokinase receptor (uPAR) and the uPAR-associated protein (uPARAP/ Endo180): membrane proteins engaged in matrix turnover during tissue remodeling. Biol Chem 385(2):103–136

Bergers G, Coussens LM (2000) Extrinsic regulators of epithelial tumor progression: metalloproteinases. Curr Opin Genet Dev 10(1):120–127

Bhat KP, Greer SF (2011) Proteolytic and non-proteolytic roles of ubiquitin and the ubiquitin proteasome system in transcriptional regulation. Biochim Biophys Acta 1809(2):150–155

Blasi F (1993) Urokinase and urokinase receptor: a paracrine/autocrine system regulating cell migration and invasiveness. Bioessays 15(2):105–111

Blasi F, Carmeliet P (2002) uPAR: a versatile signalling orchestrator. Nat Rev Mol Cell Biol 3 (12):932–943

Blobel CP (2000a) Remarkable roles of proteolysis on and beyond the cell surface. Curr Opin Cell Biol 12(5):606–612

Blobel CP (2000b) Functional processing of fertilin: evidence for a critical role of proteolysis in sperm maturation and activation. Rev Reprod 5(2):75–83

Blum G (2008) Use of fluorescent imaging to investigate pathological protease activity. Curr Opin Drug Discov Devel 11(5):708–716

Blum G, Mullins SR, Keren K, Fonovic M, Jedeszko C, Rice MJ, Sloane BF, Bogyo M (2005) Dynamic imaging of protease activity with fluorescently quenched activity-based probes. Nat Chem Biol 1(4):203–209

Bode W, Huber R (1992) Natural protein proteinase inhibitors and their interaction with proteinases. Eur J Biochem 204(2):433–451

Bode W, Huber R (2000) Structural basis of the endoproteinase-protein inhibitor interaction. Biochim Biophys Acta 1477(1–2):241–252

Bode W, Maskos K (2001) Structural studies on MMPs and TIMPs. Methods Mol Biol 151:45–77

Bode W, Maskos K (2003) Structural basis of the matrix metalloproteinases and their physiological inhibitors, the tissue inhibitors of metalloproteinases. Biol Chem 384(6):863–872

Bode W, Renatus M (1997) Tissue-type plasminogen activator: variants and crystal/solution structures demarcate structural determinants of function. Curr Opin Struct Biol 7(6):865–872

Bode W, Grams F, Reinemer P, Gomis-Ruth FX, Baumann U, McKay DB, Stocker W (1996) The metzincin-superfamily of zinc-peptidases. Adv Exp Med Biol 389:1–11

Bode W, Brandstetter H, Mather T, Stubbs MT (1997) Comparative analysis of haemostatic proteinases: structural aspects of thrombin, factor Xa, factor IXa and protein C. Thromb Haemost 78(1):501–511

Bode W, Fernandez-Catalan C, Nagase H, Maskos K (1999) Endoproteinase-protein inhibitor interactions. APMIS 107(1):3–10

Bonifacino JS, Weissman AM (1998) Ubiquitin and the control of protein fate in the secretory and endocytic pathways. Annu Rev Cell Dev Biol 14:19–57

Brandstetter H, Kim JS, Groll M, Huber R (2001) Crystal structure of the tricorn protease reveals a protein disassembly line. Nature 414(6862):466–470

Bresciani R, Von Figura K (1996) Dephosphorylation of the mannose-6-phosphate recognition marker is localized in later compartments of the endocytic route. Identification of purple acid phosphatase (uteroferrin) as the candidate phosphatase. Eur J Biochem 238(3):669–674
Brew K, Nagase H (2010) The tissue inhibitors of metalloproteinases (TIMPs): an ancient family with structural and functional diversity. Biochim Biophys Acta 1803(1):55–71
Brinckerhoff CE, Matrisian LM (2002) Matrix metalloproteinases: a tail of a frog that became a prince. Nat Rev Mol Cell Biol 3(3):207–214
Brix K (2005) Lysosomal proteases: revival of the sleeping beauty, Chap. 5. In: Saftig P (ed) Lysosomes. Landes Bioscience/Eurekah.com/Springer, Georgetown, TX
Brix K, Herzog V (1994) Extrathyroidal release of thyroid hormones from thyroglobulin by J774 mouse macrophages. J Clin Invest 93(4):1388–1396
Brix K, Jordans S (2005) Watching proteases in action. Nat Chem Biol 1(4):186–187
Brix K, Lemansky P, Herzog V (1996) Evidence for extracellularly acting cathepsins mediating thyroid hormone liberation in thyroid epithelial cells. Endocrinology 137(5):1963–1974
Brix K, Wirtz R, Herzog V (1997) Paracrine interaction between hepatocytes and macrophages after extrathyroidal proteolysis of thyroglobulin. Hepatology 26(5):1232–1240
Brix K, Linke M, Tepel C, Herzog V (2001) Cysteine proteinases mediate extracellular prohormone processing in the thyroid. Biol Chem 382(5):717–725
Brix K, Dunkhorst A, Mayer K, Jordans S (2008) Cysteine cathepsins: cellular roadmap to different functions. Biochimie 90(2):194–207
Brix K, Fuhrer D, Biebermann H (2011) Molecules important for thyroid hormone synthesis and action – known facts and future perspectives. Thyroid Res 4(Suppl 1):S9
Bromme D, Okamoto K (1995) Human cathepsin O2, a novel cysteine protease highly expressed in osteoclastomas and ovary molecular cloning, sequencing and tissue distribution. Biol Chem Hoppe Seyler 376(6):379–384
Brooks P, Fuertes G, Murray RZ, Bose S, Knecht E, Rechsteiner MC, Hendil KB, Tanaka K, Dyson J, Rivett J (2000a) Subcellular localization of proteasomes and their regulatory complexes in mammalian cells. Biochem J 346(Pt 1):155–161
Brooks P, Murray RZ, Mason GG, Hendil KB, Rivett AJ (2000b) Association of immunoproteasomes with the endoplasmic reticulum. Biochem J 352(Pt 3):611–615
Brown MS, Ye J, Rawson RB, Goldstein JL (2000) Regulated intramembrane proteolysis: a control mechanism conserved from bacteria to humans. Cell 100(4):391–398
Brunner G, Preissner KT (1994) Pericellular enzymatic hydrolysis: implications for the regulation of cell proliferation in the vessel wall and the bone marrow. Blood Coagul Fibrinolysis 5 (4):625–639
Budihardjo I, Oliver H, Lutter M, Luo X, Wang X (1999) Biochemical pathways of caspase activation during apoptosis. Annu Rev Cell Dev Biol 15:269–290
Bugge TH, List K, Szabo R (2007) Matriptase-dependent cell surface proteolysis in epithelial development and pathogenesis. Front Biosci 12:5060–5070
Burden RE, Snoddy P, Jefferies CA, Walker B, Scott CJ (2007) Inhibition of cathepsin L-like proteases by cathepsin V propeptide. Biol Chem 388(5):541–545
Buth H, Wolters B, Hartwig B, Meier-Bornheim R, Veith H, Hansen M, Sommerhoff CP, Schaschke N, Machleidt W, Fusenig NE, Boukamp P, Brix K (2004) HaCaT keratinocytes secrete lysosomal cysteine proteinases during migration. Eur J Cell Biol 83(11–12):781–795
Buth H, Luigi Buttigieg P, Ostafe R, Rehders M, Dannenmann SR, Schaschke N, Stark HJ, Boukamp P, Brix K (2007) Cathepsin B is essential for regeneration of scratch-wounded normal human epidermal keratinocytes. Eur J Cell Biol 86(11–12):747–761
Buttle DJ (2007) Factors controlling matrix turnover in health and disease. Biochem Soc Trans 35 (Pt 4):643–646
Carlson EE, Cravatt BF (2007) Chemoselective probes for metabolite enrichment and profiling. Nat Methods 4(5):429–435
Caughey GH (2007) Mast cell tryptases and chymases in inflammation and host defense. Immunol Rev 217:141–154

Cauwe B, Opdenakker G (2010) Intracellular substrate cleavage: a novel dimension in the biochemistry, biology and pathology of matrix metalloproteinases. Crit Rev Biochem Mol Biol 45(5):351–423
Cavallo-Medved D, Sloane BF (2003) Cell-surface cathepsin B: understanding its functional significance. Curr Top Dev Biol 54:313–341
Cawston TE, Young DA (2010) Proteinases involved in matrix turnover during cartilage and bone breakdown. Cell Tissue Res 339(1):221–235
Ceru S, Konjar S, Maher K, Repnik U, Krizaj I, Bencina M, Renko M, Nepveu A, Zerovnik E, Turk B, Kopitar-Jerala N (2010) Stefin B interacts with histones and cathepsin L in the nucleus. J Biol Chem 285(13):10078–10086
Chakraborti S, Mandal M, Das S, Mandal A, Chakraborti T (2003) Regulation of matrix metalloproteinases: an overview. Mol Cell Biochem 253(1–2):269–285
Chambers RC, Laurent GJ (2002) Coagulation cascade proteases and tissue fibrosis. Biochem Soc Trans 30(2):194–200
Chapman HA (2004) Cathepsins as transcriptional activators? Dev Cell 6(5):610–611
Chapman HA Jr, Munger JS, Shi GP (1994) The role of thiol proteases in tissue injury and remodeling. Am J Respir Crit Care Med 150(6 Pt 2):S155–S159
Chapman HA, Riese RJ, Shi GP (1997) Emerging roles for cysteine proteases in human biology. Annu Rev Physiol 59:63–88
Chen WT (1996) Proteases associated with invadopodia, and their role in degradation of extracellular matrix. Enzyme Protein 49(1–3):59–71
Chen Y, Klionsky DJ (2011) The regulation of autophagy – unanswered questions. J Cell Sci 124 (Pt 2):161–170
Choi SY, Bertram S, Glowacka I, Park YW, Pohlmann S (2009) Type II transmembrane serine proteases in cancer and viral infections. Trends Mol Med 15(7):303–312
Ciechanover A (2003) The ubiquitin proteolytic system and pathogenesis of human diseases: a novel platform for mechanism-based drug targeting. Biochem Soc Trans 31(2):474–481
Ciechanover A (2005) Proteolysis: from the lysosome to ubiquitin and the proteasome. Nat Rev Mol Cell Biol 6(1):79–87
Ciechanover A, Iwai K (2004) The ubiquitin system: from basic mechanisms to the patient bed. IUBMB Life 56(4):193–201
Clegg JS (1991) Metabolic organization and the ultrastructure of animal cells. Biochem Soc Trans 19(4):986–991
Cobbe N, Marshall KM, Gururaja Rao S, Chang CW, Di Cara F, Duca E, Vass S, Kassan A, Heck MM (2009) The conserved metalloprotease invadolysin localizes to the surface of lipid droplets. J Cell Sci 122(Pt 18):3414–3423
Collins GA, Tansey WP (2006) The proteasome: a utility tool for transcription? Curr Opin Genet Dev 16(2):197–202
Conner SD, Schmid SL (2003) Regulated portals of entry into the cell. Nature 422(6927):37–44
Coussens LM, Fingleton B, Matrisian LM (2002) Matrix metalloproteinase inhibitors and cancer: trials and tribulations. Science 295(5564):2387–2392
Creemers JW, Khatib AM (2008) Knock-out mouse models of proprotein convertases: unique functions or redundancy? Front Biosci 13:4960–4971
Dano K, Behrendt N, Hoyer-Hansen G, Johnsen M, Lund LR, Ploug M, Romer J (2005) Plasminogen activation and cancer. Thromb Haemost 93(4):676–681
Das R, Pluskota E, Plow EF (2010) Plasminogen and its receptors as regulators of cardiovascular inflammatory responses. Trends Cardiovasc Med 20(4):120–124
Dash C, Kulkarni A, Dunn B, Rao M (2003) Aspartic peptidase inhibitors: implications in drug development. Crit Rev Biochem Mol Biol 38(2):89–119
Dauth S, Arampatzidou M, Rehders M, Yu DMT, Tedelind S, Brix K (2011a) Thyroid cathepsin K – roles in physiology and thyroid disease. Clin Rev Bone Miner Metab 9:94–106

Dauth S, Sirbulescu RF, Jordans S, Rehders M, Avena L, Oswald J, Lerchl A, Saftig P, Brix K (2011b) Cathepsin K deficiency in mice induces structural and metabolic changes in the central nervous system that are associated with learning and memory deficits. BMC Neurosci 12(1):74
Dauth S, Schmidt MM, Rehders M, Dietz F, Kelm S, Dringen R, Brix K (2012) Characterization and metabolism of astroglia-rich primary cultures from cathepsin K-deficient mice. Biol Chem 393:959–970
Davies KJ (2001) Degradation of oxidized proteins by the 20S proteasome. Biochimie 83 (3–4):301–310
De Mot R, Nagy I, Walz J, Baumeister W (1999) Proteasomes and other self-compartmentalizing proteases in prokaryotes. Trends Microbiol 7(2):88–92
De Strooper B, Annaert W (2000) Proteolytic processing and cell biological functions of the amyloid precursor protein. J Cell Sci 113(Pt 11):1857–1870
Declercq W, Vanden Berghe T, Vandenabeele P (2009) RIP kinases at the crossroads of cell death and survival. Cell 138(2):229–232
Degli Esposti M (2008) Organelle intermixing and membrane scrambling in cell death. Methods Enzymol 442:421–438
Desmarais S, Masse F, Percival MD (2009) Pharmacological inhibitors to identify roles of cathepsin K in cell-based studies: a comparison of available tools. Biol Chem 390(9):941–948
Deveraux QL, Stennicke HR, Salvesen GS, Reed JC (1999) Endogenous inhibitors of caspases. J Clin Immunol 19(6):388–398
Di Cara F, Duca E, Dunbar DR, Cagney G, Heck MM (2013) Invadolysin, a conserved lipid droplet-associated metalloprotease, is required for mitochondrial function in Drosophila. J Cell Sci, in press, doi:10.1242/jcs.133306
Dice JF (1987) Molecular determinants of protein half-lives in eukaryotic cells. FASEB J 1 (5):349–357
Dice JF (1990) Peptide sequences that target cytosolic proteins for lysosomal proteolysis. Trends Biochem Sci 15(8):305–309
Dice JF (2007) Chaperone-mediated autophagy. Autophagy 3(4):295–299
Dice JF, Walker CD (1979) Protein degradation in metabolic and nutritional disorders. Ciba Found Symp 75:331–350
Dikic I, Schmidt MH (2010) Notch: implications of endogenous inhibitors for therapy. Bioessays 32(6):481–487
Dobson CM (2003) Protein folding and misfolding. Nature 426(6968):884–890
Docherty K, Steiner DF (1982) Post-translational proteolysis in polypeptide hormone biosynthesis. Annu Rev Physiol 44:625–638
Doucet A, Butler GS, Rodriguez D, Prudova A, Overall CM (2008) Metadegradomics: toward in vivo quantitative degradomics of proteolytic post-translational modifications of the cancer proteome. Mol Cell Proteomics 7(10):1925–1951
Drag M, Salvesen GS (2010) Emerging principles in protease-based drug discovery. Nat Rev Drug Discov 9(9):690–701
Driscoll JJ, Dechowdhury R (2010) Therapeutically targeting the SUMOylation, ubiquitination and proteasome pathways as a novel anticancer strategy. Target Oncol 5(4):281–289
Dunn BM, Oda K, Kay J, Rao-Naik C, Lowther WT, Beyer BM, Scarborough PE, Bukhtiyarova M (1998) Comparison of the specificity of the aspartic proteinases towards internally consistent sets of oligopeptide substrates. Adv Exp Med Biol 436:133–138
Earnshaw WC, Martins LM, Kaufmann SH (1999) Mammalian caspases: structure, activation, substrates, and functions during apoptosis. Annu Rev Biochem 68:383–424
Edwards DR, Handsley MM, Pennington CJ (2008) The ADAM metalloproteinases. Mol Aspects Med 29(5):258–289
Ehrmann M, Clausen T (2004) Proteolysis as a regulatory mechanism. Annu Rev Genet 38:709–724
Ellis V (2003) Plasminogen activation at the cell surface. Curr Top Dev Biol 54:263–312

Farquhar MG, Palade GE (1981) The Golgi apparatus (complex)-(1954-1981)-from artifact to center stage. J Cell Biol 91(3 Pt 2):77s–103s
Farquhar MG, Palade GE (1998) The Golgi apparatus: 100 years of progress and controversy. Trends Cell Biol 8(1):2–10
Felberbaum-Corti M, Van Der Goot FG, Gruenberg J (2003) Sliding doors: clathrin-coated pits or caveolae? Nat Cell Biol 5(5):382–384
Fineschi B, Miller J (1997) Endosomal proteases and antigen processing. Trends Biochem Sci 22 (10):377–382
Freeman M (2008) Rhomboid proteases and their biological functions. Annu Rev Genet 42:191–210
Fricker LD (1988) Carboxypeptidase E. Annu Rev Physiol 50:309–321
Friedl P, Wolf K (2003) Proteolytic and non-proteolytic migration of tumour cells and leucocytes. Biochem Soc Symp 70:277–285
Friedl P, Wolf K (2009) Proteolytic interstitial cell migration: a five-step process. Cancer Metastasis Rev 28(1–2):129–135
Friedrichs B, Tepel C, Reinheckel T, Deussing J, von Figura K, Herzog V, Peters C, Saftig P, Brix K (2003) Thyroid functions of mouse cathepsins B, K, and L. J Clin Invest 111(11):1733–1745
Frittoli E, Palamidessi A, Disanza A, Scita G (2011) Secretory and endo/exocytic trafficking in invadopodia formation: the MT1-MMP paradigm. Eur J Cell Biol 90(2–3):108–114
Fritz H, Schiessler H, Schleuning WD (1973) Proteinases and proteinase inhibitors in the fertilization process: new concepts of control? Adv Biosci 10:271–286
Fu X, Parks WC, Heinecke JW (2008) Activation and silencing of matrix metalloproteinases. Semin Cell Dev Biol 19(1):2–13
Funkelstein L, Beinfeld M, Minokadeh A, Zadina J, Hook V (2010) Unique biological function of cathepsin L in secretory vesicles for biosynthesis of neuropeptides. Neuropeptides 44 (6):457–466
Gagescu R, Gruenberg J, Smythe E (2000) Membrane dynamics in endocytosis: structure—function relationship. Traffic 1(1):84–88
Gahmberg CG, Tolvanen M (1996) Why mammalian cell surface proteins are glycoproteins. Trends Biochem Sci 21(8):308–311
Galivan J, Ryan TJ, Chave K, Rhee M, Yao R, Yin D (2000) Glutamyl hydrolase. Pharmacological role and enzymatic characterization. Pharmacol Ther 85(3):207–215
Garten W, Hallenberger S, Ortmann D, Schafer W, Vey M, Angliker H, Shaw E, Klenk HD (1994) Processing of viral glycoproteins by the subtilisin-like endoprotease furin and its inhibition by specific peptidylchloroalkylketones. Biochimie 76(3–4):217–225
Gerasimenko JV, Gerasimenko OV, Petersen OH (2001) Membrane repair: Ca(2+)-elicited lysosomal exocytosis. Curr Biol 11(23):R971–R974
Gerlt JA, Babbitt PC (2001) Divergent evolution of enzymatic function: mechanistically diverse superfamilies and functionally distinct suprafamilies. Annu Rev Biochem 70:209–246
Gocheva V, Joyce JA (2007) Cysteine cathepsins and the cutting edge of cancer invasion. Cell Cycle 6(1):60–64
Goldberg AL, Dice JF (1974) Intracellular protein degradation in mammalian and bacterial cells. Annu Rev Biochem 43:835–869
Gomez-Lazaro M, Rinn C, Aroso M, Amado F, Schrader M (2010) Proteomic analysis of zymogen granules. Expert Rev Proteomics 7(5):735–747
Gomis-Ruth FX (2008) Structure and mechanism of metallocarboxypeptidases. Crit Rev Biochem Mol Biol 43(5):319–345
Gomis-Ruth FX, Botelho TO, Bode W (2012) A standard orientation for metallopeptidases. Biochim Biophys Acta 1824(1):157–163
Goulet B, Nepveu A (2004) Complete and limited proteolysis in cell cycle progression. Cell Cycle 3(8):986–989

Goulet B, Baruch A, Moon NS, Poirier M, Sansregret LL, Erickson A, Bogyo M, Nepveu A (2004) A cathepsin L isoform that is devoid of a signal peptide localizes to the nucleus in S phase and processes the CDP/Cux transcription factor. Mol Cell 14(2):207–219
Goulet B, Sansregret L, Leduy L, Bogyo M, Weber E, Chauhan SS, Nepveu A (2007) Increased expression and activity of nuclear cathepsin L in cancer cells suggests a novel mechanism of cell transformation. Mol Cancer Res 5(9):899–907
Goulet B, Markovic Y, Leduy L, Nepveu A (2008) Proteolytic processing of cut homeobox 1 by neutrophil elastase in the MV4;11 myeloid leukemia cell line. Mol Cancer Res 6(4):644–653
Greenbaum D, Baruch A, Hayrapetian L, Darula Z, Burlingame A, Medzihradszky KF, Bogyo M (2002) Chemical approaches for functionally probing the proteome. Mol Cell Proteomics 1 (1):60–68
Groettrup M, Soza A, Kuckelkorn U, Kloetzel PM (1996) Peptide antigen production by the proteasome: complexity provides efficiency. Immunol Today 17(9):429–435
Groettrup M, Khan S, Schwarz K, Schmidtke G (2001a) Interferon-gamma inducible exchanges of 20S proteasome active site subunits: why? Biochimie 83(3–4):367–372
Groettrup M, van den Broek M, Schwarz K, Macagno A, Khan S, de Giuli R, Schmidtke G (2001b) Structural plasticity of the proteasome and its function in antigen processing. Crit Rev Immunol 21(4):339–358
Groettrup M, Kirk CJ, Basler M (2010) Proteasomes in immune cells: more than peptide producers? Nat Rev Immunol 10(1):73–78
Groll M, Clausen T (2003) Molecular shredders: how proteasomes fulfill their role. Curr Opin Struct Biol 13(6):665–673
Groll M, Ditzel L, Lowe J, Stock D, Bochtler M, Bartunik HD, Huber R (1997) Structure of 20S proteasome from yeast at 2.4 A resolution. Nature 386(6624):463–471
Gruenberg J (2001) The endocytic pathway: a mosaic of domains. Nat Rev Mol Cell Biol 2 (10):721–730
Gruenberg J, Howell KE (1989) Membrane traffic in endocytosis: insights from cell-free assays. Annu Rev Cell Biol 5:453–481
Gruenberg J, Maxfield FR (1995) Membrane transport in the endocytic pathway. Curr Opin Cell Biol 7(4):552–563
Gruenberg J, Stenmark H (2004) The biogenesis of multivesicular endosomes. Nat Rev Mol Cell Biol 5(4):317–323
Gu F, Gruenberg J (1999) Biogenesis of transport intermediates in the endocytic pathway. FEBS Lett 452(1–2):61–66
Gu F, Aniento F, Parton RG, Gruenberg J (1997) Functional dissection of COP-I subunits in the biogenesis of multivesicular endosomes. J Cell Biol 139(5):1183–1195
Gumy LF, Tan CL, Fawcett JW (2010) The role of local protein synthesis and degradation in axon regeneration. Exp Neurol 223(1):28–37
Ha Y (2007) Structural principles of intramembrane proteases. Curr Opin Struct Biol 17 (4):405–411
Hajjar KA, Acharya SS (2000) Annexin II and regulation of cell surface fibrinolysis. Ann NY Acad Sci 902:265–271
Halangk W, Lerch MM, Brandt-Nedelev B, Roth W, Ruthenbuerger M, Reinheckel T, Domschke W, Lippert H, Peters C, Deussing J (2000) Role of cathepsin B in intracellular trypsinogen activation and the onset of acute pancreatitis. J Clin Invest 106(6):773–781
Hansen TH, Bouvier M (2009) MHC class I antigen presentation: learning from viral evasion strategies. Nat Rev Immunol 9(7):503–513
Haugen MH, Johansen HT, Pettersen SJ, Solberg R, Brix K, Flatmark K, Maelandsmo GM (2013) Nuclear legumain activity in colorectal cancer. PLoS One 8(1):e52980
He B, Lu N, Zhou Z (2009) Cellular and nuclear degradation during apoptosis. Curr Opin Cell Biol 21(6):900–912
Heck MM (1997) Condensins, cohesins, and chromosome architecture: how to make and break a mitotic chromosome. Cell 91(1):5–8

Hendrickx AP, Budzik JM, Oh SY, Schneewind O (2011) Architects at the bacterial surface – sortases and the assembly of pili with isopeptide bonds. Nat Rev Microbiol 9(3):166–176
Henrich S, Lindberg I, Bode W, Than ME (2005) Proprotein convertase models based on the crystal structures of furin and kexin: explanation of their specificity. J Mol Biol 345 (2):211–227
Herget M, Tampe R (2007) Intracellular peptide transporters in human–compartmentalization of the "peptidome". Pflugers Arch 453(5):591–600
Hicke L, Dunn R (2003) Regulation of membrane protein transport by ubiquitin and ubiquitin-binding proteins. Annu Rev Cell Dev Biol 19:141–172
Hille A, Klumperman J, Geuze HJ, Peters C, Brodsky FM, von Figura K (1992) Lysosomal acid phosphatase is internalized via clathrin-coated pits. Eur J Cell Biol 59(1):106–115
Hiraiwa M (1999) Cathepsin A/protective protein: an unusual lysosomal multifunctional protein. Cell Mol Life Sci 56(11–12):894–907
Holmbeck K, Bianco P, Caterina J, Yamada S, Kromer M, Kuznetsov SA, Mankani M, Robey PG, Poole AR, Pidoux I, Ward JM, Birkedal-Hansen H (1999) MT1-MMP-deficient mice develop dwarfism, osteopenia, arthritis, and connective tissue disease due to inadequate collagen turnover. Cell 99(1):81–92
Hook V, Yasothornsrikul S, Greenbaum D, Medzihradszky KF, Troutner K, Toneff T, Bundey R, Logrinova A, Reinheckel T, Peters C, Bogyo M (2004) Cathepsin L and Arg/Lys aminopeptidase: a distinct prohormone processing pathway for the biosynthesis of peptide neurotransmitters and hormones. Biol Chem 385(6):473–480
Hook V, Schechter I, Demuth HU, Hook G (2008) Alternative pathways for production of beta-amyloid peptides of Alzheimer's disease. Biol Chem 389(8):993–1006
Hooper NM (2002) Proteases: a primer. Essays Biochem 38:1–8
Hooper JD, Clements JA, Quigley JP, Antalis TM (2001) Type II transmembrane serine proteases. Insights into an emerging class of cell surface proteolytic enzymes. J Biol Chem 276 (2):857–860
Howell KE, Devaney E, Gruenberg J (1989) Subcellular fractionation of tissue culture cells. Trends Biochem Sci 14(2):44–47
Hudson DF, Marshall KM, Earnshaw WC (2009) Condensin: architect of mitotic chromosomes. Chromosome Res 17(2):131–144
Ichihara A, Kaneshiro Y, Suzuki F (2006) Prorenin receptor blockers: effects on cardiovascular complications of diabetes and hypertension. Expert Opin Investig Drugs 15(10):1137–1139
Ishidoh K, Kominami E (2002) Processing and activation of lysosomal proteinases. Biol Chem 383(12):1827–1831
Itoh Y (2006) MT1-MMP: a key regulator of cell migration in tissue. IUBMB Life 58 (10):589–596
Ivanova S, Repnik U, Bojic L, Petelin A, Turk V, Turk B (2008) Lysosomes in apoptosis. Methods Enzymol 442:183–199
Izuhara K, Ohta S, Kanaji S, Shiraishi H, Arima K (2008) Recent progress in understanding the diversity of the human ov-serpin/clade B serpin family. Cell Mol Life Sci 65(16):2541–2553
Jacques LB (1979) Heparin: an old drug with a new paradigm. Science 206(4418):528–533
Jedeszko C, Sameni M, Olive MB, Moin K, Sloane BF (2008) Visualizing protease activity in living cells: from two dimensions to four dimensions. Curr Protoc Cell Biol Chapter 4:Unit 4.20
Jiang A, Lehti K, Wang X, Weiss SJ, Keski-Oja J, Pei D (2001) Regulation of membrane-type matrix metalloproteinase 1 activity by dynamin-mediated endocytosis. Proc Natl Acad Sci U S A 98(24):13693–13698
Jordans S, Jenko-Kokalj S, Kuhl NM, Tedelind S, Sendt W, Bromme D, Turk D, Brix K (2009) Monitoring compartment-specific substrate cleavage by cathepsins B, K, L, and S at physiological pH and redox conditions. BMC Biochem 10:23
Joyce JA, Hanahan D (2004) Multiple roles for cysteine cathepsins in cancer. Cell Cycle 3 (12):1516–1619

Kaiser CA, Chen EJ, Losko S (2002) Subcellular fractionation of secretory organelles. Methods Enzymol 351:325–338
Kaiserman D, Whisstock JC, Bird PI (2006) Mechanisms of serpin dysfunction in disease. Expert Rev Mol Med 8(31):1–19
Katzmann DJ, Odorizzi G, Emr SD (2002) Receptor downregulation and multivesicular-body sorting. Nat Rev Mol Cell Biol 3(12):893–905
Kennedy RD, D'Andrea AD (2005) The Fanconi Anemia/BRCA pathway: new faces in the crowd. Genes Dev 19(24):2925–2940
Kenny AJ, Maroux S (1982) Topology of microvillar membrance hydrolases of kidney and intestine. Physiol Rev 62(1):91–128
Kersse K, Verspurten J, Berghe TV, Vandenabeele P (2011) The death-fold superfamily of homotypic interaction motifs. Trends Biochem Sci 36(10):541–552
Kessenbrock K, Dau T, Jenne DE (2011) Tailor-made inflammation: how neutrophil serine proteases modulate the inflammatory response. J Mol Med 89(1):23–28
Khan AR, James MN (1998) Molecular mechanisms for the conversion of zymogens to active proteolytic enzymes. Protein Sci 7(4):815–836
Kidd D, Liu Y, Cravatt BF (2001) Profiling serine hydrolase activities in complex proteomes. Biochemistry 40(13):4005–4015
Kim J, Hajjar KA (2002) Annexin II: a plasminogen-plasminogen activator co-receptor. Front Biosci 7:d341–d348
Kirschke H, Wiederanders B (1994) Cathepsin S and related lysosomal endopeptidases. Methods Enzymol 244:500–511
Kirschke H, Barrett AJ, Rawlings ND (1995) Proteinases 1: lysosomal cysteine proteinases. Protein Profile 2(14):1581–1643
Klionsky DJ (2007) Autophagy: from phenomenology to molecular understanding in less than a decade. Nat Rev Mol Cell Biol 8(11):931–937
Kloetzel PM, Ossendorp F (2004) Proteasome and peptidase function in MHC-class-I-mediated antigen presentation. Curr Opin Immunol 16(1):76–81
Koohmaraie M (1992) The role of Ca(2+)-dependent proteases (calpains) in post mortem proteolysis and meat tenderness. Biochimie 74(3):239–245
Kopito RR (2000) Aggresomes, inclusion bodies and protein aggregation. Trends Cell Biol 10 (12):524–530
Korkmaz B, Moreau T, Gauthier F (2008) Neutrophil elastase, proteinase 3 and cathepsin G: physicochemical properties, activity and physiopathological functions. Biochimie 90 (2):227–242
Kornfeld S (1992) Structure and function of the mannose 6-phosphate/insulinlike growth factor II receptors. Annu Rev Biochem 61:307–330
Koster A, Saftig P, Matzner U, von Figura K, Peters C, Pohlmann R (1993) Targeted disruption of the M(r) 46,000 mannose 6-phosphate receptor gene in mice results in misrouting of lysosomal proteins. EMBO J 12(13):5219–5223
Krysko DV, Agostinis P, Krysko O, Garg AD, Bachert C, Lambrecht BN, Vandenabeele P (2011) Emerging role of damage-associated molecular patterns derived from mitochondria in inflammation. Trends Immunol 32(4):157–164
Kveiborg M, Albrechtsen R, Couchman JR, Wewer UM (2008) Cellular roles of ADAM12 in health and disease. Int J Biochem Cell Biol 40(9):1685–1702
Kwak J, Workman JL, Lee D (2011) The proteasome and its regulatory roles in gene expression. Biochim Biophys Acta 1809(2):88–96
Lamkanfi M, Festjens N, Declercq W, Vanden Berghe T, Vandenabeele P (2007) Caspases in cell survival, proliferation and differentiation. Cell Death Differ 14(1):44–55
Lamond AI, Earnshaw WC (1998) Structure and function in the nucleus. Science 280 (5363):547–553
Lecaille F, Weidauer E, Juliano MA, Bromme D, Lalmanach G (2003) Probing cathepsin K activity with a selective substrate spanning its active site. Biochem J 375(Pt 2):307–312

Lemansky P, Brix K, Herzog V (1998) Iodination of mature cathepsin D in thyrocytes as an indicator for its transport to the cell surface. Eur J Cell Biol 76(1):53–62

Lenarcic B, Bevec T (1998) Thyropins–new structurally related proteinase inhibitors. Biol Chem 379(2):105–111

Liaudet-Coopman E, Beaujouin M, Derocq D, Garcia M, Glondu-Lassis M, Laurent-Matha V, Prebois C, Rochefort H, Vignon F (2006) Cathepsin D: newly discovered functions of a long-standing aspartic protease in cancer and apoptosis. Cancer Lett 237(2):167–179

Lichtenthaler SF, Haass C, Steiner H (2011) Regulated intramembrane proteolysis–lessons from amyloid precursor protein processing. J Neurochem 117(5):779–796

Lin SX, Mallet WG, Huang AY, Maxfield FR (2004) Endocytosed cation-independent mannose 6-phosphate receptor traffics via the endocytic recycling compartment en route to the trans-Golgi network and a subpopulation of late endosomes. Mol Biol Cell 15(2):721–733

Linder S (2007) The matrix corroded: podosomes and invadopodia in extracellular matrix degradation. Trends Cell Biol 17(3):107–117

Linke M, Herzog V, Brix K (2002a) Trafficking of lysosomal cathepsin B-green fluorescent protein to the surface of thyroid epithelial cells involves the endosomal/lysosomal compartment. J Cell Sci 115(Pt 24):4877–4889

Linke M, Jordans S, Mach L, Herzog V, Brix K (2002b) Thyroid stimulating hormone upregulates secretion of cathepsin B from thyroid epithelial cells. Biol Chem 383(5):773–784

Lipowsky R (1995) The morphology of lipid membranes. Curr Opin Struct Biol 5(4):531–540

List K, Bugge TH, Szabo R (2006) Matriptase: potent proteolysis on the cell surface. Mol Med 12 (1–3):1–7

Liu H, Lazarus SC, Caughey GH, Fahy JV (1999) Neutrophil elastase and elastase-rich cystic fibrosis sputum degranulate human eosinophils in vitro. Am J Physiol 276(1 Pt 1):L28–L34

Lopez-Otin C, Bond JS (2008) Proteases: multifunctional enzymes in life and disease. J Biol Chem 283(45):30433–30437

Lopez-Otin C, Matrisian LM (2007) Emerging roles of proteases in tumour suppression. Nat Rev Cancer 7(10):800–808

Ludwig T, Munier-Lehmann H, Bauer U, Hollinshead M, Ovitt C, Lobel P, Hoflack B (1994) Differential sorting of lysosomal enzymes in mannose 6-phosphate receptor-deficient fibroblasts. EMBO J 13(15):3430–3437

Mach L (2002) Biosynthesis of lysosomal proteinases in health and disease. Biol Chem 383 (5):751–756

Mach L, Mort JS, Glossl J (1994) Maturation of human procathepsin B. Proenzyme activation and proteolytic processing of the precursor to the mature proteinase, in vitro, are primarily unimolecular processes. J Biol Chem 269(17):13030–13035

Mackie EJ, Pagel CN, Smith R, de Niese MR, Song SJ, Pike RN (2002) Protease-activated receptors: a means of converting extracellular proteolysis into intracellular signals. IUBMB Life 53(6):277–281

Magdolen V, Arroyo de Prada N, Sperl S, Muehlenweg B, Luther T, Wilhelm OG, Magdolen U, Graeff H, Reuning U, Schmitt M (2000) Natural and synthetic inhibitors of the tumor-associated serine protease urokinase-type plasminogen activator. Adv Exp Med Biol 477:331–341

Mai J, Waisman DM, Sloane BF (2000) Cell surface complex of cathepsin B/annexin II tetramer in malignant progression. Biochim Biophys Acta 1477(1–2):215–230

Manon-Jensen T, Itoh Y, Couchman JR (2010) Proteoglycans in health and disease: the multiple roles of syndecan shedding. FEBS J 277(19):3876–3889

Marfany G, Denuc A (2008) To ubiquitinate or to deubiquitinate: it all depends on the partners. Biochem Soc Trans 36(Pt 5):833–838

Matrisian LM (1999) Cancer biology: extracellular proteinases in malignancy. Curr Biol 9(20): R776–R778

Matteucci E, Giampietro O (2009) Dipeptidyl peptidase-4 (CD26): knowing the function before inhibiting the enzyme. Curr Med Chem 16(23):2943–2951

Mayer K, Schwartz S, Lentze MJ, Kalff JC, Brix K (2006) Extracellular localization of intestinal cathepsins: implications of their actions during post-operative ileus. In: XLI congress of the European society for surgical research. Medimond International Proceedings, Rostock, Germany

Mayer K, Iolyeva ME, Meyer-Grahle U, Brix K (2008) Intestine-specific expression of green fluorescent protein-tagged cathepsin B: proof-of-principle experiments. Biol Chem 389 (8):1085–1096

Mayer K, Vreemann A, Qu H, Brix K (2009) Release of endo-lysosomal cathepsins B, D, and L from IEC6 cells in a cell culture model mimicking intestinal manipulation. Biol Chem 390 (5–6):471–480

McGrath ME (1999) The lysosomal cysteine proteases. Annu Rev Biophys Biomol Struct 28:181–204

McHugh B, Krause SA, Yu B, Deans AM, Heasman S, McLaughlin P, Heck MM (2004) Invadolysin: a novel, conserved metalloprotease links mitotic structural rearrangements with cell migration. J Cell Biol 167(4):673–686

McIntyre GF, Godbold GD, Erickson AH (1994) The pH-dependent membrane association of procathepsin L is mediated by a 9-residue sequence within the propeptide. J Biol Chem 269 (1):567–572

McNeil PL, Steinhardt RA (1997) Loss, restoration, and maintenance of plasma membrane integrity. J Cell Biol 137(1):1–4

Meister T, Niehues R, Hahn D, Domschke W, Sendler M, Lerch MM, Schnekenburger J (2010) Missorting of cathepsin B into the secretory compartment of CI-MPR/IGFII-deficient mice does not induce spontaneous trypsinogen activation but leads to enhanced trypsin activity during experimental pancreatitis–without affecting disease severit. J Physiol Pharmacol 61 (5):565–575

Mentlein R (2004) Cell-surface peptidases. Int Rev Cytol 235:165–213

Meusser B, Hirsch C, Jarosch E, Sommer T (2005) ERAD: the long road to destruction. Nat Cell Biol 7(8):766–772

Meyer-Hoffert U (2009) Neutrophil-derived serine proteases modulate innate immune responses. Front Biosci 14:3409–3418

Mizushima N, Levine B, Cuervo AM, Klionsky DJ (2008) Autophagy fights disease through cellular self-digestion. Nature 451(7182):1069–1075

Mockaitis K, Estelle M (2008) Auxin receptors and plant development: a new signaling paradigm. Annu Rev Cell Dev Biol 24:55–80

Mohamed MM, Sloane BF (2006) Cysteine cathepsins: multifunctional enzymes in cancer. Nat Rev Cancer 6(10):764–775

Mondino A, Blasi F (2004) uPA and uPAR in fibrinolysis, immunity and pathology. Trends Immunol 25(8):450–455

Morita M, Kurochkin IV, Motojima K, Goto S, Takano T, Okamura S, Sato R, Yokota S, Imanaka T (2000) Insulin-degrading enzyme exists inside of rat liver peroxisomes and degrades oxidized proteins. Cell Struct Funct 25(5):309–315

Mort JS, Buttle DJ (1997) Cathepsin B. Int J Biochem Cell Biol 29(5):715–720

Mort JS, Recklies AD, Poole AR (1984) Extracellular presence of the lysosomal proteinase cathepsin B in rheumatoid synovium and its activity at neutral pH. Arthritis Rheum 27 (5):509–515

Mott JD, Werb Z (2004) Regulation of matrix biology by matrix metalloproteinases. Curr Opin Cell Biol 16(5):558–564

Mukherjee S, Ghosh RN, Maxfield FR (1997) Endocytosis. Physiol Rev 77(3):759–803

Munshi HG, Stack MS (2006) Reciprocal interactions between adhesion receptor signaling and MMP regulation. Cancer Metastasis Rev 25(1):45–56

Murphy G (2009) Regulation of the proteolytic disintegrin metalloproteinases, the ‘Sheddases’. Semin Cell Dev Biol 20(2):138–145

Murphy G, Gavrilovic J (1999) Proteolysis and cell migration: creating a path? Curr Opin Cell Biol 11(5):614–621
Murphy G, Nagase H (2011) Localizing matrix metalloproteinase activities in the pericellular environment. FEBS J 278(1):2–15
Murphy JE, Padilla BE, Hasdemir B, Cottrell GS, Bunnett NW (2009) Endosomes: a legitimate platform for the signaling train. Proc Natl Acad Sci U S A 106(42):17615–17622
Nagase H, Suzuki K, Itoh Y, Kan CC, Gehring MR, Huang W, Brew K (1996) Involvement of tissue inhibitors of metalloproteinases (TIMPS) during matrix metalloproteinase activation. Adv Exp Med Biol 389:23–31
Nakagawa T, Roth W, Wong P, Nelson A, Farr A, Deussing J, Villadangos JA, Ploegh H, Peters C, Rudensky AY (1998) Cathepsin L: critical role in Ii degradation and CD4 T cell selection in the thymus. Science 280(5362):450–453
Nakanishi H (2003) Microglial functions and proteases. Mol Neurobiol 27(2):163–176
Nakayama K (1997) Furin: a mammalian subtilisin/Kex2p-like endoprotease involved in processing of a wide variety of precursor proteins. Biochem J 327(Pt 3):625–635
Nasmyth K, Peters JM, Uhlmann F (2000) Splitting the chromosome: cutting the ties that bind sister chromatids. Science 288(5470):1379–1385
Navarro A, Anand-Apte B, Parat MO (2004) A role for caveolae in cell migration. FASEB J 18 (15):1801–1811
Nichols BJ, Lippincott-Schwartz J (2001) Endocytosis without clathrin coats. Trends Cell Biol 11 (10):406–412
Nicotra G, Castino R, Follo C, Peracchio C, Valente G, Isidoro C (2010) The dilemma: does tissue expression of cathepsin D reflect tumor malignancy? The question: does the assay truly mirror cathepsin D mis-function in the tumor? Cancer Biomark 7(1):47–64
Nixon RA, Cataldo AM (1993) The lysosomal system in neuronal cell death: a review. Ann NY Acad Sci 679:87–109
Noel A, Maillard C, Rocks N, Jost M, Chabottaux V, Sounni NE, Maquoi E, Cataldo D, Foidart JM (2004) Membrane associated proteases and their inhibitors in tumour angiogenesis. J Clin Pathol 57(6):577–584
Norman LP, Matters GL, Crisman JM, Bond JS (2003) Expression of meprins in health and disease. Curr Top Dev Biol 54:145–166
Novinec M, Kordis D, Turk V, Lenarcic B (2006) Diversity and evolution of the thyroglobulin type-1 domain superfamily. Mol Biol Evol 23(4):744–755
O'Brien P, O'Connor BF (2008) Seprase: an overview of an important matrix serine protease. Biochim Biophys Acta 1784(9):1130–1145
Olson TS, Dice JF (1989) Regulation of protein degradation rates in eukaryotes. Curr Opin Cell Biol 1(6):1194–1200
Ong PC, McGowan S, Pearce MC, Irving JA, Kan WT, Grigoryev SA, Turk B, Silverman GA, Brix K, Bottomley SP, Whisstock JC, Pike RN (2007) DNA accelerates the inhibition of human cathepsin V by serpins. J Biol Chem 282(51):36980–36986
Ovaere P, Lippens S, Vandenabeele P, Declercq W (2009) The emerging roles of serine protease cascades in the epidermis. Trends Biochem Sci 34(9):453–463
Overall CM, Blobel CP (2007) In search of partners: linking extracellular proteases to substrates. Nat Rev Mol Cell Biol 8(3):245–257
Owen CA (2008) Leukocyte cell surface proteinases: regulation of expression, functions, and mechanisms of surface localization. Int J Biochem Cell Biol 40(6–7):1246–1272
Owen CA, Campbell EJ (1995) Neutrophil proteinases and matrix degradation. The cell biology of pericellular proteolysis. Semin Cell Biol 6(6):367–376
Pagano M, Reboud-Ravaux M (2011) Cryptic activities of fibronectin fragments, particularly cryptic proteases. Front Biosci 16:698–706
Palade GE (1964) The organization of living matter. Proc Natl Acad Sci U S A 52:613–634
Palade GE (1966) Structure and function at the cellular level. JAMA 198(8):815–825

Palade G (1975) Intracellular aspects of the process of protein synthesis. Science 189 (4200):347–358
Pappot H, Gardsvoll H, Romer J, Pedersen AN, Grondahl-Hansen J, Pyke C, Brunner N (1995) Plasminogen activator inhibitor type 1 in cancer: therapeutic and prognostic implications. Biol Chem Hoppe Seyler 376(5):259–267
Parkin ET, Turner AJ, Hooper NM (2004) Secretase-mediated cell surface shedding of the angiotensin-converting enzyme. Protein Pept Lett 11(5):423–432
Paulick MG, Bogyo M (2008) Application of activity-based probes to the study of enzymes involved in cancer progression. Curr Opin Genet Dev 18(1):97–106
Pelkmans L, Helenius A (2002) Endocytosis via caveolae. Traffic 3(5):311–320
Pellman D, Christman MF (2001) Separase anxiety: dissolving the sister bond and more. Nat Cell Biol 3(9):E207–E209
Peters C, von Figura K (1994) Biogenesis of lysosomal membranes. FEBS Lett 346(1):108–114
Peters C, Braun M, Weber B, Wendland M, Schmidt B, Pohlmann R, Waheed A, von Figura K (1990) Targeting of a lysosomal membrane protein: a tyrosine-containing endocytosis signal in the cytoplasmic tail of lysosomal acid phosphatase is necessary and sufficient for targeting to lysosomes. EMBO J 9(11):3497–3506
Pham CT (2006) Neutrophil serine proteases: specific regulators of inflammation. Nat Rev Immunol 6(7):541–550
Pham CT (2008) Neutrophil serine proteases fine-tune the inflammatory response. Int J Biochem Cell Biol 40(6–7):1317–1333
Pickart CM (1997) Targeting of substrates to the 26S proteasome. FASEB J 11(13):1055–1066
Pickart CM (2001) Mechanisms underlying ubiquitination. Annu Rev Biochem 70:503–533
Pickart CM, Eddins MJ (2004) Ubiquitin: structures, functions, mechanisms. Biochim Biophys Acta 1695(1–3):55–72
Pickart CM, Fushman D (2004) Polyubiquitin chains: polymeric protein signals. Curr Opin Chem Biol 8(6):610–616
Pillay CS, Elliott E, Dennison C (2002) Endolysosomal proteolysis and its regulation. Biochem J 363(Pt 3):417–429
Piper RC, Katzmann DJ (2007) Biogenesis and function of multivesicular bodies. Annu Rev Cell Dev Biol 23:519–547
Platta HW, Erdmann R (2007) Peroxisomal dynamics. Trends Cell Biol 17(10):474–484
Podgorski I (2009) Future of anticathepsin K drugs: dual therapy for skeletal disease and atherosclerosis? Future Med Chem 1(1):21–34
Pohlmann R, Boeker MW, von Figura K (1995) The two mannose 6-phosphate receptors transport distinct complements of lysosomal proteins. J Biol Chem 270(45):27311–27318
Pop C, Salvesen GS (2009) Human caspases: activation, specificity, and regulation. J Biol Chem 284(33):21777–21781
Potempa J, Sroka A, Imamura T, Travis J (2003) Gingipains, the major cysteine proteinases and virulence factors of Porphyromonas gingivalis: structure, function and assembly of multidomain protein complexes. Curr Protein Pept Sci 4(6):397–407
Predescu SA, Predescu DN, Palade GE (2001) Endothelial transcytotic machinery involves supramolecular protein-lipid complexes. Mol Biol Cell 12(4):1019–1033
Pungercar JR, Caglic D, Sajid M, Dolinar M, Vasiljeva O, Pozgan U, Turk D, Bogyo M, Turk V, Turk B (2009) Autocatalytic processing of procathepsin B is triggered by proenzyme activity. FEBS J 276(3):660–668
Rachner TD, Khosla S, Hofbauer LC (2011) Osteoporosis: now and the future. Lancet 377 (9773):1276–1287
Ramsay AJ, Hooper JD, Folgueras AR, Velasco G, Lopez-Otin C (2009) Matriptase-2- (TMPRSS6): a proteolytic regulator of iron homeostasis. Haematologica 94(6):840–849
Rao JS (2003) Molecular mechanisms of glioma invasiveness: the role of proteases. Nat Rev Cancer 3(7):489–501
Rawlings ND (2010) Peptidase inhibitors in the MEROPS database. Biochimie 92(11):1463–1483

Rawlings ND, Tolle DP, Barrett AJ (2004) Evolutionary families of peptidase inhibitors. Biochem J 378(Pt 3):705–716

Rechsteiner M (1990) PEST sequences are signals for rapid intracellular proteolysis. Semin Cell Biol 1(6):433–440

Rechsteiner M, Hill CP (2005) Mobilizing the proteolytic machine: cell biological roles of proteasome activators and inhibitors. Trends Cell Biol 15(1):27–33

Reed SI (2003) Ratchets and clocks: the cell cycle, ubiquitylation and protein turnover. Nat Rev Mol Cell Biol 4(11):855–864

Reinheckel T, Grune T, Davies KJ (2000) The measurement of protein degradation in response to oxidative stress. Methods Mol Biol 99:49–60

Reinheckel T, Deussing J, Roth W, Peters C (2001) Towards specific functions of lysosomal cysteine peptidases: phenotypes of mice deficient for cathepsin B or cathepsin L. Biol Chem 382(5):735–741

Reiser J, Adair B, Reinheckel T (2010) Specialized roles for cysteine cathepsins in health and disease. J Clin Invest 120(10):3421–3431

Reiss K, Saftig P (2009) The "a disintegrin and metalloprotease" (ADAM) family of sheddases: physiological and cellular functions. Semin Cell Dev Biol 20(2):126–137

Reudelhuber TL, Brechler V, Jutras I, Mercure C, Methot D (1998) Proteolytic and non-proteolytic activation of prorenin. Adv Exp Med Biol 436:229–238

Reverter D, Sorimachi H, Bode W (2001) The structure of calcium-free human m-calpain: implications for calcium activation and function. Trends Cardiovasc Med 11(6):222–229

Reyes-Turcu FE, Ventii KH, Wilkinson KD (2009) Regulation and cellular roles of ubiquitin-specific deubiquitinating enzymes. Annu Rev Biochem 78:363–397

Riedl SJ, Salvesen GS (2007) The apoptosome: signalling platform of cell death. Nat Rev Mol Cell Biol 8(5):405–413

Riedl SJ, Shi Y (2004) Molecular mechanisms of caspase regulation during apoptosis. Nat Rev Mol Cell Biol 5(11):897–907

Riese RJ, Chapman HA (2000) Cathepsins and compartmentalization in antigen presentation. Curr Opin Immunol 12(1):107–113

Rivett AJ (1998) Intracellular distribution of proteasomes. Curr Opin Immunol 10(1):110–114

Rivett AJ, Hearn AR (2004) Proteasome function in antigen presentation: immunoproteasome complexes, Peptide production, and interactions with viral proteins. Curr Protein Pept Sci 5 (3):153–161

Rochefort H, Garcia M, Glondu M, Laurent V, Liaudet E, Rey JM, Roger P (2000) Cathepsin D in breast cancer: mechanisms and clinical applications, a 1999 overview. Clin Chim Acta 291 (2):157–170

Rockwell NC, Thorner JW (2004) The kindest cuts of all: crystal structures of Kex2 and furin reveal secrets of precursor processing. Trends Biochem Sci 29(2):80–87

Romisch K (2005) Endoplasmic reticulum-associated degradation. Annu Rev Cell Dev Biol 21:435–456

Rosenberg GA (2009) Matrix metalloproteinases and their multiple roles in neurodegenerative diseases. Lancet Neurol 8(2):205–216

Rosenblum JS, Kozarich JW (2003) Prolyl peptidases: a serine protease subfamily with high potential for drug discovery. Curr Opin Chem Biol 7(4):496–504

Rothman JE, Wieland FT (1996) Protein sorting by transport vesicles. Science 272(5259):227–234

Ruchaud S, Carmena M, Earnshaw WC (2007) Chromosomal passengers: conducting cell division. Nat Rev Mol Cell Biol 8(10):798–812

Sachse M, Urbe S, Oorschot V, Strous GJ, Klumperman J (2002) Bilayered clathrin coats on endosomal vacuoles are involved in protein sorting toward lysosomes. Mol Biol Cell 13 (4):1313–1328

Saftig P, Reiss K (2011) The "A Disintegrin And Metalloproteases" ADAM10 and ADAM17: novel drug targets with therapeutic potential? Eur J Cell Biol 90(6–7):527–535

Saftig P, Hunziker E, Wehmeyer O, Jones S, Boyde A, Rommerskirch W, Moritz JD, Schu P, von Figura K (1998) Impaired osteoclastic bone resorption leads to osteopetrosis in cathepsin-K-deficient mice. Proc Natl Acad Sci U S A 95(23):13453–13458

Salvesen GS (2002) Caspases and apoptosis. Essays Biochem 38:9–19

Salvesen GS, Dixit VM (1997) Caspases: intracellular signaling by proteolysis. Cell 91 (4):443–446

Salvesen GS, Duckett CS (2002) IAP proteins: blocking the road to death's door. Nat Rev Mol Cell Biol 3(6):401–410

Sameni M, Cavallo-Medved D, Dosescu J, Jedeszko C, Moin K, Mullins SR, Olive MB, Rudy D, Sloane BF (2009) Imaging and quantifying the dynamics of tumor-associated proteolysis. Clin Exp Metastasis 26(4):299–309

Sato K, Kawashima S (2001) Calpain function in the modulation of signal transduction molecules. Biol Chem 382(5):743–751

Sato H, Takino T (2010) Coordinate action of membrane-type matrix metalloproteinase-1 (MT1-MMP) and MMP-2 enhances pericellular proteolysis and invasion. Cancer Sci 101 (4):843–847

Scamuffa N, Calvo F, Chretien M, Seidah NG, Khatib AM (2006) Proprotein convertases: lessons from knockouts. FASEB J 20(12):1954–1963

Schaller A, Ryan CA (1996) Systemin–a polypeptide defense signal in plants. Bioessays 18 (1):27–33

Schaschke N, Assfalg-Machleidt I, Machleidt W, Moroder L (1998) Substrate/propeptide-derived endo-epoxysuccinyl peptides as highly potent and selective cathepsin B inhibitors. FEBS Lett 421(1):80–82

Schechter I (2005) Mapping of the active site of proteases in the 1960s and rational design of inhibitors/drugs in the 1990s. Curr Protein Pept Sci 6(6):501–512

Schenk S, Quaranta V (2003) Tales from the crypt[ic] sites of the extracellular matrix. Trends Cell Biol 13(7):366–375

Schmid SL (1993) Toward a biochemical definition of the endosomal compartment. Studies using free flow electrophoresis. Subcell Biochem 19:1–28

Schwartz DC, Hochstrasser M (2003) A superfamily of protein tags: ubiquitin, SUMO and related modifiers. Trends Biochem Sci 28(6):321–328

Schweitzer K, Naumann M (2010) Control of NF-kappaB activation by the COP9 signalosome. Biochem Soc Trans 38(Pt 1):156–161

Scott CC, Gruenberg J (2011) Ion flux and the function of endosomes and lysosomes: pH is just the start: the flux of ions across endosomal membranes influences endosome function not only through regulation of the luminal pH. Bioessays 33(2):103–110

Scott CJ, Taggart CC (2010) Biologic protease inhibitors as novel therapeutic agents. Biochimie 92(11):1681–1688

Seglen PO, Berg TO, Blankson H, Fengsrud M, Holen I, Stromhaug PE (1996) Structural aspects of autophagy. Adv Exp Med Biol 389:103–111

Seidah NG (2011) What lies ahead for the proprotein convertases? Ann NY Acad Sci 1220:149–161

Seidah NG, Chretien M (1999) Proprotein and prohormone convertases: a family of subtilases generating diverse bioactive polypeptides. Brain Res 848(1–2):45–62

Seidah NG, Prat A (2002) Precursor convertases in the secretory pathway, cytosol and extracellular milieu. Essays Biochem 38:79–94

Seidah NG, Day R, Marcinkiewicz M, Benjannet S, Chretien M (1991) Mammalian neural and endocrine pro-protein and pro-hormone convertases belonging to the subtilisin family of serine proteinases. Enzyme 45(5–6):271–284

Seiki M (2002) The cell surface: the stage for matrix metalloproteinase regulation of migration. Curr Opin Cell Biol 14(5):624–632

Selkoe DJ, Wolfe MS (2007) Presenilin: running with scissors in the membrane. Cell 131 (2):215–221

Shi Z, Stack MS (2007) Urinary-type plasminogen activator (uPA) and its receptor (uPAR) in squamous cell carcinoma of the oral cavity. Biochem J 407(2):153–159
Silverman GA, Bird PI, Carrell RW, Church FC, Coughlin PB, Gettins PG, Irving JA, Lomas DA, Luke CJ, Moyer RW, Pemberton PA, Remold-O'Donnell E, Salvesen GS, Travis J, Whisstock JC (2001) The serpins are an expanding superfamily of structurally similar but functionally diverse proteins. Evolution, mechanism of inhibition, novel functions, and a revised nomenclature. J Biol Chem 276(36):33293–33296
Simionescu N, Simionescu M, Palade GE (1972) Permeability of intestinal capillaries. Pathway followed by dextrans and glycogens. J Cell Biol 53(2):365–392
Singhvi A, Garriga G (2009) Asymmetric divisions, aggresomes and apoptosis. Trends Cell Biol 19(1):1–7
Sloane BF, Yan S, Podgorski I, Linebaugh BE, Cher ML, Mai J, Cavallo-Medved D, Sameni M, Dosescu J, Moin K (2005) Cathepsin B and tumor proteolysis: contribution of the tumor microenvironment. Semin Cancer Biol 15(2):149–157
Sloane BF, Sameni M, Podgorski I, Cavallo-Medved D, Moin K (2006) Functional imaging of tumor proteolysis. Annu Rev Pharmacol Toxicol 46:301–315
Smirnova IV, Ho GJ, Fenton JW II, Festoff BW (1994) Extravascular proteolysis and the nervous system: serine protease/serpin balance. Semin Thromb Hemost 20(4):426–432
Smith HW, Marshall CJ (2010) Regulation of cell signalling by uPAR. Nat Rev Mol Cell Biol 11 (1):23–36
Sohail A, Sun Q, Zhao H, Bernardo MM, Cho JA, Fridman R (2008) MT4-(MMP17) and MT6-MMP (MMP25), a unique set of membrane-anchored matrix metalloproteinases: properties and expression in cancer. Cancer Metastasis Rev 27(2):289–302
Sol-Church K, Picerno GN, Stabley DL, Frenck J, Xing S, Bertenshaw GP, Mason RW (2002) Evolution of placentally expressed cathepsins. Biochem Biophys Res Commun 293(1):23–29
Song L, Rape M (2008) Reverse the curse–the role of deubiquitination in cell cycle control. Curr Opin Cell Biol 20(2):156–163
Song Z, Steller H (1999) Death by design: mechanism and control of apoptosis. Trends Cell Biol 9 (12):M49–M52
Sorimachi H, Ishiura S, Suzuki K (1997) Structure and physiological function of calpains. Biochem J 328(Pt 3):721–732
Sorimachi H, Hata S, Ono Y (2010) Expanding members and roles of the calpain superfamily and their genetically modified animals. Exp Anim 59(5):549–566
Sounni NE, Noel A (2005) Membrane type-matrix metalloproteinases and tumor progression. Biochimie 87(3–4):329–342
Squier TC (2006) Redox modulation of cellular metabolism through targeted degradation of signaling proteins by the proteasome. Antioxid Redox Signal 8(1–2):217–228
Stadtmueller BM, Hill CP (2011) Proteasome activators. Mol Cell 41(1):8–19
Stamenkovic I (2003) Extracellular matrix remodelling: the role of matrix metalloproteinases. J Pathol 200(4):448–464
Steiner DF (1969) Proinsulin and the biosynthesis of insulin. N Engl J Med 280(20):1106–1113
Steiner DF (1998) The proprotein convertases. Curr Opin Chem Biol 2(1):31–39
Steiner H, Haass C (2000) Intramembrane proteolysis by presenilins. Nat Rev Mol Cell Biol 1 (3):217–224
Steiner DF, Quinn PS, Chan SJ, Marsh J, Tager HS (1980) Processing mechanisms in the biosynthesis of proteins. Ann NY Acad Sci 343:1–16
Sterchi EE, Stocker W, Bond JS (2008) Meprins, membrane-bound and secreted astacin metalloproteinases. Mol Aspects Med 29(5):309–328
Sternlicht MD, Werb Z (2001) How matrix metalloproteinases regulate cell behavior. Annu Rev Cell Dev Biol 17:463–516
Stetler-Stevenson WG (2008) The tumor microenvironment: regulation by MMP-independent effects of tissue inhibitor of metalloproteinases-2. Cancer Metastasis Rev 27(1):57–66

Storr SJ, Carragher NO, Frame MC, Parr T, Martin SG (2011) The calpain system and cancer. Nat Rev Cancer 11(5):364–374
Strongin AY (2010) Proteolytic and non-proteolytic roles of membrane type-1 matrix metalloproteinase in malignancy. Biochim Biophys Acta 1803(1):133–141
Stubbs MT, Bode W (1994) Coagulation factors and their inhibitors. Curr Opin Struct Biol 4 (6):823–832
Sun Y, Kucej M, Fan HY, Yu H, Sun QY, Zou H (2009) Separase is recruited to mitotic chromosomes to dissolve sister chromatid cohesion in a DNA-dependent manner. Cell 137 (1):123–132
Szabo R, Bugge TH (2008) Type II transmembrane serine proteases in development and disease. Int J Biochem Cell Biol 40(6–7):1297–1316
Tai HC, Schuman EM (2008) Ubiquitin, the proteasome and protein degradation in neuronal function and dysfunction. Nat Rev Neurosci 9(11):826–838
Takahashi T, Tang J (1981) Cathepsin D from porcine and bovine spleen. Methods Enzymol 80 (Pt C):565–581
Tallant C, Marrero A, Gomis-Ruth FX (2010) Matrix metalloproteinases: fold and function of their catalytic domains. Biochim Biophys Acta 1803(1):20–28
Tanaka Y, Tanaka R, Kawabata T, Noguchi Y, Himeno M (2000) Lysosomal cysteine protease, cathepsin B, is targeted to lysosomes by the mannose 6-phosphate-independent pathway in rat hepatocytes: site-specific phosphorylation in oligosaccharides of the proregion. J Biochem 128 (1):39–48
Tatnell PJ, Fowler SD, Bur D, Lees WE, Kay J (1998) Cathepsin E. The best laid plans of mice and men. Adv Exp Med Biol 436:147–152
Tedelind S, Poliakova K, Valeta A, Hunegnaw R, Yemanaberhan EL, Heldin NE, Kurebayashi J, Weber E, Kopitar-Jerala N, Turk B, Bogyo M, Brix K (2010) Nuclear cysteine cathepsin variants in thyroid carcinoma cells. Biol Chem 391(8):923–935
Tedelind S, Jordans S, Resemann H, Blum G, Bogyo M, Fuhrer D, Brix K (2011) Cathepsin B trafficking in thyroid carcinoma cells. Thyroid Res 4(Suppl 1):S2
Tepel C, Bromme D, Herzog V, Brix K (2000) Cathepsin K in thyroid epithelial cells: sequence, localization and possible function in extracellular proteolysis of thyroglobulin. J Cell Sci 113 (Pt 24):4487–4498
Thomas G (2002) Furin at the cutting edge: from protein traffic to embryogenesis and disease. Nat Rev Mol Cell Biol 3(10):753–766
Tooze J, Hollinshead M, Hensel G, Kern HF, Hoflack B (1991) Regulated secretion of mature cathepsin B from rat exocrine pancreatic cells. Eur J Cell Biol 56(2):187–200
Travis J (1988) Structure, function, and control of neutrophil proteinases. Am J Med 84(6A):37–42
Travis J, Fritz H (1991) Potential problems in designing elastase inhibitors for therapy. Am Rev Respir Dis 143(6):1412–1415
Travis J, Salvesen GS (1983) Human plasma proteinase inhibitors. Annu Rev Biochem 52:655–709
Trombetta ES, Mellman I (2005) Cell biology of antigen processing in vitro and in vivo. Annu Rev Immunol 23:975–1028
Tschopp J, Martinon F, Burns K (2003) NALPs: a novel protein family involved in inflammation. Nat Rev Mol Cell Biol 4(2):95–104
Tsukuba T, Okamoto K, Yasuda Y, Morikawa W, Nakanishi H, Yamamoto K (2000) New functional aspects of cathepsin D and cathepsin E. Mol Cells 10(6):601–611
Turk V, Bode W (1991) The cystatins: protein inhibitors of cysteine proteinases. FEBS Lett 285 (2):213–219
Turk B, Turk V (2009) Lysosomes as "suicide bags" in cell death: myth or reality? J Biol Chem 284(33):21783–21787
Turk B, Turk V, Turk D (1997) Structural and functional aspects of papain-like cysteine proteinases and their protein inhibitors. Biol Chem 378(3–4):141–150

Turk B, Turk D, Turk V (2000) Lysosomal cysteine proteases: more than scavengers. Biochim Biophys Acta 1477(1–2):98–111
Turk V, Turk B, Turk D (2001) Lysosomal cysteine proteases: facts and opportunities. EMBO J 20 (17):4629–4633
Turk D, Turk B, Turk V (2003) Papain-like lysosomal cysteine proteases and their inhibitors: drug discovery targets? Biochem Soc Symp 70:15–30
Turner AJ, Nalivaeva NN (2007) New insights into the roles of metalloproteinases in neurodegeneration and neuroprotection. Int Rev Neurobiol 82:113–135
Uddin MN, Nabi AH, Nakagawa T, Ichihara A, Inagami T, Suzuki F (2008) Non-proteolytic activation of prorenin: activation by (pro)renin receptor and its inhibition by a prorenin prosegment, "decoy peptide". Front Biosci 13:745–753
Uhlmann F (2003) Separase regulation during mitosis. Biochem Soc Symp 70:243–251
Urban S (2010) Taking the plunge: integrating structural, enzymatic and computational insights into a unified model for membrane-immersed rhomboid proteolysis. Biochem J 425 (3):501–512
Urban S, Freeman M (2002) Intramembrane proteolysis controls diverse signalling pathways throughout evolution. Curr Opin Genet Dev 12(5):512–518
Urban S, Lee JR, Freeman M (2001) Drosophila rhomboid-1 defines a family of putative intramembrane serine proteases. Cell 107(2):173–182
Van den Eynde BJ, Morel S (2001) Differential processing of class-I-restricted epitopes by the standard proteasome and the immunoproteasome. Curr Opin Immunol 13(2):147–153
van Goor H, Melenhorst WB, Turner AJ, Holgate ST (2009) Adamalysins in biology and disease. J Pathol 219(3):277–286
van Meer G, de Kroon AI (2011) Lipid map of the mammalian cell. J Cell Sci 124(Pt 1):5–8
Vandenabeele P, Galluzzi L, Vanden Berghe T, Kroemer G (2010) Molecular mechanisms of necroptosis: an ordered cellular explosion. Nat Rev Mol Cell Biol 11(10):700–714
Vanlangenakker N, Vanden Berghe T, Krysko DV, Festjens N, Vandenabeele P (2008) Molecular mechanisms and pathophysiology of necrotic cell death. Curr Mol Med 8(3):207–220
Varki A (1993) Biological roles of oligosaccharides: all of the theories are correct. Glycobiology 3 (2):97–130
Varki A (1998) Factors controlling the glycosylation potential of the Golgi apparatus. Trends Cell Biol 8(1):34–40
Vasiljeva O, Reinheckel T, Peters C, Turk D, Turk V, Turk B (2007) Emerging roles of cysteine cathepsins in disease and their potential as drug targets. Curr Pharm Des 13(4):387–403
Victor BC, Sloane BF (2007) Cysteine cathepsin non-inhibitory binding partners: modulating intracellular trafficking and function. Biol Chem 388(11):1131–1140
Vierstra RD (2009) The ubiquitin-26S proteasome system at the nexus of plant biology. Nat Rev Mol Cell Biol 10(6):385–397
Villadangos JA, Bryant RA, Deussing J, Driessen C, Lennon-Dumenil AM, Riese RJ, Roth W, Saftig P, Shi GP, Chapman HA, Peters C, Ploegh HL (1999) Proteases involved in MHC class II antigen presentation. Immunol Rev 172:109–120
Vreemann A, Qu H, Mayer K, Andersen LB, Stefana MI, Wehner S, Lysson M, Farcas AM, Peters C, Reinheckel T, Kalff J, Brix K (2009) Cathepsin B release from rodent intestine mucosa due to mechanical injury results in extracellular matrix damage in early post-traumatic phases. Biol Chem 390(5–6):481–492
Watts C (2012) The endosome-lysosome pathway and information generation in the immune system. Biochim Biophys Acta 1824(1):14–21
Watts C, Matthews SP, Mazzeo D, Manoury B, Moss CX (2005) Asparaginyl endopeptidase: case history of a class II MHC compartment protease. Immunol Rev 207:218–228
Weihofen A, Martoglio B (2003) Intramembrane-cleaving proteases: controlled liberation of proteins and bioactive peptides. Trends Cell Biol 13(2):71–78
Whisstock JC, Bottomley SP (2006) Molecular gymnastics: serpin structure, folding and misfolding. Curr Opin Struct Biol 16(6):761–768

Whisstock JC, Bottomley SP (2008) Structural biology: serpins' mystery solved. Nature 455 (7217):1189–1190
White JM (2003) ADAMs: modulators of cell-cell and cell-matrix interactions. Curr Opin Cell Biol 15(5):598–606
Wiederanders B, Kaulmann G, Schilling K (2003) Functions of propeptide parts in cysteine proteases. Curr Protein Pept Sci 4(5):309–326
Wiertz EJ, Jones TR, Sun L, Bogyo M, Geuze HJ, Ploegh HL (1996a) The human cytomegalovirus US11 gene product dislocates MHC class I heavy chains from the endoplasmic reticulum to the cytosol. Cell 84(5):769–779
Wiertz EJ, Tortorella D, Bogyo M, Yu J, Mothes W, Jones TR, Rapoport TA, Ploegh HL (1996b) Sec61-mediated transfer of a membrane protein from the endoplasmic reticulum to the proteasome for destruction. Nature 384(6608):432–438
Wojcik C, DeMartino GN (2003) Intracellular localization of proteasomes. Int J Biochem Cell Biol 35(5):579–589
Wolf DH, Hilt W (2004) The proteasome: a proteolytic nanomachine of cell regulation and waste disposal. Biochim Biophys Acta 1695(1–3):19–31
Wolfe MS, Kopan R (2004) Intramembrane proteolysis: theme and variations. Science 305 (5687):1119–1123
Woodward JK, Holen I, Coleman RE, Buttle DJ (2007) The roles of proteolytic enzymes in the development of tumour-induced bone disease in breast and prostate cancer. Bone 41 (6):912–927
Wyllie AH, Kerr JF, Currie AR (1980) Cell death: the significance of apoptosis. Int Rev Cytol 68:251–306
Yamamoto K (1995) Cathepsin E and cathepsin D: biosynthesis, processing and subcellular location. Adv Exp Med Biol 362:223–229
Yanagida M (2009) Clearing the way for mitosis: is cohesin a target? Nat Rev Mol Cell Biol 10 (7):489–496
Yao C (2010) Major surface protease of trypanosomatids: one size fits all? Infect Immun 78 (1):22–31
Yu DM, Wang XM, McCaughan GW, Gorrell MD (2006) Extraenzymatic functions of the dipeptidyl peptidase IV-related proteins DP8 and DP9 in cell adhesion, migration and apoptosis. FEBS J 273(11):2447–2460
Yu DM, Yao TW, Chowdhury S, Nadvi NA, Osborne B, Church WB, McCaughan GW, Gorrell MD (2010) The dipeptidyl peptidase IV family in cancer and cell biology. FEBS J 277 (5):1126–1144
Zaidi N, Hermann C, Herrmann T, Kalbacher H (2008) Emerging functional roles of cathepsin E. Biochem Biophys Res Commun 377(2):327–330
Zhao JH, Yang CT, Wu JW, Tsai WB, Lin HY, Fang HW, Ho Y, Liu HL (2008) RING domains functioning as E3 ligases reveal distinct structural features: a molecular dynamics simulation study. J Biomol Struct Dyn 26(1):65–74
Zhou A, Webb G, Zhu X, Steiner DF (1999) Proteolytic processing in the secretory pathway. J Biol Chem 274(30):20745–20748
Zhou A, Carrell RW, Murphy MP, Wei Z, Yan Y, Stanley PL, Stein PE, Broughton Pipkin F, Read RJ (2010) A redox switch in angiotensinogen modulates angiotensin release. Nature 468 (7320):108–111
Zogg T, Brandstetter H (2011) Complex assemblies of factors IX and X regulate the initiation, maintenance, and shutdown of blood coagulation. Prog Mol Biol Transl Sci 99:51–103

Chapter 4
Cathepsins: Getting in Shape for Lysosomal Proteolysis

Ann H. Erickson, Ciro Isidoro, Lukas Mach, and John S. Mort

4.1 Introduction

Besides their pivotal functions in general cellular protein turnover, cathepsins play important roles in a diverse range of other physiological processes which include tissue remodelling during embryogenesis and development, programmed cell death, autophagy, prohormone and neuropeptide processing, antigen presentation, wound healing and bone resorption. Furthermore, substantial experimental evidence has been accumulated that cathepsins are of pathological relevance in disease states such as cancer, arthritis, osteopetrosis, pancreatitis, cholestatic liver disease, and epilepsy (Mohamed and Sloane 2006; Vasiljeva et al. 2007; Turk and Turk 2009; Reiser et al. 2010). To prevent tissue damage due to unwanted proteolysis, the activities of cathepsins have to be strictly controlled *in situ*. The main regulatory pathways rely on restricting the subcellular localization of these proteases to lysosomes, the presence of specific cathepsin inhibitors in other cellular compartments, and their initial synthesis as latent proenzymes (Cygler and Mort 1997; Mort and Buttle 1997;

A.H. Erickson
Department of Biochemistry and Biophysics, University of North Carolina, Chapel Hill, NC, USA
e-mail: annherickson@gmail.com

C. Isidoro
Department of Medical Sciences, Amedeo Avogadro University, Novara, Italy
e-mail: ciro.isidoro@med.unipmn.it

L. Mach (✉)
Department of Applied Genetics and Cell Biology, University of Natural Resources and Life Sciences, Muthgasse 18, 1190 Vienna, Austria
e-mail: lukas.mach@boku.ac.at

J.S. Mort
Genetics Unit, Shriners Hospital for Children, Montreal, Canada
e-mail: jmort@shriners.mcgill.ca

K. Brix and W. Stöcker (eds.), *Proteases: Structure and Function*,
DOI 10.1007/978-3-7091-0885-7_4, © Springer-Verlag Wien 2013

Turk et al. 2001b). Interference with any of these control mechanisms can lead to pathological consequences. It is therefore crucial to understand the molecular basis of cathepsin biosynthesis and intracellular transport as well as the mechanisms leading to activation of their precursors in a cellular context.

4.2 Mannose 6-Phosphate Receptors: Key Cellular Interaction Partners of Lysosomal Cathepsins

The extracellular accumulation of various cathepsins in cancer, arthritis and other human disorders indicates disease-associated changes in the biosynthesis and intracellular transport of these proteinases. However, extensive studies on the biosynthesis of individual cathepsins have been performed only for three of them: cathepsin B, cathepsin D, and cathepsin L. These proteinases are synthesized and targeted to their final intracellular destination in a similar manner as most other soluble lysosomal proteins, involving the following general steps: synthesis as latent preproenzymes by ribosomes associated with the rough endoplasmic reticulum (ER), translocation into the lumen of this compartment, delivery to and passage through the different stacks of the Golgi apparatus, and finally receptor-mediated transport from the *trans*-Golgi network (TGN) to lysosomal compartments followed by receptor-ligand dissociation and proteolytic maturation of the proteinase precursors *in situ* (Erickson 1989; Hasilik 1992; Mach 2002).

Like most other soluble lysosomal enzymes, cathepsins are glycoproteins which are modified in their *N*-glycosidically linked oligosaccharide chains with mannose 6-phosphate (M6P) residues. This unique post-translational modification is critical for the correct intracellular targeting of the proteinases. Two enzymes are responsible for the creation of the M6P recognition marker: UDP-*N*-acetylglucosamine-1-phosphotransferase (phosphotransferase) and *N*-acetylglucosamine-1-phosphodiester α-*N*-acetyl-glucosaminidase (uncovering enzyme). Phosphotransferase, a *cis*-Golgi enzyme, attaches GlcNAc-1-phosphate residues from UDP-GlcNAc to the C-6 hydroxyl group of selected mannoses in the high mannose-type oligosaccharides of newly synthesized lysosomal hydrolases (Reitman and Kornfeld 1981; Tiede et al. 2005). This generates a phosphodiester, Man-P-GlcNAc. The sugar moiety masking the phosphate group is then removed by the uncovering enzyme localized in the TGN (Waheed et al. 1981; Rohrer and Kornfeld 2001). This exposes the phosphomonoester residue that is then recognized by specific M6P receptors residing in the same compartment, which then deliver the newly synthesized enzymes to lysosomes (Braulke and Bonifacino 2009; Saftig and Klumperman 2009).

Phosphotransferase is unique in its requirement for specific oligosaccharide acceptor sites only present in lysosomal proteins. The fundamental importance of this enzyme for lysosome biogenesis is documented by the fact that its deficiency leads to the inherited disorder mucolipidosis II, also referred to as I-cell disease

(Tiede et al. 2005). It is still unclear how phosphotransferase distinguishes between lysosomal and secretory proteins. However, it has been proposed that its interaction with cathepsins B, D and L involves the recognition of a structural motif based on distinct spatial positioning of certain lysine residues (Cuozzo et al. 1998; Lukong et al. 1999). At least for fibroblast cathepsin B, modification with M6P residues seems mandatory for delivery of the enzyme to lysosomes, since cells from I-cell disease patients almost quantitatively fail to retain their newly-synthesized procathepsin B (Hanewinkel et al. 1987). The latter is also observed in the case of murine fibroblasts lacking both M6P receptors (Probst et al. 2006).

Two distinct M6P-binding proteins occur in mammalian cells, the 300-kDa mannose 6-phosphate/insulin-like growth factor II receptor (M6P/IGF2R) and the 46-kDa cation-dependent mannose 6-phosphate receptor (MPR46). The comparison of their cDNA sequences has revealed that the two receptors are related proteins. While M6P/IGF2R has an extracytoplasmic domain consisting of 15 repeating segments, the entire extracytoplasmic domain of MPR46 is similar to each of the repeating units of the former protein. This suggests that the two receptors may be derived from a common ancestor (Ghosh et al. 2003; Dahms et al. 2008). Fibroblasts devoid of both receptors have a phenotype similar to cells from patients with mucolipidosis II, quantitatively secreting most of their newly synthesized lysosomal enzymes. In cells lacking only one of the two proteins, the remaining receptor cannot fully compensate for the absence of the other (Ludwig et al. 1994; Pohlmann et al. 1995). Interestingly, hypersecretion of lysosomal enzymes by cells lacking M6P/IGF2R can be only partially rescued by overexpression of MPR46 (Watanabe et al. 1990).

Besides sorting newly synthesized lysosomal enzymes from the Golgi apparatus to lysosomes, M6P receptors also function in the endocytosis of extracellular lysosomal enzymes. It is remarkable that M6P/IGF2R, but not MPR46, is capable of mediating this process. Thus, mammalian cells have two means to deliver M6P-tagged proteins to lysosomes: (a) the biosynthetic route via vesicles derived from the Golgi apparatus, and (b) recapture of mistargeted and hence secreted lysosomal enzymes via the endocytic route, with the latter process being strictly dependent on M6P/IGF2R. It has been proposed that the secretion-recapture pathway contributes significantly to the efficiency of lysosomal enzyme sorting in mammalian cells (Kasper et al. 1996). Taken together, this indicates that two distinct M6P receptors are engaged in cathepsin trafficking to lysosomes, with M6P/IGF2R being the major protein responsible for this important cellular process.

The displacement of M6P-modified cathepsins from their sorting receptors occurs upon reaching endosomes, followed by rapid M6P hydrolysis due to the action of phosphatases. Receptor-ligand dissociation is thought to be due to the reduced affinities of M6P receptors for their ligands at the low pH of the endosomal lumen. In cells expressing both M6P receptors, agents that perturb endosomal acidification have been shown to interfere with cathepsin trafficking to lysosomes (Braulke et al. 1987). However, such lysosomotropic amines do not noticeably affect the residual intracellular retention of cathepsins B, D and L in fibroblasts and epithelial cells lacking M6P/IGF2R. Intriguingly, intracellular transport of cathepsin B in M6P/IGF2R-deficient cells is still entirely dependent on the presence of the M6P recognition marker, indicating that it is mediated by MPR46. These findings strongly suggest that at least dissociation of MPR46 and its ligands can also occur in the absence of endosomal acidification (Probst et al. 2006).

Although the M6P receptor system is the main pathway for lysosomal enzyme sorting, it has become evident that in specialised cell types intracellular transport of soluble acid hydrolases to lysosomes can also occur in an M6P-independent manner. The strongest support for the existence of M6P-independent lysosomal trafficking pathways comes from studies involving cells of lymphocytic origin. In lymphoblasts from I-cell disease patients, transport of cathepsin D to lysosomal compartments is not dependent on *N*-glycosylation of the protein (Glickman and Kornfeld 1993). Cathepsin B was also localized in lysosomal compartments of I-cell disease lymphoblastoid cell lines (Griffiths and Isaaz 1993). Moreover, murine thymocytes deficient in both M6P receptors retain their newly-synthesized procathepsin D as efficiently as their normal counterparts (Dittmer et al. 1999). Recently, evidence has been provided that sortilin could be involved in M6P-independent targeting of cathepsin D to lysosomes (Canuel et al. 2008). Alternatively, M6P-independent transfer of cathepsins to lysosomes could rely on a secretion-recapture pathway, possibly engaging cell-surface lectins such as the asialoglycoprotein receptor or the mannose receptor. M6P-independent uptake of secreted cathepsins might also occur by means of direct cell-to-cell contact, involving pinocytotic microinvaginations and non-coated vesicles (Dittmer et al. 1999).

Cathepsins are delivered to endosomes as latent proenzymes. It is generally believed that proteolytic maturation of the precursors is initiated once receptor-mediated delivery to endosomal compartments has been achieved. Proteinase activation is then triggered by the local acidic environment. It has been demonstrated that purified procathepsins B, D and L can undergo autocatalytic activation in acidic conditions (Hasilik et al. 1982; Mach et al. 1994a; Ménard et al. 1998). Activation is quickly followed by cleavage of the latent proenzymes into the single-chain forms of the respective proteinases (see chapter 4.5 for details). Finally, terminal processing of the single-chain enzymes into the corresponding double-chain forms occurs as a late biosynthetic event in the lysosomes.

4.3 Cysteine Cathepsins: Endopeptidases and Exopeptidases

Protein degradation in lysosomes and related compartments is thought to involve two main phases. The initial digestion of the substrates is achieved by the action of cathepsins with endopeptidase activity. The fragments thus generated are then converted by exopeptidases into small peptides and free amino acids. Three endopeptidases appear to be present in all mammalian lysosomes: the aspartic proteinase cathepsin D, and the cysteine proteinases cathepsin B and cathepsin

L. Cathepsin B also displays dipeptidylcarboxypeptidase activity. In contrast, cathepsin D and cathepsin L are classical proteinases lacking any exopeptidolytic potential (Brix et al. 2008; Masson et al. 2010).

Cathepsin B and cathepsin L are members of the so-called papain superfamily of lysosomal cysteine proteinases, which includes another nine human enzymes: cathepsins C, F, H, K, O, S, V, W and X. Two ubiquitously expressed genuine exopeptidases without endopeptidolytic activity belong to this family, the dipeptidylaminopeptidase cathepsin C and the carboxypeptidase cathepsin X. Cathepsin H exhibits both aminopeptidase and endopeptidase activity. Some specialised cell types express other family members with close homology to cathepsin L. The best studied of these tissue-specific lysosomal cysteine proteinases are cathepsins K and S (Turk et al. 2000; Brix et al. 2008).

Of the 11 human cysteine cathepsins, 7 appear to be obligate endopeptidases. Three of these—cathepsins K, L and S—have been studied extensively whereas cathepsins F, O, V and W have received less attention. The mature forms of these proteases consist of a two-lobed papain-like structure, the junction between which forms the substrate-binding cleft. This channel is usually capable of housing three amino acid residues on the unprimed side of the peptide linkage targeted for cleavage and at least two residues on the primed side (Turk et al. 1998; Stern et al. 2004). The major selectivity of the cysteine cathepsins lies in the P2 position. As endopeptidases, cathepsins K, L and S are very efficient protein dismantling agents.

Building on the basic papain model, additional structural elements have evolved to restrict the active site cleft so that the resulting enzyme becomes one of four exopeptidase types. Cathepsins B and C remove dipeptides from the C- or N-terminus, while cathepsins X and H are monocarboxy- and aminopeptidases, respectively. The carboxypeptidase activities are mediated by the inclusion of an extra loop in the basic structure of the mature protease. Termed the occluding loop in the case of cathepsin B (Musil et al. 1991) and the mini loop in the case of cathepsin X (Nägler et al. 1999b), these elements block off the primed side of the active site cleft and position a histidine side chain to accept the negative charge on the substrate C-terminus. The size of the loop allows for the positioning of two or one residues, respectively, at the primed side of the site of cleavage. In cathepsin B the occluding loop is held in position by an ionic interaction between a second histidine residue in the loop and an aspartate residue located in the main body of the enzyme. Disruption of this interaction as occurs with increasing pH permits the loop to reorient (Nägler et al. 1997), and the enzyme then acts as an endopeptidase.

The molecular basis for aminopeptidase activity is more complex and depends on the positioning of remnants of the propeptides which restrict the unprimed side of the active-site cleft so that it can only accept one or two residues. In the case of cathepsin H an octapeptide mini-chain is disulfide bonded in place providing space for only a single residue before the cleavage site. The positive charge on the substrate N-terminus is matched by the carboxylate of the C-terminus of the mini-chain (Guncar et al. 1998). As would be expected, cathepsin H lacking the mini-chain is a functional endopeptidase (Vasiljeva et al. 2003). In cathepsin C a

119-residue β-barrel unit termed the 'exclusion domain' is employed for positioning of an aspartate side chain as the acceptor for the substrate N-terminus in the S2 pocket of the enzyme (Turk et al. 2001a). With this diversity of activities, the combined actions of the cysteine cathepsins can recycle a substrate protein to its constituent amino acids.

4.4 Structures of Cysteine Cathepsin Precursors

Proteolytic activity is inherently destructive and thus potentially hazardous to the cell. Proteases are therefore usually synthesized as inactive precursors which require processing to release functional enzymes at cellular locations where their action is required. Various strategies are employed by different protease families to realize this self-defense mechanism (Khan et al. 1998). In most cases the proenzyme contains an N-terminal propeptide extension which folds back onto the active site of the protease, thus blocking access to the substrate. The propeptides can be as small as a dipeptide as in the case of some serine proteases such as neutrophil elastase (Salvesen and Enghild 1990) and granzymes, or domains with over 90 residues as for various cysteine cathepsins.

In 1993, based on an analysis of the proenzyme sequences, the cysteine cathepsins were divided into two classes based on the presence of a highly conserved interspersed motif termed 'ERFNIN' (Karrer et al. 1993) which represents a series of residues predicted to lie along one surface of a large α-helix. Variants of this motif are present in the cathepsin L-like class members but absent in the cathepsin B-like enzymes. The three-dimensional structures of procathepsins K (LaLonde et al. 1999; Sivaraman et al. 1999), L (Coulombe et al. 1996) and S (Kaulmann et al. 2006) show a distinct α-helix-rich domain which is stabilized by a conserved tripartite tryptophan motif located at the interaction site between the two major α-helices (Kreusch et al. 2000), the second of which contains the ERFNIN motif. This double-helical domain is linked to the body of the catalytic unit by the insertion of an aromatic side chain into a hydrophobic pocket formed by a region termed the propeptide binding loop (Fig. 4.1). The structures of the proenzymes immediately explain their inability to cleave protein substrates since the propeptide extends through the active site blocking access to the catalytic machinery. Critically, the sense of the propeptide strand is reversed relative to that required for normal substrate binding, thus hampering processing of the propeptide (Cygler and Mort 1997). The inhibitory nature of the propeptide regions of these enzymes has been demonstrated by their production as independent modules. These can fold into helix-rich domains and were shown to be tight-binding inhibitors of their cognate enzymes (Schilling et al. 2009).

With 38 residues, the propeptide of cathepsin X is the shortest in the whole cysteine cathepsin family (Fig. 4.2), barely reaching beyond the active-site cleft (Sivaraman et al. 2000). However, its position is stabilized by a disulfide bond located between the prosegment and the active-site cysteine residue.

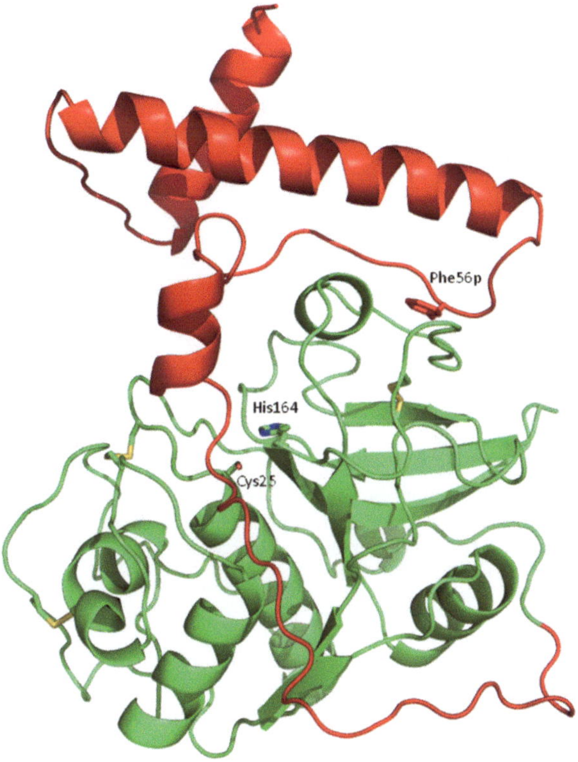

Fig. 4.1 Three-dimensional structure of human procathepsin L. The propeptide is illustrated in *red* while the mature enzyme is in *green*. Disulfide bridges are displayed in *yellow*. The side chains of the catalytic cysteine and histidine residues are shown in *stick* representation, as is the side chain of the phenylalanine residue in the propeptide which anchors the propeptide to the mature enzyme through hydrophobic interactions (PDB 1CJL)

The propeptide of cathepsin B is intermediate in length between that of cathepsin X and those of the cathepsin L-like class. As in procathepsin L, an α-helical structural element is present which is bound to the protease through a hydrophobic interaction (Cygler et al. 1996; Turk et al. 1996). In mature cathepsin B, the occluding loop is positioned to accept the carboxylate of the P2′ substrate residue, as illustrated in the structure of a complex between the protease and the irreversible inhibitor CA-074 (Yamamoto et al. 2000) (Fig. 4.3b, c). In the proenzyme however, the position of the occluding loop is deflected by the propeptide (Fig. 4.3a). In fact, the occluding loop appears to be also adaptable in other situations such as the binding of the inhibitors chagasin (Redzynia et al. 2008) and cystatin A (Renko et al. 2010).

The large variation in the size of the cysteine cathepsin proregions suggests that the α-helical structural element may have additional functional roles. Some evidence for this is provided by the characterization of a receptor for procathepsin L (McIntyre and Erickson 1993) and binding of procathepsin B to annexin II (Mai et al. 2000). These interactions may play a role in the processing of the proenzyme forms or the targeting of the enzymes to lysosomes or other subcellular locations. Procathepsin F is a unique member of this family of enzymes since it contains a cystatin-like domain N-terminal to the cathepsin L-like propeptide region (Nägler

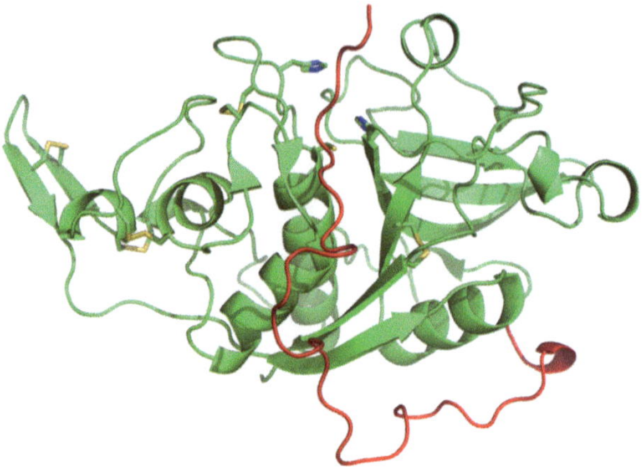

Fig. 4.2 Three-dimensional structure of human procathepsin X. The propeptide is illustrated in *red* while the mature enzyme is in *green*. Disulfide bridges are displayed in *yellow*. The side chains of the catalytic cysteine and histidine residues are shown in *stick* representation, as is the side chain of the histidine residue of the mini-loop which is the acceptor for the C-terminal carboxylate of the substrate (PDB 1DEU)

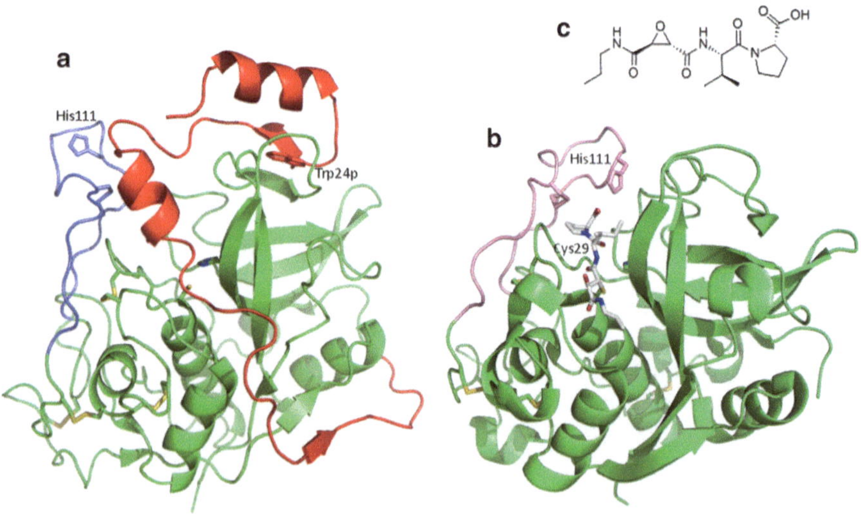

Fig. 4.3 Three-dimensional structures of human procathepsin B and the mature enzyme. (**a**) Procathepsin B (PDB 3PBH) showing the propeptide in *red* and the mature enzyme in *green*. Disulfide bridges are displayed in *yellow*. The side chains of the catalytic cysteine and histidine residues are shown in *stick* representation, as is the side chain of the tryptophan residue in the proregion which anchors the propeptide to the mature enzyme through hydrophobic interactions. The occluding loop is illustrated in *blue*. Note that the loop is deflected by the presence of the propeptide. (**b**) A complex (PDB 1QDQ) between the mature protease and the irreversible inhibitor CA-074 (*trans*-epoxysuccinyl(propylamide)-Ile-Pro, chemical structure shown in **c**). The Ile-Pro portion of the inhibitor occupies the S1′ and S2′ subsites. The occluding loop is indicated in *pink*. The carboxylate of the Pro residue of the inhibitor interacts with the side chain of His111 located in the occluding loop

et al. 1999a; Wex et al. 1999). The functional significance of this domain is, however, still unclear.

The proregions of the exopeptidases follow the same overall strategy of using the propeptide to obstruct the active site cleft, but the N-terminal regions of the propeptides differ widely. No structural studies have been published so far for the proforms of cathepsins H and C. However, their sequences indicate that a similar crossed α-helix structure is present. This is also supported by a model of the procathepsin H structure (Horn et al. 2005).

4.5 Activation and Maturation of Cysteine Cathepsin Precursors

Activation of cysteine cathepsin precursors generally accompanies acidification of their environment as they are transported to the lysosome. Proteolytic cleavage of the proregion then occurs, and as was demonstrated in kinetic studies on the isolated proregions of cathepsins B (Fox et al. 1992), K (Billington et al. 2000) and L (Carmona et al. 1996), the binding affinity of the propeptide for its cognate enzyme decreases dramatically as the pH is lowered. Thus, once cleaved the propeptide dissociates from the catalytic unit and becomes a substrate for further proteolysis. Autoprocessing of several proenzymes has been observed in many cases and this can be facilitated by interactions with surfaces and with anionic polymers (Mason and Massey 1992; Caglic et al. 2007). Autoprocessing can occur through two possible mechanisms, intra-molecular or inter-molecular. Early studies using a procathepsin B in which the active-site cysteine residue had been mutated into serine demonstrated that processing can occur in *trans* with the propeptide being liberated by several lysosomal proteases including cathepsin B itself acting as an endopeptidase (Rowan et al. 1992). Claims of intra-molecular processing have been controversial (Rozman et al. 1999). Studies on the concentration dependence of autoprocessing of procathepsin B (Mach et al. 1994a), procathepsin L (Ménard et al. 1998) and procathepsin S (Quraishi and Storer 2001) clearly showed, by extrapolation of the observed rates to zero proenzyme concentration, that intra-molecular processing occurs, which probably requires the transient formation of short-lived processing intermediates (Quraishi and Storer 2001; Pungercar et al. 2009). While the propeptide passes through the active site in the reverse orientation to that required for normal substrate hydrolysis, the carbonyl residue of the peptide bond closest to the active-site cysteine is close enough to form a tetrahedral intermediate although this cannot be stabilized by the canonical oxyanion-hole mechanism. However, slow peptide-bond hydrolysis still occurs and this is followed by removal of the residual propeptide segment by inter-molecular processing. In contrast to cathepsin B and the members of the cathepsin L class, maturation of procathepsins X (Nägler et al. 1999b) and C (Dahl et al. 2001) requires the action of cathepsin L-like proteases.

4.6 Unconventional Cysteine Cathepsin Gene Products

For cathepsins B and L, isoforms have been described which lack parts of their respective propeptides. These variants are also devoid of a signal peptide for translocation into the ER and are therefore not transported to lysosomes. In the case of cathepsin B, an alternate transcript lacking exons two and three is produced in various tissues (Mehtani et al. 1998). In addition to placing the translation start site within the propeptide region, this deletion leads to the generation of a mitochondrial import targeting sequence so that the newly synthesized protein product is located in that organelle (Müntener et al. 2004). The functional significance of the presence of the truncated proenzyme in this location and its status as a protease are still not clear. It has been shown that such a shortened propeptide reduces the folding competence of cathepsin B (Müntener et al. 2005). However, evidence has been provided that expression of truncated cathepsin B can provoke cell death in a manner independent of its enzymatic activity (Müntener et al. 2003; Baici et al. 2006).

Truncated procathepsin L gene products have also been reported. Here it was shown that initiation of translation occurs at methionine codons 3′ to the conventional initiation codon (Goulet et al. 2004). The translation product lacks the signal peptide and several residues of the propeptide. The resulting protein is cytoplasmic and the activity of such a product has been implicated in the processing of dynamin in proteinuric kidney disease (Sever et al. 2007). It has also been detected in the nucleus and is believed to process the transcription factor CPD/Cux (Goulet et al. 2006). However, it is still unclear whether proper disulfide bond formation and processing of the truncated proenzyme is possible in this cellular environment.

4.7 Biosynthesis and Molecular Forms of Cathepsin B

Cathepsin B usually occurs in human tissues and cell lines as a mixture of single-chain and double-chain variants. The ratio between the two forms can differ considerably between individual cell types. While conversion into the double-chain form is essentially complete in HepG2 cells (Mach et al. 1992), the single-chain enzyme is only partially processed in skin fibroblasts (Hanewinkel et al. 1987). In both cell types, cathepsin B is initially synthesized as a latent proenzyme of 45 kDa. Upon delivery to the lysosomal pathway, procathepsin B is first converted into the mature single-chain form of the proteinase (33 kDa). Further processing of this protein leads to the generation of the double-chain enzyme, consisting of subunits of 27 kDa (heavy chain) and 5 kDa (light chain). This endoproteolytic cleavage is accompanied by the excision of a dipeptide that connects the N-terminal light chain with the C-terminal heavy chain. The large fragment carries one *N*-glycan that is gradually lost due to the action of lysosomal glycosidases, ultimately giving rise to a carbohydrate-free, 24-kDa polypeptide (Mach et al. 1992). Furthermore, cathepsin B is subject to exopeptidolytic trimming in the lysosomes, resulting in the removal of N- and C-terminal extensions each consisting of six amino acids. However, the extent of C-terminal cathepsin B processing varies between cell types. Removal of the C-terminal extension is quantitative in HepG2 cells (Mach et al. 1993), while C-terminal processing of cathepsin B is incomplete in human fibroblasts (Schmid et al. 1999).

In the case of cathepsin B, the subunits of the double-chain enzyme are held together by disulfide bridges. It has been shown that single-chain and double-chain forms are equivalent in their enzymatic properties. However, double-chain cathepsin B appears less stable than its single-chain counterpart (Hasnain et al. 1992). This is consistent with the hypothesis that intralysosomal generation of double-chain cathepsins represents a first step in the autocatalytic degradation of these proteinases (Erickson 1989).

Detailed information is also available on the proteolytic maturation of cathepsin B in rodent cells. In murine fibroblasts, cathepsin B is first synthesised as a 42-kDa precursor, which is then converted into the mature, single-chain enzyme (35 kDa). Further processing into the double-chain form of the proteinase is not observed. Therefore, at least two proteolytic steps occur during the biosynthesis of human fibroblast cathepsin B, while mouse procathepsin B is processed only at one site in these cells (Lorenzo et al. 2000; Probst et al. 2006). Similar results have been reported for rat hepatocytes, where single-chain cathepsin B also resists further proteolytic processing (Nishimura et al. 1988). However, a small amount of double-chain cathepsin B is generated during the biosynthesis of this enzyme in rat macrophages (Kominami et al. 1988).

4.7.1 Proteolytic Maturation of Cathepsin B: A Lysosomal Proteinase as Catalyst and Substrate

In most mammalian cells, the generation of the mature single-chain forms of lysosomal proteinases including cathepsin B is believed to take place in acidic post-Golgi compartments. The local acidic environment in endosomes could be sufficient to trigger cathepsin B self-activation, as demonstrated for recombinant procathepsin B *in vitro* (Mach et al. 1994a). Presumably, activation is then quickly followed by both intra-molecular (Mach et al. 1994a; Quraishi and Storer 2001) and inter-molecular (Rowan et al. 1992; Rozman et al. 1999) autocatalytic cleavage events yielding mature single-chain cathepsin B. Terminal processing into the double-chain form is a much slower event and hence occurs most likely in lysosomes (Fig. 4.4).

In human and rodent cells, generation of double-chain cathepsin B can be abolished by treatment with synthetic cysteine proteinase inhibitors such as leupeptin, E-64d and Z-Phe-Ala-CHN_2 (Hanewinkel et al. 1987; Hara et al. 1988; Mach et al. 1992). This provided support for the proposal that processing of human

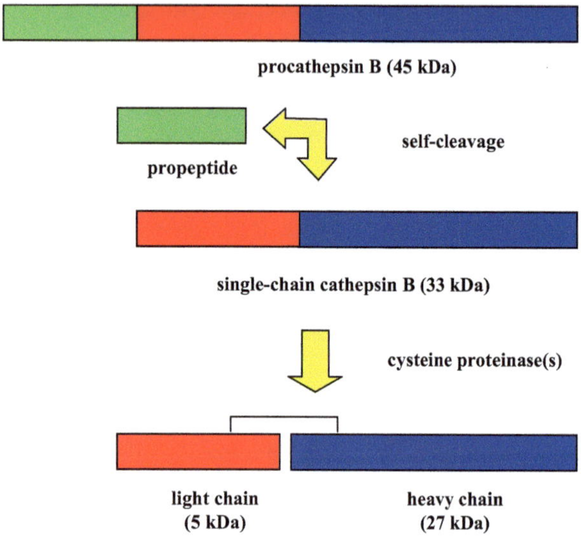

Fig. 4.4 Proteolytic maturation of cathepsin B. The key steps in the proteolytic maturation of human cathepsin B are illustrated. Cathepsin B from other species is processed in an analogous manner. Cysteine proteinases of different families have been implicated in the intralysosomal generation of the double-chain form of the enzyme

cathepsin B in lysosomes is executed by cysteine cathepsins such as cathepsins B and L (Mach et al. 1992; Ishidoh and Kominami 2002). However, gene disruption studies have indicated that an unrelated cysteine proteinase, asparaginyl endopeptidase (legumain), plays a role in lysosomal processing of cathepsin B in murine kidney cells (Shirahama-Noda et al. 2003). Apart from internal fragmentation, cathepsin B appears to be also subject of processing by lysosomal exopeptidases (Mach et al. 1993). It has been proposed that cathepsin C accounts for the progressive trimming of three dipeptides from the N-terminus of the enzyme (Rowan et al. 1992), whereas the C-terminal extension of the proteinase can be removed by the dipeptidylcarboxypeptidase activity of cathepsin B itself (Rowan et al. 1993).

4.7.2 Biosynthesis of Cathepsin B in Cancer Cells

Various investigators have reported the elevated synthesis and secretion of cathepsin B in response to viral transformation or malignant dedifferentiation (Sloane et al. 1981; Joyce et al. 2004). In the case of Moloney-murine-sarcoma-virus-transformed mouse fibroblasts, enhanced secretion of procathepsin B was attributed to the lack of functional M6P receptors on the cell surface (Achkar et al. 1990). The biosynthesis and molecular forms of cathepsin B have been studied in various human and rodent carcinoma cell lines. In most cases, no major differences to normal cells were detected (Mach et al. 1992; Braulke et al. 1992). However, unique glycoforms of cathepsin B as well as enhanced secretion of the proenzyme were observed in M6P/IGF2R-deficient SCC-VII murine squamous carcinoma cells. Interestingly, SCC-VII cells lack mature dense lysosomes, with the bulk of intracellular acid hydrolases residing in immature granules with properties reminiscent of late endosomes (Lorenzo et al. 2000). Reconstitution of functional M6P/IGF2R expression in SCC-VII cells was sufficient to restore dense lysosome formation and cathepsin B retention (Probst et al. 2009). In contrast to SCC-VII cells, all other carcinoma cell lines studied so far were found to be M6P/IGF2R-positive. Accordingly, these tumour cells secrete normal amounts of procathepsin B (Mach et al. 1992; Braulke et al. 1992). Interestingly, this also applies to SW1116 human colon carcinoma cells which are unable to internalise M6P-modified proteins despite normal M6P/IGF2R expression (Braulke et al. 1992). These findings suggest that receptor-mediated endocytosis of secreted procathepsin B does not contribute significantly to lysosomal sorting of the enzyme in tumour cells.

4.7.3 Extracellular Forms of Cathepsin B and the Mechanisms of Its Release and Activation

Under normal conditions, lysosomal proteinases occur in the pericellular environment only as their latent precursors. However, enzymatically active extracellular forms of cathepsin B have been found in tumours, either in a soluble state (Poole et al. 1978) or bound to the plasma membrane (Sloane et al. 1986). Possible explanations for these phenomena include autoactivation of secreted latent procathepsin B and/or regulated exocytosis of the mature enzyme from lysosomes. It has been demonstrated that secreted procathepsin B may undergo self-activation triggered by the acidic microenvironment around tumour cells, leading to the transient formation of non-covalent complexes between cathepsin B and its autoinhibitory propeptide and thus stabilisation of the mature enzyme which would be otherwise short-lived in body fluids (Mach et al. 1994b). Interestingly, it has been observed that self-activation of procathepsin B is accelerated in the presence of negatively charged macromolecules like glycosaminoglycan chains, which are major constituents of the extracellular matrix (Mach et al. 1994a). Recently, evidence has been provided that the proregion accounts for the interaction of the cathepsin B precursor with negatively charged molecules as found on cellular surfaces (Caglic et al. 2007). Cathepsin B has been detected at the plasma membrane of transformed breast epithelial cells and colon carcinoma cells (Sloane et al. 1994; Cavallo-Medved et al. 2005). In addition, procathepsin B was found to interact with the small subunit of the peripheral membrane protein annexin II, which could contribute to the presence of cathepsin B on the surface of tumour cells (Mai et al. 2000).

An alternative hypothesis to explain the extracellular occurrence of enzymatically active cathepsin B in tumours relies on the observation that this enzyme is frequently redistributed to peripheral vesicles in tumour cells (Rozhin et al. 1994). It was proposed that this could promote exocytosis of cathepsin B due to retrograde transport from lysosomes along the endocytic pathway followed by fusion of endosomal compartments with the plasma membrane. Such processes also occur under physiological conditions, as for instance in the case of the degranulation of the highly specialised secretory lysosomes of lymphocytes (Blott and Griffiths 2002). However, other cell types are also able to secrete their lysosomal contents. For instance, such a regulated release of cysteine cathepsins seems to occur in the process of the generation of thyroid hormones. It was observed that thyroid epithelial cells secrete mature cathepsin B, thus initiating limited extracellular proteolysis of the hormone precursor thyroglobulin which ultimately culminates in the generation of the thyroid hormone thyroxine (Friedrichs et al. 2003). It was shown that regulated secretion of cathepsin B by thyroid epithelial cells is linked to the redistribution of cathepsin B-containing vesicles from the perinuclear region to the cell periphery, indicating that the enzyme is first delivered to endosomal/ lysosomal compartments and then secreted (Linke et al. 2002).

4.8 Biosynthesis of Cathepsin L

Cathepsin L was initially reported to be 'the most active endopeptidase from rat liver lysosomes acting at pH 6–7' (Kirschke et al. 1977). Later, a protein secreted by various transformed cells (Gottesman 1978) was determined to be procathepsin L (Gal and Gottesman 1986; Joseph et al. 1988), suggesting that the physiological functions of this cathepsin are not limited to general protein turnover within lysosomes. A cDNA cloned from mouse macrophages (Portnoy et al. 1986) was predicted to encode procathepsin L based on alignment with sequences of human liver cathepsin L (Mason et al. 1986). The cDNA of the rat enzyme was sequenced a year later (Ishidoh et al. 1987), and the sequence of the cDNA encoding the human enzyme was reported in 1988 (Gal and Gottesman 1988; Joseph et al. 1988). The first crystal structure of human procathepsin L (Fig. 4.1) was reported in 1996 (Coulombe et al. 1996).

Procathepsin L was initially predicted to reach lysosomes via the classic secretory pathway based on the fact that the enzyme possesses high-mannose *N*-linked oligosaccharides. Thus the presence of a 17-amino acid signal peptide mediating ER import was predicted (Portnoy et al. 1986) and subsequently confirmed experimentally by radiosequence analysis of immunoprecipitated proenzyme (Erickson 1989). Disc electrophoretograms of the purified enzyme first identified multiple forms of cathepsin L in rat liver (Kirschke et al. 1977), suggesting that enzyme maturation requires post-translational proteolysis. This was confirmed and the relationship of the various cathepsin L isoforms was established by pulse-chase analysis of the mouse enzyme (Gal et al. 1985; Portnoy et al. 1986). Preprocathepsin L from all species loses a signal peptide co-translationally, an activation peptide in late endosomes, and undergoes chain cleavage in lysosomes (Fig. 4.5a).

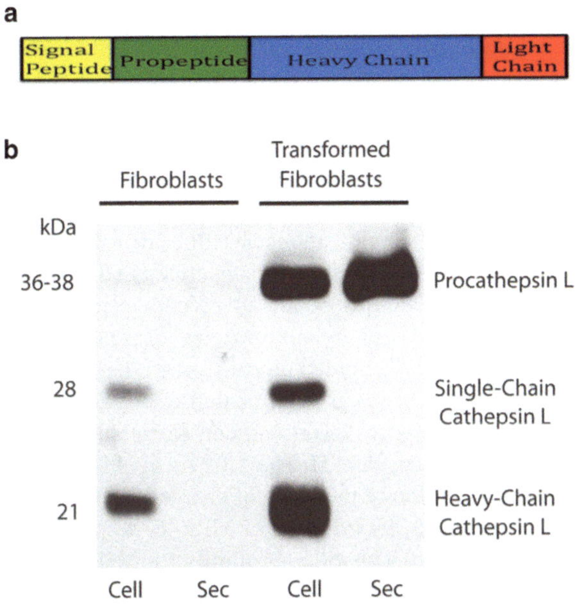

Fig. 4.5 Procathepsin L synthesis, storage and secretion are increased in transformed fibroblasts. (**a**) Diagram of preprocathepsin L. The N-terminal signal peptide is cleaved off co-translationally, while the propeptide is removed to activate the enzyme once the protein reaches late endosomes. In late endosomes or lysosomes single-chain cathepsin L is cleaved into a heavy and a light chain held together by a disulfide bridge. (**b**) These biosynthetic forms of cathepsin L can be visualized in Kirsten virus-transformed KNIH mouse fibroblasts and wild-type mouse fibroblasts incubated in serum-free medium for 2 h. Equal amounts of cellular protein (Cell) and proteins secreted to the culture medium (Sec) were resolved by polyacrylamide gel electrophoresis and visualized by western blotting. Cell transformation results in increased synthesis of cathepsin L, increased secretion of the proenzyme, and cellular storage of the proenzyme. The majority of the cathepsin L protein synthesized by the KNIH cells (46 %) is present in the 2 h culture medium as the 38-kDa proform of the protease. Significant procathepsin L remains in the transformed fibroblasts (23 %), avoiding activation in lysosomes. Figure published in Traffic 1: 724–737 (2000)

Co-translationally, before synthesis of the protein is complete, high-mannose carbohydrate chains are added to one and sometimes two selected asparagine residues of procathepsin L. For the mouse proenzyme, the sugar chain is added at Asn221 (numbering from the first residue of the signal peptide). Mouse Asn268 is normally not modified, but high expression levels resulting in chaperone shortage and thus slow folding, point mutations in the propeptide, or addition of a C-terminal epitope tag can modify the conformation sufficiently to expose this site to oligosaccharyltransferase (Chapman et al. 1997). Glycosylation of cathepsin L is not essential for enzymatic function and has little effect on protein folding or stability (Smith et al. 1989). Procathepsin L that lacks carbohydrate and thus M6P residues is secreted (Kane 1993; Smith et al. 1989). Procathepsin L that acquires two high-mannose carbohydrate chains subsequently undergoes modification with complex sugars, which are detected by increased molecular mass and resistance of the sugar chain to endoglycosidase H (Chapman et al. 1997; Collette et al. 2004a). This form of the protease is also rapidly secreted and is not detectable within cells unless a reagent such as brefeldin A is utilized to block constitutive secretion (Chapman et al. 1997).

Modification of the high-mannose carbohydrate chain(s) on lysosomal proteases with phosphate is critical for segregation out of the secretory pathway. For mouse cathepsin L, the key residues for recognition by phosphotransferase are Lys54 and Lys99, but low levels of M6P modification can still be detected when these residues are ablated, indicating that other lysines can substitute (Cuozzo et al. 1998; Warner et al. 2002). The oligosaccharides bound to procathepsin L are thought to be uniformly processed to diphosphorylated species (Lazzarino and Gabel 1990; Stearns et al. 1990). As correctly folded procathepsin L has only a single carbohydrate chain and as phosphorylated high-mannose oligosaccharides cannot be converted into complex *N*-glycans (Kornfeld and Kornfeld 1985), procathepsin L remains sensitive to endoglycosidase H. The M6P recognition marker generated by phosphotransferase recognition of the patch of charged lysine residues is bound by one of the two M6P receptors, M6P/IGF2R or MPR46. As procathepsin L has only a single phosphorylated oligosaccharide, its affinity for M6P/IGF2R is low relative to other lysosomal enzymes (Dong and Sahagian 1990). Additionally, uncharacterized protein determinants are thought to impair its interaction with M6P/IGF2R (Lazzarino and Gabel 1990).

4.8.1 Proteolytic Processing and Activation of Procathepsin L

Most cysteine cathepsin precursors contain an autoinhibitory propeptide to avoid premature enzymatic activity during intracellular targeting. As for other cathepsins, it is believed that proteolytic removal of the cathepsin L propeptide/activation peptide only occurs once the protein reaches the acidic endolysosomal membrane system. Pulse-chase analysis in mouse fibroblasts (Gal et al. 1985) and mouse macrophages (Portnoy et al. 1986) confirmed that a propeptide is removed from a 36–38 kDa proenzyme precursor (Fig. 4.5b). The site of the cleavage that releases the 96-amino-acid mouse macrophage propeptide was determined by radiosequencing of immunoprecipitated protein (Erickson 1989). The cleavage site can vary, however, suggesting that either multiple sites are utilized and/or that exopeptidase cleavage occurs. Intermediates have been detected (Ishidoh and Kominami 1994; Ishidoh et al. 1998; Ménard et al. 1998; Salminen and Gottesman 1990), but these are not necessarily fully proteolytically active (Ishidoh and Kominami 2002). Structural studies of the procathepsin L propeptide show that the N-terminal portion of the propeptide occludes the active site (Carmona et al. 1996), while the C-terminal portion serves as a chaperone that initiates correct folding of the protein within the ER (Schilling et al. 2001; Tao et al. 1994). The propeptide avoids degradation, although positioned in the active site, because it lies in the pocket in opposite orientation to cleavable substrates (Coulombe et al. 1996).

Procathepsin L can autoactivate *in vitro* (Mason et al. 1987; Ménard et al. 1998) and *in vivo* (Nomura and Fujisawa 1997; Salminen and Gottesman 1990) if the pH drops below 5. Self-activation can also occur at higher pH (5.5–6.0) if the proenzyme is in the presence of a negatively charged surface, such as that presented by dextran sulfate (Mason and Massey 1992) or glycosaminoglycans, e.g. those of heparan sulfate or chondroitin sulfate proteoglycans (Ishidoh and Kominami 1995; Kihara et al. 2002). Cathepsin D has also been shown to be capable of activating procathepsin L (Nishimura et al. 1989; Wiederanders and Kirschke 1989). The activation of procathepsin L is a regulated event; thus the amount of proenzyme in cells varies with the cell type and its physiological state (Fig. 4.5b). While the proenzyme is barely detectable in normal fibroblasts, it is commonly the major form of cathepsin L in transformed fibroblasts (Ahn et al. 2002; Collette et al. 2004a). Thus, measuring enzyme activity alone does not reveal the total amount of the protease in cells, while assaying mRNA levels alone does not provide a reliable measure of active cathepsin L levels present in cells or tissues.

Once the enzyme reaches acidic compartments, additional proteolysis leads to generation of light and heavy cathepsin L chains. As initially detected by pulse-chase analysis, conversion of the mouse 36–38 kDa proform to a 28-kDa single-chain enzyme is followed by cleavage to a 21-kDa heavy chain derived from the N-terminus of the single chain and a 6–7 kDa light chain derived from the C-terminus of the single chain (Fig. 4.5a). The small light chain is not routinely detected on standard polyacrylamide gels (Gal et al. 1985; Portnoy et al. 1986). The sites of cleavage in human and rat cathepsin L have been determined by sequencing of the purified double-chain enzymes, which revealed that 2–3 amino acids are missing at the light-heavy chain boundary due to exopeptidase cleavage (Mason et al. 1986; Ishidoh et al. 1987; Towatari and Katunuma 1988; Ritonja et al. 1988). Recently, asparaginyl endopeptidase has been implicated in double-chain cathepsin L formation as mouse cells lacking the former enzyme accumulate the single-chain form (Shirahama-Noda et al. 2003; Maehr et al. 2005). However, inhibitor studies have also provided evidence for the involvement of cysteine cathepsins, including cathepsin L itself, in this process (Hara et al. 1988; Salminen and Gottesman 1990; Nishimura et al. 1995; Ahn et al. 2002). The two chains remain connected through a disulfide bridge. Active-site labelling experiments established that both the single and the double-chain forms of the enzyme are proteolytically active (Mason et al. 1989). Like removal of the activation peptide, the efficiency of this secondary cleavage depends on the cell type. The amount of heavy chain detected in cell extracts relative to the single-chain form is thus variable (Erickson 1989).

4.8.2 Non-lysosomal Localization of Cathepsin L

Cathepsin L activity has been detected at multiple intra- and extracellular sites, suggesting the protease has specific physiological functions in addition to general protein turnover in lysosomes. In certain cell types, enzyme activity is found in endosomes, in secretion granules and/or secretory lysosomes, in the cytoplasm, and in the nucleus. Thus, targeting mechanisms must exist to mediate transport to sites other than lysosomes.

An initial indication that procathepsin L can function extracellularly, contributing to cancer cell metastasis, came from the observation that transformed mouse fibroblasts secrete large quantities of a protein (Fig. 4.5b) initially named the 'major excreted polypeptide' or MEP (Gottesman 1978). Expression of this protein is induced by oncogenic Ras (Joseph et al. 1987). MEP was shown to be modified with M6P moieties (Sahagian and Gottesman 1982), localized to lysosomes (Gal et al. 1985), and eventually found to be identical to procathepsin L (Joseph et al. 1987; Mason et al. 1987; Troen et al. 1987). Support for the idea that secreted cathepsin L might serve physiological functions extracellularly came from the demonstration that MEP could degrade extracellular matrix (Gal and Gottesman 1986), and that purified enzyme could efficiently degrade kidney glomerular basement membrane (Baricos et al. 1988). Additional evidence that cathepsin L can serve a physiological function outside cells comes, for example, from demonstration that secreted cathepsin L can liberate thyroid hormone (Brix et al. 1996) and from the extensive studies of cathepsin L-like proteases secreted by parasites (Robinson et al. 2008).

Most lysosomal proteases are constitutively secreted to some minor extent, presumably because of failure to bind to M6P receptors in the TGN, but relative to procathepsin L, other endogenous lysosomal enzymes are efficiently targeted to lysosomes in transformed fibroblasts. Thus this secretion is selective for procathepsin L. The upregulation of gene expression that is characteristic of transformation (Ishidoh and Kominami 1998) would lead to increased synthesis of protein that could swamp M6P receptors, causing the excess proenzyme to be secreted by the constitutive secretory pathway. This, however, should concomitantly result in increased secretion of other lysosomal enzymes synthesized at lower levels, which must compete for binding to M6P receptors; surprisingly, this is not detected. Consistent with this, while ectopic expression of Ras leads to increased expression of procathepsin L in both fibroblasts and epithelial cells, high secretion is only detected in fibroblasts, which secrete 50 % of their total cathepsin L, compared to 16 % for epithelial cells (Collette et al. 2004b). Recently, procathepsin L secretion by human melanoma cells has been correlated with Rab4 expression and/or function (Barbarin and Frade 2011). However, the exact nature of the molecular mechanisms underlying these unique properties of procathepsin L within the secretory pathway remains unknown.

The primary form of cathepsin L detected outside transformed fibroblasts is the proteolytically inactive proenzyme, leading to questions as to the physiological relevance of the secreted protease, as the extracellular pH is normally assumed to be neutral and cathepsin L has been reported to be the most unstable of the lysosomal cysteine proteases at neutral or alkaline pH (Turk et al. 1993). The demonstration that extracellular matrix-like molecules rich in negative charge can induce self-activation provides a mechanism for activation of the secreted protease (Ishidoh and Kominami 1995; Kihara et al. 2002). Additionally, numerous studies have established that the pericellular pH of tumour cells can be sufficiently low to enable cathepsins to be active (Stubbs et al. 2000). Finally, the activity of cathepsin L has been shown to be modulated by the extracellular matrix proteoglycan testican-1 (Bocock et al. 2003) and by purified thyroglobulin type-1 domain (Meh et al. 2005) found not only in secreted proteins such as thyroglobulin (Brix et al. 1996) and testican-1, but also in the p41 isoform of the invariant chain associated with MHCII molecules (Bevec et al. 1996; Hitzel et al. 2000). The p41 protein not only stabilizes mature cathepsin L in endocytic compartments of antigen-processing endosomes (Turk et al. 1999), but like testican-1 (Bocock et al. 2003), stabilizes active cathepsin L in the neutral extracellular environment, possibly potentiating its role in the inflammatory response (Fiebiger et al. 2002). Together these observations argue that the secreted proenzyme can impact extracellular events.

Active, mature cathepsin L can also be detected in cell culture medium under certain physiological conditions. Activated thioglycollate-elicited mouse macrophages abundantly secrete single-chain and double-chain forms of cathepsin L, in contrast to macrophages resident in the peritoneum, which primarily secrete proenzyme (Erickson 1989; Collette et al. 2004a). The cellular pathway mediating secretion of active cathepsin L has not been elucidated but may relate to the recent finding that almost all proteins in human primary macrophages lack M6P and thus utilize an M6P-independent pathway for targeting proteases to lysosomes (Pohl et al. 2010).

4.8.3 M6P-Independent Intracellular Transport of Cathepsin L

Early evidence that cathepsin L could be packaged in regulated secretory vesicles came from the detection of the protease in sperm acrosomes (McDonald and Kadkhodayan 1988) and in melanosomes (Diment et al. 1995). Cathepsin L has more recently been documented to cleave perforin in cytotoxic granules of natural killer cells and cytotoxic T lymphocytes (Konjar et al. 2010) and to participate in neuropeptide production in neuroendocrine cells (Yasothornsrikul et al. 2003). Incorporation into vesicles capable of regulated secretion allows directed delivery to specific sites on the membrane, such as an immunological synapse (Griffiths et al. 2010), while constitutive secretion from the TGN would deliver the protease to basolateral surfaces. These observations suggest that, at least in certain cell types, a mechanism exists to target procathepsin L to secretory granules, as well as to lysosomes.

Procathepsin L has been detected in dense-core vesicles of transformed mouse fibroblasts (Yeyeodu et al. 2000; Ahn et al. 2002), suggesting that the protease can be stored in regulated secretory vesicles even in a non-secretory cell type (Fig. 4.6). The use of an antibody specific for the propeptide established that it is the latent proenzyme that is concentrated in these vesicles, explaining the stability of the large amount of proenzyme detectable in these cells under steady-state conditions (Yeyeodu et al. 2000; Ahn et al. 2002), in contrast to untransformed fibroblasts such as NIH3T3 or L fibroblasts which contain primarily mature single-chain and double-chain cathepsin L (Fig. 4.5b). Colocalization with the transmembrane protein CD63 (Ahn et al. 2002) indicates that this site of storage is a multivesicular endosome, an endosome that can undergo fusion with the plasma membrane in response to signaling (Piper and Katzmann 2007). This is consistent with the recent demonstration that fibroblasts contain specialized endosomes which fuse with the plasma membrane in response to changes in cytosolic calcium levels (Laulagnier et al. 2011).

Procathepsin L has been observed to undergo self-association in yeast two-hybrid assays (Yeyeodu et al. 2000), suggesting that high synthesis levels could lead to formation of procathepsin L aggregates, possibly as early as in the ER. These aggregates might react less efficiently with M6P receptors than the monomeric proenzyme. The low affinity of procathepsin L for M6P/IGF2R (Dong and Sahagian 1990; Lazzarino and Gabel 1990) may also contribute to storage granule targeting. Alternatively, an unidentified alternate targeting receptor which could mediate the transport of these proteases to secretory vesicles could be selectively expressed or upregulated upon cell transformation.

The existence of M6P-independent or alternate targeting pathways for lysosomal proteins has been suggested since the early observation that I-cell hepatocytes, Kupffer cells and leukocytes that lack phosphotransferase activity possess functional lysosomes (Owada and Neufeld 1982; Waheed et al. 1982). Sortilin has recently been reported to bind several lysosomal enzymes, explaining their M6P-independent targeting to lysosomes in I-cell fibroblasts, but cathepsin L reaches lysosomes in these cells without binding sortilin (Canuel et al. 2009), suggesting the existence of additional, yet unidentified targeting receptor(s). Procathepsin L, but not the active protease, has been observed to undergo pH-dependent M6P-independent association with a 43-kDa mouse fibroblast membrane protein (McIntyre and Erickson 1991, 1993), an interaction inhibited by a 9-residue fragment of the N-terminal activation peptide (McIntyre et al. 1994). Procathepsin D undergoes similar M6P-independent membrane association in these and other cells (McIntyre and Erickson 1991; Zhu and Conner 1994; Godbold et al. 1998). A receptor mediating this interaction has not been isolated so far. Precedence for M6P-independent, concentration-dependent sorting of proteases to storage vesicles comes from studies of plant cells that contain protein storage vacuoles, which likewise store enzymes in dense cores comprised of internal membranes (Wang et al. 2011). Protein targeting to these vacuoles is mediated by 43-kDa homology-transmembrane RING-H2 (RMR) proteins (Jiang et al. 2000),

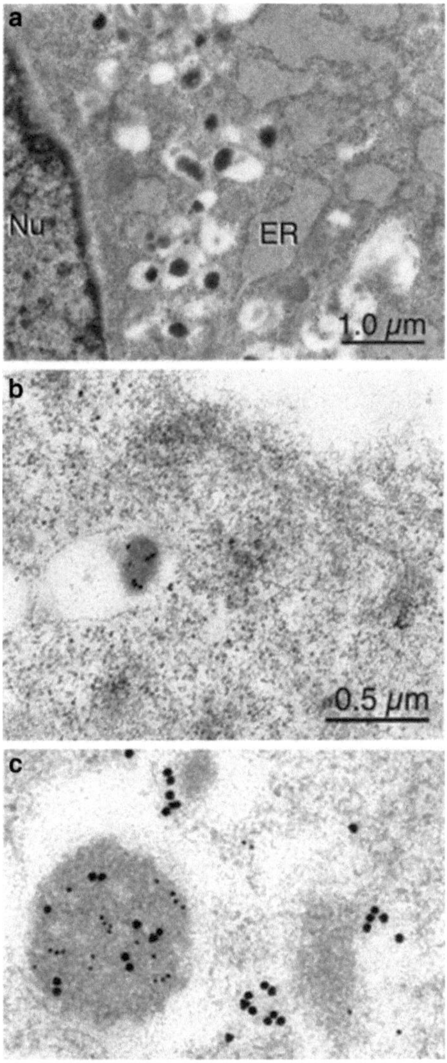

Fig. 4.6 Procathepsin L of transformed fibroblasts is stored in dense cores of multivesicular endosomes. Procathepsin L was localized in Kirsten virus-transformed KNIH mouse fibroblasts (Panel **a**) using polyclonal rabbit antibodies specific for the 96-amino acid propeptide of mouse procathepsin L detected with goat anti-rabbit Fab fragments conjugated with 15-nm gold (Panel **b**). In Panel (**c**), the multivesicular endosome marker protein CD63 was colocalized in the same section using guinea pig anti-rat CD63 antibodies that were detected with biotinylated goat anti-guinea pig IgG, followed by avidin conjugated to 20-nm gold. Procathepsin L was detected with antibodies specific for the propeptide followed by goat anti-rabbit IgG conjugated with 10-nm gold. Figure published in Traffic 3: 147–159 (2002). *ER* endoplasmic reticulum, *Nu* nucleus

integral endosomal membrane PA-TM-RING proteins (Erickson 2011) which have an N-terminal protease-associated (PA) domain that recognizes C-terminal sequences on ligands (Wang et al. 2011).

4.9 Biosynthesis of Other Cysteine Cathepsins

Cathepsin H is one of the two other cysteine cathepsins whose biosynthesis and intracellular transport has been thoroughly characterized. This enzyme is synthesized in rat hepatocytes and rat macrophages as a 41-kDa proenzyme.

The latent precursor is rapidly converted into the mature single-chain form of the proteinase (28 kDa). A double-chain form of cathepsin H composed of an N-terminal large subunit (22 kDa) and a C-terminal light chain (6 kDa) is generated in macrophages (Kominami et al. 1988), while single-chain cathepsin H persists in hepatocytes even upon prolonged incubation (Nishimura et al. 1988). Generation of the double-chain form can be inhibited by treatment with the cysteine cathepsin inhibitor E-64d (Hara et al. 1988). Interestingly, genetic ablation of asparaginyl endopeptidase expression exerts a similar inhibitory effect on cathepsin H processing in mice (Shirahama-Noda et al. 2003). In contrast to cathepsin B, mature cathepsin H is not subject of N- and C-terminal trimming by exopeptidases. A glycosylated octapeptide derived from the prosegment is linked to the mature enzyme through a disulfide bridge, binding to the active-site cleft of the proteinase in a substrate-like manner. Cathepsin H is also *N*-glycosylated within its catalytic domain, carrying high-mannose *N*-glycans as typical for soluble lysosomal proteins. Evidence has been provided that the presence of these oligosaccharides is required for the delivery of the enzyme to lysosomes (Nishimura et al. 1988). Recently, it has been shown that cathepsin H binds to M6P/IGF2R in an M6P-dependent manner (Sleat et al. 2006). These findings indicate that cathepsin H can be transported to lysosomes via the M6P receptor pathway. However, a recent study has highlighted that lysosomal targeting of cathepsin H could also involve interactions with alternate sorting receptors such as sortilin (Canuel et al. 2008).

The biosynthesis and intracellular transport of cathepsin C has been studied in rat hepatocytes and hepatoma cells. This oligomeric enzyme is first synthesized as a precursor with an apparent molecular mass of 55 kDa. Procathepsin C is *N*-glycosylated and carries the M6P recognition marker in its carbohydrate moiety. Transport to lysosomes has been shown to depend on its interaction with M6P receptors (Mainferme et al. 1985; Muno et al. 1993). Upon delivery to lysosomes, the cathepsin C precursor is proteolytically processed into 25-kDa and 8-kDa fragments, corresponding to the heavy and light chains of the mature enzyme (Ishidoh et al. 1991). In contrast to cathepsins B and L, procathepsin C maturation is not affected by treatment of the cells with the cysteine proteinase inhibitor Z-Phe-Ala-CHN_2 (Mainferme et al. 1985).

Comparatively little information is available on the biosynthesis and intracellular transport of other cysteine cathepsins. Evidence has been provided that procathepsins F, K, S and X are modified with M6P residues (Czupalla et al. 2006), and their mature counterparts have been localized in lysosomes (Wiederanders et al. 1992; Tepel et al. 2000; Kos et al. 2005; Tang et al. 2006; van Meel et al. 2011). None of these cathepsins seems to undergo processing into a double-chain species. Two variants of mature cathepsin V have been detected in lysosomal fractions of human thyroid carcinoma cells (Tedelind et al. 2010), which suggests intralysosomal conversion of the single-chain enzyme into a double-chain form as in the case of its close relative cathepsin L. Opposite to all other cysteine cathepsins, the precursor of cathepsin W is retained in the endoplasmic reticulum and appears to be resistant to proteolytic maturation (Wex et al. 2001; Ondr and Pham 2004).

4.10 Biosynthesis and Intracellular Transport of Cathepsin D

The major non-cysteine cathepsin is the aspartyl proteinase cathepsin D. The open reading frame of human cathepsin D cDNA (Faust et al. 1985) encodes a polypeptide of 412 amino acids with N-terminal pre- and prosegments of 20 and 44 residues, as judged by comparison with the N-terminal sequences of the biosynthetic precursors of porcine cathepsin D (Erickson and Blobel 1979; Erickson et al. 1981). From a structural point of view, mature human cathepsin D exists as a single-chain protein, which is relatively more abundant in endosomes, and as a double-chain form, which is relatively more abundant in lysosomes (Hasilik and Neufeld 1980; Follo et al. 2007). The initial product of its biosynthesis is preprocathepsin D, which upon entry into the lumen of the rough ER loses its signal peptide, leading to the formation of procathepsin D consisting of 392 amino acids. While synthesis is still ongoing, procathepsin D is *N*-glycosylated on both Asn134 and Asn263, belonging to the N-terminal and C-terminal lobes, respectively, in the double-chain mature polypeptide (Faust et al. 1985). These Asn residues are located in a peptide region specifically involved in the recognition of cathepsin D by phosphotransferase (Baranski et al. 1990, 1991, 1992; Cantor et al. 1992; Dustin et al. 1995) and thus can be decorated with *N*-linked oligosaccharides harbouring M6P residues. The last step of M6P biosynthesis occurs in a compartment beyond the site of action of brefeldin A (Radons et al. 1990), a fungal antibiotic that disrupts the organization of the Golgi stacks and causes retrograde transport of Golgi proteins to the ER (Lippincott-Schwartz et al. 1989). The facts that ammonium chloride impairs the uncovering of phosphorylated procathepsin D and stimulates the secretion of procathepsin D bearing masked M6P residues (Isidoro et al. 1990) further support the view that the two reactions are spatially separated, in agreement with the different subcellular localizations of the two enzymes involved in the generation of the M6P recognition marker (Rohrer and Kornfeld 2001; Tiede et al. 2005).

The molecular weight of doubly glycosylated procathepsin D is about 53 kDa. Procathepsin D is then entrapped within vesicles budding from the TGN and transported to the endosomal compartments by M6P-dependent and M6P-independent routes. In the TGN, M6P receptors sequester procathepsin D into transport vesicles that travel along microtubules and then fuse with pre-lysosomal organelles, whereupon procathepsin D is discharged and activated due to the mild acidic milieu of these compartments. Both M6P/IGF2R and MPR46 are involved in this transport (Braulke and Bonifacino 2009), though the former receptor exhibits a greater affinity for procathepsin D (Pohlmann et al. 1995).

Nonetheless, procathepsin D devoid of M6P residues may reach endosomal compartments (Glickman and Kornfeld 1993) by an M6P-independent route that possibly involves the transient interaction with the lysosomal protein prosaposin (Zhu and Conner 1994; Gopalakrishnan et al. 2004). Very recently, it has been shown that transport of procathepsin D to lysosomes at least partially depends on its interaction with sortilin (Canuel et al. 2008).

Once procathepsin D is released into the lumen of late endosomes, the propeptide segment of 44 amino acids is proteolytically removed and an enzymatically active, single-chain enzyme is formed (Fig. 4.7). This molecular form has a molecular weight of about 48 kDa and consists of 348 amino acids. The removal of the propeptide is not purely an autoproteolytic process, as it occurs also in a mutant form (D295N) of human procathepsin D in which the active site of the enzyme has been inactivated. Rather, cysteine cathepsins such as cathepsins B and L are probably involved in this processing event (Laurent-Matha et al. 2006). The predominant molecular form of human cathepsin D is in fact the mature double-chain protein. It is composed of an N-terminal light chain of 14 kDa and a C-terminal heavy chain of 34 kDa (Hasilik and Neufeld 1980; Gieselmann et al. 1983). The light and heavy chains of cathepsin D are not covalently linked by inter-chain disulfide bonds, as previously believed; rather, their association arises from hydrophobic and steric interactions between the two chains, which keep their proper conformations due to intra-chain disulfide bridges. The conversion of single-chain cathepsin D into the double-chain form is a lysosomal event and depends on the activity of cysteine cathepsins (Gieselmann et al. 1985; Samarel et al. 1989). Cathepsin D accumulating in lysosomes undergoes progressive C-terminal trimming that removes the two residues Leu412 and Arg411, and probably others, from the heavy chain. The trimming at the heavy-chain C-terminus was first shown for the double-chain form of porcine cathepsin D (Erickson and Blobel 1983). It has been proposed that this late processing initiates the turnover of cathepsin D (Erickson 1989).

4.10.1 Biosynthesis of Rodent Cathepsin D

Purified rat liver cathepsin D is a mixture of a single-chain form (approximately 95 %) and two different double-chain proteins. The rat cathepsin D cDNA encodes a polypeptide containing a 20-residue signal sequence followed by a 44-residue propeptide (Fujita et al. 1991). A plethora of molecular forms of rat cathepsin D, comprising the 53 kDa precursor, the 47 kDa intermediate, the 43 kDa mature single-chain proteinase, two 34/30 kDa heavy chains and two 9/14 kDa light chains has been described in rat basophilic leukemia cells (Dragonetti et al. 2000). This study confirmed that the most abundant rat cathepsin D form is the mature single-chain enzyme, and revealed that the predominant double-chain isoform is composed of the 34-kDa and 9-kDa chains.

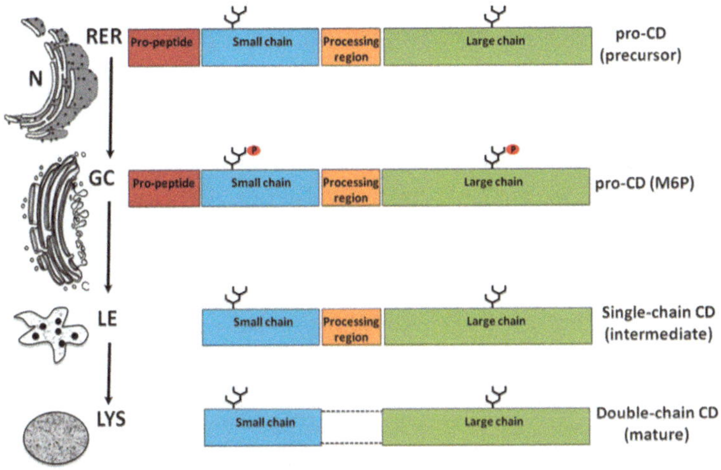

Fig. 4.7 Proteolytic maturation of cathepsin D. The key steps in the proteolytic maturation of human cathepsin D (CD) are illustrated. The intralysosomal generation of the double-chain form of the enzyme does not occur in all species. *GC* Golgi complex, *LE* late endosome, *LYS* lysosome, *N* nucleus, *RER* rough endoplasmic reticulum

The complete cDNA sequence coding for mouse cathepsin D has also been reported (Grusby et al. 1990). The molecular forms of cathepsin D found in mouse 3T3 fibroblasts include the precursor (with a molecular weight of about 52 kDa) and mature polypeptides mostly represented by a 48 kDa single-chain isoform and, to a minor extent, by a double-chain protein composed of 31-kDa and 16-kDa heavy and light chains (Isidoro et al. 1995). Rat and mouse cathepsin D both contain two potential *N*-glycosylation sites, one in each chain.

4.10.2 Biosynthesis and Molecular Forms of Cathepsin D in Non-mammalian Species

Chicken (*Gallus gallus*) cathepsin D has been purified from oocytic yolk, pre-ovulatory follicle and liver homogenates by affinity chromatography. Based on peptide sequencing of the N-terminus, a full-length cDNA clone encoding chicken cathepsin D was isolated from a chicken follicle cDNA library. It is predicted that chicken cathepsin D exists as a double-chain isoform (Retzek et al. 1992). In 1997, Gerhartz and colleagues purified and characterized cathepsin D from the yolk-sac membrane of quail (*Coturnix coturnix japonica*) eggs. The molecular forms of mature quail cathepsin D include a double-chain isoform composed of a light chain of 14 kDa and a heavy chain of 30 kDa, which is the most abundant molecular species, and a single-chain isoform of about 44 kDa

(Gerhartz et al. 1997). The entire amino-acid sequence of quail cathepsin D is not known, and its cDNA has not been cloned yet.

The molecular features of cathepsin D isolated from *Chionodraco hamatus* (Antarctic icefish) and *Clupaea harengus* (Atlantic herring) are known. The mature enzyme purified from the liver of Antarctic icefish shows a molecular weight of 40 kDa (Capasso et al. 1999), whereas that isolated from herring muscle shows a molecular weight of 38–39 kDa (Nielsen and Nielsen 2001). In both fishes cathepsin D has been found to occur exclusively as a single-chain isoform. The N-termini of icefish and herring cathepsin D have been determined by automated Edman degradation (Capasso et al. 1999; Nielsen and Nielsen 2001). No data are available about the glycosylation status of icefish cathepsin D. Glycosidase treatments (with endoglycosidase H or *N*-glycosidase F) and binding to the lectin concanavalin A indicate that herring cathepsin D contains only one *N*-linked carbohydrate moiety of the high-mannose type (Nielsen and Nielsen 2001). Very recently, the biosynthesis of *Danio rerio* (zebrafish) cathepsin D has been characterized. The protein is synthesized as a mono-glycosylated precursor of 43 kDa that is transported in an M6P-dependent manner to endosomal-lysosomal compartments where it matures and accumulates as a single-chain peptide of 41 kDa (Follo et al. 2011).

Xenopus laevis is a frog commonly used as a model organism for studies on embryonic development. Mature cathepsin D of *Xenopus laevis* shows different isoforms, depending on the examined tissue: in fact, two different isoforms of 43 and 36 kDa can be isolated from ovarian extracts, while only the 36-kDa isoform is present in liver extracts (Nakamura et al. 1996).

Aedes aegypti is a mosquito that can spread the Dengue and Yellow fever viruses. A cDNA coding for a mosquito cathepsin D-like lysosomal aspartic proteinase was cloned and sequenced. The cDNA encodes a polypeptide of 387 amino acids starting with an 18-residue signal sequence and a 35-residue pro-segment. Mosquito cathepsin D purified by pepstatin A-agarose displayed an apparent molecular weight of 80 kDa or 40 kDa under non-denaturing or denaturing conditions, respectively. Two-dimensional gel electrophoresis revealed a predominant spot of 40 kDa with an isoelectric point of 5.4. The mature enzyme is a single-chain polypeptide with only one *N*-glycosylation site in the N-terminal part of the protein (Cho and Raikhel 1992). Cathepsin D cDNA from the silkworm *Bombyx mori* was cloned in 2006 (Gui et al. 2006) and encodes a single-chain protein of 385 amino acids bearing one potential *N*-glycosylation site, with an apparent molecular weight of 40–44 kDa. *B. mori* cathepsin D contains the two active-site aspartic acid residues as well as the six cysteine residues characteristic of aspartic proteinases, and displays substantial amino-acid identity with the aspartic proteinases of the mosquitoes *Anopheles gambiae* (64 % identity) and *Aedes aegypti* (63 % identity). Three cDNAs coding for preprocathepsin D-like proteinases have been cloned from *Musca domestica*, but these await further characterization on the protein level (Padilha et al. 2009).

Caenorhabditis elegans is a small unsegmented nematode. Cathepsin D purified by affinity chromatography from this worm is a mixture of several enzymatically active single-chain isoforms, which differ in their glycosylation status, with an apparent molecular weight of 36 and 34 kDa (the most abundant forms) and of 32, 38 and 44 kDa (Jacobson et al. 1988). Cathepsin D has been also isolated from the trematode *Schistosoma japonicum* (commonly known as blood fluke). The precursor and the active single-chain enzyme showed an apparent molecular weight of 47 kDa and 41 kDa, respectively (Verity et al. 1999).

The slime mold *Dictyostelium discoideum* is a eukaryotic amoeba phylogenetically very distant from man. Slime mold cathepsin D, Ddp 44, was first identified in 1999 by peptide sequence analysis of soluble endosomal/lysosomal proteins. *D. discoideum* preprocathepsin D is composed of 383 amino acids. The short signal sequence (18 amino acids) is co-translationally removed upon translocation into the ER lumen, while the propeptide segment (from position 19 to 48) is removed later in endosomal compartments. Mature *D. discoideum* cathepsin D is a single-chain enzyme that shows a molecular weight of about 44 kDa. It mainly presents with only one *N*-linked oligosaccharide (on Asn118), although two other potential *N*-glycosylation sites (Asn238 and Asn310) are present in the protein (Journet et al. 1999).

4.10.3 Maturation of Cathepsin D

The initial biogenetic events leading to the biosynthesis of *N*-glycosylated procathepsin D and its transport from the TGN to endosomes are substantially similar in all mammals, whereas the processing of procathepsin D into the mature enzyme(s) may present species-specific peculiarities in the generation of the single-chain and of the double-chain isoforms. Thus, in human and porcine cells, the single-chain polypeptide is rapidly processed and the double-chain enzyme is the most predominant molecular form of mature cathepsin D (Hasilik and Neufeld 1980; Erickson et al. 1981). The most abundant isoform of purified porcine cathepsin D is a double-chain protein formed by a 15-kDa light chain and a 31-kDa heavy chain, although a 43-kDa single-chain intermediate can also be isolated (Barth and Afting 1984). It was shown that porcine double-chain cathepsin D originated from the latter through the removal of five amino acids (Yonezawa et al. 1988). By contrast, in ovine, rat, hamster and mouse cells cathepsin D accumulates mainly as a single-chain isoform (Fujita et al. 1991; Isidoro et al. 1991, 1995; Tyynelä et al. 2000). In these mammalian species, the processing of the single-chain isoform into the double-chain isoform proceeds very slowly and, as a consequence, at steady state the latter represents less than 10–15 % of the whole cathepsin D isolated from acidic compartments. Of note, rodent cells possess the enzymatic machinery for the single- to double-chain processing of cathepsin D. In fact, human cathepsin D heterologously expressed in hamster cells is efficiently maturated into the double-chain isoform (Conner et al. 1989; Isidoro et al. 1991).

4.10.4 Mechanism and Physiological Significance of the Single- to Double-Chain Processing of Cathepsin D

Based on the observation that single-chain bovine cathepsin D converts into the double-chain form upon incubation at pH 3.5, it was suggested that the active intermediate molecule can process itself through an autocatalytic cleavage once it reaches the lysosome (Lah and Turk 1982). A more recent study proposes that human cathepsin D is processed into the double-chain form by cathepsins B and/or L (Laurent-Matha et al. 2006), which is consistent with earlier studies employing different cysteine cathepsin inhibitors (Gieselmann et al. 1985; Samarel et al. 1989). However, other cysteine cathepsins might also be involved in this process since cathepsin B/L double-knockout mice still contain normal levels of mature double-chain cathepsin D (Felbor et al. 2002; Stahl et al. 2007). Whether this processing occurs in endosomes before the single-chain intermediate is translocated into lysosomes or within the lysosome soon after the intermediate reaches this compartment is not yet fully clarified (Hasilik 1992). Recently, in an attempt to clarify the molecular mechanism and the nature of the acidic compartment(s) involved in this maturation step, a human cathepsin D mutant was studied in which the beta-hairpin loop excised during double-chain processing had been deleted (Follo et al. 2007). While wild-type human cathepsin D was efficiently maturated, thus establishing the lysosomal processing capacity of the recipient mouse cells, the cathepsin D mutant was not converted into the mature double-chain form although it reached the lysosome (Follo et al. 2007). These data argue against a quality control mechanism that would impair the transport of unprocessed intermediate cathepsin D from endosomes to lysosomes, and are rather compatible with the view that the last maturation step occurs within lysosomes.

What about the physiological relevance of the single- to double-chain processing? It can be hypothesized that these two cathepsin D variants have different substrate specificity. In general, single-chain cathepsin D seems more prone to act on small substrates and to be active at neutral or slightly acidic pH. Endosomal single-chain rabbit cathepsin D was shown to be able to process parathyroid hormone into bioactive peptides (Diment et al. 1989). Rat single-chain cathepsin D secreted at the basolateral site of mammary acini was shown to be able to mediate, at neutral pH, the cleavage of the hormone prolactin into bioactive peptides (Lkhider et al. 2004). During apoptosis associated with lysosomal membrane disruption, endosomal/lysosomal cathepsin D translocates into the cytosol and cleaves a so-far unknown substrate followed by activation of Bax (Bidère et al. 2003; Castino et al. 2007). *In vitro* cathepsin D activity is classically assayed using bovine haemoglobin, a rather complex substrate, at very acidic pH. Single- and double-chain cathepsin D isolated from bovine spleen were shown to possess different activity in such an assay, the double-chain form being more active towards haemoglobin at pH 3.5 (Lah et al. 1984). Similarly, Tanji et al. (1991) demonstrated that two different isoenzymes of cathepsin D purified from skeletal muscle of Japanese monkey differ in their specificity towards oxidized insulin β-chain.

Human double-chain cathepsin D also shows a higher affinity towards haemoglobin than its single-chain counterpart (Kawada et al. 1997). It is interesting to note that these authors found that single-chain cathepsin D was the most abundant form of the enzyme in epithelial cells of healthy skin, whereas the double-chain form was the major cathepsin D variant present in psoriatic skin biopsies. This observation would indicate that the extent of the cathepsin D maturation process likely depends on the tissue, the microenvironment and on particular pathophysiological conditions.

4.10.5 Structure and pH-Dependent Activation of Cathepsin D

In vivo, procathepsin D is activated by acid-dependent proteolytic removal of the 44-residue propeptide culminating in generation of an active single-chain intermediate (see above). However, at least *in vitro*, procathepsin D can undergo another type of acid-dependent proteolytic activation that leads to 'pseudo'-cathepsin D which misses only the first 26 amino acids (Conner and Richo 1992). Pseudo-cathepsin D is enzymatically active (Gopalakrishnan et al. 2004) and retains the last 18 residues of the prosegment (amino acids 27–44). Interestingly, this residual fragment of the propeptide has been claimed to possess mitogenic properties (Vetvicka et al. 1998). Whether generation of pseudo-cathepsin D also occurs during pro-cathepsin D maturation *in vivo* is still debated (Richo and Conner 1994; Wittlin et al. 1999). It is also unclear whether this process is purely autocatalytic (Wittlin et al. 1999; Laurent-Matha et al. 2006). Although complete removal of the propeptide seems obligatory for cathepsin D to gain full enzymatic activity, *in vitro* studies indicated that at acid pH the propeptide does not occupy the active site, implying that procathepsin D could be active under these conditions (Wittlin et al. 1998). A better understanding of the pH-dependent mechanism of cathepsin D activation has been made possible by the elucidation of the three-dimensional structures of native and pepstatin-inhibited cathepsin D in combination with modelling studies (Baldwin et al. 1993; Metcalf and Fusek 1993; Lee et al. 1998; Goldfarb et al. 2005). By comparing the crystal structures obtained at pH 5.1 and 7.5, it was observed that an N-terminal segment (residues 3–7) relocated into the active site at neutral pH, thus explaining its inaccessibility for substrates or inhibitors at pH above 7.0 (Lee et al. 1998). A recent study proposes that the pH-dependent conformational change occurring in the molecule is initiated by charge repulsion between the two carboxylates of Glu180 and Asp187 (Goldfarb et al. 2005). Such a pH-dependent conformational switching may have physiological relevance for general and restricted cathepsin D-mediated proteolysis in terms of substrate specificity, compartmentalized activity, as well as for procathepsin D function as mitogenic ligand (Beyer and Dunn 1996; Berchem et al. 2002).

4.10.6 Cathepsin D in Biological Fluids and Pathological Conditions

Environmental stimuli (e.g. cytokines, growth factors, hormones, pH, hypoxia) or genetic factors (e.g. mutations in proteins involved in lysosomal protein traffic) may affect lysosomal targeting of cathepsin D or induce the exocytosis of the enzyme from endosomal-lysosomal compartments (Chiarpotto et al. 1999; Carini et al. 2004; Koo et al. 2008; Takenouchi et al. 2011). As a consequence, a substantial portion of procathepsin D and/or mature cathepsin D may be located extracellularly. This event is likely to occur for instance in cancer cells (Vashishta et al. 2005; Liaudet-Coopman et al. 2006), but also under certain physiological conditions such as allergen stimulation of mast cells (Dragonetti et al. 2000; Puri and Roche 2008), activation of dendritic cells following interaction with cytotoxic lympocytes (Gardella et al. 2001), bone remodelling (Czupalla et al. 2006), lactation (Lkhider et al. 2004; Castino et al. 2008) and luteal regression (Erdmann et al. 2007). Therefore, it is not surprising that cathepsin D, either as inactive precursor or as mature enzyme, can be found in biological fluids. Cathepsin D has been found in serum or plasma (Zühlsdorf et al. 1983; Naseem et al. 2005), milk (Větvicka et al. 1993; Larsen and Petersen 1995; Benes et al. 2002; Christensen et al. 2010), urine (Zühlsdorf et al. 1983) and even sweat (Baechle et al. 2006) or gastric juice (Ruan et al. 2011). The basal presence of cathepsin D in biological fluids probably merely reflects default lysosomal spillage into extracellular fluids and/or limited tissue damage. However, during abnormal 'leakage' from pathological tissues the plasma level of cathepsin D rises far above its physiological concentration and thus often represents a disease marker. Thus, for instance, abnormal levels of cathepsin D are found in the serum of cancer patients (Abbott et al. 2010; Szajda et al. 2008; Fukuda et al. 2005; Nicotra et al. 2010) and in the plasma of patients after myocardial infarction (Naseem et al. 2005). The physiological role of cathepsin D in milk is still obscure, but it is known that this protease is involved in coagulation and degradation of milk proteins (Larsen et al. 1996; Hurley et al. 2000). More recently, cathepsin D has been shown to process osteopontin in milk (Christensen et al. 2010). Cathepsin D found in urine presents with a glycosylation pattern different from that of its counterpart found in serum which suggests a local origin from renal tissue, and not from blood, as also supported by the elevated level of urinary cathepsin D in patients with nephritic syndrome (Zühlsdorf et al. 1983). The presence of active mature cathepsin D in eccrine sweat has been related to antimicrobial activity and immune defense on the skin surface (Baechle et al. 2006).

Abnormalities in the expression and secretion of cathepsin D have been reported in a variety of diseases, including psoriasis (Chen et al. 2000), atherosclerosis (Li and Yuan 2004), Alzheimer's disease (Cataldo et al. 1995; Zhou et al. 2006; Hamano et al. 2008), Parkinson's disease (Qiao et al. 2008; Cullen et al. 2009) and cancer (Nicotra et al. 2010). Cathepsin D appears implicated in all critical steps of cancer development and progression (Garcia et al. 1990; Isidoro et al. 1995; Liaudet et al. 1995; Berchem et al. 2002; Benes et al. 2008; Ohri et al. 2008), including chemoresistance (Wu et al. 1998; Sagulenko et al. 2008; Castino et al. 2009) and formation of metastases (Liaudet et al. 1994; Glondu et al. 2002; Vashishta et al. 2007). It has been noticed that the role of cathepsin D in tumourigenesis is not only attributable to its proteolytic activity. In fact, an enzymatically inactive mutant of cathepsin D was shown to still possess mitogenic properties (Glondu et al. 2001). This mitogenic ability seems to reside on a fragment of the propeptide (amino acids 27–44) that interacts with a so far unknown membrane receptor (Fusek and Vetvicka 1994; Vetvicka et al. 1997, 2004).

4.11 Cathepsins in Lysosomal Storage Disorders

Lysosomal storage diseases comprise a heterogeneous group of about 50 inherited metabolic disorders. Many of these disease states can be attributed to the selective loss of a single hydrolase involved in the strictly ordered breakdown of glycoproteins, proteoglycans and sphingolipids. These deficiencies result in the gradual accumulation of the respective substrate within lysosomes, ultimately blocking the functions of these organelles (Neufeld 1991; Futerman and van Meer 2004). Surprisingly, only one lysosomal storage disease, pycnodysostosis, is due to a cathepsin deficiency. This disorder, manifested by osteosclerosis and short stature, is caused by mutations in the cathepsin K gene (Gelb et al. 1996). This is consistent with cathepsin K being the main collagenolytic proteinase produced by osteoclasts, as is evident from the analysis of cathepsin K-deficient mice (Saftig et al. 1998).

Another congenital disorder associated with a cathepsin deficiency is Papillon-Lefevbre syndrome, a hereditary disease characterised by hyperkeratosis and severe periodontitis. This disorder is due to genetic inactivation of cathepsin C (Toomes et al. 1999). However, as yet it is not clear which metabolic pathways are affected by the absence of the enzyme. In mice, disruption of the cathepsin C gene abolishes the cytolytic activity of T-lymphocytes, owing to interference with the activation of the destructive and pro-apoptotic serine proteinases granzymes A and B (Pham and Ley 1999) which requires cathepsin C-mediated removal of their N-terminal dipeptide proregions.

The neuronal ceroid lipofuscinoses constitute a group of neurodegenerative lysosomal storage diseases characterised by progressive psychomotor retardation, blindness and premature death. Interestingly, an inactivating mutation of cathepsin D accounts for a related disease in sheep (Tyynelä et al. 2000). Furthermore, disruption of the murine cathepsin F gene causes neuronal lipofuscinosis and neuropathological symptoms (Tang et al. 2006). Intriguingly, cathepsin F is the only cysteine cathepsin whose individual inactivation leads to such a lysosomal storage defect in mice. Animals deficient in both cathepsin B and cathepsin L die soon after birth due to massive brain atrophy and neuronal degeneration whereas mice lacking either cathepsin B or cathepsin L are relatively healthy, indicating that these proteinases may functionally compensate for each other (Felbor et al. 2002).

4.12 Concluding Remarks

Cathepsins play a fundamental role in a wide range of important physiological processes. However, these enzymes also contribute to various devastating disease states. The latter is frequently associated with changes in cathepsin biosynthesis and intracellular transport. This chapter summarizes our current knowledge of the molecular mechanisms governing cathepsin trafficking and *in vivo* activation in the context of the pathophysiological significance of these potent proteases.

Acknowledgements This article is based on work funded by the United States National Science Foundation (MCB-8908842, 9204834, 9604139 and 0235680; A.H.E.), Fondazione Compagnia S. Paolo (Neuroscienze, 2008.2395; C.I.) and Polo di Innovazione Tecnologica (POR-FESR, BANP 2010; C.I.). Molecular structures were prepared using the program PyMOL (http://sourceforge.net/projects/pymol/).

References

Abbott DE, Margaryan NV, Jeruss JS, Khan S, Kaklamani V, Winchester DJ, Hansen N, Rademaker A, Khalkhali-Ellis Z, Hendrix MJ (2010) Reevaluating cathepsin D as a biomarker for breast cancer: serum activity levels versus histopathology. Cancer Biol Ther 9:23–30

Achkar C, Gong Q, Frankfater A, Bajkowski AS (1990) Differences in targeting and secretion of cathepsins B and L by BALB/3T3 fibroblasts and Moloney murine sarcoma virus-transformed BALB/3T3 fibroblasts. J Biol Chem 265:13650–13654

Ahn K, Yeyeodu S, Collette J, Madden V, Arthur J, Li L, Erickson AH (2002) An alternate targeting pathway for procathepsin L in mouse fibroblasts. Traffic 3:147–159

Baechle D, Flad T, Cansier A, Steffen H, Schittek B, Tolson J, Herrmann T, Dihazi H, Beck A, Mueller GA, Mueller M, Stevanovic S, Garbe C, Mueller CA, Kalbacher H (2006) Cathepsin D is present in human eccrine sweat and involved in the postsecretory processing of the antimicrobial peptide DCD-1L. J Biol Chem 281:5406–5415

Baici A, Müntener K, Willimann A, Zwicky R (2006) Regulation of human cathepsin B by alternative mRNA splicing: homeostasis, fatal errors and cell death. Biol Chem 387:1017–1021

Baldwin ET, Bhat TN, Gulnik S, Hosur MV, Sowder RC II, Cachau RE, Collins J, Silva AM, Erickson JW (1993) Crystal structures of native and inhibited forms of human cathepsin D: implications for lysosomal targeting and drug design. Proc Natl Acad Sci U S A 90:6796–6800

Baranski TJ, Faust PL, Kornfeld S (1990) Generation of a lysosomal enzyme targeting signal in the secretory protein pepsinogen. Cell 63:281–291

Baranski TJ, Koelsch G, Hartsuck JA, Kornfeld S (1991) Mapping and molecular modeling of a recognition domain for lysosomal enzyme targeting. J Biol Chem 266:23365–23372

Baranski TJ, Cantor AB, Kornfeld S (1992) Lysosomal enzyme phosphorylation. I. Protein recognition determinants in both lobes of procathepsin D mediate its interaction with UDP-GlcNAc:lysosomal enzyme N-acetylglucosamine-1-phosphotransferase. J Biol Chem 267:23342–23348

Barbarin A, Frade R (2011) Procathepsin L secretion, which triggers tumour progression, is regulated by Rab4a in human melanoma cells. Biochem J 437:97–107
Baricos WH, Zhou Y, Mason RW, Barrett AJ (1988) Human kidney cathepsins B and L. Characterization and potential role in degradation of glomerular basement membrane. Biochem J 252:301–304
Barth R, Afting EG (1984) Cathepsin D from pig myometrium. Characterization of the proteinase. Biochem J 219:899–904
Benes P, Koelsch G, Dvorak B, Fusek M, Vetvicka V (2002) Detection of procathepsin D in rat milk. Comp Biochem Physiol B Biochem Mol Biol 133:113–118
Benes P, Vetvicka V, Fusek M (2008) Cathepsin D-many functions of one aspartic protease. Crit Rev Oncol Hematol 68:12–28
Berchem G, Glondu M, Gleizes M, Brouillet JP, Vignon F, Garcia M, Liaudet-Coopman E (2002) Cathepsin-D affects multiple tumor progression steps in vivo: proliferation, angiogenesis and apoptosis. Oncogene 21:5951–5955
Bevec T, Stoka V, Pungercic G, Dolenc I, Turk V (1996) Major histocompatibility complex class II-associated p41 invariant chain fragment is a strong inhibitor of lysosomal cathepsin L. J Exp Med 183:1331–1338
Beyer BM, Dunn BM (1996) Self-activation of recombinant human lysosomal procathepsin D at a newly engineered cleavage junction, "short" pseudocathepsin D. J Biol Chem 271:15590–15596
Bidère N, Lorenzo HK, Carmona S, Laforge M, Harper F, Dumont C, Senik A (2003) Cathepsin D triggers Bax activation, resulting in selective apoptosis-inducing factor (AIF) relocation in T lymphocytes entering the early commitment phase to apoptosis. J Biol Chem 278:31401–31411
Billington CJ, Mason P, Magny M-C, Mort JS (2000) The slow-binding inhibition of cathepsin K by its propeptide. Biochem Biophys Res Commun 276:924–929
Blott EJ, Griffiths GM (2002) Secretory lysosomes. Nat Rev Mol Cell Biol 3:122–131
Bocock JP, Edgell CJ, Marr HS, Erickson AH (2003) Human proteoglycan testican-1 inhibits the lysosomal cysteine protease cathepsin L. Eur J Biochem 270:4008–4015
Braulke T, Bonifacino JS (2009) Sorting of lysosomal proteins. Biochim Biophys Acta 1793:605–614
Braulke T, Geuze HJ, Slot JW, Hasilik A, von Figura K (1987) On the effects of weak bases and monensin on sorting and processing of lysosomal enzymes in human cells. Eur J Cell Biol 43:316–321
Braulke T, Mach L, Hoflack B, Glössl J (1992) Biosynthesis and endocytosis of lysosomal enzymes in human colon carcinoma SW 1116 cells: impaired internalization of plasma membrane-associated cation-independent mannose 6-phosphate receptor. Arch Biochem Biophys 298:176–181
Brix K, Lemansky P, Herzog V (1996) Evidence for extracellularly acting cathepsins mediating thyroid hormone liberation in thyroid epithelial cells. Endocrinology 137:1963–1974
Brix K, Dunkhorst A, Mayer K, Jordans S (2008) Cysteine cathepsins: cellular roadmap to different functions. Biochimie 90:194–207
Caglic D, Pungercar JR, Pejler G, Turk V, Turk B (2007) Glycosaminoglycans facilitate procathepsin B activation through disruption of propeptide-mature enzyme interactions. J Biol Chem 282:33076–33085
Cantor AB, Baranski TJ, Kornfeld S (1992) Lysosomal enzyme phosphorylation. II. Protein recognition determinants in either lobe of procathepsin D are sufficient for phosphorylation of both the amino and carboxyl lobe oligosaccharides. J Biol Chem 267:23349–23356
Canuel M, Korkidakis A, Konnyu K, Morales CR (2008) Sortilin mediates the lysosomal targeting of cathepsins D and H. Biochem Biophys Res Commun 373:292–297
Canuel M, Libin Y, Morales CR (2009) The interactomics of sortilin: an ancient lysosomal receptor evolving new functions. Histol Histopathol 24:481–492

Capasso C, Lees WE, Capasso A, Scudiero R, Carginale V, Kille P, Kay J, Parisi E (1999) Cathepsin D from the liver of the antarctic icefish Chionodraco hamatus exhibits unusual activity and stability at high temperatures. Biochim Biophys Acta 1431:64–73

Carini R, Castino R, De Cesaris MG, Splendore R, Démoz M, Albano E, Isidoro C (2004) Preconditioning-induced cytoprotection in hepatocytes requires Ca(2+)-dependent exocytosis of lysosomes. J Cell Sci 117:1065–1077

Carmona E, Dufour É, Plouffe C, Takebe S, Mason P, Mort JS, Ménard R (1996) Potency and selectivity of the cathepsin L propeptide as an inhibitor of cysteine proteases. Biochemistry 35:8149–8157

Castino R, Bellio N, Nicotra G, Follo C, Trincheri NF, Isidoro C (2007) Cathepsin D-Bax death pathway in oxidative stressed neuroblastoma cells. Free Radic Biol Med 42:1305–1316

Castino R, Delpal S, Bouguyon E, Demoz M, Isidoro C, Ollivier-Bousquet M (2008) Prolactin promotes the secretion of active cathepsin D at the basal side of rat mammary acini. Endocrinology 149:4095–4105

Castino R, Peracchio C, Salini A, Nicotra G, Trincheri NF, Démoz M, Valente G, Isidoro C (2009) Chemotherapy drug response in ovarian cancer cells strictly depends on a cathepsin D-Bax activation loop. J Cell Mol Med 13:1096–1109

Cataldo AM, Barnett JL, Berman SA, Li J, Quarless S, Bursztajn S, Lippa C, Nixon RA (1995) Gene expression and cellular content of cathepsin D in Alzheimer's disease brain: evidence for early up-regulation of the endosomal-lysosomal system. Neuron 14:671–680

Cavallo-Medved D, Mai J, Dosescu J, Sameni M, Sloane BF (2005) Caveolin-1 mediates the expression and localization of cathepsin B, pro-urokinase plasminogen activator and their cell-surface receptors in human colorectal carcinoma cells. J Cell Sci 118:1493–1503

Chapman RL, Kane SE, Erickson AH (1997) Abnormal glycosylation of procathepsin L due to N-terminal point mutations correlates with failure to sort to lysosomes. J Biol Chem 272:8808–8816

Chen SH, Arany I, Apisarnthanarax N, Rajaraman S, Tyring SK, Horikoshi T, Brysk H, Brysk MM (2000) Response of keratinocytes from normal and psoriatic epidermis to interferon-gamma differs in the expression of zinc-alpha(2)-glycoprotein and cathepsin D. FASEB J 14:565–571

Chiarpotto E, Domenicotti C, Paola D, Vitali A, Nitti M, Pronzato MA, Biasi F, Cottalasso D, Marinari UM, Dragonetti A, Cesaro P, Isidoro C, Poli G (1999) Regulation of rat hepatocyte protein kinase C beta isoenzymes by the lipid peroxidation product 4-hydroxy-2,3-nonenal: a signaling pathway to modulate vesicular transport of glycoproteins. Hepatology 29:1565–1572

Cho WL, Raikhel AS (1992) Cloning of cDNA for mosquito lysosomal aspartic protease. Sequence analysis of an insect lysosomal enzyme similar to cathepsins D and E. J Biol Chem 267:21823–21829

Christensen B, Schack L, Kläning E, Sørensen ES (2010) Osteopontin is cleaved at multiple sites close to its integrin-binding motifs in milk and is a novel substrate for plasmin and cathepsin D. J Biol Chem 285:7929–7937

Collette J, Bocock JP, Ahn K, Chapman RL, Godbold G, Yeyeodu S, Erickson AH (2004a) Biosynthesis and alternate targeting of the lysosomal cysteine protease cathepsin L. Int Rev Cytol 241:1–51

Collette J, Ulku AS, Der CJ, Jones A, Erickson AH (2004b) Enhanced cathepsin L expression is mediated by different Ras effector pathways in fibroblasts and epithelial cells. Int J Cancer 112:190–199

Conner GE, Richo G (1992) Isolation and characterization of a stable activation intermediate of the lysosomal aspartyl protease cathepsin D. Biochemistry 31:1142–1147

Conner GE, Udey JA, Pinto C, Sola J (1989) Nonhuman cells correctly sort and process the human lysosomal enzyme cathepsin D. Biochemistry 28:3530–3533

Coulombe R, Grochulski P, Sivaraman J, Ménard R, Mort JS, Cygler M (1996) Structure of human procathepsin L reveals the molecular basis of inhibition by the prosegment. EMBO J 15:5492–5503

Cullen V, Lindfors M, Ng J, Paetau A, Swinton E, Kolodziej P, Boston H, Saftig P, Woulfe J, Feany MB, Myllykangas L, Schlossmacher MG, Tyynelä J (2009) Cathepsin D expression level affects alpha-synuclein processing, aggregation, and toxicity in vivo. Mol Brain 2:5
Cuozzo JW, Tao K, Cygler M, Mort JS, Sahagian GG (1998) Lysine-based structure responsible for selective mannose phosphorylation of cathepsin D and cathepsin L defines a common structural motif for lysosomal enzyme targeting. J Biol Chem 273:21067–21076
Cygler M, Mort JS (1997) Proregion structure of members of the papain superfamily. Mode of inhibition of enzymatic activity. Biochimie 79:645–652
Cygler M, Sivaraman J, Grochulski P, Coulombe R, Storer AC, Mort JS (1996) Structure of rat procathepsin B. Model for inhibition of cysteine protease activity by the proregion. Structure 4:405–416
Czupalla C, Mansukoski H, Riedl T, Thiel D, Krause E, Hoflack B (2006) Proteomic analysis of lysosomal acid hydrolases secreted by osteoclasts: implications for lytic enzyme transport and bone metabolism. Mol Cell Proteomics 5:134–143
Dahl SW, Halkier T, Lauritzen C, Dolenc I, Pedersen J, Turk V, Turk B (2001) Human recombinant pro-dipeptidyl peptidase I (cathepsin C) can be activated by cathepsins L and S but not by autocatalytic processing. Biochemistry 40:1671–1678
Dahms NM, Olson LJ, Kim JJ (2008) Strategies for carbohydrate recognition by the mannose 6-phosphate receptors. Glycobiology 18:664–678
Diment S, Martin KJ, Stahl PD (1989) Cleavage of parathyroid hormone in macrophage endosomes illustrates a novel pathway for intracellular processing of proteins. J Biol Chem 264:13403–13406
Diment S, Eidelman M, Rodriguez GM, Orlow SJ (1995) Lysosomal hydrolases are present in melanosomes and are elevated in melanizing cells. J Biol Chem 270:4213–4215
Dittmer F, Ulbrich EJ, Hafner A, Schmahl W, Meister T, Pohlmann R, von Figura K (1999) Alternative mechanisms for trafficking of lysosomal enzymes in mannose 6-phosphate receptor-deficient mice are cell type-specific. J Cell Sci 112:1591–1597
Dong JM, Sahagian GG (1990) Basis for low affinity binding of a lysosomal cysteine protease to the cation-independent mannose 6-phosphate receptor. J Biol Chem 265:4210–4217
Dragonetti A, Baldassarre M, Castino R, Démoz M, Luini A, Buccione R, Isidoro C (2000) The lysosomal protease cathepsin D is efficiently sorted to and secreted from regulated secretory compartments in the rat basophilic/mast cell line RBL. J Cell Sci 113:3289–3298
Dustin ML, Baranski TJ, Sampath D, Kornfeld S (1995) A novel mutagenesis strategy identifies distantly spaced amino acid sequences that are required for the phosphorylation of both the oligosaccharides of procathepsin D by N-acetylglucosamine 1-phosphotransferase. J Biol Chem 270:170–179
Erdmann S, Ricken A, Merkwitz C, Struman I, Castino R, Hummitzsch K, Gaunitz F, Isidoro C, Martial J, Spanel-Borowski K (2007) The expression of prolactin and its cathepsin D-mediated cleavage in the bovine corpus luteum vary with the estrous cycle. Am J Physiol Endocrinol Metab 293:E1365–E1377
Erickson AH, Blobel G (1979) Early events in the biosynthesis of the lysosomal enzyme cathepsin D. J Biol Chem 254:11771–11774
Erickson AH (1989) Biosynthesis of lysosomal endopeptidases. J Cell Biochem 40:31–41
Erickson AH (2011) PA-TM-RING proteins: a new family of endosomal membrane proteins. FEBS J 278:46
Erickson AH, Blobel G (1983) Carboxyl-terminal proteolytic processing during biosynthesis of the lysosomal enzymes beta-glucuronidase and cathepsin D. Biochemistry 22:5201–5205
Erickson AH, Conner GE, Blobel G (1981) Biosynthesis of a lysosomal enzyme. Partial structure of two transient and functionally distinct NH2-terminal sequences in cathepsin D. J Biol Chem 256:11224–11231
Faust PL, Kornfeld S, Chirgwin JM (1985) Cloning and sequence analysis of cDNA for human cathepsin D. Proc Natl Acad Sci U S A 82:4910–4914

Felbor U, Kessler B, Mothes W, Goebel HH, Ploegh HL, Bronson RT, Olsen BR (2002) Neuronal loss and brain atrophy in mice lacking cathepsins B and L. Proc Natl Acad Sci U S A 99:7883–7888

Fiebiger E, Maehr R, Villadangos J, Weber E, Erickson A, Bikoff E, Ploegh HL, Lennon-Duménil AM (2002) Invariant chain controls the activity of extracellular cathepsin L. J Exp Med 196:1263–1269

Follo C, Castino R, Nicotra G, Trincheri NF, Isidoro C (2007) Folding, activity and targeting of mutated human cathepsin D that cannot be processed into the double-chain form. Int J Biochem Cell Biol 39:638–649

Follo C, Ozzano M, Mugoni V, Castino R, Santoro M, Isidoro C (2011) Knock-down of cathepsin D affects the retinal pigment epithelium, impairs swim-bladder ontogenesis and causes premature death in zebrafish. PLoS One 6:e21908

Fox T, de Miguel E, Mort JS, Storer AC (1992) Potent slow-binding inhibition of cathepsin B by its propeptide. Biochemistry 31:12571–12576

Friedrichs B, Tepel C, Reinheckel T, Deussing J, von Figura K, Herzog V, Peters C, Saftig P, Brix K (2003) Thyroid functions of mouse cathepsins B, K, and L. J Clin Invest 111:1733–1745

Fujita H, Tanaka Y, Noguchi Y, Kono A, Himeno M, Kato K (1991) Isolation and sequencing of a cDNA clone encoding rat liver lysosomal cathepsin D and the structure of three forms of mature enzymes. Biochem Biophys Res Commun 179:190–196

Fukuda ME, Iwadate Y, Machida T, Hiwasa T, Nimura Y, Nagai Y, Takiguchi M, Tanzawa H, Yamaura A, Seki N (2005) Cathepsin D is a potential serum marker for poor prognosis in glioma patients. Cancer Res 65:5190–5194

Fusek F, Vetvicka V (1994) Mitogenic function of human procathepsin D: the role of the propeptide. Biochem J 303:775–780

Futerman AH, van Meer G (2004) The cell biology of lysosomal storage disorders. Nat Rev Mol Cell Biol 5:554–565

Gal S, Gottesman MM (1986) The major excreted protein (MEP) of transformed mouse cells and cathepsin L have similar protease specificity. Biochem Biophys Res Commun 139:156–162

Gal S, Gottesman MM (1988) Isolation and sequence of a cDNA for human pro-(cathepsin L). Biochem J 253:303–306

Gal S, Willingham MC, Gottesman MM (1985) Processing and lysosomal localization of a glycoprotein whose secretion is transformation stimulated. J Cell Biol 100:535–544

Garcia M, Derocq D, Pujol P, Rochefort H (1990) Overexpression of transfected cathepsin D in transformed cells increases their malignant phenotype and metastatic potency. Oncogene 12:1809–1814

Gardella S, Andrei C, Lotti LV, Poggi A, Torrisi MR, Zocchi MR, Rubartelli A (2001) CD8(+) T lymphocytes induce polarized exocytosis of secretory lysosomes by dendritic cells with release of interleukin-1beta and cathepsin D. Blood 98:2152–2159

Gelb BD, Shi GP, Chapman HA, Desnick RJ (1996) Pycnodysostosis, a lysosomal disease caused by cathepsin K deficiency. Science 273:1236–1238

Gerhartz B, Auerswald EA, Mentele R, Fritz H, Machleidt W, Kolb HJ, Wittmann J (1997) Proteolytic enzymes in yolk-sac membrane of quail egg. Purification and enzymatic characterisation. Comp Biochem Physiol B Biochem Mol Biol 118:159–166

Ghosh P, Dahms NM, Kornfeld S (2003) Mannose 6-phosphate receptors: new twists in the tale. Nat Rev Mol Cell Biol 4:202–212

Gieselmann V, Pohlmann R, Hasilik A, Von Figura K (1983) Biosynthesis and transport of cathepsin D in cultured human fibroblasts. J Cell Biol 97:1–5

Gieselmann V, Hasilik A, von Figura K (1985) Processing of human cathepsin D in lysosomes in vitro. J Biol Chem 260:3215–3220

Glickman JN, Kornfeld S (1993) Mannose 6-phosphate-independent targeting of lysosomal enzymes in I-cell disease B lymphoblasts. J Cell Biol 123:99–108

Glondu M, Coopman P, Laurent-Matha V, Garcia M, Rochefort H, Liaudet-Coopman E (2001) A mutated cathepsin-D devoid of its catalytic activity stimulates the growth of cancer cells. Oncogene 20:6920–6929

Glondu M, Liaudet-Coopman E, Derocq D, Platet N, Rochefort H, Garcia M (2002) Down regulation of cathepsin-D expression by antisense gene transfer inhibits tumor growth and experimental lung metastasis of human breast cancer cells. Oncogene 21:5127–5134

Godbold GD, Ahn K, Yeyeodu S, Lee LF, Ting JP, Erickson AH (1998) Biosynthesis and intracellular targeting of the lysosomal aspartic proteinase cathepsin D. Adv Exp Med Biol 436:153–162

Goldfarb NE, Lam MT, Bose AK, Patel AM, Duckworth AJ, Dunn BM (2005) Electrostatic switches that mediate the pH-dependent conformational change of "short" recombinant human pseudocathepsin D. Biochemistry 44:15725–15733

Gopalakrishnan MM, Grosch HW, Locatelli-Hoops S, Werth N, Smolenová E, Nettersheim M, Sandhoff K, Hasilik A (2004) Purified recombinant human prosaposin forms oligomers that bind procathepsin D and affect its autoactivation. Biochem J 383:507–515

Gottesman MM (1978) Transformation-dependent secretion of a low molecular weight protein by murine fibroblasts. Proc Natl Acad Sci U S A 75:2767–2771

Goulet B, Baruch A, Moon NS, Poirier M, Sansregret LL, Erickson A, Bogyo M, Nepveu A (2004) A cathepsin L isoform that is devoid of a signal peptide localizes to the nucleus in S phase and processes the CDP/Cux transcription factor. Mol Cell 14:207–219

Goulet B, Truscott M, Nepveu A (2006) A novel proteolytically processed CDP/Cux isoform of 90 kDa is generated by cathepsin L. Biol Chem 387:1285–1293

Griffiths GM, Isaaz S (1993) Granzymes A and B are targeted to the lytic granules of lymphocytes by the mannose-6-phosphate receptor. J Cell Biol 120:885–896

Griffiths GM, Tsun A, Stinchcombe JC (2010) The immunological synapse: a focal point for endocytosis and exocytosis. J Cell Biol 189:399–406

Grusby MJ, Mitchell SC, Glimcher LH (1990) Molecular cloning of mouse cathepsin D. Nucleic Acids Res 18:4008

Gui ZZ, Lee KS, Kim BY, Choi YS, Wei YD, Choo YM, Kang PD, Yoon HJ, Kim I, Je YH, Seo SJ, Lee SM, Guo X, Sohn HD, Jin BR (2006) Functional role of aspartic proteinase cathepsin D in insect metamorphosis. BMC Dev Biol 6:49

Guncar G, Podobnik M, Pungercar J, Strukelj B, Turk V, Turk D (1998) Crystal structure of porcine cathepsin H determined at 2.1 Å resolution: location of the mini-chain C-terminal carboxyl group defines cathepsin H aminopeptidase function. Structure 6:51–61

Hamano T, Gendron TF, Causevic E, Yen SH, Lin WL, Isidoro C, Deture M, Ko LW (2008) Autophagic-lysosomal perturbation enhances tau aggregation in transfectants with induced wild-type tau expression. Eur J Neurosci 27:1119–1130

Hanewinkel H, Glössl J, Kresse H (1987) Biosynthesis of cathepsin B in cultured normal and I-cell fibroblasts. J Biol Chem 262:12351–12355

Hara K, Kominami E, Katunuma N (1988) Effect of proteinase inhibitors on intracellular processing of cathepsin B, H and L in rat macrophages. FEBS Lett 231:229–231

Hasilik A (1992) The early and late processing of lysosomal enzymes: proteolysis and compartmentation. Experientia 48:130–151

Hasilik A, Neufeld EF (1980) Biosynthesis of lysosomal enzymes in fibroblasts. Phosphorylation of mannose residues. J Biol Chem 255:4946–4950

Hasilik A, von Figura K, Conzelmann E, Nehrkorn H, Sandhoff K (1982) Lysosomal enzyme precursors in human fibroblasts. Activation of cathepsin D precursor *in vitro* and activity of β-hexosaminidase A precursor towards ganglioside G_{M2}. Eur J Biochem 125:317–321

Hasnain S, Hirama T, Tam A, Mort JS (1992) Characterization of recombinant rat cathepsin B and nonglycosylated mutants expressed in yeast. J Biol Chem 267:4713–4721

Hitzel C, Kanzler H, König A, Kummer MP, Brix K, Herzog V, Koch N (2000) Thyroglobulin type-I-like domains in invariant chain fusion proteins mediate resistance to cathepsin L digestion. FEBS Lett 485:67–70

Horn M, Doleckova-Maresova L, Rulisek L, Masa M, Vasiljeva O, Turk B, Gan-Erdene T, Baudys M, Mares M (2005) Activation processing of cathepsin H impairs recognition by its propeptide. Biol Chem 386:941–947
Hurley MJ, Larsen LB, Kelly AL, McSweeney PLH (2000) The milk acid proteinase cathepsin D: a review. Int Dairy J 10:673–681
Ishidoh K, Kominami E (1994) Multi-step processing of procathepsin L in vitro. FEBS Lett 352:281–284
Ishidoh K, Kominami E (1995) Procathepsin L degrades extracellular matrix proteins in the presence of glycosaminoglycans in vitro. Biochem Biophys Res Commun 217:624–631
Ishidoh K, Kominami E (1998) Gene regulation and extracellular functions of procathepsin L. Biol Chem 379:131–135
Ishidoh K, Kominami E (2002) Processing and activation of lysosomal proteinases. Biol Chem 383:1827–1831
Ishidoh K, Towatari T, Imajoh S, Kawasaki H, Kominami E, Katunuma N, Suzuki K (1987) Molecular cloning and sequencing of cDNA for rat cathepsin L. FEBS Lett 223:69–73
Ishidoh K, Muno D, Sato N, Kominami E (1991) Molecular cloning of cDNA for rat cathepsin C. Cathepsin C, a cysteine proteinase with an extremely long propeptide. J Biol Chem 266:16312–16317
Ishidoh K, Saido TC, Kawashima S, Hirose M, Watanabe S, Sato N, Kominami E (1998) Multiple processing of procathepsin L to cathepsin L in vivo. Biochem Biophys Res Commun 252:202–207
Isidoro C, Radons J, Baccino FM, Hasilik A (1990) Suppression of the 'uncovering' of mannose-6-phosphate residues in lysosomal enzymes in the presence of NH4Cl. Eur J Biochem 191:591–597
Isidoro C, Horst M, Baccino FM, Hasilik A (1991) Differential segregation of human and hamster cathepsin D in transfected baby-hamster kidney cells. Biochem J 273:363–367
Isidoro C, Demoz M, De Stefanis D, Baccino FM, Bonelli G (1995) Synthesis, maturation and extracellular release of procathepsin D as influenced by cell proliferation or transformation. Int J Cancer 63:866–871
Jacobson LA, Jen-Jacobson L, Hawdon JM, Owens GP, Bolanowski MA, Emmons SW, Shah MV, Pollock RA, Conklin DS (1988) Identification of a putative structural gene for cathepsin D in Caenorhabditis elegans. Genetics 119:355–363
Jiang L, Phillips TE, Rogers SW, Rogers JC (2000) Biogenesis of the protein storage vacuole crystalloid. J Cell Biol 150:755–770
Joseph L, Lapid S, Sukhatme V (1987) The major ras induced protein in NIH3T3 cells is cathepsin L. Nucleic Acids Res 15:3186
Joseph LJ, Chang LC, Stamenkovich D, Sukhatme VP (1988) Complete nucleotide and deduced amino acid sequences of human and murine preprocathepsin L. An abundant transcript induced by transformation of fibroblasts. J Clin Invest 81:1621–1629
Journet A, Chapel A, Jehan S, Adessi C, Freeze H, Klein G, Garin J (1999) Characterization of Dictyostelium discoideum cathepsin D. J Cell Sci 112:3833–3843
Joyce JA, Baruch A, Chehade K, Meyer-Morse N, Giraudo E, Tsai FY, Greenbaum DC, Hager JH, Bogyo M, Hanahan D (2004) Cathepsin cysteine proteases are effectors of invasive growth and angiogenesis during multistage tumorigenesis. Cancer Cell 5:443–453
Kane SE (1993) Mouse procathepsin L lacking a functional glycosylation site is properly folded, stable, and secreted by NIH 3T3 cells. J Biol Chem 268:11456–11462
Karrer KM, Peiffer SL, Ditomas ME (1993) Two distinct gene subfamilies within the family of cysteine protease genes. Proc Natl Acad Sci U S A 90:3063–3067
Kasper D, Dittmer F, von Figura K, Pohlmann R (1996) Neither type of mannose 6-phosphate receptor is sufficient for targeting of lysosomal enzymes along intracellular routes. J Cell Biol 134:615–623

Kaulmann G, Palm GJ, Schilling K, Hilgenfeld R, Wiederanders B (2006) The crystal structure of a Cys25 -> Ala mutant of human procathepsin S elucidates enzyme-prosequence interactions. Protein Sci 15:2619–2629

Kawada A, Hara K, Kominami E, Hiruma M, Noguchi H, Ishibashi A (1997) Processing of cathepsins L, B and D in psoriatic epidermis. Arch Dermatol Res 289:87–93

Khan AR, Parrish JC, Fraser ME, Smith WW, Bartlett PA, James MNG (1998) Lowering the entropic barrier for binding conformationally flexible inhibitors to enzymes. Biochemistry 37:16839–16845

Kihara M, Kakegawa H, Matano Y, Murata E, Tsuge H, Kido H, Katunuma N (2002) Chondroitin sulfate proteoglycan is a potent enhancer in the processing of procathepsin L. Biol Chem 383:1925–1929

Kirschke H, Langner J, Wiederanders B, Ansorge S, Bohley P (1977) Cathepsin L. A new proteinase from rat-liver lysosomes. Eur J Biochem 74:293–301

Kominami E, Tsukahara T, Hara K, Katunuma N (1988) Biosyntheses and processing of lysosomal cysteine proteinases in rat macrophages. FEBS Lett 231:225–228

Konjar S, Sutton VR, Hoves S, Repnik U, Yagita H, Reinheckel T, Peters C, Turk V, Turk B, Trapani JA, Kopitar-Jerala N (2010) Human and mouse perforin are processed in part through cleavage by the lysosomal cysteine proteinase cathepsin L. Immunology 131:257–267

Koo IC, Wang C, Raghavan S, Morisaki JH, Cox JS, Brown EJ (2008) ESX-1-dependent cytolysis in lysosome secretion and inflammasome activation during mycobacterial infection. Cell Microbiol 10:1866–1878

Kornfeld R, Kornfeld S (1985) Assembly of asparagine-linked oligosaccharides. Annu Rev Biochem 54:631–664

Kos J, Sekirnik A, Premzl A, Zavasnik Bergant V, Langerholc T, Turk B, Werle B, Golouh R, Repnik U, Jeras M, Turk V (2005) Carboxypeptidases cathepsins X and B display distinct protein profile in human cells and tissues. Exp Cell Res 306:103–113

Kreusch S, Fehn M, Maubach G, Nissler K, Rommerskirch W, Schilling K, Weber E, Wenz I, Wiederanders B (2000) An evolutionarily conserved tripartite tryptophan motif stabilizes the prodomains of cathepsin L-like cysteine proteases. Eur J Biochem 267:2965–2972

Lah T, Turk V (1982) Autolysis studies of cathepsin D. Hoppe Seylers Z Physiol Chem 363:247–254

Lah T, Drobnic-Kosorok M, Turk V, Pain RH (1984) Conformation, structure and activation of bovine cathepsin D. Unfolding and refolding studies. Biochem J 218:601–608

LaLonde JM, Zhao B, Janson CA, D'Alessio KJ, McQueney MS, Orsini MJ, Debouck CM, Smith WW (1999) The crystal structure of human procathepsin K. Biochemistry 38:862–869

Larsen LB, Petersen TE (1995) Identification of five molecular forms of cathepsin D in bovine milk. Adv Exp Med Biol 362:279–283

Larsen LB, Benfeldt C, Rasmussen LK, Petersen TE (1996) Bovine milk procathepsin D and cathepsin D: coagulation and milk protein degradation. J Dairy Res 63:119–130

Laulagnier K, Schieber NL, Maritzen T, Haucke V, Parton RG, Gruenberg J (2011) Role of AP1 and Gadkin in the traffic of secretory endo-lysosomes. Mol Biol Cell 22:2068–2082

Laurent-Matha V, Derocq D, Prébois C, Katunuma N, Liaudet-Coopman E (2006) Processing of human cathepsin D is independent of its catalytic function and auto-activation: involvement of cathepsins L and B. J Biochem 139:363–371

Lazzarino D, Gabel CA (1990) Protein determinants impair recognition of procathepsin L phosphorylated oligosaccharides by the cation-independent mannose 6-phosphate receptor. J Biol Chem 265:11864–11871

Lee AY, Gulnik SV, Erickson JW (1998) Conformational switching in an aspartic proteinase. Nat Struct Biol 5:866–871

Li W, Yuan XM (2004) Increased expression and translocation of lysosomal cathepsins contribute to macrophage apoptosis in atherogenesis. Ann N Y Acad Sci 1030:427–433

Liaudet E, Garcia M, Rochefort H (1994) Cathepsin D maturation and its stimulatory effect on metastasis are prevented by addition of KDEL retention signal. Oncogene 9:1145–1154

Liaudet E, Derocq D, Rochefort H, Garcia M (1995) Transfected cathepsin D stimulates high density cancer cell growth by inactivating secreted growth inhibitors. Cell Growth Differ 6:1045–1052
Liaudet-Coopman E, Beaujouin M, Derocq D, Garcia M, Glondu-Lassis M, Laurent-Matha V, Prébois C, Rochefort H, Vignon F (2006) Cathepsin D: newly discovered functions of a long-standing aspartic protease in cancer and apoptosis. Cancer Lett 237:167–179
Linke M, Herzog V, Brix K (2002) Trafficking of lysosomal cathepsin B-green fluorescent protein to the surface of thyroid epithelial cells involves the endosomal/lysosomal compartment. J Cell Sci 115:4877–4889
Lippincott-Schwartz J, Yuan LC, Bonifacino JS, Klausner RD (1989) Rapid redistribution of Golgi proteins into the ER in cells treated with brefeldin A: evidence for membrane cycling from Golgi to ER. Cell 56:801–813
Lkhider M, Castino R, Bouguyon E, Isidoro C, Ollivier-Bousquet M (2004) Cathepsin D released by lactating rat mammary epithelial cells is involved in prolactin cleavage under physiological conditions. J Cell Sci 117:5155–5164
Lorenzo K, Ton P, Clark JL, Coulibaly S, Mach L (2000) Invasive properties of murine squamous carcinoma cells: secretion of matrix-degrading cathepsins is attributable to a deficiency in the mannose 6-phosphate/insulin-like growth factor II receptor. Cancer Res 60:4070–4076
Ludwig T, Munier-Lehmann H, Bauer U, Hollinshead M, Ovitt C, Lobel P, Hoflack B (1994) Differential sorting of lysosomal enzymes in mannose 6-phosphate receptor-deficient fibroblasts. EMBO J 13:3430–3437
Lukong KE, Elsliger MA, Mort JS, Potier M, Pshezhetsky AV (1999) Identification of UDP-N-acetylglucosamine-phosphotransferase-binding sites on the lysosomal proteases, cathepsins A, B, and D. Biochemistry 38:73–80
Mach L (2002) Biosynthesis of lysosomal proteinases in health and disease. Biol Chem 383:751–756
Mach L, Stüwe K, Hagen A, Ballaun C, Glössl J (1992) Proteolytic processing and glycosylation of cathepsin B. The role of the primary structure of the latent precursor and of the carbohydrate moiety for cell-type-specific molecular forms of the enzyme. Biochem J 282:577–582
Mach L, Schwihla H, Stüwe K, Rowan AD, Mort JS, Glössl J (1993) Activation of procathepsin B in human hepatoma cells: the conversion into the mature enzyme relies on the action of cathepsin B itself. Biochem J 293:437–442
Mach L, Mort JS, Glössl J (1994a) Maturation of human cathepsin B. Proenzyme activation and proteolytic processing of the precursor to the mature proteinase, *in vitro*, are primarily unimolecular processes. J Biol Chem 269:13030–13035
Mach L, Mort JS, Glössl J (1994b) Non-covalent complexes between the lysosomal proteinase cathepsin B and its propeptide account for stable, extracellular, high molecular mass forms of the enzyme. J Biol Chem 269:13036–13040
Maehr R, Hang HC, Mintern JD, Kim YM, Cuvillier A, Nishimura M, Yamada K, Shirahama-Noda K, Hara-Nishimura I, Ploegh HL (2005) Asparagine endopeptidase is not essential for class II MHC antigen presentation but is required for processing of cathepsin L in mice. J Immunol 174:7066–7074
Mai J, Finley RL Jr, Waisman DM, Sloane BF (2000) Human procathepsin B interacts with the annexin II tetramer on the surface of tumor cells. J Biol Chem 275:12806–12812
Mainferme F, Wattiaux R, von Figura K (1985) Synthesis, transport and processing of cathepsin C in Morris hepatoma 7777 cells and rat hepatocytes. Eur J Biochem 153:211–216
Mason RW, Massey SD (1992) Surface activation of pro-cathepsin L. Biochem Biophys Res Commun 189:1659–1666
Mason RW, Walker JE, Northrop FD (1986) The N-terminal amino acid sequences of the heavy and light chains of human cathepsin L. Relationship to a cDNA clone for a major cysteine proteinase from a mouse macrophage cell line. Biochem J 240:373–377

Mason RW, Gal S, Gottesman MM (1987) The identification of the major excreted protein (MEP) from a transformed mouse fibroblast cell line as a catalytically active precursor form of cathepsin L. Biochem J 248:449–454

Mason RW, Wilcox D, Wikstrom P, Shaw EN (1989) The identification of active forms of cysteine proteinases in Kirsten-virus-transformed mouse fibroblasts by use of a specific radiolabelled inhibitor. Biochem J 257:125–129

Masson O, Bach AS, Derocq D, Prébois C, Laurent-Matha V, Pattingre S, Liaudet-Coopman E (2010) Pathophysiological functions of cathepsin D: targeting its catalytic activity versus its protein binding activity? Biochimie 92:1635–1643

McDonald JK, Kadkhodayan S (1988) Cathepsin L–a latent proteinase in guinea pig sperm. Biochem Biophys Res Commun 151:827–835

McIntyre GF, Erickson AH (1991) Procathepsins L and D are membrane-bound in acidic microsomal vesicles. J Biol Chem 266:15438–15445

McIntyre GF, Erickson AH (1993) The lysosomal proenzyme receptor that binds procathepsin L to microsomal membranes at pH 5 is a 43-kDa integral memebrane protein. Proc Natl Acad Sci U S A 90:10588–10592

McIntyre GF, Godbold GD, Erickson AH (1994) The pH-dependent membrane association of procathepsin L is mediated by a 9-residue sequence within the propeptide. J Biol Chem 269:567–572

Meh P, Pavsic M, Turk V, Baici A, Lenarcic B (2005) Dual concentration-dependent activity of thyroglobulin type-1 domain of testican: specific inhibitor and substrate of cathepsin L. Biol Chem 386:75–83

Mehtani S, Gong Q, Panella J, Subbiah S, Peffley DM, Frankfater A (1998) In vivo expression of an alternatively spliced human tumor message that encodes a truncated form of cathepsin B. Subcellular distribution of the truncated enzyme in COS cells. J Biol Chem 273:13236–13244

Ménard R, Carmona E, Takebe S, Dufour É, Plouffe C, Mason P, Mort JS (1998) Autocatalytic processing of recombinant human procathepsin L. Contribution of both intermolecular and unimolecular events in the processing of procathepsin L in vitro. J Biol Chem 273:4478–4484

Metcalf P, Fusek M (1993) Two crystal structures for cathepsin D: the lysosomal targeting signal and active site. EMBO J 12:1293–1302

Mohamed MM, Sloane BF (2006) Cysteine cathepsins: multifunctional enzymes in cancer. Nat Rev Cancer 6:764–775

Mort JS, Buttle DJ (1997) Cathepsin B. Int J Biochem Cell Biol 29:715–720

Muno D, Ishidoh K, Ueno T, Kominami E (1993) Processing and transport of the precursor of cathepsin C during its transfer into lysosomes. Arch Biochem Biophys 306:103–110

Müntener K, Zwicky R, Csucs G, Baici A (2003) The alternative use of exons 2 and 3 in cathepsin B mRNA controls enzyme trafficking and triggers nuclear fragmentation in human cells. Histochem Cell Biol 119:93–101

Müntener K, Zwicky R, Csucs G, Rohrer J, Baici A (2004) Exon skipping of cathepsin B: mitochondrial targeting of a lysosomal peptidase provokes cell death. J Biol Chem 279:41012–41017

Müntener K, Willimann A, Zwicky R, Svoboda B, Mach L, Baici A (2005) Folding competence of N-terminally truncated forms of human procathepsin B. J Biol Chem 280:11973–11980

Musil D, Zucic D, Turk D, Engh RA, Mayr I, Huber R, Popovic T, Turk V, Towatari T, Katunuma N, Bode W (1991) The refined 2.15 Å x-ray crystal structure of human liver cathepsin B: the structural basis for its specificity. EMBO J 10:2321–2330

Nägler DK, Storer AC, Portaro FCV, Carmona E, Juliano L, Ménard R (1997) Major increase in endopeptidase activity of human cathepsin B upon removal of occluding loop contacts. Biochemistry 36:12608–12615

Nägler DK, Sulea T, Ménard R (1999a) Full-length cDNA of human cathepsin F predicts the presence of a cystatin domain at the N-terminus of the cysteine protease zymogen. Biochem Biophys Res Commun 257:313–318

Nägler DK, Zhang R, Tam W, Sulea T, Purisima EO, Ménard R (1999b) Human cathepsin X: a cysteine protease with unique carboxypeptidase activity. Biochemistry 38:12648–12654

Nakamura K, Yonezawa S, Yoshizaki N (1996) Vitellogenesis-related ovary cathepsin D from Xenopus laevis: purification and properties in comparison with liver cathepsin D. Comp Biochem Physiol B Biochem Mol Biol 113:835–840

Naseem RH, Hedegard W, Henry TD, Lessard J, Sutter K, Katz SA (2005) Plasma cathepsin D isoforms and their active metabolites increase after myocardial infarction and contribute to plasma renin activity. Basic Res Cardiol 100:139–146

Neufeld EF (1991) Lysosomal storage diseases. Annu Rev Biochem 60:257–280

Nicotra G, Castino R, Follo C, Peracchio C, Valente G, Isidoro C (2010) The dilemma: does tissue expression of cathepsin D reflect tumor malignancy? The question: does the assay truly mirror cathepsin D mis-function in the tumor? Cancer Biomark 7:47–64

Nielsen LB, Nielsen HH (2001) Purification and characterization of cathepsin D from herring muscle (Clupea harengus). Comp Biochem Physiol B Biochem Mol Biol 128:351–363

Nishimura Y, Amano J, Sato H, Tsuji H, Kato K (1988) Biosynthesis of lysosomal cathepsins B and H in cultured rat hepatocytes. Arch Biochem Biophys 262:159–170

Nishimura Y, Kawabata T, Furuno K, Kato K (1989) Evidence that aspartic proteinase is involved in the proteolytic processing event of procathepsin L in lysosomes. Arch Biochem Biophys 271:400–406

Nishimura Y, Kato K, Furuno K, Himeno M (1995) Inhibitory effect of leupeptin on the intracellular maturation of lysosomal cathepsin L in primary cultures of rat hepatocytes. Biol Pharm Bull 18:945–950

Nomura T, Fujisawa Y (1997) Processing properties of recombinant human procathepsin L. Biochem Biophys Res Commun 230:143–146

Ohri SS, Vashishta A, Proctor M, Fusek M, Vetvicka V (2008) The propeptide of cathepsin D increases proliferation, invasion and metastasis of breast cancer cells. Int J Oncol 32:491–498

Ondr JK, Pham CT (2004) Characterization of murine cathepsin W and its role in cell-mediated cytotoxicity. J Biol Chem 279:27525–27533

Owada M, Neufeld EF (1982) Is there a mechanism for introducing acid hydrolases into liver lysosomes that is independent of mannose 6-phosphate recognition? Evidence from I-cell disease. Biochem Biophys Res Commun 105:814–820

Padilha MH, Pimentel AC, Ribeiro AF, Terra WR (2009) Sequence and function of lysosomal and digestive cathepsin D-like proteinases of Musca domestica midgut. Insect Biochem Mol Biol 39:782–791

Pham CT, Ley TJ (1999) Dipeptidyl peptidase I is required for the processing and activation of granzymes A and B *in vivo*. Proc Natl Acad Sci U S A 96:8627–8632

Piper RC, Katzmann DJ (2007) Biogenesis and function of multivesicular bodies. Annu Rev Cell Dev Biol 23:519–547

Pohl S, Tiede S, Marschner K, Encarnação M, Castrichini M, Kollmann K, Muschol N, Ullrich K, Müller-Loennies S, Braulke T (2010) Proteolytic processing of the gamma-subunit is associated with the failure to form GlcNAc-1-phosphotransferase complexes and mannose 6-phosphate residues on lysosomal enzymes in human macrophages. J Biol Chem 285:23936–23944

Pohlmann R, Boeker MW, von Figura K (1995) The two mannose 6-phosphate receptors transport distinct complements of lysosomal proteins. J Biol Chem 270:27311–27318

Poole AR, Tiltman KJ, Recklies AD, Stoker TA (1978) Differences in secretion of the proteinase cathepsin B at the edges of human breast carcinomas and fibroadenomas. Nature 273:545–547

Portnoy DA, Erickson AH, Kochan J, Ravetch JV, Unkeless JC (1986) Cloning and characterization of a mouse cysteine proteinase. J Biol Chem 261:14697–14703

Probst OC, Ton P, Svoboda B, Gannon A, Schuhmann W, Wieser J, Pohlmann R, Mach L (2006) The 46-kDa mannose 6-phosphate receptor does not depend on endosomal acidification for delivery of hydrolases to lysosomes. J Cell Sci 119:4935–4943

Probst OC, Puxbaum V, Svoboda B, Leksa V, Stockinger H, Mikula M, Mikulits W, Mach L (2009) The mannose 6-phosphate/insulin-like growth factor II receptor restricts the tumourigenicity and invasiveness of squamous cell carcinoma cells. Int J Cancer 124:2559–2567

Pungercar JR, Caglic D, Sajid M, Dolinar M, Vasiljeva O, Pozgan U, Turk D, Bogyo M, Turk V, Turk B (2009) Autocatalytic processing of procathepsin B is triggered by proenzyme activity. FEBS J 276:660–668

Puri N, Roche PA (2008) Mast cells possess distinct secretory granule subsets whose exocytosis is regulated by different SNARE isoforms. Proc Natl Acad Sci U S A 105:2580–2585

Qiao L, Hamamichi S, Caldwell KA, Caldwell GA, Yacoubian TA, Wilson S, Xie ZL, Speake LD, Parks R, Crabtree D, Liang Q, Crimmins S, Schneider L, Uchiyama Y, Iwatsubo T, Zhou Y, Peng L, Lu Y, Standaert DG, Walls KC, Shacka JJ, Roth KA, Zhang J (2008) Lysosomal enzyme cathepsin D protects against alpha-synuclein aggregation and toxicity. Mol Brain 1:17

Quraishi O, Storer AC (2001) Identification of internal autoproteolytic cleavage sites within the prosegments of recombinant procathepsin B and procathepsin S. Contribution of a plausible unimolecular autoproteolytic event for the processing of zymogens belonging to the papain family. J Biol Chem 276:8118–8124

Radons J, Isidoro C, Hasilik A (1990) Brefeldin A prevents uncovering but not phosphorylation of the recognition marker in cathepsin D. Biol Chem Hoppe Seyler 371:567–573

Redzynia I, Ljunggren A, Abrahamson M, Mort JS, Krupa JC, Jaskolski M, Bujacz G (2008) Displacement of the occluding loop by the parasite protein, chagasin, results in efficient inhibition of human cathepsin B. J Biol Chem 283:22815–22825

Reiser J, Adair B, Reinheckel T (2010) Specialized roles for cysteine cathepsins in health and disease. J Clin Invest 120:3421–3431

Reitman ML, Kornfeld S (1981) UDP-N-acetylglucosamine:glycoprotein N-acetylglucosamine-1-phosphotransferase. Proposed enzyme for the phosphorylation of the high mannose oligosaccharide units of lysosomal enzymes. J Biol Chem 256:4275–4281

Renko M, Pozgan U, Majera D, Turk D (2010) Stefin A displaces the occluding loop of cathepsin B only by as much as required to bind to the active site cleft. FEBS J 277:4338–4345

Retzek H, Steyrer E, Sanders EJ, Nimpf J, Schneider WJ (1992) Molecular cloning and functional characterization of chicken cathepsin D, a key enzyme for yolk formation. DNA Cell Biol 11:661–672

Richo GR, Conner GE (1994) Structural requirements of procathepsin D activation and maturation. J Biol Chem 269:14806–14812

Ritonja A, Popović T, Kotnik M, Machleidt W, Turk V (1988) Amino acid sequences of the human kidney cathepsins H and L. FEBS Lett 228:341–345

Robinson MW, Dalton JP, Donnelly S (2008) Helminth pathogen cathepsin proteases: it's a family affair. Trends Biochem Sci 33:601–608

Rohrer J, Kornfeld R (2001) Lysosomal hydrolase mannose 6-phosphate uncovering enzyme resides in the trans-Golgi network. Mol Biol Cell 12:1623–1631

Rowan AD, Mason P, Mach L, Mort JS (1992) Rat procathepsin B. Proteolytic processing to the mature form in vitro. J Biol Chem 267:15993–15999

Rowan AD, Feng R, Konishi Y, Mort JS (1993) Demonstration by electrospray mass spectrometry that the peptidyldipeptidase activity of cathepsin B is capable of rat cathepsin B *C*-terminal processing. Biochem J 294:923–927

Rozhin J, Sameni M, Ziegler G, Sloane BF (1994) Pericellular pH affects distribution and secretion of cathepsin B in malignant cells. Cancer Res 54:6517–6525

Rozman J, Stojan J, Kuhelj R, Turk V, Turk B (1999) Autocatalytic processing of recombinant human procathepsin B is a bimolecular process. FEBS Lett 459:358–362

Ruan HL, Hong RT, Xie HJ, Hu NZ, Xu JM, Zhang W (2011) Significance of elevated levels of collagen type IV and hyaluronic acid in gastric juice and serum in gastric cancer and precancerous lesion. Dig Dis Sci 56:2001–2008

Saftig P, Klumperman J (2009) Lysosome biogenesis and lysosomal membrane proteins: trafficking meets function. Nat Rev Mol Cell Biol 10:623–635

Saftig P, Hunziker E, Wehmeyer O, Jones S, Boyde A, Rommerskirch W, Moritz JD, Schu P, von Figura K (1998) Impaired osteoclastic bone resorption leads to osteopetrosis in cathepsin K-deficient mice. Proc Natl Acad Sci U S A 95:13453–13458

Sagulenko V, Muth D, Sagulenko E, Paffhausen T, Schwab M, Westermann F (2008) Cathepsin D protects human neuroblastoma cells from doxorubicin-induced cell death. Carcinogenesis 29:1869–1877

Sahagian GG, Gottesman MM (1982) The predominant secreted protein of transformed murine fibroblasts carries the lysosomal mannose 6-phosphate recognition marker. J Biol Chem 257:11145–11150

Salminen A, Gottesman MM (1990) Inhibitor studies indicate that active cathepsin L is probably essential to its own processing in cultured fibroblasts. Biochem J 272:39–44

Salvesen G, Enghild JJ (1990) An unusual specificity in the activation of neutrophil serine proteinase zymogens. Biochemistry 29:5304–5308

Samarel AM, Ferguson AG, Decker RS, Lesch M (1989) Effects of cysteine protease inhibitors on rabbit cathepsin D maturation. Am J Physiol 257:C1069–C1079

Schilling K, Pietschmann S, Fehn M, Wenz I, Wiederanders B (2001) Folding incompetence of cathepsin L-like cysteine proteases may be compensated by the highly conserved, domain-building N-terminal extension of the proregion. Biol Chem 382:859–865

Schilling K, Korner A, Sehmisch S, Kreusch A, Kleint R, Benedix Y, Schlabrakowski A, Wiederanders B (2009) Selectivity of propeptide-enzyme interaction in cathepsin L-like cysteine proteases. Biol Chem 390:167–174

Schmid JA, Mach L, Paschke E, Glössl J (1999) Accumulation of sialic acid in endocytic compartments interferes with the formation of mature lysosomes. Impaired proteolytic processing of cathepsin B in fibroblasts of patients with lysosomal sialic acid storage disease. J Biol Chem 274:19063–19071

Sever S, Altintas MM, Nankoe SR, Möller CC, Ko D, Wei C, Henderson J, del Re EC, Hsing L, Erickson A, Cohen CD, Kretzler M, Kerjaschki D, Rudensky A, Nikolic B, Reiser J (2007) Proteolytic processing of dynamin by cytoplasmic cathepsin L is a mechanism for proteinuric kidney disease. J Clin Invest 117:2095–2104

Shirahama-Noda K, Yamamoto A, Sugihara K, Hashimoto N, Asano M, Nishimura M, Hara-Nishimura I (2003) Biosynthetic processing of cathepsins and lysosomal degradation are abolished in asparaginyl endopeptidase-deficient mice. J Biol Chem 278:33194–33199

Sivaraman J, Lalumière M, Ménard R, Cygler M (1999) Crystal structure of wild-type human procathepsin K. Protein Sci 8:283–290

Sivaraman J, Nägler DK, Zhang R, Ménard R, Cygler M (2000) Crystal structure of human procathepsin X: a cysteine protease with the proregion covalently linked to the active site cysteine. J Mol Biol 295:939–951

Sleat DE, Zheng H, Qian M, Lobel P (2006) Identification of sites of mannose 6-phosphorylation on lysosomal proteins. Mol Cell Proteomics 5:686–701

Sloane BF, Dunn JR, Honn KV (1981) Lysosomal cathepsin B: correlation with metastatic potential. Science 212:1151–1153

Sloane BF, Rozhin J, Johnson K, Taylor H, Crissman JD, Honn KV (1986) Cathepsin B: association with plasma membrane in metastatic tumors. Proc Natl Acad Sci U S A 83:2483–2487

Sloane BF, Moin K, Sameni M, Tait LR, Rozhin J, Ziegler G (1994) Membrane association of cathepsin B can be induced by transfection of human breast epithelial cells with c-Ha-ras oncogene. J Cell Sci 107:373–384

Smith SM, Kane SE, Gal S, Mason RW, Gottesman MM (1989) Glycosylation of procathepsin L does not account for species molecular-mass differences and is not required for proteolytic activity. Biochem J 262:931–938

Stahl S, Reinders Y, Asan E, Mothes W, Conzelmann E, Sickmann A, Felbor U (2007) Proteomic analysis of cathepsin B- and L-deficient mouse brain lysosomes. Biochim Biophys Acta 1774:1237–1246

Stearns NA, Dong JM, Pan JX, Brenner DA, Sahagian GG (1990) Comparison of cathepsin L synthesized by normal and transformed cells at the gene, message, protein, and oligosaccharide levels. Arch Biochem Biophys 283:447–457

Stern I, Schaschke N, Moroder L, Turk D (2004) Crystal structure of NS-134 in complex with bovine cathepsin B: a two-headed epoxysuccinyl inhibitor extends along the entire active-site cleft. Biochem J 381:511–517

Stubbs M, McSheehy PM, Griffiths JR, Bashford CL (2000) Causes and consequences of tumour acidity and implications for treatment. Mol Med Today 6:15–19

Szajda SD, Snarska J, Jankowska A, Roszkowska-Jakimiec W, Puchalski Z, Zwierz K (2008) Cathepsin D and carcino-embryonic antigen in serum, urine and tissues of colon adenocarcinoma patients. Hepatogastroenterology 55:388–393

Takenouchi T, Iwamaru Y, Sugama S, Tsukimoto M, Fujita M, Sekigawa A, Sekiyama K, Sato M, Kojima S, Conti B, Hashimoto M, Kitani H (2011) The activation of P2X7 receptor induces cathepsin D-dependent production of a 20-kDa form of IL-1β under acidic extracellular pH in LPS-primed microglial cells. J Neurochem 117:712–723

Tang CH, Lee JW, Galvez MG, Robillard L, Mole SE, Chapman HA (2006) Murine cathepsin F deficiency causes neuronal lipofuscinosis and late-onset neurological disease. Mol Cell Biol 26:2309–2316

Tanji M, Kageyama T, Takahashi K (1991) Occurrence of cathepsin D isozymes with different specificities in monkey skeletal muscle. Biochem Biophys Res Commun 176:798–804

Tao K, Stearns NA, Dong J, Wu QL, Sahagian GG (1994) The proregion of cathepsin L is required for proper folding, stability, and ER exit. Arch Biochem Biophys 311:19–27

Tedelind S, Poliakova K, Valeta A, Hunegnaw R, Yemanaberhan EL, Heldin NE, Kurebayashi J, Weber E, Kopitar-Jerala N, Turk B, Bogyo M, Brix K (2010) Nuclear cysteine cathepsin variants in thyroid carcinoma cells. Biol Chem 391:923–935

Tepel C, Brömme D, Herzog V, Brix K (2000) Cathepsin K in thyroid epithelial cells: sequence, localization and possible function in extracellular proteolysis of thyroglobulin. J Cell Sci 113:4487–4498

Tiede S, Storch S, Lübke T, Henrissat B, Bargal R, Raas-Rothschild A, Braulke T (2005) Mucolipidosis II is caused by mutations in GNPTA encoding the alpha/beta GlcNAc-1-phosphotransferase. Nat Med 11:1109–1112

Toomes C, James J, Wood AJ, Wu CL, McCormick D, Lench N, Hewitt C, Moynihan L, Roberts E, Woods CG, Markham A, Wong M, Widmer R, Ghaffar KA, Pemberton M, Hussein IR, Temtamy SA, Davies R, Read AP, Sloan P, Dixon MJ, Thakker NS (1999) Loss-of-function mutations in the cathepsin C gene result in periodontal disease and palmoplantar keratosis. Nat Genet 23:421–424

Towatari T, Katunuma N (1988) Amino acid sequence of rat liver cathepsin L. FEBS Lett 236:57–61

Troen BR, Gal S, Gottesman MM (1987) Sequence and expression of the cDNA for MEP (major excreted protein), a transformation-regulated secreted cathepsin. Biochem J 246:731–735

Turk B, Turk V (2009) Lysosomes as "suicide bags" in cell death: myth or reality? J Biol Chem 284:21783–21787

Turk B, Dolenc I, Turk V, Bieth JG (1993) Kinetics of the pH-induced inactivation of human cathepsin L. Biochemistry 32:375–380

Turk D, Podobnik M, Kuhelj R, Dolinar M, Turk V (1996) Crystal structures of human procathepsin B at 3.2 and 3.3 Å resolution reveal an interaction motif between a papain-like cysteine protease and its propeptide. FEBS Lett 384:211–214

Turk D, Guncar G, Podobnik M, Turk B (1998) Revised definition of substrate binding sites of papain-like cysteine proteases. Biol Chem 379:137–147

Turk D, Guncar G, Turk V (1999) The p41 fragment story. IUBMB Life 48:7–12

Turk B, Turk D, Turk V (2000) Lysosomal cysteine proteases: more than scavengers. Biochim Biophys Acta 1477:98–111
Turk D, Janjic V, Stern I, Podobnik M, Lamba D, Dahl SW, Lauritzen C, Pedersen J, Turk V, Turk B (2001a) Structure of human dipeptidyl peptidase I (cathepsin C): exclusion domain added to an endopeptidase framework creates the machine for activation of granular serine proteases. EMBO J 20:6570–6582
Turk V, Turk B, Turk D (2001b) Lysosomal cysteine proteases: facts and opportunities. EMBO J 20:4629–4633
Tyynelä J, Sohar I, Sleat DE, Gin RM, Donnelly RJ, Baumann M, Haltia M, Lobel P (2000) A mutation in the ovine cathepsin D gene causes a congenital lysosomal storage disease with profound neurodegeneration. EMBO J 19:2786–2792
van Meel E, Boonen M, Zhao H, Oorschot V, Ross FP, Kornfeld S, Klumperman J (2011) Disruption of the man-6-p targeting pathway in mice impairs osteoclast secretory lysosome biogenesis. Traffic 12:912–924
Vashishta A, Fusek M, Vetvicka V (2005) Possible role of procathepsin D in human cancer. Folia Microbiol (Praha) 50:71–76
Vashishta A, Ohri SS, Proctor M, Fusek M, Vetvicka V (2007) Ribozyme-targeting procathepsin D and its effect on invasion and growth of breast cancer cells: an implication in breast cancer therapy. Int J Oncol 30:1223–1230
Vasiljeva O, Dolinar M, Turk V, Turk B (2003) Recombinant human cathepsin H lacking the mini chain is an endopeptidase. Biochemistry 42:13522–13528
Vasiljeva O, Reinheckel T, Peters C, Turk D, Turk V, Turk B (2007) Emerging roles of cysteine cathepsins in disease and their potential as drug targets. Curr Pharm Des 13:387–403
Verity CK, McManus DP, Brindley PJ (1999) Developmental expression of cathepsin D aspartic protease in Schistosoma japonicum. Int J Parasitol 29:1819–1824
Větvicka V, Vágner J, Baudys M, Tang J, Foundling SI, Fusek M (1993) Human breast milk contains procathepsin D-detection by specific antibodies. Biochem Mol Biol Int 30:921–928
Vetvicka V, Vetvickova J, Hilgert I, Voburka Z, Fusek M (1997) Analysis of the interaction of procathepsin D activation peptide with breast cancer cells. Int J Cancer 73:403–409
Vetvicka V, Vetvickova J, Fusek M (1998) Effect of procathepsin D and its activation peptide on prostate cancer cells. Cancer Lett 129:55–59
Vetvicka V, Vetvickova J, Benes P (2004) Role of enzymatically inactive procathepsin D in lung cancer. Anticancer Res 24:2739–2743
Waheed A, Hasilik A, von Figura K (1981) Processing of the phosphorylated recognition marker in lysosomal enzymes. Characterization and partial purification of a microsomal alpha-N-acetylglucosaminyl phosphodiesterase. J Biol Chem 256:5717–5721
Waheed A, Pohlmann R, Hasilik A, von Figura K, van Elsen A, Leroy JG (1982) Deficiency of UDP-N-acetylglucosamine:lysosomal enzyme N-acetylglucosamine-1-phosphotransferase in organs of I-cell patients. Biochem Biophys Res Commun 105:1052–1058
Wang H, Rogers JC, Jiang L (2011) Plant RMR proteins: unique vacuolar sorting receptors that couple ligand sorting with membrane internalization. FEBS J 278:59–68
Warner JB, Thalhauser C, Tao K, Sahagian GG (2002) Role of N-linked oligosaccharide flexibility in mannose phosphorylation of lysosomal enzyme cathepsin L. J Biol Chem 277:41897–41905
Watanabe H, Grubb JH, Sly WS (1990) The overexpressed human 46-kDa mannose 6-phosphate receptor mediates endocytosis and sorting of β-glucuronidase. Proc Natl Acad Sci U S A 87:8036–8040
Wex T, Wex H, Bromme D (1999) The human cathepsin F gene–a fusion product between an ancestral cathepsin and cystatin gene. Biol Chem 380:1439–1442
Wex T, Bühling F, Wex H, Günther D, Malfertheiner P, Weber E, Brömme D (2001) Human cathepsin W, a cysteine protease predominantly expressed in NK cells, is mainly localized in the endoplasmic reticulum. J Immunol 167:2172–2178
Wiederanders B, Kirschke H (1989) The processing of a cathepsin L precursor in vitro. Arch Biochem Biophys 272:516–521

Wiederanders B, Brömme D, Kirschke H, von Figura K, Schmidt B, Peters C (1992) Phylogenetic conservation of cysteine proteinases. Cloning and expression of a cDNA coding for human cathepsin S. J Biol Chem 267:13708–13713

Wittlin S, Rösel J, Stover DR (1998) One-step purification of cathepsin D by affinity chromatography using immobilized propeptide sequences. Eur J Biochem 252:530–536

Wittlin S, Rösel J, Hofmann F, Stover DR (1999) Mechanisms and kinetics of procathepsin D activation. Eur J Biochem 265:384–393

Wu GS, Saftig P, Peters C, El-Deiry WS (1998) Potential role for cathepsin D in p53-dependent tumor suppression and chemosensitivity. Oncogene 16:2177–2183

Yamamoto A, Tomoo K, Hara T, Murata M, Kitamura K, Ishida T (2000) Substrate specificity of bovine cathepsin B and its inhibition by CA074, based on crystal structure refinement of the complex. J Biochem (Tokyo) 127:635–643

Yasothornsrikul S, Greenbaum D, Medzihradszky KF, Toneff T, Bundey R, Miller R, Schilling B, Petermann I, Dehnert J, Logvinova A, Goldsmith P, Neveu JM, Lane WS, Gibson B, Reinheckel T, Peters C, Bogyo M, Hook V (2003) Cathepsin L in secretory vesicles functions as a prohormone-processing enzyme for production of the enkephalin peptide neurotransmitter. Proc Natl Acad Sci U S A 100:9590–9595

Yeyeodu S, Ahn K, Madden V, Chapman R, Song L, Erickson AH (2000) Procathepsin L self-association as a mechanism for selective secretion. Traffic 1:724–737

Yonezawa S, Takahashi T, Wang XJ, Wong RN, Hartsuck JA, Tang J (1988) Structures at the proteolytic processing region of cathepsin D. J Biol Chem 263:16504–16511

Zhou W, Scott SA, Shelton SB, Crutcher KA (2006) Cathepsin D-mediated proteolysis of apolipoprotein E: possible role in Alzheimer's disease. Neuroscience 143:689–701

Zhu Y, Conner GE (1994) Intermolecular association of lysosomal protein precursors during biosynthesis. J Biol Chem 269:3846–3851

Zühlsdorf M, Imort M, Hasilik A, von Figura K (1983) Molecular forms of beta-hexosaminidase and cathepsin D in serum and urine of healthy subjects and patients with elevated activity of lysosomal enzymes. Biochem J 213:733–740

Chapter 5
Limited and Degradative Proteolysis in the Context of Posttranslational Regulatory Networks: Current Technical and Conceptional Advances

Stefan Tholen, Maria Magdalena Koczorowska, Zon Weng Lai, Joern Dengjel, and Oliver Schilling

5.1 Introduction

Posttranslational modifications (PTMs) control almost all (patho-)physiological processes in living systems by affecting protein folding, conformation, turnover, cell communication, localization and protein function in an orchestral manner. Various PTMs can modulate one single protein leading to an enormous number of protein isoforms, about three orders of magnitude higher than the number of genes encoded in the genome (Cox and Mann 2007; Witze et al. 2007). As a result, PTMs are pivotal in many diseases, such as cancer, Alzheimer's disease, and

S. Tholen
Institute for Molecular Medicine and Cell Research, University of Freiburg, 79104 Freiburg, Germany

Faculty of Biology, University of Freiburg, D-79104 Freiburg, Germany

M.M. Koczorowska • Z.W. Lai
Institute for Molecular Medicine and Cell Research, University of Freiburg, 79104 Freiburg, Germany

J. Dengjel
Freiburg Institute for Advanced Studies (FRIAS), School of Life Science-LifeNet, University of Freiburg, 79104 Freiburg, Germany

ZBSA Center for Biological Systems Analysis, University of Freiburg, 79104 Freiburg, Germany

BIOSS Centre for Biological Signaling Studies, University of Freiburg, 79104 Freiburg, Germany

O. Schilling (✉)
Institute for Molecular Medicine and Cell Research, University of Freiburg, D-79104 Freiburg, Germany

BIOSS Centre for Biological Signaling Studies, University of Freiburg, D-79104 Freiburg, Germany
e-mail: oliver.schilling@mol-med.uni-freiburg.de

K. Brix and W. Stöcker (eds.), *Proteases: Structure and Function*,
DOI 10.1007/978-3-7091-0885-7_5, © Springer-Verlag Wien 2013

diabetes. The most abundant and ubiquitous PTM is phosphorylation, which plays major roles in signal transduction and regulation of enzyme activity (Kamath et al. 2011). Due to an abnormal phosphorylation process, the protein tau aggregates in Alzheimer's disease (Martin et al. 2011). Ubiquitination decides the fate of proteins since covalent attachment of ubiquitin leads to proteasomal or lysosomal degradation of a protein. Other common and well investigated PTMs are (a) glycosylation, affecting protein stability, solubility and cell-cell interactions, (b) acetylation, controlling cell signaling processes and gene expression by modification of histones and (c) methylation, affecting gene expression by modifying of histones similar to acetylation (Kamath et al. 2011).

Proteolysis has emerged as one of the most prevalent PTMs, regulating numerous physiological processes such as development, immune response or blood clotting. Hence, it is not surprising that >560 proteases are encoded in the human genome, representing the second largest enzyme family in man (Puente et al. 2003). In general, proteases break down polypeptide chains by hydrolysis of peptide bonds (Lopez-Otin and Overall 2002). Proteolytic cleavage is of particular interest compared to most PTMs since it is an irreversible process occurring intra- and extracellularly. The "irreversible nature" associates proteolytic processing with fundamental steps in cell function. Consequently, proteolysis is a tightly regulated process that affects every protein either through limited proteolysis or terminal degradation.

Limited proteolysis describes a process in which proteins are functionally modified by a proteolytic cleavage yielding truncated, stable cleavage products. The functional consequences of limited proteolysis can be diverse. Proteins can be activated, inactivated or transferred to another cell compartment, thereby maintaining the proper operation of the cell machinery. Examples for limited proteolysis include proteases involved in blood coagulation (Turk et al. 2012) and ADAM17 (a disintegrin and metalloproteinase 17) is a membrane bound protease that cleaves cell surface proteins, such as cytokines (e.g. TNFα) and cytokine receptors (e.g. IL-6R and TNF-R) thereby regulating cell signaling (Scheller et al. 2011). Less specific proteases including cathepsins, located in the lysosomal compartment, and the proteasome are involved in both limited proteolysis and protein turnover, thereby serving as a quality control by degrading denatured, misfolded and obsolete proteins (Wickner et al. 1999). Since 3–5 % of our cellular proteins are degraded and resynthesized every day (Ciechanover 2006), the rate of synthesis/degradation becomes nowadays more important to fully appreciate cellular dynamics and bridge the gap between transcriptome and proteome data (Jayapal et al. 2010).

Disproportioned distributions of protease, protease inhibitors, and protease substrates can be deleterious and many developmental disorders and diseases, including cancer and neurodegenerative diseases, are accompanied with dysregulation of proteolysis (Doucet et al. 2008; Quesada et al. 2009; Reiser et al. 2010). Therefore proteases are suggested to be valuable and suitable drug targets (Drag and Salvesen 2010). To fully understand protease functions in physiology and pathology, and to exploit the therapeutic potential of protease inhibition, knowledge about the *in vivo* substrate repertoire of a protease is needed (auf dem Keller and Schilling 2010; Impens et al. 2010). The term "degradomics" summarizes all proteomic investigations and techniques regarding the genetic, structural and functional identification and characterization of proteases, and their substrates and inhibitors (Lopez-Otin and Overall 2002). Mass spectrometry-based proteomic techniques are applied to detect alterations in substrate abundance caused by the lack or overexpression of a protease, to identify generated neo amino- and carboxy-termini generated by a protease, and to determine the protease cleavage site specificity. Moreover activity-based probes (ABPs) specifically target active proteases and monitor their activity from cellular to organ level demonstrating colocalization of proteases and their substrates (auf dem Keller and Schilling 2010).

Degradomic approaches uncovered that proteases do not operate in isolation. They built up proteolytic systems comprising of substrates and cleavage products, protease inhibitors and proteases that dynamically interact, thereby forming cascades, regulatory circuits and pathways. These dynamics form the so-called "protease web" that underlies a constant flux (Butler and Overall 2009; Overall and Dean 2006). For instance the aspartic protease cathepsin D degrades the protease inhibitor cystatin C (Laurent-Matha et al. 2012). Cystatin C not only targets cysteine cathepsins but also acts on matrix-metalloproteinase (MMP)-2 (Butler and Overall 2009). Lower cystatin C levels potentially affect MMP-2 levels which in turn will affect other proteases and protease inhibitors. This scenario visualizes that certain proteases are important nodes in this network modulating and balancing the activity of proteases and protease inhibitors, enabling the cell to react on perturbations. The direct role of a protease in knockout or transgenic animal models is hard to elucidate, since a knockout of one protease often leads to unexpected downstream effects mediated by the protease web. The protease web gets more complex by considering the occurrence of PTMs that can control the action of proteases *in vivo* including regulation of gene expression, catalytic activity, protein interactions and degradation.

Mass spectrometry (MS) has become a powerful tool for qualitative and quantitative analysis of a large number of proteins and their PTMs in complex samples. The basis was built by the invention of soft ionization techniques called matrix-associated laser desorption ionization (MALDI) (Tanaka et al. 1988) and electrospray ionization (ESI) (Fenn et al. 1989). The MS analysis for protein quantification employs two main strategies, label-free methods, which directly compare intensity profiles from liquid-chromatography mass spectrometry (LC-MS) or usage of distinct mass tags to directly compare samples (e.g. healthy versus diseased or wildtype versus gene knockout) labeled either with the light or the heavy stable isotope in one MS run (Ong and Mann 2005). The aim of a qualitative proteomic analysis is to identify all proteins present in a sample. The coverage of a proteomic sample is dependent on the sensitivity of a mass spectrometer. Important parameters are (a) the minimum amount of analyte which can be detected, (b) the dynamic range, the signal range in which analytes can be identified and (c) the duty cycle, the number of fragmentation spectra per timeframe which can be recorded (de Godoy et al. 2006). Importantly a protein cannot be considered as absent from the sample, if it was not identified in a mass spectrometric analysis (Helsens et al. 2011). Protein modifications occur only in a subpopulation of the proteome, therefore mass spectrometry in combination with separation and enrichment techniques is used to successfully characterize peptides decorated with PTMs, e.g. for phosphorylation (McNulty and Annan 2008), glycosylation (Geng et al. 2001), lysine acetylation (Choudhary et al. 2009) or ubiquitination (Peng et al. 2003).

In the following sections we want to give an overview on emerging degradomic techniques enabling the characterization of protease cleavage sites in a large, proteome-wide scale. We show perspectives how identified cleavage sites can be validated by profiling of protease specificity and monitoring of proteolysis by activity-based probes. Recent proteomic approaches highlight important nodes of the proteolytic network, illustrate how proteolysis is modulated by other PTMs, and emphasize its role in protein turnover.

5.2 Identification of Cleavage Sites by N- and C-Terminal Degradomics

5.2.1 Overview

In the field of proteolysis research, quantitative gel-based and gel-free proteomic strategies can monitor how differential proteolytic activity affects protein abundance on a proteome wide scale (auf dem Keller and Schilling 2010). As a typical gel-based approach, two-dimensional gel electrophoresis in combination with MS is applied to determine protein abundance by comparing spot-staining intensity. Most common gel-free approaches determine alterations in protein abundance by comparing stable isotope labeled samples allowing relative quantification by one MS-measurement. Several labeling strategies have been successfully applied, like isotope-coded affinity tag (ICAT) (Gygi et al. 1999), isobaric tag for relative and absolute quantification (iTRAQ) (Ross et al. 2004) or stable isotope labeling by amino acids in cell culture (SILAC) (Ong et al. 2002). In a prototypical proteomic-degradomic experiment, a sample that was exposed to proteolysis is mixed with an equal amount of an unexposed control sample (e.g. protease inhibitor treated versus untreated, wild-type versus knockout). Meanwhile multi-tag labeling strategies have evolved, e.g. triple SILAC (Mann 2006) or iTRAQ allowing up to eight different tags (Prudova et al. 2010).

For the identification of direct cleavage sites global quantitative comparisons are not suitable due to sample complexity, because internal peptides would overshadow terminal peptides in the MS analysis. Furthermore they insufficiently profile subtle alterations caused by limited proteolysis. For identification of cleavage events terminomic techniques are applied. Since every cleavage event is leading to a neo-N- and neo-C-terminus on the cleavage product terminal proteomic strategies, "terminomics", select either N- or C-terminal peptides to enrich neo-N-terminal or neo-C-terminal peptides from complex peptide mixtures (Overall and Blobel 2007). This selection process reduces complexity by excluding the overwhelming amount of noninformative internal peptides that overshadow the terminal peptides in the MS analysis. Enrichment of terminal peptides before MS analysis simplifies the proteome by using single terminal peptides for protein identification, which increases dynamic range and proteome coverage (Gevaert et al. 2007). A drawback of single peptide identification is that MS analysis may not identify all terminal peptides, because they are too long, too short, too hydrophobic, ionize and fragment poorly (Huesgen and Overall 2012; Impens et al. 2010). Terminomic strategies targeting neo-N-termini and neo-C-termini will inevitably enrich for an increased number of so called "one-hit wonders", proteins that are only identified by one terminal peptide. Often these identifications are considered as less reliable. This situation is also faced by other PTM approaches (acetylation/phosphorylation). Numerous PTM proteomic studies have contributed to increase confidence in MS based peptide identification with "one-hit wonders". These are increasingly regarded as an invaluable and rich source for the proteomic understanding of biological systems. In fact, as early as 2007 (prior to most degradomic techniques), it has been argued that "one-hit wonders" should not be discarded, since a large amount of biologically valuable data would get lost (Higdon and Kolker 2007). To reduce the number of false positive protein identifications the use of statistical approaches for a validation of search engine based spectrum-to-peptide assignment is mandatory. Different strategies like the usage of decoy databases (Elias and Gygi 2007), a statistical validation by PeptideProphet (Keller et al. 2002) and identification by at least two different MS/MS spectra can be applied (Huesgen and Overall 2012). Moreover a customized scoring tool with high sensitivity termed Peptizer was developed to point to potential false positive protein identifications in large data sets (Helsens et al. 2008). Another challenge is to handle the high amounts of data, which is addressed by tools such as TopFind, a knowledgebase linking protein termini with function (Lange et al. 2012; Lange and Overall 2011) or MEROPS the peptidase database (Rawlings et al. 2010).

N-terminomic techniques are classified into positive and negative selection procedures. Positive selection techniques directly target terminal peptides by modifying neo-N-termini and mature N-termini with an affinity tag (e.g. through specific biotinylation) and subsequent removal of internal peptides after digestion (e.g. trypsin). By definition, positive selection schemes fail to enrich for naturally modified protein termini. This is a major caveat, given that terminal modifications are abundant in nature. Negative selection techniques overcome this problem by modification of all primary amines on all neo-N-terminal and mature N-termini in the sample. In the next step a secondary digest (e.g. trypsin) creates new primary amines selectively on internal peptides that are removed from the mixture afterwards. Finally only neo-N-termini, unmodified mature N-termini with free amines as well as naturally blocked mature N-termini remain. Negative selection procedures enrich for neo-N-termini and mature N-termini more efficiently and allow complete proteome coverage by isolating naturally blocked N-termini as well (Huesgen and Overall 2012). In general, negative selection procedures might be preferred over positive selection procedures (Impens et al. 2010).

C-termini are less accessible for analysis than N-termini and so far have rarely been targeted to determine protease derived cleavage sites. The underlying problem for enrichment of C-termini is the lack of chemical methods to specifically modify terminal carboxyl groups. C-terminomic techniques normally follow a negative selection procedure. C-terminomic strategies are required to examine C-terminal processing by carboxypeptidases. They are useful to validate cleavage sites identified in N-terminomic approaches and uncover cleavage sites that are overlooked by N-terminomic approaches, e.g. N-termini closely located to the substrate C-terminus, which are too short for LC-MS/MS analysis.

5.2.2 Isolation of N-Termini: Positive Selection Procedures

5.2.2.1 Biotinylation of Protein N-Termini Using Subtiligase

In 2008 the group of J. A. Wells engineered a subtiligase enzyme to specifically biotinylate primary amines of neo-N-termini and unprotected mature N-termini (Mahrus et al. 2008). During biotinylation procedure a biotinylated peptide ester containing a cleavage site for the highly specific tobacco etch virus (TEV) protease reacts with subtiligase forming an intermediate product (Fig. 5.1). In a second reaction subtiligase is transferred onto the N-terminal α-amines, but not onto ε-amines of lysine side chains. Following trypsin digestion, biotinylated N-termini are bound to immobilized streptavidin and subsequently eluted by digestion with TEV. After TEV cleavage the N-termini keep a Ser-Tyr-dipeptide which helps to confirm true positives during MS analysis.

This N-terminomic technique has been successfully applied to identify 333 caspase-like cleavage sites on 292 protein substrates in etoposide-treated apoptotic Jurkat cells (Mahrus et al. 2008). Moreover it has been used in combination with SILAC for the discovery of caspase-1 substrates in THP-1 human monocytic leukemia cells exposed to inflammatory stimuli (Agard et al. 2010). Through low efficiency of subtiligase N-terminal labeling, this technique requires large amounts of starting material (circa 50–100 mg) (Agard and Wells 2009) and may introduce a selection bias through subtiligase specificity (Huesgen and Overall 2012).

5.2.2.2 Biotinylation of Protein N-Termini Post Lysine Guanidation

In 2007 the group of G. S. Salvesen developed a positive selection procedure based on the selective guanidation of lysine ε-amines on protein level using *O*-methylisourea (Timmer et al. 2007). *O*-methylisourea does not react with α-amines present at neo-N-termini and mature unprotected N-termini, which are than chemically biotinylated (Fig. 5.2). Following tryptic digest, biotinylated

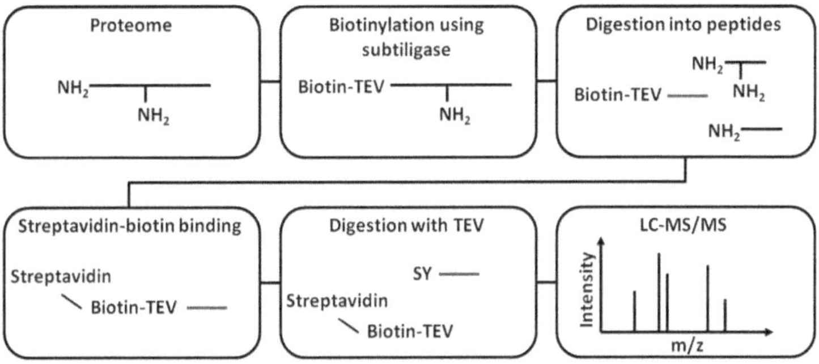

Fig. 5.1 Outline of the biotinylation of protein N-termini using subtiligase procedure

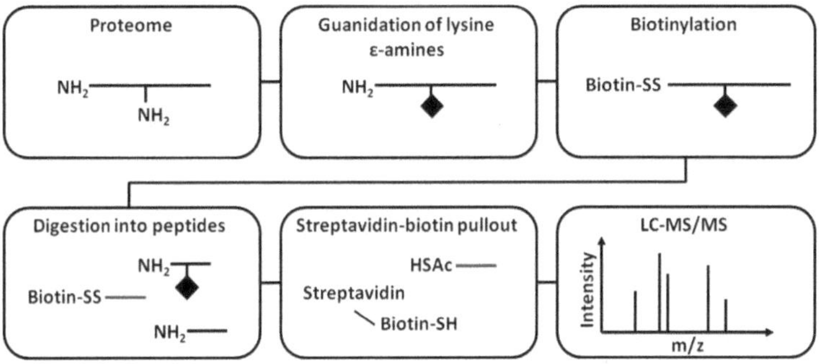

Fig. 5.2 Outline of the biotinylation of protein N-termini post lysine guanidation procedure

N-termini are captured by immobilized streptavidin and released by the reduction of disulfide bonds that link the biotin tag to the peptide. Afterwards N-termini carry a thioacyl modification and are analyzed by MS.

This method has been used for the discovery of mitochondrial transit peptides from *Escherichia coli*, yeast, mouse and human cells as well as for a study of structural and kinetic parameters that determine specificity of caspase-3 and GluC (Timmer et al. 2007, 2009).

5.2.2.3 Labeling of Protein N-Termini with iTRAQ Reagents Post Lysine Guanidation

In 2007 the group of G. S. Salvesen published a second method for identification of neo-N-termini and mature unprotected N-termini. In this approach, first lysine ε-amines are selectively guanidated and afterwards α-amines labeled with iTRAQ reagents on protein level (Enoksson et al. 2007) (Fig. 5.3). Following tryptic digest, N-termini are selected by an *in silico* analysis based on two rounds of MALDI-TOF/TOF-MS/MS. In detail, a first MS/MS analysis is performed to screen for N-termini that possess signature iTRAQ ions in their fragmentation spectra.

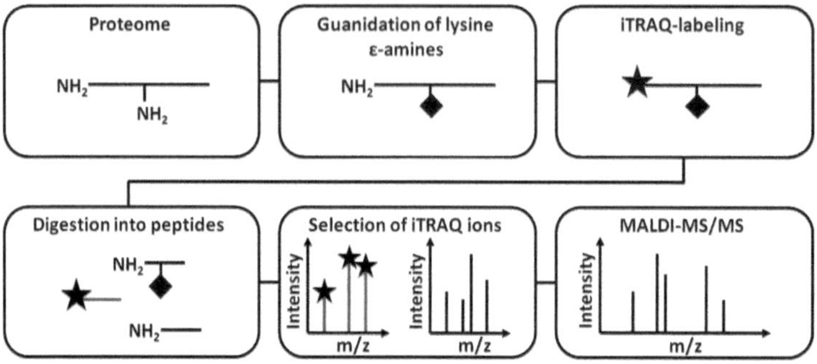

Fig. 5.3 Outline of the biotinylation of protein N-termini with iTRAQ reagents post lysine guanidation procedure

These can only be observed on MS/MS level since iTRAQ tags result in characteristic reporter ions. N-termini showing iTRAQ signatures pass through a second more efficiently in-depth MS/MS analysis identifying the actual peptide.

To validate the procedure, a set of recombinant *Escherichia coli* proteins with predicted caspase-3 cleavage sites was used. A protein mixture was treated with active or inactive caspase-3 and labeled with two different iTRAQ reagents. Ten cleavage sites, all corresponding to caspase-3 consensus, could be identified (Enoksson et al. 2007). Since this approach disregards removal of internal peptides, analyzed samples are highly complex. Moreover its usage is restricted to MALDI mass spectrometers. Identification of terminal peptides through MS2 reporter fragments was also used by Tholey and coworkers in a PICS-like approach (biochemical strategy to investigate protease specificity) (Jakoby et al. 2012).

5.2.2.4 N-CLAP: Biotinylation of Protein N-Termini Using Edman Chemistry

In 2009 the group of S. R. Jaffrey introduced a technique called N-terminalomics by chemical labeling of the alpha-amine of proteins (N-CLAP) (Xu et al. 2009a). N-CLAP takes advantage of Edman degradation chemistry to selectively label unprotected α-amines of proteins with a cleavable biotin (Fig. 5.4). In detail, phenyl isothiocyanate (PITC) is employed to block all primary amines on protein level. In the next step trifluoracetic acid (TFA) is used to initiate an intramolecular cyclization on the PITC-blocked N-terminal amino acid leading to a peptide bond breakage between the first and second amino acid. Since PITC-blocked ε-amines on lysine side chains cannot undergo this reaction, primary α-amines are generated only on N-termini, which are shortened by one amino acid. These are specifically biotinylated using sulfo-NHS-SS-biotin. Following tryptic digestion, biotinylated N-termini are bound to immobilized avidin and eluted by reducing the disulfide bond of the cleavable biotin tag.

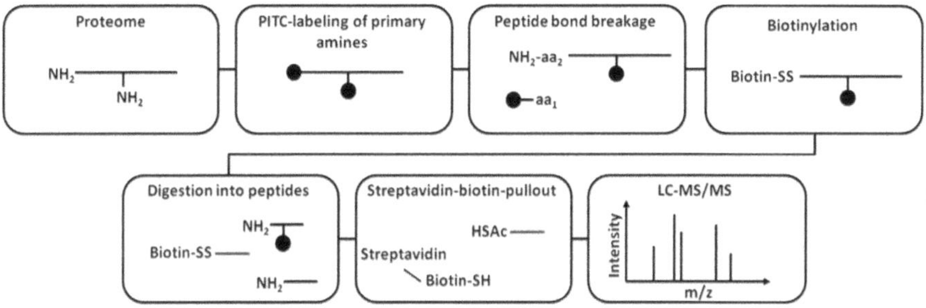

Fig. 5.4 Outline of the N-CLAP procedure

This technique was used to characterize proteolytic cleavage events associated with methionine aminopeptidases and signal peptide peptidases, as well as proteins that are proteolytically cleaved after cisplatin-induced apoptosis (Xu et al. 2009a). Disadvantage of this approach is the incompatibility with chemical stable isotope labeling. Enriched N-termini are shortened by one amino acid compared to true N-termini. Incomplete modification of α-amines by PITC can result in biotinylation of the first n-terminal amino acid leading to an unclear N-termini determination.

5.2.3 Isolation of N-Termini: Negative Selection Procedures

5.2.3.1 Combined Fractional Diagonal Chromatography

In 2003 the Gevaert et al. developed the widely applied N-terminomic technique combined fractional diagonal chromatography (COFRADIC) (Gevaert et al. 2003). During this procedure all primary amines are acetylated on protein level (Fig. 5.5). Following tryptic digestion, peptides are separated by reverse liquid chromatography (RP-HPLC), generating typically 12–15 fractions containing N-terminal and internal peptides. Only internal peptides possess primary α-amines, that are labeled in a second step with highly hydrophobic 2,4,6-trinitrobenzenesulfonyl (TNBS), which increases the hydrophobicity of these peptides. In a second RP-HPLC analysis, TNBS labeled internal peptides show a hydrophobic shift and can be discarded. N-terminal peptides elute in the same gradient concentration as in the first sorting step and can be further fractionated. The number of identified N-terminal peptides can be increased by performing strong cation exchange (SCX) chromatography at low pH (Staes et al. 2008).

This method, using a negative selection procedure, allows for identification of neo-N-termini and mature unmodified as well as modified N-termini. Note that *in vivo* acetylated N-termini can be distinguished from *in vitro* acetylated N-termini, because trideutero-acetylation can be used. Therefore COFRADIC was successfully applied to determine the N-terminal acetylation status of human

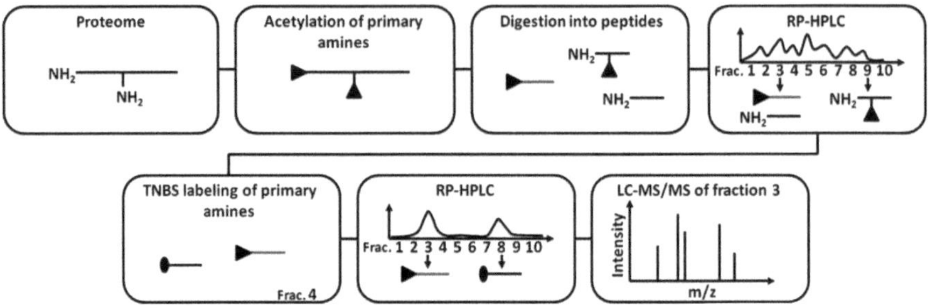

Fig. 5.5 Outline of the COFRADIC procedure

and yeast protein N-termini (Arnesen et al. 2009). Although based on a complex chromatography scheme, COFRADIC has been widely used for N-terminal analysis, also in combination with stable isotope labeling (^{18}O isotope labeling, SILAC) (Impens et al. 2008; Vande Walle et al. 2007).

5.2.3.2 Isolation of N-Termini by Phospho Tagging

In 2012 the group of Ad P.J.M. de Jong recently published a new technique to analyze the global N-proteome (Mommen et al. 2012). During this procedure all primary amines are dimethylated with formaldehyde at the protein level (Fig. 5.6). Following tryptic digest new primary amines of internal peptides are generated, which are modified using the phospho tagging (PTAG) reaction. Resulting phosphopeptides are depleted by TiO_2 affinity chromatography. The flow-through contains neo-N-termini as well as naturally modified N-termini, which are analyzed by LC-MS/MS. This recently introduced approach was used to isolate N-termini from bacteria and yeast. More than 700 N-termini were identified in *Neisseria meningitidis* and more than 900 N-termini in *Saccharomyces cerevisiae.*

5.2.3.3 Terminal Amine Isotope Labeling of Substrates

In 2010 the group of C. M. Overall developed terminal amine isotope labeling of substrates (TAILS) as a simple and highly efficient negative selection strategy (Kleifeld et al. 2010). In the first step all primary amines are labeled with different stable isotope variants of formaldehyde allowing for differential quantification between two samples (light: protease present; heavy: control) (Fig. 5.7). In the next step samples are combined and subjected to tryptic digest. Afterwards internal peptides contain primary amines and are specifically extracted from the complex peptide mixture by using a highly specific aldehyde functionalized amine reactive hyperbranched polyglycerol polymer (HPG-ALD polymer). This covalently binds internal tryptic peptides and can easily be removed by filtration with a spin filter device, caused by its high molecular weight. The flow-through contains

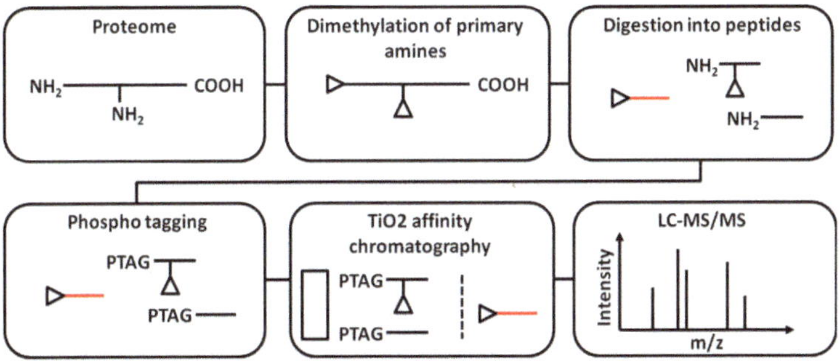

Fig. 5.6 Outline of the PTAG strategy

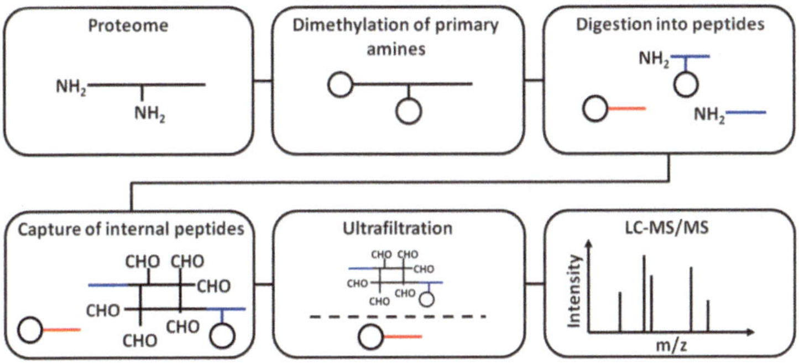

Fig. 5.7 Outline of the N-TAILS procedure

neo-N-termini, mature unmodified as well as naturally modified N-termini. The N-termini are then quantified according to their relative abundance in the protease treated and the control sample.

This N-terminomic approach has been widely applied and used for the identification of cleavage sites in complex samples. TAILS identified MMP-2-substrates in secretome samples derived from MMP-2$^{-/-}$ mouse embryonic fibroblasts (MEFs) incubated with recombinant MMP-2. Moreover it was applied for the discovery of cathepsin substrates in cell-based systems (Tholen et al. 2011, 2013a) as well as tissues from knockout and wild-type mice (Tholen et al. 2013b). Moreover, amyloid precursor protein (APP) was identified as a substrate for meprin β by TAILS (Jefferson et al. 2011). Meanwhile TAILS has been combined with iTRAQ reagents allowing the analysis of up to four or even eight samples in one experiment (Prudova et al. 2010). Additionally, a statistics-based platform for quantitative N-terminome analysis and identification of protease cleavage products has been established (auf dem Keller et al. 2010). TAILS requires low sample amounts and internal peptides are highly efficient removed by an aldehyde polymer, which is commercially available.

5.2.4 Isolation of C-Termini

5.2.4.1 COFRADIC-Based Enrichment of C-Terminal Peptides

In 2010 the group of K. Gevaert published an enrichment strategy for C-termini by SCX, based on peptide charge at low pH (Van Damme et al. 2010). In the first step all proteins are S-alkylated and all primary amines are blocked by (trideutero-) acetylation (Fig. 5.8). The proteins are digested using trypsin, which can only cleave after arginine residues. Afterwards the peptide mixture consists of α-amino-blocked N-terminal peptides, α-amino-free internal peptides and α-amino-free-C-terminal peptides, that all end with an arginine residue, except from the C-terminal peptides. In the next step these peptide mixture is passed through a SCX column using a low pH, where all internal peptides are captured caused by their positive charge at low pH whereas α-amino-blocked N-terminal peptides and α-amino-free-C-terminal peptides pass through. These enriched N- and C-terminal peptides are then separated by RP-HPLC. Afterwards C-terminal peptides are butyrylated and further separated in a second RP-HPLC run under identical conditions. Butyrylated C-termini elute later compared to the first RP-HPLC run, are collected in distinct fractions and analyzed by LC-MS/MS. This approach identified 334 neo-C-termini generated by granzyme B and 16 neo-C-termini generated by carboxypeptidase A4 in human cell lysates.

5.2.4.2 C-Terminal Amine-Based Isotope Labeling of Substrates

In 2010 Schilling et al. introduced C-terminal amine-based isotope labeling of substrates (C-TAILS), a targeted approach for the enrichment of C-terminal peptides from complex samples (Schilling et al. 2010). In this negative selection procedure first all primary amines are modified by reductive dimethylation using formaldehyde (Fig. 5.9). In the next step all carboxyl groups are protected by carbodiimide-mediated condensation of C-terminal carboxyl groups with ethanolamine. Following tryptic digestion new generated α-amines of N-terminal and internal tryptic peptides are dimethylated and removed based on their unprotected C-termini by coupling to the high-molecular-weight polymer poly-allylamine. Ultrafiltration separates the uncoupled, blocked C-terminal peptides that are subsequently analyzed by LC-MS/MS.

C-TAILS has been used for identification native protein C-termini together with neo C-termini, more than 100 cleavage sites were identified in an *Escherichia coli* proteome exposed to GluC. This approach can be used in combination with stable isotope formaldehyde-based labeling. A detailed protocol was recently published (Schilling et al. 2011c).

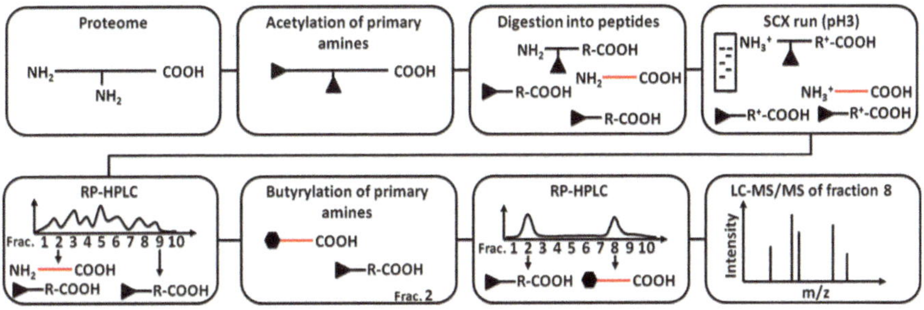

Fig. 5.8 Outline of the COFRADIC-based enrichment of C-terminal peptides procedure

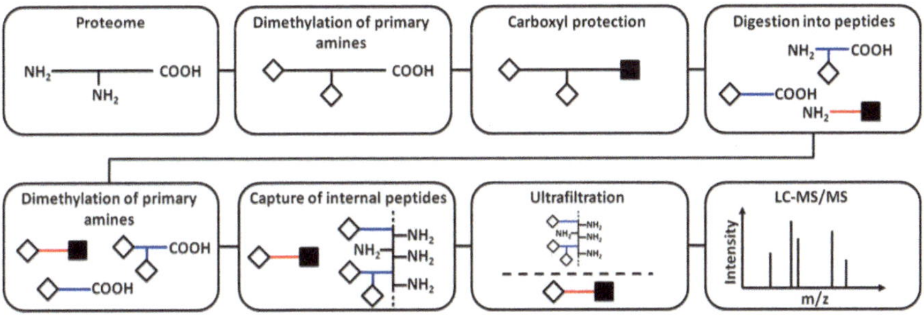

Fig. 5.9 Outline of the C-TAILS procedure

5.2.5 *Identification of Cleavage Sites by Non-selection Procedures*

5.2.5.1 The Protein Topography and Migration Platform

In 2008 the group of B. F. Cravatt developed protein topography and migration platform (PROTOMAP), an approach for the quantitative comparison of biological samples with the focus on proteolytic processing (Dix et al. 2008). This technique does not enrich for C-terminal or N-terminal peptides. In the first step an experimental and a control sample are separated independently by one-dimensional (1D) SDS-PAGE. Each lane is sliced into gel bands of fixed intervals. Peptides are released by tryptic digestion and analyzed using LC-MS/MS. Altered proteolysis of a given protein between both samples is indicated by reduced abundance at a given molecular weight and appearance of cleavage fragments at lower molecular weight. Since samples are not mixed and analyzed independently, PROTOMAP employs spectral counting for label-free quantification with peptide abundances being mapped on protein sequences to locate cleavage sites.

This technique is very powerful and has been successfully applied to establish proteolytic processing in the intrinsic apoptosis pathway in Jurkat T cells. Many known caspase mediated proteolytic events were validated and 150 additional proteins cleaved during apoptosis have been reported. In PROTOMAP, cleavage events leading to only a slight shift in molecular weight may not be detected and information on the exact cleavage site is missing but may be crucial to assign a cleavage site to a putative protease.

5.3 Validation of Cleavage Sites

5.3.1 Overview

The identification of cleavage sites by terminomic techniques results in potential protease substrates, which have to be validated in a subsequent step. Before performing post-screening experiments like biochemical and cell-biological approaches, cleavage sites that are physiologically irrelevant should be excluded. Especially if a proteome treated with a recombinant protease is compared to an untreated proteome, cleavage events can occur *in vitro* that would not have occurred *in vivo*. During sample preparation cell compartments can be disrupted, resulting in cleavage of a protein that would not meet the test protease *in vivo*. Moreover it is important to use physiological concentrations of the test protease. Higher concentrations of the test protease could increase the number of cleavage events. It is important to ensure for incubation conditions that reflect the *in vivo* situation and not induce partial unfolding of proteins thereby enhancing accessibility of cleavage sites (Huesgen and Overall 2012).

A major challenge is to distinguish cleavage sites directly generated by the protease under investigation from indirect cleavage events stemming from further proteases. Since proteases are organized in complex networks, the so-called "protease web", changes in protease activity of one protease will affect the activity of other proteases resulting in downstream effects like altered gene expression or protein turnover (Butler and Overall 2009; Overall and Dean 2006). This is of particular importance for the analysis of complex samples, for example for the comparison of protease-knockout and control samples or protease overexpression and control samples. In general, a combination of a terminomic technique and a quantitative proteome comparison is advised to distinguish direct and indirect cleavage events. In some cases, altered abundance of a proteolytically processed N-terminus reflects altered abundance of the corresponding protein rather than impaired proteolytic processing. Moreover the usage of different experimental setups like a combination of overexpression and silencing of a test-protease can prevent the identification of "false" cleavage events (Doucet et al. 2008).

To distinguish direct cleavage sites from downstream effects, knowledge about sequence specificity of the protease of interest is useful. Several techniques have been developed to characterize substrate specificity like substrate phage and bacterial display, peptide microarrays or positional scanning peptide libraries (auf dem Keller and Schilling 2010; Poreba and Drag 2010). Proteomic identification of protease cleavage sites (PICS) is a fast and effective alternative to these techniques (Schilling and Overall 2008) and will be introduced in the next section. Next to knowledge about the site specificity of a protease, knowledge about the subcellular localization can help to discriminate direct and indirect effects of altered protease activity. To prove for colocalization of a protease and its potential substrate, activity based probes (ABPs), molecules that can irreversibly bind to active proteases, are useful tools to monitor protease activity *in vivo.*

5.3.2 Proteomic Identification of Protease Cleavage Sites

Schilling et al. published in 2008 a technique called proteomic identification of protease cleavage sites (PICS) for characterization of prime- and non-prime specificity as well as subsite cooperativity in one experiment (Schilling and Overall 2008). A PICS experiment starts with generating a peptide library by digesting a proteome (e.g. cell lysate) with a specific protease such as trypsin, chymotrypsin or GluC (Fig. 5.10a). Following all primary amines and sulfhydryl groups of the resulting peptides are chemically protected and then exposed to the protease of interest (Fig. 5.10b). Newly generated primary amines are biotinylated with sulfo-NHS-SS-biotin, enriched using streptavidin, released by disulfide bond reduction and analyzed by LC-MS/MS. The corresponding non-prime side sequences are derived bioinformatically by database searches, thereby revealing the exact position of the cleavage site. PICS determines protease specificity using natural sequence diversity and enables the successful corroboration of cell contextual cleavage sites. A detailed protocol was published recently (Schilling et al. 2011b) as well as a web-based data-analysis resource termed WebPics (Schilling et al. 2011a).

PICS has been widely applied for the identification of cleavage site specificity including all protease classes. Amongst others, site specificities of MMP-2, caspases 3 and 7, cathepsins B, L, S, K and G, HIV protease 1, thrombin and elastase have been profiled. PICS has been used to validate cleavage sites identified by TAILS (Tholen et al. 2011). In this study more than 1,500 protein N-termini were identified, that mostly were contradicting cathepsin L specificity determined by PICS. This result indicates that altered cathepsin L activity affects numerous proteases and protease inhibitors leading to downstream effects. PICS revealed that cathepsin L specificity and the specificities of cathepsins B and S share some features (Biniossek et al. 2011). For instance, all three cathepsins prefer glycine residues in P1 and P1′ position. Another study revealed the cleavage site specificity for several members of the astacin metalloprotease family. A strong specificity for aspartate residues in P1′ position was observed for meprin α, meprin β and LAST_MAM (Becker-Pauly et al. 2011). Interestingly, cleavage site specificity was also influenced by proline in P2′ or P3′ position leading to an example of subsite cooperativity. Here, the obtained specificities validated TAILS data and results of other biochemical approaches revealing processing of vascular

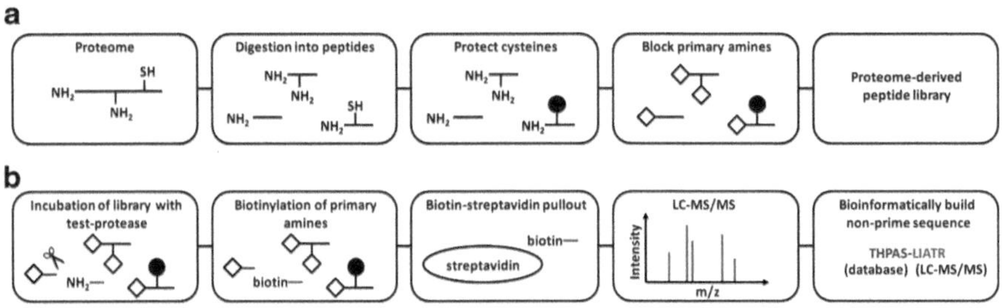

Fig. 5.10 Outline of the PICS procedure. (**a**) Library generation. (**b**) Cleavage site screen

endothelial growth factor-A (VEGF-A) by meprin α and processing of the serine protease pro-kallikrein 7 by meprin β.

5.3.3 Activity-Based Probes

Proteomic techniques determine protein abundance, but do not account for changes in protease activities. To fully understand protease action monitoring of protease activity is fundamental, but how proteases carry out their biological functions is a challenging task, since proteases can be regulated by small molecules, interactions with other proteins or PTMs (Drag and Salvesen 2010; Shen 2010). Moreover many proteases are synthesized as inactive zymogens that are activated by proteolysis or conformational changes. For instance, cysteine cathepsins are processed in a pH-dependent manner in the lysosomal compartment (Conus and Simon 2010). ABPs can help to distinguish active and inactive proteases, thereby revealing the localization of protease activity. Meanwhile ABPs are applied for biomarker discovery, drug screening and *in vivo* imaging (Deu et al. 2012; Fonovic and Bogyo 2007).

ABPs are small molecules that irreversibly bind to active proteases, but not inactive or inhibited proteases (Kidd et al. 2001; Liu et al. 1999). The general structure of ABPs consists of a chemically reactive group termed as a “warhead”, a spacer region, that targets the probe to the target protease, and a tag, normally a fluorescent dye or an affinity tag such as biotin (Deu et al. 2012) (Fig. 5.11). ABPs operate by irreversible binding of the warhead to the active site nucleophile residue of the target protease. The tag allows for visualization allowing the proof of colocalization of the protease activity in the candidate substrate, and/or for isolation of the ABP-protease-complex, identifying the active protease from complex samples (Doucet and Overall 2008).

The main application of ABPs is their use to visualize protease activity using modern imaging techniques, revealing localization and distribution of proteases in cells and *in vivo*. This becomes of special interest for cancer research and treatment,

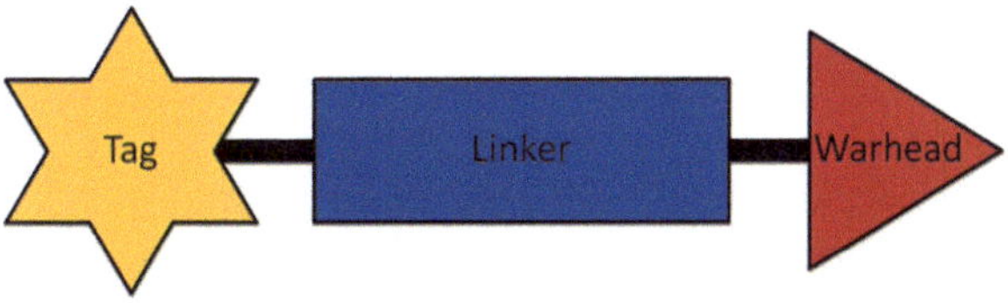

Fig. 5.11 Schematic structure of activity-based probes (ABPs); adapted from Paulick and Bogyo (2008)

since almost all human cancer tissues show increased protease activity (Blum 2008). In mice developing pancreatic cancer, but also other cancer tissues, ABPs targeting cysteine cathepsins have been used to noninvasively detect cathepsin activity (Blum et al. 2005, 2007; Joyce et al. 2004). Moreover to study the kinetics of apoptosis caspase activity has been monitored in multiple mouse models (Edgington et al. 2009, 2012). All these studies also highlight the potential of ABPs to monitor the response of protease activity to a drug treatment, which could be a helpful tool to keep tumor growth under surveillance. Most ABPs target cysteine or serine proteases, since the development of ABPs for metalloproteases or aspartic proteases is more challenging. During their action an activated water molecule conducts the nucleophilic attack towards the amide bond preventing the ABP from covalent binding to the active site nucleophile (Doucet and Overall 2008). Meanwhile ABPs for metalloproteases are available that contain an additional photoactivable handle anchoring the ABP to the protease via an amino acid outside the active site (Saghatelian et al. 2004). In summary, there are ABPs targeting almost all protease classes such as metalloproteases (Chan et al. 2004; Saghatelian et al. 2004), threonine proteases (Kessler et al. 2001; Wang et al. 2003), cysteine proteases (Greenbaum et al. 2000; Thornberry et al. 1994) and serine proteases (Kidd et al. 2001; Williams et al. 1989). A major challenge remains the development of probes that selectively bind to one member of a protein class, but some new approaches have emerged (Blair et al. 2007; Hagel et al. 2011). Another limitation of ABPs is that their production requires extensive knowledge of organic chemistry and synthesis efforts, because the probes have to be stabile, not toxic and suitable for *in vivo* use.

Besides *in vivo* imaging applications ABPs have been combined with mass spectrometry to isolate active proteases from complex proteomes for identification and quantification, a technique known as ABPP-MudPIT (activity-based protein profiling—multi-dimensional protein identification technology) (Speers and Cravatt 2009). In these approaches often lysates of cells or tissues are incubated with a probe targeting a certain protease class, afterwards the ABP-protease complexes are isolated by affinity purification and then identified by MS analysis. ABPs have been used to profile metalloprotease activity in several cell lines (Saghatelian et al. 2004). More than 20 metalloproteases were identified by applying a cocktail of metalloprotease-directed probes to cell proteomes followed by LC-MS/MS (Sieber et al. 2006). Additionally, the role of cysteine proteases in tumorigenesis, microbial pathogenesis or apoptosis was investigated using ABPs (Berger et al. 2006; Paulick and Bogyo 2008; Puri and Bogyo 2009).

5.4 Proteolysis Regulation Networks: Exemplary Studies

Proteases are involved a multitude of physiological reactions from simple digestion of food proteins to highly regulated systems. Maintaining homeostasis in living systems requires strict control of all physiological processes. Importantly, most of these processes are directly or indirectly interconnected. Also, sequential effects are interdependent, meaning that diverse initial events can result in the same downstream process. Thus, the term networking of proteases and inhibitors better depicts the actual state of our knowledge on proteolytic regulation.

Proteases create a complex network of sequential activation and inhibition episodes. The main agents in this system are: (i) zymogens, (ii) activated proteases, and (iii) protease inhibitors. On the most basic level, zymogen activation is a proteolytic process, hence representing an initial layer of functional protease interconnectedness.

Maintaining the balance between active and non-active forms of proteases as well as their inhibitors is crucial for appropriate cellular processes. Many proteases act in cascades, which integrate and amplify primary proteolytic "signals". Thus, they yield dominant downstream effects that often overshadow the initiating proteolytic events. Most cascades are tightly controlled to sustain an appropriate homeostasis within the proteolytic system (Garcia-Verdugo et al. 2010; Le Magueresse-Battistoni 2007).

The most prominent example of a zymogen activation cascade is the coagulation system, which sequentially recruits factors required for blood clotting. The coagulation factors are mostly serine proteases (Morrissey 2012; Ott 2011; Zogg and Brandstetter 2009). Two parallel pathways are involved in the cascade depending on the initiation source. In the extrinsic pathway the blood clotting process begins when the vessel wall is injured and membrane bound tissue factor forms a complex with the zymogen FVIIa in circulating blood, finally leading to activation of factor X. In the intrinsic pathway of coagulation a sequence of reactions leading to fibrin formation begins with the contact activation of factor XII, and also results in the activation of factor X. As an example, this protease activation cascade is shown in Fig. 5.12.

In addition to these prototypical cases, recent data indicates widespread networking and regulation of proteolytic activities of proteases of different kinds.

Protease networking is illustrated by the interplay between members of the papain subfamily of cysteine proteases, which consists of 11 human cysteine cathepsins (Rawlings et al. 2006), and their physiological inhibitors. In the present compilation, a detailed review on cysteine cathepsins is provided in the sections "Cathepsins: getting in shape for lysosomal proteolysis" and "Exploring systemic functions of lysosomal cysteine proteases: the perspective of genetically modified mouse models". As for most other protease families, cathepsin activity is regulated on several levels. Cathepsins are strictly regulated by endogenous inhibitors and impaired regulation leads to various malignancies (Glondu et al. 2002; Hu et al. 2008; Sevenich et al. 2010). Endogenous cysteine cathepsin inhibitors include

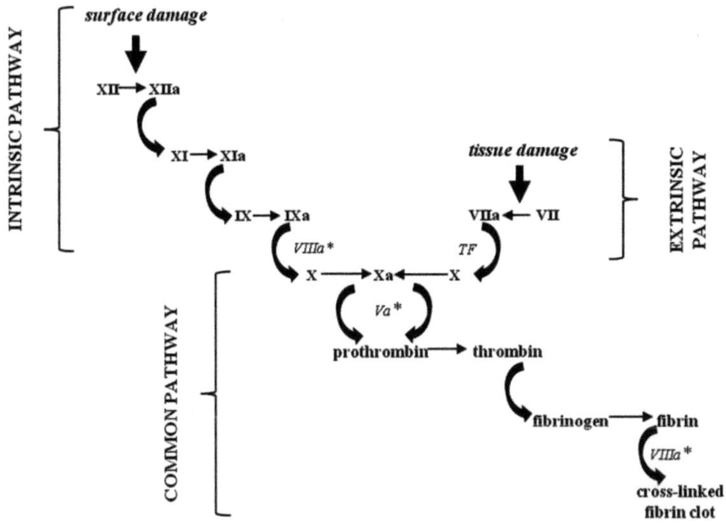

Fig. 5.12 Blood clotting cascade (blood coagulation pathway); *Roman numerals* represent coagulation factors, a—active form, *Asterisk*—activated by thrombin. *TF* tissue factor

extracellular cystatin and intracellular stefins (Dubin 2005). Overexpression of cystatin C, cystatin M and stefin/cystatin A in tumor cell lines has shown that cysteine cathepsins have functional roles in growth, invasion and metastasis of tumor cells of both epithelial and mesenchymal origins. On the other hand, cystatin M inhibition significantly increased the enzymatic activities of cathepsins B and L and legumain (Li et al. 2005; Sokol and Schiemann 2004; Vigneswaran et al. 2006; Zhang et al. 2004). Other studies showed that cystatin M as well as cystatin C also inhibit legumain (Rawlings et al. 2006). Regulation, and hence interplay on the enzyme-inhibitor axis was further shown in studies using the mouse model of multistage pancreatic islet cell carcinogenesis. As an example selective cathepsin S deficiency impaired angiogenesis and tumor cell proliferation, angiogenic islet formation, and the growth of solid tumors; whereas the absence of its endogenous inhibitor cystatin C had the opposite result (Wang et al. 2006).

Compensatory effects are also found for cysteine cathepsins and constitute another aspect of protease networking. Cathepsins B and Z are the only carboxypeptidases among the cysteine cathepsins (Klemencic et al. 2000). During breast cancer development in the PymT mouse model both cathepsins B and Z exert synergistic anticancer effects. Single deficiencies show partial or full reciprocal compensation (Sevenich et al. 2010) and deletion of either cathepsin B or Z alone produces considerably mild phenotypes. In fact, a more substantial anti-cancer phenotype occurred in the double knock-out, in comparison to the single knockout or wild type mice (Sevenich et al. 2010; Vasiljeva et al. 2006). Thus, only the combined loss of both proteases led to significant reductions in tumor and metastatic burden, while single deficiencies resulted in reciprocal compensation.

Additionally, migratory and invasive properties of tumor cells with different cathepsin B and Z genotypes were investigated *in vitro*. Notably, cathepsin B or cathepsin Z deficient tumor cells showed less invasive phenotypes compared to wild type controls, However, the invasiveness of the cathepsin B and Z double-deficient cells was most impaired, indicating a synergistic effect of cathepsin B and Z on cancer cell invasion by proteolytic matrix remodeling (Sevenich et al. 2010).

These results demonstrate that although cathepsins B and Z are critical for early tumor formation and affect this process independently, the loss of cathepsin Z function in promotion of established tumors is completely compensated by cathepsin B alone or in combination with proteases other than cysteine cathepsins, whereas cathepsin Z at least partially compensates for a lack of cathepsin B (Sevenich et al. 2010).

A recently discovered example of protease networking is degradation of cystatin C by cathepsin D, a ubiquitously expressed aspartic endoprotease. Yeast 2-hybrid screening using human cathepsin D as a bait and a cDNA library isolated from normal human breast tissue identified cystatin C as a putative cathepsin D interaction partner. It has been confirmed that both cathepsin D precursor and cystatin C interact extracellularly at neutral pH. Transcriptome analysis showed that the cystatin C extracellular levels are cathepsin D dependent. Degradomic analysis demonstrated that cathepsin D cleaves cystatin C at acidic pH at multiple sites. Additionally, system-wide effects of cathepsin D depletion were observed. Upon cathepsin D silencing, cystatin C was not degraded and the extracellular activity of cysteine cathepsins was decreased (Laurent-Matha et al. 2012).

Cystatin M/E is the endogenous inhibitor of cathepsin L and legumain. In mice, cystatin M/E deficiency is embryonically lethal. Lethality is rescued by additional depletion of cathepsin L. Cystatin E/M can regulate both intracellular and extracellular processing and activation of legumain. It has been observed that prolegumain was highly secreted into the conditioned media of legumain overexpressing cells. Additionally the increased processing of cathepsin L to the two-chain form in legumain expressing cells has been showed, whereas this processing was inhibited in the cystatin E/M overexpressing cells (Smith et al. 2012).

Interestingly, loss of legumain does not rescue the cystatin M/E-knockout phenotype. By generating mice lacking both cystatin M/E and cathepsin L, it was shown that cathepsin L deficiency rescues the lethal cystatin M/E-knockout mice phenotype. Also, in the same study it was shown that cathepsin D (unlike cathepsin L) is able to process legumain, and in addition some autoactivation was observed, which makes the interplay between proteases even more complex (Zeeuwen et al. 2010). This is an example on how sophisticated *in vivo* studies untangle a protease—inhibitor network.

Networking in the regulation of protease activity is also illustrated by examples within the metalloproteinase family. Inhibitors often target multiple proteases so alteration in inhibitor levels can affect seemingly unrelated proteases. Tissue inhibitors of metalloproteases (TIMPs) are another example: four TIMPs each target a different array of metalloproteases, including different MMPs and ADAMs (Murphy 2011). Moreover, non-canonical inhibitory profiles are beginning to emerge. For example, cystatin C (the prototypical cysteine protease inhibitors) also exerts inhibitory activity on meprin metalloproteases (Hedrich et al. 2010).

Activation of MMP-2 involves a ternary complex, consisting of another metalloprotease (typically MT1-MMP) and TIMP2 (Butler et al. 1998). A trimeric complex is formed on the cell surface by an interaction between the carboxyterminal domain of TIMP-2 and the hemopexin domain of proMMP-2(Strongin et al. 1993). Notably, the local concentration of TIMP-2 in the tissue is crucial for the complex formation and MMP-2 cleavage regulation; if it is too low, insufficient proMMP is brought to the cell surface, whereas too high TIMP-2 concentration inhibits all of the MT1-MMP and thus blocks the initial cleavage. Interestingly, the experiments on TIMP-2 knock-down cells, with different expression levels of MT1-MMP did not exhibit any MMP-2 active form production, which indicates that MT1-MMP absolutely requires TIMP-2 to catalyze the conversion of MMP-2 to the fully active form (Butler et al. 1998). Notably, in this context a protease inhibitor augments protease activity rather than repress it, thereby adding further complexity to proteolytic systems *in vivo*.

Impaired control of protease signaling leads to severe abnormalities, defects, and even lethality. The SPINK5 gene encodes the serine protease inhibitor LEKTI, lympho-epithelial Kazal-type related inhibitor (Magert et al. 1999). LEKTI is expressed in the differentiated viable layers of stratified epithelium. In the epidermis, it is mainly restricted to the granular layer (Bitoun et al. 2003). LEKTI is comprised of 15 potential Kazal-type serine proteinase inhibitory domains. The full-length recombinant protein was shown to inhibit serine proteases such as trypsin, plasmin, subtilisin A, cathepsin G, and elastase (Mitsudo et al. 2003); however, kallikrein proteases are thought to constitute the primary in vivo targets of LEKTI (Jayakumar et al. 2004).

Due to its versatile activity, LEKTI is involved in multiple biological pathways relevant to tissue homeostasis, inflammation and antimicrobial defense. LEKTI is involved in the regulation of proteolytic events crucial for barrier formation and maintenance. Loss of balance between LEKTI and its target proteases leads to severe skin barrier defects with repetitive inflammations and allergic symptoms The impaired functioning of LEKTI due to mutations leads to Netherton syndrome (Chavanas et al. 2000). The group of Alain Hovnanian has shown that LEKTI-deficient mice faithfully replicate a skin phenotype reminiscent of the Netherton syndrome in humans. To a large extent, the severe skin phenotype is a result of unregulated kallikrein 5 and kallikrein 7-like activity and subsequent loss of stratum corneum adhesion through degradation of desmoglein. In mice, further depletion of matriptase, an auto-activating transmembrane serine protease, rescues the skin phenotype stemming from LEKTI deficiency. Matriptase initiates kallikrein activation, hence loss of matriptase counter-balances excessive kallikrein activity in LEKTI deficient mice. Although this is a superb example for fine-tuned interaction of proteases and protease inhibitors, it should be noted that matriptase is not a natural target of LEKTI (Chavanas et al. 2000). Deficiency of matriptase alone—in the presence of LEKTI—impairs epidermal barrier function by weakening epidermal tight junctions (List et al. 2009). In LEKTI-deficient mice, matriptase initiates Netherton syndrome by premature activation of a pro-kallikrein-related cascade. Once converted to its active form, kallikrein-related peptidase 5 is capable of proteolytically activating pro-kallikrein-related peptidase 5 and pro-kallikrein-related peptidase 7, in line with the previously proposed role of the protease in the propagation of an epidermal prokallikrein-related peptidase cascade (Sales et al. 2010).

5.5 Interactions Between Proteolytic Processing and Post Translational Modification

Limited proteolysis and protein degradation are essential processes in a diversity of (patho-) physiological responses. Proteolysis is integrated in a complex network of multiple PTMs, such as phosphorylation, acetylation or ubiquitination. While the ubiquitin proteasome system (UPS) represents a prototypical case for synergistic and co-operative action of proteolysis with another type of PTM, proteome-wide studies are now beginning to shed light on further links, e.g. between proteolysis and phosphorylation. However, while the ubiquitin proteasome system has been understood in great detail (see below), much less is known how limited proteolysis is fine-tuned by PTMs.

Although the precise mechanisms of protein degradation processes are yet to be fully defined, our knowledge in understanding the cellular and molecular mechanisms of different proteolytic pathways has increased significantly over the years (Attaix et al. 2001). Some of the major degradation pathways utilize lysosome proteases (Agarraberes et al. 1997; Cuervo et al. 1997; Franch et al. 2001), calcium-dependent proteases (Sorimachi et al. 1997), metalloproteases (Yong et al. 1998), proteases that are involved in apoptosis (Tseng et al. 2008), as well as the ubiquitin-proteasome system (Coux et al. 1996; Goldberg et al. 1995; King et al. 1996). The ubiquitin-proteasome system is perhaps the most comprehensively described degradation pathway to date (Price et al. 1996; Rock and Goldberg 1999). This complex pathway is one of the major proteolytic processes that are responsible for the removal of abnormal or damaged proteins in all living cells (Ciechanover et al. 1980; Etlinger and Goldberg 1977). It involves a cascade of enzymatic reactions including activation and attachment of ubiquitin to proteins that are targeted for degradation (Fig. 5.13). This process comprises several ubiquitin ligases with varying substrate specificity and different roles in cellular physiology.

Ubiquitin chains are linked through Lys^{11}/Lys^{48} for proteosomal degradation, however alternatively, other linkages (Lys^{11}, Lys^{29} and Lys^{63}) have also been previously described (Weissman 2001; (Komander and Rape 2012). Structure and composition of the (poly)ubiquitin chain constitutes itself a cellular signal with different functional consequences and outcomes; e.g. lysosomal or proteasomal degradation. Details of the "ubiquitin code" have been reviewed previously

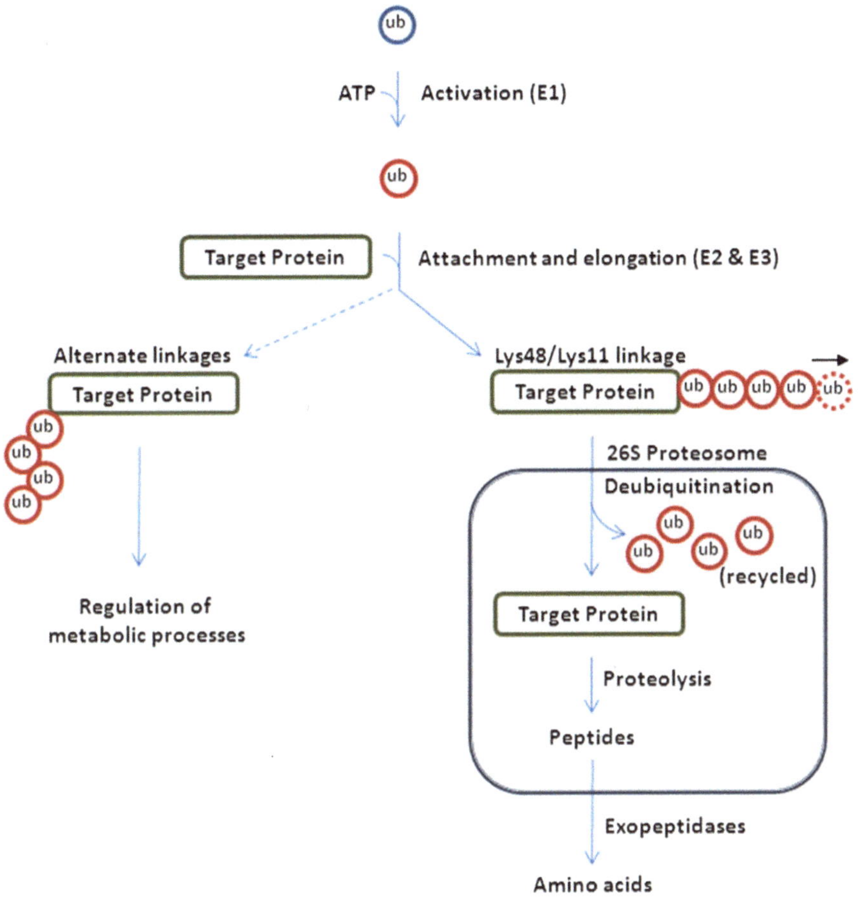

Fig. 5.13 Ubiquitin-proteasome system for proteolytic degradation of proteins into free amino acids

(Komander and Rape 2012). It is well established that when monoubiquitylation and homogenous chains of four or more ubiquitins are formed, this chain is recognized by the 26S proteasome for degradation (Komander and Rape 2012). Within the proteasome complex, the polyubiquitin chain is cleaved for recycling prior to cleavage of targeted protein. Once the proteins are degraded into peptide fragments, they are released from the proteasome for further degradation into free amino acids by exopeptidases in the cytosol. The elements that constitute the ubiquitin-proteosome system had been extensively reviewed in (Chitra et al. 2012; Debigare and Price 2003). Proteolysis that involves the ubiquitin-proteosome pathway underlies a multitude of cellular events such as cell growth and proliferation, antigen presentation and DNA repair. Further examples of cellular processes involved include the degradation of protein regulators and inhibitors. When regulators such as cyclin or some transcriptional regulators should act transiently, or when a process is initiated by the degradation of inhibitors such as cyclin-dependent kinase inhibitors (Ckis), or a class of transcriptional inhibitor proteins (IκBs), proteolysis of these regulators and inhibitors is achieved through the ubiquitin-proteosome pathway (Pagano et al. 1995).

Small ubiquitin-like modifier (SUMO) proteins are similar to ubiquitin. However unlike ubiquitin, sumoylated proteins are not targeted for destruction by the proteasome. Instead, sumoylation of proteins are associated with various cellular processes such as nuclear-transport, transcriptional regulation, cell death, protein stability, response to cellular stress, and cell cycle (Hay 2005). Studies have shown that the N-terminal sumoylation of the main ubiquitin E3 ligase for p53, MDM2, reduces the self-ubiquitination and degradation activity by proteasome (Buschmann et al. 2001; Lee et al. 2006; Miyauchi et al. 2002). Similarly, tumor the tumor suppressor protein, p53, can be modified by sumoylation with either up-regulation or down-regulation of its activity (Hock and Vousden 2010). Furthermore a recent report had shown that the protooncogene SKI is able regulate sumoylation of MDM2 and p53, which leads to the increase in MDM2 self-ubiquitination activity and enhanced degradation of p53 (Ding et al. 2012). Another known PTM that acts reversibly on lysine residues is acetylation. Interestingly, a recent proteome-wide study has shown substantial overlap of acetylation and ubiquitination site in Saccharomyces cerevisae (Henriksen et al. 2012). This acetylation site is localized within the critical regulatory domain of SAGA (Spt-Ada-Gcn5-Acetyltransferase) complex and it is involve in the Ubp8-containing histone H2B deubiquitylase complex (Henriksen et al. 2012). It is likely that the acetylation of this site reversibly shields the protein complex from UPS-based degradation.

Further studies have described protective effects of PTMs from proteolytic cleavage. Huntington's disease (HD), an autosomal dominant neurodegenerative disorder, is characterized by impaired muscle movements, psychiatric problems and cognitive decline (Rubinsztein and Carmichael 2003). It is caused the abnormal expansion of polyglutamine (polyQ) tract in the huntingtin (htt) protein. It was proposed that proteolysis of the polyQ proteins by a range of proteases such as caspases (Goldberg et al. 1996; Wellington et al. 2002), calpains (Gafni and Ellerby 2002; Kim et al. 2001), and aspartic proteases (Lunkes et al. 2002) to generate shorter, diffusible fragments that are responsible for aggregation. Studies have shown that polyQ repeats in htt protein when cleaved, yield short amino-terminal toxic fragments (Kim et al. 2001; Lunkes et al. 2002; Wellington et al. 2000), which in turn lead to protein aggregation and cellular toxicity. A number of studies have shown that htt exhibits a range of PTMs, including ubiquitination (Kalchman et al. 1996), sumoylation (Steffan et al. 2004) and phosphorylation (Schilling et al. 2006a). It has been reported that htt undergo calpain-mediated proteolytic cleavage at amino acid Ser^{536} (Gafni et al. 2004). *In vitro* study showed that this amino acid is able to undergo phosphorylation to prevent cleavage and modulate cellular cytotoxicity of the polyQ disease. In addition, another study has reported that It has been reported that Cdk5-mediated phosphorylation of Ser^{434} reduces caspase-mediated htt cleavage of the protein at residue Asp^{513} (Luo et al. 2005). In another polyQ related study of Kennedy's disease (degeneration of motor neurons), it was also shown that phosphorylation of Ser^{514} of the androgen receptor that blocks caspase-3 cleavage, thus preventing cell death (LaFevre-Bernt and Ellerby 2003).

The paradigm in which protein cleavage can reveal PTM site(s), and conversely, PTM of protein can promote proteolytic cleavage, highlights the tight regulation and diverse interactions of proteolysis and post-translational processing. Apart from PTM and proteolytic crosstalk in degradation, protease-mediated cleavage and PTM such as protein phosphorylation events play essential roles in regulating multitude biological and pathological processes including tissue development, cancer and apoptosis (Kurokawa and Kornbluth 2009; Lopez-Otin and Hunter 2010). It has been proposed that up to 5 % of the proteome is subjected to caspase-mediated proteolysis during apoptosis (Arntzen and Thiede 2012; Crawford and Wells 2011). The diverse roles of caspase include activation as well as inactivation of protein kinases that are involved in phosphorylation. For example, caspase 3 is able to cleave and inactivate focal adhesion kinase (FAK) to downregulate phosphorylation (Taylor et al. 2008). The complexity of caspase functions is highlighted in a study by Cravatt and colleagues, where more than 700 cleaved proteins were identified in Jurkat T cells that are involved in apoptotic pathway, with over 5,000 phosphorylation sites (Dix et al. 2012). Approximately 500 apoptosis-specific phosphorylation were enriched on cleaved proteins and clustered around sites of caspase proteolysis (Dix et al. 2012). The dynamic interactions between phosphorylation kinases and proteases have been extensively reviewed previously by Lopez-Otin and Hunter (Lopez-Otin and Hunter 2010). The association of proteolytic cleavage/degradation and PTM is also well characterized in Alzheimer's disease. The aggregation of Aβ peptides is influenced by different PTMs (Kuo et al. 1998) [such as peptide truncations (Hartig et al. 2010; Kuo et al. 1997; Miravalle et al. 2005; Saido et al. 1996; Tekirian et al. 1998), racemization (Mori et al. 1994; Tomiyama et al. 1994), isomerization (Murakami et al. 2008; Shimizu et al. 2000), pyroglutamination (Kuo et al. 1997; Saido et al. 1995), metal induced oxidation (Dong et al. 2003) as well as phosphorylation (Kumar et al. 2011)] which contribute to the insolubility, stability and resistance of the amyloid filaments to proteolytic degradation (Fig. 5.14). While the majority of cellular PTMs (e.g. phosphorylation and glycosylation) are actively induced by cells through dedicated enzymes (e.g. kinases), other PTMs such as oxidation and racemization rather represent accidental chemical events against which cells tend to possess protective machineries. It has been reported that some of these post-translationally modified Aβ peptides are detected in early stages of Alzheimer (Hartig et al. 2010; Kumar et al. 2011; Schilling et al. 2008; Sergeant et al. 2003), resulting in higher cellular cytotoxicity compared to non-modified counterparts (Millucci et al. 2010). It has also been proposed that these modified peptides serve as seeding species to propagate the aggregate formation of amyloid plague in vivo (Kumar et al. 2011; Schilling et al. 2006b, 2008).

Another significant PTM that involves in proteolytic degradation include protein glycosylation. Glycosylation represents the most abundant extracellular PTM in eukaryotes. It has been suggested that glycosylation can act as a protective storage

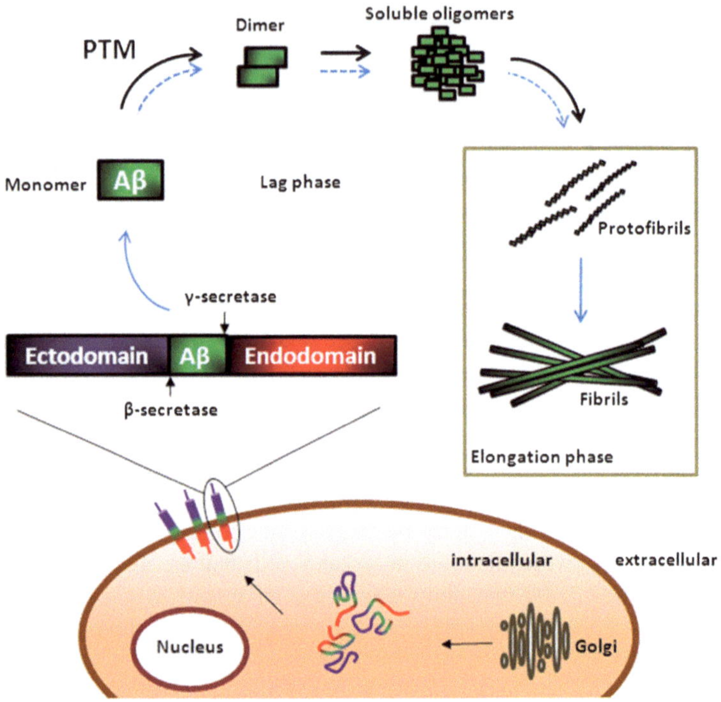

Fig. 5.14 Schematic diagram showing fibrils formation from Alzheimer's amyloid β-peptide. Post-translational modifications of amyloid β-peptide reduce lag phase promoting aggregation

depot for some important glycoproteins, such as growth factors from non-specific proteolysis in the extracellular matrix and thus prolonging their activities (Varki 1993). Another implicated role of post-translational glycosylation of extracellular proteins is the enhancement of structural stability and rigidity by sugar moieties (Imperiali and O'Connor 1999; Wyss et al. 1995) and it also assists in folding and transport processes, by protecting glycoproteins from proteolytic cleavage (Opdenakker et al. 1993). Many examples have demonstrated the protective role of glycosylation from proteolytic cleavage. For example, in Drosophila, the glycosylation of acetylcholinsterase, a membrane-anchored protein, prevents it from proteolytic cleavage by cellular processing enzyme(s) (Mutero and Fournier 1992), while in bacteria, glycosylated cellulases from Cellulomonas fimi had been shown to be protected by C. fimi protease (Langsford et al. 1987). Another example of bacterial glycoproteins has also shown a protective effect of glycosylation against proteolytic cleavage of a lipoprotein antigen from Mycobacterium tuberculosis by cellular proteases, where up to six cleaved fragments of the protein were observed when the protein was not glycosylated (Herrmann et al. 1996). However, like phosphorylation, glycosylation could also play a role in the activation and/or facilitating proteolysis. Studies have reported the involvement of glycosylation in proteolytic processing of various proteins, including CREB-H (Chan et al. 2010), CD97 (Hsiao et al. 2009), anti-inflammatory protein A20 (Shrikhande et al. 2010), ATP-binding cassette transporter ABCG2 (Nakagawa et al. 2009), and immunoglobulin A (Taylor and Wall 1988). Nevertheless, PTMs are one of the major processing events in cellular homeostasis and they are implicated in the pathogenesis of various diseases. More importantly, PTM could result in enhanced or ablation of specific proteolytic activity.

5.6 Protein Turnover

Initially proteases were not regarded as "precision tools" altering activities/characteristics of their respective substrates and protease cleavage was not regarded as posttranslational modification. The major role of proteases was thought to be protein degradation, which is indeed one of their vital and important functions. As cells have to respond to changing environmental conditions the cellular proteome is constantly being turned over with protein synthesis and degradation balancing each other under steady-state conditions (Ciechanover 2005). Mass spectrometry (MS)-based proteomics has been successfully employed to study protein turnover and underlying molecular mechanisms [for review see Engelke et al. (2012)]. Especially the development of quantitative MS methods has greatly expanded our knowledge on protein dynamics of degradative processes.

Historically, protein degradation was thought to take place solely in lysosomes (De Duve et al. 1953). However, the observation that protein half-lives differ substantially raised the question how bulk lysosomal degradation should account for this. The discovery of the ubiquitin-proteasome system (UPS) provided an unexpected answer: next to lysosomal degradation proteins can be specifically targeted to proteasomal degradation by the covalent attachment of ubiquitin (Ciechanover 2005). To date, the UPS and the autophagosomal-lysosomal system are regarded as the two major cellular degradation systems. However, autophagy, lysosomal degradation of intracellular proteins, has emerged as tightly regulated protein degradation pathways being far less "unspecific" than initially anticipated (Yang and Klionsky 2010). Several autophagy subtypes allowing specific degradation of protein complexes and organelles have been described such as mitophagy for mitochondrial degradation (Kim et al. 2007; Tolkovsky 2009), reticulophagy for ER degradation (Bernales et al. 2006), ribophagy for ribosomal (Beau et al. 2008; Kraft et al. 2008), pexophagy for peroxisomal degradation (Dunn et al. 2005). In addition, xenophagy, the selective removal of intracellular bacteria and viruses via the autophagic machinery, has attracted attention, particularly due to its important role in human health and disease (Romao and Munz 2011). Also the unspecific nature of classical macroautophagy, the stereotype of lysosomal bulk degradation, has been debated lately. Thus, it could be shown that organelles are degraded in an ordered fashion during starvation-induced macroautophagy (Kristensen et al. 2008) and that autophagosomal proteome compositions reflect the inducing stimuli and differ over time (Dengjel et al. 2012).

In protein turnover studies classic radiolabeling principles have been transferred to quantitative MS-based proteomics approaches using metabolic, stable isotope labeling strategies allowing global unbiased investigations (Hinkson and Elias 2011). Protein degradation constants can be determined by pulse-labeling approaches using the decrease of stable isotope-labeled peptide signals (Pratt et al. 2002). Compared to studies using radiolabeled tracers proteomics studies allow the use of more than one isotopic label, which enables the simultaneous recording of protein turnover, synthesis and degradation values (Boisvert et al. 2012). Also absolute quantification approaches have been successfully employed to study both, protein synthesis and degradation (Schwanhausser et al. 2011). MS-based proteomics experiments did not only lead to new insights with respect to protein turnover, also mechanistic insights on protein degradation have been generated.

5.6.1 Ubiquitin-Proteasome System

The 26S proteasome has been investigated in detail and MS-based proteomics approaches highlighted the heterogeneity of the multi-protein complex. Hence, proteasomes containing two different activator subunits could be described, as well as proteasomes differing in their posttranslational modification status (Drews et al. 2007; Schmidt et al. 2005; Wang et al. 2007). Another field of active research is the role of proteasomal activity in the generation of MHC peptides and their implication on human immune responses (Mester et al. 2011).

Next to proteasomal composition and activity the role of ubiquitin and ubiquitin-like modifiers has been extensively studied by MS-based proteomics. The existence of branched ubiquitin chains, the use of all seven ubiquitin lysine residues for branching, and their involvement in protein degradation was outlined (Peng et al. 2003; Xu et al. 2009b). In general, ubiquitinated proteins have to be enriched prior analysis, which can be done by several ways: immunoaffinity purifications using anti-ubiquitin antibodies or tagged ubiquitin versions, as e.g. HA-, FLAG-, or His-tagged ubiquitin, and affinity purifications (AP) by ubiquitin-binding domains. The latter has been successfully employed inter alia to study ubiquitination dynamics during growth factor receptor signaling (Akimov et al. 2011). In global, bottom-up approaches a di-glycine remnant can be identified by MS, which marks ubiquitination sites after tryptic digestion. A purification strategy using a respective monoclonal antibody has been developed (Xu et al. 2010). However, as also ubiquitin-like modifiers like NEDD8 and ISG15 leave the same mark the data has to be interpreted cautiously (Kim et al. 2011).

5.6.2 Autophagosomal-Lysosomal System

MS-based proteomics approaches have focused on (I) protein-protein networks responsible for regulation of autophagosomal-lysosomal protein degradation and (II) studying of underlying organellar proteome dynamics (Zimmermann et al. 2010). In an extensive protein interaction study a complex, autophagy-related protein network was outlined having implications on vesicle trafficking, protein and lipid phosphorylation (Behrends et al. 2010). Interestingly, a sequence motif was identified as binding to the autophagosomal marker protein LC3 (LC3 interacting region; LIR) serving as a signal for specific autophagy degradation receptors (Johansen and Lamark 2011). It could be shown that in the case of the protein optineurin the interaction is phosphorylation dependent and that phosphorylation promoted selective removal of ubiquitin-coated cytosolic Salmonella bacteria by autophagy (Wild et al. 2011).

Both, lysosomes as well as autophagosomes, have been studied by proteomics means and were either enriched by AP, or profiled over gradient centrifugations. As lysosomes are relatively labile AP have been used, e.g. magnetic separation of iron-containing lysosomes and immunoprecipitation of whole organelles by anti-vacuolar H+-ATPase antibodies allowing insights into lysosomal trafficking (Cardoso et al. 2009; Nylandsted et al. 2011). Autophagosomes have been studied by both AP and profiling approaches. AP purifications are commonly performed using anti-GFP antibodies in combination with cells stably expressing the autophagosomal marker protein LC3 fused to GFP (Dengjel et al. 2012; Gao et al. 2010). In combination with time-resolved data and different autophagy inducing stimuli interesting insights into organellar dynamics have been generated identifying a close crosstalk between the UPS and the autophagosomal-lysosomal system (Dengjel et al. 2012).

5.7 Outlook

Proteomic-based protein discovery has seen rapid technical advances and enhanced our understanding of diverse and complex systems by applying various mass spectrometry based techniques. The area of degradomics has evolved in recent years and has started a new era, enabling identification of proteolytic cleavage products of natural substrates in their biological context in cell-based as well as in tissue samples. A number of key publications describing promising techniques arose in the past 5 years that build up the basement for protease substrate discovery. Now is the time to apply these techniques in sophisticated experimental settings. Degradomic techniques have already identified hundreds of potential substrate candidates and many cleavage sites have been validated by biochemical or cell-based approaches uncovering multiple roles of proteases in physiology and pathology. Even though not many *in vivo* degradomic studies have been published to date, recent presentations and discussions on scientific meetings raise the assumption of an upcoming wealth of knowledge.

A key challenge will be the discrimination between direct and downstream effects resulting from alterations in protease activity *in vivo*. Substrate candidates revealed by degradomics have to be validated by biochemical approaches, such as immunoblotting, *in vitro* cleavage assays, imaging techniques to establish co-localization of protease and substrate, and cell-based assays to evaluate protease function in a more physiological context. Since MS instruments constantly improve in sensitivity and accuracy, it will be equally important to create computational platforms for data analysis that are accessible and easy to handle even for non-experts. It is desirable to integrate proteomics data with phenomic and functional genomic information to fully reconstruct proteolytic networks. Present degradomic studies already provided impressive insights into the function of proteases, protease inhibitors and their interactions among each other (Butler and Overall 2009; Overall and Dean 2006). The question still remains if the protease network is guided through direct compensatory mechanisms or rather represents "accidental" accumulations of downstream and secondary effects. Independently of this debate, present studies highlight that the proteomic phenotypes, stemming from loss- or gain of protease function, often represent strong connections between different proteolytic systems.

To fully understand the biological role of protease networks in cancer or other diseases it is important to not only focus on substrates alone but also include other proteases, protease inhibitors, cofactors and receptors to analyze their relationships. This has to be extended to the analysis of further PTMs, since different PTMs act in concert to control gene expression, control cell signaling processes, and affect protein stability and degradation.

The mass spectrometry-based analysis and identification of PTMs is still challenging and methodological improvements are required. Due to the complexity of PTM regulation typical approaches follow a discovery based strategy detecting a mixture of hundreds to thousands of proteins including their PTMs through mass spectrometric analysis. In degradomic approaches targeted proteomics is playing an increasingly important role as it provides a sensitive and specific way to measure selectively proteins and peptides. Techniques like multiple reaction monitoring (MRM) (Picotti et al. 2007, 2009) or multiplexed selected ion monitoring (Gallien et al. 2012) will be increasingly used to specifically monitor protein processing by measuring several proteins with high sensitivity and increasing throughput. With the present proteomic and degradomic, we now possess valuable tools to dissect the protease web in both health and disease states with the aim to better understand protease biology.

Acknowledgements O.S. is supported by the Deutsche Forschungsgemeinschaft (DFG, SCHI 871/2 (Emmy-Noether) and SCHI 871/5), a starting grant of the European Research Council (Programme "Ideas"—Call identifier: ERC-2011-StG 282111-ProteaSys), and the Excellence Initiative of the German Federal and State Governments (EXC 294, BIOSS). J.D. is supported by the Excellence Initiative of the German Federal and State Governments through Freiburg Institute for Advanced Studies (FRIAS), School of Life Sciences—LifeNet and the Center for

Biological Signalling Studies (BIOSS), by grants DE 1757/2-1 from the German Research Foundation, DFG, and through GerontoSys II—NephAge (031 5896 A) from the German Ministry for Education and Research, BMBF.

References

Agard NJ, Wells JA (2009) Methods for the proteomic identification of protease substrates. Curr Opin Chem Biol 13:503–509

Agard NJ, Maltby D, Wells JA (2010) Inflammatory stimuli regulate caspase substrate profiles. Mol Cell Proteomics 9:880–893

Agarraberes FA, Terlecky SR, Dice JF (1997) An intralysosomal hsp70 is required for a selective pathway of lysosomal protein degradation. J Cell Biol 137:825–834

Akimov V, Rigbolt KT, Nielsen MM, Blagoev B (2011) Characterization of ubiquitination dependent dynamics in growth factor receptor signaling by quantitative proteomics. Mol Biosyst 7(12):3223–3233

Arnesen T, Van Damme P, Polevoda B, Helsens K, Evjenth R, Colaert N, Varhaug JE, Vandekerckhove J, Lillehaug JR, Sherman F et al (2009) Proteomics analyses reveal the evolutionary conservation and divergence of N-terminal acetyltransferases from yeast and humans. Proc Natl Acad Sci U S A 106:8157–8162

Arntzen MO, Thiede B (2012) ApoptoProteomics, an integrated database for analysis of proteomics data obtained from apoptotic cells. Mol Cell Proteomics 11(M111):010447

Attaix D, Combaret L, Pouch MN, Taillandier D (2001) Regulation of proteolysis. Curr Opin Clin Nutr Metab Care 4:45–49

auf dem Keller U, Schilling O (2010) Proteomic techniques and activity-based probes for the system-wide study of proteolysis. Biochimie 92:1705–1714

auf dem Keller U, Prudova A, Gioia M, Butler GS, Overall CM (2010) A statistics-based platform for quantitative N-terminome analysis and identification of protease cleavage products. Mol Cell Proteomics 9:912–927

Beau I, Esclatine A, Codogno P (2008) Lost to translation: when autophagy targets mature ribosomes. Trends Cell Biol 18:311–314

Becker-Pauly C, Barre O, Schilling O, auf dem Keller U, Ohler A, Broder C, Schutte A, Kappelhoff R, Stocker W, Overall CM (2011) Proteomic analyses reveal an acidic prime side specificity for the astacin metalloprotease family reflected by physiological substrates. Mol Cell Proteomics 10(M111):009233

Behrends C, Sowa ME, Gygi SP, Harper JW (2010) Network organization of the human autophagy system. Nature 466:68–76

Berger AB, Witte MD, Denault JB, Sadaghiani AM, Sexton KM, Salvesen GS, Bogyo M (2006) Identification of early intermediates of caspase activation using selective inhibitors and activity-based probes. Mol Cell 23:509–521

Bernales S, McDonald KL, Walter P (2006) Autophagy counterbalances endoplasmic reticulum expansion during the unfolded protein response. PLoS Biol 4:e423

Biniossek ML, Nagler DK, Becker-Pauly C, Schilling O (2011) Proteomic identification of protease cleavage sites characterizes prime and non-prime specificity of cysteine cathepsins B, L, and S. J Proteome Res 10:5363–5373

Bitoun E, Micheloni A, Lamant L, Bonnart C, Tartaglia-Polcini A, Cobbold C, Al Saati T, Mariotti F, Mazereeuw-Hautier J, Boralevi F et al (2003) LEKTI proteolytic processing in human primary keratinocytes, tissue distribution and defective expression in Netherton syndrome. Hum Mol Genet 12:2417–2430

Blair JA, Rauh D, Kung C, Yun CH, Fan QW, Rode H, Zhang C, Eck MJ, Weiss WA, Shokat KM (2007) Structure-guided development of affinity probes for tyrosine kinases using chemical genetics. Nat Chem Biol 3:229–238

Blum G (2008) Use of fluorescent imaging to investigate pathological protease activity. Curr Opin Drug Discov Devel 11:708–716

Blum G, Mullins SR, Keren K, Fonovic M, Jedeszko C, Rice MJ, Sloane BF, Bogyo M (2005) Dynamic imaging of protease activity with fluorescently quenched activity-based probes. Nat Chem Biol 1:203–209

Blum G, von Degenfeld G, Merchant MJ, Blau HM, Bogyo M (2007) Noninvasive optical imaging of cysteine protease activity using fluorescently quenched activity-based probes. Nat Chem Biol 3:668–677

Boisvert FM, Ahmad Y, Gierlinski M, Charriere F, Lamond D, Scott M, Barton G, Lamond AI (2012) A quantitative spatial proteomics analysis of proteome turnover in human cells. Mol Cell Proteomics 11(3):M111.011429

Buschmann T, Lerner D, Lee CG, Ronai Z (2001) The Mdm-2 amino terminus is required for Mdm2 binding and SUMO-1 conjugation by the E2 SUMO-1 conjugating enzyme Ubc9. J Biol Chem 276:40389–40395

Butler GS, Overall CM (2009) Updated biological roles for matrix metalloproteinases and new "intracellular" substrates revealed by degradomics. Biochemistry 48:10830–10845

Butler GS, Butler MJ, Atkinson SJ, Will H, Tamura T, Schade van Westrum S, Crabbe T, Clements J, d'Ortho MP, Murphy G (1998) The TIMP2 membrane type 1 metalloproteinase "receptor" regulates the concentration and efficient activation of progelatinase A. A kinetic study. J Biol Chem 273:871–880

Cardoso CM, Groth-Pedersen L, Hoyer-Hansen M, Kirkegaard T, Corcelle E, Andersen JS, Jaattela M, Nylandsted J (2009) Depletion of kinesin 5B affects lysosomal distribution and stability and induces peri-nuclear accumulation of autophagosomes in cancer cells. PLoS One 4:e4424

Chan EW, Chattopadhaya S, Panicker RC, Huang X, Yao SQ (2004) Developing photoactive affinity probes for proteomic profiling: hydroxamate-based probes for metalloproteases. J Am Chem Soc 126:14435–14446

Chan CP, Mak TY, Chin KT, Ng IO, Jin DY (2010) N-linked glycosylation is required for optimal proteolytic activation of membrane-bound transcription factor CREB-H. J Cell Sci 123:1438–1448

Chavanas S, Bodemer C, Rochat A, Hamel-Teillac D, Ali M, Irvine AD, Bonafe JL, Wilkinson J, Taieb A, Barrandon Y et al (2000) Mutations in SPINK5, encoding a serine protease inhibitor, cause Netherton syndrome. Nat Genet 25:141–142

Chitra S, Nalini G, Rajasekhar G (2012) The ubiquitin proteasome system and efficacy of proteasome inhibitors in diseases. Int J Rheum Dis 15:249–260

Choudhary C, Kumar C, Gnad F, Nielsen ML, Rehman M, Walther TC, Olsen JV, Mann M (2009) Lysine acetylation targets protein complexes and co-regulates major cellular functions. Science 325:834–840

Ciechanover A (2005) Proteolysis: from the lysosome to ubiquitin and the proteasome. Nat Rev Mol Cell Biol 6:79–87

Ciechanover A (2006) The ubiquitin proteolytic system: from an idea to the patient bed. Proc Am Thorac Soc 3:21–31

Ciechanover A, Elias S, Heller H, Ferber S, Hershko A (1980) Characterization of the heat-stable polypeptide of the ATP-dependent proteolytic system from reticulocytes. J Biol Chem 255:7525–7528

Conus S, Simon HU (2010) Cathepsins and their involvement in immune responses. Swiss Med Wkly 140:w13042

Coux O, Tanaka K, Goldberg AL (1996) Structure and functions of the 20S and 26S proteasomes. Annu Rev Biochem 65:801–847

Cox J, Mann M (2007) Is proteomics the new genomics? Cell 130:395–398

Crawford ED, Wells JA (2011) Caspase substrates and cellular remodeling. Annu Rev Biochem 80:1055–1087
Cuervo AM, Dice JF, Knecht E (1997) A population of rat liver lysosomes responsible for the selective uptake and degradation of cytosolic proteins. J Biol Chem 272:5606–5615
De Duve C, Gianetto R, Appelmans F, Wattiaux R (1953) Enzymic content of the mitochondria fraction. Nature 172:1143–1144
de Godoy LM, Olsen JV, de Souza GA, Li G, Mortensen P, Mann M (2006) Status of complete proteome analysis by mass spectrometry: SILAC labeled yeast as a model system. Genome Biol 7:R50
Debigare R, Price SR (2003) Proteolysis, the ubiquitin-proteasome system, and renal diseases. Am J Physiol Renal Physiol 285:F1–F8
Dengjel J, Hoyer-Hansen M, Nielsen MO, Eisenberg T, Harder LM, Schandorff S, Farkas T, Kirkegaard T, Becker AC, Schroeder S et al (2012) Identification of autophagosome-associated proteins and regulators by quantitative proteomic analysis and genetic screens. Mol Cell Proteomics 11(M111):014035
Deu E, Verdoes M, Bogyo M (2012) New approaches for dissecting protease functions to improve probe development and drug discovery. Nat Struct Mol Biol 19:9–16
Ding B, Sun Y, Huang J (2012) Overexpression of SKI oncoprotein leads to p53 degradation through regulation of MDM2 protein sumoylation. J Biol Chem 287:14621–14630
Dix MM, Simon GM, Cravatt BF (2008) Global mapping of the topography and magnitude of proteolytic events in apoptosis. Cell 134:679–691
Dix MM, Simon GM, Wang C, Okerberg E, Patricelli MP, Cravatt BF (2012) Functional interplay between caspase cleavage and phosphorylation sculpts the apoptotic proteome. Cell 150:426–440
Dong J, Atwood CS, Anderson VE, Siedlak SL, Smith MA, Perry G, Carey PR (2003) Metal binding and oxidation of amyloid-beta within isolated senile plaque cores: Raman microscopic evidence. Biochemistry 42:2768–2773
Doucet A, Overall CM (2008) Protease proteomics: revealing protease in vivo functions using systems biology approaches. Mol Aspects Med 29:339–358
Doucet A, Butler GS, Rodriguez D, Prudova A, Overall CM (2008) Metadegradomics: toward in vivo quantitative degradomics of proteolytic post-translational modifications of the cancer proteome. Mol Cell Proteomics 7:1925–1951
Drag M, Salvesen GS (2010) Emerging principles in protease-based drug discovery. Nat Rev Drug Discov 9:690–701
Drews O, Wildgruber R, Zong C, Sukop U, Nissum M, Weber G, Gomes AV, Ping P (2007) Mammalian proteasome subpopulations with distinct molecular compositions and proteolytic activities. Mol Cell Proteomics 6:2021–2031
Dubin G (2005) Proteinaceous cysteine protease inhibitors. Cell Mol Life Sci 62:653–669
Dunn WA Jr, Cregg JM, Kiel JA, van der Klei IJ, Oku M, Sakai Y, Sibirny AA, Stasyk OV, Veenhuis M (2005) Pexophagy: the selective autophagy of peroxisomes. Autophagy 1:75–83
Edgington LE, Berger AB, Blum G, Albrow VE, Paulick MG, Lineberry N, Bogyo M (2009) Noninvasive optical imaging of apoptosis by caspase-targeted activity-based probes. Nat Med 15:967–973
Edgington LE, van Raam BJ, Verdoes M, Wierschem C, Salvesen GS, Bogyo M (2012) An optimized activity-based probe for the study of caspase-6 activation. Chem Biol 19:340–352
Elias JE, Gygi SP (2007) Target-decoy search strategy for increased confidence in large-scale protein identifications by mass spectrometry. Nat Methods 4:207–214
Engelke R, Becker AC, Dengjel J (2012) The degradative inventory of the cell: proteomic insights. Antioxid Redox Signal 17:803–812
Enoksson M, Li J, Ivancic MM, Timmer JC, Wildfang E, Eroshkin A, Salvesen GS, Tao WA (2007) Identification of proteolytic cleavage sites by quantitative proteomics. J Proteome Res 6:2850–2858

Etlinger JD, Goldberg AL (1977) A soluble ATP-dependent proteolytic system responsible for the degradation of abnormal proteins in reticulocytes. Proc Natl Acad Sci U S A 74:54–58

Fenn JB, Mann M, Meng CK, Wong SF, Whitehouse CM (1989) Electrospray ionization for mass spectrometry of large biomolecules. Science 246:64–71

Fonovic M, Bogyo M (2007) Activity based probes for proteases: applications to biomarker discovery, molecular imaging and drug screening. Curr Pharm Des 13:253–261

Franch HA, Sooparb S, Du J, Brown NS (2001) A mechanism regulating proteolysis of specific proteins during renal tubular cell growth. J Biol Chem 276:19126–19131

Gafni J, Ellerby LM (2002) Calpain activation in Huntington's disease. J Neurosci 22:4842–4849

Gafni J, Hermel E, Young JE, Wellington CL, Hayden MR, Ellerby LM (2004) Inhibition of calpain cleavage of huntingtin reduces toxicity: accumulation of calpain/caspase fragments in the nucleus. J Biol Chem 279:20211–20220

Gallien S, Duriez E, Crone C, Kellmann M, Moehring T, Domon B (2012) Targeted proteomic quantification on quadrupole-orbitrap mass spectrometer. Mol Cell Proteomics 11 (12):1709–1723

Gao W, Kang JH, Liao Y, Ding WX, Gambotto AA, Watkins SC, Liu YJ, Stolz DB, Yin XM (2010) Biochemical isolation and characterization of the tubulovesicular LC3-positive autophagosomal compartment. J Biol Chem 285:1371–1383

Garcia-Verdugo I, Descamps D, Chignard M, Touqui L, Sallenave JM (2010) Lung protease/anti-protease network and modulation of mucus production and surfactant activity. Biochimie 92:1608–1617

Geng M, Zhang X, Bina M, Regnier F (2001) Proteomics of glycoproteins based on affinity selection of glycopeptides from tryptic digests. J Chromatogr B Biomed Sci Appl 752:293–306

Gevaert K, Goethals M, Martens L, Van Damme J, Staes A, Thomas GR, Vandekerckhove J (2003) Exploring proteomes and analyzing protein processing by mass spectrometric identification of sorted N-terminal peptides. Nat Biotechnol 21:566–569

Gevaert K, Van Damme P, Ghesquiere B, Impens F, Martens L, Helsens K, Vandekerckhove J (2007) A la carte proteomics with an emphasis on gel-free techniques. Proteomics 7:2698–2718

Glondu M, Liaudet-Coopman E, Derocq D, Platet N, Rochefort H, Garcia M (2002) Down-regulation of cathepsin-D expression by antisense gene transfer inhibits tumor growth and experimental lung metastasis of human breast cancer cells. Oncogene 21:5127–5134

Goldberg AL, Stein R, Adams J (1995) New insights into proteasome function: from archaebacteria to drug development. Chem Biol 2:503–508

Goldberg YP, Nicholson DW, Rasper DM, Kalchman MA, Koide HB, Graham RK, Bromm M, Kazemi-Esfarjani P, Thornberry NA, Vaillancourt JP et al (1996) Cleavage of huntingtin by apopain, a proapoptotic cysteine protease, is modulated by the polyglutamine tract. Nat Genet 13:442–449

Greenbaum D, Medzihradszky KF, Burlingame A, Bogyo M (2000) Epoxide electrophiles as activity-dependent cysteine protease profiling and discovery tools. Chem Biol 7:569–581

Gygi SP, Rist B, Gerber SA, Turecek F, Gelb MH, Aebersold R (1999) Quantitative analysis of complex protein mixtures using isotope-coded affinity tags. Nat Biotechnol 17:994–999

Hagel M, Niu D, St Martin T, Sheets MP, Qiao L, Bernard H, Karp RM, Zhu Z, Labenski MT, Chaturvedi P et al (2011) Selective irreversible inhibition of a protease by targeting a noncatalytic cysteine. Nat Chem Biol 7:22–24

Hartig W, Goldhammer S, Bauer U, Wegner F, Wirths O, Bayer TA, Grosche J (2010) Concomitant detection of beta-amyloid peptides with N-terminal truncation and different C-terminal endings in cortical plaques from cases with Alzheimer's disease, senile monkeys and triple transgenic mice. J Chem Neuroanat 40:82–92

Hay RT (2005) SUMO: a history of modification. Mol Cell 18:1–12

Hedrich J, Lottaz D, Meyer K, Yiallouros I, Jahnen-Dechent W, Stocker W, Becker-Pauly C (2010) Fetuin-A and cystatin C are endogenous inhibitors of human meprin metalloproteases. Biochemistry 49:8599–8607

Helsens K, Timmerman E, Vandekerckhove J, Gevaert K, Martens L (2008) Peptizer, a tool for assessing false positive peptide identifications and manually validating selected results. Mol Cell Proteomics 7:2364–2372

Helsens K, Martens L, Vandekerckhove J, Gevaert K (2011) Mass spectrometry-driven proteomics: an introduction. Methods Mol Biol 753:1–27

Henriksen P, Wagner SA, Weinert BT, Sharma S, Bacinskaja G, Rehman M, Juffer AH, Walther TC, Lisby M, Choudhary C (2012) Proteome-wide analysis of lysine acetylation suggests its broad regulatory scope in Saccharomyces cerevisiae. Mol Cell Proteomics 11(11):1510–1522

Herrmann JL, O'Gaora P, Gallagher A, Thole JE, Young DB (1996) Bacterial glycoproteins: a link between glycosylation and proteolytic cleavage of a 19 kDa antigen from Mycobacterium tuberculosis. EMBO J 15:3547–3554

Higdon R, Kolker E (2007) A predictive model for identifying proteins by a single peptide match. Bioinformatics 23:277–280

Hinkson IV, Elias JE (2011) The dynamic state of protein turnover: it's about time. Trends Cell Biol 21:293–303

Hock A, Vousden KH (2010) Regulation of the p53 pathway by ubiquitin and related proteins. Int J Biochem Cell Biol 42:1618–1621

Hsiao CC, Cheng KF, Chen HY, Chou YH, Stacey M, Chang GW, Lin HH (2009) Site-specific N-glycosylation regulates the GPS auto-proteolysis of CD97. FEBS Lett 583:3285–3290

Hu L, Roth JM, Brooks P, Luty J, Karpatkin S (2008) Thrombin up-regulates cathepsin D which enhances angiogenesis, growth, and metastasis. Cancer Res 68:4666–4673

Huesgen PF, Overall CM (2012) N- and C-terminal degradomics: new approaches to reveal biological roles for plant proteases from substrate identification. Physiol Plant 145:5–17

Impens F, Van Damme P, Demol H, Van Damme J, Vandekerckhove J, Gevaert K (2008) Mechanistic insight into taxol-induced cell death. Oncogene 27:4580–4591

Impens F, Colaert N, Helsens K, Plasman K, Van Damme P, Vandekerckhove J, Gevaert K (2010) MS-driven protease substrate degradomics. Proteomics 10:1284–1296

Imperiali B, O'Connor SE (1999) Effect of N-linked glycosylation on glycopeptide and glycoprotein structure. Curr Opin Chem Biol 3:643–649

Jakoby T, van den Berg BH, Tholey A (2012) Quantitative protease cleavage site profiling using tandem-mass-tag labeling and LC-MALDI-TOF/TOF MS/MS analysis. J Proteome Res 11:1812–1820

Jayakumar A, Kang Y, Mitsudo K, Henderson Y, Frederick MJ, Wang M, El-Naggar AK, Marx UC, Briggs K, Clayman GL (2004) Expression of LEKTI domains 6-9' in the baculovirus expression system: recombinant LEKTI domains 6-9' inhibit trypsin and subtilisin A. Protein Expr Purif 35:93–101

Jayapal KP, Sui S, Philp RJ, Kok YJ, Yap MG, Griffin TJ, Hu WS (2010) Multitagging proteomic strategy to estimate protein turnover rates in dynamic systems. J Proteome Res 9:2087–2097

Jefferson T, Causevic M, auf dem Keller U, Schilling O, Isbert S, Geyer R, Maier W, Tschickardt S, Jumpertz T, Weggen S et al (2011) Metalloprotease meprin beta generates nontoxic N-terminal amyloid precursor protein fragments in vivo. J Biol Chem 286:27741–27750

Johansen T, Lamark T (2011) Selective autophagy mediated by autophagic adapter proteins. Autophagy 7:279–296

Joyce JA, Baruch A, Chehade K, Meyer-Morse N, Giraudo E, Tsai FY, Greenbaum DC, Hager JH, Bogyo M, Hanahan D (2004) Cathepsin cysteine proteases are effectors of invasive growth and angiogenesis during multistage tumorigenesis. Cancer Cell 5:443–453

Kalchman MA, Graham RK, Xia G, Koide HB, Hodgson JG, Graham KC, Goldberg YP, Gietz RD, Pickart CM, Hayden MR (1996) Huntingtin is ubiquitinated and interacts with a specific ubiquitin-conjugating enzyme. J Biol Chem 271:19385–19394

Kamath KS, Vasavada MS, Srivastava S (2011) Proteomic databases and tools to decipher post-translational modifications. J Proteomics 75:127–144

Keller A, Nesvizhskii AI, Kolker E, Aebersold R (2002) Empirical statistical model to estimate the accuracy of peptide identifications made by MS/MS and database search. Anal Chem 74:5383–5392

Kessler BM, Tortorella D, Altun M, Kisselev AF, Fiebiger E, Hekking BG, Ploegh HL, Overkleeft HS (2001) Extended peptide-based inhibitors efficiently target the proteasome and reveal overlapping specificities of the catalytic beta-subunits. Chem Biol 8:913–929

Kidd D, Liu Y, Cravatt BF (2001) Profiling serine hydrolase activities in complex proteomes. Biochemistry 40:4005–4015

Kim YJ, Yi Y, Sapp E, Wang Y, Cuiffo B, Kegel KB, Qin ZH, Aronin N, DiFiglia M (2001) Caspase 3-cleaved N-terminal fragments of wild-type and mutant huntingtin are present in normal and Huntington's disease brains, associate with membranes, and undergo calpain-dependent proteolysis. Proc Natl Acad Sci U S A 98:12784–12789

Kim I, Rodriguez-Enriquez S, Lemasters JJ (2007) Selective degradation of mitochondria by mitophagy. Arch Biochem Biophys 462:245–253

Kim W, Bennett EJ, Huttlin EL, Guo A, Li J, Possemato A, Sowa ME, Rad R, Rush J, Comb MJ et al (2011) Systematic and quantitative assessment of the ubiquitin-modified proteome. Mol Cell 44:325–340

King RW, Deshaies RJ, Peters JM, Kirschner MW (1996) How proteolysis drives the cell cycle. Science 274:1652–1659

Kleifeld O, Doucet A, Prudova A, Schilling O, Kainthan RK, Starr AE, Foster LJ, Kizhakkedathu JN, Overall CM (2010) Isotopic labeling of terminal amines in complex samples identifies protein N-termini and protease cleavage products. Nat Biotechnol 28:281–288

Klemencic I, Carmona AK, Cezari MH, Juliano MA, Juliano L, Guncar G, Turk D, Krizaj I, Turk V, Turk B (2000) Biochemical characterization of human cathepsin X revealed that the enzyme is an exopeptidase, acting as carboxymonopeptidase or carboxydipeptidase. Eur J Biochem 267:5404–5412

Komander D, Rape M (2012) The ubiquitin code. Annu Rev Biochem 81:203–229

Kraft C, Deplazes A, Sohrmann M, Peter M (2008) Mature ribosomes are selectively degraded upon starvation by an autophagy pathway requiring the Ubp3p/Bre5p ubiquitin protease. Nat Cell Biol 10:602–610

Kristensen AR, Schandorff S, Hoyer-Hansen M, Nielsen MO, Jaattela M, Dengjel J, Andersen JS (2008) Ordered organelle degradation during starvation-induced autophagy. Mol Cell Proteomics 7:2419–2428

Kumar S, Rezaei-Ghaleh N, Terwel D, Thal DR, Richard M, Hoch M, Mc Donald JM, Wullner U, Glebov K, Heneka MT et al (2011) Extracellular phosphorylation of the amyloid beta-peptide promotes formation of toxic aggregates during the pathogenesis of Alzheimer's disease. EMBO J 30:2255–2265

Kuo YM, Emmerling MR, Woods AS, Cotter RJ, Roher AE (1997) Isolation, chemical characterization, and quantitation of A beta 3-pyroglutamyl peptide from neuritic plaques and vascular amyloid deposits. Biochem Biophys Res Commun 237:188–191

Kuo YM, Webster S, Emmerling MR, De Lima N, Roher AE (1998) Irreversible dimerization/tetramerization and post-translational modifications inhibit proteolytic degradation of A beta peptides of Alzheimer's disease. Biochim Biophys Acta 1406:291–298

Kurokawa M, Kornbluth S (2009) Caspases and kinases in a death grip. Cell 138:838–854

LaFevre-Bernt MA, Ellerby LM (2003) Kennedy's disease. Phosphorylation of the polyglutamine-expanded form of androgen receptor regulates its cleavage by caspase-3 and enhances cell death. J Biol Chem 278:34918–34924

Lange PF, Overall CM (2011) TopFIND, a knowledgebase linking protein termini with function. Nat Methods 8:703–704

Lange PF, Huesgen PF, Overall CM (2012) TopFIND 2.0–linking protein termini with proteolytic processing and modifications altering protein function. Nucleic Acids Res 40:D351–D361

Langsford ML, Gilkes NR, Singh B, Moser B, Miller RC Jr, Warren RA, Kilburn DG (1987) Glycosylation of bacterial cellulases prevents proteolytic cleavage between functional domains. FEBS Lett 225:163–167

Laurent-Matha V, Huesgen PF, Masson O, Derocq D, Prebois C, Gary-Bobo M, Lecaille F, Rebiere B, Meurice G, Orear C et al (2012) Proteolysis of cystatin C by cathepsin D in the breast cancer microenvironment. FASEB J 26(12):5172–5181

Le Magueresse-Battistoni B (2007) Serine proteases and serine protease inhibitors in testicular physiology: the plasminogen activation system. Reproduction 134:721–729

Lee MH, Lee SW, Lee EJ, Choi SJ, Chung SS, Lee JI, Cho JM, Seol JH, Baek SH, Kim KI et al (2006) SUMO-specific protease SUSP4 positively regulates p53 by promoting Mdm2 self-ubiquitination. Nat Cell Biol 8:1424–1431

Li W, Ding F, Zhang L, Liu Z, Wu Y, Luo A, Wu M, Wang M, Zhan Q, Liu Z (2005) Overexpression of stefin A in human esophageal squamous cell carcinoma cells inhibits tumor cell growth, angiogenesis, invasion, and metastasis. Clin Cancer Res 11:8753–8762

List K, Kosa P, Szabo R, Bey AL, Wang CB, Molinolo A, Bugge TH (2009) Epithelial integrity is maintained by a matriptase-dependent proteolytic pathway. Am J Pathol 175:1453–1463

Liu Y, Patricelli MP, Cravatt BF (1999) Activity-based protein profiling: the serine hydrolases. Proc Natl Acad Sci U S A 96:14694–14699

Lopez-Otin C, Hunter T (2010) The regulatory crosstalk between kinases and proteases in cancer. Nat Rev Cancer 10:278–292

Lopez-Otin C, Overall CM (2002) Protease degradomics: a new challenge for proteomics. Nat Rev Mol Cell Biol 3:509–519

Lunkes A, Lindenberg KS, Ben-Haiem L, Weber C, Devys D, Landwehrmeyer GB, Mandel JL, Trottier Y (2002) Proteases acting on mutant huntingtin generate cleaved products that differentially build up cytoplasmic and nuclear inclusions. Mol Cell 10:259–269

Luo S, Vacher C, Davies JE, Rubinsztein DC (2005) Cdk5 phosphorylation of huntingtin reduces its cleavage by caspases: implications for mutant huntingtin toxicity. J Cell Biol 169:647–656

Magert HJ, Standker L, Kreutzmann P, Zucht HD, Reinecke M, Sommerhoff CP, Fritz H, Forssmann WG (1999) LEKTI, a novel 15-domain type of human serine proteinase inhibitor. J Biol Chem 274:21499–21502

Mahrus S, Trinidad JC, Barkan DT, Sali A, Burlingame AL, Wells JA (2008) Global sequencing of proteolytic cleavage sites in apoptosis by specific labeling of protein N termini. Cell 134:866–876

Mann M (2006) Functional and quantitative proteomics using SILAC. Nat Rev Mol Cell Biol 7:952–958

Martin L, Latypova X, Terro F (2011) Post-translational modifications of tau protein: implications for Alzheimer's disease. Neurochem Int 58:458–471

McNulty DE, Annan RS (2008) Hydrophilic interaction chromatography reduces the complexity of the phosphoproteome and improves global phosphopeptide isolation and detection. Mol Cell Proteomics 7:971–980

Mester G, Hoffmann V, Stevanovic S (2011) Insights into MHC class I antigen processing gained from large-scale analysis of class I ligands. Cell Mol Life Sci 68:1521–1532

Millucci L, Ghezzi L, Bernardini G, Santucci A (2010) Conformations and biological activities of amyloid beta peptide 25-35. Curr Protein Pept Sci 11:54–67

Miravalle L, Calero M, Takao M, Roher AE, Ghetti B, Vidal R (2005) Amino-terminally truncated Abeta peptide species are the main component of cotton wool plaques. Biochemistry 44:10810–10821

Mitsudo K, Jayakumar A, Henderson Y, Frederick MJ, Kang Y, Wang M, El-Naggar AK, Clayman GL (2003) Inhibition of serine proteinases plasmin, trypsin, subtilisin A, cathepsin G, and elastase by LEKTI: a kinetic analysis. Biochemistry 42:3874–3881

Miyauchi Y, Yogosawa S, Honda R, Nishida T, Yasuda H (2002) Sumoylation of Mdm2 by protein inhibitor of activated STAT (PIAS) and RanBP2 enzymes. J Biol Chem 277:50131–50136

Mommen GP, van de Waterbeemd B, Meiring HD, Kersten G, Heck AJ, de Jong AP (2012) Unbiased selective isolation of protein N-terminal peptides from complex proteome samples using phospho tagging (PTAG) and TiO2-based depletion. Mol Cell Proteomics 11:832–842

Mori H, Ishii K, Tomiyama T, Furiya Y, Sahara N, Asano S, Endo N, Shirasawa T, Takio K (1994) Racemization: its biological significance on neuropathogenesis of Alzheimer's disease. Tohoku J Exp Med 174:251–262

Morrissey JH (2012) Polyphosphate: a link between platelets, coagulation and inflammation. Int J Hematol 95:346–352

Murakami K, Uno M, Masuda Y, Shimizu T, Shirasawa T, Irie K (2008) Isomerization and/or racemization at Asp23 of Abeta42 do not increase its aggregative ability, neurotoxicity, and radical productivity in vitro. Biochem Biophys Res Commun 366:745–751

Murphy G (2011) Tissue inhibitors of metalloproteinases. Genome Biol 12:233

Mutero A, Fournier D (1992) Post-translational modifications of Drosophila acetylcholinesterase. In vitro mutagenesis and expression in Xenopus oocytes. J Biol Chem 267:1695–1700

Nakagawa H, Wakabayashi-Nakao K, Tamura A, Toyoda Y, Koshiba S, Ishikawa T (2009) Disruption of N-linked glycosylation enhances ubiquitin-mediated proteasomal degradation of the human ATP-binding cassette transporter ABCG2. FEBS J 276:7237–7252

Nylandsted J, Becker AC, Bunkenborg J, Andersen JS, Dengjel J, Jaattela M (2011) ErbB2-associated changes in the lysosomal proteome. Proteomics 11:2830–2838

Ong SE, Mann M (2005) Mass spectrometry-based proteomics turns quantitative. Nat Chem Biol 1:252–262

Ong SE, Blagoev B, Kratchmarova I, Kristensen DB, Steen H, Pandey A, Mann M (2002) Stable isotope labeling by amino acids in cell culture, SILAC, as a simple and accurate approach to expression proteomics. Mol Cell Proteomics 1:376–386

Opdenakker G, Rudd PM, Ponting CP, Dwek RA (1993) Concepts and principles of glycobiology. FASEB J 7:1330–1337

Ott I (2011) Inhibitors of the initiation of coagulation. Br J Clin Pharmacol 72:547–552

Overall CM, Blobel CP (2007) In search of partners: linking extracellular proteases to substrates. Nat Rev Mol Cell Biol 8:245–257

Overall CM, Dean RA (2006) Degradomics: systems biology of the protease web. Pleiotropic roles of MMPs in cancer. Cancer Metastasis Rev 25:69–75

Pagano M, Tam SW, Theodoras AM, Beer-Romero P, Del Sal G, Chau V, Yew PR, Draetta GF, Rolfe M (1995) Role of the ubiquitin-proteasome pathway in regulating abundance of the cyclin-dependent kinase inhibitor p27. Science 269:682–685

Paulick MG, Bogyo M (2008) Application of activity-based probes to the study of enzymes involved in cancer progression. Curr Opin Genet Dev 18:97–106

Peng J, Schwartz D, Elias JE, Thoreen CC, Cheng D, Marsischky G, Roelofs J, Finley D, Gygi SP (2003) A proteomics approach to understanding protein ubiquitination. Nat Biotechnol 21:921–926

Picotti P, Aebersold R, Domon B (2007) The implications of proteolytic background for shotgun proteomics. Mol Cell Proteomics 6:1589–1598

Picotti P, Bodenmiller B, Mueller LN, Domon B, Aebersold R (2009) Full dynamic range proteome analysis of S. cerevisiae by targeted proteomics. Cell 138:795–806

Poreba M, Drag M (2010) Current strategies for probing substrate specificity of proteases. Curr Med Chem 17:3968–3995

Pratt JM, Petty J, Riba-Garcia I, Robertson DH, Gaskell SJ, Oliver SG, Beynon RJ (2002) Dynamics of protein turnover, a missing dimension in proteomics. Mol Cell Proteomics 1:579–591

Price SR, Bailey JL, Wang X, Jurkovitz C, England BK, Ding X, Phillips LS, Mitch WE (1996) Muscle wasting in insulinopenic rats results from activation of the ATP-dependent, ubiquitin-proteasome proteolytic pathway by a mechanism including gene transcription. J Clin Invest 98:1703–1708

Prudova A, auf dem Keller U, Butler GS, Overall CM (2010) Multiplex N-terminome analysis of MMP-2 and MMP-9 substrate degradomes by iTRAQ-TAILS quantitative proteomics. Mol Cell Proteomics 9:894–911

Puente XS, Sanchez LM, Overall CM, Lopez-Otin C (2003) Human and mouse proteases: a comparative genomic approach. Nat Rev Genet 4:544–558

Puri AW, Bogyo M (2009) Using small molecules to dissect mechanisms of microbial pathogenesis. ACS Chem Biol 4:603–616

Quesada V, Ordonez GR, Sanchez LM, Puente XS, Lopez-Otin C (2009) The Degradome database: mammalian proteases and diseases of proteolysis. Nucleic Acids Res 37:D239–D243

Rawlings ND, Morton FR, Barrett AJ (2006) MEROPS: the peptidase database. Nucleic Acids Res 34:D270–D272

Rawlings ND, Barrett AJ, Bateman A (2010) MEROPS: the peptidase database. Nucleic Acids Res 38:D227–D233

Reiser J, Adair B, Reinheckel T (2010) Specialized roles for cysteine cathepsins in health and disease. J Clin Invest 120:3421–3431

Rock KL, Goldberg AL (1999) Degradation of cell proteins and the generation of MHC class I-presented peptides. Annu Rev Immunol 17:739–779

Romao S, Munz C (2011) Autophagy of pathogens alarms the immune system and participates in its effector functions. Swiss Med Wkly 141:w13198

Ross PL, Huang YN, Marchese JN, Williamson B, Parker K, Hattan S, Khainovski N, Pillai S, Dey S, Daniels S et al (2004) Multiplexed protein quantitation in Saccharomyces cerevisiae using amine-reactive isobaric tagging reagents. Mol Cell Proteomics 3:1154–1169

Rubinsztein DC, Carmichael J (2003) Huntington's disease: molecular basis of neurodegeneration. Expert Rev Mol Med 5:1–21

Saghatelian A, Jessani N, Joseph A, Humphrey M, Cravatt BF (2004) Activity-based probes for the proteomic profiling of metalloproteases. Proc Natl Acad Sci U S A 101:10000–10005

Saido TC, Iwatsubo T, Mann DM, Shimada H, Ihara Y, Kawashima S (1995) Dominant and differential deposition of distinct beta-amyloid peptide species, A beta N3(pE), in senile plaques. Neuron 14:457–466

Saido TC, Yamao-Harigaya W, Iwatsubo T, Kawashima S (1996) Amino- and carboxyl-terminal heterogeneity of beta-amyloid peptides deposited in human brain. Neurosci Lett 215:173–176

Sales KU, Masedunskas A, Bey AL, Rasmussen AL, Weigert R, List K, Szabo R, Overbeek PA, Bugge TH (2010) Matriptase initiates activation of epidermal pro-kallikrein and disease onset in a mouse model of Netherton syndrome. Nat Genet 42:676–683

Scheller J, Chalaris A, Garbers C, Rose-John S (2011) ADAM17: a molecular switch to control inflammation and tissue regeneration. Trends Immunol 32:380–387

Schilling O, Overall CM (2008) Proteome-derived, database-searchable peptide libraries for identifying protease cleavage sites. Nat Biotechnol 26:685–694

Schilling B, Gafni J, Torcassi C, Cong X, Row RH, LaFevre-Bernt MA, Cusack MP, Ratovitski T, Hirschhorn R, Ross CA et al (2006a) Huntingtin phosphorylation sites mapped by mass spectrometry. Modulation of cleavage and toxicity. J Biol Chem 281:23686–23697

Schilling S, Lauber T, Schaupp M, Manhart S, Scheel E, Bohm G, Demuth HU (2006b) On the seeding and oligomerization of pGlu-amyloid peptides (in vitro). Biochemistry 45:12393–12399

Schilling S, Zeitschel U, Hoffmann T, Heiser U, Francke M, Kehlen A, Holzer M, Hutter-Paier B, Prokesch M, Windisch M et al (2008) Glutaminyl cyclase inhibition attenuates pyroglutamate Abeta and Alzheimer's disease-like pathology. Nat Med 14:1106–1111

Schilling O, Barre O, Huesgen PF, Overall CM (2010) Proteome-wide analysis of protein carboxy termini: C terminomics. Nat Methods 7:508–511

Schilling O, auf dem Keller U, Overall CM (2011a) Factor Xa subsite mapping by proteome-derived peptide libraries improved using WebPICS, a resource for proteomic identification of cleavage sites. Biol Chem 392:1031–1037

Schilling O, Huesgen PF, Barre O, auf dem Keller U, Overall CM (2011b) Characterization of the prime and non-prime active site specificities of proteases by proteome-derived peptide libraries and tandem mass spectrometry. Nat Protoc 6:111–120
Schilling O, Huesgen PF, Barre O, Overall CM (2011c) Identification and relative quantification of native and proteolytically generated protein C-termini from complex proteomes: C-terminome analysis. Methods Mol Biol 781:59–69
Schmidt M, Haas W, Crosas B, Santamaria PG, Gygi SP, Walz T, Finley D (2005) The HEAT repeat protein Blm10 regulates the yeast proteasome by capping the core particle. Nat Struct Mol Biol 12:294–303
Schwanhausser B, Busse D, Li N, Dittmar G, Schuchhardt J, Wolf J, Chen W, Selbach M (2011) Global quantification of mammalian gene expression control. Nature 473:337–342
Sergeant N, Bombois S, Ghestem A, Drobecq H, Kostanjevecki V, Missiaen C, Wattez A, David JP, Vanmechelen E, Sergheraert C et al (2003) Truncated beta-amyloid peptide species in pre-clinical Alzheimer's disease as new targets for the vaccination approach. J Neurochem 85:1581–1591
Sevenich L, Schurigt U, Sachse K, Gajda M, Werner F, Muller S, Vasiljeva O, Schwinde A, Klemm N, Deussing J et al (2010) Synergistic antitumor effects of combined cathepsin B and cathepsin Z deficiencies on breast cancer progression and metastasis in mice. Proc Natl Acad Sci U S A 107:2497–2502
Shen A (2010) Allosteric regulation of protease activity by small molecules. Mol Biosyst 6:1431–1443
Shimizu T, Watanabe A, Ogawara M, Mori H, Shirasawa T (2000) Isoaspartate formation and neurodegeneration in Alzheimer's disease. Arch Biochem Biophys 381:225–234
Shrikhande GV, Scali ST, da Silva CG, Damrauer SM, Csizmadia E, Putheti P, Matthey M, Arjoon R, Patel R, Siracuse JJ et al (2010) O-glycosylation regulates ubiquitination and degradation of the anti-inflammatory protein A20 to accelerate atherosclerosis in diabetic ApoE-null mice. PLoS One 5:e14240
Sieber SA, Niessen S, Hoover HS, Cravatt BF (2006) Proteomic profiling of metalloprotease activities with cocktails of active-site probes. Nat Chem Biol 2:274–281
Smith R, Johansen HT, Nilsen H, Haugen MH, Pettersen SJ, Maelandsmo GM, Abrahamson M, Solberg R (2012) Intra- and extracellular regulation of activity and processing of legumain by cystatin E/M. Biochimie 94(12):2590–2599
Sokol JP, Schiemann WP (2004) Cystatin C antagonizes transforming growth factor beta signaling in normal and cancer cells. Mol Cancer Res 2:183–195
Sorimachi H, Ishiura S, Suzuki K (1997) Structure and physiological function of calpains. Biochem J 328(Pt 3):721–732
Speers AE, Cravatt BF (2009) Activity-based protein profiling (ABPP) and click chemistry (CC)-ABPP by MudPIT mass spectrometry. Curr Protoc Chem Biol 1:29–41
Staes A, Van Damme P, Helsens K, Demol H, Vandekerckhove J, Gevaert K (2008) Improved recovery of proteome-informative, protein N-terminal peptides by combined fractional diagonal chromatography (COFRADIC). Proteomics 8:1362–1370
Steffan JS, Agrawal N, Pallos J, Rockabrand E, Trotman LC, Slepko N, Illes K, Lukacsovich T, Zhu YZ, Cattaneo E et al (2004) SUMO modification of Huntingtin and Huntington's disease pathology. Science 304:100–104
Strongin AY, Marmer BL, Grant GA, Goldberg GI (1993) Plasma membrane-dependent activation of the 72-kDa type IV collagenase is prevented by complex formation with TIMP-2. J Biol Chem 268:14033–14039
Tanaka K, Waki H, Ido Y, Akita S, Yoshida Y (1988) Protein and polymer analyses up to m/z 100,000 by laser ionization time-of-flight mass spectrometry. Rapid Commun Mass Spectrom 2:151–153
Taylor AK, Wall R (1988) Selective removal of alpha heavy-chain glycosylation sites causes immunoglobulin A degradation and reduced secretion. Mol Cell Biol 8:4197–4203

Taylor RC, Cullen SP, Martin SJ (2008) Apoptosis: controlled demolition at the cellular level. Nat Rev Mol Cell Biol 9:231–241

Tekirian TL, Saido TC, Markesbery WR, Russell MJ, Wekstein DR, Patel E, Geddes JW (1998) N-terminal heterogeneity of parenchymal and cerebrovascular Abeta deposits. J Neuropathol Exp Neurol 57:76–94

Tholen S, Biniossek ML, Gessler AL, Muller S, Weisser J, Kizhakkedathu JN, Reinheckel T, Schilling O (2011) Contribution of cathepsin L to secretome composition and cleavage pattern of mouse embryonic fibroblasts. Biol Chem 392:961–971

Tholen S, Biniossek ML, Gansz M, Ahrens TD, Schlimpert M, Kizhakkedathu JN, Reinheckel T, Schilling O (2013a) Double deficiency of cathepsins B and L results in massive secretome alterations and suggests a degradative cathepsin-MMP axis. Cell Mol Life Sci. doi:10.1007/s00018-013-1406-1

Tholen S, Biniossek ML, Gansz M, Gomez-Auli A, Bengsch F, Noel A, Kizhakkedathu JN, Boerries M, Busch H, Reinheckel T, Schilling O (2013b) Deletion of cysteine cathepsins B or L yields differential impacts on murine skin proteome and degradome. Mol Cell Proteomics 12:611–625

Thornberry NA, Peterson EP, Zhao JJ, Howard AD, Griffin PR, Chapman KT (1994) Inactivation of interleukin-1 beta converting enzyme by peptide (acyloxy)methyl ketones. Biochemistry 33:3934–3940

Timmer JC, Enoksson M, Wildfang E, Zhu W, Igarashi Y, Denault JB, Ma Y, Dummitt B, Chang YH, Mast AE et al (2007) Profiling constitutive proteolytic events in vivo. Biochem J 407:41–48

Timmer JC, Zhu W, Pop C, Regan T, Snipas SJ, Eroshkin AM, Riedl SJ, Salvesen GS (2009) Structural and kinetic determinants of protease substrates. Nat Struct Mol Biol 16:1101–1108

Tolkovsky AM (2009) Mitophagy. Biochim Biophys Acta 1793:1508–1515

Tomiyama T, Asano S, Furiya Y, Shirasawa T, Endo N, Mori H (1994) Racemization of Asp23 residue affects the aggregation properties of Alzheimer amyloid beta protein analogues. J Biol Chem 269:10205–10208

Tseng A, Inuzuka H, Gao D, Singh A, Wei W (2008) Experimental approaches to investigate the proteasomal degradation pathways involved in regulation of apoptosis. Methods Enzymol 446:205–223

Turk B, du Turk SA, Turk V (2012) Protease signalling: the cutting edge. EMBO J 31:1630–1643

Van Damme P, Staes A, Bronsoms S, Helsens K, Colaert N, Timmerman E, Aviles FX, Vandekerckhove J, Gevaert K (2010) Complementary positional proteomics for screening substrates of endo- and exoproteases. Nat Methods 7:512–515

Vande Walle L, Van Damme P, Lamkanfi M, Saelens X, Vandekerckhove J, Gevaert K, Vandenabeele P (2007) Proteome-wide Identification of HtrA2/Omi Substrates. J Proteome Res 6:1006–1015

Varki A (1993) Biological roles of oligosaccharides: all of the theories are correct. Glycobiology 3:97–130

Vasiljeva O, Papazoglou A, Kruger A, Brodoefel H, Korovin M, Deussing J, Augustin N, Nielsen BS, Almholt K, Bogyo M et al (2006) Tumor cell-derived and macrophage-derived cathepsin B promotes progression and lung metastasis of mammary cancer. Cancer Res 66:5242–5250

Vigneswaran N, Wu J, Nagaraj N, James R, Zeeuwen P, Zacharias W (2006) Silencing of cystatin M in metastatic oral cancer cell line MDA-686Ln by siRNA increases cysteine proteinases and legumain activities, cell proliferation and in vitro invasion. Life Sci 78:898–907

Wang CC, Bozdech Z, Liu CL, Shipway A, Backes BJ, Harris JL, Bogyo M (2003) Biochemical analysis of the 20 S proteasome of Trypanosoma brucei. J Biol Chem 278:15800–15808

Wang B, Sun J, Kitamoto S, Yang M, Grubb A, Chapman HA, Kalluri R, Shi GP (2006) Cathepsin S controls angiogenesis and tumor growth via matrix-derived angiogenic factors. J Biol Chem 281:6020–6029

Wang X, Guerrero C, Kaiser P, Huang L (2007) Proteomics of proteasome complexes and ubiquitinated proteins. Expert Rev Proteomics 4:649–665

Weissman AM (2001) Themes and variations on ubiquitylation. Nat Rev Mol Cell Biol 2:169–178
Wellington CL, Singaraja R, Ellerby L, Savill J, Roy S, Leavitt B, Cattaneo E, Hackam A, Sharp A, Thornberry N et al (2000) Inhibiting caspase cleavage of huntingtin reduces toxicity and aggregate formation in neuronal and nonneuronal cells. J Biol Chem 275:19831–19838
Wellington CL, Ellerby LM, Gutekunst CA, Rogers D, Warby S, Graham RK, Loubser O, van Raamsdonk J, Singaraja R, Yang YZ et al (2002) Caspase cleavage of mutant huntingtin precedes neurodegeneration in Huntington's disease. J Neurosci 22:7862–7872
Wickner S, Maurizi MR, Gottesman S (1999) Posttranslational quality control: folding, refolding, and degrading proteins. Science 286:1888–1893
Wild P, Farhan H, McEwan DG, Wagner S, Rogov VV, Brady NR, Richter B, Korac J, Waidmann O, Choudhary C et al (2011) Phosphorylation of the autophagy receptor optineurin restricts Salmonella growth. Science 333:228–233
Williams EB, Krishnaswamy S, Mann KG (1989) Zymogen/enzyme discrimination using peptide chloromethyl ketones. J Biol Chem 264:7536–7545
Witze ES, Old WM, Resing KA, Ahn NG (2007) Mapping protein post-translational modifications with mass spectrometry. Nat Methods 4:798–806
Wyss DF, Choi JS, Li J, Knoppers MH, Willis KJ, Arulanandam AR, Smolyar A, Reinherz EL, Wagner G (1995) Conformation and function of the N-linked glycan in the adhesion domain of human CD2. Science 269:1273–1278
Xu G, Shin SB, Jaffrey SR (2009a) Global profiling of protease cleavage sites by chemoselective labeling of protein N-termini. Proc Natl Acad Sci U S A 106:19310–19315
Xu P, Duong DM, Seyfried NT, Cheng D, Xie Y, Robert J, Rush J, Hochstrasser M, Finley D, Peng J (2009b) Quantitative proteomics reveals the function of unconventional ubiquitin chains in proteasomal degradation. Cell 137:133–145
Xu G, Paige JS, Jaffrey SR (2010) Global analysis of lysine ubiquitination by ubiquitin remnant immunoaffinity profiling. Nat Biotechnol 28:868–873
Yang Z, Klionsky DJ (2010) Mammalian autophagy: core molecular machinery and signaling regulation. Curr Opin Cell Biol 22:124–131
Yong VW, Krekoski CA, Forsyth PA, Bell R, Edwards DR (1998) Matrix metalloproteinases and diseases of the CNS. Trends Neurosci 21:75–80
Zeeuwen PL, van Vlijmen-Willems IM, Cheng T, Rodijk-Olthuis D, Hitomi K, Hara-Nishimura I, John S, Smyth N, Reinheckel T, Hendriks WJ et al (2010) The cystatin M/E-cathepsin L balance is essential for tissue homeostasis in epidermis, hair follicles, and cornea. FASEB J 24:3744–3755
Zhang J, Shridhar R, Dai Q, Song J, Barlow SC, Yin L, Sloane BF, Miller FR, Meschonat C, Li BD et al (2004) Cystatin m: a novel candidate tumor suppressor gene for breast cancer. Cancer Res 64:6957–6964
Zimmermann AC, Zarei M, Eiselein S, Dengjel J (2010) Quantitative proteomics for the analysis of spatio-temporal protein dynamics during autophagy. Autophagy 6:1009–1016
Zogg T, Brandstetter H (2009) Activation mechanisms of coagulation factor IX. Biol Chem 390:391–400

Chapter 6
Exploring Systemic Functions of Lysosomal Proteases: The Perspective of Genetically Modified Mouse Models

Martina Gansz, Ursula Kern, Christoph Peters, and Thomas Reinheckel

6.1 Introduction

The development of techniques for random and targeted mutagenesis of the mouse genome represents a milestone for the functional analysis of genes in the context of a living organism (Capecchi 2005; Gondo et al. 2009). The investigation of proteases employing genetically engineered mice (GEM) started soon after the methods for genome modification had been inaugurated in the second half of the 1980ies (Frels et al. 1985; Thomas and Capecchi 1987). The first mouse line with targeted inactivation of a gene for a lysosomal peptidase was that for the aspartic endoprotease cathepsin D (Saftig et al. 1995), with more 'knock-out' mouse lines for the aspartic and cysteine cathepsins to follow (Buhling et al. 2011; D'Angelo et al. 2010; Halangk et al. 2000; Nakagawa et al. 1999; Ondr and Pham 2004; Pham and Ley 1999; Roth et al. 2000; Saftig et al. 1998; Sevenich et al. 2010; Tang et al. 2006; Tsukuba et al. 2003). However, it soon became evident that generation of protease GEM does not only require mastering the introduction of the genome modification into the mouse germ line—it also required systematic phenotyping of

M. Gansz
Institute of Molecular Medicine and Cell Research and Faculty of Biology, Albert-Ludwigs-University Freiburg, Freiburg, Germany

U. Kern
Institue of Molecular Medicine and Cell Research, Spemann Graduate School of Biology and Medicine (SGBM) and Faculty of Biology, Albert-Ludwigs-University Freiburg, Freiburg, Germany

C. Peters
Institue of Molecular Medicine and Cell Research, and BIOSS Centre for Biological Signalling Studies, Albert-Ludwigs-University Freiburg, Freiburg, Germany

T. Reinheckel (✉)
Institute of Molecular Medicine and Cell Research and BIOSS Centre for Biological Signalling Studies, Albert-Ludwigs-University Freiburg, Stefan Meier Str. 17, Freiburg 79104, Germany
e-mail: thomas.reinheckel@uniklinik-freiburg.de

K. Brix and W. Stöcker (eds.), *Proteases: Structure and Function*,
DOI 10.1007/978-3-7091-0885-7_6, © Springer-Verlag Wien 2013

the GEM on molecular, biochemical, cellular, and tissue levels as well as functional analysis of the organ systems. In 2001 we published a review entitled 'Towards Specific Functions of Lysosomal Cysteine Peptidases: Phenotypes of mice deficient for Cathepsin B or Cathepsin L', which summarized the data that had emerged from the analysis of GEM with loss of function mutations of these cysteine cathepsins (Reinheckel et al. 2001). The paper was discussing the periodic hair loss and the epidermal hyperproliferation of cathepsin L deficient mice, the role of cathepsins in MHC class II mediated antigen presentation, the role of cathepsin B in the early pathogenesis of acute pancreatitis, and the involvement of cathepsin B in cell death processes. To date one can state that much progress in understanding the cathepsin functions in all the topics mentioned above has been obtained during the past decade. This has been addressed by a number of recent in-depth reviews (Brix et al. 2008; Hsing and Rudensky 2005; Reiser et al. 2010; Turk and Turk 2009; Zeeuwen et al. 2009). Over the past 10 years of 'cathepsin mouse genetics' a clear trend extending the analysis of GEM with a single gene mutation to analysis of double- and triple mutants that have deficiencies for multiple proteases, express the human enzyme in the context of the respective gene knock-out, or carry genetic modifications of protease genes simultaneously with their inhibitors, emerged. This enabled stringent gain- and loss-of-function studies of lysosomal proteases and their inhibitors in a complex living system, i.e. *mus musculus*. Here we illustrate these developments by addressing four areas in which considerable progress has been made by employing GEM: Firstly we will revisit the role of endolysosomal cysteine proteases and their inhibitors in skin homeostasis and skin cancer. Secondly, the involvement of cathepsins in atherosclerosis and a previously unknown role of cathepsin L in the heart will be discussed. Finally, we focus on the interplay of cathepsins and their inhibitors in neurodegeneration. Throughout these diverse subjects we show how the normal proteolytic balance has been disturbed by the genetic alternations introduced in the GEM.

6.2 Cathepsin L in Epidermal Homeostasis and Regulation of Hair Cycling

6.2.1 Cathepsin L Deficiency as Molecular Cause of the Furless Phenotype

The balance of proteolytic activity is detrimental for skin and hair follicle homeostasis (Zeeuwen et al. 2009). A well known example is the complex phenotype of periodical hair loss and epidermal hyperproliferation of cathepsin L deficient mice (Roth et al. 2000). This skin phenotype is identical to the classical spontaneous mouse mutant *furless* (*fs*) (Green 1954) and the radiation induced mutation nackt (*nkt*) (Benavides et al. 1999). By genetic complementation experiments it turned out that *furless* and *nkt* are caused by a point mutation and a deletion of the cathepsin L gene, respectively (Benavides et al. 2001; Roth et al. 2000).

How the loss of cathepsin L disturbs the differentiation and cycling of the hair follicle is still not fully understood. However, the cell type in which cathepsin L is critically required to maintain regular hair growth has been identified by crossing transgenic mice that expressed mouse cathepsin L (Ctsl) under control of the keratin 14 promotor (K14) with cathepsin L deficient ($Ctsl^{-/-}$) mice. The resulting K14-Ctsl/$Ctsl^{-/-}$ mice express the protease only in stratified epithelia, i.e. in keratinocytes of the epidermis and hair follicles, but not in the mesenchymal parts of the skin (Reinheckel et al. 2005). The result of this genetic experiment was a rescue of the *furless* phenotype indicating a keratinocyte specific function of mouse cathepsin L. Notably, GEM expressing human cathepsin V, the orthologue to mouse cathepsin L, controlled by the K14-promotor and crossed to cathepsin L null mice showed the same normalization of the fur suggesting identical functions of mouse cathepsin L and human cathepsin V in keratinocytes (Hagemann et al. 2004). Human cathepsin V and the highly homologous 'classical' human cathepsin L are both expressed in human skin keratinocytes (Bernard et al. 2003; Zeeuwen et al. 2007). Interestingly, crossing mice transgenic for human cathepsin L with the cathepsin L knock-out did not affect the periodic hair loss (Sevenich et al. 2006). However, these GEM express human cathepsin L not controlled by the K14-promotor but regulated by the genuine human cathepsin L promoter elements (Houseweart et al. 2003). Although these GEM express active human cathepsin L in various organs such as kidneys, heart, and brain we were not able to detect the human cathepsin L in the epidermis of these mice (TR unpublished data). Hence, the functional analysis of human cathepsin L in mouse epidermis requires further investigation.

On cellular level the disturbed hair cycle in cathepsin L deficient mice is caused by impaired apoptosis and increased proliferation of hair follicle keratinocytes during the physiological regression phase (the catagen) of the hair follicle (Tobin et al. 2002). The prolonged catagen results in profoundly disturbed differentiation of the hair follicle inner root sheet that normally anchors the hair shaft. Loss of this anchor in cathepsin L deficient mice causes the typical periodic hair loss at the end of the catagen and during the resting phase (telogen) of the hair cycle. The massive regression of the hair follicle in the catagen is required for anchoring of the hair shaft and failure to regress properly leads to the abnormal shedding observed in cathepsin L deficient mice. In this regard, trichohyalin the major structural protein of the inner root sheet, which is subjected to protease dependent processing, is a possible substrate protein of cathepsin L (Tobin et al. 2002). Trichohyalin is associated with filaggrin, another structural protein expressed in the inner root sheet and in differentiated epidermal keratinocytes. The final step in processing of filaggrin is impaired in the cathepsin L deficient nkt mice (Benavides et al. 2002). Since hair follicles are extremely difficult to address by biochemical methods the detailed analysis of the proteolytic system of these 'mini-organs' remains a considerable challenge. However, novel generations of GEM provide excellent tools to address these fascinating biological questions (Rendl et al. 2005).

6.2.2 Cathepsin L Regulating Epidermal Proliferation and Carcinogenesis

In addition to the hair loss phenotype, cathepsin L deficiency causes hyperproliferation of basal keratinocytes resulting in epidermal hyperproliferation, acanthosis, and hyperkeratosis (Roth et al. 2000). This epidermal phenotype of cathepsin L knock-out mice is very likely to be caused by a keratinocyte specific defect, because keratin 14 promoter controlled expression of mouse cathepsin L or human cathepsin V in Ctsl-null mice results in a normal state of epidermal proliferation and histology (Hagemann et al. 2004; Reinheckel et al. 2005). In primary mouse keratinocyte cultures the lack of cathepsin L resulted in increased recycling of the epidermal growth factor (EGF) and higher levels of the EGF-receptor (Reinheckel et al. 2005). In consequence of this impaired growth signal termination, cathepsin L deficient keratinocytes show higher proliferation rates in response to EGF as compared to wild-type keratinocytes. Work with keratinocyte conditioned media showed that this concept is not restricted to EGF but seems to apply for various proliferative signals (Dennemarker et al. 2010a; Reinheckel et al. 2005). The relatively mild epidermal phenotype of cathepsin L null mice is exaggerated in the context of skin cancer as shown in transgenic K14-HPV16 skin cancer mice crossed to congenic cathepsin L knock-out mice (Dennemarker et al. 2010a). The K14-HPV16/$Ctsl^{-/-}$ mice showed early tumor onset, anaplastic squamous cell cancers, and a highly increased frequency of lymph node metastasis. On molecular level the concentration of active GTP-Ras was increased in cathepsin L deficient epidermis. Furthermore there was an increased activation of the protein kinase B/Akt and mitogen-activated protein kinase pathways in cathepsin L knock-out cells, especially if treated with media conditioned by cathepsin L deficient keratinocytes (Dennemarker et al. 2010a). Taken together, cathepsin L deficiency seems to result in sustained oncogenic signaling in keratinocytes due to impaired termination of growth factor stimulation. In line with these findings, the formation of chemically (DMBA/TPA) induced skin cancer is enhanced in the radiation-induced cathepsin L mutant *nkt* (Benavides et al. 2011). Moreover, transgenic expression of the human cathepsin L inhibitor hurpin in mice increased their susceptibility to develop chemically induced skin carcinomas and reduced apoptosis in UVB-irradiated skin (Walz et al. 2007).

However, cathepsins are often thought to act as tumor promoters by enhancing invasiveness and angiogenesis facilitated by degradation of extracellular matrix components and processing of proangiogenic peptides (Mason and Joyce 2011; Mohamed and Sloane 2006). In support of this idea the depletion of one of the extracellular inhibitors of cysteine cathepsins, i.e. cystatin C, in the K14-HPV16 model leads to an accelerated tumor phenotype with concomitant increase in general cathepsin activity (Yu et al. 2010). In this work, the facilitated tumor progression is ascribed to enhanced proliferation and reduction of cell death as well as improved angiogenesis enabled by increased production of angiogenic peptides. These contradictory observations obtained by cathepsin L deficient and cystatin C deficient mice may be explained as the ablation of cystatin C increases extracellular activity of cysteine cathepsins whereas in the critical function of cathepsin L in keratinocytes appears to be in growth factor termination within the endolysosome. Regarding the widely discussed issue of cathepsin inhibition as potential cancer therapy, it is important to realize that cathepsin L is a multifunctional protease whose inhibition has the potential to promote cancer rather than blocking it.

6.2.3 Balanced Cathepsin Activity Required for Skin Barrier Function

The formation of the stratum corneum is essential for the barrier function of the skin protecting against mechanical stress, infections and water loss (Proksch et al. 2008). The process of cornification, as the final step in epidermal differentiation, is mainly fulfilled by protein- and lipid-crosslinking that is carried out by transglutaminases. Ichtyosis is the general term for disturbances in cornification and is caused by different genetic defects affecting either structural proteins like keratin or the transglutaminases. Additionally, the balance between protease activity and inhibition turned out to be important in the cornification process; not at least because transglutaminases need to be activated by limited proteolysis (Zeeuwen et al. 2009).

Mice that are deficient for cystatin M/E, an extracellular inhibitor of cysteine proteases, show severe defects in cornification and desquamation of the epidermis and die shortly after birth due to dehydration (Zeeuwen et al. 2002). This was at first ascribed to overshooting activity of transglutaminase-3 due to deregulated activation by the cysteine protease legumain. However, it turned out that cystatin M/E has an additional protease binding site recognizing cathepsin L (Cheng et al. 2006). Impressively, cystatin M/E deficient mice crossed with $Ctsl^{-/-}$ mice do no longer develop ichtyosis and are viable, while the ablation of legumain or transglutaminase-3 did not rescue the phenotype of cystatin M/E deficient mice (Zeeuwen et al. 2011). Hence excessive cathepsin L activity appears to be the key pathomechanism of ichtyosis in cystatin M/E deficient mice. The increase in cathepsin L due to loss of cystatin M/E leads to accelerated activation of cathepsin D which in turn might overactivate transglutaminase-1. Although the cystatin M/E-Ctsl double deficient mice are rescued from ichtyosis, skin and hair homeostasis is not restored completely as these mice show in addition to the periodical hair loss of cathepsin L deficient mice a complete loss of fur after 2–3 hair cycles (Zeeuwen et al. 2011). In these mice the dermal connective tissue strongly resembles fibrotic scar tissue with parallel organization of collagen fibers. Moreover, cystatin M/E deficient mice expressing cathepsin L from one allele develop inflammatory alterations of the eye cornea. This eye phenotype, which is 100 % persistent, is again caused by cathepsin L dependent imbalanced proteolysis as this is not observed in the double deficient animals. An ablation of cystatin B (also named stefin B), which causes the Unverricht-Lundborg epilepsy (EPM1), leads to a

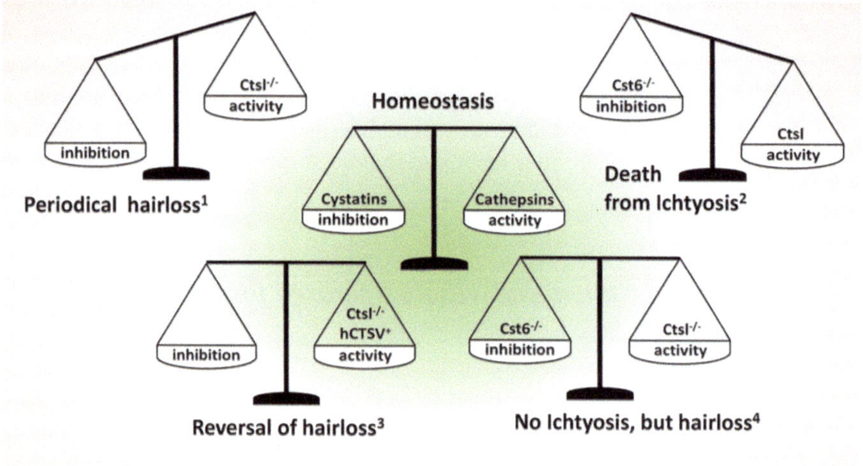

Fig. 6.1 The balance in proteolytic activity of cathepsin L is essential for skin and hair homeostasis. As soon as this balance is disturbed by a deficiency in cathepsin L ($Ctsl^{-/-}$) or its inhibitor cystatin M/E ($Cst6^{-/-}$) functions in hair and skin development and maintenance are severely impaired. The combination of both deficiencies or the expression of a human cathepsin L homologue ($hCTSV^{+}$) redresses the proteolytic balance which manifests in moderate or completely reversed phenotypes. (1)—Roth et al. (2000); (2)—Zeeuwen et al. (2002); (3)—Hagemann et al. (2004); (4)—Zeeuwen et al. (2010)

comparable eye phenotype (Pennacchio et al. 1998). However, the responsible protease in this case was has not been identified yet, because the phenotype was not rescued by additional ablation of cathepsin L (Houseweart et al. 2003). Figure 6.1 summarizes the GEM findings regarding the role of cathepsin L in epidermis and hair follicles. Although there are many unsolved questions regarding the biochemical and cellular mechanisms of cathepsin L function, the genetic evidence strongly suggests that tightly balanced cathepsin L activity is essential for epidermal homeostasis.

6.3 Cysteine Cathepsins and Their Inhibitors in the Cardiovascular System

6.3.1 Atherosclerosis and Abdominal Aortic Aneurysm

Atherosclerosis and abdominal aortic aneurysm are life threatening inflammatory diseases of blood vessel walls. They involve major remodeling of the extracellular matrix in different layers of vessel walls (Garcia-Touchard et al. 2005). Collagens and elastin are the most abundant macromolecular constituents of the extracellular matrix of the intima and media layers. Elevated collagenolytic and elastinolytic activities have been detected in atherosclerotic vessels. They are involved in the reduction of collagens and elastin in the final stages of both diseases ultimately resulting in rupture of atherosclerotic plaques with e.g. subsequent thrombosis of coronary arteries and myocardial infarction or rupture of the entire aortic wall in the case of abdominal aortic aneurysm. Multiple proteolytic enzymes belonging to the classes of matrix metallo-, serine- and cysteine proteases contribute to the extracellular degradation of collagens and elastin in the course of these devastating pathophysiological processes (Newby 2005; Nicholl et al. 2006; Sukhova and Shi 2006; Sukhova et al. 1998; Sun et al. 2009, 2011). Since the contribution of the metallo- and serine-proteases are not within the immediate scope of this chapter and have been reviewed recently, we are going to focus on the role of cysteine proteases and their inhibitors. Enzymatic activities of three major collagenolytic and elastinolytic cysteine cathepsins S, K and L have been shown to be significantly elevated in atherosclerotic as well as aneurismal lesions in humans (Sukhova et al. 1998). On one hand this is due to an elevated expression of these enzymes under these pathological conditions but on the other hand a significant reduction of the expression of their major inhibitor cystatin C has been described contributing to the imbalance towards enhanced collagenolytic and elastinolytic activities in the diseased tissue (Shi et al. 1999; Sukhova and Shi 2006). For experimental evaluation GEMs developing diet-induced atherosclerosis after consuming a 'Western' high-cholesterol diet were made deficient for one of the three cysteine cathepsins S, K or L by crossing with the respective knock-out strains. In the absence of any one of these three cathepsins an attenuation of the developing atherosclerosis has been observed (Kitamoto et al. 2007; Lutgens et al. 2006; Sukhova et al. 2003). On the other hand knocking out cystatin C—a major extracellular inhibitor of cysteine cathepsins—in the atherosclerosis model and thereby elevating the extracellular activities of the relevant cysteine cathepsins results in a significant reduction of the elastin content of the vessel media and an increased collagen as well as smooth muscle cell content (Sukhova et al. 2005). In the course of the development of atherosclerotic and aneurismal lesions elevated numbers of different cell types, e.g. endothelial cells, smooth muscle cells and macrophages have been described in and around these lesions. Elevated levels of one of the most potent elastinolytic and collagenolytic enzymes, cathepsin L, have been detected in these cells (Liu et al. 2006).

Macrophages receive increasing attention as key players in the pathogenesis of atherosclerosis and possibly target cells for therapy (Wilson et al. 2009). Macrophages may also be the main source for proteolytic enzymes within the atherosclerotic lesions. Recent evidence on an exciting intracellular function of cathepsins added to the picture. Macrophages deficient for cathepsin B or cathepsin L show significantly reduced inflammasome activation and secretion of the proinflammatory interleukine-1β after phagocytosis of crystalline cholesterol (Duewell et al. 2010). As shown also for other crystalline agents, the cathepsins enter the cytosol caused by mechanical disruption of the phagolysosomal membrane and activate the NALP3 inflammasome by a yet unknown mechanism (Halle et al. 2008; Hornung et al. 2008). Subsequently activated caspase 1, as part of the inflammasome complex, cleaves pro- interleukine-1β to produce its active form that attracts further immune cells; thus enhancing inflammation in the atherosclerotic vessel wall.

Together these data suggest that cysteine cathepsins in vessel walls and inflammatory cells aggravate the course of atherosclerosis as well as abdominal aortic aneurysm.

6.3.2 Cardiac Homeostasis and Cardiomyopathy

Upon necropsy of 1-year-old cathepsin L deficient mice enlarged hearts have been found. A systematic investigation of this phenotype revealed that the relative heart weight of cathepsin L deficient mice of this age group is elevated in comparison to wild type controls and that approximately 25 % of these animals presented with grossly enlarged hearts closely resembling human late onset dilated cardiomyopathy. Echocardiographic investigations confirmed an increase in size of the left ventricle, which was accompanied by an enlargement of the left atrium in the severe cases. Functionally, the fractional shortening—a measure of the contractility of the left ventricle—was significantly reduced in mutant mice. In the severely dilated hearts the pressure gradient across the aortic and mitral valves was elevated more than threefold in comparison to the respective controls. In these dilated hearts valve insufficiencies and changes in electrophysiology characteristic of clinical cardiomyopathies were seen. However, between weaning and 12 months of age the mortality of cathepsin L deficient animals was not elevated in comparison to wild type controls (Petermann et al. 2006; Stypmann et al. 2002). On the microscopic level these pathologies are reflected by a progressive and extensive interstitial fibrosis of the myocardium first observed at an age of 4 months. Cardiomyocytes presented with an elevated number of pleomorphic nuclei characteristic of cardiomyopathies. Infiltration of immune cells into the myocardium was not observed. Ultrastructurally, cathepsin L deficient cardiomyocytes revealed numerous vacuoles and end-stage lysosomes—also termed residual bodies—not observed in age-matched controls as early as 3 days post partum. The enlargement of the endosomal/lysosomal compartment in cardiomyocytes devoid of cathepsin L was also reflected by an elevated staining with an acidophilic dye exhibiting a patchy pattern of the lysosomes as compared to a reticular staining in wild type controls (Petermann et al. 2006). The early appearance of these alterations points at an essential role of cathepsin L in the endosomal/lysosomal compartment of cardiomyocytes. The alterations of the endosomal/lysosomal compartment in cardiomyocytes, which were also found in thyroid epithelial cells and keratinocytes of cathepsin L deficient animals, are characteristic of a defective macroautophagy (Friedrichs et al. 2003; Tobin et al. 2002). Further studies of *Ctsl*$^{-/-}$ mouse embryonic fibroblasts expressing the autophagy marker GFP-LC3 as a model system revealed no defects in the initiation of autophagy, formation of autophagosomes and fusion of these structures with lysosomes but rather a defect in degradation of contents of the autophagolysosomes in the absence of cathepsin L resulting in an enlargement and accumulations of the organelles as observed

(Dennemarker et al. 2010b). Notably, the cardiac phenotype of cathepsin L deficient mice was rescued by cardiomyocyte-specific expression of cathepsin L under control of the myosin heavy chain promoter in otherwise cathepsin L deficient mice (Spira et al. 2007). The heart weight and the size of the left ventricle were normal in the latter animals. Atrial enlargement was not observed and fractional shortening had improved. Expression of the natriuretic peptide, ANP, a marker of sheer stress in the myocardial wall, which is elevated in cathepsin L deficient mice, was normalized in these rescue transgenic animals indicating normalized heart function. The interstitial fibrosis, however, was not rescued (Spira et al. 2007). Interestingly, in transgenic mice constitutively overexpressing human cathepsin L in cardiomyocytes cardiac hypertrophy induced by aortic banding was reduced in comparison to wild type controls due to an attenuation of the Akt/GSK3β cascade (Tang et al. 2009).

Taken together, late onset dilated cardiomyopathy in mice can be caused by a deficiency of the lysosomal endoproteinase cathepsin L triggering a defect in macroautophagy with disturbed degradation of the autophagolysosomal content in cardiomyocytes.

6.4 Cathepsins in Lysosomal Neurodegenerative Disorders

Neurodegenerative disorders are often caused by altered turnover of proteins that form plaques and fibrils. Genetic defects in lysosomal catabolic enzymes cause lysosomal storage disorders and may also lead to pathogenic protein accumulation in the brain (Bellettato and Scarpa 2010; Neufeld 1991). As discussed below, the aspartic protease cathepsin D and the combined deficiency of the cysteine cathepsins B and L have been associated with neurodegeneration mainly connected with their role in bulk protein degradation and cell death (Fig. 6.2).

6.4.1 Cathepsins in Neuronal Storage Disorders

Cathepsin D deficient mice die approximately 26 days after birth (Saftig et al. 1995). Undigested lipoprotein, mainly the subunit c of mitochondrial ATP synthase accumulates within lysosomes in affected neurons and a neurodegenerative phenotype similar to neuronal ceroid lipofuscinosis (NCL) was described in the cathepsin D deficient mouse brain (Koike et al. 2000, 2005). Interestingly, several spontaneous cathepsin D mutations in sheep and bulldog that show an NCL-like disease phenotype have been identified (Awano et al. 2006; Tyynela et al. 2000). Most importantly mutations in the human cathepsin D gene are the cause for a severe juvenile form of NCL (Siintola et al. 2006; Steinfeld et al. 2006). Hence, cathepsin D is essential for neuronal homeostasis in the mammalian organism. In contrast, mice lacking either cathepsin B or L do not show any spontaneous neurological phenotypes. However, combined deficiency is lethal and leads to rapid loss of cerebral cortical neurons and cerebellar Purkinje and granule cells

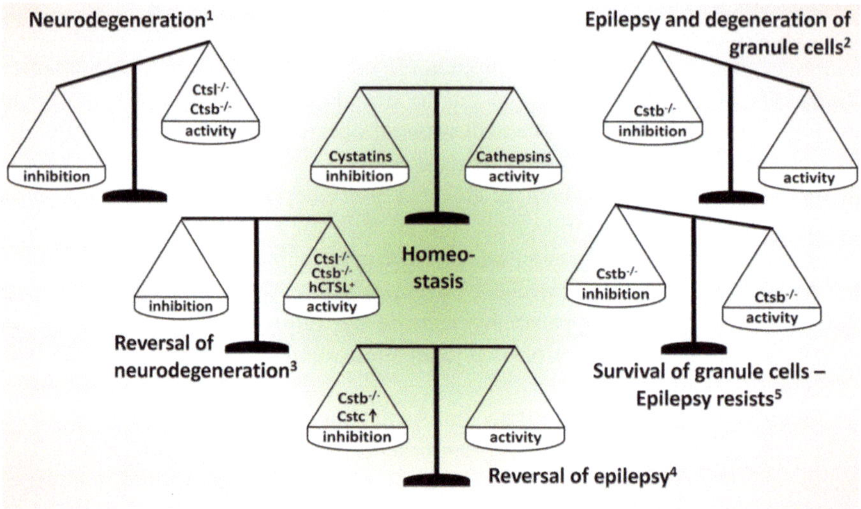

Fig. 6.2 Cellular homeostasis of neurons depends on the balance in proteolytic activity of cysteine cathepsins and their inhibitors. In case of Cathepsin L and B double deficient mice (Ctsl$^{-/-}$; Ctsb$^{-/-}$) the lack protease activity leads to neurodegeneration which is reversed if the human cathepsin L homologue (hCTSL^{+}) is expressed. Epilepsy and degeneration of granule cells is caused by an increase in proteolytic activity due to depletion of the cathepsin inhibitor cystatin B (Cstb$^{-/-}$). This phenotype is ameliorated by an additional knock out of cathepsin B and completely reversed by overexpression of cystatin C (Cstc). (1)—Felbor et al. (2002); (2)—Pennacchio et al. (1998); (3)—Sevenich et al. (2006); (4)—Kaur et al. (2010); (5)—Houseweart et al. (2003)

(Felbor et al. 2002). Again neurons show substantial accumulation of lysosomes and double-membrane vesicles in the perikarya of neurons and uncommon swelling of axons. In how far the composition of the accumulated material within the vesicles differs from those observed in cathepsin D deficiency is still not fully understood. The comparison of brain lysosomes of cathepsin B/L double deficient mice with wild-type controls by proteome analysis revealed increased abundance of proteins either associated with lysosomal recycling or neural growth (Stahl et al. 2007). This suggests a potential role for cathepsin B and L in recycling during postnatal neuronal development, which may influence axon outgrowth and synapse formation. These functions are shared by cathepsin B and L which is a typical example for the functional redundancy among cysteine cathepsins. Accordingly, human cathepsin L is able to overtake brain associated functions of murine cathepsin L and B, as transgene expression in double deficient mice prevents neurodegeneration (Sevenich et al. 2006).

6.4.2 Cathepsins as Executor Proteases in Neuronal Macroautophagy

In the current understanding of how cells are protected from aggregate formation the induction and efficient turnover of autophagosomes is essential (Banerjee et al. 2010). Degradation of aggregate-prone proteins and damaged organelles via the autophagy-lysosomal pathway is possibly even more efficient than via the ubiquitin-proteasome pathway (Korolchuk et al. 2010). In several common neurodegenerative disorders the neurons show an accumulation of autophagosomes, which is most likely due to impaired turnover, rather than enforced induction of autophagy. Thereby a mechanism which is carried out to protect the cell in the end promotes neuronal cell death due to accumulation of autophagosomes (overview in Banerjee et al. 2010). This concept has been proven in Atg-7 deficient autophagy impaired mice that die from massive loss of cerebral and cerebellar neurons (Komatsu et al. 2006). Detailed investigation of the accumulated vesicles in brains of cathepsin D single deficient and cathepsin B and L double deficient mice revealed that these carry the autophagosome marker LC3 (Koike et al. 2005). The processed membrane-bound version of LC3, LC-3 II, is highly increased in cathepsin deficient brains and correlates with disease progression which is a clear sign for massive autophagy. Defects in turnover of autophago-lysosomes have been described even in cathepsin L single deficient mice (Dennemarker et al. 2010b). It turned out that neither the initiation of autophagy nor formation of autophagolysosomes is impaired. This implies a non redundant function of cathepsin L in the final degradation process of the incorporated material. Is cathepsin L missing the digestion cannot be completed and these autophagolysosomes accumulate. In conclusion neurodegenerative disorders induced by loss of lysosomal enzymes may be due to insufficient and deregulated autophagy. This concept may be extended to other forms of NCL as autophagy is also accelerated in initiation but incomplete in CLN3 mutated mice, a mouse model for juvenile NCL (Cao et al. 2006). In general enforcing autophagy as a neuroprotective program could be a therapeutic approach in several common neurodegenerative disorders. This may be achieved by improving cathepsin function in the autophagy-lysosomal pathway. This has been shown in mice transgenic for the human amyloid precursor protein as the accumulation of this aggregate-prone protein in autophagosomes is rescued by enhancement of lysosomal cathepsin activity by depletion of cystatin B (Yang et al. 2011).

6.4.3 Cystatin B a Critical Regulator of Neuronal Proteolytic Balance

As mentioned above, cystatin B depletion can be neuroprotective in some conditions. However, the cystatin B example illustrates delicate the balance of proteases and their inhibitors as loss of cystatin B causes the familial Unverricht-Lundborg epilepsy (Lalioti et al. 1997; Pennacchio et al. 1996). Cystatin B knock-out mice also suffer from seizures and ataxia (Pennacchio et al. 1998). In these mice lots of granule cells in the cerebellum die by apoptosis, which is most likely the reason for the neurological phenotype. There is no accumulation of vesicles or plaques found in this kind of disease. To investigate whether the deregulated activity of cathepsins is responsible for this phenotype, cathepsins B, L or S have been deleted in cystatin B deficient mice (Houseweart et al. 2003). Dying of granule cells in the cerebellum was remarkably reduced by ablation of cathepsin B, whereas deficiency in cathepsin L or S did not affect cell death (Houseweart et al. 2003). However, also ablation of cathepsin B did not rescue the phenotype of seizures and ataxia indicating that either other proteases compensate for cathepsin B function or other targets of cystatin B contribute to disease progression. Accordingly, the overexpression of cystatin C, another cysteine protease inhibitor, is able to rescue the phenotype of cystatin B deficiency, most likely because it targets more than one protease and is therefore not affected by compensatory effects (Kaur et al. 2010; Lieuallen et al. 2001).

Figure 6.2 summarizes the findings on lysosomal proteases and their endogenous inhibitors, namely cystatin B, in the nervous system of protease or inhibitor deficient GEM. Again, cellular homeostasis depends on the fine tuned balance of the proteolytic system.

6.5 Conclusions

The—still incomplete—list of examples presented in the previous sections illustrates how GEM models have been extensively employed in order to study lysosomal proteases and their inhibitors in a complex mammalian organism. What are the most important general lessons learned from these studies? Firstly, there are highly conserved *in vivo* functions of the orthologous mouse and human cathepsins illustrated by the rescue of knock-out phenotypes by expression of the respective human cathepsin orthologue. Secondly, the phenotypic effects caused by genetic deletion or a transgenic expression of cathepsins or their endogenous inhibitors are strikingly dependent on the cell-, tissue- and process-context. This is even true for the ubiquitously expressed proteins, such as cathepsins B and L or the cystatins B and C. Finally, these recent investigations provided strong *in vivo* evidence for the intriguing balance of proteolytic and anti-proteolytic processes. This demonstrates that cathepsins are embedded in a complex proteolytic network as a part of the entire signaling network of cells and tissues. In future, tools and methods of systems biology will help us to further understand these interactions.

Acknowledgement The work of the laboratory is supported by the Deutsche Forschungsgemeinschaft SFB 850 Project B7, the Centre of Chronic Immunodeficiency (CCI) Freiburg grant TP8, the Excellence Initiative of the German Federal and State Governments (EXC 294), and the European Union Framework Program (FP7 "MICROENVIMET" No 201279).

References

Awano T, Katz ML, O'Brien DP, Taylor JF, Evans J, Khan S, Sohar I, Lobel P, Johnson GS (2006) A mutation in the cathepsin D gene (CTSD) in American Bulldogs with neuronal ceroid lipofuscinosis. Mol Genet Metab 87:341–348

Banerjee R, Beal MF, Thomas B (2010) Autophagy in neurodegenerative disorders: pathogenic roles and therapeutic implications. Trends Neurosci 33:541–549

Bellettato CM, Scarpa M (2010) Pathophysiology of neuropathic lysosomal storage disorders. J Inherit Metab Dis 33:347–362

Benavides F, Giordano M, Fiette L, Bueno Brunialti AL, Martin Palenzuela N, Vanzulli S, Baldi P, Schmidt R, Dosne Pasqualini C, Guenet J (1999) Nackt (nkt), a new hair loss mutation of the mouse with associated CD4 deficiency. Immunogenetics 49:413–419

Benavides F, Venables A, Poetschke Klug H, Glasscock E, Rudensky A, Gomez M, Martin Palenzuela N, Guenet JL, Richie ER, Conti CJ (2001) The CD4 T cell-deficient mouse mutation nackt (nkt) involves a deletion in the cathepsin L (Ctsl) gene. Immunogenetics 53: 233–242

Benavides F, Starost MF, Flores M, Gimenez-Conti IB, Guenet JL, Conti CJ (2002) Impaired hair follicle morphogenesis and cycling with abnormal epidermal differentiation in nackt mice, a cathepsin L-deficient mutation. Am J Pathol 161:693–703

Benavides F, Perez C, Blando J, Contreras O, Shen J, Coussens LM, Fischer SM, Kusewitt DF, Digiovanni J, Conti CJ (2012) Protective role of cathepsin L in mouse skin carcinogenesis. Mol Carcinog 51:352–361

Bernard D, Mehul B, Thomas-Collignon A, Simonetti L, Remy V, Bernard MA, Schmidt R (2003) Analysis of proteins with caseinolytic activity in a human stratum corneum extract revealed a yet unidentified cysteine protease and identified the so-called "stratum corneum thiol protease" as cathepsin l2. J Invest Dermatol 120:592–600

Brix K, Dunkhorst A, Mayer K, Jordans S (2008) Cysteine cathepsins: cellular roadmap to different functions. Biochimie 90:194–207

Buhling F, Kouadio M, Chwieralski CE, Kern U, Hohlfeld JM, Klemm N, Friedrichs N, Roth W, Deussing JM, Peters C et al (2011) Gene targeting of the cysteine peptidase cathepsin H impairs lung surfactant in mice. PLoS One 6:e26247

Cao Y, Espinola JA, Fossale E, Massey AC, Cuervo AM, MacDonald ME, Cotman SL (2006) Autophagy is disrupted in a knock-in mouse model of juvenile neuronal ceroid lipofuscinosis. J Biol Chem 281:20483–20493

Capecchi MR (2005) Gene targeting in mice: functional analysis of the mammalian genome for the twenty-first century. Nat Rev Genet 6:507–512

Cheng T, Hitomi K, van Vlijmen-Willems IM, de Jongh GJ, Yamamoto K, Nishi K, Watts C, Reinheckel T, Schalkwijk J, Zeeuwen PL (2006) Cystatin M/E is a high affinity inhibitor of cathepsin V and cathepsin L by a reactive site that is distinct from the legumain-binding site. A novel clue for the role of cystatin M/E in epidermal cornification. J Biol Chem 281: 15893–15899

D'Angelo ME, Bird PI, Peters C, Reinheckel T, Trapani JA, Sutton VR (2010) Cathepsin H is an additional convertase of pro-granzyme B. J Biol Chem 285:20514–20519

Dennemarker J, Lohmuller T, Mayerle J, Tacke M, Lerch MM, Coussens LM, Peters C, Reinheckel T (2010a) Deficiency for the cysteine protease cathepsin L promotes tumor progression in mouse epidermis. Oncogene 29:1611–1621

Dennemarker J, Lohmuller T, Muller S, Aguilar SV, Tobin DJ, Peters C, Reinheckel T (2010b) Impaired turnover of autophagolysosomes in cathepsin L deficiency. Biol Chem 391:913–922

Duewell P, Kono H, Rayner KJ, Sirois CM, Vladimer G, Bauernfeind FG, Abela GS, Franchi L, Nunez G, Schnurr M et al (2010) NLRP3 inflammasomes are required for atherogenesis and activated by cholesterol crystals. Nature 464:1357–1361

Felbor U, Kessler B, Mothes W, Goebel HH, Ploegh HL, Bronson RT, Olsen BR (2002) Neuronal loss and brain atrophy in mice lacking cathepsins B and L. Proc Natl Acad Sci U S A 99: 7883–7888

Frels WI, Bluestone JA, Hodes RJ, Capecchi MR, Singer DS (1985) Expression of a microinjected porcine class I major histocompatibility complex gene in transgenic mice. Science 228: 577–580

Friedrichs B, Tepel C, Reinheckel T, Deussing J, von Figura K, Herzog V, Peters C, Saftig P, Brix K (2003) Thyroid functions of mouse cathepsins B, K, and L. J Clin Invest 111:1733–1745

Garcia-Touchard A, Henry TD, Sangiorgi G, Spagnoli LG, Mauriello A, Conover C, Schwartz RS (2005) Extracellular proteases in atherosclerosis and restenosis. Arterioscler Thromb Vasc Biol 25:1119–1127

Gondo Y, Fukumura R, Murata T, Makino S (2009) Next-generation gene targeting in the mouse for functional genomics. BMB Rep 42:315–323

Green EL (1954) The genetics of a new hair deficiency, furless, in the house mouse. J Hered 45: 115–118

Hagemann S, Gunther T, Dennemarker J, Lohmuller T, Bromme D, Schule R, Peters C, Reinheckel T (2004) The human cysteine protease cathepsin V can compensate for murine cathepsin L in mouse epidermis and hair follicles. Eur J Cell Biol 83:775–780

Halangk W, Lerch MM, Brandt-Nedelev B, Roth W, Ruthenbuerger M, Reinheckel T, Domschke W, Lippert H, Peters C, Deussing J (2000) Role of cathepsin B in intracellular trypsinogen activation and the onset of acute pancreatitis. J Clin Invest 106:773–781

Halle A, Hornung V, Petzold GC, Stewart CR, Monks BG, Reinheckel T, Fitzgerald KA, Latz E, Moore KJ, Golenbock DT (2008) The NALP3 inflammasome is involved in the innate immune response to amyloid-beta. Nat Immunol 9:857–865

Hornung V, Bauernfeind F, Halle A, Samstad EO, Kono H, Rock KL, Fitzgerald KA, Latz E (2008) Silica crystals and aluminum salts activate the NALP3 inflammasome through phagosomal destabilization. Nat Immunol 9:847–856

Houseweart MK, Pennacchio LA, Vilaythong A, Peters C, Noebels JL, Myers RM (2003) Cathepsin B but not cathepsins L or S contributes to the pathogenesis of Unverricht-Lundborg progressive myoclonus epilepsy (EPM1). J Neurobiol 56:315–327

Hsing LC, Rudensky AY (2005) The lysosomal cysteine proteases in MHC class II antigen presentation. Immunol Rev 207:229–241

Kaur G, Mohan P, Pawlik M, DeRosa S, Fajiculay J, Che S, Grubb A, Ginsberg SD, Nixon RA, Levy E (2010) Cystatin C rescues degenerating neurons in a cystatin B-knockout mouse model of progressive myoclonus epilepsy. Am J Pathol 177:2256–2267

Kitamoto S, Sukhova GK, Sun J, Yang M, Libby P, Love V, Duramad P, Sun C, Zhang Y, Yang X et al (2007) Cathepsin L deficiency reduces diet-induced atherosclerosis in low-density lipoprotein receptor-knockout mice. Circulation 115:2065–2075

Koike M, Nakanishi H, Saftig P, Ezaki J, Isahara K, Ohsawa Y, Schulz-Schaeffer W, Watanabe T, Waguri S, Kametaka S et al (2000) Cathepsin D deficiency induces lysosomal storage with ceroid lipofuscin in mouse CNS neurons. J Neurosci 20:6898–6906

Koike M, Shibata M, Waguri S, Yoshimura K, Tanida I, Kominami E, Gotow T, Peters C, von Figura K, Mizushima N et al (2005) Participation of autophagy in storage of lysosomes in neurons from mouse models of neuronal ceroid-lipofuscinoses (Batten disease). Am J Pathol 167:1713–1728

Komatsu M, Waguri S, Chiba T, Murata S, Iwata J, Tanida I, Ueno T, Koike M, Uchiyama Y, Kominami E et al (2006) Loss of autophagy in the central nervous system causes neurodegeneration in mice. Nature 441:880–884

Korolchuk VI, Menzies FM, Rubinsztein DC (2010) Mechanisms of cross-talk between the ubiquitin-proteasome and autophagy-lysosome systems. FEBS Lett 584:1393–1398

Lalioti MD, Mirotsou M, Buresi C, Peitsch MC, Rossier C, Ouazzani R, Baldy-Moulinier M, Bottani A, Malafosse A, Antonarakis SE (1997) Identification of mutations in cystatin B, the

gene responsible for the Unverricht-Lundborg type of progressive myoclonus epilepsy (EPM1). Am J Hum Genet 60:342–351
Lieuallen K, Pennacchio LA, Park M, Myers RM, Lennon GG (2001) Cystatin B-deficient mice have increased expression of apoptosis and glial activation genes. Hum Mol Genet 10: 1867–1871
Liu J, Sukhova GK, Yang JT, Sun J, Ma L, Ren A, Xu WH, Fu H, Dolganov GM, Hu C et al (2006) Cathepsin L expression and regulation in human abdominal aortic aneurysm, atherosclerosis, and vascular cells. Atherosclerosis 184:302–311
Lutgens E, Lutgens SP, Faber BC, Heeneman S, Gijbels MM, de Winther MP, Frederik P, van der Made I, Daugherty A, Sijbers AM et al (2006) Disruption of the cathepsin K gene reduces atherosclerosis progression and induces plaque fibrosis but accelerates macrophage foam cell formation. Circulation 113:98–107
Mason SD, Joyce JA (2011) Proteolytic networks in cancer. Trends Cell Biol 21:228–237
Mohamed MM, Sloane BF (2006) Cysteine cathepsins: multifunctional enzymes in cancer. Nat Rev Cancer 6:764–775
Nakagawa TY, Brissette WH, Lira PD, Griffiths RJ, Petrushova N, Stock J, McNeish JD, Eastman SE, Howard ED, Clarke SR et al (1999) Impaired invariant chain degradation and antigen presentation and diminished collagen-induced arthritis in cathepsin S null mice. Immunity 10: 207–217
Neufeld EF (1991) Lysosomal storage diseases. Annu Rev Biochem 60:257–280
Newby AC (2005) Dual role of matrix metalloproteinases (matrixins) in intimal thickening and atherosclerotic plaque rupture. Physiol Rev 85:1–31
Nicholl SM, Roztocil E, Davies MG (2006) Plasminogen activator system and vascular disease. Curr Vasc Pharmacol 4:101–116
Ondr JK, Pham CT (2004) Characterization of murine cathepsin W and its role in cell-mediated cytotoxicity. J Biol Chem 279:27525–27533
Pennacchio LA, Lehesjoki AE, Stone NE, Willour VL, Virtaneva K, Miao J, D'Amato E, Ramirez L, Faham M, Koskiniemi M et al (1996) Mutations in the gene encoding cystatin B in progressive myoclonus epilepsy (EPM1). Science 271:1731–1734
Pennacchio LA, Bouley DM, Higgins KM, Scott MP, Noebels JL, Myers RM (1998) Progressive ataxia, myoclonic epilepsy and cerebellar apoptosis in cystatin B-deficient mice. Nat Genet 20:251–258
Petermann I, Mayer C, Stypmann J, Biniossek ML, Tobin DJ, Engelen MA, Dandekar T, Grune T, Schild L, Peters C et al (2006) Lysosomal, cytoskeletal, and metabolic alterations in cardiomyopathy of cathepsin L knockout mice. FASEB J 20:1266–1268
Pham CT, Ley TJ (1999) Dipeptidyl peptidase I is required for the processing and activation of granzymes A and B in vivo. Proc Natl Acad Sci U S A 96:8627–8632
Proksch E, Brandner JM, Jensen JM (2008) The skin: an indispensable barrier. Exp Dermatol 17:1063–1072
Reinheckel T, Deussing J, Roth W, Peters C (2001) Towards specific functions of lysosomal cysteine peptidases: phenotypes of mice deficient for cathepsin B or cathepsin L. Biol Chem 382:735–741
Reinheckel T, Hagemann S, Dollwet-Mack S, Martinez E, Lohmuller T, Zlatkovic G, Tobin DJ, Maas-Szabowski N, Peters C (2005) The lysosomal cysteine protease cathepsin L regulates keratinocyte proliferation by control of growth factor recycling. J Cell Sci 118:3387–3395
Reiser J, Adair B, Reinheckel T (2010) Specialized roles for cysteine cathepsins in health and disease. J Clin Invest 120:3421–3431
Rendl M, Lewis L, Fuchs E (2005) Molecular dissection of mesenchymal-epithelial interactions in the hair follicle. PLoS Biol 3:e331
Roth W, Deussing J, Botchkarev VA, Pauly-Evers M, Saftig P, Hafner A, Schmidt P, Schmahl W, Scherer J, Anton-Lamprecht I et al (2000) Cathepsin L deficiency as molecular defect of furless: hyperproliferation of keratinocytes and pertubation of hair follicle cycling. FASEB J 14:2075–2086

Saftig P, Hetman M, Schmahl W, Weber K, Heine L, Mossmann H, Koster A, Hess B, Evers M, von Figura K et al (1995) Mice deficient for the lysosomal proteinase cathepsin D exhibit progressive atrophy of the intestinal mucosa and profound destruction of lymphoid cells. EMBO J 14:3599–3608

Saftig P, Hunziker E, Wehmeyer O, Jones S, Boyde A, Rommerskirch W, Moritz JD, Schu P, von Figura K (1998) Impaired osteoclastic bone resorption leads to osteopetrosis in cathepsin-K-deficient mice. Proc Natl Acad Sci U S A 95:13453–13458

Sevenich L, Pennacchio LA, Peters C, Reinheckel T (2006) Human cathepsin L rescues the neurodegeneration and lethality in cathepsin B/L double-deficient mice. Biol Chem 387: 885–891

Sevenich L, Schurigt U, Sachse K, Gajda M, Werner F, Muller S, Vasiljeva O, Schwinde A, Klemm N, Deussing J et al (2010) Synergistic antitumor effects of combined cathepsin B and cathepsin Z deficiencies on breast cancer progression and metastasis in mice. Proc Natl Acad Sci U S A 107:2497–2502

Shi GP, Sukhova GK, Grubb A, Ducharme A, Rhode LH, Lee RT, Ridker PM, Libby P, Chapman HA (1999) Cystatin C deficiency in human atherosclerosis and aortic aneurysms. J Clin Invest 104:1191–1197

Siintola E, Partanen S, Stromme P, Haapanen A, Haltia M, Maehlen J, Lehesjoki AE, Tyynela J (2006) Cathepsin D deficiency underlies congenital human neuronal ceroid-lipofuscinosis. Brain 129:1438–1445

Spira D, Stypmann J, Tobin DJ, Petermann I, Mayer C, Hagemann S, Vasiljeva O, Gunther T, Schule R, Peters C et al (2007) Cell type-specific functions of the lysosomal protease cathepsin L in the heart. J Biol Chem 282:37045–37052

Stahl S, Reinders Y, Asan E, Mothes W, Conzelmann E, Sickmann A, Felbor U (2007) Proteomic analysis of cathepsin B- and L-deficient mouse brain lysosomes. Biochim Biophys Acta 1774: 1237–1246

Steinfeld R, Reinhardt K, Schreiber K, Hillebrand M, Kraetzner R, Bruck W, Saftig P, Gartner J (2006) Cathepsin D deficiency is associated with a human neurodegenerative disorder. Am J Hum Genet 78:988–998

Stypmann J, Glaser K, Roth W, Tobin DJ, Petermann I, Matthias R, Monnig G, Haverkamp W, Breithardt G, Schmahl W et al (2002) Dilated cardiomyopathy in mice deficient for the lysosomal cysteine peptidase cathepsin L. Proc Natl Acad Sci U S A 99:6234–6239

Sukhova GK, Shi GP (2006) Do cathepsins play a role in abdominal aortic aneurysm pathogenesis? Ann N Y Acad Sci 1085:161–169

Sukhova GK, Shi GP, Simon DI, Chapman HA, Libby P (1998) Expression of the elastolytic cathepsins S and K in human atheroma and regulation of their production in smooth muscle cells. J Clin Invest 102:576–583

Sukhova GK, Zhang Y, Pan JH, Wada Y, Yamamoto T, Naito M, Kodama T, Tsimikas S, Witztum JL, Lu ML et al (2003) Deficiency of cathepsin S reduces atherosclerosis in LDL receptor-deficient mice. J Clin Invest 111:897–906

Sukhova GK, Wang B, Libby P, Pan JH, Zhang Y, Grubb A, Fang K, Chapman HA, Shi GP (2005) Cystatin C deficiency increases elastic lamina degradation and aortic dilatation in apolipoprotein E-null mice. Circ Res 96:368–375

Sun J, Zhang J, Lindholt JS, Sukhova GK, Liu J, He A, Abrink M, Pejler G, Stevens RL, Thompson RW et al (2009) Critical role of mast cell chymase in mouse abdominal aortic aneurysm formation. Circulation 120:973–982

Sun J, Sukhova GK, Zhang J, Chen H, Sjoberg S, Libby P, Xiang M, Wang J, Peters C, Reinheckel T et al (2011) Cathepsin L activity is essential to elastase perfusion-induced abdominal aortic aneurysms in mice. Arterioscler Thromb Vasc Biol 31:2500–2508

Tang CH, Lee JW, Galvez MG, Robillard L, Mole SE, Chapman HA (2006) Murine cathepsin f deficiency causes neuronal lipofuscinosis and late-onset neurological disease. Mol Cell Biol 26:2309–2316

Tang Q, Cai J, Shen D, Bian Z, Yan L, Wang YX, Lan J, Zhuang GQ, Ma WZ, Wang W (2009) Lysosomal cysteine peptidase cathepsin L protects against cardiac hypertrophy through blocking AKT/GSK3beta signaling. J Mol Med 87:249–260

Thomas KR, Capecchi MR (1987) Site-directed mutagenesis by gene targeting in mouse embryo-derived stem cells. Cell 51:503–512

Tobin DJ, Foitzik K, Reinheckel T, Mecklenburg L, Botchkarev VA, Peters C, Paus R (2002) The lysosomal protease cathepsin L is an important regulator of keratinocyte and melanocyte differentiation during hair follicle morphogenesis and cycling. Am J Pathol 160:1807–1821

Tsukuba T, Okamoto K, Okamoto Y, Yanagawa M, Kohmura K, Yasuda Y, Uchi H, Nakahara T, Furue M, Nakayama K et al (2003) Association of cathepsin E deficiency with development of atopic dermatitis. J Biochem 134:893–902

Turk B, Turk V (2009) Lysosomes as "suicide bags" in cell death: myth or reality? J Biol Chem 284:21783–21787

Tyynela J, Sohar I, Sleat DE, Gin RM, Donnelly RJ, Baumann M, Haltia M, Lobel P (2000) A mutation in the ovine cathepsin D gene causes a congenital lysosomal storage disease with profound neurodegeneration. EMBO J 19:2786–2792

Walz M, Kellermann S, Bylaite M, Andree B, Ruther U, Paus R, Kloepper JE, Reifenberger J, Ruzicka T (2007) Expression of the human Cathepsin L inhibitor hurpin in mice: skin alterations and increased carcinogenesis. Exp Dermatol 16:715–723

Wilson HM, Barker RN, Erwig LP (2009) Macrophages: promising targets for the treatment of atherosclerosis. Curr Vasc Pharmacol 7:234–243

Yang DS, Stavrides P, Mohan PS, Kaushik S, Kumar A, Ohno M, Schmidt SD, Wesson D, Bandyopadhyay U, Jiang Y et al (2011) Reversal of autophagy dysfunction in the TgCRND8 mouse model of Alzheimer's disease ameliorates amyloid pathologies and memory deficits. Brain 134:258–277

Yu W, Liu J, Shi MA, Wang J, Xiang M, Kitamoto S, Wang B, Sukhova GK, Murphy GF, Orasanu G et al (2010) Cystatin C deficiency promotes epidermal dysplasia in K14-HPV16 transgenic mice. PLoS One 5:e13973

Zeeuwen PL, van Vlijmen-Willems IM, Hendriks W, Merkx GF, Schalkwijk J (2002) A null mutation in the cystatin M/E gene of ichq mice causes juvenile lethality and defects in epidermal cornification. Hum Mol Genet 11:2867–2875

Zeeuwen PL, Ishida-Yamamoto A, van Vlijmen-Willems IM, Cheng T, Bergers M, Iizuka H, Schalkwijk J (2007) Colocalization of cystatin M/E and cathepsin V in lamellar granules and corneodesmosomes suggests a functional role in epidermal differentiation. J Invest Dermatol 127:120–128

Zeeuwen PL, Cheng T, Schalkwijk J (2009) The biology of cystatin M/E and its cognate target proteases. J Invest Dermatol 129:1327–1338

Zeeuwen PL, van Vlijmen-Willems IM, Cheng T, Rodijk-Olthuis D, Hitomi K, Hara-Nishimura I, John S, Smyth N, Reinheckel T, Hendriks WJ et al (2010) The cystatin M/E-cathepsin L balance is essential for tissue homeostasis in epidermis, hair follicles, and cornea. FASEB J 24:3744–3755

Zeeuwen PL, van Vlijmen-Willems IM, Cheng T, Rodijk-Olthuis D, Hitomi K, Hara-Nishimura I, John S, Smyth N, Reinheckel T, Hendriks WJ et al (2011) The cystatin M/E-cathepsin L balance is essential for tissue homeostasis in epidermis, hair follicles, and cornea. FASEB J 24:3744–3755

Chapter 7
Astacins: Proteases in Development and Tissue Differentiation

Walter Stöcker and F. Xavier Gomis-Rüth

7.1 Introduction

The astacins form a versatile family of multi-domain zinc peptidases within the metzincin superfamily (Bond and Beynon 1995; Gomis-Rüth 2003, 2009; Stöcker et al. 1995). To date, the MEROPS database (http://merops.sanger.ac.uk/) of proteolytic enzymes and their inhibitors lists more than a thousand astacin proteases of both prokaryotic and eukaryotic origin. Most of them are secreted proteins; only the members of the meprin subfamily are translated with a membrane anchor. Astacins are generally synthesized as inactive zymogens (pro-enzymes). Thus, their activity relies on the post-translational removal of amino terminal pro-peptides. Thereafter, biological protein inhibitors control their activity.

Eukaryotic astacin proteases are composed of amino terminal signal peptides and pro-segments, zinc-binding protease modules, various carboxy terminal domains, trans-membrane anchors, and cytosolic domains. The X-ray crystal structures of four astacins and one complete zymogen have been solved. The catalytic domain comprises about 200 residues and contains in its center the conserved zinc-binding sequence, HEXXHXXGXXH, typical for metzincin peptidases.

The first family member to be discovered was the digestive metalloprotease astacin from the European fresh water crayfish *Astacus astacus* L.—originally termed 'low molecular weight protease' or '*Astacus* protease' (Stöcker and Zwilling 1995). Generally, the genomes of lower vertebrates and invertebrates

W. Stöcker (✉)
Institute of Zoology, Cell and Matrix Biology, Johannes Gutenberg-University, Johannes-von-Müller-Weg 6, 55128 Mainz, Germany
e-mail: stoecker@uni-mainz.de

F.X. Gomis-Rüth
Proteolysis Lab, Molecular Biology Institute of Barcelona, CSIC, Barcelona Science Park, Helix Building, c/Baldiri Reixac, 15-21, E-08028 Barcelona, Spain
e-mail: fxgr@ibmb.csic.es

K. Brix and W. Stöcker (eds.), *Proteases: Structure and Function*,
DOI 10.1007/978-3-7091-0885-7_7, © Springer-Verlag Wien 2013

contain more astacin genes than mammalian genomes. MEROPS lists 7–18 astacins in amphibians and fishes, at least four in cnidarians, 13–25 in insects, and up to 40 in nematodes such as *Caenorhabditis elegans*. The large number in the latter example can at least partially be attributed to the parasitic lifestyle of nematodes, which require an array of proteases to break down host connective tissue. Characterized examples of this phylum are enzymes from *Trichinella spiralis* (Lun et al. 2003) and *Onchocerca volvolus* (Borchert et al. 2007). Astacins with digestive function have been observed in several invertebrate (mostly decapod crustacean) species (Möhrlen et al. 2003; Stöcker and Zwilling 1995). However, most astacins are not involved in digestion, but rather in proteolytic processing of precursors of extracellular matrix constituents and growth factors, including their antagonists and cell surface receptors, during embryonic patterning and cell differentiation.

The first astacin reported to participate in development was bone morphogenetic protein 1 (BMP1, also known as procollagen C-protease), which is co-expressed with TGFβ-like, non-proteolytic growth factors termed BMPs due to their capability to induce ectopic bone formation in mice (Wozney et al. 1988). Since then a variety of astacins have been discovered, which are key players in developmental processes, tissue differentiation, and embryonic hatching, as exemplified by UVS.2 from clawed frog (Sato and Sargent 1990), tolloid from the fruit fly (Shimell et al. 1991), the low (LCE) and high (HCE) choriolytic enzymes from medaka fish (Yasumasu et al. 1992), and SPAN and blastula protein BP10 from sea urchin (Lepage et al. 1992; Reynolds et al. 1992). Consequently, due to these basic regulatory functions, astacins have found to be linked to diseases like cancer, connective tissue disorders, neuro-degenerative disorders like Alzheimer's disease, etc.

This review will focus on the structure and function of astacin proteases involved in various aspects of development, morphogenesis and tissue differentiation.

7.2 Structure of Astacin Proteases

7.2.1 Modular Composition of Astacins

The smallest prokaryote astacins consist only of a catalytic domain of approximately 200 amino acid residues. Eukaryote astacins are larger, at least elongated by an amino terminal pro-peptide, which renders the enzyme inactive (Yiallouros et al. 2002) (Fig. 7.1).

A distinct subfamily of astacins are the BMP-1/tolloid-like proteases (BTPs), named after bone morphogenetic protein 1 and the dorso-ventral patterning protein tolloid, first described in *Drosophila* embryos. The BTPs contain additional downstream domains such as CUB modules (named according to their occurrence in

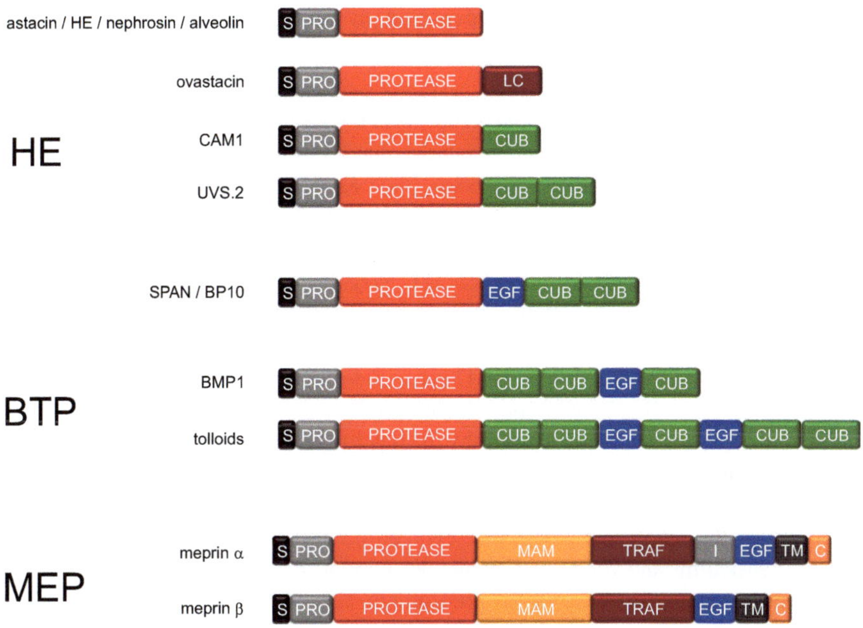

Fig. 7.1 Domain composition of selected astacin proteases. S = (signal peptide); PRO = (pro-peptide); PROTEASE = (catalytic domain); EGF (epidermal growth factor; PFAM PF00008); CUB (domain found in complement C1r/1s, sea urchin Uegf, BMP1; PF00431); MAM (domain found in meprins, A5 receptor protein, and tyrosine phosphatase μ; PF00629); TRAF (tumor necrosis factor receptor-associated factor; PF00917); LC (low complexity domain); C (cytosolic domain); I (inserted domain); TM (transmembrane anchor); HE = (hatching enzymes); BTP = (BMP1/Tolloid-subfamily); MEP = (meprin subfamily)

complement C1s and C1r, sea urchin UEGF and BMP1), EGF (epidermal growth factor like) modules (Bond and Beynon 1995; Stöcker et al. 1995). Some of these domains have been shown to bind calcium and facilitate protein-protein interactions. Especially, they seem to be important for the selective substrate specificity of BTPs (Garrigue-Antar et al. 2004; Hintze et al. 2006; Sieron et al. 2000; Wermter et al. 2007). Sea urchin astacins like SPAN and BP10 are structurally related to tolloids (Lepage et al. 1992; Reynolds et al. 1992), because they also contain CUB and EGF domains.

Astacins of the meprin subfamily are distinguished by TRAF domains (from tumor necrosis factor receptor-associated factor) (Rothe et al. 1994; Zapata et al. 2001) and MAM domains initially identified in meprins, A5 protein, and receptor tyrosine phosphatase μ (Beckmann and Bork 1993). MAM domains alone are also present in HMP2 from hydra (Yan et al. 2000a, b), LAST-MAM from the horseshoe crab (Becker-Pauly et al. 2009) and in the so-called myosinases from squids (Tamori et al. 1999). A considerable number of astacin proteases contain less characterized additional modules often termed LC (regions of low complexity) with little similarity to other protein modules. Such regions have been identified in *Caenorhabditis elegans* astacins, sea urchin astacins SPAN/BP10 (Lepage et al. 1992; Reynolds et al. 1992) and ovastacin (Quesada et al. 2004). Likewise, there are other uncharacterized protein domains such as the EB module found in some *C. elegans* proteins and the TT domain of bacterial astacins, which has been named after the viral ORF2 (open reading frame) of the TT virus. Several *C. elegans* astacins have thrombospondin type 1 repeats, which are also found in other metzincins, such as the ADAMTS peptidases (Apte 2009). Further extracellular domains seen in astacins are ShK toxin moieties, i.e. six-cysteine (SXC) domains, which were originally observed in metridin, a toxin from sea anemone, and several hypothetical *C. elegans* proteins. For more information on astacin domain confer to http://merops.sanger.ac.uk (Rawlings et al. 2010).

Based on the sequences of their catalytic domains, the astacins can be grouped in several major subfamilies (Fig. 7.1). There is the distinct BTP cluster present in all animal phyla. On the other hand, the meprin cluster is as yet only present in vertebrates. Another cluster is formed by the hatching enzymes, which in amphibians and fishes have evolved in several divergent lineages. A different scenario becomes evident in cnidarians, nematodes, insects, and mollusks, where specific clusters not seen outside the respective phyla have evolved (Gomis-Rüth et al. 2012b).

Most structural and functional details are known of the pro-peptide regions and catalytic domains (Fig. 7.2). In the pro-domain, there is a unique conserved motif termed 'aspartate-switch'; in the middle of the catalytic domain there is the zinc-binding consensus sequence; 40 residues downstream, the 'Met-turn' is found; a further 25 residues later, the absolutely crucial S_1'-subsite shaping "170-loop" is found engaged in substrate recognition and cleavage specificity within the catalytic domain.

7.2.2 Structure of Catalytic Domains and Metal Binding Sites

The three dimensional structures of five astacins have been solved by X-ray crystallography so far. These are crayfish astacin (Bode et al. 1992; Gomis-Rüth et al. 1993; Grams et al. 1996), human BMP1 and Tll1 (Mac Sweeney et al. 2008), the zebrafish hatching enzyme ZHE1 (Okada et al. 2010) and the human meprin beta dimer (Arolas et al. 2012) (Fig. 7.2). Astacin catalytic domains are compact kidney-shaped ellipsoids with dimensions of about 55 × 45 × 35 Å (Fig. 7.3). If viewed in standard orientation (Gomis-Rüth et al. 2012a), the active-site cleft divides the catalytic domain into an upper N-terminal and a lower C-terminal sub-domain. Superposition of the catalytic domains of BMP1, ZHE1, meprin and astacin reveals high topological equivalence.

In all cases, the catalytic domain is composed of a twisted five-stranded β-sheet of four parallel strands and one antiparallel strand, which forms the upper edge of the active site cleft. After passing the sheet, the polypeptide chain enters the active-site

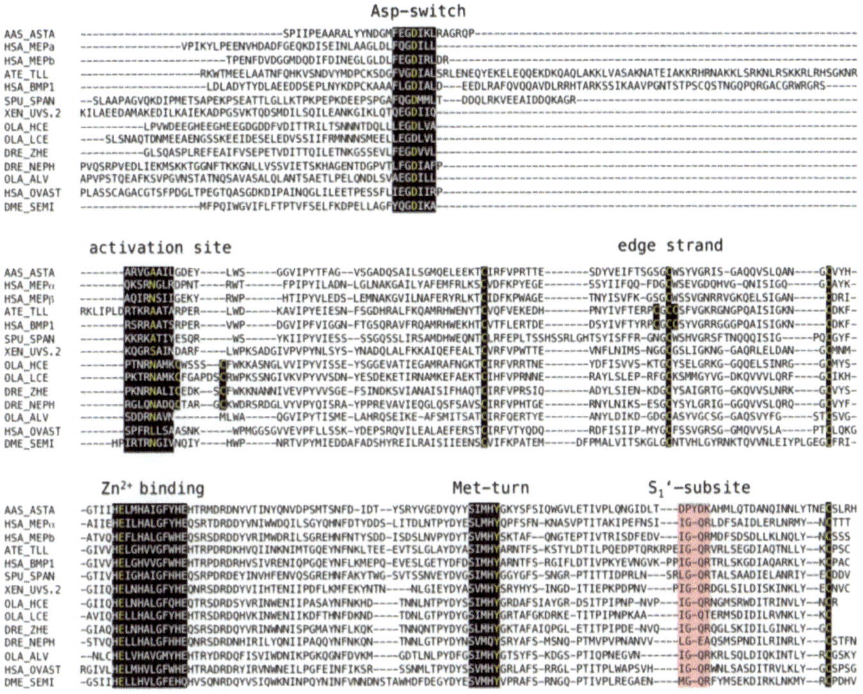

Fig. 7.2 Alignment of pro- and catalytic domains of selected astacins. On *black* background: the aspartate-switch residue in the pro-peptide, the activation site, the zinc binding motif, the Met-turn, and disulfide-forming cysteines; in *pink*: residues shaping the S_1'-subsite. Proteins and UniProt database accession numbers: AAS_AST (P07584) astacin from the crayfish *Astacus astacus* (Titani et al. 1987); ATE_TLL (Q75UQ6) tolloid from house spider *Achaearanea tepidariorum*; CJA_CAM1 (P42662) hatching enzyme from Japanese quail *Coturnix japonica* (Elaroussi and DeLuca 1994); DME_SEMI (CG11864, Q9VJN9) seminase from the fruitfly *Drosophila melanogaster* (LaFlamme et al. 2012); DRE_ZHE (Q75NR9) zebrafish hatching enzyme from *Danio rerio* (Okada et al. 2010); DRE_NEPH (Q8AYF4) zebrafish nephrosin from *Danio rerio* (Hung et al. 1997); HSA_BMP1 (P13497) human BMP1 (Wozney et al. 1988); HSA_MEPa (Q16819) human meprin α (Dumermuth et al. 1991); HSA_MEPb (Q16820) human meprin β (Dumermuth et al. 1991); HSA_OVAST (Q6HA08) human ovastacin (Quesada et al. 2004); OLA_HCE (P31580) medaka fish high choriolytic enzyme from *Oryzias latipes* (Yasumasu et al. 1992); OLA_LCE (P31579) medaka fish low choriolytic enzyme from *Oryzias latipes* (Yasumasu et al. 1992); OLA_ALV (Q9VJN9) alveolin from *Oryzias latipes* (Shibata et al. 2003); SPU_SPAN (P98068) sea urchin blastula protease from *Strongylocentrotus purpuratus* (Reynolds et al. 1992); and XLA_UVS.2 (P42664) hatching enzyme from the clawed frog *Xenopus laevis* (Sato and Sargent 1990)

helix, which provides the first two His residues of the typical metzincin-type zinc-binding motif, $H^{92}EXXHXXG^{99}XXH^{102}$ (mature astacin numbering, single letter amino-acid code; X = any residue) (Bode et al. 1993; Gomis-Rüth 2003, 2009; Stöcker and Bode 1995; Stöcker et al. 1993, 1995). At G^{99} the active-site helix is terminated, and the chain is bent sharply thereby placing the third zinc-binding residue

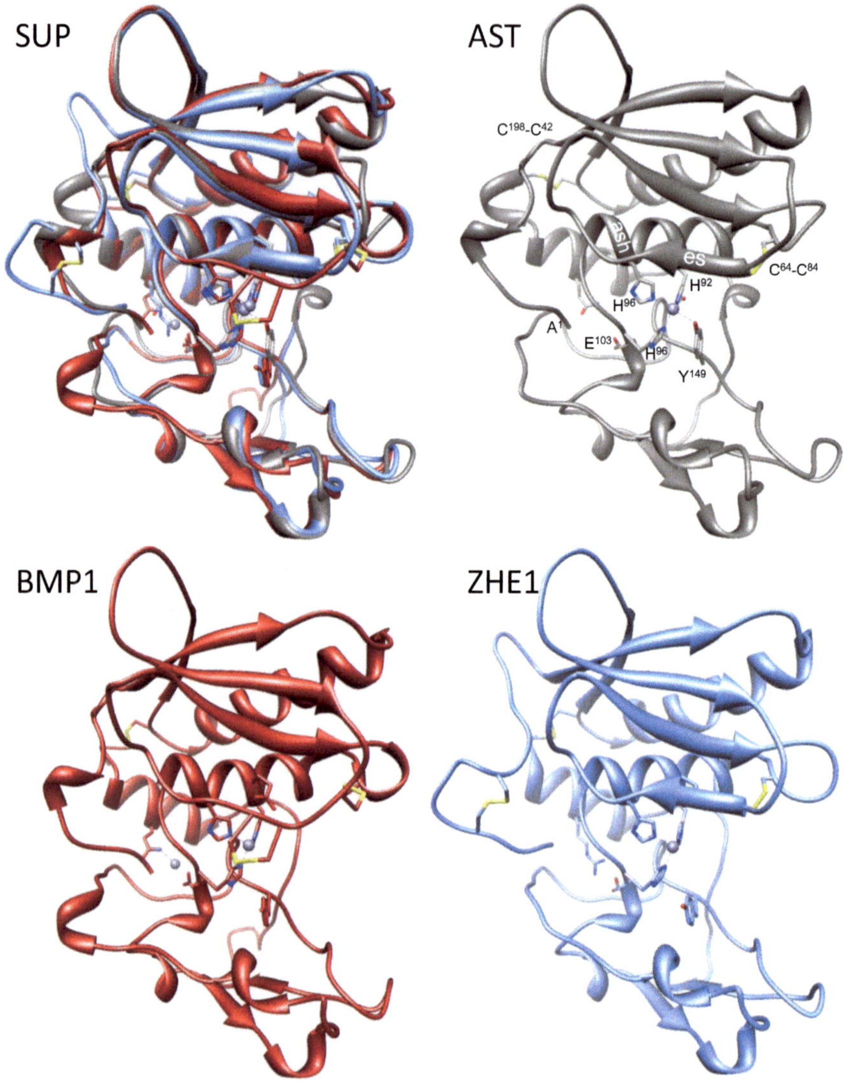

Fig. 7.3 Structure of astacin proteases. Shown are ribbon representations based on the X-ray crystal structures of the mature catalytic domains of crayfish astacins in *grey* (AST) (PDB accession code: 1AST), human BMP1 (PDB accession code 3LQB) in *red* and zebrafish ZHE1 (PDB accession code 3EDH) modeled with CHIMERA (http://www.cgl.ucsf.edu/chimera/) in *blue*. Labeled in AST are the side chains of the three zinc-binding histidines, Y^{149} of the Met-turn, the two conserved disulfide bonds (between C^{42}–C^{198} and C^{64}–C^{84}) and the zinc-bound water symbolized by a *red dot*. (SUP) superposition of BMP1 and ZHE1 onto the structure of astacin (*es* edge strand, *ash* active site helix)

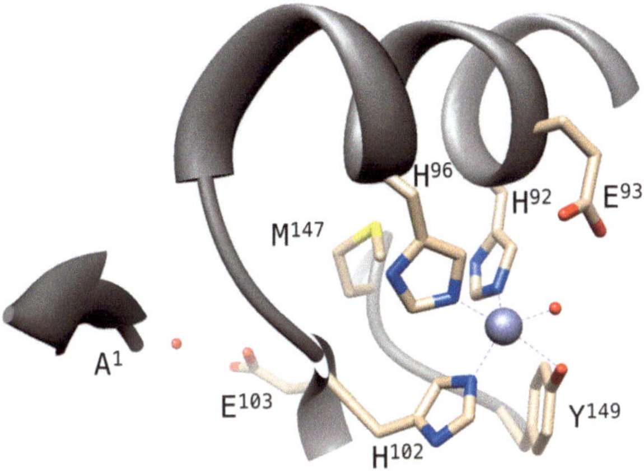

Fig. 7.4 Zinc-binding region of astacin. Shown are the zinc ligands H^{92}, H^{96}, H^{102}, Y^{149}, the catalytic base E^{93}, the water-mediated (*red dot*) salt bridge between A^{1} and E^{103}, and the methionine residue M^{147} of the Met-turn backing the zinc site. Modeling of PDB 1AST was performed with CHIMERA (http://www.cgl.ucsf.edu/chimera/)

H^{102} in a competent position, and the astacin family-specific E^{103} (Stöcker et al. 1993), which is salt bridged to the mature amino terminus (see below).

The subsequent part of the lower sub-domain contains little regular secondary structure except for a short 3_{10}-helix and two short strands before the chain enters the prominent C-terminal helix. However, this rather unstructured C-terminal half contains a unique 1,4-β-turn immediately beneath the zinc-site. This turn is characterized by a methionine residue (M^{147}), shown in a close-up view of the zinc site in Fig. 7.4, which is conserved even for its side-chain conformation in all metzincins of known structure (Gomis-Rüth 2009; Goulas et al. 2010; Waltersperger et al. 2010), although its ultimate functional and structural implications are still under debate (Boldt et al. 2001; Butler et al. 2004; Hege and Baumann 2001; Oberholzer et al. 2009; Perez et al. 2007; Pieper et al. 1997; Tallant et al. 2010; Walasek and Honek 2005). Also located in the Met-turn is Y^{149}, which is engaged in zinc binding and catalysis.

The astacin catalytic domains contain two or three internal disulfide bonds. Conserved in most astacins are C^{42}–C^{198} and C^{64}–C^{84} first observed in crayfish astacin (Figs. 7.2 and 7.3). The first connects the end of the C-terminal helix to the body of the protein. The second links the so-called edge strand (es) to the active-site helix (ash) and thus contributes to shaping the active-site cleft at its primed side and to substrate binding (Figs. 7.2 and 7.5) (Stöcker et al. 1993). The hatching enzymes like ZHE1 contain an additional pair of cysteines within the amino terminal segment (Fig. 7.2). As seen in ZHE1 (Fig. 7.3), these cysteines are cross-linked, thus locking the amino terminus distant from E^{103} (Fig. 7.3). BMP1 and TLL1, likewise contain an additional disulfide bond, albeit in a different position compared to the hatching enzymes. Interestingly, this additional pair of cysteines rearranges the conserved disulfide pattern (Figs. 7.2 and 7.3) by introducing another unique link between the side chains of two consecutive cysteines of the edge strand (C^{64}–C^{65}; BMP1 numbering), whereas the remaining C^{62} in the edge strand is now linked to C^{84}. This 'cysteine-rich loop', is unique for the tolloid subfamily (Fig. 7.2) and has implications for substrate binding, because the two consecutive cysteines

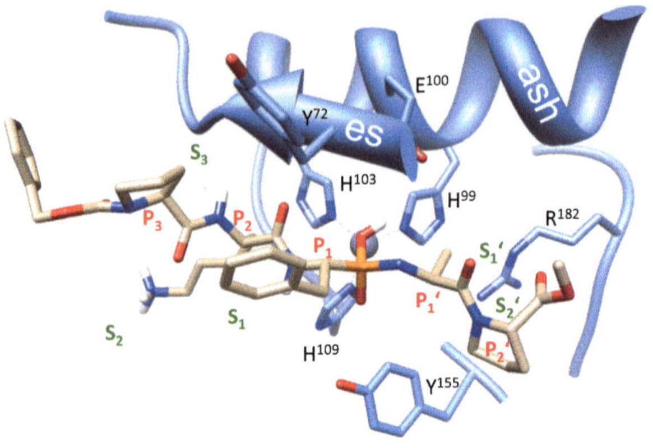

Fig. 7.5 Active site of astacins. Superposition of the structure of the zebrafish hatching enzyme ZHE1 (PDB accession code 3LQB, in *blue*) (Okada et al. 2010) onto the complex of astacin with the phosphinic pseudopeptide inhibitor, BOC-PKRΨ(PO_2CH_2)AP-OCH_3 (PDB accession code 1QJI) (BOC = benzoyl-oxycarbonyl) (K_i = 14 μM against astacin) (Grams et al. 1996; Yiallouros et al. 1998). Shown are the inhibitor in *grey*, the zinc-binding histidine imidazoles, Y^{155} (ZHE1 numbering) in hydrogen-bonding distance to the lower phosphinyl oxygen, the catalytic base E^{100} and, finally, R^{182}, which is thought to be responsible for the specificity of ZHE1 for acidic side chains in P_1' and P_2' of the substrate. The substrate positions from P_3 to P_2' and their corresponding binding sites in the enzyme, S_3 to S_2', are labeled according to (Schechter and Berger 1967). The metal-chelating phosphinyl group adopts the geometry of a tetrahedrally coordinated carbon as visualized with CHIMERA, http://www.cgl.ucsf.edu/chimera/, during peptide bond cleavage

generate a flap occluding the active site. Substrate binding requires opening of this flap (Mac Sweeney et al. 2008).

7.2.3 Buried N-Terminal Region in Mature Astacins

In mature astacin, the three N-terminal residues A^1-A^2-I^3 are plugged into a cavity and the A^1-ammonium integrates in a buried hydrogen-bonding network together with eight solvent molecules (one of them shown as a red dot in Fig. 7.4) and the carboxylate of E^{103}, the direct neighbor of zinc ligand H^{102} (Fig. 7.4) (Stöcker et al. 1993, 1995). The solvent molecule is further hydrogen bonded to the side chain of Q^{189}. These interactions are crucial for the structural stability in mature astacin catalytic domains. Replacement of E^{103} by Q or A did not diminish catalytic efficiency, but caused thermal instability (Yiallouros et al. 2002). In both BMP1 and TLL1, the amino terminal alanine is acetylated and the carbonyl oxygen of the acetyl group is integrated in the aforementioned network of hydrogen bonds. There is also electron density indicating a second metal ion (Fig. 7.3), which is inserted between Q^{189} of the terminal helix and E^{103} (Mac Sweeney et al. 2008).

A different scenario is seen in ZHE1, where the amino terminus is more than 8 Å apart from the conserved family-specific glutamate, which, instead, is linked to

R^{189} (rather than Q^{189}) provided by the carboxy terminal helix (Fig. 7.3). This distinct arrangement can in part be attributed to the additional disulfide bond, which locks the amino terminus in a glutamate-distal position (see above). Alignment and modeling of the mature amino termini of representative astacin family members (Figs. 7.2 and 7.3) (Stöcker et al. 1993) suggests that most of them are arranged in structures similar to astacin/BMP1/TLL1 or the hatching enzymes like ZHE1. This also implicates that variability in the length of the amino terminal segment is restricted. This is a structural analogy to the salt bridged amino terminus of trypsin-like serine proteinases to an aspartate that neighbors the catalytic serine residue (Fehlhammer et al. 1977; Huber and Bode 1978).

7.2.4 Active-Site Cleft and Substrate Specificity

The catalytic mechanism suggested for astacin-like zinc-peptidases relies on the polarization of the metal-bound water between the zinc(II)-ion, which acts as a Lewis-acid, and the glutamic acid residue (E^{93}), which acts as the general base (Grams et al. 1996). Since the catalytic water is also bound to Y^{149} (Figs. 7.3, 7.4, and 7.5), both residues were mutated to assay their significance in catalysis. The $Y^{149}F$ mutant still retained low activity, whereas the $E^{93}A$ mutant was completely inactive (Yiallouros et al. 2000). This supports the role of E^{93} as a general base and proton shuttle to the leaving amino group during catalysis in analogy to E^{143} of thermolysin (Matthews 1988) or E^{270} of carboxypeptidase A (Christianson and Lipscomb 1989). Y^{149} on the other hand seems to stabilize the transition state similarly to H^{231} in thermolysin and Y^{248} of carboxypeptidase A.

The specific interactions of astacin with the reaction-intermediate analog inhibitor, BOC-PKRΨ(PO_2CH_2)AP-OCH_3, have provided insight into the catalytic mechanism of astacin-like enzymes (Grams et al. 1996) (Fig. 7.5). In this mimic, the scissile peptide bond is replaced by a phosphinic pseudo peptide bond Ψ(PO_2CH_2), in which the PO_2 mimicks the tetrahedrally coordinated carbon of the intermediate during peptide bond cleavage.

Astacins bind protein substrates in an extended conformation by antiparallel alignment to their upper-rim edge β-strand (es) (Figs. 7.3 and 7.5). The non-primed part (Schechter and Berger 1967) is hydrogen bonded via backbone carbonyl and amide groups (Fig. 7.5). The two phosphinyl oxygens of the PO_2 group chelate the metal ion. Most importantly, upon inhibitor binding, the tyrosine side chain moves into a position about 5.0 Å removed from the metal, and becomes hydrogen-bonded with the PO_2 group, which mimics a water-attacked peptide bond (Fig. 7.5). This 'tyrosine switch' is a unique feature of the astacin-like proteinases and also of serralysins. In the structures of BMP1 and ZHE1 dimethyl sulfoxide and sulfate were found, respectively, to bind the zinc ion. This is probably the reason, why in these structures the tyrosine is shifted into this remote position even in the absence of an inhibitor, for sterical reasons. The overlay of ZHE1 and BOC-PKRΨ

$(PO_2CH_2)AP\text{-}OCH_3$ depicted in Fig. 7.5 therefore most likely resembles the 'inhibited' conformation of ZHE1.

The side chain carboxy terminally of the cleavage site is an alanine methyl group in BOC-PKRΨ$(PO_2CH_2)AP\text{-}OCH_3$ (Fig. 7.5). For the subsite specificity of zinc endopeptidases, this is the most important side chain, since it points directly into the specificity pocket. According to Schechter and Berger (1967), this is the P_1' position of the substrate, which is harbored in the corresponding S_1' subsite of the enzyme. Remarkably, the great majority of astacin proteases in the MEROPS database are very similar in their S_1' regions (see Fig. 7.2). Particularly, the consensus motif highlighted in Fig. 7.2 contains a conserved arginine residue, which forms the bottom of the S_1'-pocket in the structures of ZHE1, TLL1, and BMP1 as seen in Fig. 7.5. Crayfish astacin belongs to the minority of astacins, which do not share this arginine but instead have a rather shallow S_1' subsite. This is probably the reason for the preference of acidic side chains of aspartate and glutamate in P_1' and P_2' by most astacins, whereas astacin itself prefers small aliphatic residues, as shown recently on a quantitative basis in a proteomics approach covering cleavage sites in complete cellular proteomes employing several astacins. This preference is even more pronounced in meprin β and BTP astacins (e.g. BMP1), which have additional basic side chains for recognition of acidic substrates in their S_2' subsites (Becker-Pauly et al. 2011).

In astacin and ZHE1 the upper-rim strand lines the top of the cleft on its primed side, together with the conserved disulfide bond between Cys^{64} and Cys^{84} (astacin numbering; Figs. 7.2 and 7.3). By contrast, in BMP1 and TLL1 this disulfide bond is slightly displaced and a further, unique SS-bridge is found between two consecutive cysteines within a cysteine-rich loop that replaces the upper-rim strand in astacin and ZHE1. This gives rise to an eight-membered, largely hydrophobic ring above the S_1 pocket (Fig. 7.3), which prevents substrate binding to the cleft and causes the upper rim to deviate from a regular β-strand. This cysteine-rich loop is disordered in the inhibitor free structures, and it has been proposed to act as a mobile flap clamping substrates into a competent position for a Michaelis complex (Mac Sweeney et al. 2008). On its non-primed side, the cleft is limited in astacins by the end of the edge strand. At its bottom, the cleft is constrained on its non-primed side by I^4-G^5 and the loops following the active-site helix and connecting the 3_{10}-helix with the Met-turn, and, on its primed side, by the Met-turn and the downstream segment up to W^{158}.

7.2.5 *Zymogen Structure and Activation Mechanism*

Eukaryotic astacins are synthesized as inactive pro-proteases (zymogens), which require proteolytic removal of an amino terminal pro-peptide (Guevara et al. 2010). The pro-segments of distinct family members are aligned in Fig. 7.2. They differ in length (from 34 to 393 residues), but they share a unique consensus sequence $F^{18P}XGD^{21P}$ (pro-peptide residues are labeled with the suffix 'P'). The only

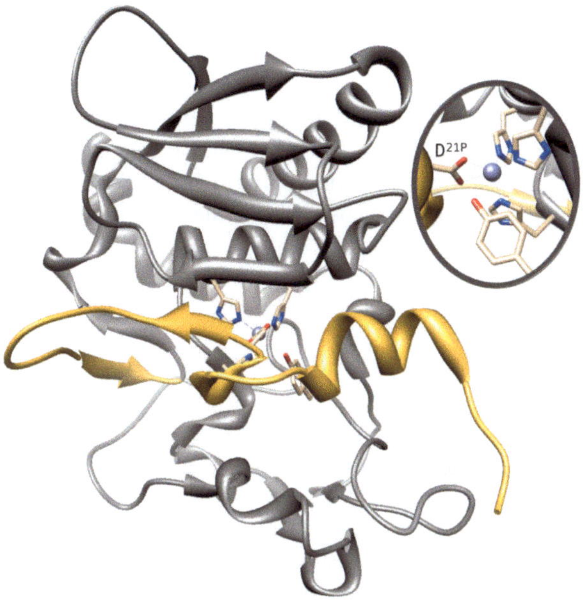

Fig. 7.6 Structure of pro-astacin. Ribbon model of pro-astacin (standard view) shown in *grey* with the pro-peptide in *yellow*. The ellipsoid shows the interaction of D^{21P} with the metal. Modeling of PDB 3LQ0 was performed with CHIMERA (http://www.cgl.ucsf.edu/chimera/)

structurally characterized zymogen of the astacin family is that of crayfish pro-astacin (Guevara et al. 2010) (Fig. 7.6). Pro-astacin could be purified from *Escherichia coli* inclusion bodies and correctly folded as a potentially activatable zymogen (Reyda et al. 1999; Yiallouros et al. 2000, 2002). It contains a short pro-peptide of 34 residues, which is not required for folding as an intramolecular chaperon (Reyda et al. 1999). However, this possibility cannot been excluded for pro-domains, which are even larger than the catalytic domain, as it is the case in *Drosophila* tolloid-related (Nguyen et al. 1994).

The pro-segment of the astacin zymogen runs across the front surface of the catalytic domain in the inverse direction of a substrate (Fig. 7.6). This prevents self-cleavage, as observed in cysteine-protease and matrix metalloproteinase zymogens (Khan and James 1998). Most remarkable is a Z-shaped loop directly in front of the zinc site, which posts D^{21P} as a bidentate chelator of the catalytic zinc ion (Figs. 7.2 and 7.6). This loop contains two tight 1,4-β-turns, which also explains the conservation of a glycine next to the aspartate, and it also contains conserved hydrophobic residues that facilitate the formation of a compact globular moiety backing D^{21P} (Fig. 7.5).

The activation site (Figs. 7.2 and 7.6), $G^{34P}*A^{1}$, is buried like the amino-terminus of the mature form and is located at the tip of a sharp turn. Activation occurs through successive cleavages by trypsin and mature astacin, which liberate the mature N-terminus at A^{1} (Guevara et al. 2010; Yiallouros et al. 2002) and enable formation of the salt bridge to E^{103} and the release of the pro-peptide's D^{21P} from the zinc ion. In a similar fashion, the amino terminus of matrix metalloproteinases (MMPs) is trimmed by successive cleavage events to become finally engaged in a hydrogen bond with a conserved aspartate of the C-terminal helix (Nagase 1997; Reinemer et al. 1994). In analogy to the cysteine-switch of MMPs the removal of the zinc-blocking aspartate has been termed 'aspartate-switch' (Guevara et al. 2010). This causes major rearrangement of the 'activation domain' beneath the mature amino terminus, which adopts a rigid and competent conformation in the mature enzyme only (Fig. 7.6). This increase in rigidity and stability is another analogy to trypsin-like serine proteinases (Bode and Huber 1978; Khan and James 1998).

7.2.6 Protein Inhibitors and Enhancers of Astacins

Astacins are resistant to inhibition by tissue inhibitors of metalloproteinases (TIMPs), which are effective against MMPs ad ADAMs. However, the general protein scavenger and regulator of vascular and interstitial proteolysis, α_2-macroglobulin, is a potent inactivator of members of the astacin family. This holds true not only for small single domain astacins, but also for middle-sized multi-domain proteins like BMP1 (Marrero et al. 2012; Meier et al. 1994; Stöcker et al. 1991; Zhang et al. 2006). However, α_2-macroglobulin does not inhibit larger oligomeric astacins such as meprins (Kruse et al. 2004).

Another natural protein inhibitor of astacins was discovered in complex with nephrosin, an astacin-like protease from the carp, *Cyprinus carpio* (Hung et al. 1997; Tsai et al. 2004). The enzyme was termed nephrosin after its site of biosynthesis, the head kidney, which is a hematopoietic organ in fishes. The nephrosin inhibitor turned out to be the fish homolog of fetuin, a large mammalian plasma protein with many functions (Jahnen-Dechent et al. 2011; Schäfer et al. 2003). These fetuins contain cystatin-like domains and are related to cystatin C-like inhibitors of papain-like cysteine proteases. It has been shown recently that the plasma proteins fetuin and cystatin C act as physiological inhibitors of human astacin proteases such as ovastacin and meprins (Hedrich et al. 2010; Dietzel et al. 2013).

Other protein inhibitors of astacin proteases have been discovered in the context of body axis formation during early embryogenesis in amphibians and fishes. In these lower vertebrates secreted frizzled like proteins (sFRPs) are potent antagonists of BTPs (Lee et al. 2006). sFRPs are composed of a carboxy terminal netrin-like domain and an amino terminal frizzled domain. Frizzled is a cell surface receptor of the Wnt-signalling pathway. Interestingly, there are several sFRPs in *Xenopus*, which are expressed in different regions of the embryo. One of these, the dorsally expressed crescent protein enables cross talk between Wnt- and BMP-pathways, since it is an inhibitor of BTPs and can also trigger Wnt signalling (Ploper et al. 2011). By contrast, the mammalian sFRPs do not inhibit BTS proteases (Bijakowski et al. 2012) but may act as BTS protease enhancers (Kobayashi et al. 2009), albeit this enhancing function has yet to be corroborated (von Marschall and Fisher 2010b).

However, there are true enhancers, which regulate BTPs in a highly substrate specific manner. These enhancers have been termed procollagen C-proteinase enhancers (PCPE1 and PDCPE2), because they bind selectively to the C-propeptides of fibrillar collagens and increase the activity of BTS proteases by about 20-fold. PCPEs are composed of two CUB domains and a netrin-like domain (Kessler et al. 1990; Kronenberg et al. 2009; Moali et al. 2005; Steiglitz et al. 2002; Vadon-Le Goff et al. 2011). Studies with PCPE$^{-/-}$ mice also indicate a role of these proteins for proper collagen fiber assembly (Steiglitz et al. 2006).

7.3 Distribution and Physiological Role of Astacins

In the human and mouse genomes (see http://degradome.uniovi.es/met.html), there are six genes encoding astacin proteases, namely *bmp1*, *tll1*, *tll2*, *mep1a*, *mep1b* and *astl*. The first three code for the BTPs, which include protein BMP1 and its major splice variant, mammalian tolloid, and the mammalian tolloid-like proteins mTll1 and mTll2 (Muir and Greenspan 2011). Genes *mep1a* and *mep1b* encode the multi-domain proteins meprin α and meprin β, respectively (Sterchi et al. 2008). The third subgroup of astacins in vertebrates and invertebrates comprises the so-called hatching enzymes, which degrade embryonic envelopes during the free water developmental stage of crustaceans, echinoderms, fishes, and frogs (Kawaguchi et al. 2010a).

7.3.1 BMP1/Tolloid Proteases (BTPs)

BTPs cleave precursors of fibrillar procollagens for proper matrix assembly. They also process other matrix proteins including proteoglycans, laminins, and anchoring fibrils. In addition, tolloids also cleave growth factors and their antagonists, which are crucial for dorso-ventral patterning during gastrulation in the embryo (for reviews, see Ge and Greenspan 2006b; Hopkins et al. 2007; Muir and Greenspan 2011). In vertebrates, the four major BTPs are all expressed during embryonic development in the gastrula (Ploper et al. 2011). They are crucial for dorso-ventral patterning through cleavage of chordin, an antagonist of the transforming growth factor-like bone morphogenetic proteins BMP2 and BMP4. In later development and in the adult, BMP1, mTLD, and mTLL1 are further required for bone formation and connective tissue differentiation, because they specifically activate and trim a variety of procollagens and therefore are also termed 'procollagen C-proteases' (Kessler et al. 1996; Li et al. 1996). mTLL2, by contrast, localizes to skeletal muscle in later development (Scott et al. 1999). However, it has been found in bmp1$^{-/-}$ mice that it also may be involved in procollagen VII processing (Rattenholl et al. 2002). Validated BTP substrates are listed in Table 7.1.

Table 7.1 Cleavage sites of BMP1 in extracellular matrix proteins

ProCOLL α_1(I)	YYRA*DDAN	Kessler et al. (1996), Li et al. (1996)
ProCOLL α_2(I)	FYRAA*DQPR	Kessler et al. (1996), Li et al. (1996)
ProCOLL α_1(II)	YMRAA*DQAA	Kessler et al. (1996), Li et al. (1996)
ProCOLL α_1(III)	PYYGA*DEPM	Kessler et al. (1996), Li et al. (1996)
ProCOLL α_1(V)C-Pro	QLLDA*DGNG	Pappano et al. (2003), Unsöld et al. (2002)
ProCOLL α_2(V)C-Pro	EFTEA*DQAA	Pappano et al. (2003), Unsöld et al. (2002)
ProCOLL α_1(VII)	SYAAA*DTAG	Rattenholl et al. (2002)
ProCOLL α_3(V)N-Pro	SFQQA*AQAQ	Gopalakrishnan et al. (2004)
ProCOLL α_1(V)N-Pro	TPQSA*QDPN	Pappano et al. (2003), Unsöld et al. (2002)
IGFBP3	ESQSA*TDTQ	Kim et al. (2011)
α_1(XI)COLL	AAQAA*QEPQ	Pappano et al. (2003), Unsöld et al. (2002)
α_2(XI)COLL	RPQNA*QQPH	Pappano et al. (2003), Unsöld et al. (2002)
Pro-LOX	RMVGA*DDPY	Borel et al. (2001), Uzel et al. (2001)
Pro-LOX-like	VAVGA*DSTG	Borel et al. (2001), Uzel et al. (2001)
Pro-LOX-like	VRSSA*DAPP	Borel et al. (2001), Uzel et al. (2001)
Laminin 332 γ_2	CYSGA*DENP	Veitch et al. (2003)
Laminin 332 α_3	QEPKA*DSSP	Veitch et al. (2003)
Probiglycan	FMMNA*DEEA	Scott et al. (2000a)
Decorin	FLMEA*DEAS	von Marschall and Fisher (2010a)
DMP1	EMQSA*DDPE	Steiglitz et al. (2004)
DSPP	SMQGA*DDPN	Tsuchiya et al. (2011), von Marschall and Fisher (2010b)
Myostatin	DVQRA*DDSS	Wolfman et al. (2003)
GDF11	DFQGA*DALQ	Ge et al. (2005)
LTBP		
N-term	IPSLA*DQEK	Ge and Greenspan (2006a)
C-term	YFIQA*DRFL	Ge and Greenspan (2006a)
Perlecan	SGGNA*DAPG	Gonzalez et al. (2005)
Chordin		
N-term	RSYSA*DRGE	Piccolo et al. (1997)
C-term	PMQAA*DGPR	Piccolo et al. (1997)
Osteoglycin	QLQKA*DEVI	Ge et al. (2004)
Gliomedin	AIPNA*DDTL	Maertens et al. (2007)

Cleavage sites with four adjacent amino acid residues (single-letter code) on either side of the scissile bond (marked by an asterisk *) are indicated

Procoll procollagen, *coll* collagen, *IGFBP3* insulin-like growth factor binding protein III, *LOX* lysyloxidase, *DMP1* dentin matrix protein 1, *DSPP* dentin sialophosphoprotein, *GDF11* growth and differentiation factor 11, *LTBP* latent TGFβ binding protein

In lower vertebrates BTPs have been studied in the zebrafish *Danio rerio* (Blader et al. 1997) and the frog *Xenopus laevis* (Holley et al. 1996; Ploper et al. 2011). Basic insight on BTPs has been gained for *Drosophila* tolloid, which cleaves the chordin homolog short gastrulation, the antagonist of the BMP2/4 homolog decapentaplegic (DPP) in fly embryos (Shimell et al. 1991), and for tolloid related (tolkin) from *Drosophila* larvae (Serpe and O'Connor 2006). BTPs are not only expressed in the ECM, but also in the developing and adult nervous system of both invertebrates as the sea hare *Aplysia* (Liu et al. 1997) or the fruitfly *Drosophila* (Serpe and O'Connor 2006) and mammals (Clark et al. 1999; Scott et al. 2000b; Takahara et al. 1994, 1996).

Related to BTPs are the sea urchin proteases SPAN from *Strongylocentrotus purpuratus* (Reynolds et al. 1992) and blastula protein 10 (BP10) from *Paracentrotus lividus* (Lepage et al. 1992). They are expressed in the blastula stage and as BTPs they play roles in early gastrulation. For example they are involved in spicule formation, the larval skeleton of echinoderms. Their domain composition is similar to that of vertebrate BTPs (Fig. 7.1).

bmp1$^{-/-}$ mice have skeletal abnormalities and incomplete ventral body closure due to abnormal collagen fibrils; however, they do have bony skeleton because of the partially compensating effect of mTLL1 (Suzuki et al. 1996). mTLL1 deficiency, on the other hand, causes defects in heart compartmentalization (Clark et al. 1999). *mtll2*$^{-/-}$ mice have less muscle tissue (Lee 2008). mTld, the longer splice form derived from the *bmp1* gene is thought to circulate in the blood and to play a role in bone fracture healing (Grgurevic et al. 2011) and kidney fibrosis (Grgurevic et al. 2007).

Other matrix proteins like lysyl oxidase and lysyl oxidase like enzymes (Borel et al. 2001; Uzel et al. 2001), dentin matrix protein 1, and dentin sialophosphoprotein, osteoglycin and leucin-rich proteoglycans like biglycan (Scott et al. 2000a) and decorin (von Marschall and Fisher 2010a) are also activated by BTS proteases, which thereby promote extracellular matrix assembly in manifold ways. Likewise, BTPs proteolytically modify basement membrane-laminin-332 (Veitch et al. 2003) and perlecan (Gonzalez et al. 2005) (Table 7.1).

Transforming growth factor β (TGFβ, an important regulator of cell functions is released from latent TGFβ-binding proteins by BTPs (Ge and Greenspan 2006a). TGFβ-related factors like GDF8 and 11 (Ge et al. 2005) (Table 7.1) as well as myostatin acting as negative regulators of skeletal muscle growth are also activated by BTS proteases, which adds an intriguing facet to BTP function (Wolfman et al. 2003).

Other BTP substrates are IGFBPs (insulin growth factor binding proteins), as shown for IGFBP3, which regulate cell differentiation and development (Kim et al. 2011).

7.3.2 Meprin Proteases

Meprins are membrane-bound or soluble astacin proteases (Barnes et al. 1989; Beynon et al. 1981; Bond and Beynon 1995; Corbeil et al. 1993; Dumermuth et al. 1991; Johnson and Hersh 1992, 1994; Milhiet et al. 1994; Sterchi et al. 1982, 1988a, 2008; Broder and Becker-Pauly 2013). They are composed of an amino terminal pro-domain, followed by the astacin-like catalytic domain. Typical meprins also contain a MAM domain (meprin, A5 protein, and receptor protein tyrosine phosphatase μ (Beckmann and Bork 1993) (Sterchi et al. 1988b), a

TRAF domain (tumor necrosis factor receptor-associated factor) (Rothe et al. 1994; Zapata et al. 2001), an EGF-like domain, a transmembrane segment, and a cytosolic domain (Fig. 7.1).

There is a significant difference between the α subunit and the β subunit. Meprin α contains an additional I-domain inserted between the TRAF and the EGF-like domain, which can be cleaved by furin like enzymes (Fig. 7.1) during the passage through the endoplasmic reticulum and Golgi (Dumermuth et al. 1993; Gorbea et al. 1993; Grünberg et al. 1992; Hahn et al. 2000; Jiang et al. 1992; Kounnas et al. 1991; Milhiet et al. 1995; Sterchi et al. 2008). For this reason, meprin α is secreted as a soluble enzyme and forms high-molecular weight multimers up to megadalton size (Becker et al. 2003; Ishmael et al. 2001). Therefore it appears to be the largest extracellular protease reported so far. By contrast, meprin β homodimers and αβ heterodimers remain cell-surface-bound unless shed proteolytically (Hahn et al. 2003). Meprins are important for tissue differentiation and pericellular signaling. In this context, a variety of meprin substrates have been reported (Bertenshaw et al. 2001) including biologically active peptides such as gastrin and cholecystokinin, substance P, cytokines, and chemokines (for a review see Sterchi et al. 2008). Of special interest is the fact that meprins cleave components of the extracellular matrix (Kruse et al. 2004; Walker et al. 1998), in particular the basal lamina but also adhesion proteins at the cell-cell interface (Ambort et al. 2010; Huguenin et al. 2008; Sterchi et al. 2008; Vazeille et al. 2011).

Recent proteomics approaches have identified previously known and new physiologically relevant *in vivo* substrates such as vascular endothelial growth factor (Schütte et al. 2010), amyloid precursor protein (Jefferson et al. 2011), procollagens I and III (Kronenberg et al. 2010), interleukin-1β (Herzog et al. 2005), interleukin 18 (Banerjee and Bond 2008), pro-kallikrein 7 (Ohler et al. 2010), and fibroblast growth factor 19 (Becker-Pauly et al. 2011). Their ability to cleave procollagens at exactly the same sites as BTPs sheds additional light on the differential function of meprins in health and disease (Becker-Pauly et al. 2007) (Kronenberg et al. 2010; Broder et al. 2013).

Meprins have been found to be implicated in various pathological situations. The processing of interleukins by meprins might explain the observation that meprin$^{-/-}$ mice have deficiencies in their immune system (Bylander et al. 2008; Crisman et al. 2004; Sun et al. 2009). Meprins are also linked to intestinal disorders like inflammatory bowel disease (Banerjee et al. 2009), Crohn's disease (Vazeille et al. 2011), and celiac disease (Lottaz et al. 2007). Furthermore they are linked to atherosclerosis (Gao et al. 2009), kidney disorders (Bylander et al. 2008; Carmago et al. 2002; Herzog et al. 2007; Mathew et al. 2005; Oneda et al. 2008; Red Eagle et al. 2005; Takayama et al. 2008; Yura et al. 2009) and tumor metastasis (Heinzelmann-Schwarz et al. 2007; Matters et al. 2005; Rösmann et al. 2002).

Recent observations of meprins in Alzheimer's disease support their basic regulatory signaling functions. The amyloid precursor protein is obviously processed by meprin β at several sites including the β-secretase site (Jefferson et al. 2011; Bien et al. 2012).

7.3.3 Hatching Enzymes

The embryonic development of many aquatic invertebrates, e.g. crustaceans, fishes, amphibians and reptiles, takes place in free ambient water. The eggs and embryos of these organisms are protected by a durable envelope of extracellular matrix (also termed *zona pellucida* or vitelline membrane), which has to be cleaved off after a certain stage of embryogenesis to release the young larva. This process is called 'hatching' and is frequently triggered by proteolytic enzymes.

Among the members of the astacin family there is a considerable number of so-called 'hatching enzymes' (Figs. 7.1 and 7.2). Some of these are composed only of a pro-domain and an astacin-like catalytic domain, others additionally contain carboxy terminal CUB domains or other protein modules. Examples include crayfish embryonic astacin (Geier and Zwilling 1998), and the 'low choriolytic enzymes' (LCE) and 'high chorolytic enzymes' (HCE), first described in teleost fishes (Yasumasu et al. 1996), or corresponding proteases from amphibians like the UVS.2 gene product from the frog *Xenopus laevis* (Fan and Katagiri 2001; Katagiri et al. 1997; Sato and Sargent 1990).

The hatching process has been studied most intensively in the medaka fish *Oryzias latipes* and other euteleosts. In these organisms, the hatching seems to be organized in a conserved fashion. Enzymes of the HCE type cleave *zona pellucida* proteins at specific sites, which allow for swelling of the envelope. In a second step, enzymes of the LCE type then completely digest the egg envelope (Hiroi et al. 2004; Kawaguchi et al. 2006, 2010a, b; Sano et al. 2010; Yasumasu et al. 2010a, b). LCE and HCE hatching enzymes are distinguished by an additional pair of cysteine residues in the amino terminal segments of their catalytic astacin-like domains (Figs. 7.2 and 7.3).

Closely related to hatching enzymes is nephrosin from carp head kidney (Hung et al. 1997). The physiological function of this protease is not clarified yet. However, since it is expressed in the head kidney, which is an ancient hematopoetic organ, it might be involved in immune defense and/or general blood cell differentiation.

Also sequentially related to hatching enzymes is bird CAM1 (chorio allantoic membrane protein 1) from the Japanese quail, which is important for calcification of the egg shell (Elaroussi and DeLuca 1994).

Other hatching-type enzymes are alveolin fom medaka fish (Shibata et al. 2000) and ovastacin from mammals (Quesada et al. 2004). Both are involved in *zona pellucida* hardening (ZPH). Ovastacin cleaves *zona pellucida* protein 2 at a highly conserved diacidic site (Burkart et al. 2012), typical for astacins (Becker-Pauly et al. 2011), which results in ZPH immediately after sperm penetration in order to prevent polyspermy; this is supported by the presence of ovastacin in cortical granules of the egg, which are released into the *zona pellucida* upon sperm penetration (Burkart et al. 2012). As observed recently, ovastacin is controlled by fetuin B, a cystatin-like plasma protein, in order to prevent premature ZPH (Dietzel et al. 2013). Another report on ovastacin suggested also a role in sperm-egg interaction (Sachdev et al. 2012).

7.3.4 Seminase Activated Astacin-Like Protease

The seminal fluid of many animal taxa contains proteases and protease inhibitors. In mammals these are important for regulating seminal clot liquefaction, for example. An intriguing protolytic network has recently been discovered in the seminal fluid of the fruitfly *Drosophila melanogaster*. During mating of fruit flies, seminal fluid proteases are transferred from males to females (Ravi Ram et al. 2006). These enzymes form a cascade and one of them is an astacin-like metalloprotease (Ayroles et al. 2011; Sirot et al. 2009). Obviously, this protease cascade is required for post-mating responses in the female. In the first step, a serine protease termed 'seminase' (CG10586) activates the astacin protease CG11864, which subsequently cleaves the sex peptide ovulin and other seminal plasma proteins. These observations indicate an important role for this protease cascade to temporally regulate many responses in females after mating (LaFlamme et al. 2012).

References

Ambort D, Brellier F, Becker-Pauly C, Stöcker W, Andrejevic-Blant S, Chiquet M, Sterchi EE (2010) Specific processing of tenascin-C by the metalloprotease meprinbeta neutralizes its inhibition of cell spreading. Matrix Biol 29:31–42

Apte S (2009) A disintegrin-like and metalloprotease (reprolysin-type) with thrombospondin type 1 motif (ADAMTS) superfamily: functions and mechanisms. J Biol Chem 284:31493–31497

Arolas JL, Broder C, Jefferson T, Guevara T, Sterchi EE, Bode W, Stöcker W, Becker-Pauly C, Gomis-Rüth FX (2012) Structural basis for the sheddase function of human meprin β metalloproteinase at the plasma membrane. Proc Natl Acad Sci USA 109:16131–16136

Ayroles JF, Laflamme BA, Stone EA, Wolfner MF, Mackay TF (2011) Functional genome annotation of Drosophila seminal fluid proteins using transcriptional genetic networks. Genet Res 93:387–395

Banerjee S, Bond J (2008) Prointerleukin-18 is activated by meprin beta in vitro and in vivo in intestinal inflammation. J Biol Chem 283:31371–31377

Banerjee S, Oneda B, Yap LM, Jewell DP, Matters GL, Fitzpatrick LR, Seibold F, Sterchi EE, Ahmad T, Lottaz D, Bond JS (2009) MEP1A allele for meprin A metalloprotease is a susceptibility gene for inflammatory bowel disease. Mucosal Immunol 2:220–231

Barnes K, Ingram J, Kenny AJ (1989) Proteins of the kidney microvillar membrane. Structural and immunochemical properties of rat endopeptidase-2 and its immunohistochemical localization in tissues of rat and mouse. Biochem J 264:335–346

Becker C, Kruse MN, Slotty KA, Köhler D, Harris JR, Rösmann S, Sterchi EE, Stöcker W (2003) Differences in the activation mechanism between the alpha and beta subunits of human meprin. Biol Chem 384:825–831

Becker-Pauly C, Höwel M, Walker T, Vlad A, Aufenvenne K, Oji V, Lottaz D, Sterchi EE, Debela M, Magdolen V, Traupe H, Stöcker W (2007) The alpha and beta subunits of the metalloprotease meprin are expressed in separate layers of human epidermis, revealing different functions in keratinocyte proliferation and differentiation. J Invest Dermatol 127:1115–1125

Becker-Pauly C, Bruns BC, Damm O, Schütte A, Hammouti K, Burmester T, Stöcker W (2009) News from an ancient world: two novel astacin metalloproteases from the horseshoe crab. J Mol Biol 385:236–248

Becker-Pauly C, Barre O, Schilling O, Auf dem Keller U, Ohler A, Broder C, Schütte A, Kappelhoff R, Stöcker W, Overall CM (2011) Proteomic analyses reveal an acidic prime side specificity for the astacin metalloprotease family reflected by physiological substrates. Mol Cell Proteomics 10(9):M111.009233

Beckmann G, Bork P (1993) An adhesive domain detected in functionally diverse receptors. Trends Biochem Sci 18:40–41

Bertenshaw GP, Turk BE, Hubbard SJ, Matters GL, Bylander JE, Crisman JM, Cantley LC, Bond JS (2001) Marked differences between metalloproteases meprin A and B in substrate and peptide bond specificity. J Biol Chem 276:13248–13255

Beynon RJ, Shannon JD, Bond JS (1981) Purification and characterization of a metallo-endoproteinase from mouse kidney. Biochem J 199:591–598

Bien J, Jefferson T, Causevic M, Jumpertz T, Munter L, Multhaup G, Weggen S, Becker-Pauly C, Pietrzik CU (2012) The metalloprotease meprin β generates amino terminal-truncated amyloid β peptide species. J Biol Chem 287:33304–33313

Bijakowski C, Vadon-Le Goff S, Delolme F, Bourhis J-M, Lécorché P, Ruggiero F, Becker-Pauly C, Yiallouros I, Stöcker W, Dive V, Hulmes DJS, Moali C (2012) Sizzled is unique among secreted frizzled-related proteins for its ability to specifically inhibit BMP-1/tolloid-like proteinases. J Biol Chem 287:33581–33593

Blader P, Rastegar S, Fischer N, Strähle U (1997) Cleavage of the BMP-4 antagonist chordin by zebrafish tolloid. Science 278:1937–1940

Bode W, Huber R (1978) Crystal structure analysis and refinement of two variants of trigonal trypsinogen: trigonal trypsin and PEG (polyethylene glycol) trypsinogen and their comparison with orthorhombic trypsin and trigonal trypsinogen. FEBS Lett 90:265–269

Bode W, Gomis-Rüth FX, Huber R, Zwilling R, Stöcker W (1992) Structure of astacin and implications for activation of astacins and zinc-ligation of collagenases. Nature 358:164–167

Bode W, Gomis-Rüth FX, Stöcker W (1993) Astacins, serralysins, snake venom and matrix metalloproteinases exhibit identical zinc-binding environments (HEXXHXXGXXH and Met-turn) and topologies and should be grouped into a common family, the 'metzincins'. FEBS Lett 331:134–140

Boldt HB, Overgaard MT, Laursen LS, Weyer K, Sottrup-Jensen L, Oxvig C (2001) Mutational analysis of the proteolytic domain of pregnancy-associated plasma protein-A (PAPP-A): classification as a metzincin. Biochem J 358:359–367

Bond JS, Beynon RJ (1995) The astacin family of metalloendopeptidases. Protein Sci 4: 1247–1261

Borchert N, Becker-Pauly C, Wagner A, Fischer P, Stöcker W, Brattig NW (2007) Identification and characterization of onchoastacin, an astacin-like metalloproteinase from the filaria Onchocerca volvulus. Microbes Infect 9:498–506

Borel A, Eichenberger D, Farjanel J, Kessler E, Gleyzal C, Hulmes DJ, Sommer P, Font B (2001) Lysyl oxidase-like protein from bovine aorta. Isolation and maturation to an active form by bone morphogenetic protein-1. J Biol Chem 276:48944–48949

Broder C, Becker-Pauly C (2013) The metalloproteases meprin α and meprin β: unique enzymes in inflammation, neurodegeneration, cancer and fibrosis. Biochem J 450:253–264

Broder C, Arnold P, Vadon-Le Goff S, Konerding MA, Bahr K, Müller S, Overall CM, Bond JS, Koudelka T, Tholey A, Hulmes DJS, Moali C, Becker-Pauly C (2013) Metalloproteases meprin α and meprin β are C- and N-procollagen proteinases important for collagen assembly and tensile strength. Proc Natl Acad Sci USA 110(35):14219–14224. doi:10.1073/pnas.1305464110

Burkart AD, Xiong B, Baibakov B, Jimenez-Movilla M, Dean J (2012) Ovastacin, a cortical granule protease, cleaves ZP2 in the zona pellucida to prevent polyspermy. J Cell Biol 197: 37–44

Butler GS, Tam EM, Overall CM (2004) The canonical methionine 392 of matrix metalloproteinase 2 (gelatinase A) is not required for catalytic efficiency or structural integrity: probing the role of the methionine-turn in the metzincin metalloprotease superfamily. J Biol Chem 279:15615–15620

Bylander J, Li Q, Ramesh G, Zhang B, Reeves W, Bond J (2008) Targeted disruption of the meprin metalloproteinase beta gene protects against renal ischemia/reperfusion injury in mice. Am J Physiol Renal Physiol 294(3):F480–F490

Carmago S, Shah SV, Walker PD (2002) Meprin, a brush-border enzyme, plays an important role in hypoxic/ischemic acute renal tubular injury in rats. Kidney Int 61:959–966

Christianson DW, Lipscomb WN (1989) Carboxypeptidase A. Acc Chem Res 22:62–69

Clark TG, Conway SJ, Scott IC, Labosky PA, Winnier G, Bundy J, Hogan BL, Greenspan DS (1999) The mammalian Tolloid-like 1 gene, Tll1, is necessary for normal septation and positioning of the heart. Development 126:2631–2642

Corbeil D, Milhiet PE, Simon V, Ingram J, Kenny AJ, Boileau G, Crine P (1993) Rat endopeptidase-24.18 alpha subunit is secreted into the culture medium as a zymogen when expressed by COS-1 cells. FEBS Lett 335:361–366

Crisman JM, Zhang B, Norman LP, Bond JS (2004) Deletion of the mouse meprin beta metalloprotease gene diminishes the ability of leukocytes to disseminate through extracellular matrix. J Immunol 172:4510–4519

Dietzel E, Wessling J, Floehr J, Schäfer C, Ensslen S, Denecke B, Rösing B, Neulen J, Veitinger T, Spehr M, Tropartz T, Tolba R, Renné T, Egert A, Schorle H, Gottenbusch Y, Hildebrand A, Yiallouros I, Stöcker W, Weiskirchen R, Jahnen-Dechent W (2013) Fetuin-B, a liver-derived plasma protein is essential for fertilization. Dev Cell 25:106–112

Dumermuth E, Sterchi EE, Jiang WP, Wolz RL, Bond JS, Flannery AV, Beynon RJ (1991) The astacin family of metalloendopeptidases. J Biol Chem 266:21381–21385

Dumermuth E, Eldering JA, Grünberg J, Jiang WP, Sterchi EE (1993) Cloning of the paba peptide-hydrolase alpha-subunit (Pph-Alpha) from human small-intestine and its expression in Cos-1 cells. FEBS Lett 335:367–375

Elaroussi MA, DeLuca HF (1994) A new member to the astacin family of metalloendopeptidases: a novel 1,25-dihydroxyvitamin D-3-stimulated mRNA from chorioallantoic membrane of quail. Biochim Biophys Acta 1217:1–8

Fan TJ, Katagiri C (2001) Properties of the hatching enzyme from Xenopus laevis. Eur J Biochem 268:4892–4898

Fehlhammer H, Bode W, Huber R (1977) Crystal structure of bovine trypsinogen at 1-8 A resolution. II. Crystallographic refinement, refined crystal structure and comparison with bovine trypsin. J Mol Biol 111:415–438

Gao P, Guo R, Chen J, Chen Y, Wang H, Yu Y, Huang L (2009) A meprin inhibitor suppresses atherosclerotic plaque formation in ApoE(-/-) mice. Atherosclerosis 207(1):84–92

Garrigue-Antar L, Francois V, Kadler KE (2004) Deletion of epidermal growth factor-like domains converts mammalian tolloid into a chordinase and effective procollagen C-proteinase. J Biol Chem 279:49835–49841

Ge G, Greenspan D (2006a) BMP1 controls TGFbeta1 activation via cleavage of latent TGFbeta-binding protein. J Cell Biol 175:111–120

Ge G, Greenspan DS (2006b) Developmental roles of the BMP1/TLD metalloproteinases. Birth Defects Res C Embryo Today 78:47–68

Ge G, Seo NS, Liang X, Hopkins DR, Hook M, Greenspan DS (2004) Bone morphogenetic protein-1/tolloid-related metalloproteinases process osteoglycin and enhance its ability to regulate collagen fibrillogenesis. J Biol Chem 279:41626–41633

Ge G, Hopkins DR, Ho WB, Greenspan DS (2005) GDF11 forms a bone morphogenetic protein 1-activated latent complex that can modulate nerve growth factor-induced differentiation of PC12 cells. Mol Cell Biol 25:5846–5858

Geier G, Zwilling R (1998) Cloning and characterization of a cDNA coding for Astacus embryonic astacin, a member of the astacin family of metalloproteases from the crayfish Astacus astacus. Eur J Biochem 253:796–803

Gomis-Rüth FX (2003) Structural aspects of the metzincin clan of metalloendopeptidases. Mol Biotechnol 24:157–202

Gomis-Rüth FX (2009) Catalytic domain architecture of metzincin metalloproteases. J Biol Chem 284:15353–15357

Gomis-Rüth FX, Stöcker W, Huber R, Zwilling R, Bode W (1993) Refined 1.8 A X-ray crystal structure of astacin, a zinc-endopeptidase from the crayfish Astacus astacus L. Structure determination, refinement, molecular structure and comparison with thermolysin. J Mol Biol 229:945–968

Gomis-Rüth FX, Botelho TO, Bode W (2012a) A standard orientation for metallopeptidases. Biochim Biophys Acta 1824:157–163

Gomis-Rüth FX, Trillo-Muyo S, Stöcker W (2012b) Functional and structural insights into astacin metallopeptidases. Biol Chem. doi:10.1515/bc-2012-0149

Gonzalez EM, Reed CC, Bix G, Fu J, Zhang Y, Gopalakrishnan B, Greenspan DS, Iozzo RV (2005) BMP-1/Tolloid-like metalloproteases process endorepellin, the angiostatic C-terminal fragment of perlecan. J Biol Chem 280:7080–7087

Gopalakrishnan B, Wang W, Greenspan D (2004) Biosynthetic processing of the Pro-alpha1(V) Pro-alpha2(V)Pro-alpha3(V) procollagen heterotrimer. J Biol Chem 279:30904–30912

Gorbea CM, Marchand P, Jiang W, Copeland NG, Gilbert DJ, Jenkins NA, Bond JS (1993) Cloning, expression, and chromosomal localization of the mouse meprin beta subunit. J Biol Chem 268:21035–21043

Goulas V, Papoti VT, Exarchou V, Tsimidou MZ, Gerothanassis IP (2010) Contribution of flavonoids to the overall radical scavenging activity of olive (Olea europaea L.) leaf polar extracts. J Agric Food Chem 58:3303–3308

Grams F, Dive V, Yiotakis A, Yiallouros I, Vassiliou S, Zwilling R, Bode W, Stöcker W (1996) Structure of astacin with a transition-state analogue inhibitor. Nat Struct Biol 3:671–675

Grgurevic L, Macek B, Durdevic D, Vukicevic S (2007) Detection of bone and cartilage-related proteins in plasma of patients with a bone fracture using liquid chromatography-mass spectrometry. Int Orthop 31:743–751

Grgurevic L, Macek B, Mercep M, Jelic M, Smoljanovic T, Erjavec I, Dumic-Cule I, Prgomet S, Durdevic D, Vnuk D, Lipar M, Stejskal M, Kufner V, Brkljacic J, Maticic D, Vukicevic S (2011) Bone morphogenetic protein (BMP)1-3 enhances bone repair. Biochem Biophys Res Commun 408:25–31

Grünberg J, Luginbuhl U, Sterchi EE (1992) Proteolytic processing of human intestinal lactase-phlorizin hydrolase precursor is not a prerequisite for correct sorting in Madin Darby canine kidney (MDCK) cells. FEBS Lett 314:224–228

Guevara T, Yiallouros I, Kappelhoff R, Bissdorf S, Stöcker W, Gomis-Rüth FX (2010) Proenzyme structure and activation of astacin metallopeptidase. J Biol Chem 285:13958–13965

Hahn D, Illisson R, Metspalu A, Sterchi EE (2000) Human N-benzoyl-L-tyrosyl-p-aminobenzoic acid hydrolase (human meprin): genomic structure of the alpha and beta subunits. Biochem J 346(Pt 1):83–91

Hahn D, Pischitzis A, Roesmann S, Hansen MK, Leuenberger B, Luginbuehl U, Sterchi EE (2003) Phorbol 12-myristate 13-acetate-induced ectodomain shedding and phosphorylation of the human meprinbeta metalloprotease. J Biol Chem 278:42829–42839

Hedrich J, Lottaz D, Meyer K, Yiallouros I, Jahnen-Dechent W, Stocker W, Becker-Pauly C (2010) Fetuin-A and cystatin C are endogenous inhibitors of human meprin metalloproteases. Biochemistry 49:8599–8607

Hege T, Baumann U (2001) The conserved methionine residue of the metzincins: a site-directed mutagenesis study. J Mol Biol 314:181–186

Heinzelmann-Schwarz VA, Scolyer RA, Scurry JP, Smith AN, Gardiner-Garden M, Biankin AV, Baron-Hay S, Scott C, Ward RL, Fink D, Hacker NF, Sutherland RL, O'Brien PM (2007) Low meprin alpha expression differentiates primary ovarian mucinous carcinoma from gastrointestinal cancers that commonly metastasise to the ovaries. J Clin Pathol 60:622–626

Herzog C, Kaushal GP, Haun RS (2005) Generation of biologically active interleukin-1beta by meprin B. Cytokine 31:394–403

Herzog C, Seth R, Shah SV, Kaushal GP (2007) Role of meprin A in renal tubular epithelial cell injury. Kidney Int 71:1009–1018

Hintze V, Höwel M, Wermter C, Grosse Berkhoff E, Becker-Pauly C, Beermann B, Yiallouros I, Stöcker W (2006) The interaction of recombinant subdomains of the procollagen C-proteinase with procollagen I provides a quantitative explanation for functional differences between the two splice variants, mammalian tolloid and bone morphogenetic protein 1. Biochemistry 45: 6741–6748

Hiroi J, Maruyama K, Kawazu K, Kaneko T, Ohtani-Kaneko R, Yasumasu S (2004) Structure and developmental expression of hatching enzyme genes of the Japanese eel Anguilla japonica: an aspect of the evolution of fish hatching enzyme gene. Dev Genes Evol 214:176–184

Holley S, Neul J, Attisano L, Wrana J, Sasai Y, O'Connor M, De Robertis E, Ferguson E (1996) The Xenopus dorsalizing factor noggin ventralizes Drosophila embryos by preventing DPP from activating its receptor. Cell 86:607–617

Hopkins DR, Keles S, Greenspan DS (2007) The bone morphogenetic protein 1/Tolloid-like metalloproteinases. Matrix Biol 26:508–523

Huber R, Bode W (1978) Structural basis for the activation and action of trypsin. Acc Chem Res 11:114–122

Huguenin M, Müller EJ, Trachsel-Rösmann S, Oneda B, Ambort D, Sterchi EE, Lottaz D (2008) The metalloprotease meprinbeta processes E-cadherin and weakens intercellular adhesion. PLoS One 3:e2153

Hung CH, Huang HR, Huang CJ, Huang FL, Chang GD (1997) Purification and cloning of carp nephrosin, a secreted zinc endopeptidase of the astacin family. J Biol Chem 272:13772–13778

Ishmael FT, Norcum MT, Benkovic SJ, Bond JS (2001) Multimeric structure of the secreted meprin A metalloproteinase and characterization of the functional protomer. J Biol Chem 276: 23207–23211

Jahnen-Dechent W, Heiss A, Schäfer C, Ketteler M (2011) Fetuin-A regulation of calcified matrix metabolism. Circ Res 108:1494–1509

Jefferson T, Causevic M, Auf dem Keller U, Schilling O, Isbert S, Geyer R, Maier W, Tschickardt S, Jumpertz T, Weggen S, Bond JS, Overall CM, Pietrzik CU, Becker-Pauly C (2011) Metalloprotease meprin beta generates nontoxic N-terminal amyloid precursor protein fragments in vivo. J Biol Chem 286:27741–27750

Jiang W, Gorbea CM, Flannery AV, Beynon RJ, Grant GA, Bond JS (1992) The alpha subunit of meprin A. Molecular cloning and sequencing, differential expression in inbred mouse strains, and evidence for divergent evolution of the alpha and beta subunits. J Biol Chem 267:9185–9193 [published erratum appears in J Biol Chem 1992 Jul 5;267(19):13779]

Johnson GD, Hersh LB (1992) Cloning a rat meprin cDNA reveals the enzyme is a heterodimer. J Biol Chem 267:13505–13512 [published erratum appears in J Biol Chem 1993 Aug 15;268 (23):17647]

Johnson GD, Hersh LB (1994) Expression of meprin subunit precursors. Membrane anchoring through the beta subunit and mechanism of zymogen activation. J Biol Chem 269:7682–7688

Katagiri C, Maeda R, Yamashika C, Mita K, Sargent TD, Yasumasu S (1997) Molecular cloning of Xenopus hatching enzyme and its specific expression in hatching gland cells. Int J Dev Biol 41:19–25

Kawaguchi M, Yasumasu S, Hiroi J, Naruse K, Inoue M, Iuchi I (2006) Evolution of teleostean hatching enzyme genes and their paralogous genes. Dev Genes Evol 216:769–784

Kawaguchi M, Hiroi J, Miya M, Nishida M, Iuchi I, Yasumasu S (2010a) Intron-loss evolution of hatching enzyme genes in Teleostei. BMC Evol Biol 10:260

Kawaguchi M, Yasumasu S, Shimizu A, Sano K, Iuchi I, Nishida M (2010b) Conservation of the egg envelope digestion mechanism of hatching enzyme in euteleostean fishes. FEBS J 277: 4973–4987

Kessler E, Mould AP, Hulmes DJ (1990) Procollagen type I C-proteinase enhancer is a naturally occurring connective tissue glycoprotein. Biochem Biophys Res Commun 173:81–86

Kessler E, Takahara K, Biniaminov L, Brusel M, Greenspan DS (1996) Bone morphogenetic protein-1: the type I procollagen C-proteinase. Science 271:360–362

Khan AR, James MN (1998) Molecular mechanisms for the conversion of zymogens to active proteolytic enzymes. Protein Sci 7:815–836

Kim B, Huang G, Ho WB, Greenspan DS (2011) Bone morphogenetic protein-1 processes insulin-like growth factor-binding protein 3. J Biol Chem 286:29014–29025

Kobayashi K, Luo M, Zhang Y, Wilkes DC, Ge G, Grieskamp T, Yamada C, Liu TC, Huang G, Basson CT, Kispert A, Greenspan DS, Sato TN (2009) Secreted Frizzled-related protein 2 is a procollagen C proteinase enhancer with a role in fibrosis associated with myocardial infarction. Nat Cell Biol 11:46–55

Kounnas MZ, Wolz RL, Gorbea CM, Bond JS (1991) Meprin-A and -B. Cell surface endopeptidases of the mouse kidney. J Biol Chem 266:17350–17357

Kronenberg D, Vadon-Le Goff S, Bourhis JM, Font B, Eichenberger D, Hulmes DJ, Moali C (2009) Strong cooperativity and loose geometry between CUB domains are the basis for procollagen c-proteinase enhancer activity. J Biol Chem 284:33437–33446

Kronenberg D, Bruns BC, Moali C, Vadon-Le Goff S, Sterchi EE, Traupe H, Böhm M, Hulmes DJ, Stöcker W, Becker-Pauly C (2010) Processing of procollagen III by meprins: new players in extracellular matrix assembly? J Invest Dermatol 130:2727–2735

Kruse MN, Becker C, Lottaz D, Köhler D, Yiallouros I, Krell HW, Sterchi EE, Stöcker W (2004) Human meprin alpha and beta homo-oligomers: cleavage of basement membrane proteins and sensitivity to metalloprotease inhibitors. Biochem J 378:383–389

LaFlamme BA, Ram KR, Wolfner MF (2012) The Drosophila melanogaster seminal fluid protease "seminase" regulates proteolytic and post-mating reproductive processes. PLoS Genet 8: e1002435

Lee S-J (2008) Genetic analysis of the role of proteolysis in the activation of latent myostatin. PLoS One 3:e1628

Lee HX, Ambrosio AL, Reversade B, De Robertis EM (2006) Embryonic dorsal-ventral signaling: secreted frizzled-related proteins as inhibitors of tolloid proteinases. Cell 124:147–159

Lepage T, Ghiglione C, Gache C (1992) Spatial and temporal expression pattern during sea urchin embryogenesis of a gene coding for a protease homologous to the human protein BMP-1 and to the product of the Drosophila dorsal-ventral patterning gene tolloid. Development 114: 147–163

Li SW, Sieron AL, Fertala A, Hojima Y, Arnold WV, Prockop DJ (1996) The C-proteinase that processes procollagens to fibrillar collagens is identical to the protein previously identified as bone morphogenic protein-1. Proc Natl Acad Sci U S A 93:5127–5130

Liu QR, Hattar S, Endo S, MacPhee K, Zhang H, Cleary LJ, Byrne JH, Eskin A (1997) A developmental gene (Tolloid/BMP-1) is regulated in Aplysia neurons by treatments that induce long-term sensitization. J Neurosci 17:755–764

Lottaz D, Buri C, Monteleone G, Rösmann S, Macdonald TT, Sanderson IR, Sterchi EE (2007) Compartmentalised expression of meprin in small intestinal mucosa: enhanced expression in lamina propria in coeliac disease. Biol Chem 388:337–341

Lun HM, Mak CH, Ko RC (2003) Characterization and cloning of metallo-proteinase in the excretory/secretory products of the infective-stage larva of Trichinella spiralis. Parasitol Res 90:27–37

Mac Sweeney A, Gil-Parrado S, Vinzenz D, Bernardi A, Hein A, Bodendorf U, Erbel P, Logel C, Gerhartz B (2008) Structural basis for the substrate specificity of bone morphogenetic protein 1/tolloid-like metalloproteases. J Mol Biol 384:228–239

Maertens B, Hopkins D, Franzke CW, Keene DR, Bruckner-Tuderman L, Greenspan DS, Koch M (2007) Cleavage and oligomerization of gliomedin, a transmembrane collagen required for node of ranvier formation. J Biol Chem 282:10647–10659

Marrero A, Duquerroy S, Trapani S, Goulas T, Guevara T, Andersen GR, Navaza J, Sottrup-Jensen L, Gomis-Ruth FX (2012) The crystal structure of human alpha(2)-macroglobulin reveals a unique molecular cage. Angew Chem Int Ed Engl. doi:10.1002/anie.201108015

Mathew R, Futterweit S, Valderrama E, Tarectecan AA, Bylander JE, Bond JS, Trachtman H (2005) Meprin-alpha in chronic diabetic nephropathy: interaction with the renin-angiotensin axis. Am J Physiol 289:F911–F921

Matters GL, Manni A, Bond JS (2005) Inhibitors of polyamine biosynthesis decrease the expression of the metalloproteases meprin alpha and MMP-7 in hormone-independent human breast cancer cells. Clin Exp Metastasis 22:331–339

Matthews BW (1988) Structural basis for the action of thermolysin and related zinc peptidases. Acc Chem Res 21:333–340

Meier UC, Boetzel J, Kellermann J, Mann K, Billich A, Stöcker W, Schramm HJ (1994) The cleavage of the bait region of alpha 2-macroglobulin by human immunodeficiency virus proteinases and by astacin. Ann NY Acad Sci 737:431–433

Milhiet PE, Corbeil D, Simon V, Kenny AJ, Crine P, Boileau G (1994) Expression of rat endopeptidase-24.18 in COS-1 cells: membrane topology and activity. Biochem J 300(Pt 1): 37–43

Milhiet PE, Chevallier S, Corbeil D, Seidah NG, Crine P, Boileau G (1995) Proteolytic processing of the alpha-subunit of rat endopeptidase-24.18 by furin. Biochem J 309(Pt 2):683–688

Moali C, Font B, Ruggiero F, Eichenberger D, Rousselle P, Francois V, Oldberg A, Bruckner-Tuderman L, Hulmes DJ (2005) Substrate-specific modulation of a multisubstrate proteinase. C-terminal processing of fibrillar procollagens is the only BMP-1-dependent activity to be enhanced by PCPE-1. J Biol Chem 280:24188–24194

Möhrlen F, Hutter H, Zwilling R (2003) The astacin protein family in Caenorhabditis elegans. Eur J Biochem 270:4909–4920

Muir A, Greenspan DS (2011) Metalloproteinases in Drosophila to humans that are central players in developmental processes. J Biol Chem 286:41905–41911

Nagase H (1997) Activation mechanisms of matrix metalloproteinases. Biol Chem 378:151–160

Nguyen T, Jamal J, Shimell MJ, Arora K, O'Connor MB (1994) Characterization of tolloid-related-1: a BMP-1-like product that is required during larval and pupal stages of Drosophila development. Dev Biol 166:569–586

Oberholzer AE, Bumann M, Hege T, Russo S, Baumann U (2009) Metzincin's canonical methionine is responsible for the structural integrity of the zinc-binding site. Biol Chem 390:875–881

Ohler A, Debela M, Wagner S, Magdolen V, Becker-Pauly C (2010) Analyzing the protease web in skin: meprin metalloproteases are activated specifically by KLK4, 5 and 8 vice versa leading to processing of proKLK7 thereby triggering its activation. Biol Chem 391:455–460

Okada A, Sano K, Nagata K, Yasumasu S, Ohtsuka J, Yamamura A, Kubota K, Iuchi I, Tanokura M (2010) Crystal structure of zebrafish hatching enzyme 1 from the zebrafish Danio rerio. J Mol Biol 402:865–878

Oneda B, Lods N, Lottaz D, Becker-Pauly C, Stöcker W, Pippin J, Huguenin M, Ambort D, Marti HP, Sterchi EE (2008) Metalloprotease meprin beta in rat kidney: glomerular localization and differential expression in glomerulonephritis. PLoS One 3:e2278

Pappano W, Steiglitz B, Scott I, Keene D, Greenspan D (2003) Use of Bmpl/Tll1 doubly homozygous null mice and proteomics to identify and validate in vivo substrates of bone morphogenetic protein 1/tolloid-like metalloproteinases. Mol Cell Biol 23:4428–4438

Perez L, Kerrigan JE, Li X, Fan H (2007) Substitution of methionine 435 with leucine, isoleucine, and serine in tumor necrosis factor alpha converting enzyme inactivates ectodomain shedding activity. Biochem Cell Biol 85:141–149

Piccolo S, Agius E, Lu B, Goodman S, Dale L, De Robertis EM (1997) Cleavage of Chordin by Xolloid metalloprotease suggests a role for proteolytic processing in the regulation of Spemann organizer activity. Cell 91:407–416

Pieper M, Betz M, Budisa N, Gomis-Rüth F-X, Bode W, Tschesche H (1997) Expression, purification, characterization, and X-ray analysis of selenomethionine 215 variant of leukocyte collagenase. J Protein Chem 16:637–650

Ploper D, Lee HX, De Robertis EM (2011) Dorsal-ventral patterning: crescent is a dorsally secreted Frizzled-related protein that competitively inhibits Tolloid proteases. Dev Biol 352: 317–328

Quesada V, Sanchez LM, Alvarez J, Lopez-Otin C (2004) Identification and characterization of human and mouse ovastacin: a novel metalloproteinase similar to hatching enzymes from arthropods, birds, amphibians, and fish. J Biol Chem 279:26627–26634

Rattenholl A, Pappano WN, Koch M, Keene DR, Kadler KE, Sasaki T, Timpl R, Burgeson RE, Greenspan DS, Bruckner-Tuderman L (2002) Proteinases of the bone morphogenetic protein-1 family convert procollagen VII to mature anchoring fibril collagen. J Biol Chem 277: 26372–26378

Ravi Ram K, Sirot LK, Wolfner MF (2006) Predicted seminal astacin-like protease is required for processing of reproductive proteins in Drosophila melanogaster. Proc Natl Acad Sci U S A 103:18674–18679

Rawlings ND, Barrett AJ, Bateman A (2010) MEROPS: the peptidase database. Nucleic Acids Res 38:D227–D233

Red Eagle AR, Hanson RL, Jiang W, Han X, Matters GL, Imperatore G, Knowler WC, Bond JS (2005) Meprin beta metalloprotease gene polymorphisms associated with diabetic nephropathy in the Pima Indians. Hum Genet 118:12–22

Reinemer P, Grams F, Huber R, Kleine T, Schnierer S, Piper M, Tschesche H, Bode W (1994) Structural implications for the role of the N terminus in the 'superactivation' of collagenases. A crystallographic study. FEBS Lett 338:227–233

Reyda S, Jacob E, Zwilling R, Stöcker W (1999) cDNA cloning, bacterial expression, in vitro renaturation and affinity purification of the zinc endopeptidase astacin. Biochem J 344: 851–857

Reynolds SD, Angerer LM, Palis J, Nasir A, Angerer RC (1992) Early mRNAs, spatially restricted along the animal-vegetal axis of sea urchin embryos, include one encoding a protein related to tolloid and BMP-1. Development 114:769–786

Rösmann S, Hahn D, Lottaz D, Kruse MN, Stöcker W, Sterchi EE (2002) Activation of human meprin-alpha in a cell culture model of colorectal cancer is triggered by the plasminogen-activating system. J Biol Chem 277:40650–40658

Rothe M, Wong SC, Henzel WJ, Goeddel DV (1994) A novel family of putative signal transducers associated with the cytoplasmic domain of the 75 kDa tumor necrosis factor receptor. Cell 78:681–692

Sachdev M, Mandal A, Mulders S, Digilio LC, Panneerdoss S, Suryavathi V, Pires E, Klotz KL, Hermens L, Herrero MB, Flickinger CJ, van Duin M, Herr JC (2012) Oocyte specific oolemmal SAS1B involved in sperm binding through intra-acrosomal SLLP1 during fertilization. Dev Biol 363:40–51

Sano K, Kawaguchi M, Yoshikawa M, Iuchi I, Yasumasu S (2010) Evolution of the teleostean zona pellucida gene inferred from the egg envelope protein genes of the Japanese eel, Anguilla japonica. FEBS J 277:4674–4684

Sato SM, Sargent TD (1990) Molecular approach to dorsoanterior development in Xenopus laevis. Dev Biol 137:135–141

Schäfer C, Heiss A, Schwarz A, Westenfeld R, Ketteler M, Floege J, Müller-Esterl W, Schinke T, Jahnen-Dechent W (2003) The serum protein alpha 2-Heremans-Schmid glycoprotein/fetuin-A is a systemically acting inhibitor of ectopic calcification. J Clin Invest 112:357–366

Schechter I, Berger A (1967) On the size of the active site in proteases. I. Papain. Biochem Biophys Res Commun 27:157–162

Schütte A, Hedrich J, Stöcker W, Becker-Pauly C (2010) Let it flow: morpholino knockdown in zebrafish embryos reveals a pro-angiogenic effect of the metalloprotease meprin alpha2. PLoS One 5:e8835

Scott IC, Clark TG, Takahara K, Hoffman GG, Eddy RL, Haley LL, Shows TB, Greenspan DS (1999) Assignment of TLL1 and TLL2, which encode human BMP-1/Tolloid-related metalloproteases, to chromosomes 4q32–>q33 and 10q23–>q24 and assignment of murine Tll2 to chromosome 19. Cytogenet Cell Genet 86:64–65

Scott IC, Imamura Y, Pappano WN, Troedel JM, Recklies AD, Roughley PJ, Greenspan DS (2000a) Bone morphogenetic protein-1 processes probiglycan. J Biol Chem 275:30504–30511

Scott IC, Steiglitz BM, Clark TG, Pappano WN, Greenspan DS (2000b) Spatiotemporal expression patterns of mammalian chordin during postgastrulation embryogenesis and in postnatal brain. Dev Dyn 217:449–456

Serpe M, O'Connor MB (2006) The metalloprotease tolloid-related and its TGF-beta-like substrate Dawdle regulate Drosophila motoneuron axon guidance. Development 133:4969–4979

Shibata Y, Iwamatsu T, Oba Y, Kobayashi D, Tanaka M, Nagahama Y, Suzuki N, Yoshikuni M (2000) Identification and cDNA cloning of alveolin, an extracellular metalloproteinase, which induces chorion hardening of medaka (Oryzias latipes) eggs upon fertilization. J Biol Chem 275:8349–8354

Shibata Y, Nagahama Y, Yoshikuni M (2003) An astacin-like protease, alveolin, is released from cortical vesicles and induces changes in egg envelope proteins during fertilization of medaka (Oryzias latipes) eggs. Fish Physiol Biochem 28:365

Shimell MJ, Ferguson EL, Childs SR, O'Connor MB (1991) The Drosophila dorsal-ventral patterning gene tolloid is related to human bone morphogenetic protein 1. Cell 67:469–481

Sieron AL, Tretiakova A, Jameson BA, Segall ML, Lund-Katz S, Khan MT, Sw L, Stöcker W (2000) Structure and function of procollagen C-proteinase (mTolloid) domains determined by protease digestion, circular dichroism, binding to procollagen type I, and computer modeling. Biochemistry 39:3231–3239

Sirot LK, LaFlamme BA, Sitnik JL, Rubinstein CD, Avila FW, Chow CY, Wolfner MF (2009) Molecular social interactions: Drosophila melanogaster seminal fluid proteins as a case study. Adv Genet 68:23–56

Steiglitz BM, Keene DR, Greenspan DS (2002) PCOLCE2 encodes a functional procollagen C-proteinase enhancer (PCPE2) that is a collagen-binding protein differing in distribution of expression and post-translational modification from the previously described PCPE1. J Biol Chem 277:49820–49830

Steiglitz BM, Ayala M, Narayanan K, George A, Greenspan DS (2004) Bone morphogenetic protein-1/Tolloid-like proteinases process dentin matrix protein-1. J Biol Chem 279:980–986

Steiglitz BM, Kreider JM, Frankenburg EP, Pappano WN, Hoffman GG, Meganck JA, Liang X, Hook M, Birk DE, Goldstein SA, Greenspan DS (2006) Procollagen C proteinase enhancer 1 genes are important determinants of the mechanical properties and geometry of bone and the ultrastructure of connective tissues. Mol Cell Biol 26:238–249

Sterchi EE, Green JR, Lentze MJ (1982) Non-pancreatic hydrolysis of N-benzoyl-L-tyrosyl-p-aminobenzoic acid (PABA-peptide) in the human small intestine. Clin Sci 62:557–560

Sterchi EE, Naim HY, Lentze MJ (1988a) Biosynthesis of N-benzoyl-L-tyrosyl-p-aminobenzoic acid hydrolase: disulfide-linked dimers are formed at the site of synthesis in the rough endoplasmic reticulum. Arch Biochem Biophys 265:119–127

Sterchi EE, Naim HY, Lentze MJ, Hauri HP, Fransen JA (1988b) N-benzoyl-L-tyrosyl-p-aminobenzoic acid hydrolase: a metalloendopeptidase of the human intestinal microvillus membrane which degrades biologically active peptides. Arch Biochem Biophys 265:105–118

Sterchi EE, Stöcker W, Bond JS (2008) Meprins, membrane-bound and secreted astacin metalloproteinases. Mol Aspects Med 29:309–328

Stöcker W, Bode W (1995) Structural features of a superfamily of zinc-endopeptidases: the metzincins. Curr Opin Struct Biol 5:383–390

Stöcker W, Zwilling R (1995) Astacin. Methods Enzymol 248:305–325

Stöcker W, Breit S, Sottrup-Jensen L, Zwilling R (1991) alpha$_2$-Macroglobulin from hemolymph of the freshwater crayfish Astacus astacus. Comp Biochem Physiol B 98:501–509

Stöcker W, Gomis-Rüth F-X, Bode W, Zwilling R (1993) Implications of the 3-dimensional structure of astacin for the structure and function of the astacin family of zinc- endopeptidases. Eur J Biochem 214:215–231

Stöcker W, Grams F, Baumann U, Reinemer P, Gomis-Rüth FX, McKay DB, Bode W (1995) The metzincins–topological and sequential relations between the astacins, adamalysins, serralysins, and matrixins (collagenases) define a superfamily of zinc-peptidases. Protein Sci 4:823–840

Sun Q, Jin HJ, Bond JS (2009) Disruption of the meprin alpha and beta genes in mice alters homeostasis of monocytes and natural killer cells. Exp Hematol 37:346–356

Suzuki N, Labosky PA, Furuta Y, Hargett L, Dunn R, Fogo AB, Takahara K, Peters DM, Greenspan DS, Hogan BL (1996) Failure of ventral body wall closure in mouse embryos lacking a procollagen C-proteinase encoded by Bmp1, a mammalian gene related to Drosophila tolloid. Development 122:3587–3595

Takahara K, Kessler E, Biniaminov L, Brusel M, Eddy RL, Jani-Sait S, Shows TB, Greenspan D (1994) Type I procollagen COOH-terminal proteinase enhancer protein: identification, primary structure, and chromosomal localization of the cognate human gene (PCOLCE). J Biol Chem 269:26280–26285

Takahara K, Brevard R, Hoffman GG, Suzuki N, Greenspan DS (1996) Characterization of a novel gene product (mammalian tolloid-like) with high sequence similarity to mammalian tolloid/ bone morphogenetic protein-1. Genomics 34:157–165

Takayama J, Takaoka M, Yamamoto S, Nohara A, Ohkita M, Matsumura Y (2008) Actinonin, a meprin inhibitor, protects ischemic acute kidney injury in male but not in female rats. Eur J Pharmacol 581:157–163

Tallant C, Garcia-Castellanos R, Baumann U, Gomis-Rüth FX (2010) On the relevance of the Met-turn methionine in metzincins. J Biol Chem 285:13951–13957

Tamori J, Kanzawa N, Tajima T, Tamiya T, Tsuchiya T (1999) Purification and characterization of a novel isoform of myosinase from spear squid liver. J Biochem 126:969–974

Titani K, Torff HJ, Hormel S, Kumar S, Walsh KA, Rödl J, Neurath H, Zwilling R (1987) Amino acid sequence of a unique protease from the crayfish Astacus fluviatilis. Biochemistry 26: 222–226

Tsai PL, Chen CH, Huang CJ, Chou CM, Chang GD (2004) Purification and cloning of an endogenous protein inhibitor of carp nephrosin, an astacin metalloproteinase. J Biol Chem 279:11146–11155

Tsuchiya S, Simmer JP, Hu JC, Richardson AS, Yamakoshi F, Yamakoshi Y (2011) Astacin proteases cleave dentin sialophosphoprotein (Dspp) to generate dentin phosphoprotein (Dpp). J Bone Miner Res 26:220–228

Unsöld C, Pappano W, Imamura Y, Steiglitz B, Greenspan D (2002) Biosynthetic processing of the pro-alpha 1(V)2pro-alpha 2(V) collagen heterotrimer by bone morphogenetic protein-1 and furin-like proprotein convertases. J Biol Chem 277:5596–5602

Uzel MI, Scott IC, Babakhanlou-Chase H, Palamakumbura AH, Pappano WN, Hong HH, Greenspan DS, Trackman PC (2001) Multiple bone morphogenetic protein 1-related mammalian metalloproteinases process pro-lysyl oxidase at the correct physiological site and control lysyl oxidase activation in mouse embryo fibroblast cultures. J Biol Chem 276:22537–22543

Vadon-Le Goff S, Kronenberg D, Bourhis JM, Bijakowski C, Raynal N, Ruggiero F, Farndale RW, Stocker W, Hulmes DJ, Moali C (2011) Procollagen C-proteinase enhancer stimulates procollagen processing by binding to the C-propeptide region only. J Biol Chem 286: 38932–38938

Vazeille E, Bringer MA, Gardarin A, Chambon C, Becker-Pauly C, Pender SL, Jakob C, Muller S, Lottaz D, Darfeuille-Michaud A (2011) Role of meprins to protect ileal mucosa of Crohn's disease patients from colonization by adherent-invasive E. coli. PLoS One 6:e21199

Veitch DP, Nokelainen P, McGowan KA, Nguyen TT, Nguyen NE, Stephenson R, Pappano WN, Keene DR, Spong SM, Greenspan DS, Findell PR, Marinkovich MP (2003) Mammalian tolloid metalloproteinase, and not matrix metalloprotease 2 or membrane type 1 metalloprotease, processes laminin-5 in keratinocytes and skin. J Biol Chem 278:15661–15668

von Marschall Z, Fisher LW (2010a) Decorin is processed by three isoforms of bone morphogenetic protein-1 (BMP1). Biochem Biophys Res Commun 391:1374–1378

von Marschall Z, Fisher LW (2010b) Dentin sialophosphoprotein (DSPP) is cleaved into its two natural dentin matrix products by three isoforms of bone morphogenetic protein-1 (BMP1). Matrix Biol 29:295–303

Walasek P, Honek JF (2005) Nonnatural amino acid incorporation into the methionine 214 position of the metzincin Pseudomonas aeruginosa alkaline protease. BMC Biochem 6:21

Walker PD, Kaushal GP, Shah SV (1998) Meprin A, the major matrix degrading enzyme in renal tubules, produces a novel nidogen fragment in vitro and in vivo. Kidney Int 53:1673–1680

Waltersperger S, Widmer C, Wang M, Baumann U (2010) Crystal structure of archaemetzincin AmzA from Methanopyrus kandleri at 1.5 A resolution. Proteins 78:2720–2723

Wermter C, Höwel M, Hintze V, Bombosch B, Aufenvenne K, Yiallouros I, Stöcker W (2007) The protease domain of procollagen C-proteinase (BMP1) lacks substrate selectivity, which is conferred by non-proteolytic domains. Biol Chem 388:513–521

Wolfman N, McPherron A, Pappano W, Davies M, Song K, Tomkinson K, Wright J, Zhao L, Sebald S, Greenspan D, Lee S (2003) Activation of latent myostatin by the BMP-1/tolloid family of metalloproteinases. Proc Natl Acad Sci U S A 100:15842–15846

Wozney JM, Rosen V, Celeste AJ, Mitsock LM, Whitters MJ, Kriz RW, Hewick RM, Wang EA (1988) Novel regulators of bone formation: molecular clones and activities. Science 242: 1528–1534

Yan L, Fei K, Zhang J, Dexter S, Sarras MP Jr (2000a) Identification and characterization of hydra metalloproteinase 2 (HMP2): a meprin-like astacin metalloproteinase that functions in foot morphogenesis. Development 127:129–141

Yan L, Leontovich A, Fei K, Sarras MP Jr (2000b) Hydra metalloproteinase 1: a secreted astacin metalloproteinase whose apical axis expression is differentially regulated during head regeneration. Dev Biol 219:115–128

Yasumasu S, Yamada K, Akasaka K, Mitsunaga K, Iuchi I, Shimada H, Yamagami K (1992) Isolation of cDNAs for LCE and HCE, two constituent proteases of the hatching enzyme of Oryzias latipes, and concurrent expression of their mRNAs during development. Dev Biol 153: 250–258

Yasumasu S, Shimada H, Inohaya K, Yamazaki K, Iuchi I, Yasumasu I, Yamagami K (1996) Different exon-intron organizations of the genes for two astacin-like proteases, high choriolytic enzyme (choriolysin H) and low choriolytic enzyme (choriolysin L), the constituents of the fish hatching enzyme. Eur J Biochem 237:752–758

Yasumasu S, Kawaguchi M, Ouchi S, Sano K, Murata K, Sugiyama H, Akema T, Iuchi I (2010a) Mechanism of egg envelope digestion by hatching enzymes, HCE and LCE in medaka, Oryzias latipes. J Biochem 148:439–448

Yasumasu S, Uzawa M, Iwasawa A, Yoshizaki N (2010b) Hatching mechanism of the Chinese soft-shelled turtle Pelodiscus sinensis. Comp Biochem Physiol B 155:435–441

Yiallouros I, Vassiliou S, Yiotakis A, Zwilling R, Stöcker W, Dive V (1998) Phosphinic peptides, the first potent inhibitors of astacin, behave as extremely slow-binding inhibitors. Biochem J 331:375–379

Yiallouros I, Grosse Berkhoff E, Stöcker W (2000) The roles of Glu93 and Tyr149 in astacin-like zinc peptidases. FEBS Lett 484:224–228

Yiallouros I, Kappelhoff R, Schilling O, Wegmann F, Helms MW, Auge A, Brachtendorf G, Berkhoff EG, Beermann B, Hinz HJ, König S, Peter-Katalinic J, Stöcker W (2002) Activation mechanism of pro-astacin: role of the pro-peptide, tryptic and autoproteolytic cleavage and importance of precise amino-terminal processing. J Mol Biol 324:237–246

Yura R, Bradley S, Ramesh G, Reeves W, Bond J (2009) Meprin A metalloproteases enhance renal damage and bladder inflammation after LPS challenge. Am J Physiol Renal Physiol 296(1): F135–F144

Zapata JM, Pawlowski K, Haas E, Ware CF, Godzik A, Reed JC (2001) A diverse family of proteins containing tumor necrosis factor receptor-associated factor domains. J Biol Chem 276:24242–24252

Zhang Y, Ge G, Greenspan DS (2006) Inhibition of bone morphogenetic protein 1 by native and altered forms of alpha2-macroglobulin. J Biol Chem 281:39096–39104

Chapter 8
Proteases in Death Pathways

Andreas Flütsch and Markus G. Grütter

Abbreviations

Aβ	Amyloid-β
AIF	Apoptosis-inducing factor
Apaf-1	Apoptotic protease activating factor 1
APP	Amyloid precursor protein
BH	Bcl-2 homology
Bid	BB3-interacting domain
BIR	Baculoviral IAP repeat
C. elegans	Caenorhabditis elegans
CAD	Caspase-activated DNAse
CARD	Caspase activation and recruitment domain
DAMP	danger associated molecular pattern
DARPin	Designed ankyrin repeat protein
DD	Death domain
DED	Death effector domain
DFF40	DNA fragmentation factor 40 also termed CAD
DIABLO	Direct IAP binding protein with low pI, also termed Smac
DISC	Death-inducing signaling complex
ENDOG	Endonuclease g
FADD	Fas-associated death domain
FLICE	FADD-like interleukin-1 beta-converting enzyme, today known as caspase-8
FLIP	FLICE-inhibitory protein
HTRA2	High temperature requirement protein 2
htt	Huntingtin

A. Flütsch • M.G. Grütter (✉)
Department of Biochemistry, University of Zurich, Winterthurerstrasse 190, 8057 Zürich, Switzerland
e-mail: gruetter@bioc.uzh.ch

K. Brix and W. Stöcker (eds.), *Proteases: Structure and Function*,
DOI 10.1007/978-3-7091-0885-7_8, © Springer-Verlag Wien 2013

IAP	Inhibitor of apoptosis protein
ICAD	Inhibitor of CAD, also known as DFF45
ICE	Interleukin-1β-converting enzyme
IMS	Inter-membrane space
LRR	Leucine-rich repeat
MOMP	Mitochondrial outer membrane permeabilization
NC-IUBMB	Nomenclature Committee of the International Union of Biochemistry and Molecular Biology
NCCD	Nomenclature Committee on Cell Death
NFT	Neurofibrillary tangle
NLR	NOD-like receptor
NOD	Nucleotide binding and oligomerization domain
PARP	Poly(ADP-ribose) polymerase
PIDD	p53-induced protein with a death domain
PS	Phosphatidylserine
RIP	Receptor-interacting protein
ROCK	Rho-associated kinase
Smac	Second mitochondria-derived activator of caspase, also termed DIABLO
tBid	Truncated Bid
TNF	Tumor necrosis factor
TRAIL	TNF-related apoptosis-inducing ligand
UBA	Ubiquitin associated
XIAP	X-linked inhibitor of apoptosis protein

8.1 Introduction

The theory of the cell as the basic unit of life was developed by the German scientists Schleiden and Schwann in 1839. Shortly afterwards, cellular demise was observed by researches in different tissues and cell types. A landmark was in 1885, when Walther Flemming morphologically described the process of dying ovarian follicles containing "chromatin chunks", which he termed chromatolysis (Flemming 1885). In fact, the condensation of chromatin is nowadays well known as a morphological feature of apoptosis.

In 1972, Kerr et al. (1972) described the morphology of cell deletion in tissues with a reproducible and different mechanism compared to the known process of necrosis. Disappearing cells showed aggregation of nuclear chromatin similar to Flemmings observations. The mechanism could be dissected into two stages: A step of nuclear and cytoplasmic condensation into membrane-bound cellular fragments and a degradation step performed by other cells after phagocytosis. Kerr proposed the term apoptosis for this programmed cell death derived from the Greek word for "falling off".

Studies on embryonic development of *Caenorhabditis elegans* (*C. elegans*) by Ellis and Horvitz (1986) yielded in the discovery of the cell death controlling genes ced-3 and ced-4. A fundamental starting point in death pathway research was in 1993, when this *C. elegans* gene ced-3 was recognized to encode a protein similar to the mammalian interleukin-1β-converting enzyme (ICE) (Yuan et al. 1993). Several homologues of ICE were identified in mammals forming a proteolytic network as the molecular basis for apoptosis. The increasing amount of identified cysteine-aspartic-dependent proteases with inconsistent and multiple names were unified for simplicity with the name caspase in 1996 (Alnemri et al. 1996).

Caspases were further investigated and revealed a complex network of regulation and activation during apoptosis. They are known as the key-players in apoptosis and besides of this crucial role, caspases are involved in inflammatory pathways and immunity and can regulate proliferation and differentiation.

Besides the caspase-mediated apoptosis, other death pathways like necroptosis, autophagy and pyroptosis were discovered and expanded the research field of death pathways with caspase-dependent and -independent mechanisms.

The numerous different ways of cell removal and destruction underlines the significance of cell deletion. It is crucial for multicellular organisms to maintain the correct numbers of cells in tissues and to shape the architecture of organs. In addition, unspecific cell proliferation and tumor development is related to down-regulation of apoptosis whereas its up-regulation is associated to degenerative diseases like Parkinson and Alzheimer.

8.2 Death Pathways

8.2.1 Nomenclature

Various distinctive death pathways have been discovered till date and were historically classified by cellular morphology. Since 2012, the Nomenclature Committee on Cell Death (NCCD) suggests a classification based not only on morphological observations but also on biochemical data analysis (Galluzzi et al. 2012). It has been revealed that similar morphological features in cell death show biochemical, functional and immunological heterogeneity. This renders the nomenclature based on morphology as not sufficient and the exact identification of a specific pathway as a rather challenging task. The suggested classification on quantitative biochemical analysis is less prone to misinterpretations, since morphological observations require an experienced operator. Thus, the following pathway definitions will consider the NCCD recommendations.

8.2.2 Extrinsic Apoptosis

The extrinsic apoptotic pathway is a caspase-dependent mechanism, which is initiated by extracellular stress signals. Extracellular death ligands bind to death receptors like FAS, tumor necrosis factor (TNF) receptor 1 or TNF-related apoptosis inducing ligand (TRAIL) receptor 1 and 2 leading to receptor trimerization. The activated receptor conformation triggers the assembly of an intracellular death inducing signaling complex (DISC) involving the adaptor protein Fas-associated death domain (FADD) and the initiator pro-caspases-8 or -10.

The DISC recruitment of pro-caspases leads to activation by dimerization and subsequent autocatalytic processing. Then, initiator caspases catalyze the maturation of executioner caspases to start the apoptotic removal of the cell. In particular cells like hepatocytes (type II cells), caspase-8 cleaves the death agonist BB3-interacting domain (Bid) to a truncated form (tBid) instead of activating executioner caspases (Yin et al. 1999). Bid-cleavage leads to mitochondrial outer membrane permeabilization (MOMP) that activates apoptosis in a similar manner as the intrinsic apoptosis driven by caspase-9. Bid-cleavage can thus substitute a direct caspase-8 mediated activation of executioner caspases and can provide an amplification loop of extrinsic apoptosis through the intrinsic apoptotic pathway in selected cells (Yin et al. 1999).

8.2.3 Intrinsic Apoptosis

Intrinsic apoptosis is triggered by intracellular stress signals such as DNA damage, oxidative stress and accumulation of unfolded protein. The NCCD suggests a classification into two different mechanisms due to highly heterogeneous signaling cascades with the mitochondrion as the main control mechanism. Predominance of pro-apoptotic proteins like Bak or Bax can lead to MOMP due to their pore forming activity. Consequently, inter-membrane space (IMS) residential proteins are released into the cytosol where they mediate two apoptotic pathways:

- ***Caspase-dependent***
  The release of cytochrome c upon MOMP triggers the caspase-dependent mechanism of intrinsic apoptosis. Cytochrome c assembles together with the apoptotic protease activating factor 1 (Apaf-1) to the recruitment platform for pro-caspase-9, known as the apoptosome. After the activation by dimerization, caspase-9 facilitates downstream cleavage of executioner caspases leading to cellular demise.
- ***Caspase-independent***
  The release of IMS proteins, such as apoptosis-inducing factor (AIF) and endonuclease g (ENDOG), leads to caspase-independent fragmentation of DNA after relocation into the nucleus. The IMS serine protease high temperature requirement protein A2 (HTRA2) cleaves numerous cytoplasmic substrates

including cytoskeleton proteins and contributes to the caspase-independent mechanism of intrinsic apoptosis.

8.2.4 Pyroptosis

Pyroptosis is a regulated death pathway highly connected to caspase-1 activation and inflammatory responses. Brennan and Cookson (2000) described the death of macrophages that were infected by *Salmonella typhimurium* and named this reproducible mechanism pyroptosis (Cookson and Brennan 2001). Indeed, several other bacteria are able to induce pyroptosis like *Shigella flexneri*, *Bacillus anthracis toxin* and *Pseudomona aeruginosa*.

It became clear that activation of caspase-1 is a crucial step, which is promoted by a multi-protein complex known as the inflammasome. In contrast to the apoptotic pathways, executioner caspases are not involved during pyroptosis.

The caspase-1-activating inflammasome is a large protein complex, which is formed upon infection with bacteria, viruses or parasites, as well as by host-derived danger associated molecular patterns (DAMP). NOD-like receptor (NLR) proteins recognize danger signals and bind via the adaptor protein ASC and its caspase activation and recruitment domain (CARD) to pro-caspase-1. Proximity-induced oligomerization and activation of caspase-1 leads to the maturation of pro-interleukins. The activated cytokines interleukin-1 and -18 are secreted to recruit and activate immune cells for an inflammatory response. The activation of caspase-1 results in cleavage of several other substrates, however the complex mechanism of inflammasome activation and pyroptotic cell death is still not completely understood but has recently been well reviewed by Rathinam et al. (2012).

8.2.5 Regulated Necrosis

Necrosis is commonly known as an accidental cell death with a morphotype of disrupted membranes and release of intracellular content. Hitomi et al. (2008) recently pointed out that there is as well a regulated form of necrosis, which can be triggered under specific energy dependent circumstances. Key mediators of the regulated necrotic pathway are ATP-dependent enzymes like poly(ADP-ribose) polymerase 1 (PARP-1) or receptor-interacting protein kinase 1 (RIP1). Alkylated DNA damage can act as an inducing signal leading to PARP-1 hyperactivation and cytosolic ATP reduction. The depletion of the cellular energy storage results in activation of RIP1 or its homolog RIP3, which triggers the execution of necrosis.

8.2.6 Autophagy

Morphological features like massive cytoplasmic vacuolization were first described in the 1950s. In most cases, autophagy is activated upon cell stress and as a cytoprotective response of dying cells. It is a conserved, kinase-dependent cell degradation pathway in eukaryotes with autophagy related proteins and Beclin-1 as the key players of autophagosome biogenesis. A clear elucidation of autophagosome formation and its regulation is beyond the scope of this chapter but reviewed by Yang and Klionsky (2009) and Rubinsztein et al. (2012). Like apoptosis, the autophagic cell degradation is non-inflammatory due to vacuolization of cytoplasmic content.

8.2.7 Other Modalities

Numerous additional death pathways exist that are in some cases overlapping or bypassing other mechanisms. Thus, a variety of cell death pathways can be observed in cultures induced by the same stimulus. This makes a clear identification of the exact molecular mechanism even more important. In addition to apoptosis as the most important non-inflammatory pathway and pyroptosis as the inflammatory pathway, the following mechanisms are known:

- **Mitotic catastrophe** is initiated in an arrested M-phase in the cell cycle due to mitotic machinery aberrations. It finally leads to cell death or senescence.
- **Anoikis** is an apoptotic response due to loss of cell-to-matrix interactions
- **Entosis** or "cell-cannibalism" describes the complete engulfment of a cell into another cell, which is never released and finally degraded.
- **Parthanatos** is initiated by PARP-1 hyperactivation and its resulting ATP-depletion. It is a particular form of regulated necrosis.
- **Netosis** is a cell death mechanism restricted to granulocytic cells, sharing biochemical features with autophagy and regulated necrosis.
- **Cornification** is important for epidermis formation and restricted to keratinocytes.

8.3 Caspases: The Essential Proteases in Death Pathways

The caspase family members are well known as the key players in apoptosis. They play important roles in initiation and execution of the extrinsic and caspase-dependent intrinsic apoptotic pathways as well as in pyroptotic cell deletion. The first hint of these pro-apoptotic proteins was found in *C. elegans* in 1986, when Ellis and Horvitz (1986) identified that the genes ced-3 and ced-4 are required for the initiation of apoptosis. Deletion of these two genes leads to the survival of cells, which otherwise usually died during embryonic development. Using the simple model organism *C. elegans*, more apoptosis-related genes and proteins like ced-9 (Hengartner et al. 1992) and EGL-1 (Conradt and Horvitz 1998) were unraveled decomposing the complex molecular system of apoptosis into its relevant proteins. CED-4 dimeric proteins are blocked in an inactive form by CED-9 in living cells. Apoptotic stimuli induce the production of EGL-1, which binds to CED-9 and thus liberates the CED-4 dimers. The released dimers build the apoptosome by tetramerization of the asymmetric dimers and recruitment of proCED-3, where CED-3 is activated by proteolysis or conformational changes and then released to trigger apoptosis.

Apoptotic research on mammalian cells and human proteins was boosted after the discovery of ICE as a unique cysteine protease (Thornberry et al. 1992) and its homology to the nematodic CED-3 (Yuan et al. 1993). The unified name caspase was finally introduced after the identification of several ICE homologues and their confusing nomenclature (Alnemri et al. 1996). In fact, the mammalian genomes encode numerous ced-like genes that may have arisen by gene duplication during evolution. Compared to nematodes, this emphasizes a much more complex apoptotic system with more redundancy and regulatory effectors.

In 1994, the first high-resolution crystal structure of human ICE was reported at a resolution of 2.6 Å and gave first insights into the catalytic mechanism and its regulation (Wilson et al. 1994). The tetrameric enzyme consists of two identical heterodimers made by a small (p10) and a large (p20) subunit. Two substrate-binding pockets are shaped between the p10 and p20 of the individual heterodimer. The active sites are formed by a catalytic diad (Cys-285 and His-237) and cleave substrates specifically after aspartate residues. Both, the conserved catalytic diad residues Cys-285 and His-237 as well as the requirement of a substrate with an aspartic acid at P_1 are common for all caspase family members.

Based on the structure and selective mutagenesis, an allosteric regulation of ICE activity was proposed (Wilson et al. 1994). Mutations in the dimer interface of the heterodimeric subunits, far away from the active site cleft, resulted in an inactive ICE suggesting the tetrameric enzyme as a prerequisite for activity.

During the last two decades, 12 different human caspases have been identified and biochemically characterized. Seven of them could be crystallized or studied by NMR spectroscopy till date. Even though this progress led to one of the best-characterized protease families among proteolytic enzymes, open questions are still remaining.

8.3.1 Classification, Nomenclature and Family Members

The Nomenclature Committee of the International Union of Biochemistry and Molecular Biology (NC-IUBMB) classifies the family of caspases as cysteine endopeptidases with the enzyme category (EC) number 3.4.22 where today caspase-1 to caspase-11 are listed in the IUBMB database with individual EC numbers. The MEROPS database assigned these enzymes to the clan CD with the family C14 [http://merops.sanger.ac.uk/ (Rawlings et al. 2012)].

The term caspase was introduced in 1996 with "c" abbreviating the catalytic residue cysteine and "aspase" referring to the required aspartic acid at position P_1 of the peptidase (Alnemri et al. 1996). Other similar terminologies for caspase can be found in the literature like "cysteine-dependent aspartate-directed proteases" (Eckhart et al. 2008).

The number of each family member is assigned to the date of publication and reflects the history of caspase research. Caspases are expressed as zymogens, indicated by the prefix "pro" (e.g. pro-caspase-1). The polypeptide chain consists of either a N-terminal pro-domain or a short pro-peptide and a catalytic domain that can be proteolytically separated from the former.

The architecture of the catalytic domain is made by a large and a small subunit, connected by a short linker. The individual subunits are particularly named by their exact molecular weight (e.g. caspase-1-p10 for the small subunit of caspase-1). Complete caspase activation is achieved by proteolytic removal of the N-terminal pro-domain or short pro-peptide as well as the inter-subunit linker. In addition, two catalytic domains dimerize and form a heterotetrameric structure also known as the caspase-fold.

A total of 16 different mammalian caspases could be identified by comparative genomics and evolutionary analysis (Eckhart et al. 2008). The individual family members can be classified according to their function, their length and structure of the N-terminal pro-domain, their optimal cleavage motives or their phylogenetic analysis of the amino acid sequence. The existing human homologues are generally grouped according to sequence identity and functionality.

8.3.1.1 Human Caspases

Homology based analysis of the human caspases divides the family into three subgroups that can be assigned as well to their function (Fig. 8.1).

Caspase-1, -4, -5 and -12 are inflammatory caspases according to their roles in cytokine activation. For instance, caspase-1 gets activated by the inflammasome formation and triggers pyroptosis, an inflammatory death pathway (Rathinam et al. 2012). The role of caspase-5, a close homologue to caspase-1, could be assigned to cytokine maturation and inflammasome formation (reviewed by Martinon and Tschopp 2004) whereas caspase-4 was recently shown to be involved in inflammasome activation (Sollberger et al. 2012). A special case is the full-length caspase-12, known for increased susceptibility to severe sepsis. It is expressed only in a small population of African descent, while the human majority expresses a truncated, pro-domain-only form due to a single nucleotide polymorphism (Saleh et al. 2004).

Caspase-2, -8, -9 and -10 are known as apoptotic-inducing peptidases, which are activated in response to either extra- or intracellular apoptotic stimuli. The characteristic N-terminal CARD or two death effector domains (DED) are a prerequisite

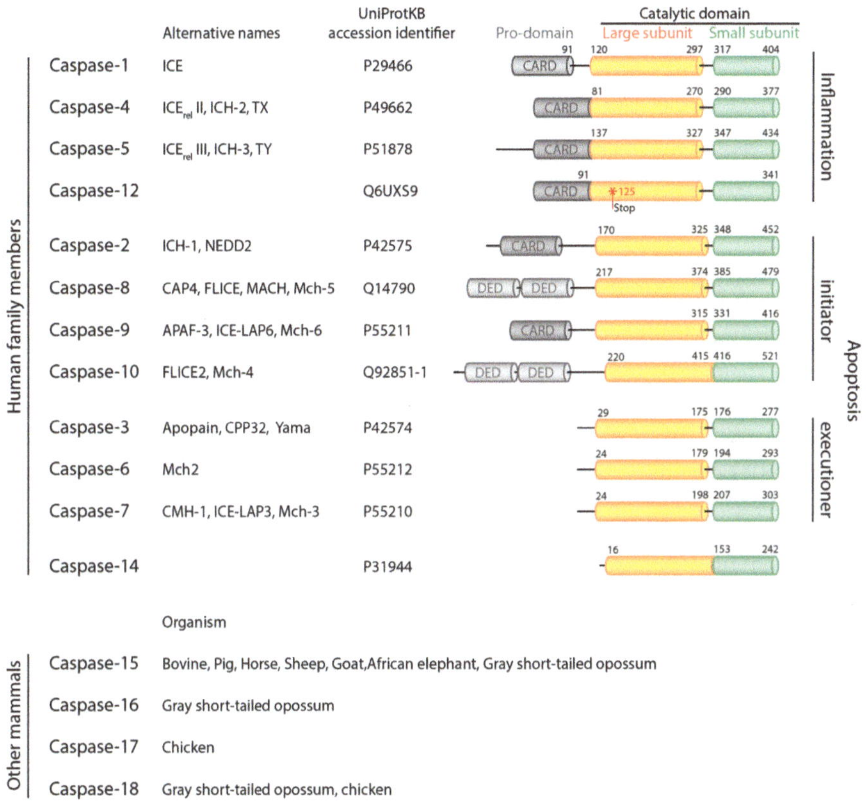

Fig. 8.1 The caspase family. The human caspase family members are ordered by their function into inflammatory and apoptotic initiator and executioner caspases. Caspase-14 is mainly expressed in keratinocytes and assigned to cornification processes. A human majority expresses a truncated form of caspase-12 due to a single nucleotide polymorphism, which leads to a stop codon at position 125 (indicated by *asterisk*). Numbers indicate terminal residues of domains. UniProtKB accession numbers and alternative names are listed for the human members. More mammalian caspase family members have been identified in other species. Caspase-11 and caspase-13 are orthologues of human caspase-4 and thus not listed

for initiator caspase activation. These domains control the oligomerization at activation platforms like the apoptosome and provoke autocatalytic activation by induced-proximity (Salvesen and Dixit 1999). In this manner, a pro-apoptotic signal can be translated into a proteolytic response.

Activated initiator caspases are responsible for the proteolytic maturation of the executioner caspases-3, -6 and -7. Executioner caspases are lacking the structured N-terminal domains and contain only a short N-terminal peptide sequence, which is thought to be a regulatory signal (Meergans et al. 2000; Denault and Salvesen 2003). In cells, these enzymes exist as heterotetrameric pro-forms and are activated by proteolysis of the N-terminal peptide and the inter-subunit linker. Once activated, they provoke the majority of proteolytic events leading to the known morphology of apoptotic cells.

Caspase-14 is a special member and mainly expressed in keratinocytes (reviewed by Denecker et al. 2008). Its function is assigned to cornification processes and mice experiments recently showed a relationship between caspase-14-deficiency and UVB-induced photodamage (Hoste et al. 2013).

8.3.1.2 Caspases in Mammals

Besides the well-characterized human members, numerous other homologues have been discovered in different mammalian species increasing the family members from caspase-1 to caspase-18 (Eckhart et al. 2008). Notably, murine caspase-11 and bovine caspase-13 are orthologues of human caspase-4 (Koenig et al. 2001; Eckhart et al. 2008) and thus named inconsistently with respect to the nomenclature guidelines of Alnemri et al. (1996).

A remarkable finding was published in 2011, when Kayagaki et al. (2011) discovered that the generally used caspase-1 knockout mouse (strain 129) is particularly a double knockout of caspase-1 and -11. A new caspase-1-only-deficient mouse strain depicts a crucial role of caspase-11 in the non-canonical activation of inflammation. Translated to humans, this suggests a similar role for caspase-4 and -5 and thus a specific function in incidents such as sepsis. The findings suggest a careful interpretation of the role of caspase-1 versus -11, which so far only has been determined experimentally using the mouse strain 129.

The evolutionary relationship of mammalian caspases has recently been studied. This revealed that the caspase family has gone through extensive changes with numerous gene deletions and duplications (Eckhart et al. 2008) (Fig. 8.1).

8.3.1.3 Non-mammalian Caspases and Metacaspases

Identification of the cell death related gene ced-3 in *C. elegans* was a landmark in the discovery of caspases (Ellis and Horvitz 1986; Yuan et al. 1993). CED-3 gets activated upon binding to the nematodic apoptosome formed by a CED-4 octamer and then executes the apoptotic pathway in C. elegans. A crystal structure of the formed complex between CED-3 and CED-4 revealed a funnel shaped octamer formed by CED-4, on which CED-3 is supposed to become activated, although no electron density could be observed for CED-3 in the structure (Qi et al. 2010).

Caspase homologues exist as well in insects. The initiator caspase Dronc is a caspase-9 ortholog in *Drosophila* and involved in apoptosis and cell proliferation (Steller 2008). Additional initiator (Dredd, Strica) and executioner (Drice, Dcp-1, Decay, Damm) caspases have been discovered in *Drosophila* (reviewed by Kumar and Doumanis 2000; Steller 2008). Recently, another Dredd homologue was characterized from an insect vector of human disease: *Aedis* Dredd (AeDredd) from the yellow fever mosquito *Aedis aegypti* (Cooper et al. 2007). This finding could be a first step to unravel the role of apoptosis in innate immune response of insects and in the particular case of mosquitoes to understand their exceptional vectorial capacity.

While caspases can be found in metazoan, distant homologues could be identified as well in protozoa, fungi and plants. These distantly related cysteine-dependent peptidases cleave after arginine or lysine and are called metacaspases (Tsiatsiani et al. 2011).

8.3.2 The Caspase Fold

The tetrameric caspase structure can be described as a globular fold made by two equivalent and symmetry-related catalytic domains with a α/β topology.

Each subdomain derives from a single polypeptide chain, encoding a N-terminal pro-domain or -peptide, and a large (17–20 kDa) and a small subunit (10–12 kDa). The N-terminal domains are separated from the caspase structure by an inter-domain linker, which can be cleaved during activation processes. Due to this flexible linker, the crystallization of a full-length caspase with its pro-domain was not successful and thus their hypothetical interactions remain unclear. However, several pro-domain homologues have been separately crystallized or investigated by NMR spectroscopy. These dead domains revealed a globular fold of six antiparallel α-helices with $\alpha1$–$\alpha5$ forming a conserved Greek key motive (Vaughn et al. 1999; reviewed by Kersse et al. 2011).

A tightly packed caspase catalytic domain roughly forms a cuboid with the approximate dimension of 25 Å × 35 Å × 42 Å. It is composed of a β-sheet, which is sandwiched between two layers of overall five α-helices. The twisted β-sheet involves five parallel strands ($\beta1$–$\beta5$) and one antiparallel strand ($\beta6$). The latter aligns with the $\beta6$ strand of the second catalytic domain in antiparallel manner. This results in a continuous β-sheet with 12 strands and together with ten α-helices in a twofold symmetry related structure (Fig. 8.2a).

8.3.3 The Active Site, Substrate Recognition and Cleavage Mechanism

The numbering of a specific amino acid in different caspase family members can diverge due to residue insertions or deletions. In general, a facilitated numbering based on caspase-1 sequence is used as proposed by Fuentes-Prior and Salvesen (2004).

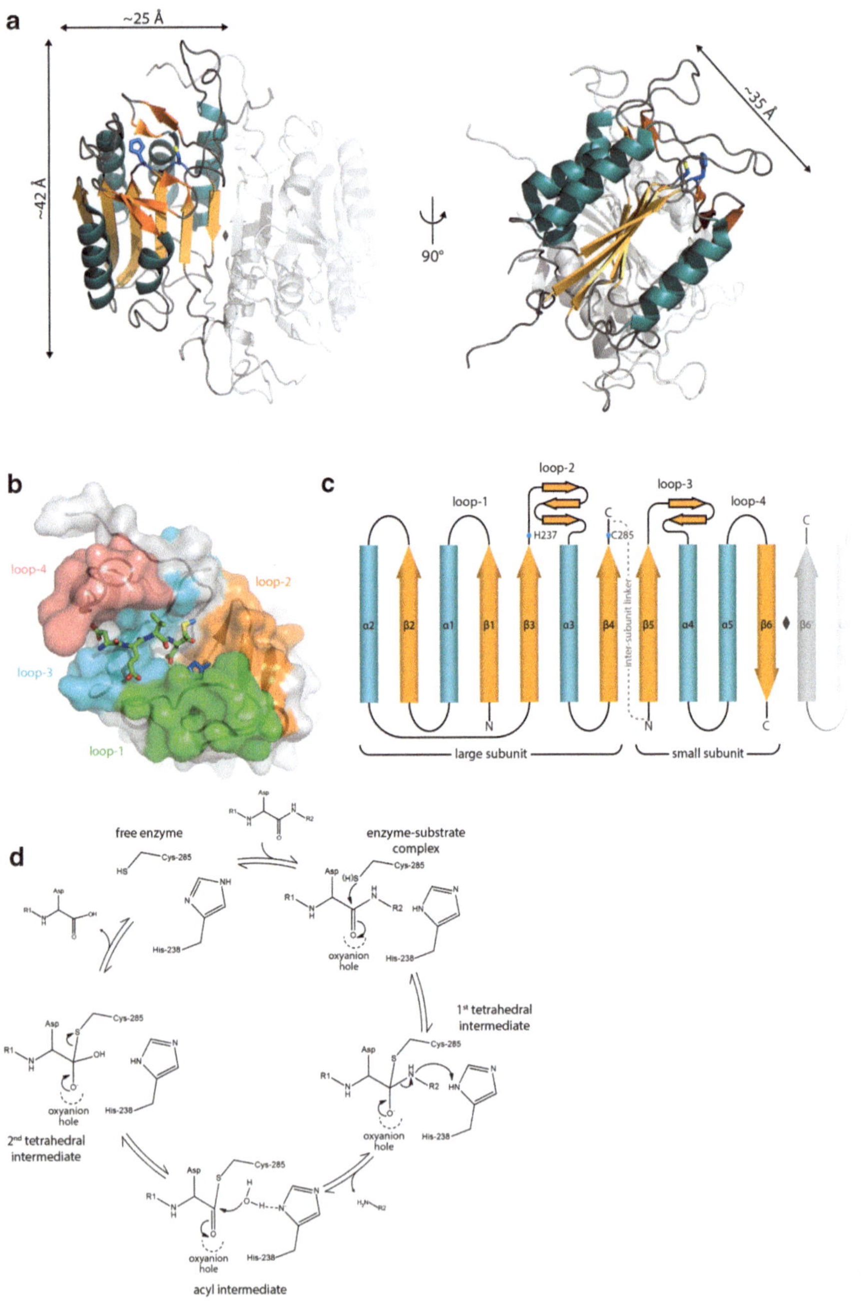

Fig. 8.2 The caspase fold. (**a**) Cartoon representation of caspase-3 crystal structure (2DKO) representing a dimer of two catalytic subunits (colored in *green/orange* and *gray*). Twelve central β-strands are sandwiched between ten α-helices. Active site residues His-237 and Cys-285 are colored in *blue* and located after strand β3 and β4, respectively. The five parallel β-strands of a catalytic domain points towards the substrate-binding cleft. A 90° rotation reveals a square-like arrangement of the α-helices. (**b**) Substrate-binding groove of caspase-3 (2DKO) covalently bound

8.3.3.1 Active Site Architecture and Substrate Recognition

The active site and binding groove of the catalytic domain is constructed by loops of its large and small subunits (Fig. 8.2b). The catalytic Cys-285 resides at the end of the central β-strand 4 whereas His-237 is located at the end of β-strand 3 (Fig. 8.2c). In a cartoon representation, the five parallel β-strands point towards the substrate-binding groove, with the tip of β-strand 3 and 4 indicating the position of the active site residues.

A strictly conserved Arg-179 positioned in front of helix α1 in loop-1 is crucial for the P_1 specificity. It shapes the aspartate specific binding pocket S_1 together with the conserved residues Gln-283 and Arg-341. In addition, loop-3 residue Arg-341 is involved in the formation of the binding subsite S_3 mediating main-chain-main-chain hydrogen bonds and interacting with the carboxylate group of a preferred glutamate at P_3 (Fuentes-Prior and Salvesen 2004).

The S_2 and S_4 pocket residues are less conserved. This leads to more substrate variability at P_2 and P_4 that can be observed directly by comparing small, peptide-based substrates (Roschitzki-Voser et al. 2012).

In inflammatory and initiator caspases, Val-338 of loop-3 shapes a large hydrophobic S_2 pocket that tolerates bulky side chains like His or Thr as P_2 residue. In contrast, in executioner caspases Val-338 is substituted by Tyr leading to a preference for small aliphatic substrate residues such as Val or Ala. A similar effect is observed at the S_4 subsite, where a bulky Trp-348 in apoptotic caspases shapes a small S_4 pocket with an increased affinity for branched aliphatic residues or aspartate at position P_4 (Fuentes-Prior and Salvesen 2004). A substitution in caspase-1 (Trp-348 by Val) or caspase-4 and -5 (Trp-348 by Ile) reshapes the S_4 subsite that favors large and hydrophobic residues.

A S_5 pocket could only be characterized in caspase-2 with its preferential accommodation of a small hydrophobic substrate residue (Fuentes-Prior and Salvesen 2004). Furthermore, residues after the scissile bond do not require special properties to be accommodated in the less restrictive primed subsite. An exception is the S_1' pocket that mildly discriminates for charged or bulky residues (Timmer and Salvesen 2007).

In summary, a consensus caspase substrate sequence can be defined as (D or W)-E-X-D ↓ ϕ with P_4-P_3-P_2-P_1 ↓ P_1' where Asp at position P4 is preferred by caspases of the apoptotic subfamily, X can be substituted by any amino acid and ϕ stands for a small uncharged residue (Timmer and Salvesen 2007). Today, a large number of caspase substrates are commercially available with a proposed high specificity for a single caspase family member. When looking at the consensus substrate sequence,

Fig. 8.2 (continued) to substrate-analogue inhibitor DEVD-chloromethylketone. Binding cleft forming loops are colored in *green* (loop-1), *orange* (loop-2), *blue* (loop-3) and *red* (loop-4). (**c**) Topology map of caspases colored according to crystal structure shown in (**a**). Active site residues His-237 and Cys-285 are indicated. Loop-2 is also known as 240-loop. (**d**) Catalytic mechanism according to Fuentes-Prior and Salvesen (2004) it is evident that such specificity is difficult to achieve and in fact could not be observed experimentally (McStay et al. 2008). This effect is not only observed for specific caspase substrates but also for peptide inhibitors, which are designed based on the consensus sequence and thus are not feasible for a highly specific caspase inhibition.

8.3.3.2 Catalytic Mechanism

Fuentes-Prior and Salvesen (2004) proposed a first catalytic mechanism in 2004. The authors suggested a mechanism with close similarity to the proteolytic reaction of serine peptidases (Fig. 8.2d). Their mechanism is based on crystal structures with active site covalently bound inhibitors mimicking a tetrahedral intermediate.

In a hypothetical enzyme-substrate complex, the P_1 carbonyl oxygen is oriented by hydrogen bonds towards the oxyanion hole, which is formed by backbone amine groups of the conserved residues Gly-238 and Cys-285. A nucleophilic attack of the Cys-285 sulfur atom on the carbonyl carbon forms a tetrahedral intermediate with covalently bound substrate. This state is stabilized by the imidaloze moiety of His-237, which protonates the α-amino group of the P_1' residue and thus triggers the release of the C-terminal peptide product from the S_1' subsite. The formed acyl intermediate is subsequently hydrolytically processed. The deprotonated π-nitrogen of His-237 abstracts a proton from a water molecule, which itself attacks the thioester carbonyl. This forms a second tetrahedral intermediate followed by a cleavage of the covalent bond formed between the carbonyl carbon and the sulfur of Cys-285. The non-covalently bound N-terminal peptide product is released and the regenerated enzyme starts a new catalytic cycle.

8.3.4 Activation in Death Pathways

Caspases are expressed as zymogens that undergo proteolytic activation in a highly regulated manner. Monomeric initiator and inflammatory caspases are activated at macromolecular platforms formed by adapter proteins. According to the induced-proximity model, the recruitment to these platforms results in a dimerization of two catalytic subunits followed by activation and autocatalytic processing (Salvesen and Dixit 1999). In contrast, the executioner caspases exist as intracellular dimers of two catalytic subunits and are activated upon cleavage by inducer or as well active executioner caspases without the requirement of additional adapter proteins.

8.3.4.1 Caspase-Dependent Intrinsic Apoptosis: The Apoptosome

Formation of the macromolecular complex known as the apoptosome is the pivotal point of caspase activation in intrinsic apoptosis (Fig. 8.3a). MOMP-mediated release of cytochrome c from the mitochondrial IMS into the cytosol triggers the assembly of Apaf-1 into a wheel-like heptamer (Yuan et al. 2010).

The adapter protein Apaf-1 consists of a N-terminal CARD, a nucleotide binding and oligomerization domain (NOD) and a C-terminal regulatory region with two domains (WD1 and WD2). Its fundamental role in apoptosome formation relies on the Apaf-1 oligomerization with a molecular mechanism proposed based on the crystal structure of a full-length murine Apaf-1 protein (Reubold et al. 2011). Binding of cytochrome c to the regulatory region of monomeric Apaf-1 releases the attached NOD, which rotates upon binding of ATP to expose its oligomerization areas and thereby relocates the now accessible CARD. This conformational change leads to the assembly into a circular structure of seven Apaf-1 subunits with central NOD and propeller-like regulatory regions (Reubold et al. 2011). The N-terminal CARDs form a recruitment disk on top of the protomer, accessible for the targeted pro-caspase-9.

Recruitment of pro-caspase-9 to the apoptosome via homotypic interactions between the Apaf-1 and the pro-caspase-9 CARDs results in dimerization of two catalytic domains and activation of the enzyme by cleavage (Pop et al. 2006). Notably, inter-subunit linker cleavage is not necessary for activity (Stennicke et al. 1999) but may provide stability to the active dimer (Fuentes-Prior and Salvesen 2004). Thus, the Apaf-1 based apoptosome can be defined as a cofactor of caspase-9 (Pop et al. 2006). As a last step, limited proteolysis of downstream executioner caspases by caspase-9 promotes the execution of apoptosis.

8.3.4.2 Extrinsic Apoptosis: Death Inducing Signaling Complex (DISC)

Numerous extracellular death stimuli provoke the intracellular formation of a macromolecular recruitment platform for caspase-8 or -10, known as the death inducing signaling complex (DISC) (Kischkel et al. 1995). Initially, a death ligand binds to its receptor and induces oligomerization (Fig. 8.3b). A prominent and well-studied example is the receptor Fas (also known as APO-1 or CD95), which binds to Fas ligand and thereby forms a trimeric complex (Kischkel et al. 1995). TRAIL receptors (death receptor 4 and 5) are other examples that form a homotrimer upon binding to their ligand (Chaudhary et al. 1997; Pan et al. 1997; Walczak et al. 1997). Both receptor types are members of the death receptor family with its superfamily of TNF receptors and trigger the intracellular DISC formation.

Death receptor oligomerization results in a conformational change that exposes the intracellular death domains (DD) of the receptor as structurally shown for Fas (Scott et al. 2009). Then, homotypic interactions of the accessible DDs with the DDs of the adaptor protein FADD build the recruitment platform for the apical caspases.

FADD is a cytosolic protein with a N-terminal DED and a C-terminal DD (Chinnaiyan et al. 1995). The crystal structure of the protein-complex between DDs from Fas and FADD revealed a circular assembly of 5–7 Fas-DDs and 5 FADD-DDs, which optimally orients the FADD-DEDs for caspase recruitment

(Wang et al. 2010). Additionally, both recent structural studies on DISC formation described the observation of higher oligomeric clusters and denote at least two trimeric Fas receptors as a prerequisite for signal transduction (Scott et al. 2009; Wang et al. 2010).

The complete DISC is formed upon the recruitment of pro-caspase-8 or -10 via homotypic interactions of their N-terminal DED to FADD-DEDs. This leads to an increased local pro-caspase concentration, a dimerization of two catalytic domains by induced proximity and enzymatic activity of the caspase dimer (Salvesen and Dixit 1999; Pop et al. 2007).

Caspase-8 is autocatalytically processed in the inter-subunit linker to stabilize the active dimeric conformation (Pop et al. 2007) and cleavage of the pro-domain linker allows the release of the peptidase into the cytosole (Martin et al. 1998).

Structural investigations on the unprocessed catalytic domain of caspase-8 by NMR spectroscopy visualized a highly flexible inter-subunit linker and an activation induced rearrangement of loop-1 and -3 in comparison to the X-ray structure of an active dimer (Keller et al. 2009). Included mutational studies stressed the combination of proteolytic cleavage and dimerization as a prerequisite for full activity.

Further studies on caspase-8 using a reconstituted DISC revealed a substrate specificity switch between pro- and active caspase-8 (Hughes et al. 2009). Initial dimerization of pro-caspase-8 induces activity for a limited substrate range, including autoproteolysis of pro-caspase-8 or cleavage of c-FLIP, a caspase-related regulatory protein involved in death survival signaling (Scaffidi et al. 1999; Micheau et al. 2002). Autocatalytic processing then forms the fully active enzyme with the complete substrate repertoire, including the executioner caspases and Bid that are responsible for the downstream signaling in apoptosis.

The activation mechanism for caspase-10 is similar to caspase-8 initiation (Wachmann et al. 2010). Induced proximity results in the formation of an active dimer and the further autocatalytic processing accelerates the cleavage of downstream proteins. However, the authors could show that Bid cleavage does not require processed caspase-10 in contrast to tBid production by caspase-8. This finding suggested a pro-apoptotic role of this enzyme in the uncleaved form and further research will be needed to clarify this function.

8.3.4.3 Pyroptosis: The Inflammasome

The activation of inflammatory caspases remained elusive until 2002 with the discovery of a molecular activation platform termed the inflammasome (Martinon et al. 2002). This intracellular protein complex triggers the activation of caspase-1 and -5 and is formed by NLRP1 and an adapter protein ASC (Fig. 8.3c). Subsequently, researchers identified more NLR family members that are involved in inflammasome formation, like NLRP3 (Agostini et al. 2004), NLRC4 (Mariathasan et al. 2004) and others (reviewed by Rathinam et al. 2012).

NLR proteins act as the danger-sensing molecule in the inflammasome and contain three important domains: A leucine-reach repeat (LRR) with an assumed function of danger signal recognition similar to Toll like receptors (Martinon et al. 2002), a nucleotide-binding and oligomerization domain NACHT and a death mediating domain, which is either a PYD in NLRP1 and NLRP3 or a CARD in NLRC4. Although the initiation of inflammasomes remains unclear, Martinon and Tschopp (2005) proposed a possible mechanism: Binding of a danger-associated ligand to LRR induces a conformational change that exposes the NACHT domain and thereby triggers oligomerization and recruitment of adapter proteins and inflammatory caspases.

The activation of pro-caspase-1 by NLRP1- or NLRP3-inflammasomes involves the adapter protein ASC, which contains a N-terminal CARD and a C-terminal PYD. The solution structure of full-length ASC visualized the two isolated domains connected via a flexible linker in a back-to-back orientation (de Alba 2009). This may prevent steric hindrance between the two domains and raises the capture radius for pro-caspases and NRL molecules. In addition, NLRP1 contains a C-terminal CARD that is able to directly recruit pro-caspase-1 although the presence of ASC increases caspase-1 activation (Faustin et al. 2007). A similar observation has been reported for the NLRC4 inflammasome, which can directly recruit pro-caspase-1 via its N-terminal CARD (Miao et al. 2010) but showed an ASC-dependent increase of activation (Mariathasan et al. 2004; Miao et al. 2010).

Insights into the activation mechanism of inflammatory caspases have been obtained with the crystal structure of pro-caspase-1 (Elliott et al. 2009). The inter-subunit linker of one chain occupies the central cavity of the zymogen dimer and locates its residue Asp-297 in close proximity to the active site of the other chain. After a possible interdimeric cleavage at Asp-297, the peptide chain is released and the central cavity can then harbor the other inter-subunit linker to trigger a second interdimeric processing. Additional autoproteolysis at a second cleavage site (Asp-316) then stabilizes the active dimer by conversion of an α-helix from one chain into an intradimer β-sheet (Elliott et al. 2009). These findings support the induced proximity model (Salvesen and Dixit 1999) for inflammatory caspases and emphasizes the crucial role of macromolecular oligomerization platforms for dimerization.

8.3.4.4 Caspase-2 Activation: The PIDDosome

The activation platform for caspase-2 was termed PIDDosome after the identification of the involved proteins by Tinel and Tschopp (2004). This large protein complex is formed by oligomerization of PIDD (p53-induced protein with a death domain) and the adapter protein RAIDD (Fig. 8.3d).

PIDD consists of a N-terminal LRR domain, two central ZU-5 domains and a C-terminal DD. Autoproteolysis between or after the two ZU-5 domains has been observed as a switch for pro-survival or pro-death signaling (Tinel et al. 2007). The adapter protein RAIDD, a N-terminal CARD and a C-terminal DD, is responsible

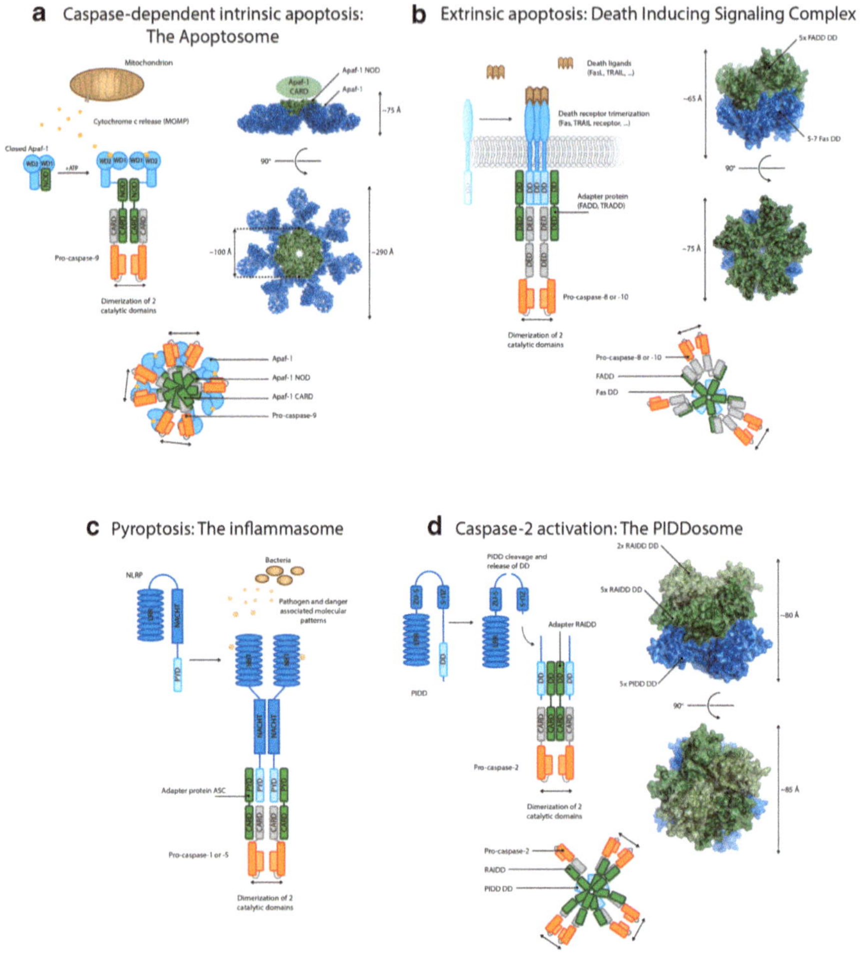

Fig. 8.3 Caspase activation platforms. Inflammatory and apoptotic initator caspases are activated at macromolecular complexes, which induces dimerization and autoproteolytic cleavage. (**a**) Apoptosome formation: released cytochrome c from the mitochondrion binds to Apaf-1 and induces a conformational change that reorients its NOD and CARD. Seven Apaf-1 molecules then form a wheel-like heptamer, which have been analyzed by electron microscopy (3IZA). The structure revealed a CARD-CARD interaction between the Apaf-1 and caspase-9 that form a disc above the central hub. (**b**) Death inducing signaling complex: activation of caspase-8 or -10 occurs at the DISC, which is initiated by trimerization of death receptors. Trimeric receptors recruit pro-caspase-8 or -10 via adapter proteins (FADD, TRADD). The assembly of FAS/FADD DDs have been structurally characterized (3OQ9) and showed a pentameric arrangement of 5–7 Fas DD and 5 FADD DDs. First, this structure shows a ring-like arrangement of the DDs similar to the Apaf-1 arrangement. Second, it indicates a requirement of more than one trimeric Fas receptor to initiate the DISC formation. (**c**) Proposed mechanism of inflammasome formation: NLRP as well as NLRC proteins exist in a conformation, where its NACHT domain is blocked. Pathogen and danger associated molecular patterns can be recognized by the LRR, which triggers a conformational change leading to a freed NACHT domain and subsequent oligomerization. Adapter proteins like ASC are then recruiting caspase-1 or -5 for activation. (**d**) PIDDosome formation: sequential cleavage of PIDD releases its DD, which recruits pro-caspase-2 via the adapter protein RAIDD. The crystal structure of PIDD-RAIDD DD assembly (2OF5) revealed a ring like structure

for caspase-2 recruitment via CARD-CARD interactions. Besides its involvement in the PIDDosome, RAIDD has been observed in specific binding of RIP1, a serine/threonine protein kinase involved in death pathway (Duan and Dixit 1997).

The crystal structure of a complex of the DDs that are involved in PIDDosome formation discovered a stoichiometry of seven RAIDD DDs and five PIDD DD molecules (Park et al. 2007). Five RAIDD DDs are stacked in a ring-like structure on top of an asymmetric, pentameric ring of PIDD DDs. Two additional RAIDD DD molecules are located in a third level on top of the RAIDD DD ring-like structure. In principal, this complete assembly is able to recruit seven pro-caspase-2 molecules for activation although only three active enzymes can be formed.

8.3.4.5 Activation of Executioner Caspases

Executioner caspases exist as stable dimers in solution in contrast to the zymogens of the inflammatory and initiator related family members. Activation occurs upon partial proteolysis by initiator or executioner caspases and results in a rearrangement of the inter-subunit linker (Riedl et al. 2001a).

The pro-caspase-7 crystal structures revealed an asymmetric blocking of the central cavity by the inter-subunit linker (Chai et al. 2001b; Riedl et al. 2001a). Linker cleavage at Asp-297 promotes its release from the central cavity. It then stabilizes the active conformation via bundle formation, which is promoted between the new C-terminus of the large subunit of one catalytic domain and the newly formed N-terminus of the small subunit of the other adjacent catalytic domain. This bundle is further interacting with loop-4 and enables the formation of the active site as shown in peptide-bound caspase-7 structures (Riedl et al. 2001a; Fuentes-Prior and Salvesen 2004).

The important role of the central cavity in caspase activation is not only depicted by the zymogen structures of caspase-1 and -7 (Chai et al. 2001b; Riedl et al. 2001a; Elliott et al. 2009) but also by the identification of allosteric inhibitors of caspase-3 and -7 (Hardy et al. 2004). These inhibitors bind in the central cavity and trap the protease in a zymogen-like conformation.

Fig. 8.3 (continued) with five RAIDD DD (*dark green*) stacked on top of five PIDD DD (*blue*). Two additional RAIDD DD (*light green*) are located on a third level. (**a–d**) The available complex structures of these oligomeric activation platforms point out a stoichiometric asymmetry: in principal, one of the pro-caspase monomers will not dimerize due to the odd-numbered adapter proteins

8.3.5 Caspase Substrates

The commercially available short peptide substrates have been derived from a combinatorial approach by scanning of large peptide libraries (Thornberry et al. 1997) unraveling the caspase substrate consensus. These small molecule substrates are coupled to a fluorophore reporter and have generally been used for caspase characterization *in-vitro* (Roschitzki-Voser et al. 2012).

Diverse methods to identify *in-vivo* cleavage sites have been developed in the field of degradomics, like scanning of the transcriptome using bioinformatics (Boyd et al. 2005) as well as cell-based degradation methods, recently reviewed by Agard and Wells (2009). A growing list of identified *in-vivo* caspase substrates is available in the MEROPS database [http://merops.sanger.ac.uk/ (Rawlings et al. 2012)] with almost 900 physiological relevant entries. Undoubtedly, some of these entries are cleaved only in specific cell types or not explicitly in cell death pathways while others are well known for being processed during apoptosis. Proteolysis of those proteins typically leads to a loss-of-function or a gain-of-function. While the executioner caspases are prominent examples for the latter, several other important substrates have been described.

8.3.5.1 Rho-Associated Kinase I (ROCK I)

The family of Rho GTPases is known as an actin cytoskeleton regulator with the Rho-associated kinases (ROCK) as their effectors (Hall 1998). These serine/threonine kinase isoforms are activated upon interaction with GTP-bound Rho proteins.

During apoptosis, caspases cleave ROCK 1 at a DETD ↓ G motive near the C-terminus and thereby remove a putative autophosphorylation/auto-inhibitory domain (Coleman et al. 2001). The result is an intrinsically more active kinase, which promotes membrane blebbing in apoptotic cells by phosphorylation of downstream targets. It was further shown that a truncated ROCK 1 is independent of Rho GTPases and triggers membrane blebbing without activated caspases.

8.3.5.2 Caspase-Activated DNAse (CAD)

A morphological feature of apoptosis is the degradation of chromosomal DNA observable as a characteristic DNA ladder. A crucial nuclease for this process has been identified as the caspase-activated DNAse (CAD, also known as DNA fragmentation factor DFF40) (Nagata et al. 2003). In cells, this enzyme forms an inactive dimeric complex with the inhibitor of CAD (ICAD, also known as DFF45).

The cleavage sites DETD ↓ S and DAVD ↓ T are located in two flexible interdomain linkers of CAD and get cleaved by caspase-3. This triggers the release from ICAD and a formation of active CAD dimers. Notably, CAD is not only responsible for apoptotic degradation of nucleosomal DNA, it also promotes cell differentiation induced by DNA strand breaks (Larsen et al. 2010). Thus, CAD is an excellent example for the caspase-3 functional dualism in cell death execution and cell proliferation.

8.3.5.3 Poly(ADP-Ribose) Polymerase (PARP)

First notices of cleavage of poly(ADP-ribose) polymerase (PARP) have been described in chicken cells in 1994 (Lazebnik et al. 1994). PARP is involved in repair mechanisms of single and double strand DNA breaks. It forms branched poly(ADP-ribose) molecules, attaches them to acceptor proteins, including itself, leading to the recruitment of repair proteins to the DNA breakage site (Javle and Curtin 2011).

The executioner caspase-3 and -7 cleave the 116 kDa PARP after a canonical sequence DEVD ↓ G into an 85 kDa fragment having no enzymatic activity. This loss-of-function prevents the repair of DNA fragments induced by CAD. Remarkably, hyperactivation of PARP can lead to a low ATP level inside cells that is a known trigger of regulated necrosis. In this prospect, the loss-of-function supports not only the apoptotic pathway but also prevents the activation of necrotic cell death and inflammatory responses.

Furthermore, PARP cleavage is a good example for exosite interactions between a protease and its target protein. It could be shown that PARP, which has been modified with long branched poly(ADP-ribose) molecules, is cleaved with higher efficiency by caspase-7 than by caspase-3 (Germain et al. 1999) although the latter cleaves short peptide substrates more efficiently. The enhanced affinity to caspase-7 was assigned to exosite interactions between the caspase-7-p20 and the branched poly(ADP-ribose) molecule.

8.3.5.4 Human Telomerase Reverse Transcriptase (hTERT)

The human telomerase hTERT maintains the length of DNA telomeres by its reverse transcriptase activity and ensures chromosomal stability. The reverse transcriptase as the catalytic domain of the protein was recently identified as a unique caspase-6 and -7 substrate that is cleaved during apoptosis (Soares et al. 2011). Four cleavage sites have been identified with TVTD ↓ A and VNMD ↓ Y cleaved by caspase-6 and TSLE ↓ G and PKPD ↓ G cleaved by caspase-7. These cleavage sites are conserved in old world monkeys and apes although the function of the persistent fragments during apoptosis remains unclear. hTERT is a recent example for a non-canonical substrate recognition of caspases.

8.3.5.5 Pro-interleukins and Interleukins

The starting point of caspase research was the identification of the ICE, today known as caspase-1 (Thornberry et al. 1992). Indeed, interleukin precursors and their active forms are caspase substrates that either provoke inflammatory responses or prevent inflammation during apoptosis.

For example, bacterial infections of cells can lead to caspase-mediated processing of pro-interleukin-1β (YVHD ↓ A) and -18 (LESD ↓ Y) with followed secretion of these activated cytokines. Caspases-1 and -4 are able to process the pro-forms (Thornberry et al. 1992; Kamens et al. 1995; Akita et al. 1997; Gu et al. 1997) and have been reported to be involved in the inflammasome-mediated response (Sahoo et al. 2011; Sollberger et al. 2012).

Remarkably, pro-interleukins are not only activated by inflammatory caspases but also by apoptotic members as reported for pro-interleukin-16 activation (PKPD ↓ G) by caspase-3 (Zhang et al. 1998). Pro-interleukin-16 is expressed in more than 90 % of all T cells (Chupp et al. 1998), is located in the cytoplasm and the nucleus, while a nuclear presence inhibits progression (Center et al. 2004). Thus, proteolytic modification of such a cell cycle regulator by caspase-3 illustrates the important role of these proteases in other molecular mechanisms.

In contrast to other cytokines, interleukin-33 is expressed in its active form and is not secreted after inflammatory stimuli (Lüthi et al. 2009). It is more likely to be released after membrane disruption during necrosis, while apoptotic cell death leads to interleukin-33 deactivation mediated by caspase-3 and -7 after cleavage at DGVD ↓ G (Lüthi et al. 2009). In principal, the proteolysis of interleukin-33 by apoptotic executioner caspases acts as an intracellular proinflammatory extinguisher in a tissue protective manner.

The enormous variety of caspase substrates illustrates the important roles of these enzymes in cell death, proliferation and inflammatory pathways. Although, a consensus sequence has been defined (Timmer and Salvesen 2007), natural substrates can as well be cleaved at non-canonical sites as demonstrated in rare cases (Krippner-Heidenreich et al. 2001; Soares et al. 2011).

8.3.6 Morphological Changes upon Activation

Activation of cellular death pathways results in major morphological changes that can be observed under the light microscope. Biochemical techniques have been used to characterize death pathways (Galluzzi et al. 2012). Particular morphological features (reviewed by Ziegler and Groscurth 2004) are well characterized and especially the apoptotic morphological hallmarks can be explained by caspase-mediated activation of effector proteins.

Nuclear condensation and DNA fragmentation during apoptosis relies on executioner caspases. On one hand, the cleavage of nuclear lamins by caspase-3 and -6 induces a loss of structural integrity and a nuclear shrinkage, which promotes a collapse into smaller nuclear particles in a late apoptotic state (Rao et al. 1996) (Fig. 8.4). On the other hand, CAD is activated by caspase-3 and produces DNA fragments of 180- to 200-bp length, visible by gel electrophoresis (Nagata et al. 2003; Larsen et al. 2010). In addition, cleavage of the DNA repair enzyme

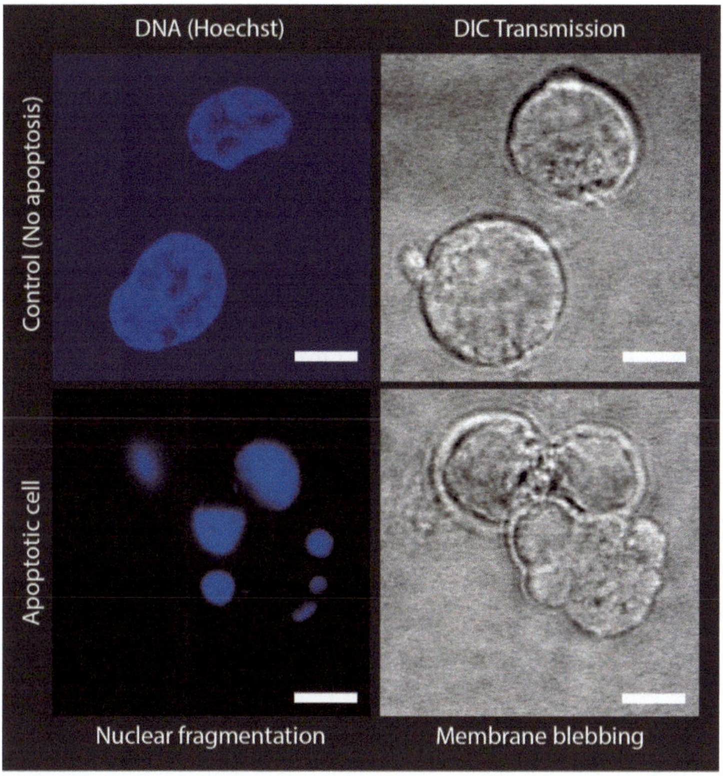

Fig. 8.4 Morphology of apoptosis. Jurkat cells were growing in media containing etoposide (100 μM, *lower panels*) or without etoposide (*upper panels*). Paraformaldehyde fixation was performed after 6 h. Etoposide treated cells show condensed and fragmented nuclei while untreated cells possess round shaped nuclei (*left panels*, DNA staining with Hoechst33342). Transmission images show round shaped healthy cells, whereas etoposide treatment leads to apoptotic cells displaying membrane blebbing (*right panels*)

PARP by caspase-3 and -7 (Germain et al. 1999) prevents the repair response thus further provoking apoptosis.

ROCK 1 cleavage by executioner caspases leads to an intrinsically more active kinase that phosphorylates downstream targets in the actin-myosin system (Coleman et al. 2001). Actin-myosin related cell contractility initiates the formation of membrane protrusions commonly known as membrane blebs (Fig. 8.4). Further progression of cell shrinkage separates these blebs into apoptotic bodies, packed with organelles and nuclear fragments, which are subsequently engulfed by surrounding cells via phagocytosis (Ziegler and Groscurth 2004).

Other phosphorylation targets in the Rho-ROCK signaling pathway are flippases, which maintain the membrane asymmetry of phosphatidylserine (PS) that is located within the inner membrane leaflet of a normal cell (Krijnen et al. 2010). In an early stage of apoptosis, PS is presented at the outer membrane as an "eat-me" signal for phagocytes (Grimsley and Ravichandran 2003). First studies on oxidatively stressed erythrocytes indicated caspase-related PS externalization (Mandal et al. 2002) while research on cardiomyocytes identified a Rho-ROCK signaling to be crucial for PS exposure (Krijnen et al. 2010).

8.3.7 Regulation and Specific Inhibition

A tight regulation of caspase activation is essential to avoid accidental induction of cellular demise. Thus, a complex network involving numerous proteins has been evolved, which in particular cases guarantees the survival or decease of a cell (Fig. 8.5). Most of those proteins belong to the Bcl-2 family or to the inhibitor of apoptosis protein (IAP) family. In addition, released mitochondrial proteins upon MOMP can interact with members of these two families and facilitate the progression of apoptosis.

Members of the Bcl-2 family are known to have cell regulatory functions and often possess either pro- or anti-apoptotic activity, depending on the environmental conditions (reviewed by Hardwick et al. 2012). Bcl-2 proteins (e.g. Bcl-2, Bcl-B, Bax, Bak,. . .) consist of 3–4 Bcl-2 homology (BH) domains enumerated from BH1 to BH4. Members of this family (e.g. Bim, Bid, Puma,. . .) contain only one BH3 domain, are thus called BH3-only proteins and known as promoters of apoptosis. Certain family members contain an additional C-terminal transmembrane domain (e.g. Bcl-2, Bax, Bim,. . .) that localizes them at membranes. Bcl-2 proteins exist in the cytosol, at outer mitochondrial membranes and supposedly inside mitochondria, however this submitochondrial localization remains controversial (Hardwick et al. 2012).

The IAP family members are multi domain proteins that directly suppress apoptosis by caspase inhibition (Deveraux et al. 1997) or indirectly via interaction with pro-apoptotic signaling complexes (Fulda and Vucic 2012). In addition, several members contain a C-terminal RING domain that functions as an E3 ubiquitin ligase with increasing significance in cell survival (Vaux and Silke 2005; Vucic et al. 2011).

Intracellular apoptotic stimuli can activate BH3-only proteins, which promotes Bax and Bak oligomerization and pore formation at the outer mitochondrial membrane (Kuwana et al. 2002). This leads to the release of cytochrome c and other mitochondrial proteins that function in a pro-apoptotic manner. The second mitochondria-derived activator of caspases (Smac, also termed DIABLO) is known to bind and inhibit IAP (Verhagen et al. 2000) but can itself be antagonized by IAPs (Vucic et al. 2005). Another released protein is HtrA2 (also known as Omi), a serine protease that is involved in apoptosis and suppression of IAP (Verhagen et al. 2002). In addition, caspase-8 can cleave the BH3-only protein Bid, forming a truncated form tBid and thus integrates the extracellular apoptotic stimulus by inducing MOMP.

Furthermore, phosphorylation of caspases and upstream adapter proteins has been reported as regulatory mechanisms in cell death pathways (Kitazumi and

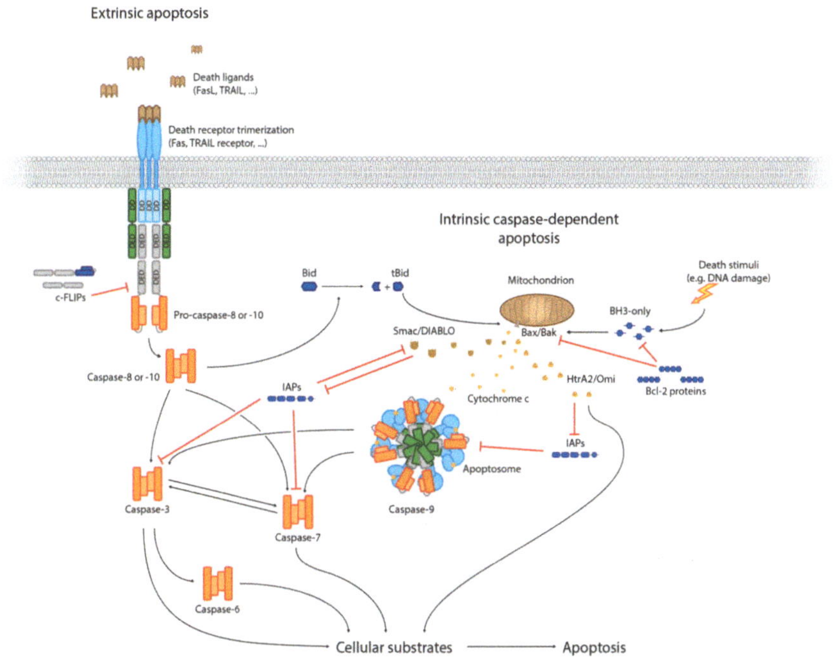

Fig. 8.5 Regulatory network. Activation of caspase-8 or -10 at the DISC can be repressed by cFLIPs. Once activated, these caspases cleave downstream executioner caspases that trigger the apoptotic pathway. In addition, caspase-8 or -10 can produce tBid that stimulates the intrinsic caspase-dependent apoptotic pathway. This pathway is also triggered by intracellular death stimuli like DNA damage and activates pro-apoptotic Bcl-2 proteins, known as the subfamily BH3-only proteins. These proteins are regulated by anti-apoptotic Bcl-2 family members that provide a first threshold for death stimuli. The BH3-only proteins Bax and Bak possess a pore forming activity to induce MOMP. Released mitochondrial proteins then trigger the progression of the intrinsic apoptotic pathway: cytochrome c binds to Apaf-1 and induces apoptosome formation and caspase-9 activation. Released Smac/DIABLO is binding and inhibiting IAPs and HtrA2/Omi not only antagonizes IAPs but also cleave cellular substrates. Additional important regulators of apoptosis are the IAPs: they specifically inhibit caspases and are able to counteract with Smac/DIABLO in anti-apoptotic manner

Tsukahara 2011). For instance, caspase-9 can be phosphorylated at the catalytic domain residue Ser-196 and in the CARD at Thr-125, which both prevents caspase-9 activation (Cardone et al. 1998; Allan et al. 2003). Phosphorylation of caspase-8 at Ser-364 and caspase-3 at Ser-150 has been reported as survival signals during neutrophil apoptosis (Alvarado-Kristensson et al. 2004). Other phosphorylation sites on caspase-8 have been discovered (Tyr-310, Tyr-397, Tyr-465) and revealed a dynamic post-translational regulation of apoptosis in neutrophils (Jia et al. 2008). These findings strongly emphasize the important control functions of kinases and phosphatases in death pathways.

8.3.7.1 X-Linked Inhibitor of Apoptosis Protein (XIAP)

A well-studied IAP family member is the X-linked IAP (XIAP) that is able to directly inhibit caspases-3, -7 and -9 (Deveraux et al. 1997, 1998). It is characterized by three baculoviral IAP repeat (BIR1–3) modules, an ubiquitin associated (UBA) domain and a C-terminal RING domain. Structural investigations on XIAP binding to caspase-3, -7 and -9 unraveled two different binding and inhibition mechanisms, which are mainly related to binding of linker regions near the BIR2 and BIR3 domains, respectively (Chai et al. 2001a; Riedl et al. 2001b; Shiozaki et al. 2003). In addition, XIAP can simultaneously bind and inhibit both, initiator caspase-9 and executioner caspase-3 (Bratton et al. 2002) at a tightly packed "holo-apoptosome" that not only recruits pro-caspase-9 but also binds active caspase-3 (Yuan et al. 2011).

Crystal structures of caspase-3 and -7 have been reported with a truncated XIAP containing a short N-terminal linker peptide and the BIR2 domain (Fig. 8.6a). Remarkably, both homologue caspases showed the same principle of inhibition: The XIAP residues involved in binding interactions are located in the peptide linker, which lies inside the substrate-binding cleft of the enzyme in opposite direction when compared to the observed natural binding of substrate (Riedl et al. 2001b). Furthermore, the S_4 pocket is occupied by Asp-148 of XIAP whereas upstream residues sterically block the access to the S1 and S1′ subsites and thus competitively inhibit the enzyme.

The complex of caspase-9 with the XIAP-BIR3 domain revealed a completely different inhibition mechanism: The BIR3 domain binds at the caspase-9 dimer interface of the catalytic domain and consequently traps caspase-9 in an inactive, monomeric form thus preventing active site formation (Shiozaki et al. 2003) (Fig. 8.6b). This interaction is further provoked by the N-terminal peptide of the small subunit of caspase-9, which interacts with BIR3 at the so-called Smac pocket (Shiozaki et al. 2003).

As other IAPs, XIAP can regulate the apoptotic pathway via its C-terminal RING domain. Ubiquitination of caspase-3 mediates proteosomal degradation (Suzuki et al. 2001) while RING domain truncations of XIAP lead to elevated caspase-3 activity (Schile et al. 2008). In contrast, ubiquitination of AIF is nondegradative but also attenuates its death-inducing activity (Lewis et al. 2011).

8.3.7.2 c-FLIP: A Structural Homologue of Caspase-8

The family of FLICE-inhibitory proteins (FLIPs) has been discovered in viral genomes (termed vFLIPs) and supports viral infections by down regulation of cellular death responses (Thome et al. 1997). In humans, numerous splice variants of cellular FLIPs (c-FLIPs) have been reported whereas three isoforms are expressed (Krueger et al. 2001; Golks et al. 2005; Safa 2012). The two known short forms, c-FLIP$_S$ and c-FLIP$_R$, consist of two DEDs that can be recruited to the

DISC where they block binding of caspase-8 and thus prevent a death-receptor mediated apoptosis (Krueger et al. 2001; Golks et al. 2005).

A longer isoform c-FLIP$_L$ is structurally similar to caspase-8 and comprises two DED and a C-terminal caspase-like domain lacking enzymatic activity (Yu et al. 2009). Its regulation of the extrinsic-apoptotic pathway is more adaptive and concentration dependent compared to its short isoforms. On one hand, FLIP$_L$ is recruited to the DISC where it can form heterodimers with catalytic domains of caspase-8 at physiological conditions and thus mediates its activation (Chang et al. 2002). In contrast, a high expression level of FLIP$_L$ induces partial autocatalytic processing of caspase-8 (Krueger et al. 2001), which then remains at the DISC-platform, cleaves local substrates like RIP1 and prevents apoptosis (Micheau et al. 2002; Kavuri et al. 2011). Furthermore, c-FLIP$_L$ is also known as a limiting factor of kinase-dependent necrosis (reviewed by Green et al. 2011; Oberst et al. 2011).

The highly variable signal transduction from death receptors to caspase-8 and c-FLIP$_L$ strongly emphasizes the critical interplay of these proteins in regulation of pro- or anti-apoptotic responses.

8.3.7.3 Non-natural Inhibition: Designed Ankyrin Repeat Proteins (DARPins)

Gene knockouts or artificial inhibition of a specific protein are general methods used in cell biology to target its distinct role in diverse cellular processes. These methods have also been applied to unravel the known functions of the caspase family members, which have been broadly studied in cells derived from gene knockout mice. However, the production of a knockout mouse is a time-consuming process, not always successful or occasionally inaccurate, as shown for the caspase-1 deficient strain 129 that also harbors a deficient caspase-11 (Kayagaki et al. 2011).

A protein knock-out by specific inhibition is a second approach to study the role of a specific protein in cellular processes. In the case of caspases this is rather difficult to achieve due to the high structural homology of these enzymes. The commercially available substrate-based inhibitors target the conserved active site and are thus not very specific (McStay et al. 2008). In contrast, selected binding proteins like anti- and nanobodies or designed ankyrin repeat proteins (DARPins) bind via an increased protein-protein interaction surface that may enhance the specificity. Furthermore, the binding may induce functional modification of the target protein, as shown in several examples like a Fas receptor agonistic antibody (Chodorge et al. 2012) or an ABC transporter modulating DARPin (Seeger et al. 2012).

Highly specific inhibition of caspase-2 and -3 has been achieved with two different DARPins (Schweizer et al. 2007; Schroeder et al. 2013). These binding proteins were designed based on the ankyrin protein scaffold and usually contain two or three internal repeats that are sandwiched between a N- and C-terminal capping repeat (Binz et al. 2004). Each internal repeat obtains six randomized positions that are potentially involved in interactions, forming a diverse library with at least 10^{10} different molecules that can be selected by phage- or ribosome

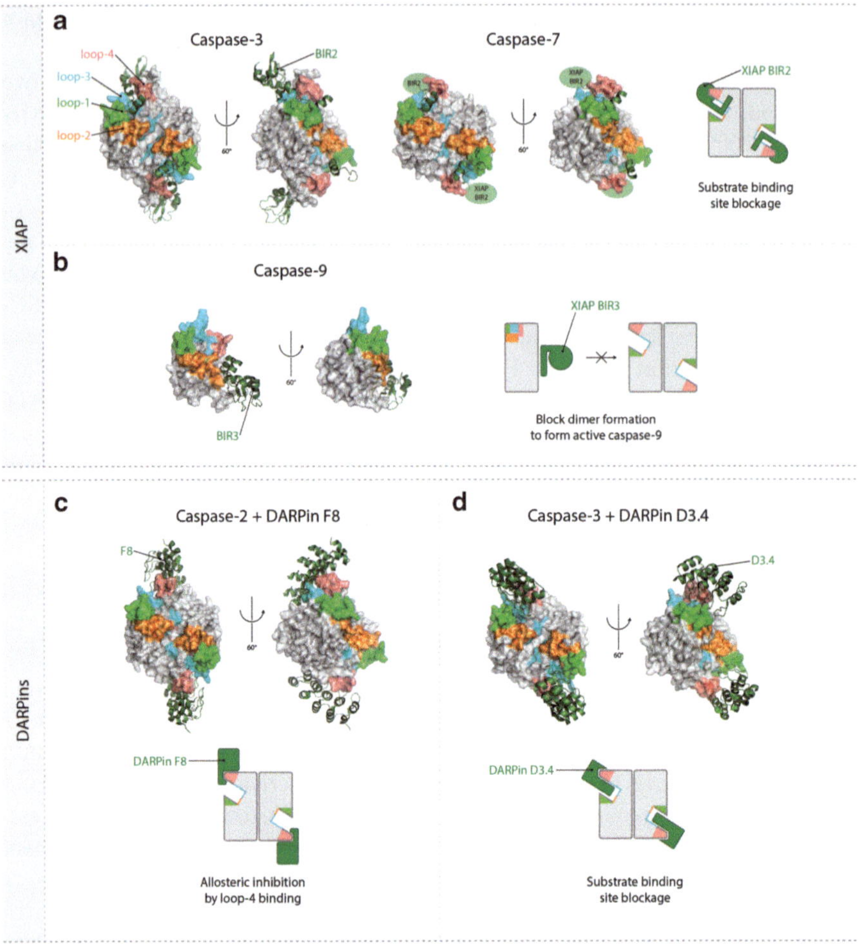

Fig. 8.6 Specific inhibition. (**a**) XIAP inhibition of caspase-3 and -7. Crystal structures of XIAP-BIR2 domain in complex with either caspase-3 (1I3O) or caspase-7 (1I51) revealed a competitive inhibition mechanism by direct binding into the substrate-binding groove. The BIR2 domain was crystallized in both complexes, although no electron density for the domain could be observed in the complex with caspase-7. (**b**) The crystal structure of XIAP-BIR3 domain in complex with caspase-9 (1NW9) unraveled a different inhibition mechanism: BIR3 interacts with residues in the dimer interface and thus prevents a dimerization of caspase-9. (**c**) Complex structure of caspase-2 with DARPin F8 (2P2C) showed a binding of the DARPin to loop-4. This binding stabilizes the loop in a distinct conformation compared to substrate bound caspase-2 and thus inhibits the enzyme allosteric. (**d**) DARPin D3.4 binds to caspase-3 (2XZD) at the substrate-binding cleft and inhibits the enzyme in a linear competitive mechanism

display (Hanes and Plückthun 1997; Steiner et al. 2008). Selected DARPins combine the advantages of high affinity and stability with high purity and expression yields in a cysteine less framework.

The caspase-2 specific DARPin (AR_F8) binds and inhibits the enzyme with a tight binding mixed-type inhibition mechanism (Schweizer et al. 2007). AR_F8 binding to caspase-2 occurs primarily at the binding groove forming loop-4 and stabilizes a distinct conformation of this loop compared to substrate analogue bound caspase-2 (Fig. 8.6c). Additional structural rearrangements have been observed N-terminal at the small subunit of the adjacent catalytic domain that finally results in a displacement of the catalytic residue Cys-285. Altogether, the structural and kinetic analysis of this DARPin clearly revealed an allosteric inhibition of caspase-2 with high specificity.

Another selected DARPin (D3.4) has been reported to specifically target caspase-3 (Schroeder et al. 2013). This binder inhibits the protease in a tight binding and purely competitive mechanism (Fig. 8.6d). Structural data uncovered a complex formation of D3.4 and caspase-3 at the active site cleft leading to obstructed binding pockets. D3.4 positions its residue Asp-45 inside the S_4 pocket mimicking a natural peptide binding. Further interactions of the second internal repeat lock the caspase-3 residue Tyr-204 in a closed state that occupies the subsite S_2 as seen in the XIAP caspase-3 complex structure (Riedl et al. 2001b). Although the structural mechanism of inhibition displays similarities to XIAP, the active site occupying peptide is not in a reversal orientation. Furthermore, this binder is a decent example for active site directed inhibition that is combined with an enlarged target-inhibitor interaction surface to achieve high specificity.

8.4 Caspase-Related Diseases

The important roles that caspases mediate in cell death and differentiation pathways are highly regulated and often include redundant mechanisms. However, their importance in cellular processes also increases the susceptibility for severe diseases in case of misregulation. For instance, alteration of the apoptotic pathway is related to cancer, neurological and cardiovascular disorders as well as autoimmune diseases (reviewed by Favaloro et al. 2012).

Cancer development and progression is highly connected to overexpression of regulatory Bcl-2 proteins, down-regulation of apoptosome forming Apaf-1 and modulation of death receptor pathways by reduced or increased expression of death receptors or death ligands, respectively (Favaloro et al. 2012). Direct mutations of the caspase genes can lead to a reduction or loss of enzymatic activity, while other studies reported dominant negative caspase variants that prevent activation of the wild type form (reviewed by Olsson and Zhivotovsky 2011). This results in a decrease of apoptotic capacity that may favor cancer manifestation. However, there is little evidence for tumorigenesis induced by an individual caspase-mutant that indeed emphasizes a redundant network with tumor suppressor functionalities (Olsson and Zhivotovsky 2011).

Caspases have been reported as major conductors in the development of Alzheimer's disease (reviewed by Rohn 2010). Two of the major pathological markers in Alzheimer's disease are the formation of neurofibrillary tangles (NFTs), consisting of hyperphosphorylated microtubule-associated protein tau and the extracellular deposition of amyloid-β (Aβ) in plaques. Aβ is created after sequential proteolysis of the amyloid precursor protein (APP) by β- and γ-secretases. However, caspase involvement of APP cleavage has also been observed (Zhang et al. 2011).

In addition, it has been proposed that the presence of Aβ stimulates caspase activation that leads to tau cleavage, predominantly mediated by caspase-3 and -6, and a following tau hyperphosphorylation (Rohn 2010). Moreover, this hypothesis of Aβ-linked tau cleavage has been verified using a transgenic mouse model that overexpresses the antiapoptotic protein Bcl-2 (Rohn et al. 2008). Thus, the role of caspases in Alzheimer's disease lays assumably not only terminal in the execution of neuronal cell death but also proximal in the initiation of the disease mechanism.

Caspase involvement has also been shown in Huntington's disease, a neuronal disorder that is caused by mutation of the huntingtin (htt) protein (reviewed by Ehrnhoefer et al. 2011). Disease related htt mutations lead to an abnormal elongation of a polyglutamine stretch in the N-terminal region, which is supposed to prevent aggregation in wildtype proteins by turnover regulation (Ehrnhoefer et al. 2011). Both, either wildtype or mutant proteins can be processed by proteases, including the caspase family.

N-terminal cleavage fragments of the htt mutant have been reported to increase the apoptotic susceptibility in dependence of fragment and polyglutamine extension length (Hackam et al. 1998). Removal of the five predicted caspase cleavage site in htt could reduce its toxicity and furthermore unraveled a specific caspase-6 cleavage site that is crucial in the development of Huntington's disease symptoms (Graham et al. 2006). Active caspase-6 has been proposed as an early disease marker due to its activation before the manifestation of the first motor abnormalities (Graham et al. 2010).

Activation of caspase-6 and downstream cleavage of this enzyme has been shown to be a fundamental step in the development of neurodegenerative diseases and highlights its potential as a novel therapeutic target (reviewed by Graham et al. 2011).

Alterations in the inflammasome formation mechanisms can lead to hyperactivation of caspase-1 and extensive production of IL-1β and IL-18, which can be related to autoimmune diseases (reviewed by Lamkanfi et al. 2011). For instance, enhanced expression of caspase-1 and IL-18 has been found in patients suffering from multiple sclerosis (Huang et al. 2004).

8.5 Concluding Remarks

Death pathways are fundamental mechanisms of multicellular organisms to ensure the correct development, tissue maintenance and homeostasis. The diverse processes are tightly regulated to prevent accidental cellular demise but also to facilitate a fast removal of cells upon cell death activation. Its redundant system of overlapping, distinctive pathways secures the multicellular integrity upon various cytotoxic stimuli in an exceptional manner.

The particular death pathways of apoptosis and pyroptosis are primarily driven by caspases. These enzymes are classified according to their activity related to inflammation and apoptosis. Active caspases cleave specific substrates, which results in either activation of demise effectors or deactivation of survival proteins that both accelerate cell destruction. Thus, a precise regulation is indispensable and occurs at numerous steps. Although this highly specific and minimal error-prone regulatory system has been evolved with redundancy, imbalances result in severe diseases including cancer and neurodegenerative disorders.

In conclusion, modern research unraveled a broad contribution of caspases not only in cell death mechanism but also in inflammatory responses and tissue protection by anti-inflammatory activity during apoptosis. Thus, researchers began to reconsider the general classification of inflammatory and apoptotic caspases. Martin et al. (2012) recently suggested a new classification of positive and negative regulators of inflammation: Besides the pro-inflammatory caspases-1, -4 and -5, the nowadays known apoptotic caspases are categorized as anti-inflammatory proteases. Overall, it further underlines the remarkable role of this protease family in numerous distinctive cellular processes besides death pathways.

Acknowledgement This work was supported by the Swiss National Science Foundation grant 310030-122342 to Markus G. Grütter.

References

Agard NJ, Wells JA (2009) Methods for the proteomic identification of protease substrates. Curr Opin Chem Biol 13(5–6):503–509

Agostini L, Martinon F, Burns K, McDermott MF, Hawkins PN, Tschopp J (2004) NALP3 forms an IL-1beta-processing inflammasome with increased activity in Muckle-Wells autoinflammatory disorder. Immunity 20(3):319–325

Akita K, Ohtsuki T, Nukada Y, Tanimoto T, Namba M, Okura T, Takakura-Yamamoto R, Torigoe K, Gu Y, Su MS, Fujii M, Satoh-Itoh M, Yamamoto K, Kohno K, Ikeda M, Kurimoto M (1997) Involvement of caspase-1 and caspase-3 in the production and processing of mature human interleukin 18 in monocytic THP.1 cells. J Biol Chem 272(42):26595–26603

Allan LA, Morrice N, Brady S, Magee G, Pathak S, Clarke PR (2003) Inhibition of caspase-9 through phosphorylation at Thr 125 by ERK MAPK. Nat Cell Biol 5(7):647–654

Alnemri ES, Livingston DJ, Nicholson DW, Salvesen G, Thornberry NA, Wong WW, Yuan J (1996) Human ICE/CED-3 protease nomenclature. Cell 87(2):171

Alvarado-Kristensson M, Melander F, Leandersson K, Rönnstrand L, Wernstedt C, Andersson T (2004) p38-MAPK signals survival by phosphorylation of caspase-8 and caspase-3 in human neutrophils. J Exp Med 199(4):449–458

Binz HK, Amstutz P, Kohl A, Stumpp MT, Briand C, Forrer P, Grütter MG, Plückthun A (2004) High-affinity binders selected from designed ankyrin repeat protein libraries. Nat Biotechnol 22(5):575–582

Boyd SE, Pike RN, Rudy GB, Whisstock JC, Garcia de la Banda M (2005) PoPS: a computational tool for modeling and predicting protease specificity. J Bioinform Comput Biol 3(3):551–585

Bratton SB, Lewis J, Butterworth M, Duckett CS, Cohen GM (2002) XIAP inhibition of caspase-3 preserves its association with the Apaf-1 apoptosome and prevents CD95- and Bax-induced apoptosis. Cell Death Differ 9(9):881–892

Brennan MA, Cookson BT (2000) Salmonella induces macrophage death by caspase-1-dependent necrosis. Mol Microbiol 38(1):31–40
Cardone MH, Roy N, Stennicke HR, Salvesen GS, Franke TF, Stanbridge E, Frisch S, Reed JC (1998) Regulation of cell death protease caspase-9 by phosphorylation. Science 282(5392): 1318–1321
Center DM, Cruikshank WW, Zhang Y (2004) Nuclear pro-IL-16 regulation of T cell proliferation: p27(KIP1)-dependent G0/G1 arrest mediated by inhibition of Skp2 transcription. J Immunol 172(3):1654–1660
Chai J, Shiozaki E, Srinivasula SM, Wu Q, Datta P, Alnemri ES, Shi Y, Dataa P (2001a) Structural basis of caspase-7 inhibition by XIAP. Cell 104(5):769–780
Chai J, Wu Q, Shiozaki E, Srinivasula SM, Alnemri ES, Shi Y (2001b) Crystal structure of a procaspase-7 zymogen: mechanisms of activation and substrate binding. Cell 107(3):399–407
Chang DW, Xing Z, Pan Y, Algeciras-Schimnich A, Barnhart BC, Yaish-Ohad S, Peter ME, Yang X (2002) c-FLIP(L) is a dual function regulator for caspase-8 activation and CD95-mediated apoptosis. EMBO J 21(14):3704–3714
Chaudhary PM, Eby M, Jasmin A, Bookwalter A, Murray J, Hood L (1997) Death receptor 5, a new member of the TNFR family, and DR4 induce FADD-dependent apoptosis and activate the NF-kappaB pathway. Immunity 7(6):821–830
Chinnaiyan AM, O'Rourke K, Tewari M, Dixit VM (1995) FADD, a novel death domain-containing protein, interacts with the death domain of Fas and initiates apoptosis. Cell 81 (4):505–512
Chodorge M, Züger S, Stirnimann C, Briand C, Jermutus L, Grütter MG, Minter RR (2012) A series of Fas receptor agonist antibodies that demonstrate an inverse correlation between affinity and potency. Cell Death Differ 19(7):1187–1195
Chupp GL, Wright EA, Wu D, Vallen-Mashikian M, Cruikshank WW, Center DM, Kornfeld H, Berman JS (1998) Tissue and T cell distribution of precursor and mature IL-16. J Immunol 161 (6):3114–3119
Coleman ML, Sahai EA, Yeo M, Bosch M, Dewar A, Olson MF (2001) Membrane blebbing during apoptosis results from caspase-mediated activation of ROCK I. Nat Cell Biol 3(4): 339–345
Conradt B, Horvitz HR (1998) The C. elegans protein EGL-1 is required for programmed cell death and interacts with the Bcl-2-like protein CED-9. Cell 93(4):519–529
Cookson BT, Brennan MA (2001) Pro-inflammatory programmed cell death. Trends Microbiol 9 (3):113–114
Cooper DM, Pio F, Thi EP, Theilmann D, Lowenberger C (2007) Characterization of Aedes Dredd: a novel initiator caspase from the yellow fever mosquito, Aedes aegypti. Insect Biochem Mol Biol 37(6):559–569
de Alba E (2009) Structure and interdomain dynamics of apoptosis-associated speck-like protein containing a CARD (ASC). J Biol Chem 284(47):32932–32941
Denault J-B, Salvesen GS (2003) Human caspase-7 activity and regulation by its N-terminal peptide. J Biol Chem 278(36):34042–34050
Denecker G, Ovaere P, Vandenabeele P, Declercq W (2008) Caspase-14 reveals its secrets. J Cell Biol 180(3):451–458
Deveraux QL, Takahashi R, Salvesen GS, Reed JC (1997) X-linked IAP is a direct inhibitor of cell-death proteases. Nature 388(6639):300–304
Deveraux QL, Roy N, Stennicke HR, Van Arsdale T, Zhou Q, Srinivasula SM, Alnemri ES, Salvesen GS, Reed JC (1998) IAPs block apoptotic events induced by caspase-8 and cytochrome c by direct inhibition of distinct caspases. EMBO J 17(8):2215–2223
Duan H, Dixit VM (1997) RAIDD is a new 'death' adaptor molecule. Nature 385(6611):86–89
Eckhart L, Ballaun C, Hermann M, VandeBerg JL, Sipos W, Uthman A, Fischer H, Tschachler E (2008) Identification of novel mammalian caspases reveals an important role of gene loss in shaping the human caspase repertoire. Mol Biol Evol 25(5):831–841

Ehrnhoefer DE, Sutton L, Hayden MR (2011) Small changes, big impact: posttranslational modifications and function of huntingtin in Huntington disease. Neuroscientist 17(5):475–492

Elliott JM, Rouge L, Wiesmann C, Scheer JM (2009) Crystal structure of procaspase-1 zymogen domain reveals insight into inflammatory caspase autoactivation. J Biol Chem 284(10): 6546–6553

Ellis HM, Horvitz HR (1986) Genetic control of programmed cell death in the nematode C. elegans. Cell 44(6):817–829

Faustin B, Lartigue L, Bruey J-M, Luciano F, Sergienko E, Bailly-Maitre B, Volkmann N, Hanein D, Rouiller I, Reed JC (2007) Reconstituted NALP1 inflammasome reveals two-step mechanism of caspase-1 activation. Mol Cell 25(5):713–724

Favaloro B, Allocati N, Graziano V, Di Ilio C, De Laurenzi V (2012) Role of apoptosis in disease. Aging 4(5):330–349

Flemming W (1885) Ueber die Bildung von Richtungsfiguren in Säugethiereiern beim Untergang Graaf'scher Follikel. Arch Anat Entw Gesch (Veit & Comp). 221–244

Fuentes-Prior P, Salvesen GS (2004) The protein structures that shape caspase activity, specificity, activation and inhibition. Biochem J 384(Pt 2):201–232

Fulda S, Vucic D (2012) Targeting IAP proteins for therapeutic intervention in cancer. Nat Rev Drug Discov 11(2):109–124

Galluzzi L, Vitale I, Abrams JM, Alnemri ES, Baehrecke EH, Blagosklonny MV, Dawson TM, Dawson VL, El-Deiry WS, Fulda S, Gottlieb E, Green DR, Hengartner MO, Kepp O, Knight RA, Kumar S, Lipton SA, Lu X, Madeo F, Malorni W, Mehlen P, Nuñez G, Peter ME, Piacentini M, Rubinsztein DC, Shi Y, Simon H-U, Vandenabeele P, White E, Yuan J, Zhivotovsky B, Melino G, Kroemer G (2012) Molecular definitions of cell death subroutines: recommendations of the Nomenclature Committee on Cell Death 2012. Cell Death Differ 19(1):107–120

Germain M, Affar EB, D'Amours D, Dixit VM, Salvesen GS, Poirier GG (1999) Cleavage of automodified poly(ADP-ribose) polymerase during apoptosis. Evidence for involvement of caspase-7. J Biol Chem 274(40):28379–28384

Golks A, Brenner D, Fritsch C, Krammer PH, Lavrik IN (2005) c-FLIPR, a new regulator of death receptor-induced apoptosis. J Biol Chem 280(15):14507–14513

Graham RK, Deng Y, Slow EJ, Haigh B, Bissada N, Lu G, Pearson J, Shehadeh J, Bertram L, Murphy Z, Warby SC, Doty CN, Roy S, Wellington CL, Leavitt BR, Raymond LA, Nicholson DW, Hayden MR (2006) Cleavage at the caspase-6 site is required for neuronal dysfunction and degeneration due to mutant huntingtin. Cell 125(6):1179–1191

Graham RK, Deng Y, Carroll J, Vaid K, Cowan C, Pouladi MA, Metzler M, Bissada N, Wang L, Faull RLM, Gray M, Yang XW, Raymond LA, Hayden MR (2010) Cleavage at the 586 amino acid caspase-6 site in mutant huntingtin influences caspase-6 activation in vivo. J Neurosci 30 (45):15019–15029

Graham RK, Ehrnhoefer DE, Hayden MR (2011) Caspase-6 and neurodegeneration. Trends Neurosci 34(12):646–656

Green DR, Oberst A, Dillon CP, Weinlich R, Salvesen GS (2011) RIPK-dependent necrosis and its regulation by caspases: a mystery in five acts. Mol Cell 44(1):9–16

Grimsley C, Ravichandran KS (2003) Cues for apoptotic cell engulfment: eat-me, don't eat-me and come-get-me signals. Trends Cell Biol 13(12):648–656

Gu Y, Kuida K, Tsutsui H, Ku G, Hsiao K, Fleming MA, Hayashi N, Higashino K, Okamura H, Nakanishi K, Kurimoto M, Tanimoto T, Flavell RA, Sato V, Harding MW, Livingston DJ, Su MS (1997) Activation of interferon-gamma inducing factor mediated by interleukin-1beta converting enzyme. Science 275(5297):206–209

Hackam AS, Singaraja R, Wellington CL, Metzler M, McCutcheon K, Zhang T, Kalchman M, Hayden MR (1998) The influence of huntingtin protein size on nuclear localization and cellular toxicity. J Cell Biol 141(5):1097–1105

Hall A (1998) Rho GTPases and the actin cytoskeleton. Science 279(5350):509–514

Hanes J, Plückthun A (1997) In vitro selection and evolution of functional proteins by using ribosome display. Proc Natl Acad Sci U S A 94(10):4937–4942

Hardwick JM, Chen Y-b, Jonas EA (2012) Multipolar functions of BCL-2 proteins link energetics to apoptosis. Trends Cell Biol 22(6):318–328
Hardy JA, Lam J, Nguyen JT, O'Brien T, Wells JA (2004) Discovery of an allosteric site in the caspases. Proc Natl Acad Sci U S A 101(34):12461–12466
Hengartner MO, Ellis RE, Horvitz HR (1992) Caenorhabditis elegans gene ced-9 protects cells from programmed cell death. Nature 356(6369):494–499
Hitomi J, Christofferson DE, Ng A, Yao J, Degterev A, Xavier RJ, Yuan J (2008) Identification of a molecular signaling network that regulates a cellular necrotic cell death pathway. Cell 135 (7):1311–1323
Hoste E, Denecker G, Gilbert B, Van Nieuwerburgh F, van der Fits L, Asselbergh B, De Rycke R, Hachem J-P, Deforce D, Prens EP, Vandenabeele P, Declercq W (2013) Caspase-14-deficient mice are more prone to the development of parakeratosis. J Invest Dermatol 133(3):742–750
Huang W-X, Huang P, Hillert J (2004) Increased expression of caspase-1 and interleukin-18 in peripheral blood mononuclear cells in patients with multiple sclerosis. Mult Scler 10(5): 482–487
Hughes MA, Harper N, Butterworth M, Cain K, Cohen GM, Macfarlane M (2009) Reconstitution of the death-inducing signaling complex reveals a substrate switch that determines CD95-mediated death or survival. Mol Cell 35(3):265–279
Javle M, Curtin NJ (2011) The role of PARP in DNA repair and its therapeutic exploitation. Br J Cancer 105(8):1114–1122
Jia SH, Parodo J, Kapus A, Rotstein OD, Marshall JC (2008) Dynamic regulation of neutrophil survival through tyrosine phosphorylation or dephosphorylation of caspase-8. J Biol Chem 283 (9):5402–5413
Kamens J, Paskind M, Hugunin M, Talanian RV, Allen H, Banach D, Bump N, Hackett M, Johnston CG, Li P (1995) Identification and characterization of ICH-2, a novel member of the interleukin-1 beta-converting enzyme family of cysteine proteases. J Biol Chem 270(25): 15250–15256
Kavuri SM, Geserick P, Berg D, Dimitrova DP, Feoktistova M, Siegmund D, Gollnick H, Neumann M, Wajant H, Leverkus M (2011) Cellular FLICE-inhibitory protein (cFLIP) isoforms block CD95- and TRAIL death receptor-induced gene induction irrespective of processing of caspase-8 or cFLIP in the death-inducing signaling complex. J Biol Chem 286 (19):16631–16646
Kayagaki N, Warming S, Lamkanfi M, Vande Walle L, Louie S, Dong J, Newton K, Qu Y, Liu J, Heldens S, Zhang J, Lee WP, Roose-Girma M, Dixit VM (2011) Non-canonical inflammasome activation targets caspase-11. Nature 479(7371):117–121
Keller N, Mares J, Zerbe O, Grütter MG (2009) Structural and biochemical studies on procaspase-8: new insights on initiator caspase activation. Structure 17(3):438–448
Kerr JF, Wyllie AH, Currie AR (1972) Apoptosis: a basic biological phenomenon with wide-ranging implications in tissue kinetics. Br J Cancer 26(4):239–257
Kersse K, Verspurten J, Vanden Berghe T, Vandenabeele P (2011) The death-fold superfamily of homotypic interaction motifs. Trends Biochem Sci 36(10):541–552
Kischkel FC, Hellbardt S, Behrmann I, Germer M, Pawlita M, Krammer PH, Peter ME (1995) Cytotoxicity-dependent APO-1 (Fas/CD95)-associated proteins form a death-inducing signaling complex (DISC) with the receptor. EMBO J 14(22):5579–5588
Kitazumi I, Tsukahara M (2011) Regulation of DNA fragmentation: the role of caspases and phosphorylation. FEBS J 278(3):427–441
Koenig U, Eckhart L, Tschachler E (2001) Evidence that caspase-13 is not a human but a bovine gene. Biochem Biophys Res Commun 285(5):1150–1154
Krijnen PAJ, Sipkens JA, Molling JW, Rauwerda JA, Stehouwer CDA, Muller A, Paulus WJ, van Nieuw Amerongen GP, Hack CE, Verhoeven AJ, van Hinsbergh VWM, Niessen HWM (2010) Inhibition of Rho-ROCK signaling induces apoptotic and non-apoptotic PS exposure in cardiomyocytes via inhibition of flippase. J Mol Cell Cardiol 49(5):781–790

Krippner-Heidenreich A, Talanian RV, Sekul R, Kraft R, Thole H, Ottleben H, Lüscher B (2001) Targeting of the transcription factor Max during apoptosis: phosphorylation-regulated cleavage by caspase-5 at an unusual glutamic acid residue in position P1. Biochem J 358(Pt 3):705–715
Krueger A, Schmitz I, Baumann S, Krammer PH, Kirchhoff S (2001) Cellular FLICE-inhibitory protein splice variants inhibit different steps of caspase-8 activation at the CD95 death-inducing signaling complex. J Biol Chem 276(23):20633–20640
Kumar S, Doumanis J (2000) The fly caspases. Cell Death Differ 7(11):1039–1044
Kuwana T, Mackey MR, Perkins G, Ellisman MH, Latterich M, Schneiter R, Green DR, Newmeyer DD (2002) Bid, Bax, and lipids cooperate to form supramolecular openings in the outer mitochondrial membrane. Cell 111(3):331–342
Lamkanfi M, Walle LV, Kanneganti T-D (2011) Deregulated inflammasome signaling in disease. Immunol Rev 243(1):163–173
Larsen BD, Rampalli S, Burns LE, Brunette S, Dilworth FJ, Megeney LA (2010) Caspase 3/caspase-activated DNase promote cell differentiation by inducing DNA strand breaks. Proc Natl Acad Sci U S A 107(9):4230–4235
Lazebnik YA, Kaufmann SH, Desnoyers S, Poirier GG, Earnshaw WC (1994) Cleavage of poly (ADP-ribose) polymerase by a proteinase with properties like ICE. Nature 371(6495):346–347
Lewis EM, Wilkinson AS, Davis NY, Horita DA, Wilkinson JC (2011) Nondegradative ubiquitination of apoptosis inducing factor (AIF) by X-linked inhibitor of apoptosis at a residue critical for AIF-mediated chromatin degradation. Biochemistry 50(51):11084–11096
Lüthi AU, Cullen SP, Mcneela EA, Duriez PJ, Afonina IS, Sheridan C, Brumatti G, Taylor RC, Kersse K, Vandenabeele P, Lavelle EC, Martin SJ (2009) Suppression of interleukin-33 bioactivity through proteolysis by apoptotic caspases. Immunity 31(1):84–98
Mandal D, Moitra PK, Saha S, Basu J (2002) Caspase 3 regulates phosphatidylserine externalization and phagocytosis of oxidatively stressed erythrocytes. FEBS Lett 513(2–3):184–188
Mariathasan S, Newton K, Monack DM, Vucic D, French DM, Lee WP, Roose-Girma M, Erickson S, Dixit VM (2004) Differential activation of the inflammasome by caspase-1 adaptors ASC and Ipaf. Nature 430(6996):213–218
Martin DA, Siegel RM, Zheng L, Lenardo MJ (1998) Membrane oligomerization and cleavage activates the caspase-8 (FLICE/MACHalpha1) death signal. J Biol Chem 273(8):4345–4349
Martin SJ, Henry CM, Cullen SP (2012) A perspective on mammalian caspases as positive and negative regulators of inflammation. Mol Cell 46(4):387–397
Martinon F, Tschopp J (2004) Inflammatory caspases: linking an intracellular innate immune system to autoinflammatory diseases. Cell 117(5):561–574
Martinon F, Tschopp J (2005) NLRs join TLRs as innate sensors of pathogens. Trends Immunol 26 (8):447–454
Martinon F, Burns K, Tschopp J (2002) The inflammasome: a molecular platform triggering activation of inflammatory caspases and processing of proIL-beta. Mol Cell 10(2):417–426
McStay GP, Salvesen GS, Green DR (2008) Overlapping cleavage motif selectivity of caspases: implications for analysis of apoptotic pathways. Cell Death Differ 15(2):322–331
Meergans T, Hildebrandt AK, Horak D, Haenisch C, Wendel A (2000) The short prodomain influences caspase-3 activation in HeLa cells. Biochem J 349(Pt 1):135–140
Miao EA, Mao DP, Yudkovsky N, Bonneau R, Lorang CG, Warren SE, Leaf IA, Aderem A (2010) Innate immune detection of the type III secretion apparatus through the NLRC4 inflammasome. Proc Natl Acad Sci U S A 107(7):3076–3080
Micheau O, Thome M, Schneider P, Holler N, Tschopp J, Nicholson DW, Briand C, Grütter MG (2002) The long form of FLIP is an activator of caspase-8 at the Fas death-inducing signaling complex. J Biol Chem 277(47):45162–45171
Nagata S, Nagase H, Kawane K, Mukae N, Fukuyama H (2003) Degradation of chromosomal DNA during apoptosis. Cell Death Differ 10(1):108–116
Oberst A, Dillon CP, Weinlich R, Mccormick LL, Fitzgerald P, Pop C, Hakem R, Salvesen GS, Green DR (2011) Catalytic activity of the caspase-8-FLIP(L) complex inhibits RIPK3-dependent necrosis. Nature 471(7338):363–367

Olsson M, Zhivotovsky B (2011) Caspases and cancer. Cell Death Differ 18(9):1441–1449
Pan G, O'Rourke K, Chinnaiyan AM, Gentz R, Ebner R, Ni J, Dixit VM (1997) The receptor for the cytotoxic ligand TRAIL. Science 276(5309):111–113
Park HH, Logette E, Raunser S, Cuenin S, Walz T, Tschopp J, Wu H (2007) Death domain assembly mechanism revealed by crystal structure of the oligomeric PIDDosome core complex. Cell 128 (3):533–546
Pop C, Timmer J, Sperandio S, Salvesen GS (2006) The apoptosome activates caspase-9 by dimerization. Mol Cell 22(2):269–275
Pop C, Fitzgerald P, Green DR, Salvesen GS (2007) Role of proteolysis in caspase-8 activation and stabilization. Biochemistry 46(14):4398–4407
Qi S, Pang Y, Hu Q, Liu Q, Li H, Zhou Y, He T, Liang Q, Liu Y, Yuan X, Luo G, Li H, Wang J, Yan N, Shi Y (2010) Crystal structure of the Caenorhabditis elegans apoptosome reveals an octameric assembly of CED-4. Cell 141(3):446–457
Rao L, Perez D, White E (1996) Lamin proteolysis facilitates nuclear events during apoptosis. J Cell Biol 135(6 Pt 1):1441–1455
Rathinam VAK, Vanaja SK, Fitzgerald KA (2012) Regulation of inflammasome signaling. Nat Immunol 13(4):333–342
Rawlings ND, Barrett AJ, Bateman A (2012) MEROPS: the database of proteolytic enzymes, their substrates and inhibitors. Nucleic Acids Res 40:D343–D350
Reubold TF, Wohlgemuth S, Eschenburg S (2011) Crystal structure of full-length Apaf-1: how the death signal is relayed in the mitochondrial pathway of apoptosis. Structure 19(8):1074–1083
Riedl SJ, Fuentes-Prior P, Renatus M, Kairies N, Krapp S, Huber R, Salvesen GS, Bode W (2001a) Structural basis for the activation of human procaspase-7. Proc Natl Acad Sci U S A 98(26): 14790–14795
Riedl SJ, Renatus M, Schwarzenbacher R, Zhou Q, Sun C, Fesik SW, Liddington RC, Salvesen GS (2001b) Structural basis for the inhibition of caspase-3 by XIAP. Cell 104(5):791–800
Rohn TT (2010) The role of caspases in Alzheimer's disease; potential novel therapeutic opportunities. Apoptosis 15(11):1403–1409
Rohn TT, Vyas V, Hernandez-Estrada T, Nichol KE, Christie L-A, Head E (2008) Lack of pathology in a triple transgenic mouse model of Alzheimer's disease after overexpression of the anti-apoptotic protein Bcl-2. J Neurosci 28(12):3051–3059
Roschitzki-Voser H, Schroeder T, Lenherr ED, Frölich F, Schweizer A, Donepudi M, Ganesan R, Mittl PRE, Baici A, Grütter MG (2012) Human caspases in vitro: expression, purification and kinetic characterization. Protein Expr Purif 84(2):236–246
Rubinsztein DC, Shpilka T, Elazar Z (2012) Mechanisms of autophagosome biogenesis. Curr Biol 22(1):R29–R34
Safa AR (2012) c-FLIP, a master anti-apoptotic regulator. Exp Oncol 34(3):176–184
Sahoo M, Ceballos-Olvera I, Del Barrio L, Re F (2011) Role of the inflammasome, IL-1β, and IL-18 in bacterial infections. Scientific World J 11:2037–2050
Saleh M, Vaillancourt JP, Graham RK, Huyck M, Srinivasula SM, Alnemri ES, Steinberg MH, Nolan V, Baldwin CT, Hotchkiss RS, Buchman TG, Zehnbauer BA, Hayden MR, Farrer LA, Roy S, Nicholson DW (2004) Differential modulation of endotoxin responsiveness by human caspase-12 polymorphisms. Nature 429(6987):75–79
Salvesen GS, Dixit VM (1999) Caspase activation: the induced-proximity model. Proc Natl Acad Sci U S A 96(20):10964–10967
Scaffidi C, Schmitz I, Krammer PH, Peter ME (1999) The role of c-FLIP in modulation of CD95-induced apoptosis. J Biol Chem 274(3):1541–1548
Schile AJ, García-Fernández M, Steller H (2008) Regulation of apoptosis by XIAP ubiquitin-ligase activity. Genes Dev 22(16):2256–2266
Schroeder T, Barandun J, Flütsch A, Briand C, Mittl PRE, Grütter MG (2013) Specific inhibition of caspase-3 by a competitive DARPin: molecular mimicry between native and designed inhibitors. Structure 21(2):277–289

Schweizer A, Roschitzki-Voser H, Amstutz P, Briand C, Gulotti-Georgieva M, Prenosil E, Binz HK, Capitani G, Baici A, Plückthun A, Grütter MG (2007) Inhibition of caspase-2 by a designed ankyrin repeat protein: specificity, structure, and inhibition mechanism. Structure 15(5):625–636

Scott FL, Stec B, Pop C, Dobaczewska MK, Lee JJ, Monosov E, Robinson H, Salvesen GS, Schwarzenbacher R, Riedl SJ (2009) The Fas-FADD death domain complex structure unravels signalling by receptor clustering. Nature 457(7232):1019–1022

Seeger MA, Mittal A, Velamakanni S, Hohl M, Schauer S, Salaa I, Grütter MG, Van Veen HW (2012) Tuning the drug efflux activity of an ABC transporter in vivo by in vitro selected DARPin binders. PLoS One 7(6):e37845

Shiozaki EN, Chai J, Rigotti DJ, Riedl SJ, Li P, Srinivasula SM, Alnemri ES, Fairman R, Shi Y (2003) Mechanism of XIAP-mediated inhibition of caspase-9. Mol Cell 11(2):519–527

Soares J, Lowe MM, Jarstfer MB (2011) The catalytic subunit of human telomerase is a unique caspase-6 and caspase-7 substrate. Biochemistry 50(42):9046–9055

Sollberger G, Strittmatter GE, Kistowska M, French LE, Beer H-D (2012) Caspase-4 is required for activation of inflammasomes. J Immunol 188(4):1992–2000

Steiner D, Forrer P, Plückthun A (2008) Efficient selection of DARPins with sub-nanomolar affinities using SRP phage display. J Mol Biol 382(5):1211–1227

Steller H (2008) Regulation of apoptosis in Drosophila. Cell Death Differ 15(7):1132–1138

Stennicke HR, Deveraux QL, Humke EW, Reed JC, Dixit VM, Salvesen GS (1999) Caspase-9 can be activated without proteolytic processing. J Biol Chem 274(13):8359–8362

Suzuki Y, Nakabayashi Y, Takahashi R (2001) Ubiquitin-protein ligase activity of X-linked inhibitor of apoptosis protein promotes proteasomal degradation of caspase-3 and enhances its anti-apoptotic effect in Fas-induced cell death. Proc Natl Acad Sci U S A 98(15):8662–8667

Thome M, Schneider P, Hofmann K, Fickenscher H, Meinl E, Neipel F, Mattmann C, Burns K, Bodmer JL, Schröter M, Scaffidi C, Krammer PH, Peter ME, Tschopp J (1997) Viral FLICE-inhibitory proteins (FLIPs) prevent apoptosis induced by death receptors. Nature 386(6624): 517–521

Thornberry NA, Bull HG, Calaycay JR, Chapman KT, Howard AD, Kostura MJ, Miller DK, Molineaux SM, Weidner JR, Aunins J (1992) A novel heterodimeric cysteine protease is required for interleukin-1 beta processing in monocytes. Nature 356(6372):768–774

Thornberry NA, Rano TA, Peterson EP, Rasper DM, Timkey T, Garcia-Calvo M, Houtzager VM, Nordstrom PA, Roy S, Vaillancourt JP, Chapman KT, Nicholson DW (1997) A combinatorial approach defines specificities of members of the caspase family and granzyme B. Functional relationships established for key mediators of apoptosis. J Biol Chem 272(29):17907–17911

Timmer JC, Salvesen GS (2007) Caspase substrates. Cell Death Differ 14(1):66–72

Tinel A, Tschopp J (2004) The PIDDosome, a protein complex implicated in activation of caspase-2 in response to genotoxic stress. Science 304(5672):843–846

Tinel A, Janssens S, Lippens S, Cuenin S, Logette E, Jaccard B, Quadroni M, Tschopp J (2007) Autoproteolysis of PIDD marks the bifurcation between pro-death caspase-2 and pro-survival NF-kappaB pathway. EMBO J 26(1):197–208

Tsiatsiani L, Van Breusegem F, Gallois P, Zavialov A, Lam E, Bozhkov PV (2011) Metacaspases. Cell Death Differ 18(8):1279–1288

Vaughn DE, Rodriguez J, Lazebnik Y, Joshua-Tor L (1999) Crystal structure of Apaf-1 caspase recruitment domain: an alpha-helical Greek key fold for apoptotic signaling. J Mol Biol 293 (3):439–447

Vaux DL, Silke J (2005) IAPs, RINGs and ubiquitylation. Nat Rev Mol Cell Biol 6(4):287–297

Verhagen AM, Ekert PG, Pakusch M, Silke J, Connolly LM, Reid GE, Moritz RL, Simpson RJ, Vaux DL (2000) Identification of DIABLO, a mammalian protein that promotes apoptosis by binding to and antagonizing IAP proteins. Cell 102(1):43–53

Verhagen AM, Silke J, Ekert PG, Pakusch M, Kaufmann H, Connolly LM, Day CL, Tikoo A, Burke R, Wrobel C, Moritz RL, Simpson RJ, Vaux DL (2002) HtrA2 promotes cell death through its serine protease activity and its ability to antagonize inhibitor of apoptosis proteins. J Biol Chem 277(1):445–454

Vucic D, Franklin MC, Wallweber HJA, Das K, Eckelman BP, Shin H, Elliott LO, Kadkhodayan S, Deshayes K, Salvesen GS, Fairbrother WJ (2005) Engineering ML-IAP to produce an extraordinarily potent caspase 9 inhibitor: implications for Smac-dependent anti-apoptotic activity of ML-IAP. Biochem J 385(Pt 1):11–20

Vucic D, Dixit VM, Wertz IE (2011) Ubiquitylation in apoptosis: a post-translational modification at the edge of life and death. Nat Rev Mol Cell Biol 12(7):439–452

Wachmann K, Pop C, Van Raam BJ, Drag M, Mace PD, Snipas SJ, Zmasek C, Schwarzenbacher R, Salvesen GS, Riedl SJ (2010) Activation and specificity of human caspase-10. Biochemistry 49(38):8307–8315

Walczak H, Degli-Esposti MA, Johnson RS, Smolak PJ, Waugh JY, Boiani N, Timour MS, Gerhart MJ, Schooley KA, Smith CA, Goodwin RG, Rauch CT (1997) TRAIL-R2: a novel apoptosis-mediating receptor for TRAIL. EMBO J 16(17):5386–5397

Wang L, Yang JK, Kabaleeswaran V, Rice AJ, Cruz AC, Park AY, Yin Q, Damko E, Jang SB, Raunser S, Robinson CV, Siegel RM, Walz T, Wu H (2010) The Fas-FADD death domain complex structure reveals the basis of DISC assembly and disease mutations. Nat Struct Mol Biol 17(11):1324–1329

Wilson KP, Black JA, Thomson JA, Kim EE, Griffith JP, Navia MA, Murcko MA, Chambers SP, Aldape RA, Raybuck SA (1994) Structure and mechanism of interleukin-1 beta converting enzyme. Nature 370(6487):270–275

Yang Z, Klionsky DJ (2009) An overview of the molecular mechanism of autophagy. Curr Top Microbiol Immunol 335:1–32

Yin XM, Wang K, Gross A, Zhao Y, Zinkel S, Klocke B, Roth KA, Korsmeyer SJ (1999) Bid-deficient mice are resistant to Fas-induced hepatocellular apoptosis. Nature 400 (6747):886–891

Yu JW, Jeffrey PD, Shi Y (2009) Mechanism of procaspase-8 activation by c-FLIPL. Proc Natl Acad Sci U S A 106(20):8169–8174

Yuan J, Shaham S, Ledoux S, Ellis HM, Horvitz HR (1993) The C. elegans cell death gene ced-3 encodes a protein similar to mammalian interleukin-1 beta-converting enzyme. Cell 75(4): 641–652

Yuan S, Yu X, Topf M, Ludtke SJ, Wang X, Akey CW (2010) Structure of an apoptosome-procaspase-9 CARD complex. Structure 18(5):571–583

Yuan S, Yu X, Asara JM, Heuser JE, Ludtke SJ, Akey CW (2011) The holo-apoptosome: activation of procaspase-9 and interactions with caspase-3. Structure 19(8):1084–1096

Zhang Y-w, Thompson R, Zhang H, Xu H (2011) APP processing in Alzheimer's disease. Mol Brain 4:3

Zhang Y, Center DM, Wu DM, Cruikshank WW, Yuan J, Andrews DW, Kornfeld H (1998) Processing and activation of pro-interleukin-16 by caspase-3. J Biol Chem 273(2):1144–1149

Ziegler U, Groscurth P (2004) Morphological features of cell death. News Physiol Sci 19:124–128

Chapter 9
ADAM Proteases in Physiology and Pathophysiology: Cleave to Function in Health or to Cause Disease

Joachim Grötzinger and Stefan Rose-John

9.1 Introduction

It is believed that 10 % of all cell surface proteins are cleaved from the cell surface (Fong et al. 2011). Shedding of membrane proteins is common for type I and type II transmembrane or GPI-anchored proteins. The cleavage site is generally located in close proximity to the transmembrane region of the protein. Ectodomain shedding of transmembrane proteins does not only provide a mechanism for protein down-regulation from the cell surface but also for initiating or inhibiting autocrine and paracrine signaling via soluble functional proteins. Such proteins include agonistic cytokines as well as agonistic and antagonistic soluble cytokine receptors.

Members of the ADAM family have been recognized as important shedding proteinases. ADAMs are type I transmembrane proteins typically comprising 750 amino acids, with an N-terminal signal sequence followed by a prodomain, a metalloproteinase (catalytic)-domain, a disintegrin-domain, and/or a EGF-like-domain, and/or a Cystein-rich domain, a single transmembrane region and a cytoplasmic region (Fig. 9.1). Prodomain cleavage is performed by furin proteinases in the secretory pathway. The prodomain has inhibitory functions by a mechanism most likely different from the conventional cysteine switch mechanism where a conserved cysteine residue interacts with the zinc in the active site and thereby prevents binding and cleavage of the substrate (see below) (Moss et al. 2007). Some disintegrin-domains of ADAM family members have been shown to interact with integrins, which influenced cell adhesion and cell-cell interactions (Edwards et al. 2008). The disintegrin- and/or EGF-like-domains are thought to play a role in substrate recognition and/or in (hetero)-dimerization of ADAM proteases. For the majority of shedding events analyzed so far, ADAM17 [TNFα converting enzyme (TACE)] and its closest relative ADAM10 were

J. Grötzinger • S. Rose-John (✉)
Institute of Biochemistry, Christian-Albrechts-University Kiel, Olshausenstraße 40, 24098 Kiel, Germany
e-mail: jgroetzinger@biochem.uni-kiel.de; rosejohn@biochem.uni-kiel.de

K. Brix and W. Stöcker (eds.), *Proteases: Structure and Function*,
DOI 10.1007/978-3-7091-0885-7_9, © Springer-Verlag Wien 2013

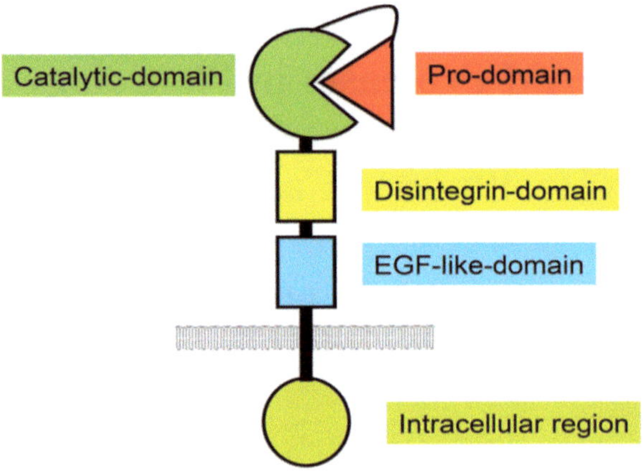

Fig. 9.1 Domain structure of ADAM10 and ADAM17

identified as responsible proteinases (Scheller et al. 2011). To date, 75 substrates of ADAM17 have been identified (Table 9.1). ADAM17 influences inflammatory responses via regulated release of TNFα, L-Selectin and sIL-6R whereas cell growth is affected via the activation of ligands of the EGF-R family. It has been shown for several substrates that ADAM10 and ADAM17 overlap in their specificity and that they can compensate of each other (Le Gall et al. 2009). There are several recent reviews of ADAM17 (Arribas and Esselens 2009; Gooz 2010; Murphy 2008; Murphy et al. 2008; Scheller et al. 2011).

This overview will summarize the role of ADAM17 as essential molecular switch in the regulation of immune responses and cancer development and its potential as future therapeutic target will be discussed.

9.2 ADAM17 Expression

At the transcriptional level, ADAM17 is ubiquitously expressed and activation seems to be under post-translational control (Soond et al. 2005). Activity and cell surface expression of ADAM17 is strongly upregulated during inflammation and cancer. The protein tissue inhibitor of metalloproteinase-3 (TIMP3) directly inhibits ADAM17 (Guinea-Viniegra et al. 2009). A recent study revealed that TIMP3 expression is part of a physiological regulation mechanism of TNF-α-activity via inhibition of ADAM17 (Guinea-Viniegra et al. 2009).

The molecular set-up of ADAM proteins is shown in Fig. 9.1. Although the prodomain is cleaved by furin within the Golgi compartment, it has been shown that the prodomain after cleavage remains attached to the mature protease and is detectable in cell supernatants. The catalytic domain contains the typical HEXXHXXGXXH motif found in all zinc dependent metalloproteases together with the conserved M-loop (Stöcker and Bode 1995) (Fig. 9.2). The zinc ion is coordinated by the three histidines and a water molecule, which is held in position

Table 9.1 Known substrates of ADAM17

Cytokines, growth factors	Receptors	Adhesion molecules	Other molecules
TNF[a,b]	p55 TNFa RI[a,b]	ICAM-1[a]	APP[a]
TGFa[a,c]	p75 TNF-a RII[a,b]	VCAM-1[a]	GP[a]
AREG[a,c]	p75 NTR[a]	NCAM[a]	CA9[a]
EREG[a]	IL-6Ra[a]	ALCAM[a]	PRNP, PrPc[a]
EPGN, Epigen[a]	IL-1R2[a]	L1-CAM[a]	KL[a]
NRG1,Heregulin[a]	NTRK1[a]	EpCAM[t]	MUC-1[a]
HB-EGF[a,c]	TrkA[a]	DSG2[a]	LYPD3, C4.4A[a]
Pref1[a]	GHR[a]	CD62L[u]	VASN[w]
Fractalkine/CX3CL1[a]	CSF1R[a]	Collagen XVII[a]	CD163[x]
TRANCE/RANKL[d]	M-CSFR[a]	PVRL4, Nectin-4[a]	PMEL17[y]
CSF-1[e]	SORL1[a]	CD44[a]	
SEMA4D[a]	SORLA[a]	F11R, JAM-A[v]	
LAG-3[a]	SORCS1[a]		
DLL1[a]	SORCS3[a]		
KL-1[a]	SORT1[a]		
KL-2[a]	CD91/APOER[j]		
MICA[a]	PTPRF, PTP-LAR[a]		
MICB[e]	EPCR[a]		
Jagged[f]	ACE2[a]		
LTA[g]	LOX-1[k]		
TMEFF2[h]	NPR[a]		
LT-3L[i]	HER4/ErbB4[a,k]		
	Notch1[a]		
	TNFRSF8, CD30[a]		
	TNFRSF5, CD40[a]		
	GPIba[a,l]		
	GPV[m]		
	GPVI[n]		
	SDC1[o]		
	SDC4[o]		
	KDR, VEGFR2[p]		
	CD89[q]		
	Ptprz[r]		
	IGF2-R[s]		
	M6P/IGF2R[s]		

[a]Arribas and Esselens (2009); [b]Bell et al. (2007); [c]Murthy et al. (2010); [d]Lum et al. (1999); [e]Boutet et al. (2009); [f]Parr-Sturgess et al. (2010); [g]Young et al. (2010); [h]Ali and Knäuper (2008); [i]Horiuchi et al. (2009b); [j]Liu et al. (2009); [k]Zhao et al. (2011); [l]Bergmeier et al. (2004); [m]Rabie et al. (2005); [n]Bender et al. (2010); [o]Pruessmeyer et al. (2010); [p]Swendeman et al. (2008); [q]Peng et al. (2010); [r]Chow et al. (2008); [s]Leksa et al. (2011); [t]Maetzel et al. (2009); [u]Arribas and Esselens (2009), Li et al. (2006); [v]Koenen et al. (2009); [w]Malapeira et al. (2011); [x]Etzerodt et al. (2010); [y]Kummer et al. (2009)

by the glutamic acid. The water molecule is in fact the nucleophile, which is responsible for the initial attack on the cleaved peptide bond (Stöcker and Bode 1995).

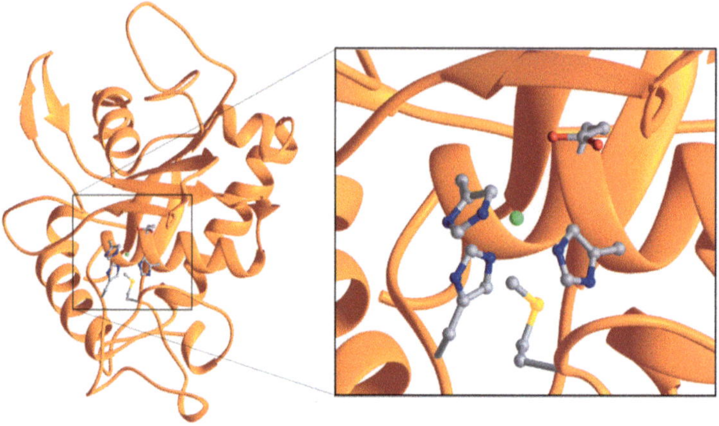

Fig. 9.2 Structure of the metalloprotease domain of ADAM17. The three-dimensional structure of the catalytic domain of ADAM17 (*left*) with close-up of the active center of ADAM17 (*right*). The side chains of the three histidines and the glutamic acid coordinating the Zn^{2+} ion (*green*) are depicted

9.3 In Vivo Function of the Major Sheddases

Activation of ADAM17 is mediated by MAP kinases, PKCs and activated oncogenes. Such signals are known to be involved in cellular activation but also in the induction of apoptosis. Activation of ADAM17, however, leads to the activation of very different pathways. On the one hand, cleavage of substrates including TNFα, IL-6R and L-selectin can be considered pro-inflammatory since they lead to a stimulation of the innate and acquired immune system. On the other hand, profound activation of the EGF-R pathway via trans-activation and via Notch cleavage has different consequences. Activation of EGF-R responses are critically involved in the regenerative response of the body to wounding but also in the growth of tumors. In studies with ADAM17 conditional knock-out animals (Horiuchi et al. 2007, 2009a, b), this has been less clear than in studies, in which ADAM17 was downregulated in all tissues (Brandl et al. 2010; Chalaris et al. 2010). Further, these mouse studies revealed the dominant role of the pro-inflammatory TNFα and of pro-regeneratory EGF-R-ligands as major ADAM17 substrates.

In the future, the molecular mechanisms leading to activation of ADAM17 in inflammation and cancer need to be understood in full detail to evaluate the feasibility of therapeutic intervention. Moreover, the available animal models have taught us that inhibition of ADAM17 can be expected to provoke negative side effects e.g. in the intestine and in the liver.

In pathophysiology, ADAM17 activation therefore governs a balance of pro- and anti-inflammatory responses leading to the appropriate response of the body to damage and stress. It should be noted that this orchestrating role of ADAM17 make it likely that this protease is deeply involved in driving inflammation-associated cancer, which is caused by a chronic activation of the immune system but which often culminates in the over-activation of members of the EGF-R family. In this respect it is noteworthy that in ADAM17 hypomorphic mice inhibition of ADAM17 led to a nearly complete loss of EGF-R signaling (Chalaris et al. 2010), which might prove to be an alternative to monoclonal antibodies or small molecules directly blocking EGF-R activity, which have been observed to elicit resistance in human tumors.

9.4 Structure

In the case of ADAM10, the prodomain contains the typically cystein-switch which binds to the active center of the catalytic domain and thereby inactivates the enzyme. Interestingly, ADAM17 does not contain this motif and it is unclear how the prodomain of ADAM17 interferes with protease activity. This difference might be responsible for a major difference in the activation step of the two proteases; whereas ADAM10 often is constitutively active, the activity of ADAM17 is normally low and needs to be stimulated by several so far incompletely understood mechanisms (see below).

There has been a long debate whether ADAM10 and ADAM17 contain a membrane-proximal EGF-like or cystein-rich domain. The structure of the two membrane-proximal domains of ADAM10 clearly demonstrated that, besides the disintegrin-domain, it contains an EGF-like domain (Janes et al. 2005). Due to the high homology of ADAM10 and ADAM17 it can be assumed that this will also hold true for ADAM17. The membrane proximal domain of ADAM17 differs in the disulfide pattern from the canonical EGF-fold, and, in addition, there is a C-terminal extension. Nevertheless, this domain belongs to the EGF-like superfamily, wherein it constitutes a new family, which should be called ADAM-EGF-like family.

For many members of the matrix-metallo-proteases (MMPs) it has been shown that membrane proximal domains with different folds (e.g. hemopexin domains) are involved in substrate recognition as so called exo-sites. In case of ADAM10 it has been shown that the EGF-like domain is responsible for the interaction with EpH, one substrate of this enzyme (Janes et al. 2005). At this point it can be speculated that also the disintegrin domain might be involved in substrate recognition by ADAM10 and ADAM17.

The role of the intracellular region of ADAM10 and ADAM17 is still unclear. It has been demonstrated that in the absence of the cytoplasmic domain, PMA stimulated shedding mediated by ADAM17 (Reddy et al. 2000). On the other hand, it has recently been shown that phosphorylation of ADAM17 at threonine 735 within the cytoplasmic portion was required for activation of shedding of TGFα by p38 MAP kinase (Peng et al. 2010).

9.5 Shedding in Health and Disease

ADAM17-deficient mice die between embryonic day 17.5 and the first days after birth (Peschon et al. 1998) and present defects in mammary epithelium, vascular system, lung, eye, hair, heart and skin. Such a phenotype was reminiscent of mice deficient for EGFR ligands and immediately pointed to the importance of the ADAM17-EGFR axis during development (Peschon et al. 1998). A background dependent subpopulation of surviving ADAM17-deficient mice revealed a reduction in lymphocyte numbers, a defect in T- and B-cell development, reduced body weight and a hypermetabolic phenotype (Gelling et al. 2008; Li et al. 2007).

More recently, conditional ADAM17 deficient mice have been developed. These mice were used to analyze the role of ADAM17 in inflammation and cancer. When mice were deficient for ADAM17 in leukocytes, monocytes and granulocytes, they were completely protected from LPS- and *E. coli*-induced septic shock due to the failure of TNFα release (Horiuchi et al. 2007). ADAM17 also affected the recruitment of neutrophils to inflamed areas by shedding of the adhesion protein L-selectin (CD62L) (Long et al. 2010).

The role of ADAM17 in the skeletal development was studied in mice with a targeted deletion of ADAM17 in all osteochondroprogenitor cells (Weskamp et al. 2010). These mice had a reduced life span and showed increased osteoblasts and osteoclasts activation and early onset of bone loss, shorter long bones and osteoporosis. Mice with a deletion of ADAM17 in endothelial cells exhibited resistance in the murine B16 melanoma tumor model and reduced pathological neovascularization in the oxygen induced retinopathy model, probably by inhibition of HB-EGF shedding. In a recent study, Murthy et al. demonstrated that ADAM17 protected hepatocytes from apoptosis. Deletion of ADAM17 in hepatocytes led to increased sensitivity to Fas-induced liver failure. In addition, adenoviral ADAM17 gene delivery prevented acute liver failure in drug-induced toxicity by blockade of apoptosis (Murthy et al. 2010).

When working with conditional ADAM17 knock-out mice—due to ubiquitous expression of ADAM17—it is often difficult to choose the appropriate transgenic *Cre* line for complex disease models, which depend on activation of several cell types. To study ADAM17 biology *in vivo* a mouse model with ADAM17 expression downregulated in all tissues seemed to be desirable. We developed a novel gene targeting strategy called exon induced translational stop (EXITS) and generated hypomorphic ADAM17$^{ex/ex}$ mice with dramatically reduced ADAM17 levels in all tissues analyzed. ADAM17$^{ex/ex}$ were viable, showed reduced shedding of ADAM17 substrates and developed very similar defects as ADAM17 knock-out mice (Chalaris et al. 2010; Peschon et al. 1998). ADAM17$^{ex/ex}$ mice showed increased susceptibility to inflammation in dextran sulfate sodium colitis, due to reduced shedding of EGF-R ligands and mucosal regeneration. A mouse strain with impaired ADAM17 activity recently generated by random mutagenesis exhibited increased inflammation upon administration of DSS confirming the protective role of ADAM17 in experimental colitis (Brandl et al. 2010; Chalaris et al. 2010).

EGF-R stimulation is involved in the growth of many tumors (Arribas and Esselens 2009; Kenny and Bissell 2007). ADAM17 is upregulated in most tumor cells (Kenny and Bissell 2007). In a process called transactivation, shedding of EGF-R ligands by ADAM17 (Table 9.1) is necessary for appropriate stimulation of EGF-R (Rodland et al. 2008). ADAM17-mediated shedding of EGF-R ligands is required for growth of lung carcinoma cells. Surprisingly, on the same cells, ADAM17 cleaved Notch and the subsequent γ-secretase mediated release of the Notch-intracellular domain led to transcriptional upregulation of EGF-R mRNA and increased EGF-R expression on the cell surface of the carcinoma cells (Baumgart et al. 2010).

Besides transcriptional upregulation, ADAM17 activity is also regulated post-translationally in cancer cells by the oncogenes ras, src and v-src leading to increased shedding of the EGF-R ligand TGFα (Maretzky et al. 2010; Van Schaeybroeck et al. 2011). In addition, activated ADAM17 and subsequent EGF-R trans-activation was shown to mediate resistance to chemotherapy (Kyula et al. 2010). Moreover, the antibody Herceptin (trastuzumab), which is used in patients with HER2–positive breast cancer, was shown to increase ADAM17 expression and activity via PKB activation. Interestingly, Herceptin does not inhibit tyrosine phosphorylation of HER2. As a consequence, shed soluble EGF-R ligands activated members of the EGF-R family (Gijsen et al. 2010). These data show that the involvement of ADAM17 in tumor growth is more complex than previously anticipated. It remains to be seen whether the role of ADAM17 during cancer development is mainly restricted to shedding of EGF-R ligands.

9.6 Substrate Recognition

The fact that more than 75 substrates for ADAM17 are described (Table 9.1) raises the question how so many substrates can be recognized. It is remarkably that type-I as well as type-II trans-membrane proteins can be cleaved. Even cleavage in trans, where the substrate is located on a neighboring cell, has been observed for ADAM10 (Janes et al. 2005).

Since there is no specific consensus cleavage sequence that is recognized by the catalytic domain, other regions or domains in the ADAMs as well in the substrates must contribute to substrate binding and recognition. Amino-acid residues lining the pocket in the structure of ADAM17 and ADAM10 to which amino-acid residues of the substrates close to the cleavage site bind are very similar (Fig. 9.3), supporting this notion. For many MMPs it has been described that other domains, e.g. the hemopexin domains of MMP-9, serve as so called exo-sites which are responsible for the specificity of substrate recognition. The role of the disintegrin- and EGF-like-domain of ADAM10 and ADAM17 in substrate recognition has been studied only partially with respect to substrates specificity. By exchanging domains of ADAM10 and ADAM17 the inability of the catalytic domain of ADAM-10 to substitute for the catalytic domain of ADAM17 in shedding the membrane-bound

```
              346        401       437
ADAM17 ....GTLGL.....LVTT....PIAVSG...
ADAM10 ....GVLGL.....ITFA....ARATSG...
              326        379       418
               *  *       *          **
```

Fig. 9.3 Sequence comparison between ADAM10 and ADAM17. Sequence segments of ADAM10 and ADAM10 lining the substrate binding pocket. *Asterisks* denote amino-acid residues, the side-chains of which are exposed and might therefore be in contact with the substrate peptide

TNFα has been shown (Reddy et al. 2000). The change in inhibitor sensitivity of TNFα shedding with one of the chimeras, namely catalytic-ADAM17/dis-EGF-ADAM-10 may arise from an increased affinity for TNFα or from a change in the subcellular localization of the TNFα proteolytic cleavage (Reddy et al. 2000). Nevertheless, the involvement of the EGF-like domain of ADAM17 in IL-$1R_{II}$ shedding has been clearly demonstrated (Reddy et al. 2000). In case of ADAM10 is has been shown by X-ray crystallography and site-directed mutagenesis that its EGF-like domain is in direct contact with the substrate Eph/ephrin (Janes et al. 2005).

The exchange of cleavage sides between molecules showed that both recognition sequences and structural elements contribute to shedding susceptibility of membrane proteins. By exchanging small peptide sequences of gp130 for cleavage-site peptides of TNFα, TGFα and IL-6R it has been shown that these short sequences conferred susceptibility to spontaneous and phorbol-ester-induced shedding of gp130 (Althoff et al. 2001). The introduction of truncated cleavage-site peptides of IL-6R into gp130 resulted in chimeras, which showed the same cleavage pattern as observed in chimeras containing the complete IL-6R cleavage site. Exchange of the juxtamembrane sequence of gp130 for the corresponding region of the leukemia inhibitory factor receptor, a protein that like gp130 is not shed, led to the generation of cleavable chimeras, supporting the notion that structural changes allow the access of the protease to a membrane-proximal region, leading to shedding of the protein (Althoff et al. 2001).

One common parameter is that most if not all cleavage sides are in close proximity to the membrane, and are very often located in flexible stalk-regions of the substrates. This property points to a confirmation of the enzymes in which the catalytic domain is bend down to the membrane. It has been suggested that the catalytic domain of ADAM17 bends back to the membrane and is bound to the EGF-like domain. Upon activation a domain or structural element of the substrate replaces this intramolecular interaction, and thereby releasing the catalytic domain. The extracellular part of mature ADAM17 comprising the catalytic-, the distintegrin- and the EGF-like domain, most probably form a C-shape structure (Janes et al. 2005; Takeda et al. 2006). This arrangement allows for the localization of the so called hyper variable region in close proximity to the active site of the catalytic domain, which supports the idea that the two membrane proximal domains are involved in substrate recognition (Janes et al. 2005; Reddy et al. 2000; Takeda et al. 2006) as well as in activation mechanism (Milla et al. 1999; Wang et al. 2009; Willems et al. 2010).

The great number of ADAM10 and ADAM17 substrates and their overlapping proteolytical activity raises the question how such multi-substrate proteases can be specifically inhibited. The three-dimensional structure of the proteolytic domain of ADAM17 and ADAM10 served as the basis for the development of many inhibitors with various degrees of specificity (Arribas and Esselens 2009; Moss et al. 2008). Some compounds block both proteases while others seems to be more specific. ADAM17 inhibitors have been studied in different preclinical models of different diseases as well as in clinical trials (Arribas and Esselens 2009; Moss et al. 2008). All these small molecule inhibitors target the active center of the protease. Given the very high structural similarity of the active centers of the proteolytic active ADAMs and other MMPs, a single molecule specificity can probably not be achieved by small molecules. In some pathological situations it might be desirable to inhibit shedding of one substrate only but not blocking ADAM10 or ADAM10 completely. Therefore the identification of exo-sites for these particular substrates has to be identified. These exo-sites then might be blocked e.g. by specific monoclonal antibodies. Interestingly, in a recent report, selective neutralizing antibodies against ADAM17 have been developed (Tape et al. 2011).

9.7 Mechanisms of Activation

ADAM10 and ADAM17 are expressed as zymogens, containing an N-terminal prodomain that is processed by furin in the trans-Golgi compartment by cleavage at the consensus RX(R/K)R motif. The prodomain lock the proteases in an inactive state by binding to the catalytic site. The isolated pro-domains from different ADAMs, including ADAM17, can act as potent, selective inhibitors of the mature forms of the proteinases. The mechanism by which the prodomain interferes with the proteolytic activity is the replacement of the zinc-bound water molecule by a cysteine residue of the prodomain and is known as the cysteine-switch and was first described for members of the MMP family. However, mutation of the corresponding cysteine residue in the corresponding region of ADAM17 and ADAM10 does not change the inhibitory potency of this prodomain (Milla et al. 2006). Therefore, the cysteine switch mechanism is not valid for these molecules. In addition to the inhibitory function, ADAM10 and ADAM17 require their prodomains for efficient transport to the cell membrane by acting as a chaperone, which protects the molecules during transport through the secretory pathway. Although cleavage of the prodomain is necessary, it is not sufficient for activation of the protease. Other natural inhibitor for ADAM10 and ADAM17 are the extracellular inhibitory proteins Tissue Inhibitor of Metallo-Proteinases (TIMPs). Whereas TIMP1 inhibits ADAM10, but not ADAM17, TIMP3 efficiently blocks ADAM17 and with a lower efficiency ADAM10 (Amour et al. 1998, 2000). The mechanism by which TIMP3 inhibits ADAM17 has been clarified by the

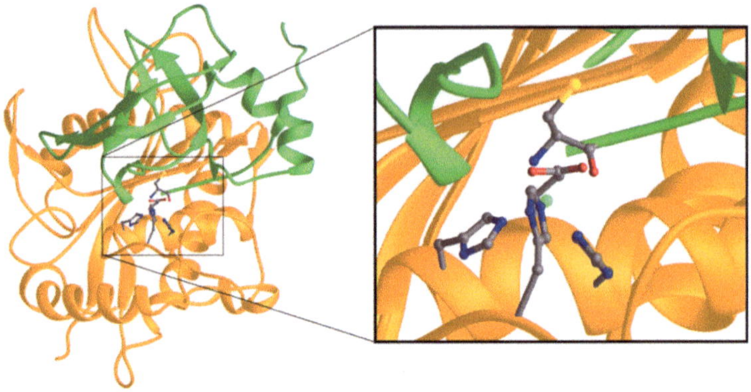

Fig. 9.4 TIMP3 is an inhibitor of ADAM17. The three-dimensional structure of the catalytic domain of ADAM17 (*yellow*) in complex with the inhibitor TIMP3 (*green*). Close-up of the active center of ADAM17 bound to TIMP3. The side chains of the three histidines and the glutamic acid of ADAM17 and the N-terminal cysteine residue of TIPM3 are depicted

three-dimensional structure of the catalytic ADAM17 domain in complex with TIMP3 (Fig. 9.4). TIMP3 binds with its four N-terminal amino-acid residues into the binding pocket of ADAM17 close to the active center thereby directing the N-terminal amino-group to the Zn-ion of the catalytic center (Wisniewska et al. 2008).

The mechanisms by which ADAM10 and ADAM17 are activated are manifold and are only partially understood. Most shedding events are induced by phorbol-esters, which are believed to mimic physiological stimuli (Blobel 2005) but little is known about the underlying mechanism. The cytoplasmic region of ADAM17 can be phosphorylated (Diaz-Rodriguez et al. 2002; Fan et al. 2003), which enhances maturation of ADAM17 (Soond et al. 2005), but this region is dispensable for phorbol-ester stimulated shedding (Doedens et al. 2003; Reddy et al. 2000). In contrast, ADAM10 is only weakly stimulated by phorbol-esters (Sahin et al. 2004; Sanderson et al. 2005) but ADAM10-dependent shedding of CD44, BTC, and N-cadherin is induced by ionophores (Nagano et al. 2004; Reiss et al. 2005; Sanderson et al. 2005).

Disturbing membrane homeostasis can induce activation of both proteases. Cellular depletion of cholesterol, breakdown of plasma membrane sphingomyelin or enrichment of the plasma membrane with ceramide induces shedding of the IL-6R by ADAM10 as well as ADAM17 (Matthews et al. 2003). Disturbing the physical integrity of the membranes by pore-forming toxins, like streptolysin O and Escherichia coli hemolysin, leads to a massive shedding of the IL-6R and the lipopolysaccharide receptor CD14 (Walev et al. 1996). Interestingly, the toxin-dependent cleavage site of the IL-6R was mapped to a position close to, but definitely distinct from that observed after stimulation with phorbol-esters (Walev et al. 1996). Also the acute phase protein C-reactive protein (CRP) activates shedding of the sIL-6R, suggesting that CRP may affect IL-6-mediated inflammatory events by enabling the formation of the IL-6/sIL-6R complex and thereby initiating trans-signaling (Jones et al. 1999).

The intracellular region of ADAM17 can also contribute to activation. Besides activation of the protease itself, the sub-cellular localization of ADAM17 plays an important role. Phosphorylation of the cytoplasmic region of ADAM17 has been described to induce the MAP-kinase pathway. ADAM17 is stored in perinuclear compartments and upon phosphorylation translocates to the cell surface. ADAM17 contains a proline rich PXXP consensus motif in its cytoplasmic region, which is able to interact with SH3 domains. In response to stress activated protein kinases by the antibiotic anisomycin, which inhibits protein synthesis, TGFα is shedd and subsequently ErbB is activated. Using an ADAM17 mutant that cannot be phosphorylated it was shown that ADAM17 has a lower basal shedding activity and that shedding was not inducible by p38 MAPK, whereas the phorbol-ester induced shedding was normal compared to the wild-type molecule (Peng et al. 2010).

The disintegrin- and EGF-like domains of ADAM10 and ADAM17 contain 28 and 26 cysteine residues, respectively. Recently it has been shown that thiol isomerases, specifically protein disulfide Isomerase (PDI) is responsible for arresting ADAM17 in its inactive form (Willems et al. 2010). The redox environment in a cell can be changed by phorbol-esters by the modulation of mitochondrial reactive oxygen species. It has been proposed that disulfide isomerization in ADAM17 may lead to a conformational change. This idea is supported by a number of adamalysin three-dimensional structures, with exhibit different disulfide bond patterns resulting in a change of the relative orientation of its domains (Guan et al. 2010).

9.8 ADAM Proteases as Therapeutic Targets

The involvement of ADAM17 into the cleavage of TNFα and IL-6R immediately stimulated the idea to block ADAM17 activity in inflammatory conditions and septic shock (Feldmann and Steinman 2005). Later it turned out that more than 70 substrates are cleaved by ADAM17 (Table 9.1). Since ADAM17 regulates the activity of the EGFR, ADAM17 activity is involved in many regenerative responses. It was shown that the regeneration of the intestinal epithelium in inflammatory bowel disease models was severely compromised in the absence of ADAM17 activity (Brandl et al. 2010; Chalaris et al. 2010), questioning the usefulness of ADAM17 blockade in inflammation. On the other hand, EGFR activity is known to be instrumental in the growth of many tumors (Kenny and Bissell 2007). Therefore, blockade of ADAM17 might be useful in late stage cancer patients, especially when tumor cells developed resistance against direct EGFR blockade by monoclonal antibodies or small molecules. More data will be needed before the usefulness of ADAM17 blockade can be fully appreciated.

References

Ali N, Knäuper V (2008) Phorbol ester-induced shedding of the prostate cancer marker transmembrane protein with epidermal growth factor and two follistatin motifs 2 is mediated by the disintegrin and metalloproteinase-17. J Biol Chem 282:37378–37388

Althoff K, Müllberg J, Aasland D, Voltz N, Kallen K-J, Grötzinger J, Rose-John S (2001) Recognition sequences and structural elements contribute to shedding susceptibility of membrane proteins. Biochem J 353:663–672

Amour A, Slocombe PM, Webster A, Butler M, Knight CG, Smith BJ, Stephens PE, Shelley C, Hutton M, Knauper V, Docherty AJ, Murphy G (1998) TNF-alpha converting enzyme (TACE) is inhibited by TIMP-3. FEBS Lett 435:39–44

Amour A, Knight CG, Webster A, Slocombe PM, Stephens PE, Knauper V, Docherty AJ, Murphy G (2000) The in vitro activity of ADAM-10 is inhibited by TIMP-1 and TIMP-3. FEBS Lett 473:275–279

Arribas J, Esselens C (2009) ADAM17 as a therapeutic target in multiple diseases. Curr Pharm Des 15:2319–2335

Baumgart A, Seidl S, Vlachou P, Michel L, Mitova N, Schatz N, Specht K, Koch I, Schuster T, Grundler R, Kremer M, Fend F, Siveke JT, Peschel C, Duyster J, Dechow T (2010) ADAM17 regulates epidermal growth factor receptor expression through the activation of Notch1 in non-small cell lung cancer. Cancer Res 70:5368–5378

Bell JH, Herrera AH, Li Y, Walcheck B (2007) Role of ADAM17 in the ectodomain shedding of TNF-alpha and its receptors by neutrophils and macrophages. J Leukoc Biol 82:173–176

Bender M, Hofmann S, Stegner D, Chalaris A, Bösl M, Braun A, Scheller J, Rose-John S, Nieswandt B (2010) Differentially regulated GPVI ectodomain shedding by multiple platelet-expressed proteinases. Blood 116:3347–3355

Bergmeier W, Piffath CL, Cheng G, Dole VS, Zhang Y, von Andrian UH, Wagner DD (2004) Tumor necrosis factor-alpha-converting enzyme (ADAM17) mediates GPIbalpha shedding from platelets in vitro and in vivo. Circ Res 95:677–683

Blobel CP (2005) ADAMs: key components in EGFR signalling and development. Nat Rev Mol Cell Biol 6:32–43

Boutet P, Agüera-González S, Atkinson S, Pennington CJ, Edwards DR, Murphy G, Reyburn HT, Valés-Gómez M (2009) Cutting edge: the metalloproteinase ADAM17/TNF-alpha-converting enzyme regulates proteolytic shedding of the MHC class I-related chain B protein. J Immunol 182:49–53

Brandl K, Sun L, Neppl C, Siggs OM, Le Gall SM, Tomisato W, Li X, Du X, Maennel DN, Blobel CP, Beutler B (2010) MyD88 signaling in nonhematopoietic cells protects mice against induced colitis by regulating specific EGF receptor ligands. Proc Natl Acad Sci USA 107:19967–19972

Chalaris A, Adam N, Sina C, Rosenstiel P, Lehmann-Koch J, Schirmacher P, Hartmann D, Cichy J, Gavrilova O, Schreiber S, Jostock T, Matthews V, Häsler R, Becker C, Neurath MF, Reiss K, Saftig P, Scheller J, Rose-John S (2010) Critical role of the disintegrin metalloprotease ADAM17 for intestinal inflammation and regeneration in mice. J Exp Med 207:1617–1624

Chow JP, Fujikawa A, Shimizu H, Suzuki R, Noda M (2008) Metalloproteinase- and gamma-secretase-mediated cleavage of protein-tyrosine phosphatase receptor type Z. J Biol Chem 283:30879–30889

Diaz-Rodriguez E, Montero JC, Esparis-Ogando A, Yuste L, Pandiella A (2002) Extracellular signal-regulated kinase phosphorylates tumor necrosis factor alpha-converting enzyme at threonine 735: a potential role in regulated shedding. Mol Biol Cell 13:2031–2044

Doedens JR, Mahimkar RM, Black RA (2003) TACE/ADAM-17 enzymatic activity is increased in response to cellular stimulation. Biochem Biophys Res Commun 308:331–338

Edwards DR, Handsley MM, Pennington CJ (2008) The ADAM metalloproteinases. Mol Aspects Med 29:258–289

Etzerodt A, Maniecki MB, Møller K, Møller HJ, Moestrup SK (2010) Tumor necrosis factor α-converting enzyme (TACE/ADAM17) mediates ectodomain shedding of the scavenger receptor CD163. J Leukoc Biol 88:1201–1205

Fan H, Turck CW, Derynck R (2003) Characterization of growth factor-induced serine phosphorylation of tumor necrosis factor-alpha converting enzyme and of an alternatively translated polypeptide. J Biol Chem 278:18617–18627

Feldmann M, Steinman L (2005) Design of effective immunotherapy for human autoimmunity. Nature 435:612–619

Fong KP, Barry C, Tran AN, Traxler EA, Wannemacher KM, Tang HY, Speicher KD, Blair IA, Speicher DW, Grosser T, Brass LF (2011) Deciphering the human platelet sheddome. Blood 117:e15–e26

Gelling RW, Yan W, Al-Noori S, Pardini A, Morton GJ, Ogimoto K, Schwartz MW, Dempsey PJ (2008) Deficiency of TNFalpha converting enzyme (TACE/ADAM17) causes a lean, hypermetabolic phenotype in mice. Endocrinology 149:6053–6064

Gijsen M, King P, Perera T, Parker PJ, Harris AL, Larijani B, Kong A (2010) HER2 phosphorylation is maintained by a PKB negative feedback loop in response to anti-HER2 herceptin in breast cancer. PLoS Biol 8:e1000563

Gooz M (2010) ADAM-17: the enzyme that does it all. Crit Rev Biochem Mol Biol 45:146–169

Guan HH, Goh KS, Davamani F, Wu PL, Huang YW, Jeyakanthan J, Wu WG, Chen CJ (2010) Structures of two elapid snake venom metalloproteases with distinct activities highlight the disulfide patterns in the D domain of ADAMalysin family proteins. J Struc Biol 169:294–303

Guinea-Viniegra J, Zenz R, Scheuch H, Hnisz D, Holcmann M, Bakiri L, Schonthaler HB, Sibilia M, Wagner EF (2009) TNFalpha shedding and epidermal inflammation are controlled by Jun protein. Genes Dev 23:2663–2674

Horiuchi K, Kimura T, Miyamoto T, Takaishi H, Okada Y, Toyama Y, Blobel CP (2007) Cutting edge: TNF-{alpha}-converting enzyme (TACE/ADAM17) inactivation in mouse myeloid cells prevents lethality from endotoxin shock. J Immunol 179:2686–2689

Horiuchi K, Kimura T, Miyamoto T, Miyamoto K, Akiyama H, Takaishi H, Morioka H, Nakamura T, Okada Y, Blobel CP, Toyama Y (2009a) Conditional inactivation of TACE by a Sox9 promoter leads to osteoporosis and increased granulopoiesis via dysregulation of IL-17 and G-CSF. J Immunol 182:2093–2101

Horiuchi K, Morioka H, Takaishi H, Akiyama H, Blobel CP, Toyama Y (2009b) Ectodomain shedding of FLT3 ligand is mediated by TNF-alpha converting enzyme. J Immunol 182:7408–7414

Janes PW, Saha N, Barton WA, Kolev MV, Wimmer-Kleikamp SH, Nievergall E, Blobel CP, Himanen JP, Lackmann M, Nikolov DB (2005) Adam meets Eph: an ADAM substrate recognition module acts as a molecular switch for ephrin cleavage in trans. Cell 123:291–304

Jones SA, Novick D, Horiuchi S, Yamamoto N, Szalai AJ, Fuller GM (1999) C-reactive protein: a physiological activator of interleukin 6 receptor shedding. J Exp Med 189:599–604

Kenny PA, Bissell MJ (2007) Targeting TACE-dependent EGFR ligand shedding in breast cancer. J Clin Invest 117:337–345

Koenen RR, Pruessmeyer J, Soehnlein O, Fraemohs L, Zernecke A, Schwarz N, Reiss K, Sarabi A, Lindbom L, Hackeng TM, Weber C, Ludwig A (2009) Regulated release and functional modulation of junctional adhesion molecule A by disintegrin metalloproteinases. Blood 113:4799–4809

Kummer MP, Maruyama H, Huelsmann C, Baches S, Weggen S, Koo EH (2009) Formation of Pmel17 amyloid is regulated by juxtamembrane metalloproteinase cleavage, and the resulting C-terminal fragment is a substrate for gamma-secretase. J Biol Chem 284:2296–2306

Kyula JN, Van Schaeybroeck S, Doherty J, Fenning CS, Longley DB, Johnston PG (2010) Chemotherapy-induced activation of ADAM-17: a novel mechanism of drug resistance in colorectal cancer. Clin Cancer Res 16:3378–3389

Le Gall SM, Bobé P, Reiss K, Horiuchi K, Niu XD, Lundell D, Gibb DR, Conrad D, Saftig P, Blobel CP (2009) ADAMs 10 and 17 represent differentially regulated components of a general shedding machinery for membrane proteins such as transforming growth factor alpha, L-selectin, and tumor necrosis factor alpha. Mol Biol Cell 20:1785–1794

Leksa V, Loewe R, Binder B, Schiller HB, Eckerstorfer P, Forster F, Cardona AS, Ondrovicová G, Kutejová E, Steinhuber E, Breuss J, Drach J, Petzelbauer P, Binder BR, Stockinger H (2011) Soluble M6P/IGF2R released by TACE controls angiogenesis via blocking plasminogen activation. Circ Res 108:676–685

Li Y, Brazzell J, Herrera A, Walcheck B (2006) ADAM17-deficiency by mature neutrophils has differential effects on L-selectin shedding. Blood 108:2275–2279

Li N, Boyd K, Dempsey PJ, Vignali DA (2007) Non-cell autonomous expression of TNF-alpha-converting enzyme ADAM17 is required for normal lymphocyte development. J Immunol 178:4214–4221

Liu Q, Zhang J, Tran H, Verbeek MM, Reiss K, Estus S, Bu G (2009) LRP1 shedding in human brain: roles of ADAM10 and ADAM17. Mol Neurodegener 4:4–11

Long C, Wang Y, Herrera AH, Horiuchi K, Walcheck B (2010) In vivo role of leukocyte ADAM17 in the inflammatory and host responses during E. coli-mediated peritonitis. J Leukoc Biol 87:1097–1101

Lum L, Wong BR, Josien R, Becherer JD, Erdjument-Bromage H, Schlondorff J, Tempst P, Choi Y, Blobel CP (1999) Evidence for a role of a tumor necrosis factor-alpha (TNF-alpha)-converting enzyme-like protease in shedding of TRANCE, a TNF family member involved in osteoclastogenesis and dendritic cell survival. J Biol Chem 274:13613–13618

Maetzel D, Denzel S, Mack B, Canis M, Went P, Benk M, Kieu C, Papior P, Baeuerle PA, Munz M, Gires O (2009) Nuclear signalling by tumour-associated antigen EpCAM. Nat Cell Biol 11:162–171

Malapeira J, Esselens C, Bech-Serra JJ, Canals F, Arribas J (2011) ADAM17 (TACE) regulates TGFβ signaling through the cleavage of vasorin. Oncogene 30(16):1912–1922

Maretzky T, Zhou W, Huang XY, Blobel CP (2010) A transforming Src mutant increases the bioavailability of EGFR ligands via stimulation of the cell-surface metalloproteinase ADAM17. Oncogene 30:611–618

Matthews V, Schuster B, Schütze S, Bußmeyer I, Ludwig A, Hundhausen C, Sadowski T, Saftig P, Hartmann D, Kallen K-J, Rose-John S (2003) Cellular cholesterol depletion triggers shedding of the human interleukin-6 receptor by ADAM10 and ADAM17 (TACE). J Biol Chem 278:38829–38839

Milla ME, Leesnitzer MA, Moss ML, Clay WC, Carter HL, Miller AB, Su J-L, Lambert MH, Willard DH, Sheeley DM, Kost TA, Burkhart W, Moyer M, Blackburn RK, Pahel GL, Mitchell JL, Hoffman CR, Becherer JD (1999) Specific sequence determinants are required for the expression of functional tumor necrosis factor-R converting enzyme (TACE). J Biol Chem 274:30563–30570

Milla ME, Gonzales PE, Leonard JD (2006) The TACE zymogen: re-examining the role of the cysteine switch. Cell Biochem Biophys 44:342–348

Moss ML, Bomar M, Liu Q, Sage H, Dempsey P, Lenhart PM, Gillispie PA, Stoeck A, Wildeboer D, Bartsch JW, Palmisano R, Zhou P (2007) The ADAM10 prodomain is a specific inhibitor of ADAM10 proteolytic activity and inhibits cellular shedding events. J Biol Chem 282:35712–35721

Moss ML, Sklair-Tavron L, Nudelman R (2008) Drug insight: tumor necrosis factor-converting enzyme as a pharmaceutical target for rheumatoid arthritis. Nat Clin Pract Rheumatol 4:300–309

Murphy G (2008) The ADAMs: signalling scissors in the tumour microenvironment. Nat Rev Cancer 8:929–941

Murphy G, Murthy A, Khokha R (2008) Clipping, shedding and RIPping keep immunity on cue. Trends Immunol 29:75–82

Murthy A, Defamie V, Smookler DS, Di Grappa MA, Horiuchi K, Federici M, Sibilia M, Blobel CP, Khokha R (2010) Ectodomain shedding of EGFR ligands and TNFR1 dictates hepatocyte apoptosis during fulminant hepatitis in mice. J Clin Invest 120:2731–2744

Nagano O, Murakami D, Hartmann D, De Strooper B, Saftig P, Iwatsubo T, Nakajima M, Shinohara M, Saya H (2004) Cell-matrix interaction via CD44 is independently regulated by different metalloproteinases activated in response to extracellular Ca(2+) influx and PKC activation. J Cell Biol 165:893–902

Parr-Sturgess CA, Rushton DJ, Parkin ET (2010) Ectodomain shedding of the Notch ligand Jagged1 is mediated by ADAM17, but is not a lipid-raft-associated event. Biochem J 432:283–294

Peng M, Guo S, Yin N, Xue J, Shen L, Zhao Q, Zhang W (2010) Ectodomain shedding of Fcalpha receptor is mediated by ADAM10 and ADAM17. Immunology 130:83–91

Peschon JJ, Slack JL, Reddy P, Stocking KL, Sunnarborg SW, Lee DC, Russell WE, Castner BJ, Johnson RS, Fitzner JN, Boyce RW, Nelson N, Kozlosky CJ, Wolfson MF, Rauch CT, Cerretti DP, Paxton RJ, March CJ, Black RA (1998) An essential role for ectodomain shedding in mammalian development. Science 282:1281–1284

Pruessmeyer J, Martin C, Hess FM, Schwarz N, Schmidt S, Kogel T, Hoettecke N, Schmidt B, Sechi A, Uhlig S, Ludwig A (2010) A disintegrin and metalloproteinase 17 (ADAM17) mediates inflammation-induced shedding of syndecan-1 and -4 by lung epithelial cells. J Biol Chem 285:555–564

Rabie T, Strehl A, Ludwig A, Nieswandt B (2005) Evidence for a role of ADAM17 (TACE) in the regulation of platelet glycoprotein V. J Biol Chem 280:14462–14468

Reddy P, Slack JL, Davis R, Cerretti DP, Kozlosky CJ, Blanton RA, Shows D, Peschon JJ, Black RA (2000) Functional analysis of the domain structure of tumor necrosis factor-alpha converting enzyme. J Biol Chem 275:14608–14614

Reiss K, Maretzky T, Ludwig A, Tousseyn T, de Strooper B, Hartmann D, Saftig P (2005) ADAM10 cleavage of N-cadherin and regulation of cell-cell adhesion and beta-catenin nuclear signalling. EMBO J 24:742–752

Rodland KD, Bollinger N, Ippolito D, Opresko LK, Coffey RJ, Zangar R, Wiley HS (2008) Multiple mechanisms are responsible for transactivation of the epidermal growth factor receptor in mammary epithelial cells. J Biol Chem 283:31477–31487

Sahin U, Weskamp G, Kelly K, Zhou HM, Higashiyama S, Peschon J, Hartmann D, Saftig P, Blobel CP (2004) Distinct roles for ADAM10 and ADAM17 in ectodomain shedding of six EGFR ligands. J Cell Biol 164:769–779

Sanderson MP, Erickson SN, Gough PJ, Garton KJ, Wille PT, Raines EW, Dunbar AJ, Dempsey PJ (2005) ADAM10 mediates ectodomain shedding of the betacellulin precursor activated by p-aminophenylmercuric acetate and extracellular calcium influx. J Biol Chem 280:1826–1837

Scheller J, Chalaris A, Garbers C, Rose-John S (2011) ADAM17: a molecular switch to control inflammation and tissue regeneration. Trends Immunol 32:380–387

Soond SM, Everson B, Riches DW, Murphy G (2005) ERK-mediated phosphorylation of Thr735 in TNFalpha-converting enzyme and its potential role in TACE protein trafficking. J Cell Sci 118:2371–2380

Stöcker W, Bode W (1995) Structural features of a superfamily of zinc-endopeptidases: the metzincins. Curr Opin Struct Biol 5:383–390

Swendeman S, Mendelson K, Weskamp G, Horiuchi K, Deutsch U, Scherle P, Hooper A, Rafii S, Blobel CP (2008) VEGF-A stimulates ADAM17-dependent shedding of VEGFR2 and crosstalk between VEGFR2 and ERK signaling. Circ Res 103:916–918

Takeda S, Igarashi T, Mori H, Araki S (2006) Crystal structures of VAP1 reveal ADAMs' MDC domain architecture and its unique C-shaped scaffold. EMBO J 25:2388–2396

Tape CJ, Willems SH, Dombernowsky SL, Stanley PL, Fogarasi M, Ouwehand W, McCafferty J, Murphy G (2011) Cross-domain inhibition of TACE ectodomain. Proc Natl Acad Sci USA 108:5578–5583

Van Schaeybroeck S, Kyula JN, Fenton A, Fenning CS, Sasazuki T, Shirasawa S, Longley DB, Johnston PG (2011) Oncogenic Kras promotes chemotherapy-induced growth factor shedding via ADAM17. Cancer Res 71:1071–1080

Walev I, Vollmer P, Palmer M, Bhakdi S, Rose-John S (1996) Pore-forming toxins trigger shedding of receptors for interleukin 6 and lipopolysaccharide. Proc Natl Acad Sci USA 93:7882–7887

Wang Y, Herrera AH, Li Y, Belani KK, Walcheck B (2009) Regulation of mature ADAM17 by redox agents for L-selectin shedding. J Immunol 182:2449–2457

Weskamp G, Mendelson K, Swendeman S, Le Gall S, Ma Y, Lyman S, Hinoki A, Eguchi S, Guaiquil V, Horiuchi K, Blobel CP (2010) Pathological neovascularization is reduced by inactivation of ADAM17 in endothelial cells but not in pericytes. Circ Res 106:932–940

Willems SH, Tape CJ, Stanley PL, Taylor NA, Mills IG, Neal DE, McCafferty J, Murphy G (2010) Thiol isomerases negatively regulate the cellular shedding activity of ADAM17. Biochem J 428:439–450

Wisniewska M, Goettig P, Maskos K, Belouski E, Winters D, Hecht R, Black R, Bode W (2008) Structural determinants of the ADAM inhibition by TIMP-3: crystal structure of the TACE-N-TIMP-3 complex. J Mol Biol 38:1307–1319

Young J, Yu X, Wolslegel K, Nguyen A, Kung C, Chiang E, Kolumam G, Wei N, Wong WL, DeForge L, Townsend MJ, Grogan JL (2010) Lymphotoxin-alphabeta heterotrimers are cleaved by metalloproteinases and contribute to synovitis in rheumatoid arthritis. Cytokine 51:78–86

Zhao XQ, Zhang MW, Wang F, Zhao YX, Li JJ, Wang XP, Bu PL, Yang JM, Liu XL, Zhang MX, Gao F, Zhang C, Zhang Y (2011) CRP enhances soluble LOX-1 release from macrophages by activating TNF-alpha converting enzyme. J Lipid Res 52:923–933

Chapter 10
Proteases in the Nervous System

Holger Cynis, Stefan F. Lichtenthaler, Leona Wagner, and Hans-Ulrich Demuth

10.1 Proteases in Alzheimer's Disease

Alzheimer's disease (AD) is characterized by the massive accumulation of the short cleavage product (beta-amyloid, Aβ) liberated from the transmembrane amyloid precursor protein (APP). The stepwise cleavage of APP is accomplished by membrane-bound β- and γ-secretase. This pathway competes with a non-amyloidogenic pathway characterized by processing of APP by α-secretase. Although the mechanism of Aβ-toxicity is still not well understood, its liberation from APP is considered as the central event in AD pathogenesis underlined by numerous mutations around the secretase cleavage sites in APP or in leading to early-onset AD.

H. Cynis
Center for Neurologic Diseases, Brigham and Women's Hospital, Harvard Medical School, 77 Avenue Louis Pasteur, Boston, MA 02115, USA

S.F. Lichtenthaler
German Center for Neurodegenerative Diseases (DZNE), Max-Lebsche-Platz 30, 81377 Munich, Germany

Institute for Clinical Chemistry and Pathobiochemistry, Neuroproteomics, Klinikum rechts der Isar, Technische Universität München, 81675 Munich, Germany

L. Wagner
Probiodrug AG, Weinbergweg 22, 06120 Halle (Saale), Germany

H.-U. Demuth (✉)
Fraunhofer Institute for Cell Therapy and Immunology (IZI), Department of Drug Design & Target Validation, Biocenter, Weinbergweg 22, 06120 Halle (Saale), Germany

K. Brix and W. Stöcker (eds.), *Proteases: Structure and Function*,
DOI 10.1007/978-3-7091-0885-7_10, © Springer-Verlag Wien 2013

10.1.1 α-Secretase

α-secretase cleavage occurs within the Aβ domain of APP and, thus, has the potential to prevent Aβ generation. α-secretase cleavage of APP occurs constitutively (constitutive α-secretase activity) (Esch et al. 1990), but can additionally be stimulated above its constitutive level (regulated α-secretase) by different pharmacological agents (reviewed in Bandyopadhyay et al. 2007; Lichtenthaler 2011), as first shown for muscarinic agonists (Nitsch et al. 1992).

The constitutive α-secretase cleavage of APP is mediated by the metalloprotease ADAM10 (A disintegrin and metalloprotease 10) [EC 3.4.24.81]. Other proteases, such as ADAM9 [EC 3.4.24.–] and ADAM17 [EC 3.4.24.86] were recently ruled-out as constitutive α-secretases (Kuhn et al. 2010). Knock-out or knock-down of ADAM10 in primary neurons as well as in several cell lines largely prevents APP α-secretase cleavage (Jorissen et al. 2010; Kuhn et al. 2010). Conversely, overexpression of ADAM10 in cell lines as well as in mouse brain increases APP α-secretase cleavage and consequently, reduces Aβ generation (Lammich et al. 1999; Postina et al. 2004). Additionally, ADAM10 cleaves APP derived peptides in vitro at the correct peptide bond (Lammich et al. 1999). The initial protease cleavage occurs between amino acids lysine16 und leucine17 of the Aβ sequence and appears to be followed by an as yet unidentified carboxypeptidase cleavage selectively removing lysine16 (Esch et al. 1990; Lammich et al. 1999; Kuhn et al. 2010).

ADAM10 has 748 amino acids and is a member of the large ADAM protease family. ADAMs are type I membrane proteins of the metzincin family requiring a zinc ion for proteolytic activity (reviewed in Edwards et al. 2008; Reiss and Saftig 2009). The large ectodomain of ADAM10 consists of an N-terminal signal peptide, followed by the prodomain, the metalloprotease domain with the conserved zinc-binding amino acid motif HEXGHXXGXXHD, a disintegrin domain and a cysteine-rich domain. In contrast to other ADAM proteases, an EGF-like domain is missing in the ADAM10 ectodomain (Janes et al. 2005). The ectodomain is followed by the transmembrane domain and a proline-rich cytoplasmic domain, which provides binding sites for SH3 domain containing proteins, such as SAP-97 (Marcello et al. 2007). A more detailed description of the ADAM protease domains, their structure and specific functions can be found in recent reviews (Edwards et al. 2008; Reiss and Saftig 2009). ADAM10 is post-translationally modified by complex N-glycosylation (Escrevente et al. 2008), and the prodomain is removed by the proprotein convertases furin and PC7 (Anders et al. 2001). Prodomain removal results in the active protease, which mediates proteolysis in late compartments of the secretory pathway and at the plasma membrane. Interestingly, the ADAM10 ectodomain can be released from the cell by ADAM9 and ADAM15 [EC 3.4.24.–] (Parkin and Harris 2009; Tousseyn et al. 2009). The ectodomain shedding is followed by γ-secretase mediated intramembrane proteolysis and translocation of the ADAM10 intracellular domain into the nucleus, where it is found in nuclear speckles, which are assumed to be involved in gene regulation (Parkin and Harris 2009; Tousseyn et al. 2009). This raises the possibility that ADAM10 may function as a signaling protein in addition to its role as protease.

APP is not the only substrate for ADAM10. By now over 30 different ADAM10 substrates have been identified, including cell adhesion proteins (e.g. N-cadherin), receptors (e.g. Notch) and growth factors (e.g. epidermal growth factor, EGF) (Pruessmeyer and Ludwig 2009; Reiss and Saftig 2009). This demonstrates that ADAM10 has a general function in the ectodomain shedding of membrane proteins. Importantly, ADAM protease-mediated shedding occurs for substrates on the same cell surface, but can also happen in trans as shown for ephrin-Eph receptor signaling (Janes et al. 2005). In vivo, the cell surface receptor Notch is a particularly important ADAM10 substrate and requires ADAM10-mediated cleavage for its signal transduction (Bozkulak and Weinmaster 2009; van Tetering et al. 2009). ADAM10 knock-out mice die embryonically due to a loss of Notch signaling (Hartmann et al. 2002). Likewise, mice with a conditional ADAM10 knock-out in the brain die perinatally, again due to a loss of Notch signaling and a major defect in brain development (Jorissen et al. 2010). An ADAM10 knock-out specifically in B cells demonstrated an essential function of ADAM10 in Notch2-mediated B cell development (Gibb et al. 2010).

ADAM10 possesses a broad substrate specificity and can cleave after distinct amino acids (Caescu et al. 2009). In the case of APP, the cleavage site seems to be located in an α-helical structure and appears to be determined by the distance from the membrane surface (Sisodia 1992), which is in agreement with ADAM10 being a membrane-bound protease.

Expression of ADAM10 is controlled at the level of transcription and translation. Posttranslational regulation of ADAM10 activity occurs through the activation of numerous signaling pathways (reviewed in Bandyopadhyay et al. 2007; Lichtenthaler 2011). Additionally intracellular protein transport is increasingly recognized as a major mechanism to control the localization of ADAM10, the access to its substrates and consequently the substrate turnover by ADAM10 (reviewed at the example of APP in Lichtenthaler 2011).

In contrast to the constitutive α-secretase the identity of the regulated α-secretase remains to be fully clarified and may be mediated by different proteases. At least ADAM10 and ADAM17 can act as regulated α-secretases after specific stimuli, such as PACAP peptides and the phorbol ester PMA (Buxbaum et al. 1998; Kojro et al. 2006). Many other stimuli, such as neurotransmitters, growth factors and cytokines, are also known to increase APP α-secretase cleavage, but it remains unclear whether the increased α-secretase cleavage occurs through ADAM10 or ADAM17 or yet other metalloproteases. In fact, overexpression of ADAM8 [EC 3.4.24.–], ADAM9 and several matrix metalloproteases also increases α-secretase cleavage of APP. This suggests that the activation or increased expression of a variety of different metalloproteases may be tested as a means to increase α-secretase cleavage and reduce Aβ generation in AD. Indeed, muscarinic agonists increase APP α-secretase cleavage and reduce amyloid pathology in an AD mouse model, presumably due to enhanced expression of ADAM17 (Caccamo et al. 2006). The therapeutic potential of ADAM10 and other metalloproteases has been reviewed elsewhere in further detail (Endres and Fahrenholz 2010; Lichtenthaler 2011).

Taken together, constitutive α-secretase cleavage of APP is mediated by ADAM10, whereas the regulated α-secretase cleavage occurs through both ADAM10 and ADAM17 and possibly additional metalloproteases.

10.1.2 *β-Secretase*

In contrast to α-secretase, β-secretase catalyzes the first step in Aβ generation and, thus, is a major drug target for Alzheimer's disease. β-secretase was identified as the transmembrane aspartyl protease BACE1 (β-site APP cleaving enzyme) [EC 3.4.23.46] (Hussain et al. 1999; Sinha et al. 1999; Vassar et al. 1999; Yan et al. 1999; Lin et al. 2000), which is related to the pepsin and retroviral aspartic protease families. Aβ-generation is largely reduced in BACE1 knock-out mice (Cai et al. 2001; Luo et al. 2001; Roberds et al. 2001; Dominguez et al. 2005). A close homolog, called BACE2 [EC 3.4.23.45], was identified shortly thereafter (Saunders et al. 1999; Yan et al. 1999; Solans et al. 2000). Much less is known about BACE2 compared to BACE1.

BACE1 is an N-glycosylated type I membrane protein with 501 amino acids. The N-terminal signal peptide is followed by a prodomain, the protease domain, a transmembrane domain and a short cytoplasmic tail. The catalytic domain comprises the two catalytic aspartic acid residues [amino acids 93–96 (DTGS) and amino acids 289–292 (DSGT)]. In the secretory pathway, the propeptide is cleaved by furin, leading to the active BACE1 protease (Bennett et al. 2000; Capell et al. 2000; Huse et al. 2000; Creemers et al. 2001). BACE1 forms dimers, which have a higher activity than the monomers (Schmechel et al. 2004; Westmeyer et al. 2004). The crystal structure of the BACE1 ectodomain shows a conserved general folding of aspartyl proteases (Hong et al. 2000).

In contrast to the ADAM proteases, which seem to be less dependent on specific amino acid motifs around the cleavage site, BACE1 has a more pronounced substrate specificity and prefers a leucine at the P1 position (Citron et al. 1995; Gruninger-Leitch et al. 2002).

BACE1 is ubiquitously expressed. The highest BACE1 expression level is found in neurons (Vassar et al. 1999), which explains why neurons are particularly vulnerable in AD due to increased Aβ generation. In mice, BACE1 expression is very high in the nervous system shortly after birth and then decreases to much lower levels (Willem et al. 2006). BACE1 localizes to the Golgi, the trans-Golgi network and to the endosomes (Vassar et al. 1999; Capell et al. 2000; Huse et al. 2000). BACE1 has an acidic pH-optimum and seems to be specifically active within acidic cellular compartments such as the late Golgi and endosomes. BACE1 activity has also been observed in other compartments, but this seems to occur only upon overexpression of the protease (Huse et al. 2002).

BACE1 is a major drug target for AD, as its inhibition lowers Aβ generation. Potent BACE1 inhibitors have been developed, but mostly do not reach sufficiently high concentrations in the brain (Vassar et al. 2009). A new generation of BACE1 inhibitors with significantly improved pharmacokinetics has been developed recently, and is currently being tested in patients. Other preclinical strategies to inhibit BACE1 activity consist of targeting BACE1 with antibodies and with modified drugs, which are specifically delivered to endosomes, where BACE1 cleaves APP (Chang et al. 2007; Rajendran et al. 2008; Mitterreiter et al. 2010; Zhou et al. 2011).

In addition to APP several other BACE1 substrates have been identified over the past few years, namely neuregulin-1 type III (NRG1), the P-selectin glycoprotein ligand-1, the sialyltransferase ST6Gal I, β-subunits of voltage-gated sodium channels, the amyloid precursor-like proteins 1 and 2, the interleukin-1 receptor 2, and the LDL receptor-related protein (Kitazume et al. 2002; Lichtenthaler et al. 2003; Li and Sudhof 2004; von Arnim et al. 2005; Wong et al. 2005; Willem et al. 2006; Kuhn et al. 2007). Additionally, a proteomic study identified further proteins as potential BACE1 substrates (Hemming et al. 2009). However, most of these substrates were not confirmed under BACE1 knock-out or knock-down conditions and should be discussed with some care. The most prominent phenotypic change in the BACE1 knock-out is a hypomyelination in the peripheral nervous system during postnatal development, which stems from the reduced cleavage of the BACE1 substrate NRG1 (Hu et al. 2006; Willem et al. 2006). Remyelination also appears to be affected (Hu et al. 2008; Farah et al. 2011). Moreover, BACE1-deficient mice show behavioral changes related to schizophrenia, which may also be due to the reduced NRG1 cleavage (Savonenko et al. 2008). Additional functions of BACE1 in the central nervous system, such as epileptic seizures, are less well understood (Hitt et al. 2010; Hu et al. 2010).

BACE1 protein expression is controlled at the transcriptional level, for example by the transcription factors NFkB, PPARγ and YY1 (reviewed in Rossner et al. 2006). Additionally, distinct mechanisms control the translation of the BACE1 mRNA, including the 5′ non-translated region, a naturally occurring antisense transcript and microRNAs (De Pietri Tonelli et al. 2004; Lammich et al. 2004; Rogers et al. 2004; Faghihi et al. 2008; Hebert et al. 2008; O'Connor et al. 2008; Wang et al. 2008). The translational mechanisms seem to be dysregulated in AD, providing an explanation for the 2–5-fold increase in BACE1 protein levels observed in AD brains.

10.1.3 *Alternative β-Secretases*

10.1.3.1 Problems with BACE1 as Sole β-Secretase

BACE 1 was identified as major β-secretase in vivo since genetic deficiency in mice dramatically reduced the amount of generated Aβ-peptides (Vassar et al. 1999; Cai et al. 2001). However, some concerns have been raised for BACE1 being the only protease possessing β-secretase activity in humans.

A major criticism is the exceptionally low catalytic specificity of BACE1 for cleavage of wildtype APP with k_{cat}/K_M between 40 and 62 $M^{-1}\ s^{-1}$ in solution (Lin et al. 2000; Shi et al. 2001) compared to, e.g. artificial BACE1 substrates being cleaved with k_{cat}/K_M $3.42 \times 10^5\ M^{-1}\ s^{-1}$ (Turner et al. 2001).

A second concern is the spectrum of N-terminal Aβ variants generated by BACE1, which differs significantly from the Aβ-composition found in human sporadic Alzheimer's disease (Kawarabayashi et al. 2001). BACE1 dominantly generates full-length Aβ-peptides starting with an N-terminal aspartate in vitro and in vivo, whereas a heterogeneous mixture of N-terminally truncated and modified Aβ peptides has been identified in human brain extracts (Saido et al. 1996). Among others, pyroglutamate (pGlu)-modified Aβ-peptides are a major species found in human AD brain (Saido et al. 1995). The truncated Aβ peptides might be generated by subsequent aminopeptidase cleavage by, e.g. aminopeptidase A (Sevalle et al. 2009) however, other studies suggest a direct liberation of Aβ-peptides differing from the full-length variants by alternative processing (Cynis et al. 2008).

A third concern is the identification of BACE1 using screening techniques based on a rare human familial Alzheimer's disease modification, namely the "Swedish" mutation (KM595/596/NL) (Vassar et al. 1999). Introduction of the "Swedish" mutation into the APP sequence makes it a much better substrate for BACE1 (k_{cat}/K_M 1.03×10^5 M^{-1} s^{-1}) than wildtype APP (Hook et al. 2008a, b) and. From the present perspective it is not surprising that BACE1 was identified using the APPswedish variant for beta-secretase screening, however, it might have precluded the identification of other putative candidates with higher preference for wildtype APP.

Although information on putative alternative β-secretases besides BACE1 are limited and frequently contradictory, considerable evidence exists, that also other proteases might at least exert some β-secretase activity in vitro and/or in vivo.

10.1.3.2 Cathepsins

Among the first proteases suspicious for β-secretase activity, several cathepsins were studied. These investigations included the cysteine proteases cathepsin B (CatB) [EC 3.4.22.1], cathepsin L (CatL) [EC 3.4.22.15] and cathepsin S (CatS) [EC 3.4.22.27] belonging to the papain family, in addition to the aspartic protease cathepsin D (CatD) [EC 3.4.23.5] belonging to the pepsin A family of proteases (Cataldo and Nixon 1990; Cataldo et al. 1997; Ladror et al. 1994; Munger et al. 1995). The rationale for investigating lysosomal proteases was the finding of internalization of APP from the cell surface and its degradation in the endosomal-lysosomal compartment. Among the tested cathepsins only CatS showed considerably increased Aβ production in co-transfection experiments in 293 cells. Neither CatB and CatL nor CatD showed an effect on APP processing under these conditions (Munger et al. 1995). CatD was later ruled out as relevant β-secretase since the gene knock out did not change Aβ generation in mice (Saftig et al. 1996). In contrast, CatB remained a putative alternative β-secretase candidate (Hook and Reisine 2003). The findings are mainly based on the sub-cellular distribution of BACE1 and CatB in primary chromaffin cells as model system for constitutive and regulated secretion. CatB seems to be localized in the regulated secretory pathway, whereas BACE1 was found to be primarily present in vesicles of the constitutive secretory pathway. Since neurons and neuronal-like cells like bovine chromaffin cells release neurotransmitters and neuropeptides as well as Aβ in a regulated fashion, it was postulated, that CatB accounts for the majority of Aβ released from chromaffin vesicles. The approach was substantiated by application of E64d as CatB-specific inhibitor to wildtype APP and "Swedish" APP mice as well as the analysis of the respective gene knock outs crossed with APP overexpressing mice (Hook et al. 2008a, b, 2009, 2011).

However, it has to be noticed, that these findings are controversial. Mueller-Steiner et al. found CatB to be involved in Aβ degradation rather than Aβ generation (Mueller-Steiner et al. 2006) and the role of CatB as alternative β-secretase has not been confirmed by other groups so far.

10.1.3.3 Caspases

Caspases play an essential role in apoptosis by cleaving a subset of cellular polypeptides at Asp-X bonds (Nicholson and Thornberry 1997). It has been demonstrated, that apoptotic cells secrete higher amounts of Aβ compared to controls (LeBlanc 1995). Therefore, the idea was raised, that activated caspases might play a role in APP turnover under apoptotic conditions. Especially caspase 3 [EC 3.4.22.56] from peptidase family C14 was found to be able to cleave APP but not in a β-secretase like fashion (Gervais et al. 1999). Interestingly, besides the absence of direct β-secretase activity of caspase 3, increased amounts of Aβ in the cell culture supernatant have been demonstrated under apoptotic conditions suggesting a shift of substrate into the amyloidogenic pathway after caspase activation. Therefore, apoptosis in general and caspase 3 in particular might play an indirect role for Aβ generation.

In addition to caspase 3, caspase 6 [EC 3.4.22.59] from peptidase family C14 was shown to be able to cleave Aβ directly at the β-secretase site. As it has already demonstrated for BACE1 and CatD, caspase 6 preferentially cleaves the "Swedish" mutant of APP with k_{cat}/K_M 2.2×10^4 M^{-1} s^{-1}, a specificity only 4–5 times lower than the specificity constant of BACE1 for "Swedish" APP. Notably, also the wildtype APP sequence was processed by caspase 6 with a specificity constant of k_{cat}/K_M 0.8×10^4 M^{-1} s^{-1}, which makes wildtype APP a 14–20 times better substrate for caspase 6 than for BACE 1 in vitro (Hook et al. 2008a, b; Gervais et al. 1999). However, the sub-cellular localization of caspases in the cytosol and their restricted activation upon apoptosis excludes them as relevant β-secretases, since the physiological turnover of APP was found to be approx. 8 %/h (Bateman et al. 2006). Nevertheless, under certain conditions, caspases could increase the heterogeneity of N-terminal Aβ-species, since caspase-activation was found to lead preferentially to the generation of truncated Aβ(2–x) via caspase 6 (Gervais et al. 1999) or to Aβ(5–x) via an yet unidentified mechanism (Takeda et al. 2004).

Summarizing, compelling evidence exist, that BACE1 is not the sole enzyme possessing β-secretase activity. Instead it appears that a number of proteases are able to cleave APP around the β-secretase cleavage site leading to the generation of a heterogeneous mixture of N-terminal truncated Aβ variants later found in brains of patients suffering from Alzheimer's disease.

10.1.4 γ-Secretase: A Hetero-tetrameric Intramembrane Protease Complex

γ-secretase cleaves the C-terminal fragments arising through the initial α- or β-secretase cleavage and results in the secretion of the p3 and Aβ-peptides, respectively. γ-secretase is an unusual aspartyl protease, in that it cleaves its substrate proteins within the phospholipid bilayer of the membrane. γ-secretase belongs to the larger family of GXGD proteases, which also comprises signal peptide peptidase and its homologs (reviewed in Fluhrer et al. 2009). γ-secretase forms a hetero-tetrameric protein complex (Seeger et al. 1997; Capell et al. 1998; Thinakaran et al. 1998; Yu et al. 1998; Li et al. 2000a) composed of its catalytic subunit presenilin (PS) [EC 3.4.23.–] (Steiner et al. 1999; Wolfe et al. 1999; Esler et al. 2000; Kimberly et al. 2000; Li et al. 2000b; Seiffert et al. 2000) and three other proteins nicastrin (NCT), APH-1 and PEN-2 (Yu et al. 2000; Lee et al. 2002; Steiner et al. 2002; Edbauer et al. 2003; Kimberly et al. 2003; Takasugi et al. 2003). These four proteins are necessary and sufficient for γ-secretase activity, as shown by reconstitution experiments in yeast (Edbauer et al. 2003).

All four γ-secretase subunits are integral membrane proteins. NCT is a type I membrane protein and is the largest subunit of the complex (Yu et al. 2000). PS has nine transmembrane domains (TMDs) and harbors the two catalytically active aspartyl residues within its TMDs 6 and 7 (Henricson et al. 2005; Laudon et al. 2005; Oh and Turner 2005a, b; Spasic et al. 2006). APH-1 spans the membrane seven times (Fortna et al. 2004), while PEN-2 contains two TMDs and is the smallest subunit of the γ-secretase complex (Crystal et al. 2003). Mammalian cells have two homologs of PS (PS1 and PS2) as well as of APH-1 (APH-1a and APH-1b) (Francis et al. 2002). Additionally, APH-1a exists as a short and a long splice variant (Lee et al. 2002). PS1 and PS2 as well as APH-1a and APH-1b do not coexist in the same γ-secretase complex (Yu et al. 1998; Steiner et al. 2002; Hebert et al. 2004; Shirotani et al. 2004). As a consequence up to six different γ-secretase complexes can exist. Whether these differ in their proteolytic properties, specifically in their ability to produce Aβ42 remains controversial (Shirotani et al. 2007; Serneels et al. 2009). The γ-secretase complex is assumed to contain one protein of each subunit (1:1:1:1 ratio) (Sato et al. 2007), but there is also evidence for a dimeric γ-secretase complex (Schroeter et al. 2003).

Several proteins have been described as γ-secretase interactors (Wakabayashi et al. 2009; Winkler et al. 2009). They are not integral subunits of the complex, but appear to be transient interactors, which may modulate the activity or the intracellular trafficking of γ-secretase. Examples are transmembrane protein 21 (TMP21)

(Chen et al. 2006a, b) and the recently identified γ-secretase activating protein GSAP (He et al. 2010).

A high-resolution structure of the γ-secretase complex is not yet available, but biochemical experiments demonstrated that the aspartyl-containing TMDs 6 and 7 of PS face each other and form a hydrophilic pore or cavity (Sato et al. 2006; Tolia et al. 2006), in which a substrate is assumed to be cleaved. Cross-linking experiments revealed several amino acids, which are part of this cavity including residues of the catalytic GXGD motif (Sato et al. 2006; Tolia et al. 2006). Amino acids in other TMDs of PS also contribute to formation of the cavity. In addition, the evolutionarily conserved amino acids PAL, which are very close to TMD6, are part of the cavity (Sato et al. 2008; Tolia et al. 2008).

γ-secretase cleaves over 80 different type I membrane proteins (Haapasalo and Kovacs 2011), demonstrating a broad role in regulated intramembrane proteolysis. Similar to the α-secretase ADAM10, the major γ-secretase substrate during development is the Notch receptor. Mice deficient in PS1 die embryonically from a loss-of-Notch signaling phenotype (De Strooper et al. 1999). Binding of a ligand induces ADAM10-mediated ectodomain shedding of Notch. This is followed by γ-secretase-mediated intramembrane proteolysis, which results in the translocation of the Notch intracellular domain into the nucleus, where it acts as a transcriptional activator of Notch target genes. Given the broad spectrum of γ-secretase substrates, a complete inhibition of γ-secretase is associated with severe side effects in mice and men and, thus, is not longer considered a viable approach to therapeutically reduce Aβ levels in AD patients.

The γ-secretase complex assembles in the endoplasmic reticulum. After assembly is complete, PS is cleaved in its large cytoplasmic loop domain between TMDs 6 and 7 into characteristic N- and C-terminal fragments. This cleavage is considered to activate γ-secretase and occurs autocatalytically (Fukumori et al. 2010).

γ-secretase does not cleave the full-length proteins, but only after their ectodomain was shortened by ectodomain shedding (e.g. α- or β-secretase cleavage). The shortened substrates seem to first bind to an exosite, before getting access to the active site. Although evidence was presented that NCT serves as an initial substrate receptor, recognizing the free N-terminus of the substrate (Shah et al. 2005), follow-up studies have yielded both additional supporting data (Dries et al. 2009) as well as further conflicting data (Chavez-Gutierrez et al. 2008; Martin et al. 2009; Zhao et al. 2010). γ-secretase cleavage occurs in a stepwise manner. This has been elucidated in detail for APP and seems to occur in a similar manner for other substrates, such as Notch and CD44. APP is first cleaved by γ-secretase at the C-terminal end of the transmembrane domain. This initial cleavage, called the ε-cleavage site (Gu et al. 2001; Sastre et al. 2001; Yu et al. 2001; Weidemann et al. 2002), is followed by further γ-secretase cleavages, each removing three or four amino acids from the C-terminus (Takami et al. 2009), until the resulting peptide is short enough to be released from the membrane. The ε-cleavage site is located after amino acid 48 or 49 of the Aβ-sequence. This gives rise to two distinct "product lines". The major product line is Aβ49–Aβ46–Aβ43–Aβ40–Aβ37 from which Aβ40 is the principal end product (Qi-Takahara et al. 2005). The minor product line is Aβ48–Aβ45–Aβ42–Aβ38 giving rise to the pathogenic Aβ42 and the non-pathogenic Aβ38 in comparable amounts. Mutations in PS, which are linked to familial forms of AD, affect the product lines and result in more of the pathogenic Aβ42 relative to Aβ40 and Aβ38 (reviewed in Lichtenthaler et al. 2011).

γ-secretase has a broad substrate specificity (Tischer and Cordell 1996; Lichtenthaler et al. 1999, 2002; Murphy et al. 1999), but amino acids in the juxtamembrane domains and within the transmembrane domains can affect the total amount of γ-secretase cleavage as well as the sites of γ-secretase cleavage (Zhang et al. 2002; Ren et al. 2007; Hemming et al. 2008). Additionally, dimerization of the APP TMD has been suggested to modulate the final cleavage sites of γ-secretase (Munter et al. 2007, 2010). In this model, APP dimerization sterically prevents γ-secretase from further shortening the Aβ42 peptides to Aβ38. Reducing APP dimerization—by site-directed mutagenesis or by using γ-secretase modulatory drugs—allows γ-secretase to continue cleavage, resulting in more Aβ38 and less Aβ42.

Taken together, γ-secretase is an unusual aspartyl protease, which acts as a heterotetrameric complex and cleaves APP and other substrates within the membrane.

10.1.5 Aβ-Degrading Proteases

10.1.5.1 Rationale

Due to the pivotal role of β- and γ-secretase in generation of Aβ-peptides, Alzheimer's disease was considered as a result of abnormal proteolysis followed by misfolding of the excessively produced Aβ molecules. However, it has been shown, that Aβ is a product of physiological APP turnover (Haass et al. 1992), rather than a per se pathophysiological side-product. Turnover of APP was found to be approx. 8 %/h throughout the life span of a human and impaired Aβ-degradation rather than increased Aβ-generation has been identified in AD patients (Bateman et al. 2006; Mawuenyega et al. 2010). Therefore, catabolism of Aβ peptides and identification of catabolic enzymes is crucial for understanding AD. To date a number of potential Aβ-degrading proteases have been identified.

10.1.5.2 Neprilysin and Endothelin-Converting Enzymes

Neprilysin (NEP) [EC 3.4.24.11] also known as neutral endopeptidase or CD10 is a zinc-dependent, membrane-bound metalloendopeptidase from the peptidase family M13 playing a pivotal role in Aβ-degradation in vivo. NEP was identified by investigating the degradation of radio-labeled Aβ(1–42) in the rat brain (Iwata et al. 2000). In this study, it could be shown, that a thiorphan and phosphoramidon-sensitive protease later identified as NEP accounts for the majority of the Aβ-degradation potential in vivo. The important role was further underlined by investigations using NEP knockout mice showing elevated A-β-deposition (Iwata et al. 2001).

In addition, the NEP-related endothelin-converting enzymes 1 and 2 (ECE-1 and -2) [EC 3.4.24.71] are able to degrade Aβ in vitro (Eckman et al. 2001). Similar to NEP, ECEs are membrane-bound zinc-dependent metalloendopeptidases of peptidase family M13. Especially ECE-1 is present at the cell membrane and within compartments of the secretory pathway and might be able to degrade Aβ molecules in vivo. First evidence comes from ECE-1 heterozygous knockout mice showing increased Aβ-deposition (Eckman et al. 2003). In addition, NEP and ECE-1 knockouts possess a more pronounced Aβ-deposition, than the single knockouts underlining a concerted proteolysis of Aβ in vivo (Eckman et al. 2006).

10.1.5.3 Insulin-Degrading Enzyme

The insulin-degrading enzyme (IDE) [EC 3.4.24.56] is another example of an Aβ-degrading metalloendopeptidase. IDE belongs to the peptidase family M16 and was described to be primarily localized in the cytosol with a prominent function in insulin degradation (Duckworth et al. 1998). In addition, IDE was found to degrade Aβ-peptides in cell culture (Vekrellis et al. 2000), however, the localization of IDE in the cytosol raised some skepticism about the relevance for extracellular Aβ degradation. Recently, IDE was shown to be secreted via an "unconventional" pathway into the extracellular space (Zhao et al. 2009). This finding is in line with increased Aβ deposition in IDE knock-out animals underlining the importance of IDE for Aβ metabolism (Farris et al. 2003, 2004). Together with NEP and the mitochondrial presequence protease "PreP", IDE possesses a characteristic catalytic chamber, which is able to encapsulate peptides of a length of 70 amino acids or less. This might explain the selectivity of these proteases for their respective substrates including Aβ (Malito et al. 2008; de Strooper 2010).

10.1.5.4 Other Aβ-Degrading Proteases

Besides NEP and IDE, a number of different proteolytic enzymes have been shown to be able to degrade Aβ peptides in vitro. Among them the metalloendopeptidase Matrix Metalloproteinase 2 (MMP-2) [3.4.24.24], MMP-9 [EC 3.4.24.35] and the serine protease Plasmin [3.4.21.7], have been identified to possess a potential for Aβ degradation.

MMP-2 and MMP-9 belong to the peptidase family M10 and degrade large macromolecules of the extracellular matrix such as collagens (e.g. IV, V, VII, X). They are expressed at low levels in brain but can be induced by stress in cell culture. Of notice is the ability of MMP-9 to degrade fibrils in contrast to most other Aβ degrading proteases (Yan et al. 2006; de Strooper 2010).

Plasmin is a serine protease of peptidase family S1, which is activated by proteolytic cleavage with uPA or tPA from its precursor protein plasminogen. Plasmin is primarily responsible for degradation of fibrin aggregates in blood (Takada and Takada 1988) and the ability of degrading fibrin aggregates makes it an interesting protease for Aβ aggregate turnover in AD. Indeed, the expression of plasminogen and its activator proteins uPA and tPA have been linked to the central nervous system (Sappino et al. 1993). In addition, uPA and tPA can be induced under certain conditions such as ischemic insults or excitotoxicity (Tsirka et al. 1995). Plasmin has been shown to degrade fibrillar Aβ in vitro with approximately 1/10th of the rate of plasmin degrading fibrin (Tucker et al. 2000). However, the knockout of plasminogen does not show altered steady-state levels of Aβ (Tucker et al. 2004). Obviously, there is no significant contribution of plasmin to the physiological catabolism of Aβ, which is conceivable in the light of plasmin activation as prerequisite step for its proteolytic function. However, under defined experimental conditions such as traumatic brain injury, e.g. by intra-cortical application of Abeta peptides, the expression of uPA and tPA might be induced, which eventually leads to plasmin activation and, therefore, to significant turnover of Aβ peptides (Melchor et al. 2003).

In addition, Angiotensin-converting enzyme (ACE) [EC 3.4.15.1] has been shown to possess some potential of Aβ degradation. ACE is able to degrade monomeric Aβ in vitro (Hu et al. 2001; Zou et al. 2007), however, this finding could not be corroborated in vivo. Neither genetic inactivation (Eckman et al. 2006) nor treatment with anti-hypertensive drugs inhibiting ACE, such as captopril (Hemming et al. 2007) showed an effect on Aβ levels in mouse brain. Therefore, despite the genetic linkage of ACE polymorphisms to the risk of developing AD (Hu et al. 1999; Farrer et al. 2000), a direct role of ACE in Aβ-catabolism could not be provided.

Finally, in spite of its discussed role as putative alternative β-secretase, CatB also has been implicated in Aβ degradation in vivo. It has been found that CatB deficiency in mice expressing the "Swedish" and "Indiana" mutant of APP leads to an increased plaque load in hippocampus and cortex. Furthermore, the ability of CatB for degradation of monomeric and fibrillar Aβ could be demonstrated and the concentration of Aβ peptides generated from wildtype APP in CatB-deficient neurons was increased suggesting a catabolic activity of CatB for Aβ (Mueller-Steiner et al. 2006). As mentioned above, especially the role of CatB in AD is highly contradictory and needs further elucidation.

10.2 Proline-Specific Peptidases in Brain and Neurodegeneration

Nature has evolved a number of neuropeptides, neurotrophic peptide hormones and cardiovascular peptides with a proline residue determining their structural conformation and biological activity. In general, the proline peptide bond has shown to be resistant to proteolytic cleavage and therefore an exclusive number of proline specific peptidases have emerged to regulate these peptides (Yaron and Naider 1993). These include two endopeptidases, five dipeptidyl peptidases, two aminopeptidases, two carboxypeptidases and two dipeptidases. They are either serine peptidases of clan SC or metallopeptidases of clan MG or MH. Except for the three members of clan MG, all of them are post-proline peptidases, whereas soluble aminopeptidase P (sAmpP), membrane aminopeptidase P (mAmpP) and prolidase from subfamily M24B are able to hydrolyze the imide peptide bond. Interestingly, out of the 12 peptidases, 9 require proline at the penultimate position and 7 of them truncate at the N-terminus. Thus, a great number of neuropeptides contain a penultimate proline at their N-termini and their truncation results either in altered receptor selectivity or inactivation (Table 10.1). The best-characterized post-proline dipeptidyl aminopeptidase is dipeptidyl peptidase 4 (DP 4), followed by dipeptidyl peptidase 2 (DP 2) (Lambeir et al. 2003; Maes et al. 2007). However, only recently three additional post-proline dipeptidyl aminopeptidases have been discovered (Abbott and Gorrell 2002; Abbott et al. 2000; Ajami et al. 2004; Scanlan et al. 1994; Gorrell 2005). The resulting X-Pro dipeptides readily cross the cell where they are cytosolically metabolized by prolidase (Cunningham and O'Connor 1997; Mitsubuchi et al. 2008).

10.2.1 Serine Peptidases of Clan SC

Serine peptidases of the SC clan have a unique catalytic triad in the order of Ser, Asp and His located in an α/β-hydrolase fold compared to the chymotrypsin catalytic triad of His, Asp, Ser. They are comprised of both exopeptidases as well as endopeptidases and include the families S9, S10, S15, S28, S33 and S37. However, only S9 and S28 contain eukaryotic proline-specific peptidases, whereas X-Pro dipeptidyl peptidase (S15.001) of S15, prolyl aminopeptidase (S33.001) of S33 as well as prolyl tripeptidyl peptidase (S09.017) of S9 are only distributed in microbial species (http://www.merops.sanger.ac.uk).

10.2.1.1 Prolyl Oligopeptidase Family S9

The S9 family, also referred to as prolyl oligopeptidase family, consists of four subfamilies represented by their respective enzymes prolyl endopeptidase (9A),

Table 10.1 Summary of known and potential substrates of proline-specific peptidases in brain

Substrate	Sequence	Functions	Cleavage by proline-specific peptidases	Comment
Neuropeptides				
Neuropeptide Y	YPSKPDNPGEDAP-AEDMARYYSALRHYINLITRQRY-NH_2	Food-intake, stress	DP4, AmpP1, AmpP2	Alteration
Peptide YY	YPIKPEAPREDASPE-ELNRYYASLRHYL-NLVTRQRY-NH_2	Food-intake	DP4, AmpP2	Alteration
Substance P	RPKPQQFFGLM-NH_2	Nociception, stress	DP4, PEP, AmpP1, AmpP2	Alteration
GALP	APAHRGRGGWTL-NSAGYLLGPVLHLPQMGDQDGKRE-TALEILDLWKAI-DGLPYSHPPQPS	Food-intake, reproduction	DP4	Inactivation
Endomorphin-1	YPWF-NH_2	↓ Nociception	DP4, AmpP1	Inactivation
Endomorphin-2	YPFF-NH_2	↓ Nociception	DP4, AmpP1	Inactivation
PACAP	HSDGIFTDS-YSRYRKQMAV-KKYLAAVLGKRYKQRVKNK	Neurotroph, behavior, circardian	DP4	Inactivation
VIP	HADGVFTSDFS-KLLGQLSAKK-YLESLM	Circardian, sleep	DP4	Inactivation
Thyroliberin	PEHP-NH_2	↑ TSH	PEP	Inactivation
Arg-vasopressin	CYFQNCPRG-NH_2	Aggression	PEP	Inactivation
Oxytocin	CYIQNCPLG-NH_2	Neuromodular, anxiety	PEP	Inactivation
Neurotensin	pELYENKPRRPYIL	Behavior	PEP, neurolysin	Inactivation
Galanin	GWTLNSAGYLL-GPHAVGNHRS-FSDKNGLTS	Food-intake, stress	PEP	Inactivation
Human orexin B	RSGPPGLQGR-LQRLLQASGN-HAAGILTM-NH_2	Food-intake, stress	PEP	Inactivation
Rat orexin B	RPGPPGL-QGRLQRLLQANG-NHAAGILTM-NH_2	Food-intake, stress	DP4	Inactivation
b-Endorphin	YGGFMTSEK-SQTPLVTLFKNAIIK-NAYKKGE	↓ Nociception	PEP	Inactivation
CLIP	PVKVYP-NGAEDESAEAF-PLEF	Circardian rhythm	PEP, AmpP1	Inactivation
MSH-a	Ac-SYSMEHFRWGKPV-NH_2	Anorexic, behavior	PEP, PCP	Inactivation
Tyr-MIF-1	YPLG-NH_2	↓ Nociception ↓ aMSH	DP4, AmpP1	Inactivation

Morphiceptin	YPFP-NH_2	↓ Nociception	DP4, AmpP1	Inactivation
Kentsin	TPRK	↓ Nociception	DP4, AmpP1	Inactivation
b-Casomorphin	YPFVEPI	↓ Nociception	DP4, AmpP1	Inactivation
Carnosine	b-Ala-His	Neuroprotection	Prolidase	Inactivation
Neurotrophic growth factors				
GLP-1	HAEGTFTSDVS-SYLEGQAAKEFIA-WLVKGRG-NH_2	Anorexic, memory	DP4	Inactivation
IGF-1	GPETLCGAEL-VDALQFVCGDRGFYFNKPTG-YGSSSRRAPQT-GIVDECCFRSCD-LRRLEMYCAPLKPAKSA	Neurotroph	DP4	Inactivation
GPE	GPE	Neuroprotective	DP2	Inactivation
Cardiovascular peptides				
Bradykinnin	RPPGFSPFR	Vasodilator	mAmpP, sAmP, PEP	Inactivation
[DesArg] bradykinin	RPPGFSPF		PCP	
Angiotensin I	DRVYIHPFHL	Vasoconstrictor	PEP	
Angiotensin II	DRVYIHPF	Vasoconstrictor	PEP, PCP	Activation
Angiotensin III	RVYIHPF	Vasoconstrictor	PEP, PCP	Inactivation

dipeptidyl peptidase 4 (9B), acyl aminoacyl peptidase (9C) and glutamyl endopeptidase (9D) as well as several peptidases and homologues not assigned to any of the above subgroups (Abbott and Gorrell 2002; Barrett and Rawlings 1992; Gorrell 2005; Polgar 2002; Rosenblum and Kozarich 2003; Yu et al. 2010) (http://www.merops.sanger.ac.uk).

PEP Gene Family S9A

Prolyl endopeptidase (PEP) [EC 3.4.21.26] is the only true proline-specific endopeptidase, representing the S9A subfamily. It is also the only monomer, consisting of 710 amino acids with a molecular weight of approx. 80 kDa (Shirasawa et al. 1994). The human gene of PEP is located on 6q22 (Garcia-Horsman et al. 2007). PEP is located cytosolically in the perivascular space and is associated with the microtubulin cytoskeleton (Schulz et al. 2005; Rossner et al. 2005). It is ubiquitously expressed with the highest expression found in brain, kidney, testis and thymus, whereas very low levels of PEP could be detected in the liver (Myohanen et al. 2008c). In brain it is mainly expressed in neurons of the cerebral cortex, CA1 of hippocampus and Purkinje cells of cerebellum. In cerebral cortex and hippocampus, PEP was specifically expressed in glutameric pyramidal cells, while it also co-localized with gamma-aminobutyric acid (GABAergic) and cholinergic interneurons of the cortex and thalamus. However, no expression of PEP could be detected in nigrostriatal dopaminergic neurons or in astrocytes (Myohanen et al. 2008b). Altered activity and expression of PEP were found in aged wildtype mice, APP transgenic mice and human brains of patients with Alzheimer's disease (AD), revealing increased levels of PEP in the hippocampus of aged wildtype and APP transgenic mice, whereas lower PEP levels were found at the amyloid-beta plaques in brains of human AD and APP transgenic mice (Rossner et al. 2005). Application of PEP selective inhibitors in vivo suggested PEP to play an important role in learning as well as memory formation and PEP has been reported to be involved in the maturation and degradation of several peptide hormones and neuropeptides (Table 10.1), with substance P (SP), arginine vasopressin (AV), thyroliberin, gonadoliberin and alpha melanocyte stimulating hormone (α-MSH) implicated to be physiological substrates according to in vivo animal models (Garcia-Horsman et al. 2007; Perroud et al. 2009; Toide et al. 1995a, b, 1996; Shinoda et al. 1995; Morain et al. 2002; Bellemere et al. 2003, 2005; Schneider et al. 2002; Yamanaka et al. 1999). The crystal structure of PEP revealed a two-domain structure comprised of a α/β-hydrolase domain typical for the SC clan with catalytic triad Ser554, Asp641 and His680 and an open seven bladed propeller domain. Since the catalytic triad is covered by a central tunnel of an unusual beta propeller, the size of the peptide substrates is restricted up to approximately 30 amino acids (Fulop et al. 1998). Nonetheless, although the presence of a membrane-bound PEP has been reported in the brain, PEP currently generally considered to be located cytosolically associated with the perinuclear cytoskeleton, where it has no access to hydrolyze secreted or vesicular neuropeptides (O'Leary et al. 1996; O'Leary and O'Connor 1995; Schulz et al. 2005; Tenorio-Laranga et al. 2008). Recently PEP has been reported to hydrolyze fragments of α-synuclein, thereby stimulating aggregation of α-synuclein (Brandt et al. 2005, 2008). In addition, PEP has been shown to be involved in the regulation of inositol-3-phosphate (IP3) signaling by its activity and association with SP, NK1 and IP3-1R (Myohanen et al. 2008a; Schulz et al. 2002). Furthermore, PEP has also been described to bind to growth-associated protein 43 (GAP43, neuromodulin) (Di Daniel et al. 2009). Finally, PEP has been associated with various neurodegenerative diseases such as Huntington's Parkinson's, Alzheimer's and Lewis body disease (Mannisto et al. 2007).

Dipeptidyl Peptidase 4 (DP4) Gene Family S9B

DP4

DP4 [EC 3.4.14.5] belongs to the serine peptidase clan SC, subfamily 9B. Currently, four members have been identified belonging to the dipeptidyl peptidase subfamily 9B, including dipeptidyl peptidase 4 (DP4), fibroblast activation protein alpha (FAP), dipeptidyl peptidase 8 (DP8) and dipeptidyl peptidase 9 (DP9) (Lambeir et al. 2003; Scanlan et al. 1994; Abbott et al. 2000; Olsen and Wagtmann 2002; Ajami et al. 2004; Abbott and Gorrell 2002).

DP 4 as representative member of the DP4 gene family is the best-characterized post-proline-dipeptidyl peptidase with most known in vivo substrates (Table 10.1) (Lambeir et al. 2003). The human gene location of DP4 is 2q24.2, encompassing 81.8 kb, spanning 26 exons, that code for two mRNAs of 2.8 kb and 4.2 kb, respectively (Abbott et al. 1994). Interestingly, the nucleotides coding for the residues of the catalytic triad are found on three different exons. The resulting protein has 766 amino acids and the primary structure consists of a short six amino acid cytoplasmic tail, a 22 amino acid transmembrane, a 738 amino acid extracellular portion comprised of a flexible stalk, glycosylation rich region, cysteine rich region and catalytic region with the catalytic triad. Although DP4 is a type II transmembrane glycoprotein, it has also a soluble shedded form in the blood circulation as illustrated in Fig. 10.1 (De Meester et al. 1999; Mentlein 2004; Cordero et al. 2009). The human crystal structure of DP4 reveals two domains, an eight bladed propeller and a catalytic α/β-hydrolase domain. The active site is composed of the catalytic triad Ser630, Asp708 and His740, two anchoring residues Glu204, Glu205 as well as substrate stabilizing residues Arg125, Asn710 and Tyr457. The propeller is open and consists of two subdomains made up of blades II–V and VI–VIII, I, respectively. Each blade has four anti-parallel β-sheets, except for blade IV that has an additional α-helix and two β-sheets forming an extended arm. There are two openings, a side opening and a propeller tunnel (Rasmussen et al. 2003). In a crystal structure, the substrate NPY suggested an entry at the side opening (Aertgeerts et al. 2004b). DP4 is reported to be a homodimer with glycosylation contributing to 23 % of the molecular weight of 110 kDa per subunit

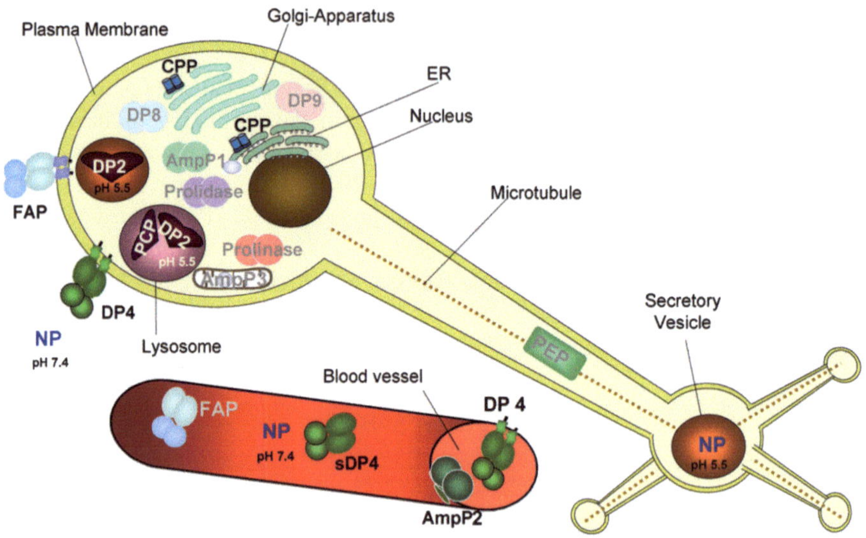

Fig. 10.1 Subcellular localization of proline-specific peptidases. Schematic representation of a neuron and a supplying blood microcapillary including pH-values in different compartiments. Further depicted are the discussed proline-specific peptidases and their sub-cellular localization. Note, that DP4 and FAP exist as membrane bound form present at the plasma membrane. In addition, a soluble form can be found in circulation. NP represents the sub-cellular localization of processed neuropeptides

(Lambeir et al. 2003). Tetramerization has also been described in porcine DP4 (Engel et al. 2003). Post-transcriptional modifications include nine N-terminal glycosylation sites and five disulfide bonds, though O-glycosylation and phosphorylation have also been reported (Rasmussen et al. 2003; Engel et al. 2003; Fan et al. 1997; Aertgeerts et al. 2004a; Alfalah et al. 2002; Tansi et al. 2010; Bilodeau et al. 2006). DP4 is known to cleave neuropeptides such as NPY, substance P, endomorphin 1 and 2; peptide hormones including GLP-1, GLP-2, GIP, glucagon, IGF-1, GHRH as well as various chemokines (Lambeir et al. 2003). According to kinetic analysis, DP4 has a very high selectivity for the neuropeptides NPY and PYY (Mentlein 1999; Medeiros and Turner 1993, 1994, 1993; Lambeir et al. 2003). Thus, it is involved in food up-take, anxiety, stress, cardiovascular, nociception, glucose homeostasis and chemotaxis. Furthermore, it functions as an extracellular adhesion molecule by binding to collagen, fibronectin and plasminogen. In addition, it is implicated in various immune responses via its interaction with several immunological molecules such as ADA or CD45 and acts as a marker for activated T-cells (De Meester et al. 1999; Ohnuma et al. 2008a, b). It is ubiquitously distributed with the highest expression in kidney, lung, liver and small intestine, whereas low expression is found in brain, heart and skeletal muscle (Lambeir et al. 2003; Gossrau 1979; Dikov et al. 1999; Frerker et al. 2007; Yu et al. 2009). However, though low levels of DP4 are found in brain parenchyma, elevated activity and expression of DP4 could be detected in the meninges, brain capillaries, choroid plexus and circumventricular organs (CVOs) (Aimes et al. 2003; Barnes et al. 1991; Bourne et al. 1989; Nagy et al. 1996; Frerker et al. 2007; Chappa et al. 2006, 2007). These results imply that DP4 is at the interphase between the CNS and the periphery via the blood circulation and CSF respectively, thereby modulating and inactivating neuropeptides and neurotrophic growth factors (Table 10.1). This already explains the possible involvement of DP4 in social and stress-related behavior as observed in DP4−/− knock-out mice, DP4-deficient rats and after in vivo application of DP4 inhibitors, since the CVO's median eminence and area postrema link the CNS with the endocrine and immune system, thereby integrating the two stress axis hypothalamic-pituitary adrenal axis (HPA) and neuro-sympathico axis and resulting in the release of stress hormones and altered cytokines (El Yacoubi et al. 2006; Frerker et al. 2009; Karl et al. 2003; Lautar et al. 2005; Sewards and Sewards 2003; Elenkov et al. 2000). Aging and neurodegeneration is associated with a dysfunctional blood brain barrier (BBB) causing it to become more leaky and eventually disintegrate (Zlokovic 2008, 2010; Pahnke et al. 2009; Altman and Rutledge 2010; Dickstein et al. 2010; Ujiie et al. 2003; Bell et al. 2010; Bell and Zlokovic 2009; Deane et al. 2009; Deane and Zlokovic 2007; Mackic et al. 1998, 2002; Martel et al. 1996; Zlokovic et al. 1990). Thus, DP4 at the altered microvascular may regulate by its enzymatic activity the bioavailability of neuropeptide, neurotrophic growth hormone, cardiovascular peptides and immunological substrates and thus contribute to the pathogenesis of neurodegeneration such as vascular dementia, brain infarcts and Alzheimer's disease.

FAP

FAP [EC 3.4.21.–] also referred to as seprase, has the highest sequence identity to DP4 and is believed to arise from gene duplication due to its gene proximity. FAP is localized on 2q24.3, encompassing 72.8 kb, spanning 26 exons and yielding a 2.8 kb mRNA, that codes for 760 amino acids. Introns are also found between the exon coding for the catalytic residues. Similar to DP4, FAP is a type II glycoprotein. In addition to a low DP4 activity, FAP also exhibits gelatinase and collagenase activity, which is collagen type I specific. Its endopeptidase activity requires the sequence Xaa-Gly-Pro-Yaa (Edosada et al. 2006a, b). Recently, the crystal structure of FAP has been elucidated and comparison with the crystal structure of DP4 points to a lower anchoring of substrates by Glu203–Glu204 due to shielding effects of surrounding hydrophobic residues and lack of Asp663. This in turn, results in a lower exopeptidase activity and enables its endopeptidase activity as confirmed by site-directed mutagenesis with subsequent kinetic studies (Aertgeerts et al. 2005). Although low levels of FAP mRNA have been detected in healthy tissue, expression at protein level is restricted to pathogenic tissues such as human meningioma and astrocytoma, in particular in close vicinity to extracellular matrix. Interestingly, while the contributions of DP8 and DP9 were higher in meningioma compared to FAP and DP4, increased levels of DP4/FAP were associated with malignancy (Stremenova et al. 2007, 2010; Mentlein et al. 2011). Thus, the expression of FAP has been investigated in several types of cancers, as to identify it as a pharmaceutical target (Garin-Chesa et al. 1990; Henry et al. 2007; Ariga et al. 2001; Iwasa et al. 2003, 2005; Jin et al. 2003; Huber et al. 2003; Mori et al. 2004; Okada et al. 2003; Goodman et al. 2003; Huang et al. 2004; Skubitz and Skubitz 2004; Wesley et al. 2004).

DP8

DP8 [EC 3.4.14.5] consists of 882 amino acids, has a molecular weight of 100 kDa and its human gene localization is 15q22, encompassing 71 kb, spanning 20 exons, encoding four alternatively spliced mRNA products between 3.8 and 4.1 kb and yielding a 882 amino acid protein (Abbott et al. 2000). Although DP8 has previously been reported to be monomeric, current data gave strong evidence for a dimeric structure with an apparent molecular weight above 200 kDa (Bjelke et al. 2006a). So far, it has been suggested to be located in the cytoplasm as a soluble protein and up to now, there has been no evidence for any secretion, though recently loose association of extra-cellular DP8 on plasma membrane was reported (Abbott et al. 2000; Chen et al. 2003a, b; Bank et al. 2011). Recent proteomic screening has revealed phosphorylation of Tyr331 and Thr334 (Yu et al. 2007). Using several chromogenic substrates, DP-8 was shown to display post-proline dipeptidyl aminopeptidase activity similar to that of DP4 (Abbott et al. 2000; Bjelke et al. 2006a). Hydrolysis of NPY, GLP-1, GLP-2, PYY, ITAC, IP-10, SDF-1α and SDF-1β, but not of MIG, Gro-β and Eotaxin could be demonstrated in vitro, though the rate of cleavage was slower compared to DP-4, in particular for PYY (Bjelke et al. 2006a, b; Ajami et al. 2008). The mRNA of DP-8 is ubiquitously distributed with its highest expression in brain and peripheral tissues, such as testis and ovaries (Abbott et al. 2000; Wagner et al. 2006; Qi et al. 2003). In baboon brains, DP 8 was shown to be highly expressed in Purkinje cells and neuronal cells in the granular layer, as well as neurites in the molecular layer, but not in cortex, forebrain or midbrain as shown by ISH. However, homogenates of mice cortex as well as pooled mid and hindbrain showed high levels of DP8 and/or DP9 activity (Yu et al. 2009). Nevertheless, its physiological function is presently unknown and still awaits further studies.

DP9

DP9 [EC 3.4.14.5] has previously been reported to be active as a cytosolic monomer, comprised of 863 amino acids with a molecular weight of approximately 100 kDa (Ajami et al. 2004). It lacks a transmembrane domain and is found intracellular near the Golgi complex, though secretion from transfected cells has not been observed yet. The gene is located on chromosome 19p13.3, encompassing 48 kb, spanning 23 exons and encoding for two alternatively spliced mRNA isoforms of 4.4 kb and 5.0 kb, respectively, with different tissue distribution (Frerker et al. 2007; Abbott and Gorrell 2002; Ajami et al. 2004; Bjelke et al. 2006a). The enzymatical active protein contains 892 amino acids (Bjelke et al. 2006a). This variant was shown to be active as homodimer with an estimated molecular weight above 200 kDa, whereas no activity could be detected for the shorter variant comprised of 863 amino acids (Olsen and Wagtmann 2002). Using several chromogenic substrates, DP9 exhibited post-proline dipeptidyl aminopeptidase activity similar to that of DP4 and it was shown to truncate NPY, GLP-1, GLP2 and to a far lesser extent PYY in vitro (Table 10.1) (Bjelke et al. 2006a, b). However, very recently the cytoplasmic proteasome-derived antigenic peptide RU134-42 could be identified as the first natural substrate of DP9, suggesting DP9 to play an important role in peptide turnover and antigen presentation (Geiss-Friedlander et al. 2009). DP9 contains an Arg-Gly-Asp cell attachment motif and two potential glycosylation sites, though SDS-analysis of expressed DP9 revealed no mass differences based on deglycosylation (Olsen and Wagtmann 2002; Qi et al. 2003). Similar to mRNAs of DP4 and DP8, mRNA of DP9 is ubiquitously distributed however, with its highest expression in liver, heart and skeletal muscle (Qi et al. 2003; Ajami et al. 2004; Olsen and Wagtmann 2002; Yu et al. 2009). Its physiological function has not been elucidated yet, though an up-regulation of DP9 mRNA was detected in human meningioma (Stremenova et al. 2010).

So far, one cannot differentiate between DP8 and DP9 enzymatic activity due to lack of selective inhibitors, however DP8/DP9 activity could be detected in rodent brain (Frerker et al. 2007). Furthermore, brain and testis have been the only organs in which DP8/DP9 activity precedes over DP4 activity (Yu et al. 2009; Frerker et al. 2007; Dubois et al. 2008). Interestingly, there seems to be a difference between the distribution of DP8 and DP9 in primates such as baboon compared to rodents, expressing low and high levels in cortex, respectively (Yu et al. 2009; Frerker et al. 2007; Ansorge et al. 2009). Similarly, histochemical enzymatic staining of brain sections obtained from DP4 deficient Fischer rats as well as their respective wildtype, revealed ubiquitous staining of DP8 and/or DP9 in both DP4-deficient as well as DP4 wildtype rats, whereas DP4 was only found at the meninges in wildtype rats. Using a DP4-selective inhibitor as well as a DP4-like inhibitor P32/98 in both fluorogenic activity assay with Ala-Pro-AMC as well as NPY hydrolysis by MALDI-TOF-MS, high levels of DP8 and/or DP9 were found in extracts of primary neuronal cells from rat compared to glial cell extracts.

Yet, the suggested cytotoxicity of DP8/DP9 inhibition is currently controversially discussed (Lankas et al. 2005; Burkey et al. 2008; Kirby et al. 2010; Wu et al. 2009; Bank et al. 2011; Ansorge et al. 2011). Non-enzymatic functions of DP8/DP9 include cell adhesion, migration and apoptosis (Yu et al. 2006). Interestingly, similar to PEP, but unlike DP4 and FAP, DP8 and DP9 are reversibly inactivated by H_2O_2 oxidation involving two cysteins in each monomer (Park et al. 2008). Up to now, there are no crystal structures of DP8 and DP9 available, although molecular modeling based on DP4 and FAP crystal structures indicate similar overall structures comprised of a β-propeller and a α/β-hydrolase domains, with the active site being located at the interphase of the two domains. However, two loops and one helix of the propeller domain extending to the interphase cavity appear to play a role at the active site. While the R125-loop and the E205/E206-helix have been described in the crystal structures of DP4, the P2-loop, containing F357 and R358, seems to be unique to DP8 and DP9 and is suggested to influence substrate and inhibitor binding to the P2-pocket (Rummey and Metz 2007; Engel et al. 2003; Rasmussen et al. 2003; Thoma et al. 2003; Oefner et al. 2003). Due to the shortest gene sizes, lowest numbers of exons, the active site being located in one exon in comparison to DP4 and FAP as well as their closest phylogenetic relationship with respect to prokaryotic members of the family, DP8 and DP9 are believed to be the ancestral genes of the DP4 gene family (Abbott and Gorrell 2002).

Although all of these enzymes described above display DP4-like activity, they exhibit distinct features with respect to cellular compartmentation and glycosylation as illustrated in Fig. 10.1.

Dipetidyl Peptidase-Like Proteins

DP-like proteins 1 (DPL1, DP6, DPP X, S09.973) and 2 (DPL2, DP10, DPP Y, S09.974) were previously assigned to the DP4-gene family S9B based on their sequence homologies to DP4 (Abbott and Gorrell 2002; Gorrell 2005). However, since they lack DP4-like activity due to mutations at the active site, they were moved to S9 family unassigned to any subfamily (http://merops.sanger.ac.uk/cgi-bin/famsum?family=S9). Both of them are type II membrane-bound glycoproteins, suggested to interact with the voltage-gated potassium channel Kv4 (Abbott and Gorrell 2002; Strop et al. 2004; Wada et al. 1992; Chen et al. 2003a, b, 2006a, b; Nadal et al. 2003; Kin et al. 2001; Zagha et al. 2005; Qi et al. 2003). While DPL1 is exclusively expressed in the brain as two variants, i.e. a short and a long form, DPL2 is found in brain, pancreas and adrenal gland (Qi et al. 2003; Chen et al. 2006a, b). The crystal structure of DPL1 resembles that of DP4 (Strop et al. 2004). DPL2 has additionally been associated with asthma.

Ex-DP4-Like Enzymes

In addition, enzymes structurally unrelated to the DP4 gene family have also been reported to display DP4 activity. These include attractin (Duke-Cohan et al. 1996) and N-acetyl alpha-linked acidic dipeptidases (NAALADases I, NAALADases II and NAALADases L) from the metalloprotease clan MH, family M28B (Pangalos et al. 1999). However, elevated DP4-like activity of the NAALADases has previously been detected only in crude extracts after cloning and expression, whereas detailed kinetic analysis of expressed and purified NAALADase I did not reveal any DP 4-like activity (Pangalos et al. 1999; Barinka et al. 2002). Likewise, the DP 4-like activity of attractin in the serum has been controversially discussed for several years (Friedrich et al. 2003; Duke-Cohan et al. 1995, 1996, 2004; Durinx et al. 2000) and only recently been disproved (Friedrich et al. 2007).

10.2.1.2 Family S28

As a member of clan SC, the S28 family has its catalytic residues in the order of Ser, Asp and His located in a α/β hydrolase fold. Peptidases of S28 contain exopeptidases that hydrolyze prolyl bonds, and are known only from eukaryotes. Human members include Prolyl carboxypeptidase (PCP), dipeptidyl peptidase 2 (DP2) and thymus-specific serine protease (TSSP), though the catalytic specificity of latter enzyme has not been elucidated yet. Intriguingly, while DP2 displays post-proline dipeptidyl aminopeptidase activity like the DP4-gene family, its high sequence homology to prolyl carboxypeptidase (PCP) designates it to family S28 (Soisson et al. 2010; Kozarich 2010).

Prolylcarboxypeptidase

PCP [EC 3.4.16.2] (prolycarboxypeptidase, Proline carboxypeptidase, lysosomal Pro-X carboxypeptidase, S28.001) hydrolyzes amino acids from the C-terminal end of oligopeptides having proline at the penultimate position (Skidgel et al. 1981). It is found soluble in lysosomes, having a pH optimum of 5.5, but its enzymatic activity is still retained at neutral pH (Odya and Erdos 1981; Odya et al. 1978). The gene location of PCP is 11q14, encompassing 147 kb covering nine exons that code for a 2.2 kb mRNA. The resulting protein contains 496 amino acids, composed of a signal peptide (1–21), propeptide (22–45) and mature PCP (46–496) (Tan et al. 1993; Watson et al. 1997; Abeywickrema et al. 2010). The enzyme is functional as a homodimer and each subunit has a molecular weight of 58 kDa (Tan et al. 1993). PCP is heavily N-glycosylated, containing six N-glycosylation sites, and the enzyme has also four disulfide bonds (Chen et al. 2009; Soisson et al. 2010). The crystal structure (pdb: 3N2Z) has only recently been elucidated, revealing the typical catalytical α/β-hydrolase domain as well as a novel SKS domain composed of five α-helices forming a novel bundle that caps the catalytical α/β-hydrolase domain. Interestingly, the active site holding the catalytic triad Ser179, Asp430 and His455 has apparently an additional charge-relay system that links the catalytic His455 with His456 and Arg460 and might even suggest a kind of dual catalytic triad bifurcated off Ser179, the ultimate nucleophile. Such His456/Arg460 diad may explain the acidic pH optimum of PCP (Soisson et al. 2010; Kozarich 2010). PCP is expressed in hypothalamus but also in lymphocytes, fibroblasts, endothelial cells, kidney and lung. It is also found soluble in CSF (Zhao et al. 2010a, b; Wardlaw 2011; Palmiter 2009; Kumamoto et al. 1981; Odya and Erdos 1981; Odya et al. 1978). Substrate hydrolysis include inactivation of Angiotensin II and III, desArg-bradykinnin and α-MSH, activation of prekallikrein and degradation of YPRPIHPA peptide fragment of human endothelin B-receptor-like protein 2 in CSF (Shariat-Madar et al. 2002, 2004; Moreira et al. 2002; Wallingford et al. 2009; Zhao et al. 2010a, b; Yang et al. 1968). Hence, it has been implicated in cardiovascular functions such as hypertension, blood coagulation, food-intake, tissue proliferation and smooth-muscle growth and proposed as a pharmaceutical target for obesity, hypertension and anti-inflammatory therapy (Adams et al. 2011; Mallela et al. 2009; Shariat-Madar et al. 2010; Ngo et al. 2009; Diano 2011; Wang et al. 2006b; Zhou et al. 2010; Zhu et al. 2010; Hagedorn 2011; Javerzat et al. 2009).

Dipeptidyl Peptidase 2 (DP2)

DP2 [EC 3.4.14.2] (DP7, quiescent proline cell dipeptidase, QPP, EC 3.4.14.2, S28.002) is a proline specific serine protease, that hydrolyzes dipeptides from the N-termini of tripeptides and small peptide fragments if proline, norisoleucine or to a lesser extent alanine are at the penultimate position (Leiting et al. 2003; Maes et al. 2005; Mentlein and Struckhoff 1989). DP2 has also been identified with quiescent proline cell dipeptidase (QPP) based on genetic homology and kinetic parameters (Araki et al. 2001; Leiting et al. 2003; Maes et al. 2005). The human gene is located on chromosome 9q34.3 corresponding to 3p13 in rat and the human gene encompasses 2,850 kb comprised of 13 exons. The soluble serine protease contains a proform and has a length of 492 amino acids with a molecular weight of 58 kDa (Underwood et al. 1999; Chiravuri et al. 2000a; Maes et al. 2007). Glycosylation and dimerization are required for the catalytic activity and latter occurs via a leucine zipper motif, which is novel for proteases (Chiravuri et al. 2000a, b). Recently, the crystal structure of DP2 was published in the Protein Data Bank (Bezerra et al. 2012), revealing a α/β-hydrolase domain as well as a novel SKS domain, comprised of 5 α-helices arranged in a helix bundle fold, capping the active site. An insertion from the SKS domain to the active site results in steric hindrance of larger substrates and contains Asp334 for anchoring N-terminus of the peptide substrate.

The homodimer is located either in cellular vesicles that are distinct from lysosomes with secretion being regulated by an increased Ca^{2+} flux, in lysosomes and parts of Golgi complex or in lipofuscin granules (Chiravuri et al. 2000a; Gorenstein and Ribak 1985; Gorenstein et al. 1985; Maes et al. 2007; McDonald et al. 1968). In addition, in some brain fractions a membrane-associated form of DP2 has also been reported and northern blot analysis of rat DP2 revealed two spliced variants in the brain (Araki et al. 2001; Mentlein and Struckhoff 1989). DP2 is ubiquitously distributed with high expression in kidney, brain, testis, heart, resting lymphocytes and differentiated macrophages (Gossrau and Lojda 1980; Maes et al. 2006, 2007; Chiravuri et al. 1999; Underwood et al. 1999). Within various brain regions, high enzyme activities were reported in rat pituitary, human cerebellum and rat human hypothalamus (Maes et al. 2005). In contrast, others found the highest level of DP2 activity in rat brain extracts from the circumventricular organs and meninges, followed by cerebellum, hypothalamus, hippocampus, striatum, subventricular zone (SVZ), amygdala, cortex and spinal cord (Wagner et al. 2008). Furthermore, real-time RT-PCR analysis revealed the highest expression of DP2 mRNA in CVOs followed by cerebellum, moderate levels in hypothalamus, hippocampus, amygdala, striatum, SVZ and spinal cord, whereas cortex had the lowest mRNA (Wagner et al. 2008). Since it was previously thought to be a lysosomal enzyme, its physiological function to date is unknown. However, altered serum activities of DP2 have been associated with various pathogenic conditions, such as Sjörgen Syndrome, rheumathoid arthritis, Lupus erythematosus, various cancers and Parkinson's disease (Maes et al. 2007). Lately, neurogenin 3-specific DP2-/- deficiency revealed a phenotype with impaired glucose tolerance, insulin resistance and visceral obesity (Danilova et al. 2009). Using chromogenic substrates, DP2 displays post-proline dipeptidyl aminopeptidase activity similar to DP4, however over a broad pH range with an acidic to neutral pH optimum (Underwood et al. 1999; Mentlein and Struckhoff 1989; Leiting et al. 2003; Maes et al. 2005, 2007). While DP2 readily hydrolyses tripeptides, its activity decreases rapidly with increasing chain length of peptide. Thus, it was shown to cleave only fragments of substance P1–4, bradykinin1–3 or bradykinin1–5 (Mentlein and Struckhoff 1989; Brandt et al. 2006; Frerker et al. 2007). Recently the N-terminal truncated tripeptide Gly-Pro-Glu (GPE) of IGF-1 was found to be a substrate of DP2. GPE has been reported to be neuroprotective by stimulating acetylcholine secretion and it has been associated with various neurodegenerative diseases such as Huntington's disease and AD. DP2 has been reported to be involved in apoptosis as a decrease of DP 2 activity caused cells to exit their G0-phase in quiescent lymphocytes and fibroblasts, resulting in an induction of apoptosis by up-regulation of p53 and c-Myc as well as a down-regulation of Bcl-2 (Chiravuri et al. 1999; Chiravuri and Huber 2000; Mele et al. 2009). Nevertheless, another study reported participation in necrosis rather than apoptosis (Maes et al. 2006). Interestingly, ADA was recently discovered to also bind to DP2 through with an order of magnitude lower potency compared to DP4 (Sharoyan et al. 2008).

10.2.2 Metallo-peptidases of Clan MG

Peptidases of the clan MG are described as being co-catalytic having water as nucleophile bound by two cobalt or manganese ions that are ligated by Asp, Asp, His, Glu, Glu. The metal ligands are pentahedrally coordinated by a Glu and the catalytic residues are three histidines. The MG clan contains only the family M24 that is divided into subfamilies M24A and M24B, respectively. The peptidases of the two subfamilies are grouped together on the basis of a common 'pitta-bread fold' comprised of both α-helices and an anti-parallel β-sheets within two structurally similar domains that are thought to be derived from an ancient gene duplication (Bazan et al. 1994). The active site is located between the two domains. The subfamily M24B holds the unique proline-specific peptidases aminopeptidase P1 (AmpP1), aminopeptidase P2 (AmpP2), aminopeptidase P3 (AmpP3) and Prolidase (PEPD), that are able to cleave the imidopeptide bond N-terminally of proline (http://merops.sanger.ac.uk/cgi-bin/famsum?family=m24b).

10.2.2.1 X-Prolyl Aminopeptidases

AmpP1

AmpP1 [EC 3.4.11.9] (X-prolyl aminopeptidases 1, XPNPEP1, M24.009) is a soluble cytosolic protein, lacking the hydrophobic signal sequence at the N-terminus and the GPI-anchor at the C-terminus (Cottrell et al. 2000a). The human gene location is 10q25.3, spans 59 kb with 19 exons and yields a 2.5 kb mRNA (Sprinkle et al. 2000). The resulting protein is a homodimer with each subunit comprised of 623 amino acids and a molecular weight of 71 kDa (Vanhoof et al. 1997a, b; Cottrell et al. 1998; Sprinkle et al. 2000). AmpP1 is not glycosylated, but has N6-acetylation at Lys304 (Choudhary et al. 2009). Recently, a 1.6 Å resolution structure of AmpP1 (pdb: 2CTZ) was elucidated, revealing an atypical three domain structure, compared to two domains in bacterial X-Pro aminopeptidase (pdb: 1A16) and prolidase (pdb: 2OKN_A). The three domains are referred to as N-terminal domain I, middle domain II and C-terminal domain III, containing the active site. While domain I and II are composed of a six-stranded mixed β-sheet, flanked by six α-helices, domain III contains three β-sheets of variable strands covered by six α-helices. The active site consists of three nucleophilic water molecules bound to two Mn^{2+} ions ligated by Asp415, Asp426, H489, Glu523 and Glu537. The side-chains of His395, His485, His498 and Glu41, surrounding the two Mn^{2+} ions are suggested to play a role in substrate recognition and catalysis (Li et al. 2008). Due to its proline specificity, AmpP1 is suggested to hydrolyze peptide hormones, neuropeptides such as tachykinins and otherwise resistant dietary protein fragments, as deficiency of AmpP1 results in excretion of large amounts of imino-oligopeptides. Thus, in vitro substrates of AmpP1 include bradykinin, NPY, β-casomorphin, substance P (SP), [Tyr1]-melanostatin, corticotropin-like intermediate lobe peptide (CLIP), IL-6, morphiceptin and kentsin (Harbeck and Mentlein 1991; Rusu and Yaron 1992; Frerker et al. 2007). Unlike the remaining proline-specific peptidases, AmpP1 is able to cleave Xaa-Pro-Pro-Yaa peptides, such as bradykinin, catalyzing it in a divalent cation-dependent manner (Griswold et al. 1996, 1999). AmpP1 is found in the brain parenchyma, particularly in astrocytes, though its physiological function in brain has not been elucidated yet. In addition, AmpP1 is ubiquitously distributed in peripheral tissues, with its highest expression in pancreas, followed by heart and muscle (Mentlein et al. 1990; Frerker et al. 2007).

AmpP2

AmpP2 [EC 3.4.11.9] (X-prolyl-aminopeptidase 2, XPNPEP2, EC 3.4.11.9, M24.005) is a GPI-anchored membrane-bound aminopeptidase encoding for 673 amino acids with a molecular mass of 75.5 kDa. The human gene localization is Xq25 encompassing 31 kb, containing 20 exons and a mRNA of 3.5 kb (Sprinkle et al. 1998; Venema et al. 1997). In addition to lipidation at Ala649 by GPI-amidation, AmpP2 is also heavily N-glycosylated, with five sugar chains contributing to about 15 % of the molecular weight, thereby increasing the predicted molecular weight to about 90 kDa per subunit. However, the functional enzyme forms a homooligomer of 217–360 kDa (Ersahin et al. 2005). Prior to GPI-anchoring to the plasma membrane, the C-terminal propeptide has to be removed from the precursor to yield the mature form of AmpP2. Currently, there is no crystal structure of AmpP2 available, however, model building of the catalytic C-terminal domain and site directed mutagenesis suggest an active site with two Mn^{2+} ions ligated to Asp450, Asp461, His524, Glu555 and Glu569 as well as the catalytic residues His430, His524 and His533, all located at the typical 'pitta-bread fold' (Cottrell et al. 2000b). AmpP2 is highly expressed in kidney, followed by lung, heart, placenta, liver, small intestine, colon as well as endothelial and smooth muscle cells of capillaries and lymphatic vessels, but not in brain, skeletal muscle, pancreas, spleen, thymus, prostate, testis, ovary, or leukocytes (Venema et al. 1997; Cottrell et al. 2000b; Mentlein and Roos 1996; Ersahin et al. 2003; Taylor-McCabe et al. 2001; Frerker et al. 2007). It hydrolyzes NPY, PYY, SP and bradykinin (Medeiros and Turner 1994, 1996; Taylor-McCabe et al. 2001; Ersahin and Simmons 1997; Orawski and Simmons 1995; Orawski et al. 1987; Maggiora et al. 1999; Abid et al. 2009; Chappa et al. 2007). Though not expressed in brain parenchyma, N-terminal truncation of NPY and SP by AmpP2 was detected in vitro BBB studies, brain perfusion and sub-cellular membrane fraction, and therefore implies expression of AmpP2 in the brain microvasculature (Frerker et al. 2007; Chappa et al. 2006, 2007). Similar to AmpP1, AmpP2 is also able to cleave Xaa-Pro-Pro-Yaa peptides such as inactivating bradykinin. Hence, it is predominantly involved in cardiovascular diseases, also associated with angioedema induced by ACE inhibitors and the selective inhibitor Apstatin has been investigated as potential drug candidate (Adam et al. 2002; Blais et al. 1999; Ersahin et al. 1999; Ersahin and Simmons 1997; Taylor-McCabe et al. 2001; Maggiora et al. 1999; Prechel et al. 1995; Wolfrum et al. 2001).

AmpP3

AmpP3 [EC 3.4.11.9] (X-prolyl aminopeptidase 3, XPNPEP3, M24.026) has been discovered by nucleotide sequencing recently. Its gene location is 22q13.2, encompasses 70.2 kb with 10 exons translating a 3 kb mRNA (O'Toole et al. 2010). Two alternative spliced isoforms have been reported, determining mitochondrial and cytosolic sub-cellular localization of AmpP3, respectively (Ersahin et al. 2005). The resulting protein is composed of 507 amino acids with a molecular weight of 51 kDa for the mature mitochondrial isoform and 57 kDa for the immature cytosolic one. AmpP3 is ubiquitously expressed both at mRNA as well as protein level with highest expression in heart followed by pancreas, kidney, and testis then T cells, B cells, and monocytes. Transcripts encoding the predicted mitochondrial protein predominated in all samples (O'Toole et al. 2010). Sequence analysis with AmpP1/2 and Prolidase points to two domains composed of an N-terminal Aminopeptidase P domain and a C-terminal Prolidase domain, which is also the catalytic domain (O'Toole et al. 2010). The enzyme is suggested to display similar activities as AmpP1 and AmpP3, having a proposed active site of two Mn^{2+} ions ligated by Asp331, Asp342, His424, Glu475 and Glu475 as determined by similarity. Substrates of AmpP3 were shown to be ciliary proteins. Mutation of AmpP3 results in nephronophthisis-like nephropathy 1 (NPHPL1), an autosomal recessive kidney disease with a phenotype of nephropathy, kidney cysts, cardiomyopathy, hypertension, seizure, tremor, mental retardation, subarchnoidal cysts, cerebellar vermis aplasia, retinal degeneration, hearing loss and liver fibrosis (O'Toole et al. 2010).

10.2.2.2 Prolidase

Prolidase [EC 3.4.13.9] (imidodipeptidase, peptidase D, PEPD, EC3.4.13.9, M24.007) catabolizes the resulting Xaa-Pro/Hyp dipeptides derived from the various DP4-like enzymes to their respective amino acids (Cunningham and O'Connor 1997). The enzyme belongs to the clan MG, subfamily M24B, requires two Mn^{2+} ions to coordinate to Asp276, Asp287, His370, Glu412 and Glu452, and has the catalytic residues H255, H366 and H377 (Besio et al. 2010). Its crystal structure revealed the typical 'pitta-bread fold' comprised of 21 α-helices and 5 anti-parallel β-sheets within two structurally similar domains, contributing 36 % and 16 % to the overall crystal structure, respectively (pdb's: 2OKN, 2IW2). It is ubiquitously distributed in the cytosol of many tissues such as brain, kidney, heart, liver, muscles, thymus, spleen, prostate, testis and placenta, where it exists as a homodimer with a molecular weight of 54.3 kDa per subunit. Posttranscriptional modifications include N-acetylation at Ala2 as well as phosphorylation at Thr188 and Thr487, respectively (Gevaert et al. 2005; Gauci et al. 2009; Beausoleil et al. 2004). Tyr-phosphorylation was shown to activate the enzyme (Surazynski et al. 2001). The gene is located at 19q12-q13.11, spans 130 kb and has 15 exons. Natural mutations of PEPD, comprised of several point mutations and deletions, result in prolidase deficiency (PD), a very rare autosomal recessive disorder associated with massive iminodipeptiduria. The clinical phenotype of PD includes skin ulcers, mental retardation, recurrent infections and abnormalities in collagenous tissues. However, these features are incompletely penetrant and highly variable in both age of onset and severity. There is a tight linkage between the polymorphisms of prolidase and the myotonic dystrophy trait as well as an increased risk factor for developing Lupus erythematosus (Tanoue et al. 1990a, b, c, 1991a, b; Endo et al. 1989a, b, 1990; Kikuchi et al. 2000; Ledoux et al. 1994, 1996; Forlino et al. 2002; Wang et al. 2006a; Falik-Zaccai et al. 2010; Lupi et al. 2004, 2006; Shrinath et al. 1997). Treatment of PD involves administration of L-proline and L-glycine containing ointments for skin, dietary supplementation with L-proline and essential amino acids, as well as erythrocytes transfusion. In addition to its Xaa-Pro dipeptidase activity, human recombinant prolidase also displays organophosphoric acid anhydrolase (OPAA) activity and is therefore suggested as a treatment for organophosphorous toxins such as soman (Wang et al. 2004, 2005, 2006c).

10.2.3 Metallo-peptidase of Clan MH

10.2.3.1 Prolinase

Prolinase [EC 3.4.13.18] (cytosol non-specific dipeptidase, carnosine dipeptidase 2, carnosinase 2, Pro-X-dipeptidase, peptidase A, EC3.4.13.18, M20.005) catabolizes Pro-Xaa dipeptides into respective amino acids, displaying a pH optimum between 8.0 and 9.5. However, amongst all the proline-specific peptidases described above, it has the broadest specificity, since P1 can be occupied by any amino acids (Teufel et al. 2003). The enzyme belongs to the clan MG, subfamily 20A, having the catalytic residues Asp101 and Glu166 and requiring two Mn^{2+} ions to coordinate to His99, Asp132, Glu167, Asp195 and His445 (Unno et al. 2008). The enzyme is strongly inhibited by p-hydroxymercurybenzoate and bestatin with IC50s of 13 μM and 7 nM, respectively (Teufel et al. 2003). A 1.7 Å crystal structure revealed a homodimer and each subunit is comprised of two domains referred to as A and B. While domain A contains the metal and bestatin-binding site, domain B provides the major interface for dimerization. Since bestatin bound to domain A of one subunit also interacts with the domain B of the other subunit, the dimer interphase is suggested to be involved in the substrate hydrolysis, which would also require structural flexibility between the domain A and B (Unno et al. 2008). The CNDP2 gene is located at 18q22.3, encompasses 27 kb containing 14 exons. The 5 kb mRNA encodes for 475 amino acids with a molecular weight of 53 kDa per subunit (McDonough et al. 2009; Wanic et al. 2008; Teufel et al. 2003). Three potential N-glycosylation sites were identified in the protein sequence and post-transcriptional modifications include N-acetylation at Ala2 as well as N6-acetylation at Lys9 (Choudhary et al. 2009). The homodimer is found in the cytosol, ubiquitously distributed in various tissues both at mRNA as well as protein levels (Teufel et al. 2003). In brain, its protein is highly expressed in the parafascicular nucleus of the thalamus, tuberomammillary nucleus of the hypothalamus and the mitral cell layer of the olfactory bulb, while in striatum only the neuronal process, but not cell bodies are stained. No expression could be detected in glial cells (Otani et al. 2005, 2008). In CNS and autonomic nervous system, the major substrate of prolinase is L-carnosine, a pseudodipeptide comprised of β-alanyl-L-histidine that has neuroprotective functions by intra-cellular buffering, chelating capabilities, anti-oxidant properties and free radical scavenger (Balion et al. 2007; Min et al. 2008; Unno et al. 2008; Stvolinsky et al. 1999; Boldyrev 1993, 1994; Boldyrev et al. 1999a, b, 2010; Klebanov et al. 1997, 1998). It is also reported to have neurotransmitter properties in the olifactory system and hypothalamic network (Unno et al. 2008; Bonfanti et al. 1999; De Marchis et al. 2000a, b;

Tabakman et al. 2002). L-carnosine is secreted via ependymal and glial cells and it was shown to protect β-amyloid induced cell death in granular neurons of the cerebellum by free radical scavenging (Bonfanti et al. 1999; De Marchis et al. 1997, 2000b; Biffo et al. 1990a, b; Boldyrev et al. 2004b). Methylation and acetylation results in the formation of = beta-Alanyl-N(pi)-methyl-L-histidine and N-acetyl carnosine, respectively, being less neuroprotective though (Boldyrev 2000; Bonfanti et al. 1999; Boldyrev and Abe 1999; Boldyrev et al. 1995). Based on its neuroprotecting properties, L-carnosine, its respective analogues as well as prolinase-resistant analogues have been investigated as pharmaceutical target for neurodegenerative disorders such as Alzheimer's disease, Parkinson's disease, ischemia and aging (Boldyrev et al. 1999a, 2001, 2004a, b, 2010; Boldyrev 2005; Pegova et al. 2000; Chasovnikova et al. 1990; Gallant et al. 2000; Stvolinsky et al. 1999, 2000, 2003, 2010; Quinn et al. 1992; Min et al. 2008; Orioli et al. 2011; Vistoli et al. 2009). Polymorphism of carnosinase alias prolinase is associated with diabetic nephropathy in type 1 diabetes, though this is currently controversially discussed (Ahluwalia et al. 2011; McDonough et al. 2009; Wanic et al. 2008).

Acknowledgements Research in our labs is funded by the Deutsche Forschungsgemeinschaft (SFB596), the European Union, the BMBF (KNDD) and the Investmentbank of Saxony-Anhalt. We apologize that not the all previous work could be included in this chapter.

References

Abbott CA, Gorrell MD (2002) The family of CD26/DP IV and related ectopeptidases. In: Langner J, Ansorge S (eds) Ectopeptidases. Kluwer, New York, pp 171–195

Abbott CA, Baker E, Sutherland GR, McCaughan GW (1994) Genomic organization, exact localization, and tissue expression of the human CD26 (dipeptidyl peptidase IV) gene. Immunogenetics 40:331–338

Abbott CA, Yu DM, Woollatt E, Sutherland GR, McCaughan GW et al (2000) Cloning, expression and chromosomal localization of a novel human dipeptidyl peptidase (DPP) IV homolog, DPP8. Eur J Biochem 267:6140–6150

Abeywickrema PD, Patel SB, Byrne NJ, Diehl RE, Hall DL et al (2010) Expression, purification and crystallization of human prolylcarboxypeptidase. Acta Crystallogr Sect F Struct Biol Cryst Commun 66:702–705

Abid K, Rochat B, Lassahn PG, Stocklin R, Michalet S et al (2009) Kinetic study of neuropeptide Y (NPY) proteolysis in blood and identification of NPY3-35, a new peptide cleaved by plasma kallikrein. J Biol Chem 284:24715–24724

Adam A, Cugno M, Molinaro G, Perez M, Lepage Y et al (2002) Aminopeptidase P in individuals with a history of angio-oedema on ACE inhibitors. Lancet 359:2088–2089

Adams GN, LaRusch GA, Stavrou E, Zhou Y, Nieman MT et al (2011) Murine prolylcarboxypeptidase depletion induces vascular dysfunction with hypertension and faster arterial thrombosis. Blood 117:3929–3937

Aertgeerts K, Ye S, Shi L, Prasad SG, Witmer D et al (2004a) N-linked glycosylation of dipeptidyl peptidase IV (CD26): effects on enzyme activity, homodimer formation, and adenosine deaminase binding. Protein Sci 13:145–154

Aertgeerts K, Ye S, Tennant MG, Kraus ML, Rogers J et al (2004b) Crystal structure of human dipeptidyl peptidase IV in complex with a decapeptide reveals details on substrate specificity and tetrahedral intermediate formation. Protein Sci 13:412–421

Aertgeerts K, Levin I, Shi L, Snell GP, Jennings A et al (2005) Structural and kinetic analysis of the substrate specificity of human fibroblast activation protein alpha. J Biol Chem 280:19441–19444

Ahluwalia TS, Lindholm E, Groop LC (2011) Common variants in CNDP1 and CNDP2, and risk of nephropathy in type 2 diabetes. Diabetologia 54:2295–2302

Aimes RT, Zijlstra A, Hooper JD, Ogbourne SM, Sit ML et al (2003) Endothelial cell serine proteases expressed during vascular morphogenesis and angiogenesis. Thromb Haemost 89:561–572

Ajami K, Abbott CA, McCaughan GW, Gorrell MD (2004) Dipeptidyl peptidase 9 has two forms, a broad tissue distribution, cytoplasmic localization and DPIV-like peptidase activity. Biochim Biophys Acta 1679:18–28

Ajami K, Pitman MR, Wilson CH, Park J, Menz R et al (2008) Stromal cell-derived factors 1alpha and 1beta, inflammatory protein-10 and interferon-inducible T cell chemo-attractant are novel substrates of dipeptidyl peptidase 8. FEBS Lett 582:819–825

Alfalah M, Jacob R, Naim HY (2002) Intestinal dipeptidyl peptidase IV is efficiently sorted to the apical membrane through the concerted action of N- and O-glycans as well as association with lipid microdomains. J Biol Chem 277:10683–10690

Altman R, Rutledge JC (2010) The vascular contribution to Alzheimer's disease. Clin Sci (Lond) 119:407–421

Anders A, Gilbert S, Garten W, Postina R, Fahrenholz F (2001) Regulation of the alpha-secretase ADAM10 by its prodomain and proprotein convertases. FASEB J 15:1837–1839

Ansorge S, Bank U, Heimburg A, Helmuth M, Koch G et al (2009) Recent insights into the role of dipeptidyl aminopeptidase IV (DPIV) and aminopeptidase N (APN) families in immune functions. Clin Chem Lab Med 47:253–261

Ansorge S, Nordhoff K, Bank U, Heimburg A, Julius H et al (2011) Novel aspects of cellular action of dipeptidyl peptidase IV/CD26. Biol Chem 392:153–168

Araki H, Li Y, Yamamoto Y, Haneda M, Nishi K et al (2001) Purification, molecular cloning, and immunohistochemical localization of dipeptidyl peptidase II from the rat kidney and its identity with quiescent cell proline dipeptidase. J Biochem (Tokyo) 129:279–288

Ariga N, Sato E, Ohuchi N, Nagura H, Ohtani H (2001) Stromal expression of fibroblast activation protein/seprase, a cell membrane serine proteinase and gelatinase, is associated with longer survival in patients with invasive ductal carcinoma of breast. Int J Cancer 95:67–72

Balion CM, Benson C, Raina PS, Papaioannou A, Patterson C et al (2007) Brain type carnosinase in dementia: a pilot study. BMC Neurol 7:38

Bandyopadhyay S, Goldstein LE, Lahiri DK, Rogers JT (2007) Role of the APP non-amyloidogenic signaling pathway and targeting alpha-secretase as an alternative drug target for treatment of Alzheimer's disease. Curr Med Chem 14:2848–2864

Bank U, Heimburg A, Wohlfarth A, Koch G, Nordhoff K et al (2011) Outside or inside: role of the subcellular localization of DP4-like enzymes for substrate conversion and inhibitor effects. Biol Chem 392:169–187

Barinka C, Rinnova M, Sacha P, Rojas C, Majer P et al (2002) Substrate specificity, inhibition and enzymological analysis of recombinant human glutamate carboxypeptidase II. J Neurochem 80:477–487

Barnes K, Bourne A, Cook PA, Turner AJ, Kenny AJ (1991) Membrane peptidases in the peripheral nervous system of the pig: their localization by immunohistochemistry at light and electron microscopic levels. Neuroscience 44:245–261

Barrett AJ, Rawlings ND (1992) Oligopeptidases, and the emergence of the prolyl oligopeptidase family. Biol Chem Hoppe-Seyler 373:353–360

Bateman RJ, Munsell LY, Morris JC, Swarm R, Yarasheski KE et al (2006) Human amyloid-beta synthesis and clearance rates as measured in cerebrospinal fluid in vivo. Nat Med 12:856–861

Bazan JF, Weaver LH, Roderick SL, Huber R, Matthews BW (1994) Sequence and structure comparison suggest that methionine aminopeptidase, prolidase, aminopeptidase P, and creatinase share a common fold. Proc Natl Acad Sci U S A 91:2473–2477

Beausoleil SA, Jedrychowski M, Schwartz D, Elias JE, Villen J et al (2004) Large-scale characterization of HeLa cell nuclear phosphoproteins. Proc Natl Acad Sci U S A 101:12130–12135

Bell RD, Zlokovic BV (2009) Neurovascular mechanisms and blood-brain barrier disorder in Alzheimer's disease. Acta Neuropathol 118:103–113

Bell RD, Winkler EA, Sagare AP, Singh I, LaRue B et al (2010) Pericytes control key neurovascular functions and neuronal phenotype in the adult brain and during brain aging. Neuron 68:409–427

Bellemere G, Morain P, Vaudry H, Jegou S (2003) Effect of S 17092, a novel prolyl endopeptidase inhibitor, on substance P and alpha-melanocyte-stimulating hormone breakdown in the rat brain. J Neurochem 84:919–929

Bellemere G, Vaudry H, Morain P, Jegou S (2005) Effect of prolyl endopeptidase inhibition on arginine-vasopressin and thyrotrophin-releasing hormone catabolism in the rat brain. J Neuroendocrinol 17:306–313

Bennett BD, Denis P, Haniu M, Teplow DB, Kahn S, Louis JC et al (2000) A furin-like convertase mediates propeptide cleavage of BACE, the Alzheimer's beta -secretase. J Biol Chem 275:37712–37717

Besio R, Alleva S, Forlino A, Lupi A, Meneghini C et al (2010) Identifying the structure of the active sites of human recombinant prolidase. Eur Biophys J 39:935–945

Bezerra GA, Dobrovetsky E, Dong A, Seitova A, Crombett L, Shewchuk LM, Hassell AM, Sweitzer SM, Sweitzer TD, McDevitt PJ, Johanson KO, Kennedy-Wilson KM, Cossar D, Bochkarev A, Gruber K, Dhe-Paganon S (2012) Structures of human DPP7 reveal the molecular basis of specific inhibition and the architectural diversity of proline-specific peptidases. PLoS One 7(8):e43019. doi:10.1371/journal.pone.0043019

Biffo S, DeLucia R, Mulatero B, Margolis F, Fasolo A (1990a) Carnosine-, calcitonin gene-related peptide- and tyrosine hydroxylase-immunoreactivity in the mouse olfactory bulb following peripheral denervation. Brain Res 528:353–357

Biffo S, Grillo M, Margolis FL (1990b) Cellular localization of carnosine-like and anserine-like immunoreactivities in rodent and avian central nervous system. Neuroscience 35:637–651

Bilodeau N, Fiset A, Poirier GG, Fortier S, Gingras MC et al (2006) Insulin-dependent phosphorylation of DPP IV in liver. Evidence for a role of compartmentalized c-Src. FEBS J 273:992–1003

Bjelke JR, Christensen J, Nielsen PF, Branner S, Kanstrup AB et al (2006a) Dipeptidyl peptidase 8 and 9 specificity and molecular characterization compared to dipeptidyl peptidase IV. Biochem J 396:391–396

Bjelke JR, Kanstrup AB, Rasmussen HB (2006b) Selectivity among dipeptidyl peptidases of the s9b family. Cell Mol Biol (Noisy-le-grand) 52:3–7

Blais CJ, Marc-Aurele J, Simmons WH, Loute G, Thibault P et al (1999) Des-Arg9-bradykinin metabolism in patients who presented hypersensitivity reactions during hemodialysis: role of serum ACE and aminopeptidase P. Peptides 20:421–430

Boldyrev AA (1993) Does carnosine possess direct antioxidant activity? Int J Biochem 25:1101–1107

Boldyrev AA (1994) Carnosine and free-radical defence mechanisms. Trends Neurosci 17:468

Boldyrev AA (2000) Problems and perspectives in studying the biological role of carnosine. Biochemistry (Mosc) 65:751–756

Boldyrev AA (2005) Protection of proteins from oxidative stress: a new illusion or a novel strategy? Ann NY Acad Sci 1057:193–205

Boldyrev A, Abe H (1999) Metabolic transformation of neuropeptide carnosine modifies its biological activity. Cell Mol Neurobiol 19:163–175

Boldyrev A, Abe H, Stvolinsky S, Tyulina O (1995) Effects of carnosine and related compounds on generation of free oxygen species: a comparative study. Comp Biochem Physiol B Biochem Mol Biol 112:481–485

Boldyrev AA, Gallant SC, Sukhich GT (1999a) Carnosine, the protective, anti-aging peptide. Biosci Rep 19:581–587

Boldyrev AA, Johnson P, Wei Y, Tan Y, Carpenter DO (1999b) Carnosine and taurine protect rat cerebellar granular cells from free radical damage. Neurosci Lett 263:169–172

Boldyrev AA, Yuneva MO, Sorokina EV, Kramarenko GG, Fedorova TN et al (2001) Antioxidant systems in tissues of senescence accelerated mice. Biochemistry (Mosc) 66:1157–1163

Boldyrev A, Bulygina E, Leinsoo T, Petrushanko I, Tsubone S et al (2004a) Protection of neuronal cells against reactive oxygen species by carnosine and related compounds. Comp Biochem Physiol B Biochem Mol Biol 137:81–88

Boldyrev A, Koudinov A, Berezov T, Carpenter DO (2004b) Amyloid-beta induced cell death is independent of free radicals. J Alzheimers Dis 6:633–638

Boldyrev AA, Stvolinsky SL, Fedorova TN, Suslina ZA (2010) Carnosine as a natural antioxidant and geroprotector: from molecular mechanisms to clinical trials. Rejuvenation Res 13:156–158

Bonfanti L, Peretto P, De Marchis S, Fasolo A (1999) Carnosine-related dipeptides in the mammalian brain. Prog Neurobiol 59:333–353

Bourne A, Barnes K, Taylor BA, Turner AJ, Kenny AJ (1989) Membrane peptidases in the pig choroid plexus and on other cell surfaces in contact with the cerebrospinal fluid. Biochem J 259:69–80

Bozkulak EC, Weinmaster G (2009) Selective use of ADAM10 and ADAM17 in activation of Notch1 signaling. Mol Cell Biol 29:5679–5695

Brandt I, De Vriendt K, Devreese B, Van Beeumen J, Van Dongen W et al (2005) Search for substrates for prolyl oligopeptidase in porcine brain. Peptides 26:2536–2546

Brandt I, Lambeir AM, Maes MB, Scharpe S, De Meester I (2006) Peptide substrates of dipeptidyl peptidases. Adv Exp Med Biol 575:3–18

Brandt I, Gerard M, Sergeant K, Devreese B, Baekelandt V et al (2008) Prolyl oligopeptidase stimulates the aggregation of alpha-synuclein. Peptides 29:1472–1478

Burkey BF, Hoffmann PK, Hassiepen U, Trappe J, Juedes M et al (2008) Adverse effects of dipeptidyl peptidases 8 and 9 inhibition in rodents revisited. Diabetes Obes Metab 10:1057–1061

Buxbaum JD, Liu KN, Luo Y, Slack JL, Stocking KL et al (1998) Evidence that tumor necrosis factor alpha converting enzyme is involved in regulated alpha-secretase cleavage of the Alzheimer amyloid protein precursor. J Biol Chem 273:27765–27767

Caccamo A, Oddo S, Billings LM, Green KN, Martinez-Coria H et al (2006) M1 receptors play a central role in modulating AD-like pathology in transgenic mice. Neuron 49:671–682

Caescu CI, Jeschke GR, Turk BE (2009) Active-site determinants of substrate recognition by the metalloproteinases TACE and ADAM10. Biochem J 424:79–88

Cai H, Wang Y, McCarthy D, Wen H, Borchelt DR et al (2001) BACE1 is the major beta-secretase for generation of Abeta peptides by neurons. Nat Neurosci 4:233–234

Capell A, Grunberg J, Pesold B, Diehlmann A, Citron M et al (1998) The proteolytic fragments of the Alzheimer's disease-associated presenilin-1 form heterodimers and occur as a 100-150-kDa molecular mass complex. J Biol Chem 273:3205–3211

Capell A, Steiner H, Willem M, Kaiser H, Meyer C et al (2000) Maturation and pro-peptide cleavage of beta-secretase. J Biol Chem 275:30849–30854

Cataldo AM, Nixon RA (1990) Enzymatically active lysosomal proteases are associated with amyloid deposits in Alzheimer brain. Proc Natl Acad Sci U S A 87:3861–3865

Cataldo AM, Barnett JL, Pieroni C, Nixon RA (1997) Increased neuronal endocytosis and protease delivery to early endosomes in sporadic Alzheimer's disease: neuropathologic evidence for a mechanism of increased beta-amyloidogenesis. J Neurosci 17:6142–6151

Chang WP, Downs D, Huang XP, Da H, Fung KM et al (2007) Amyloid-beta reduction by memapsin 2 (beta-secretase) immunization. FASEB J 21:3184–3196

Chappa AK, Audus KL, Lunte SM (2006) Characteristics of substance P transport across the blood-brain barrier. Pharm Res 23:1201–1208
Chappa AK, Cooper JD, Audus KL, Lunte SM (2007) Investigation of the metabolism of substance P at the blood-brain barrier using LC-MS/MS. J Pharm Biomed Anal 43:1409–1415
Chasovnikova LV, Formazyuk VE, Sergienko VI, Boldyrev AA, Severin SE (1990) The antioxidative properties of carnosine and other drugs. Biochem Int 20:1097–1103
Chavez-Gutierrez L, Tolia A, Maes E, Li T, Wong PC et al (2008) Glu332 in the nicastrin ectodomain is essential for γ-secretase complex maturation but not for its activity. J Biol Chem 283:20096–20105
Chen T, Ajami K, McCaughan GW, Gorrell MD, Abbott CA (2003a) Dipeptidyl peptidase IV gene family. The DPIV family. Adv Exp Med Biol 524:79–86
Chen WT, Kelly T, Ghersi G (2003b) DPPIV, seprase, and related serine peptidases in multiple cellular functions. Curr Top Dev Biol 54:207–232
Chen F, Hasegawa H, Schmitt-Ulms G, Kawarai T, Bohm C et al (2006a) TMP21 is a presenilin complex component that modulates γ-secretase but not ε-secretase activity. Nature 440:1208–1212
Chen T, Ajami K, McCaughan GW, Gai WP, Gorrell MD et al (2006b) Molecular characterization of a novel dipeptidyl peptidase like 2-short form (DPL2-s) that is highly expressed in the brain and lacks dipeptidyl peptidase activity. Biochim Biophys Acta 1764:33–43
Chen R, Jiang X, Sun D, Han G, Wang F et al (2009) Glycoproteomics analysis of human liver tissue by combination of multiple enzyme digestion and hydrazide chemistry. J Proteome Res 8:651–661
Chiravuri M, Huber BT (2000) Aminodipeptidase inhibitor-induced cell death in quiescent lymphocytes: a review. Apoptosis 5:319–322
Chiravuri M, Schmitz T, Yardley K, Underwood R, Dayal Y et al (1999) A novel apoptotic pathway in quiescent lymphocytes identified by inhibition of a post-proline cleaving aminodipeptidase: a candidate target protease, quiescent cell proline dipeptidase. J Immunol 163:3092–3099
Chiravuri M, Agarraberes F, Mathieu SL, Lee H, Huber BT (2000a) Vesicular localization and characterization of a novel post-proline-cleaving aminodipeptidase, quiescent cell proline dipeptidase. J Immunol 165:5695–5702
Chiravuri M, Lee H, Mathieu SL, Huber BT (2000b) Homodimerization via a leucine zipper motif is required for enzymatic activity of quiescent cell proline dipeptidase. J Biol Chem 275:26994–26999
Choudhary C, Kumar C, Gnad F, Nielsen ML, Rehman M et al (2009) Lysine acetylation targets protein complexes and co-regulates major cellular functions. Science 325:834–840
Citron M, Teplow DB, Selkoe DJ (1995) Generation of amyloid beta protein from its precursor is sequence specific. Neuron 14:661–670
Cordero OJ, Salgado FJ, Nogueira M (2009) On the origin of serum CD26 and its altered concentration in cancer patients. Cancer Immunol Immunother 58:1725–1749
Cottrell GS, Hyde RJ, Hooper NM, Turner AJ (1998) The cloning and functional expression of human pancreatic aminopeptidase P. Biochem Soc Trans 26:S248
Cottrell GS, Hooper NM, Turner AJ (2000a) Cloning, expression, and characterization of human cytosolic aminopeptidase P: a single manganese(II)-dependent enzyme. Biochemistry 39:15121–15128
Cottrell GS, Hyde RJ, Lim J, Parsons MR, Hooper NM et al (2000b) Identification of critical residues in the active site of porcine membrane-bound aminopeptidase P. Biochemistry 39:15129–15135
Creemers JW, Ines Dominguez D, Plets E, Serneels L, Taylor NA et al (2001) Processing of beta-secretase by furin and other members of the proprotein convertase family. J Biol Chem 276:4211–4217
Crystal AS, Morais VA, Pierson TC, Pijak DS, Carlin D et al (2003) Membrane topology of γ-secretase component PEN-2. J Biol Chem 278:20117–20123

Cunningham DF, O'Connor B (1997) Proline specific peptidases. Biochim Biophys Acta 1343:160–186

Cynis H, Scheel E, Saido TC, Schilling S, Demuth HU (2008) Amyloidogenic processing of amyloid precursor protein: evidence of a pivotal role of glutaminyl cyclase in generation of pyroglutamate-modified amyloid-beta. Biochemistry 47:7405–7413

Danilova OV, Tai AK, Mele DA, Beinborn M, Leiter AB et al (2009) Neurogenin 3-specific dipeptidyl peptidase-2 deficiency causes impaired glucose tolerance, insulin resistance, and visceral obesity. Endocrinology 150:5240–5248

De Marchis S, Melcangi RC, Modena C, Cavaretta I, Peretto P et al (1997) Identification of the glial cell types containing carnosine-related peptides in the rat brain. Neurosci Lett 237:37–40

De Marchis S, Modena C, Peretto P, Giffard C, Fasolo A (2000a) Carnosine-like immunoreactivity in the central nervous system of rats during postnatal development. J Comp Neurol 426:378–390

De Marchis S, Modena C, Peretto P, Migheli A, Margolis FL et al (2000b) Carnosine-related dipeptides in neurons and glia. Biochemistry (Mosc) 65:824–833

De Meester I, Korom S, Van Damme J, Scharpe S (1999) CD26, let it cut or cut it down. Immunol Today 20:367–375

De Pietri Tonelli D, Mihailovich M, Di Cesare A, Codazzi F, Grohovaz F et al (2004) Translational regulation of BACE-1 expression in neuronal and non-neuronal cells. Nucleic Acids Res 32:1808–1817

De Strooper B (2010) Proteases and proteolysis in Alzheimer's disease: a multifactorial view on the disease process. Physiol Rev 90:465–494

De Strooper B, Annaert W, Cupers P, Saftig P, Craessaerts K et al (1999) A presenilin-1-dependent γ-secretase-like protease mediates release of Notch intracellular domain. Nature 398:518–522

Deane R, Zlokovic BV (2007) Role of the blood-brain barrier in the pathogenesis of Alzheimer's disease. Curr Alzheimer Res 4:191–197

Deane R, Bell RD, Sagare A, Zlokovic BV (2009) Clearance of amyloid-beta peptide across the blood-brain barrier: implication for therapies in Alzheimer's disease. CNS Neurol Disord Drug Targets 8:16–30

Di Daniel E, Glover CP, Grot E, Chan MK, Sanderson TH et al (2009) Prolyl oligopeptidase binds to GAP-43 and functions without its peptidase activity. Mol Cell Neurosci 41:373–382

Diano S (2011) New aspects of melanocortin signaling: a role for PRCP in alpha-MSH degradation. Front Neuroendocrinol 32:70–83

Dickstein DL, Walsh J, Brautigam H, Stockton SD Jr, Gandy S et al (2010) Role of vascular risk factors and vascular dysfunction in Alzheimer's disease. Mt Sinai J Med 77:82–102

Dikov A, Dimitrova M, Stoineva I, Halbhuber KJ (1999) New tetrazolium method for the histochemical localization of dipeptidyl peptidase IV. Cell Mol Biol (Noisy-le-grand) 45:225–231

Dominguez D, Tournoy J, Hartmann D, Huth T, Cryns K et al (2005) Phenotypic and biochemical analyses of BACE1- and BACE2-deficient mice. J Biol Chem 280:30797–30806

Dries DR, Shah S, Han YH, Yu C, Yu S et al (2009) Glu-333 of nicastrin directly participates in γ-secretase activity. J Biol Chem 284:29714–29724

Dubois V, Lambeir AM, Van der Veken P, Augustyns K, Creemers J et al (2008) Purification and characterization of dipeptidyl peptidase IV-like enzymes from bovine testes. Front Biosci 13:3558–3568

Duckworth WC, Bennett RG, Hamel FG (1998) Insulin degradation: progress and potential. Endocr Rev 19:608–624

Duke-Cohan JS, Morimoto C, Rocker JA, Schlossman SF (1995) A novel form of dipeptidyl-peptidase IV found in human serum. Isolation, characterization, and comparison with T lymphocyte membrane dipeptidylpeptidase IV (CD26). J Biol Chem 270:14107–14114

Duke-Cohan JS, Morimoto C, Rocker JA, Schlossman SF (1996) Serum high molecular weight dipeptidyl peptidase IV (CD26) is similar to a novel antigen DPPT-L released from activated T cells. J Immunol 156:1714–1721
Duke-Cohan JS, Kim JH, Azouz A (2004) Attractin: cautionary tales for therapeutic intervention in molecules with pleiotropic functionality. J Environ Pathol Toxicol Oncol 23:1–11
Durinx C, Lambeir AM, Bosmans E, Falmagne JB, Berghmans R et al (2000) Molecular characterization of dipeptidyl peptidase activity in serum: soluble CD26/dipeptidyl peptidase IV is responsible for the release of X-Pro dipeptides. Eur J Biochem 267:5608–5613
Eckman EA, Reed DK, Eckman CB (2001) Degradation of the Alzheimer's amyloid beta peptide by endothelin-converting enzyme. J Biol Chem 276:24540–24548
Eckman EA, Watson M, Marlow L, Sambamurti K, Eckman CB (2003) Alheimer's disease beta-amyloid peptide is increased in mice deficient in endothelin-converting enzyme. J Biol Chem 278:2081–2084
Eckman EA, Adams SK, Troendle FJ, Stodola BA, Kahn MA et al (2006) Regulation of steady-state beta-amyloid levels in the brain by neprilysin and endothelin-converting enzyme but not angiotensin-converting enzyme. J Biol Chem 281:30471–30478
Edbauer D, Winkler E, Regula JT, Pesold B, Steiner H et al (2003) Reconstitution of γ-secretase activity. Nat Cell Biol 5:486–488
Edosada CY, Quan C, Tran T, Pham V, Wiesmann C et al (2006a) Peptide substrate profiling defines fibroblast activation protein as an endopeptidase of strict Gly(2)-Pro(1)-cleaving specificity. FEBS Lett 580:1581–1586
Edosada CY, Quan C, Wiesmann C, Tran T, Sutherlin D et al (2006b) Selective inhibition of fibroblast activation protein protease based on dipeptide substrate specificity. J Biol Chem 281:7437–7444
Edwards DR, Handsley MM, Pennington CJ (2008) The ADAM metalloproteinases. Mol Aspects Med 29:258–289
El Yacoubi M, Vaugeois JM, Marguet D, Sauze N, Guieu R (2006) Behavioral characterization of CD26 deficient mice in animal tests of anxiety and antidepressant-like activity. Behav Brain Res 171:279–285
Elenkov IJ, Wilder RL, Chrousos GP, Vizi ES (2000) The sympathetic nerve–an integrative interface between two supersystems: the brain and the immune system. Pharmacol Rev 52:595–638
Endo F, Tanoue A, Hata A, Kitano A, Matsuda I (1989a) Deduced amino acid sequence of human prolidase and molecular analyses of prolidase deficiency. J Inherit Metab Dis 12:351–354
Endo F, Tanoue A, Nakai H, Hata A, Indo Y et al (1989b) Primary structure and gene localization of human prolidase. J Biol Chem 264:4476–4481
Endo F, Tanoue A, Kitano A, Arata J, Danks DM et al (1990) Biochemical basis of prolidase deficiency. Polypeptide and RNA phenotypes and the relation to clinical phenotypes. J Clin Invest 85:162–169
Endres K, Fahrenholz F (2010) Upregulation of the alpha-secretase ADAM10–risk or reason for hope? FEBS J 277:1585–1596
Engel M, Hoffmann T, Wagner L, Wermann M, Heiser U et al (2003) The crystal structure of dipeptidyl peptidase IV (CD26) reveals its functional regulation and enzymatic mechanism. Proc Natl Acad Sci U S A 100:5063–5068
Ersahin C, Simmons WH (1997) Inhibition of both aminopeptidase P and angiotensin-converting enzyme prevents bradykinin degradation in the rat coronary circulation. J Cardiovasc Pharmacol 30:96–101
Ersahin C, Euler DE, Simmons WH (1999) Cardioprotective effects of the aminopeptidase P inhibitor apstatin: studies on ischemia/reperfusion injury in the isolated rat heart. J Cardiovasc Pharmacol 34:604–611
Ersahin C, Szpaderska AM, Simmons WH (2003) Rat and mouse membrane aminopeptidase P: structure analysis and tissue distribution. Arch Biochem Biophys 417:131–140

Ersahin C, Szpaderska AM, Orawski AT, Simmons WH (2005) Aminopeptidase P isozyme expression in human tissues and peripheral blood mononuclear cell fractions. Arch Biochem Biophys 435:303–310
Esch FS, Keim PS, Beattie EC, Blacher RW, Culwell AR et al (1990) Cleavage of amyloid β peptide during constitutive processing of its precursor. Science 248:1122–1124
Escrevente C, Morais VA, Keller S, Soares CM, Altevogt P et al (2008) Functional role of N-glycosylation from ADAM10 in processing, localization and activity of the enzyme. Biochim Biophys Acta 1780:905–913
Esler WP, Kimberly WT, Ostaszewski BL, Diehl TS, Moore CL et al (2000) Transition-state analogue inhibitors of γ-secretase bind directly to presenilin-1. Nat Cell Biol 2:428–433
Faghihi MA, Modarresi F, Khalil AM, Wood DE, Sahagan BG et al (2008) Expression of a noncoding RNA is elevated in Alzheimer's disease and drives rapid feed-forward regulation of beta-secretase. Nat Med 14:723–730
Falik-Zaccai TC, Khayat M, Luder A, Frenkel P, Magen D et al (2010) A broad spectrum of developmental delay in a large cohort of prolidase deficiency patients demonstrates marked interfamilial and intrafamilial phenotypic variability. Am J Med Genet B Neuropsychiatr Genet 153B:46–56
Fan H, Meng W, Kilian C, Grams S, Reutter W (1997) Domain-specific N-glycosylation of the membrane glycoprotein dipeptidylpeptidase IV (CD26) influences its subcellular trafficking, biological stability, enzyme activity and protein folding. Eur J Biochem 246:243–251
Farah MH, Pan BH, Hoffman PN, Ferraris D, Tsukamoto T (2011) Reduced BACE1 activity enhances clearance of myelin debris and regeneration of axons in the injured peripheral nervous system. J Neurosci 31:5744–5754
Farrer LA, Sherbatich T, Keryanov SA, Korovaitseva GI, Rogaeva EA et al (2000) Association between angiotensin-converting enzyme and Alzheimer disease. Arch Neurol 57:210–214
Farris W, Mansourian S, Chang Y, Lindsley L, Eckman EA et al (2003) Insulin-degrading enzyme regulates the levels of insulin, amyloid beta-protein, and the beta-amyloid precursor protein intracellular domain in vivo. Proc Natl Acad Sci U S A 100:4162–4167
Farris W, Mansourian S, Leissring MA, Eckman EA, Bertram L et al (2004) Partial loss-of-function mutations in insulin-degrading enzyme that induce diabetes also impair degradation of amyloid beta-protein. Am J Pathol 164:1425–1434
Fluhrer R, Steiner H, Haass C (2009) Intramembrane proteolysis by signal peptide peptidases – a comparative discussion of GxGD-type aspartyl proteases. J Biol Chem 284:13975–13979
Forlino A, Lupi A, Vaghi P, Icaro CA, Calligaro A et al (2002) Mutation analysis of five new patients affected by prolidase deficiency: the lack of enzyme activity causes necrosis-like cell death in cultured fibroblasts. Hum Genet 111:314–322
Fortna RR, Crystal AS, Morais VA, Pijak DS, Lee VM et al (2004) Membrane topology and nicastrin-enhanced endoproteolysis of APH-1, a component of the γ-secretase complex. J Biol Chem 279:3685–3693
Francis R, McGrath G, Zhang J, Ruddy DA, Sym M et al (2002) Aph-1 and pen-2 are required for Notch pathway signaling, γ-secretase cleavage of βAPP, and presenilin protein accumulation. Dev Cell 3:85–97
Frerker N, Wagner L, Wolf R, Heiser U, Hoffmann T et al (2007) Neuropeptide Y (NPY) cleaving enzymes: structural and functional homologues of dipeptidyl peptidase 4. Peptides 28:257–268
Frerker N, Raber K, Bode F, Skripuletz T, Nave H et al (2009) Phenotyping of congenic dipeptidyl peptidase 4 (DP4) deficient Dark Agouti (DA) rats suggests involvement of DP4 in neuro-, endocrine, and immune functions. Clin Chem Lab Med 47:275–287
Friedrich D, Kuhn-Wache K, Hoffmann T, Demuth HU (2003) Isolation and characterization of attractin-2. Adv Exp Med Biol 524:109–113
Friedrich D, Hoffmann T, Bar J, Wermann M, Manhart S et al (2007) Does human attractin have DP4 activity? Biol Chem 388:155–162

Fukumori A, Fluhrer R, Steiner H, Haass C (2010) Three-amino acid spacing of presenilin endoproteolysis suggests a general stepwise cleavage of gamma-secretase-mediated intramembrane proteolysis. J Neurosci 30:7853–7862
Fulop V, Bocskei Z, Polgar L (1998) Prolyl oligopeptidase: an unusual beta-propeller domain regulates proteolysis. Cell 94:161–170
Gallant S, Kukley M, Stvolinsky S, Bulygina E, Boldyrev A (2000) Effect of carnosine on rats under experimental brain ischemia. Tohoku J Exp Med 191:85–99
Garcia-Horsman JA, Mannisto PT, Venalainen JI (2007) On the role of prolyl oligopeptidase in health and disease. Neuropeptides 41:1–24
Garin-Chesa P, Old LJ, Rettig WJ (1990) Cell surface glycoprotein of reactive stromal fibroblasts as a potential antibody target in human epithelial cancers. Proc Natl Acad Sci U S A 87:7235–7239
Gauci S, Helbig AO, Slijper M, Krijgsveld J, Heck AJ et al (2009) Lys-N and trypsin cover complementary parts of the phosphoproteome in a refined SCX-based approach. Anal Chem 81:4493–4501
Geiss-Friedlander R, Parmentier N, Moller U, Urlaub H, Van den Eynde BJ et al (2009) The cytoplasmic peptidase DPP9 is rate-limiting for degradation of proline-containing peptides. J Biol Chem 284:27211–27219
Gervais FG, Xu D, Robertson GS, Vaillancourt JP, Zhu Y et al (1999) Involvement of caspases in proteolytic cleavage of Alzheimer's amyloid-beta precursor protein and amyloidogenic A beta peptide formation. Cell 97:395–406
Gevaert K, Staes A, Van Damme J, De Groot S, Hugelier K et al (2005) Global phosphoproteome analysis on human HepG2 hepatocytes using reversed-phase diagonal LC. Proteomics 5:3589–3599
Gibb DR, El Shikh M, Kang DJ, Rowe WJ, El Sayed R et al (2010) ADAM10 is essential for Notch2-dependent marginal zone B cell development and CD23 cleavage in vivo. J Exp Med 207:623–635
Goodman JD, Rozypal TL, Kelly T (2003) Seprase, a membrane-bound protease, alleviates the serum growth requirement of human breast cancer cells. Clin Exp Metastasis 20:459–470
Gorenstein C, Ribak CE (1985) Dendritic transport. II. Somatofugal movement of neuronal lysosomes induced by colchicine: evidence for a novel transport system in dendrites. J Neurosci 5:2018–2027
Gorenstein C, Bundman MC, Lew PJ, Olds JL, Ribak CE (1985) Dendritic transport. I. Colchicine stimulates the transport of lysosomal enzymes from cell bodies to dendrites. J Neurosci 5:2009–2017
Gorrell MD (2005) Dipeptidyl peptidase IV and related enzymes in cell biology and liver disorders. Clin Sci (Lond) 108:277–292
Gossrau R (1979) Peptidases. II. Localization of dipeptidylpeptidase IV (DPP IV). Histochemical and biochemical study. Histochemistry 60:231–248
Gossrau R, Lojda Z (1980) Study on dipeptidylpeptidase II (DPP II). Histochemistry 70:53–76
Griswold JA, Beall CV, Baker CRJ, Little DT, Little GH et al (1996) Bradykinin metabolism in the liver and lung of the rat. J Surg Res 66:12–20
Griswold JA, Baker CRJ, Little DT, Little GH, Behal FJ (1999) Bradykinin metabolism in perfused rat kidney. Shock 11:372–379
Gruninger-Leitch F, Schlatter D, Kung E, Nelbock P, Dobeli H (2002) Substrate and inhibitor profile of BACE (beta-secretase) and comparison with other mammalian aspartic proteases. J Biol Chem 277:4687–4693
Gu Y, Misonou H, Sato T, Dohmae N, Takio K et al (2001) Distinct intramembrane cleavage of the β-amyloid precursor protein family resembling γ-secretase-like cleavage of Notch. J Biol Chem 276:35235–35238
Haapasalo A, Kovacs DM (2011) The many substrates of presenilin/gamma-secretase. J Alzheimers Dis 25:3–28

Haass C, Schlossmacher MG, Hung AY, Vigo-Pelfrey C, Mellon A et al (1992) Amyloid beta-peptide is produced by cultured cells during normal metabolism. Nature 359:322–325
Hagedorn M (2011) PRCP: a key to blood vessel homeostasis. Blood 117:3705–3706
Harbeck HT, Mentlein R (1991) Aminopeptidase P from rat brain. Purification and action on bioactive peptides. Eur J Biochem 198:451–458
Hartmann D, de Strooper B, Serneels L, Craessaerts K, Herreman A et al (2002) The disintegrin/metalloprotease ADAM 10 is essential for Notch signalling but not for alpha-secretase activity in fibroblasts. Hum Mol Genet 11:2615–2624
He G, Luo W, Li P, Remmers C, Netzer WJ et al (2010) Gamma-secretase activating protein is a therapeutic target for Alzheimer's disease. Nature 467:95–98
Hebert SS, Serneels L, Dejaegere T, Horre K, Dabrowski M et al (2004) Coordinated and widespread expression of γ-secretase in vivo: evidence for size and molecular heterogeneity. Neurobiol Dis 17:260–272
Hebert SS, Horre K, Nicolai L, Papadopoulou AS, Mandemakers W et al (2008) Loss of microRNA cluster miR-29a/b-1 in sporadic Alzheimer's disease correlates with increased BACE1/beta-secretase expression. Proc Natl Acad Sci U S A 105:6415–6420
Hemming ML, Selkoe DJ, Farris W (2007) Effects of prolonged angiotensin-converting enzyme inhibitor treatment on amyloid beta-protein metabolism in mouse models of Alzheimer disease. Neurobiol Dis 26:273–281
Hemming ML, Elias JE, Gygi SP, Selkoe DJ (2008) Proteomic profiling of γ-secretase substrates and mapping of substrate requirements. PLoS Biol 6:e257
Hemming ML, Elias JE, Gygi SP, Selkoe DJ (2009) Identification of beta-secretase (BACE1) substrates using quantitative proteomics. PLoS One 4:e8477
Henricson A, Kall L, Sonnhammer EL (2005) A novel transmembrane topology of presenilin based on reconciling experimental and computational evidence. FEBS J 272:2727–2733
Henry LR, Lee HO, Lee JS, Klein-Szanto A, Watts P et al (2007) Clinical implications of fibroblast activation protein in patients with colon cancer. Clin Cancer Res 13:1736–1741
Hitt BD, Jaramillo TC, Chetkovich DM, Vassar R (2010) BACE1-/- mice exhibit seizure activity that does not correlate with sodium channel level or axonal localization. Mol Neurodegener 5:31
Hong L, Koelsch G, Lin X, Wu S, Terzyan S et al (2000) Structure of the protease domain of memapsin 2 (beta-secretase) complexed with inhibitor. Science 290:150–153
Hook VY, Reisine TD (2003) Cysteine proteases are the major bate-secretase in the regulated secretory pathway that provides most of the beta-amyloid in Alzheimer's disease: role of BACE 1 in the constitutive secretory pathway. J Neurosci Res 74:393–405
Hook V, Schechter I, Demuth HU, Hook G (2008a) Alternative pathways for production of beta-amyloid peptides of Alzheimer's disease. Biol Chem 389:993–1006
Hook VY, Kindy M, Hook G (2008b) Inhibitors of cathepsin B improve memory and reduce beta-amyloid in transgenic Alzheimer disease mice expressing the wild-type, but not the Swedish mutant, beta-secretase site of the amyloid precursor protein. J Biol Chem 283:7745–7753
Hook VY, Kindy M, Reinheckel T, Peters C, Hook G (2009) Genetic cathepsin B deficiency reduces beta-amyloid in transgenic mice expressing human wild-type amyloid precursor protein. Biochem Biophys Res Commun 386:284–288
Hook G, Hook V, Kindy M (2011) The cysteine protease inhibitor, E64d, reduces brain amyloid and improves memory deficits in Alzheimer's disease animal models by inhibiting cathepsin B, but not BACE1, β-secretase activity. J Alzheimers Dis 26:387–408
Hu J, Miyatake F, Aizu Y, Nakagawa H, Nakamura S et al (1999) Angiotensin-converting enzyme genotype is associated with Alzheimer disease in the Japanese population. Neurosci Lett 277:65–67
Hu J, Igarashi A, Kamata M, Nakagawa H (2001) Angiotensin-converting enzyme degrades Alzheimer amyloid beta-peptide (A beta) retards A beta aggregation, deposition, fibril formation; and inhibits cytotoxicity. J Biol Chem 276:47863–47868

Hu X, Hicks CW, He W, Wong P, Macklin WB et al (2006) Bace1 modulates myelination in the central and peripheral nervous system. Nat Neurosci 9:1520–1525

Hu X, He W, Diaconu C, Tang X, Kidd GJ, Macklin WB et al (2008) Genetic deletion of BACE1 in mice affects remyelination of sciatic nerves. FASEB J 22:2970–2980

Hu X, Zhou X, He W, Yang J, Xiong W et al (2010) BACE1 deficiency causes altered neuronal activity and neurodegeneration. J Neurosci 30:8819–8829

Huang Y, Wang S, Kelly T (2004) Seprase promotes rapid tumor growth and increased microvessel density in a mouse model of human breast cancer. Cancer Res 64:2712–2716

Huber MA, Kraut N, Park JE, Schubert RD, Rettig WJ et al (2003) Fibroblast activation protein: differential expression and serine protease activity in reactive stromal fibroblasts of melanocytic skin tumors. J Invest Dermatol 120:182–188

Huse JT, Pijak DS, Leslie GJ, Lee VM, Doms RW (2000) Maturation and endosomal targeting of beta-site amyloid precursor protein-cleaving enzyme. The Alzheimer's disease beta-secretase. J Biol Chem 275:33729–33737

Huse JT, Liu K, Pijak DS, Carlin D, Lee VM et al (2002) Beta -secretase processing in the trans-Golgi network preferentially generates truncated amyloid species that accumulate in Alzheimer's disease brain. J Biol Chem 14:14

Hussain I, Powell D, Howlett DR, Tew DG, Meek TD et al (1999) Identification of a novel aspartic protease (Asp 2) as beta-secretase. Mol Cell Neurosci 14:419–427

Iwasa S, Jin X, Okada K, Mitsumata M, Ooi A (2003) Increased expression of seprase, a membrane-type serine protease, is associated with lymph node metastasis in human colorectal cancer. Cancer Lett 199:91–98

Iwasa S, Okada K, Chen WT, Jin X, Yamane T et al (2005) Increased expression of seprase, a membrane-type serine protease, is associated with lymph node metastasis in human colorectal cancer. Cancer Lett 227:229–236

Iwata N, Tsubuki S, Takaki Y, Watanabe K, Sekiguchi M et al (2000) Identification of the major Abeta1-42-degrading catabolic pathway in brain parenchyma: suppression leads to biochemical and pathological deposition. Nat Med 6:143–150

Iwata N, Tsubuki S, Takaki Y, Shirotani K, Lu B et al (2001) Metabolic regulation of brain Abeta by neprilysin. Science 292:1550–1552

Janes PW, Saha N, Barton WA, Kolev MV, Wimmer-Kleikamp SH et al (2005) Adam meets Eph: an ADAM substrate recognition module acts as a molecular switch for ephrin cleavage in trans. Cell 123:291–304

Javerzat S, Franco M, Herbert J, Platonova N, Peille AL et al (2009) Correlating global gene regulation to angiogenesis in the developing chick extra-embryonic vascular system. PLoS One 4:e7856

Jin X, Iwasa S, Okada K, Mitsumata M, Ooi A (2003) Expression patterns of seprase, a membrane serine protease, in cervical carcinoma and cervical intraepithelial neoplasm. Anticancer Res 23:3195–3198

Jorissen E, Prox J, Bernreuther C, Weber S, Schwanbeck R et al (2010) The disintegrin/metalloproteinase ADAM10 is essential for the establishment of the brain cortex. J Neurosci 30:4833–4844

Karl T, Hoffmann T, Pabst R, Von Horsten S (2003) Behavioral effects of neuropeptide Y in F344 rat substrains with a reduced dipeptidyl-peptidase IV activity. Pharmacol Biochem Behav 75:869–879

Kawarabayashi T, Younkin LH, Saido TC, Shoji M, Ashe KH et al (2001) Age-dependent changes in brain, CSF, and plasma amyloid (beta) protein in the Tg2576 transgenic mouse model of Alzheimer's disease. J Neurosci 21:372–381

Kikuchi S, Tanoue A, Endo F, Wakasugi S, Matsuo N et al (2000) A novel nonsense mutation of the PEPD gene in a Japanese patient with prolidase deficiency. J Hum Genet 45:102–104

Kimberly WT, Xia W, Rahmati T, Wolfe MS, Selkoe DJ (2000) The transmembrane aspartates in presenilin 1 and 2 are obligatory for γ-secretase activity and amyloid β-protein generation. J Biol Chem 275:3173–3178

Kimberly WT, LaVoie MJ, Ostaszewski BL, Ye W, Wolfe MS et al (2003) γ-Secretase is a membrane protein complex comprised of presenilin, nicastrin, Aph-1, and Pen-2. Proc Natl Acad Sci U S A 100:6382–6387

Kin Y, Misumi Y, Ikehara Y (2001) Biosynthesis and characterization of the brain-specific membrane protein DPPX, a dipeptidyl peptidase IV-related protein. J Biochem (Tokyo) 129:289–295

Kirby M, Yu DM, O'Connor S, Gorrell MD (2010) Inhibitor selectivity in the clinical application of dipeptidyl peptidase-4 inhibition. Clin Sci (Lond) 118:31–41

Kitazume S, Tachida Y, Oka R, Kotani N, Ogawa K et al (2002) Characterization of alpha 2,6-sialyltransferase cleavage by Alzheimer's beta-secretase (BACE1). J Biol Chem 278:14865–14871

Klebanov GI, Teselkin Y, Babenkova IV, Popov IN, Levin G et al (1997) Evidence for a direct interaction of superoxide anion radical with carnosine. Biochem Mol Biol Int 43:99–106

Klebanov GI, Teselkin Y, Babenkova IV, Lyubitsky OB, Rebrova OY et al (1998) Effect of carnosine and its components on free-radical reactions. Membr Cell Biol 12:89–99

Kojro E, Postina R, Buro C, Meiringer C, Gehrig-Burger K et al (2006) The neuropeptide PACAP promotes the alpha-secretase pathway for processing the Alzheimer amyloid precursor protein. FASEB J 20:512–514

Kozarich JW (2010) S28 peptidases: lessons from a seemingly 'dysfunctional' family of two. BMC Biol 8:87

Kuhn PH, Marjaux E, Imhof A, De Strooper B, Haass C et al (2007) Regulated intramembrane proteolysis of the interleukin-1 receptor II by alpha-, beta-, and gamma-secretase. J Biol Chem 282:11982–11995

Kuhn PH, Wang H, Dislich B, Colombo A, Zeitschel U et al (2010) ADAM10 is the physiologically relevant, constitutive alpha-secretase of the amyloid precursor protein in primary neurons. EMBO J 29:3020–3032

Kumamoto K, Stewart TA, Johnson AR, Erdos EG (1981) Prolylcarboxypeptidase (angiotensinase C) in human lung and cultured cells. J Clin Invest 67:210–215

Ladror US, Snyder SW, Wang GT, Holzman TF, Krafft GA (1994) Cleavage at the amino and carboxyl termini of Alzheimer's amyloid-beta by cathepsin D. J Biol Chem 269:18422–18428

Lambeir AM, Durinx C, Scharpe S, De Meester I (2003) Dipeptidyl-peptidase IV from bench to bedside: an update on structural properties, functions, and clinical aspects of the enzyme DPP IV. Crit Rev Clin Lab Sci 40:209–294

Lammich S, Kojro E, Postina R, Gilbert S, Pfeiffer R et al (1999) Constitutive and regulated alpha-secretase cleavage of Alzheimer's amyloid precursor protein by a disintegrin metalloprotease. Proc Natl Acad Sci U S A 96:3922–3927

Lammich S, Schobel S, Zimmer AK, Lichtenthaler SF, Haass C (2004) Expression of the Alzheimer protease BACE1 is suppressed via its 5'-untranslated region. EMBO Rep 5:620–625

Lankas GR, Leiting B, Roy RS, Eiermann GJ, Beconi MG et al (2005) Dipeptidyl peptidase IV inhibition for the treatment of Type 2 diabetes: potential importance of selectivity over dipeptidyl peptidases 8 and 9. Diabetes 54:2988–2994

Laudon H, Hansson EM, Melen K, Bergman A, Farmery MR et al (2005) A nine-transmembrane domain topology for presenilin 1. J Biol Chem 280:35352–35360

Lautar SL, Rojas C, Slusher BS, Wozniak KM, Wu Y et al (2005) DPP IV inhibitor blocks mescaline-induced scratching and amphetamine-induced hyperactivity in mice. Brain Res 1048:177–184

LeBlanc A (1995) Increased production of 4 kDa amyloid peptide in serum deprived human primary neuron cultures: possible involvement of apoptosis. J Neurosci 15:7837–7846

Ledoux P, Scriver C, Hechtman P (1994) Four novel PEPD alleles causing prolidase deficiency. Am J Hum Genet 54:1014–1021

Ledoux P, Scriver CR, Hechtman P (1996) Expression and molecular analysis of mutations in prolidase deficiency. Am J Hum Genet 59:1035–1039

Lee SF, Shah S, Li H, Yu C, Han W et al (2002) Mammalian APH-1 interacts with presenilin and nicastrin and Is required for intramembrane proteolysis of amyloid-β precursor protein and Notch. J Biol Chem 277:45013–45019

Leiting B, Pryor KD, Wu JK, Marsilio F, Patel RA et al (2003) Catalytic properties and inhibition of proline-specific dipeptidyl peptidases II, IV and VII. Biochem J 371:525–532

Li Q, Sudhof TC (2004) Cleavage of amyloid-beta precursor protein and amyloid-beta precursor-like protein by BACE 1. J Biol Chem 279:10542–10550

Li YM, Lai MT, Xu M, Huang Q, DiMuzio-Mower J et al (2000a) Presenilin 1 is linked with γ-secretase activity in the detergent solubilized state. Proc Natl Acad Sci U S A 97:6138–6143

Li YM, Xu M, Lai MT, Huang Q, Castro JL et al (2000b) Photoactivated γ-secretase inhibitors directed to the active site covalently label presenilin 1. Nature 405:689–694

Li X, Lou Z, Li X, Zhou W, Ma M et al (2008) Structure of human cytosolic X-prolyl aminopeptidase: a double Mn(II)-dependent dimeric enzyme with a novel three-domain subunit. J Biol Chem 283:22858–22866

Lichtenthaler SF (2011) Alpha-secretase in Alzheimer's disease: molecular identity, regulation and therapeutic potential. J Neurochem 116:10–21

Lichtenthaler SF, Wang R, Grimm H, Uljon SN, Masters CL et al (1999) Mechanism of the cleavage specificity of Alzheimer's disease gamma- secretase identified by phenylalanine-scanning mutagenesis of the transmembrane domain of the amyloid precursor protein. Proc Natl Acad Sci U S A 96:3053–3058

Lichtenthaler SF, Beher D, Grimm HS, Wang R, Shearman MS et al (2002) The intramembrane cleavage site of the amyloid precursor protein depends on the length of its transmembrane domain. Proc Natl Acad Sci U S A 99:1365–1370

Lichtenthaler SF, Dominguez DI, Westmeyer GG, Reiss K, Haass C et al (2003) The cell adhesion protein P-selectin glycoprotein ligand-1 is a substrate for the aspartyl protease BACE1. J Biol Chem 278:48713–48719

Lichtenthaler SF, Haass C, Steiner H (2011) Regulated intramembrane proteolysis – lessons from amyloid precursor protein processing. J Neurochem 117:779–796

Lin X, Koelsch G, Wu S, Downs D, Dashti A et al (2000) Human aspartic protease memapsin 2 cleaves the beta-secretase site of beta-amyloid precursor protein. Proc Natl Acad Sci U S A 97:1456–1460

Luo Y, Bolon B, Kahn S, Bennett BD, Babu-Khan S et al (2001) Mice deficient in BACE1, the Alzheimer's beta-secretase, have normal phenotype and abolished beta-amyloid generation. Nat Neurosci 4:231–232

Lupi A, De Riso A, Torre SD, Rossi A, Campari E et al (2004) Characterization of a new PEPD allele causing prolidase deficiency in two unrelated patients: natural-occurrent mutations as a tool to investigate structure-function relationship. J Hum Genet 49:500–506

Lupi A, Rossi A, Campari E, Pecora F, Lund AM et al (2006) Molecular characterisation of six patients with prolidase deficiency: identification of the first small duplication in the prolidase gene and of a mutation generating symptomatic and asymptomatic outcomes within the same family. J Med Genet 43:e58

Mackic JB, Stins M, McComb JG, Calero M, Ghiso J et al (1998) Human blood-brain barrier receptors for Alzheimer's amyloid-beta 1–40. Asymmetrical binding, endocytosis, and transcytosis at the apical side of brain microvascular endothelial cell monolayer. J Clin Invest 102:734–743

Mackic JB, Bading J, Ghiso J, Walker L, Wisniewski T et al (2002) Circulating amyloid-beta peptide crosses the blood-brain barrier in aged monkeys and contributes to Alzheimer's disease lesions. Vascul Pharmacol 38:303–313

Maes MB, Lambeir AM, Gilany K, Senten K, Van der Veken P et al (2005) Kinetic investigation of human dipeptidyl peptidase II (DPPII)-mediated hydrolysis of dipeptide derivatives and its identification as quiescent cell proline dipeptidase (QPP)/dipeptidyl peptidase 7 (DPP7). Biochem J 386:315–324

Maes MB, Martinet W, Schrijvers DM, Van der Veken P, De Meyer GR et al (2006) Dipeptidyl peptidase II and leukocyte cell death. Biochem Pharmacol 72:70–79
Maes MB, Scharpe S, De Meester I (2007) Dipeptidyl peptidase II (DPPII), a review. Clin Chim Acta 380:31–49
Maggiora LL, Orawski AT, Simmons WH (1999) Apstatin analogue inhibitors of aminopeptidase P, a bradykinin- degrading enzyme. J Med Chem 42:2394–2402
Malito E, Ralat LA, Manolopoulou M, Tsay JL, Wadlington NL et al (2008) Molecular bases for the recognition of short peptide substrates and cystein-directed modifications of human insulin-degrading enzyme. Biochemistry 47:12822–12834
Mallela J, Yang J, Shariat-Madar Z (2009) Prolylcarboxypeptidase: a cardioprotective enzyme. Int J Biochem Cell Biol 41:477–481
Mannisto PT, Venalainen J, Jalkanen A, Garcia-Horsman JA (2007) Prolyl oligopeptidase: a potential target for the treatment of cognitive disorders. Drug News Perspect 20:293–305
Marcello E, Gardoni F, Mauceri D, Romorini S, Jeromin A et al (2007) Synapse-associated protein-97 mediates alpha-secretase ADAM10 trafficking and promotes its activity. J Neurosci 27:1682–1691
Martel CL, Mackic JB, McComb JG, Ghiso J, Zlokovic BV (1996) Blood-brain barrier uptake of the 40 and 42 amino acid sequences of circulating Alzheimer's amyloid beta in guinea pigs. Neurosci Lett 206:157–160
Martin L, Fluhrer R, Haass C (2009) Substrate requirements for SPPL2b-dependent regulated intramembrane proteolysis. J Biol Chem 284:5662–5670
Mawuenyega KG, Sigurdson W, Ovod V, Munsell L, Kasten T et al (2010) Decreased clearance of CNS beta-amyloid in Alzheimer's disease. Science 330:1774
McDonald JK, Reilly TJ, Zeitman BB, Ellis S (1968) Dipeptidyl arylamidase II of the pituitary. Properties of lysylalanyl-beta-naphthylamide hydrolysis: inhibition by cations, distribution in tissues, and subcellular localization. J Biol Chem 243:2028–2037
McDonough CW, Hicks PJ, Lu L, Langefeld CD, Freedman BI et al (2009) The influence of carnosinase gene polymorphisms on diabetic nephropathy risk in African-Americans. Hum Genet 126:265–275
Medeiros MD, Turner AJ (1993) Processing and metabolism of peptide YY. Biochem Soc Trans 21:248S
Medeiros MD, Turner AJ (1994) Processing and metabolism of peptide-YY: pivotal roles of dipeptidylpeptidase-IV, aminopeptidase-P, and endopeptidase-24.11. Endocrinology 134:2088–2094
Medeiros MD, Turner AJ (1996) Metabolism and functions of neuropeptide Y. Neurochem Res 21:1125–1132
Melchor JP, Pawlak R, Chen Z, Strickland S (2003) The possible role of tissue-type plasminogen activator (tPA) and tPA blockers in the pathogenesis and treatment of Alzheimer's disease. J Mol Neurosci 20:287–289
Mele DA, Bista P, Baez DV, Huber BT (2009) Dipeptidyl peptidase 2 is an essential survival factor in the regulation of cell quiescence. Cell Cycle 8:2425–2434
Mentlein R (1999) Dipeptidyl-peptidase IV (CD26)–role in the inactivation of regulatory peptides. Regul Pept 85:9–24
Mentlein R (2004) Cell-surface peptidases. Int Rev Cytol 235:165–213
Mentlein R, Roos T (1996) Proteases involved in the metabolism of angiotensin II, bradykinin, calcitonin gene-related peptide (CGRP), and neuropeptide Y by vascular smooth muscle cells. Peptides 17:709–720
Mentlein R, Struckhoff G (1989) Purification of two dipeptidyl aminopeptidases II from rat brain and their action on proline-containing neuropeptides. J Neurochem 52:1284–1293
Mentlein R, von Kolszynski M, Sprang R, Lucius R (1990) Proline-specific proteases in cultivated neuronal and glial cells. Brain Res 527:159–162

Mentlein R, Hattermann K, Hemion C, Jungbluth AA, Held-Feindt J (2011) Expression and role of the cell surface protease seprase/fibroblast activation protein-alpha (FAP-alpha) in astroglial tumors. Biol Chem 392:199–207

Min J, Senut MC, Rajanikant K, Greenberg E, Bandagi R et al (2008) Differential neuroprotective effects of carnosine, anserine, and N-acetyl carnosine against permanent focal ischemia. J Neurosci Res 86:2984–2991

Mitsubuchi H, Nakamura K, Matsumoto S, Endo F (2008) Inborn errors of proline metabolism. J Nutr 138:2016S–2020S

Mitterreiter S, Page RM, Kamp F, Hopson J, Winkler E et al (2010) Bepridil and amiodarone simultaneously target the Alzheimer's disease beta- and gamma-secretase via distinct mechanisms. J Neurosci 30:8974–8983

Morain P, Lestage P, De Nanteuil G, Jochemsen R, Robin JL et al (2002) S 17092: a prolyl endopeptidase inhibitor as a potential therapeutic drug for memory impairment. Preclinical and clinical studies. CNS Drug Rev 8:31–52

Moreira CR, Schmaier AH, Mahdi F, da Motta G, Nader HB et al (2002) Identification of prolylcarboxypeptidase as the cell matrix associated prekallikrein activator. FEBS Lett 523:167–170

Mori Y, Kono K, Matsumoto Y, Fujii H, Yamane T et al (2004) The expression of a type II transmembrane serine protease (Seprase) in human gastric carcinoma. Oncology 67:411–419

Mueller-Steiner S, Zhou Y, Arai H, Roberson ED, Sun B et al (2006) Antiamyloidogenic and neuroprotective functions of cathepsin B: implications for Alzheimer's disease. Neuron 51:703–714

Munger JS, Haass C, Lemere CA, Shi GP, Wong WS et al (1995) Lysosomal processing of amyloid precursor protein to A beta peptides: a distinct role for cathepsin S. Biochem J 311 (Pt 1):299–305

Munter LM, Voigt P, Harmeier A, Kaden D, Gottschalk KE et al (2007) GxxxG motifs within the amyloid precursor protein transmembrane sequence are critical for the etiology of Aβ42. EMBO J 26:1702–1712

Munter LM, Botev A, Richter L, Hildebrand PW, Althoff V et al (2010) Aberrant amyloid precursor protein (APP) processing in hereditary forms of Alzheimer disease caused by APP familial Alzheimer disease mutations can be rescued by mutations in the APP GxxxG motif. J Biol Chem 285:21636–21643

Murphy MP, Hickman LJ, Eckman CB, Uljon SN, Wang R et al (1999) gamma-Secretase, evidence for multiple proteolytic activities and influence of membrane positioning of substrate on generation of amyloid beta peptides of varying length. J Biol Chem 274:11914–11923

Myohanen TT, Venalainen JI, Garcia-Horsman JA, Mannisto PT (2008a) Spatial association of prolyl oligopeptidase, inositol 1,4,5-triphosphate type 1 receptor, substance P and its neurokinin-1 receptor in the rat brain: an immunohistochemical colocalization study. Neuroscience 153:1177–1189

Myohanen TT, Venalainen JI, Garcia-Horsman JA, Piltonen M, Mannisto PT (2008b) Cellular and subcellular distribution of rat brain prolyl oligopeptidase and its association with specific neuronal neurotransmitters. J Comp Neurol 507:1694–1708

Myohanen TT, Venalainen JI, Garcia-Horsman JA, Piltonen M, Mannisto PT (2008c) Distribution of prolyl oligopeptidase in the mouse whole-body sections and peripheral tissues. Histochem Cell Biol 130:993–1003

Nadal MS, Ozaita A, Amarillo Y, de Miera EV, Ma Y et al (2003) The CD26-related dipeptidyl aminopeptidase-like protein DPPX is a critical component of neuronal A-type K+ channels. Neuron 37:449–461

Nagy JI, Yamamoto T, Uemura H, Schrader WP (1996) Adenosine deaminase in rodent median eminence: detection by antibody to the mouse enzyme and co-localization with adenosine deaminase- complexing protein (CD26). Neuroscience 73:459–471

Ngo ML, Mahdi F, Kolte D, Shariat-Madar Z (2009) Upregulation of prolylcarboxypeptidase (PRCP) in lipopolysaccharide (LPS) treated endothelium promotes inflammation. J Inflamm (Lond) 6:3
Nicholson DW, Thornberry NA (1997) Caspases: killer proteases. Trends Biochem Sci 22:299–306
Nitsch RM, Slack BE, Wurtman RJ, Growdon JH (1992) Release of Alzheimer amyloid precursor derivatives stimulated by activation of muscarinic acetylcholine receptors. Science 258:304–307
O'Connor T, Sadleir KR, Maus E, Velliquette RA, Zhao J et al (2008) Phosphorylation of the translation initiation factor eIF2alpha increases BACE1 levels and promotes amyloidogenesis. Neuron 60:988–1009
O'Leary RM, O'Connor B (1995) Identification and localisation of a synaptosomal membrane prolyl endopeptidase from bovine brain. Eur J Biochem 227:277–283
O'Leary RM, Gallagher SP, O'Connor B (1996) Purification and characterization of a novel membrane-bound form of prolyl endopeptidase from bovine brain. Int J Biochem Cell Biol 28:441–449
O'Toole JF, Liu Y, Davis EE, Westlake CJ, Attanasio M et al (2010) Individuals with mutations in XPNPEP3, which encodes a mitochondrial protein, develop a nephronophthisis-like nephropathy. J Clin Invest 120:791–802
Odya CE and Erdos EG (1981) Human prolylcarboxypeptidase. Methods Enzymol 80 Pt C:460-466.
Odya CE, Marinkovic DV, Hammon KJ, Stewart TA, Erdos EG (1978) Purification and properties of prolylcarboxypeptidase (angiotensinase C) from human kidney. J Biol Chem 253:5927–5931
Oefner C, D'Arcy A, Mac SA, Pierau S, Gardiner R et al (2003) High-resolution structure of human apo dipeptidyl peptidase IV/CD26 and its complex with 1-[({2-[(5-iodopyridin-2-yl) amino]-ethyl}amino)-acetyl]-2-cyano-(S)-pyrrol idine. Acta Crystallogr D Biol Crystallogr 59:1206–1212
Oh YS, Turner RJ (2005a) Topology of the C-terminal fragment of human presenilin 1. Biochemistry 44:11821–11828
Oh YS, Turner RJ (2005b) Evidence that the COOH terminus of human presenilin 1 is located in extracytoplasmic space. Am J Physiol Cell Physiol 289:C576–C581
Ohnuma K, Dang NH, Morimoto C (2008a) Revisiting an old acquaintance: CD26 and its molecular mechanisms in T cell function. Trends Immunol 29:295–301
Ohnuma K, Takahashi N, Yamochi T, Hosono O, Dang NH et al (2008b) Role of CD26/dipeptidyl peptidase IV in human T cell activation and function. Front Biosci 13:2299–2310
Okada K, Chen WT, Iwasa S, Jin X, Yamane T et al (2003) Seprase, a membrane-type serine protease, has different expression patterns in intestinal- and diffuse-type gastric cancer. Oncology 65:363–370
Olsen C, Wagtmann N (2002) Identification and characterization of human DPP9, a novel homologue of dipeptidyl peptidase IV. Gene 299:185–193
Orawski AT, Simmons WH (1995) Purification and properties of membrane-bound aminopeptidase P from rat lung. Biochemistry 34:11227–11236
Orawski AT, Susz JP, Simmons WH (1987) Aminopeptidase P from bovine lung: solubilization, properties, and potential role in bradykinin degradation. Mol Cell Biochem 75:123–132
Orioli M, Vistoli G, Regazzoni L, Pedretti A, Lapolla A et al (2011) Design, synthesis, ADME properties, and pharmacological activities of beta-Alanyl-D-histidine (D-Carnosine) prodrugs with improved bioavailability. ChemMedChem 6:1269–1282
Otani H, Okumura N, Hashida-Okumura A, Nagai K (2005) Identification and characterization of a mouse dipeptidase that hydrolyzes L-carnosine. J Biochem 137:167–175
Otani H, Okumura A, Nagai K, Okumura N (2008) Colocalization of a carnosine-splitting enzyme, tissue carnosinase (CN2)/cytosolic non-specific dipeptidase 2 (CNDP2), with histidine decarboxylase in the tuberomammillary nucleus of the hypothalamus. Neurosci Lett 445:166–169

Pahnke J, Walker LC, Scheffler K, Krohn M (2009) Alzheimer's disease and blood-brain barrier function-Why have anti-beta-amyloid therapies failed to prevent dementia progression? Neurosci Biobehav Rev 33:1099–1108

Palmiter RD (2009) Reduced levels of neurotransmitter-degrading enzyme PRCP promote a lean phenotype. J Clin Invest 119:2130–2133

Pangalos MN, Neefs JM, Somers M, Verhasselt P, Bekkers M et al (1999) Isolation and expression of novel human glutamate carboxypeptidases with N-acetylated alpha-linked acidic dipeptidase and dipeptidyl peptidase IV activity. J Biol Chem 274:8470–8483

Park J, Knott HT, Navdi NA, Collyer CA, Wang XM et al (2008) Reversible inactivation of human dipeptidyl peptidases 8 and 9 by oxidation. Open Enzym Inhib J 1:52–60

Parkin E, Harris B (2009) A disintegrin and metalloproteinase (ADAM)-mediated ectodomain shedding of ADAM10. J Neurochem 108:1464–1479

Pegova A, Abe H, Boldyrev A (2000) Hydrolysis of carnosine and related compounds by mammalian carnosinases. Comp Biochem Physiol B Biochem Mol Biol 127:443–446

Perroud B, Alvarado RJ, Espinal GM, Morado AR, Phinney BS et al (2009) In vivo multiplex quantitative analysis of 3 forms of alpha melanocyte stimulating hormone in pituitary of prolyl endopeptidase deficient mice. Mol Brain 2:14

Polgar L (2002) The prolyl oligopeptidase family. Cell Mol Life Sci 59:349–362

Postina R, Schroeder A, Dewachter I, Bohl J, Schmitt U et al (2004) A disintegrin-metalloproteinase prevents amyloid plaque formation and hippocampal defects in an Alzheimer disease mouse model. J Clin Invest 113:1456–1464

Prechel MM, Orawski AT, Maggiora LL, Simmons WH (1995) Effect of a new aminopeptidase P inhibitor, apstatin, on bradykinin degradation in the rat lung. J Pharmacol Exp Ther 275:1136–1142

Pruessmeyer J, Ludwig A (2009) The good, the bad and the ugly substrates for ADAM10 and ADAM17 in brain pathology, inflammation and cancer. Semin Cell Dev Biol 20:164–174

Qi SY, Riviere PJ, Trojnar J, Junien JL, Akinsanya KO (2003) Cloning and characterization of dipeptidyl peptidase 10, a new member of an emerging subgroup of serine proteases. Biochem J 373:179–189

Qi-Takahara Y, Morishima-Kawashima M, Tanimura Y, Dolios G, Hirotani N et al (2005) Longer forms of amyloid beta protein: implications for the mechanism of intramembrane cleavage by gamma-secretase. J Neurosci 25:436–445

Quinn PJ, Boldyrev AA, Formazuyk VE (1992) Carnosine: its properties, functions and potential therapeutic applications. Mol Aspects Med 13:379–444

Rajendran L, Schneider A, Schlechtingen G, Weidlich S, Ries J et al (2008) Efficient inhibition of the Alzheimer's disease beta-secretase by membrane targeting. Science 320:520–523

Rasmussen HB, Branner S, Wiberg FC, Wagtmann N (2003) Crystal structure of human dipeptidyl peptidase IV/CD26 in complex with a substrate analog. Nat Struct Biol 10:19–25

Reiss K, Saftig P (2009) The "A Disintegrin And Metalloprotease" (ADAM) family of sheddases: physiological and cellular functions. Semin Cell Dev Biol 20:126–137

Ren Z, Schenk D, Basi GS, Shapiro IP (2007) Amyloid beta-protein precursor juxtamembrane domain regulates specificity of gamma-secretase-dependent cleavages. J Biol Chem 282:35350–35360

Roberds SL, Anderson J, Basi G, Bienkowski MJ, Branstetter DG et al (2001) BACE knockout mice are healthy despite lacking the primary beta- secretase activity in brain: implications for Alzheimer's disease therapeutics. Hum Mol Genet 10:1317–1324

Rogers GW Jr, Edelman GM, Mauro VP (2004) Differential utilization of upstream AUGs in the beta-secretase mRNA suggests that a shunting mechanism regulates translation. Proc Natl Acad Sci U S A 101:2794–2799

Rosenblum JS, Kozarich JW (2003) Prolyl peptidases: a serine protease subfamily with high potential for drug discovery. Curr Opin Chem Biol 7:496–504

Rossner S, Schulz I, Zeitschel U, Schliebs R, Bigl V et al (2005) Brain prolyl endopeptidase expression in aging, APP transgenic mice and Alzheimer's disease. Neurochem Res 30:695–702
Rossner S, Sastre M, Bourne K, Lichtenthaler SF (2006) Transcriptional and translational regulation of BACE1 expression–implications for Alzheimer's disease. Prog Neurobiol 79:95–111
Rummey C, Metz G (2007) Homology models of dipeptidyl peptidases 8 and 9 with a focus on loop predictions near the active site. Proteins 66:160–171
Rusu I, Yaron A (1992) Aminopeptidase P from human leukocytes. Eur J Biochem 210:93–100
Saftig P, Peters C, von Figura K, Craessaerts K, Van Leuven F et al (1996) Amyloidogenic processing of human amyloid precursor protein in hippocampal neurons devoid of cathepsin D. J Biol Chem 271:27241–27244
Saido TC, Iwatsubo T, Mann DM, Shimada H, Ihara Y et al (1995) Dominant and differential deposition of distinct beta-amyloid peptide species, A beta N3(pE), in senile plaques. Neuron 14:457–466
Saido TC, Yamao-Harigaya W, Iwatsubo T, Kawashima S (1996) Amino and carboxyl-terminal heterogeneity of beta-amyloid peptides deposited in human brain. Neurosci Lett 215:173–176
Sappino AP, Madani R, Huarte J, Belin D, Kiss JZ et al (1993) Extracellular proteolysis in the adult murine brain. J Clin Invest 92:679–685
Sastre M, Steiner H, Fuchs K, Capell A, Multhaup G et al (2001) Presenilin-dependent γ-secretase processing of β-amyloid precursor protein at a site corresponding to the S3 cleavage of Notch. EMBO Rep 2:835–841
Sato C, Morohashi Y, Tomita T, Iwatsubo T (2006) Structure of the catalytic pore of γ-secretase probed by the accessibility of substituted cysteines. J Neurosci 26:12081–12088
Sato T, Diehl TS, Narayanan S, Funamoto S, Ihara Y et al (2007) Active gamma-secretase complexes contain only one of each component. J Biol Chem 282:33985–33993
Sato C, Takagi S, Tomita T, Iwatsubo T (2008) The C-terminal PAL motif and transmembrane domain 9 of presenilin 1 are involved in the formation of the catalytic pore of the γ-secretase. J Neurosci 28:6264–6271
Saunders AJ, Kim TW, Tanzi RE (1999) BACE maps to chromosome 11 and a BACE homolog, BACE2, reside in the obligate down syndrome region of chromosome 21. Science 286:1255a
Savonenko AV, Melnikova T, Laird FM, Stewart KA, Price DL et al (2008) Alteration of BACE1-dependent NRG1/ErbB4 signaling and schizophrenia-like phenotypes in BACE1-null mice. Proc Natl Acad Sci U S A 105:5585–5590
Scanlan MJ, Raj BK, Calvo B, Garin-Chesa P, Sanz-Moncasi MP et al (1994) Molecular cloning of fibroblast activation protein alpha, a member of the serine protease family selectively expressed in stromal fibroblasts of epithelial cancers. Proc Natl Acad Sci U S A 91:5657–5661
Schmechel A, Strauss M, Schlicksupp A, Pipkorn R, Haass C et al (2004) Human BACE forms dimers and colocalizes with APP. J Biol Chem 279:39710–39717
Schneider JS, Giardiniere M, Morain P (2002) Effects of the prolyl endopeptidase inhibitor S 17092 on cognitive deficits in chronic low dose MPTP-treated monkeys. Neuropsychopharmacology 26:176–182
Schroeter EH, Ilagan MX, Brunkan AL, Hecimovic S, Li YM et al (2003) A presenilin dimer at the core of the γ-secretase enzyme: insights from parallel analysis of Notch 1 and APP proteolysis. Proc Natl Acad Sci U S A 100:13075–13080
Schulz I, Gerhartz B, Neubauer A, Holloschi A, Heiser U et al (2002) Modulation of inositol 1,4,5-triphosphate concentration by prolyl endopeptidase inhibition. Eur J Biochem 269:5813–5820
Schulz I, Zeitschel U, Rudolph T, Ruiz-Carrillo D, Rahfeld JU et al (2005) Subcellular localization suggests novel functions for prolyl endopeptidase in protein secretion. J Neurochem 94:970–979
Seeger M, Nordstedt C, Petanceska S, Kovacs DM, Gouras GK et al (1997) Evidence for phosphorylation and oligomeric assembly of presenilin 1. Proc Natl Acad Sci U S A 94:5090–5094

Seiffert D, Bradley JD, Rominger CM, Rominger DH, Yang F et al (2000) Presenilin-1 and -2 are molecular targets for gamma-secretase inhibitors. J Biol Chem 275:34086–34091
Serneels L, Van Biervliet J, Craessaerts K, Dejaegere T, Horre K et al (2009) gamma-Secretase heterogeneity in the Aph1 subunit: relevance for Alzheimer's disease. Science 324:639–642
Sevalle J, Amoyei A, Robert P, Fournié-Zaluski MC, Roques B et al (2009) Aminopeptidase A contributes to the N-terminal truncation of amyloid beta-peptide. J Neurochem 109:248–256
Sewards TV, Sewards MA (2003) Fear and power-dominance motivation: proposed contributions of peptide hormones present in cerebrospinal fluid and plasma. Neurosci Biobehav Rev 27:247–267
Shah S, Lee SF, Tabuchi K, Hao YH, Yu C et al (2005) Nicastrin functions as a γ-secretase-substrate receptor. Cell 122:435–447
Shariat-Madar Z, Mahdi F, Schmaier AH (2002) Identification and characterization of prolylcarboxypeptidase as an endothelial cell prekallikrein activator. J Biol Chem 277:17962–17969
Shariat-Madar Z, Mahdi F, Schmaier AH (2004) Recombinant prolylcarboxypeptidase activates plasma prekallikrein. Blood 103:4554–4561
Shariat-Madar B, Kolte D, Verlangieri A, Shariat-Madar Z (2010) Prolylcarboxypeptidase (PRCP) as a new target for obesity treatment. Diabetes Metab Syndr Obes 3:67–78
Sharoyan SG, Antonyan AA, Mardanyan SS, Lupidi G, Cuccioloni M et al (2008) Complex of dipeptidyl peptidase II with adenosine deaminase. Biochemistry (Mosc) 73:943–949
Shi XP, Chen E, Yin KC, Na S, Garsky VM et al (2001) The pro domain of beta-secretase does not confer strict zymogen-like properties but does assist proper folding of the protease domain. J Biol Chem 276:10366–10373
Shinoda M, Okamiya K, Toide K (1995) Effect of a novel prolyl endopeptidase inhibitor, JTP-4819, on thyrotropin-releasing hormone-like immunoreactivity in the cerebral cortex and hippocampus of aged rats. Jpn J Pharmacol 69:273–276
Shirasawa Y, Osawa T, Hirashima A (1994) Molecular cloning and characterization of prolyl endopeptidase from human T cells. J Biochem 115:724–729
Shirotani K, Edbauer D, Prokop S, Haass C, Steiner H (2004) Identification of distinct γ-secretase complexes with different APH-1 variants. J Biol Chem 279:41340–41345
Shirotani K, Tomioka M, Kremmer E, Haass C, Steiner H (2007) Pathological activity of familial Alzheimer's disease-associated mutant presenilin can be executed by six different gamma-secretase complexes. Neurobiol Dis 27:102–107
Shrinath M, Walter JH, Haeney M, Couriel JM, Lewis MA et al (1997) Prolidase deficiency and systemic lupus erythematosus. Arch Dis Child 76:441–444
Sinha S, Anderson JP, Barbour R, Basi GS, Caccavello R et al (1999) Purification and cloning of amyloid precursor protein beta-secretase from human brain. Nature 402:537–540
Sisodia SS (1992) Beta-amyloid precursor protein cleavage by a membrane-bound protease. Proc Natl Acad Sci U S A 89:6075–6079
Skidgel RA, Wickstrom E, Kumamoto K, Erdos EG (1981) Rapid radioassay for prolylcarboxypeptidase (angiotensinase C). Anal Biochem 118:113–119
Skubitz KM, Skubitz AP (2004) Gene expression in aggressive fibromatosis. J Lab Clin Med 143:89–98
Soisson SM, Patel SB, Abeywickrema PD, Byrne NJ, Diehl RE et al (2010) Structural definition and substrate specificity of the S28 protease family: the crystal structure of human prolylcarboxypeptidase. BMC Struct Biol 10:16
Solans A, Estivill X, de La Luna S (2000) A new aspartyl protease on 21q22.3, BACE2, is highly similar to Alzheimer's amyloid precursor protein beta-secretase. Cytogenet Cell Genet 89:177–184
Spasic D, Tolia A, Dillen K, Baert V, De Strooper B et al (2006) Presenilin-1 maintains a nine-transmembrane topology throughout the secretory pathway. J Biol Chem 281:26569–26577
Sprinkle TJ, Stone AA, Venema RC, Denslow ND, Caldwell C et al (1998) Assignment of the membrane-bound human aminopeptidase P gene (XPNPEP2) to chromosome Xq25. Genomics 50:114–116

Sprinkle TJ, Caldwell C, Ryan JW (2000) Cloning, chromosomal sublocalization of the human soluble aminopeptidase P gene (XPNPEP1) to 10q25.3 and conservation of the putative proton shuttle and metal ligand binding sites with XPNPEP2. Arch Biochem Biophys 378:51–56

Steiner H, Duff K, Capell A, Romig H, Grim MG et al (1999) A loss of function mutation of presenilin-2 interferes with amyloid β-peptide production and Notch signaling. J Biol Chem 274:28669–28673

Steiner H, Winkler E, Edbauer D, Prokop S, Basset G et al (2002) PEN-2 is an integral component of the γ-secretase complex required for coordinated expression of presenilin and nicastrin. J Biol Chem 277:39062–39065

Stremenova J, Krepela E, Mares V, Trim J, Dbaly V et al (2007) Expression and enzymatic activity of dipeptidyl peptidase-IV in human astrocytic tumours are associated with tumour grade. Int J Oncol 31:785–792

Stremenova J, Mares V, Lisa V, Hilser M, Krepela E et al (2010) Expression of dipeptidyl peptidase-IV activity and/or structure homologs in human meningiomas. Int J Oncol 36:351–358

Strop P, Bankovich AJ, Hansen KC, Garcia KC, Brunger AT (2004) Structure of a human A-type potassium channel interacting protein DPPX, a member of the dipeptidyl aminopeptidase family. J Mol Biol 343:1055–1065

Stvolinskii SL, Fedorova TN, Yuneva MO, Boldyrev AA (2003) Protective effect of carnosine on Cu, Zn-superoxide dismutase during impaired oxidative metabolism in the brain in vivo. Bull Exp Biol Med 135:130–132

Stvolinsky SL, Kukley ML, Dobrota D, Matejovicova M, Tkac I et al (1999) Carnosine: an endogenous neuroprotector in the ischemic brain. Cell Mol Neurobiol 19:45–56

Stvolinsky S, Kukley M, Dobrota D, Mezesova V, Boldyrev A (2000) Carnosine protects rats under global ischemia. Brain Res Bull 53:445–448

Stvolinsky SL, Bulygina ER, Fedorova TN, Meguro K, Sato T et al (2010) Biological activity of novel synthetic derivatives of carnosine. Cell Mol Neurobiol 30:395–404

Surazynski A, Palka J, Wolczynski S (2001) Phosphorylation of prolidase increases the enzyme activity. Mol Cell Biochem 220:95–101

Tabakman R, Lazarovici P, Kohen R (2002) Neuroprotective effects of carnosine and homocarnosine on pheochromocytoma PC12 cells exposed to ischemia. J Neurosci Res 68:463–469

Takada A, Takada Y (1988) Physiology of plasminogen: with special reference to activation and degradation. Haemostasis 18(Suppl 1):25–35

Takami M, Nagashima Y, Sano Y, Ishihara S, Morishima-Kawashima M et al (2009) gamma-Secretase: successive tripeptide and tetrapeptide release from the transmembrane domain of beta-carboxyl terminal fragment. J Neurosci 29:13042–13052

Takasugi N, Tomita T, Hayashi I, Tsuruoka M, Niimura M et al (2003) The role of presenilin cofactors in the γ-secretase complex. Nature 422:438–441

Takeda K, Araki W, Akiyama H, Tabira T (2004) Amino-truncated amyloid beta-peptide (Abeta5-40/42) produced from caspase-cleaved amyloid precursor protein is deposited in Alzheimer's disease brain. FASEB J 18:1755–1757

Tan F, Morris PW, Skidgel RA, Erdos EG (1993) Sequencing and cloning of human prolylcarboxypeptidase (angiotensinase C). Similarity to both serine carboxypeptidase and prolylendopeptidase families. J Biol Chem 268:16631–16638

Tanoue A, Endo F, Kitano A, Matsuda I (1990a) A single nucleotide change in the prolidase gene in fibroblasts from two patients with polypeptide positive prolidase deficiency. Expression of the mutant enzyme in NIH 3T3 cells. J Clin Invest 86:351–355

Tanoue A, Endo F, Matsuda I (1990b) Structural organization of the gene for human prolidase (peptidase D) and demonstration of a partial gene deletion in a patient with prolidase deficiency. J Biol Chem 265:11306–11311

Tanoue A, Endo F, Matsuda I (1990c) The human prolidase gene: structure and restriction fragment length polymorphisms. J Inherit Metab Dis 13:771–774

Tanoue A, Endo F, Akaboshi I, Oono T, Arata J et al (1991a) Molecular defect in siblings with prolidase deficiency and absence or presence of clinical symptoms. A 0.8-kb deletion with breakpoints at the short, direct repeat in the PEPD gene and synthesis of abnormal messenger RNA and inactive polypeptide. J Clin Invest 87:1171–1176

Tanoue A, Endo F, Awata H, Matsuda I (1991b) Abnormal mRNA and inactive polypeptide in a patient with prolidase deficiency. J Inherit Metab Dis 14:777–782

Tansi FL, Blanchard V, Berger M, Tauber R, Reutter W et al (2010) Interaction of human dipeptidyl peptidase IV and human immunodeficiency virus type-1 transcription transactivator in Sf9 cells. Virol J 7:267

Taylor-McCabe KJ, Ersahin C, Simmons WH (2001) Bradykinin metabolism in the isolated perfused rabbit heart. J Hypertens 19:1295–1299

Tenorio-Laranga J, Venalainen JI, Mannisto PT, Garcia-Horsman JA (2008) Characterization of membrane-bound prolyl endopeptidase from brain. FEBS J 275:4415–4427

Teufel M, Saudek V, Ledig JP, Bernhardt A, Boularand S et al (2003) Sequence identification and characterization of human carnosinase and a closely related non-specific dipeptidase. J Biol Chem 278:6521–6531

Thinakaran G, Regard JB, Bouton CM, Harris CL, Price DL et al (1998) Stable association of presenilin derivatives and absence of presenilin interactions with APP. Neurobiol Dis 4:438–453

Thoma R, Loffler B, Stihle M, Huber W, Ruf A et al (2003) Structural basis of proline-specific exopeptidase activity as observed in human dipeptidyl peptidase-IV. Structure 11:947–959

Tischer E, Cordell B (1996) Beta-amyloid precursor protein. Location of transmembrane domain and specificity of gamma-secretase cleavage. J Biol Chem 271:21914–21919

Toide K, Iwamoto Y, Fujiwara T, Abe H (1995a) JTP-4819: a novel prolyl endopeptidase inhibitor with potential as a cognitive enhancer. J Pharmacol Exp Ther 274:1370–1378

Toide K, Okamiya K, Iwamoto Y, Kato T (1995b) Effect of a novel prolyl endopeptidase inhibitor, JTP-4819, on prolyl endopeptidase activity and substance P- and arginine-vasopressin-like immunoreactivity in the brains of aged rats. J Neurochem 65:234–240

Toide K, Fujiwara T, Iwamoto Y, Shinoda M, Okamiya K et al (1996) Effect of a novel prolyl endopeptidase inhibitor, JTP-4819, on neuropeptide metabolism in the rat brain. Naunyn Schmiedebergs Arch Pharmacol 353:355–362

Tolia A, Chavez-Gutierrez L, De Strooper B (2006) Contribution of presenilin transmembrane domains 6 and 7 to a water-containing cavity in the γ-secretase complex. J Biol Chem 281:27633–27642

Tolia A, Horre K, De Strooper B (2008) Transmembrane domain 9 of presenilin determines the dynamic conformation of the catalytic site of γ-secretase. J Biol Chem 283:19793–19803

Tousseyn T, Thathiah A, Jorissen E, Raemaekers T, Konietzko U et al (2009) ADAM10, the rate-limiting protease of regulated intramembrane proteolysis of Notch and other proteins, is processed by ADAMS-9, ADAMS-15, and the gamma-secretase. J Biol Chem 284:11738–11747

Tsirka SE, Gualandris A, Amaral DG, Strickland S (1995) Excitotoxin-induced neuronal degeneration and seizure are mediated by tissue plasminogen activator. Nature 377:340–344

Tucker HM, Kihiko M, Caldwell JN, Wright S, Kawarabayashi T et al (2000) The plasmin system is induced by and degrades amyloid-beta aggregates. J Neurosci 20:2937–3946

Tucker HM, Simpson J, Kihiko-Ehmann M, Younkin LH, McGillis JP et al (2004) Plasmin deficiency does not alter endogenous murine amyloid levels in mice. Neurosci Lett 368:285–289

Turner RT III, Koelsch G, Hong L, Castanheira P, Ermolieff J et al (2001) Subsite specificity of memapsin 2 (beta-secretase): implications for inhibitor design. Biochemistry 40:10001–10006

Ujiie M, Dickstein DL, Carlow DA, Jefferies WA (2003) Blood-brain barrier permeability precedes senile plaque formation in an Alzheimer disease model. Microcirculation 10:463–470

Underwood R, Chiravuri M, Lee H, Schmitz T, Kabcenell AK et al (1999) Sequence, purification, and cloning of an intracellular serine protease, quiescent cell proline dipeptidase. J Biol Chem 274:34053–34058
Unno H, Yamashita T, Ujita S, Okumura N, Otani H et al (2008) Structural basis for substrate recognition and hydrolysis by mouse carnosinase CN2. J Biol Chem 283:27289–27299
Van Tetering G, van Diest P, Verlaan I, van der Wall E, Kopan R et al (2009) Metalloprotease ADAM10 is required for Notch1 site 2 cleavage. J Biol Chem 284:31018–31027
Vanhoof G, Goossens F, Juliano MA, Juliano L, De Meester I et al (1997a) Human lymphocyte X-prolyl aminopeptidase (aminopeptidase P)-like protein. A new member of the proline peptidase family? Adv Exp Med Biol 421:25–29
Vanhoof G, Goossens F, Juliano MA, Juliano L, Hendriks D et al (1997b) Isolation and sequence analysis of a human cDNA clone (XPNPEPL) homologous to X-prolyl aminopeptidase (aminopeptidase P). Cytogenet Cell Genet 78:275–280
Vassar R, Bennett BD, Babu-Khan S, Kahn S, Mendiaz EA et al (1999) Beta-secretase cleavage of Alzheimer's amyloid precursor protein by the transmembrane aspartic protease BACE. Science 286:735–741
Vassar R, Kovacs DM, Yan R, Wong PC (2009) The beta-secretase enzyme BACE in health and Alzheimer's disease: regulation, cell biology, function, and therapeutic potential. J Neurosci 29:12787–12794
Vekrellis K, Ye Z, Qui WQ, Walsh HD et al (2000) Neurons regulate extracellular levels of amyloid beta-protein via proteolysis by insulin-degrading enzyme. J Neurosci 20:1657–1665
Venema RC, Ju H, Zou R, Venema VJ, Ryan JW (1997) Cloning and tissue distribution of human membrane-bound aminopeptidase P. Biochim Biophys Acta 1354:45–48
Vistoli G, Orioli M, Pedretti A, Regazzoni L, Canevotti R et al (2009) Design, synthesis, and evaluation of carnosine derivatives as selective and efficient sequestering agents of cytotoxic reactive carbonyl species. Chem Med Chem 4:967–975
Von Arnim CA, Kinoshita A, Peltan ID, Tangredi MM, Herl L et al (2005) The low density lipoprotein receptor-related protein (LRP) is a novel beta-secretase (BACE1) substrate. J Biol Chem 280:17777–17785
Wada K, Yokotani N, Hunter C, Doi K, Wenthold RJ et al (1992) Differential expression of two distinct forms of mRNA encoding members of a dipeptidyl aminopeptidase family. Proc Natl Acad Sci U S A 89:197–201
Wagner L, Hoffmann T, Rahfeld JU, Demuth HU (2006) Distribution of dipeptidyl peptidase IV-like activity enzymes in canine and porcine tissue sections by RT-PCR. Adv Exp Med Biol 575:109–116
Wagner L, Gündel D, Zeitschel U, Rossner S, Frerker N et al (2008) Tissue-distribution of dipeptidyl peptidase 4-like enzymes in distinct brain areas and neuronal tissues. Clin Chem Lab Med 46:A1–A43
Wakabayashi T, Craessaerts K, Bammens L, Bentahir M, Borgions F et al (2009) Analysis of the gamma-secretase interactome and validation of its association with tetraspanin-enriched microdomains. Nat Cell Biol 11:1340–1346
Wallingford N, Perroud B, Gao Q, Coppola A, Gyengesi E (2009) Prolylcarboxypeptidase regulates food intake by inactivating alpha-MSH in rodents. J Clin Invest 119:2291–2303
Wang SH, Liu M, Chi MG, Wang QD, Sun MJ (2004) Production of human liver prolidase by Saccharomyces cerevisiae as host cells. Acta Pharmacol Sin 25:794–800
Wang SH, Zhi QW, Sun MJ (2005) Purification and characterization of recombinant human liver prolidase expressed in Saccharomyces cerevisiae. Arch Toxicol 79:253–259
Wang H, Kurien BT, Lundgren D, Patel NC, Kaufman KM et al (2006a) A nonsense mutation of PEPD in four Amish children with prolidase deficiency. Am J Med Genet A 140:580–585
Wang L, Feng Y, Zhang Y, Zhou H, Jiang S et al (2006b) Prolylcarboxypeptidase gene, chronic hypertension, and risk of preeclampsia. Am J Obstet Gynecol 195:162–171
Wang SH, Zhi QW, Sun MJ (2006c) Dual activities of human prolidase. Toxicol In Vitro 20:71–77

Wang WX, Rajeev BW, Stromberg AJ, Ren N, Tang G et al (2008) The expression of microRNA miR-107 decreases early in Alzheimer's disease and may accelerate disease progression through regulation of beta-site amyloid precursor protein-cleaving enzyme 1. J Neurosci 28:1213–1223

Wanic K, Placha G, Dunn J, Smiles A, Warram JH, Krolewski AS (2008) Exclusion of polymorphisms in carnosinase genes (CNDP1 and CNDP2) as a cause of diabetic nephropathy in type 1 diabetes: results of large case-control and follow-up studies. Diabetes 57:2547–2551

Wardlaw SL (2011) Hypothalamic proopiomelanocortin processing and the regulation of energy balance. Eur J Pharmacol 660:213–219

Watson B Jr, Nowak NJ, Myracle AD, Shows TB, Warnock DG (1997) The human angiotensinase C gene (HUMPCP) maps to 11q14 within 700 kb of D11S901: a candidate gene for essential hypertension. Genomics 44:365–367

Weidemann A, Eggert S, Reinhard FB, Vogel M, Paliga K et al (2002) A novel ε-cleavage within the transmembrane domain of the Alzheimer amyloid precursor protein demonstrates homology with Notch processing. Biochemistry 41:2825–2835

Wesley UV, Tiwari S, Houghton AN (2004) Role for dipeptidyl peptidase IV in tumor suppression of human non small cell lung carcinoma cells. Int J Cancer 109:855–866

Westmeyer GG, Willem M, Lichtenthaler SF, Lurman G, Multhaup G et al (2004) Dimerization of beta-site beta-amyloid precursor protein-cleaving enzyme. J Biol Chem 279:53205–53212

Willem M, Garratt AN, Novak B, Citron M, Kaufmann S et al (2006) Control of peripheral nerve myelination by the beta-secretase BACE1. Science 314:664–666

Winkler E, Hobson S, Fukumori A, Dumpelfeld B, Luebbers T et al (2009) Purification, pharmacological modulation, and biochemical characterization of interactors of endogenous human gamma-secretase. Biochemistry 48:1183–1197

Wolfe MS, Xia W, Ostaszewski BL, Diehl TS, Kimberly WT et al (1999) Two transmembrane aspartates in presenilin-1 required for presenilin endoproteolysis and γ-secretase activity. Nature 398:513–517

Wolfrum S, Richardt G, Dominiak P, Katus HA, Dendorfer A (2001) Apstatin, a selective inhibitor of aminopeptidase P, reduces myocardial infarct size by a kinin-dependent pathway. Br J Pharmacol 134:370–374

Wong HK, Sakurai T, Oyama F, Kaneko K, Wada K et al (2005) beta Subunits of voltage-gated sodium channels are novel substrates of beta-site amyloid precursor protein-cleaving enzyme (BACE1) and gamma-secretase. J Biol Chem 280:23009–23017

Wu JJ, Tang HK, Yeh TK, Chen CM, Shy HS et al (2009) Biochemistry, pharmacokinetics, and toxicology of a potent and selective DPP8/9 inhibitor. Biochem Pharmacol 78:203–210

Yamanaka C, Lebrethon MC, Vandersmissen E, Gerard A, Purnelle G et al (1999) Early prepubertal ontogeny of pulsatile gonadotropin-releasing hormone (GnRH) secretion: I. Inhibitory autofeedback control through prolyl endopeptidase degradation of GnRH. Endocrinology 140:4609–4615

Yan R, Bienkowski MJ, Shuck ME, Miao H, Tory MC et al (1999) Membrane-anchored aspartyl protease with Alzheimer's disease beta- secretase activity. Nature 402:533–537

Yan P, Song H, Yin K, Bateman RJ, Cirrito JR et al (2006) Matrix metalloproteinase-9 degrades amyloid-beta fibrils in vitro and compact plaques in situ. J Biol Chem 281:24566–24574

Yang HY, Erdos EG, Chiang TS (1968) New enzymatic route for the inactivation of angiotensin. Nature 218:1224–1226

Yaron A, Naider F (1993) Proline-dependent structural and biological properties of peptides and proteins. Crit Rev Biochem Mol Biol 28:31–81

Yu G, Chen F, Levesque G, Nishimura M, Zhang DM et al (1998) The presenilin 1 protein is a component of a high molecular weight intracellular complex that contains β-catenin. J Biol Chem 273:16470–16475

Yu G, Nishimura M, Arawaka S, Levitan D, Zhang L et al (2000) Nicastrin modulates presenilin-mediated notch/glp-1 signal transduction and βAPP processing. Nature 407:48–54

Yu C, Kim SH, Ikeuchi T, Xu H, Gasparini L et al (2001) Characterization of a presenilin-mediated APP carboxyl terminal fragment CTFγ: evidence for distinct mechanisms involved in γ-secretase processing of the APP and Notch1 transmembrane domains. J Biol Chem 276:43756–43760

Yu DM, Wang XM, McCaughan GW, Gorrell MD (2006) Extraenzymatic functions of the dipeptidyl peptidase IV-related proteins DP8 and DP9 in cell adhesion, migration and apoptosis. FEBS J 273:2447–2460

Yu LR, Zhu Z, Chan KC, Issaq HJ, Dimitrov DS et al (2007) Improved titanium dioxide enrichment of phosphopeptides from HeLa cells and high confident phosphopeptide identification by cross-validation of MS/MS and MS/MS/MS spectra. J Proteome Res 6:4150–4162

Yu DM, Ajami K, Gall MG, Park J, Lee CS et al (2009) The in vivo expression of dipeptidyl peptidases 8 and 9. J Histochem Cytochem 57:1025–1040

Yu DM, Yao TW, Chowdhury S, Nadvi NA, Osborne B et al (2010) The dipeptidyl peptidase IV family in cancer and cell biology. FEBS J 277:1126–1144

Zagha E, Ozaita A, Chang SY, Nadal MS, Lin U et al (2005) DPP10 modulates Kv4-mediated A-type potassium channels. J Biol Chem 280:18853–18861

Zhang J, Ye W, Wang R, Wolfe MS, Greenberg BD et al (2002) Proteolysis of chimeric beta-amyloid precursor proteins containing the Notch transmembrane domain yields amyloid beta-like peptides. J Biol Chem 277:15069–15075

Zhao J, Li L, Leissring MA (2009) Insulin-degrading enzyme is exported via an unconventional protein secretion pathway. Mol Neurodegener 4:4

Zhao G, Liu Z, Ilagan MX, Kopan R (2010) Gamma-secretase composed of PS1/Pen2/Aph1a can cleave notch and amyloid precursor protein in the absence of nicastrin. J Neurosci 30:1648–1656

Zhou C, Garcia-Calvo M, Pinto S, Lombardo M, Feng Z et al (2010) Design and synthesis of prolylcarboxypeptidase (PrCP) inhibitors to validate PrCP as a potential target for obesity. J Med Chem 53:7251–7263

Zhou L, Chavez-Gutierrez L, Bockstael K, Sannerud R, Annaert W et al (2011) Inhibition of beta-secretase in vivo via antibody binding to unique loops (D and F) of BACE1. J Biol Chem 286:8677–8687

Zhu L, Carretero OA, Liao TD, Harding P, Li H, Sumners C et al (2010) Role of prolylcarboxypeptidase in angiotensin II type 2 receptor-mediated bradykinin release in mouse coronary artery endothelial cells. Hypertension 56:384–390

Zlokovic BV (2008) New therapeutic targets in the neurovascular pathway in Alzheimer's disease. Neurotherapeutics 5:409–414

Zlokovic BV (2010) Neurodegeneration and the neurovascular unit. Nat Med 16:1370–1371

Zlokovic BV, Segal MB, Davson H, Lipovac MN, Hyman S et al (1990) Circulating neuroactive peptides and the blood-brain and blood-cerebrospinal fluid barriers. Endocrinol Exp 24:9–17

Zou K, Yamaguchi H, Akatsu H, Sakamoto T, Ko M et al (2007) Angiotensin-converting enzyme converts amyloid beta-protein 1-42 (Abeta(1-42)) to Abeta(1-40), and its inhibition enhances brain Abeta deposition. J Neurosci 27:8628–8635

Chapter 11
Proteases in the Mammalian Digestive System

S. Gaylen Bradley, Toni M. Antalis, and Judith S. Bond

11.1 Introduction

The digestive system works in concert with the other organ systems of the body, and with the external environment, in particular ingested material and microorganisms. The pathway from the point of ingestion to the point of excretion is long and tortuous, and is made up of dozens of niches with specialized physiologic roles and constellations of proteases. In each of these niches, there is a characteristic array of molecules from the lumen to the surface of the cells lining the alimentary tract (Hopfer 2011). The cellular level is a heterogeneous mosaic of resident and transient cells with special functions and arrays of proteases. This chapter will focus on the mammalian proteases found in the lumen and on the cell surface of the alimentary tract, both membrane-bound and secreted proteases. The proteases of the intestinal bacteria, mainly in the lower intestine, are dealt with in separate chapters, but structures and products of bacteria will be addressed as substrates and as factors that affect protease expression and activity. Similarly, the proteases of immune cells, which are transient members of the digestive system, are dealt with separately in other chapters, but their roles in the alimentary tract are inextricably enmeshed in normal physiologic processes of digestion and pathologic responses to injury. Accordingly, this review of proteases in the digestive system will focus on the human system beginning anatomically at the oral cavity, and proceed to the esophagus, stomach, and the niches of the intestinal tract, ending with the colon.

S.G. Bradley
Penn State University College of Medicine, Hershey, PA, USA

T.M. Antalis
University of Maryland, Baltimore, MD, USA

J.S. Bond (✉)
Department of Biochemistry and Molecular Biology, Penn State University College of Medicine, Hershey, PA 17033, USA
e-mail: jbond@psu.edu

K. Brix and W. Stöcker (eds.), *Proteases: Structure and Function*,
DOI 10.1007/978-3-7091-0885-7_11, © Springer-Verlag Wien 2013

Representative protease functions in intestinal pathobiology are discussed in the anatomical sections and in separate sections at the end of the chapter. There is detailed information on the many proteases mentioned in this chapter in the Handbook of Proteolytic Enzymes, second Edition (Barrett et al. 2004) and the third Edition (Rawlings and Salvesen 2013), and in the MEROPs database (http://www.merops.sanger.ac.uk).

11.2 Distribution of Proteases in Anatomical Sections of the Digestive Tract

11.2.1 Salivary Proteases

Over the years there has been difficulty in determining whether proteolytic activities in saliva is of bacterial or mammalian tissue origin, and whether of glandular origin or from immunocytes (Kennedy et al. 1998; Sun et al. 2009). Evidence, based upon protease inhibitors, has been found for a variety of proteases, in particular serine proteases, in secretions from submandibular glands. The concentration of trypsin in saliva of submandibular and sublingual glands is higher than that from parotid glands, which has a higher alpha-amylolytic activity (Korot'ko et al. 2002). The constellation of proteases is age dependent; for example, activity of collagenase in the saliva of juvenile periodontitis patients was low compared with that of adult periodontitis patients (Ingman et al. 1993). These investigators concluded that the salivary collagenase of localized juvenile periodontitis patients was matrix metalloproteinase-1 from fibroblasts rather than matrix metalloproteinase-8 from polymorphonuclear leukocytes. The protease activity in whole saliva appears to reflect the status of periodontal health (Nieminen et al. 1993). Gingipains, cysteine proteases secreted by bacteria, are associated with the progression of periodontal disease (Imamura 2003).

Meprin metalloproteases have been observed on the luminal surface of intercalated and striated ducts of submandibular and parotid glands and on interlobular ducts of the latter (Craig et al. 1991). The intensity of immunochemical staining was greater in males compared with females. The level of meprin protein was markedly greater in mice with high kidney meprin A activity than in low kidney meprin A mice. Meprins are expressed at high levels in kidney and intestinal tissues, and at lower levels in salivary glands, skin, immune system cells, brain and several other tissues (Sterchi et al. 2008). Meprin subunit expression is regulated at the transcriptional, post-transcriptional and post-translational level (Roff et al. 2013). The physiologic role of the meprins at a given site is dependent upon age, hormone levels, tissue distribution, and concurrent pathology (Sterchi et al. 2008).

Bacteria adhere to epithelial cells by an interaction between bacterial surface appendages and the host cell surface. The mammalian host cells utilize a number of mechanisms to prevent bacterial colonization, including secretory antibodies and glycoproteins such as fibronectin. Fibronectin, which is abundant on oropharyngeal epithelial cells, masks or competes for gram negative bacterial binding sites on the epithelial cell surface. Bacterial adherence to normal buccal epithelial cells is increased by exposure to saliva and trypsin (Hasty and Simpson 1987; Woods et al. 1981a). In clinical studies of patients with acute respiratory failure, adherence of gram-negative bacteria to buccal cells was correlated with increased protease activity in secretions, decreased cell-surface fibronectin and a poor outcome (Woods et al. 1981b). Failure to bind bacteria by the epithelial cells allows the potential pathogen to gain entry into the circulation, possibly resulting in host death. Patients with cystic fibrosis had fivefold less salivary trypsin activity than control samples (Rao and Nadler 1972), consistent with the suggestion that cystic fibrosis patients have increased adherence of gram negative bacteria by their epithelial cells, contributing to the pathophysiology of this disease.

Biofilm formation and colonization involves adhesion of microorganisms to host receptor molecules such as fibronectin and statherin. Statherin protein is a secreted phosphorylated protein consisting of 43 amino acid residues. Statherin is encoded by the *STATH* gene and is located near the histatin genes (*HIS1* and *HIS2*) with which it shows high homology. Statherin, secreted by the major salivary glands, has been implicated in the maintenance of tooth enamel homeostasis. Incubation of statherin with whole saliva generates multiple statherin-derived peptides (Helmerhorst et al. 2008). Statherin, for example, mediates adhesion of *Actinomyces* spp. *Porphyromonas gingivalis* and *Fusobacterium nucleatum*. The multiple peptides resulting from proteolysis of statherin have the potential to select which species will colonize the tooth enamel (Niemi and Johansson 2004).

Not only does saliva contain a number of proteases, it contains protease inhibitors. One example is histatin 5, a 24-residue histidine-rich salivary antimicrobial protein, which inhibits the activity of matrix metalloproteinases MMP-2 and MMP-9. These host-derived MMPs, which are elevated in the saliva of patients with periodontal disease, are considered key initiators of extracellular matrix degradation associated with periodontal diseases (Gusman et al. 2001). MMPs are also involved in the normal turnover of the extracellular matrix, which is an integral part of development, morphogenesis, and tissue remodeling. Histatin 5 is secreted by parotid, submandibular, and sublingual glands.

11.2.2 Proteases of the Esophagus

The esophagus delivers the saliva-mixed food from the mouth to the stomach and serves as an air lock between the respiratory tract and the digestive tract. The importance of the esophagus' ability to separate the mouth and stomach can be seen in the condition known as GERD (gastroesophageal reflux disease), in which the esophageal barrier is not effective, so the contents of the stomach can escape into the esophagus. The strongly acidic gastric reflux contains bile salts and pancreatic enzymes such as trypsin. Protease-activated receptor-2 (PAR-2), which is located in the lumen of the lower segment of the esophagus, can be cleaved and activated by trypsin in duodenogastric esophageal reflux. Inci et al. (2009) have suggested that the trypsin-PAR-2 pathway may be involved in the pathogenesis of gastroesphageal reflux disease (GERD).

A small percentage of people with GERD will develop Barrett's esophagus, a condition in which the cells of the lower esophagus become damaged from repeated exposure to gastric fluid. In Barrett's esophagus the color and composition of the esophagus cells change. Proenzymes of pepsinogen A (pepsinogen I) and pepsinogen C (pepsinogen II) are found consistently in biopsy specimens from Barrett's epithelium (Westerveld et al. 1987).

MMP-2 and MMP-9 were found in tumors and paired normal tissues of patients with esophageal squamous cell carcinoma (ESCC) but the levels tended to be higher in tumor tissue. The activated forms of MMP-3 and MMP-10 were strongly expressed in tumors with little or no expression in the paired normal tissue. Increased levels of MMPs occur in ESCC suggesting their up-regulation is important in esophageal tumorigenesis (Mukherjee et al. 2010).

Seprase (a serine protease), dipeptidyl peptidase IV and urokinase-type plasminogen activator (uPA) play crucial roles in the degradation of the extracellular matrix and in the progression of esophageal carcinoma. Seprase and uPA immunoreactivity have been found in dysplastic and cancer cells as well as in adjacent stromal cells, but not in normal esophageal epithelium (Goscinski et al. 2008a, b).

The esophagus is primarily a connecting valve between two important digestive chambers. Little nutritional digestion occurs within the esophagus, in part because the transient time is short. The esophagus is vulnerable to injury from gastric reflux, leading to brief episodes of discomfort and ultimately a range of lesions in which proteases play critical roles.

11.2.3 Stomach Proteases

11.2.3.1 Pepsin

Pepsin, an aspartic protease, is one of the principal protein-degrading enzymes in the digestive system. Aspartic peptidases contain aspartic acid residues at the catalytic site that activate water molecules, which act as the scissile element in cleaving substrates. Pepsin (pepsin A in *Homo sapiens*) is produced in the chief cells of the stomach as an inactive zymogen, pepsinogen. The pro-form of pepsin is activated by hydrochloric acid that is released from parietal cells of the stomach. The hormone gastrin and the vagus nerve trigger the release of both pepsinogen and HCl when food is ingested. Hydrochloric acid causes pepsinogen to unfold and cleave itself, thereby autocatalytically removing 44 amino acids from pepsinogen to generate active pepsin. Pepsin has a broad specificity with a preference for cleaving peptide bonds between hydrophobic and aromatic amino acids such as phenylalanine, tryptophan, and tyrosine (Moriyama et al. 1983).

11.2.3.2 Chymosin

Chymosin, an aspartic protease also known as rennin, is synthesized by chief cells in the stomach of ruminants. It is also expressed in the stomachs of other mammals in infancy. Its role in digestion is to curdle or coagulate milk in the stomach, an important process for young animals (Andrén 1992). If milk were not coagulated, it would rapidly flow through the stomach undigested. Chymosin secretion is maximal during the first few days after birth, and declines thereafter, replaced by secretion of pepsin as the major gastric protease. Chymosin is secreted as an inactive prochymosin that is activated by stomach acid (Richter et al. 1998).

The major milk protein is casein of which there are four types: alpha-s1, alpha-s2, beta and kappa. The alpha and beta caseins are hydrophobic proteins that are readily precipitated by calcium. Kappa casein is not calcium-precipitable. As milk is secreted, the four casein species self-associate into aggregates called micelles in which the alpha and beta caseins are kept in 'solution' by interacting with kappa casein. Chymosin proteolytically cleaves kappa casein, converting it into para-kappa-casein, a smaller protein. Para-kappa-casein does not have the ability to stabilize the micellar structure and the calcium-insoluble caseins precipitate, forming a curd.

Chymosin is an important enzyme commercially in cheesemaking (Vallejo et al. 2012). Chymosin is expensive because it is prepared from the stomachs of young cows, so it has been replaced in industrial processes by bacterial proteases that are able to coagulate milk by converting casein to paracasein.

11.2.3.3 Gastricsin

Gastricsin (Pepsinogen-C, Parapepsin-II, Pepsin-II) is an aspartic proteinase secreted as a zymogen by the gastric mucosa. Inactive gastricsin is stored in secretory granules, and released on demand to the stomach, where it acts as a relatively non-specific endopeptidase (Foster et al. 2004). Gastricsin shares 50 % identity with pepsin-A at the amino acid level, and has similar broad-ranging substrate preferences (although with a slightly more basic pH optimum than pepsin). Gastricsin is a member of the clade AA Aspartic peptidases, with the two catalytic aspartic acids at residues 91 and 276 in the peptide motif of Asp-Thr-Gly-Ser/Thr, respectively. The large A1 family of this clade also includes Pepsin-A, Chymosin, Cathepsin-D, Cathepsin-E, Napsin, Renin and Memapsins 1 and 2. Gastricsin contains a 16-residue signal sequence followed by a 43-residue propeptide domain.

The mature gastricsin molecule has a two lobed domain structure, with one catalytic residue on each lobe, the result of gene duplication from a historical precursor enzyme. The low pI of gastricsin (predicted to be 3.37 for the active enzyme) is designed to work in the acidic stomach environment, and the enzyme is thought to be irreversibly denatured at the more neutral pH outside of the stomach.

The ratio of pepsinogen-A to gastricsin has also been studied in gastric cancer patients, and there may be a correlation with the different polymorphisms of gastricsin and cancer (Hassan et al. 2010). Gastricsin has also been found in some breast cancer tissues, and in alveolar type-2 cells, where it cleaves surfactant protein-B. Elevated gastricsin levels in breast cancer seem to indicate more favorable outcomes. Gastricsin is also found in seminal fluid, where it is thought to help liquefy seminal proteins as well as decrease the immune response that would decrease fertility (Reid et al. 1984). The acidic environment in the vagina is thought to allow gastricsin to function in that environment, but it is unclear what role the enzyme has in less acidic environments.

11.2.4 Duodenum/Pancreas Proteases

The duodenum is the first and shortest section of the small intestine in mammals, about 10–12 in. long in humans. The duodenum is a secretory chamber connecting the stomach to the jejunum, and is the site of much of the enzymatic digestion of food and of iron absorption. The duodenum and pancreas together function to aid digestion. The pancreas is located behind the stomach and connects to the duodenum via two pancreatic ducts.

The pancreas is both an endocrine gland that produces hormones released into the circulatory system (such as insulin, and glucagon), and an exocrine gland, that produces digestive enzymes which are secreted via the pancreatic duct into duodenum. The ductal cells of the pancreas produce bicarbonate, which is needed to neutralize the acidity of the stomach chyme when the pylorus opens. Bicarbonate production by pancreatic ductal cells is stimulated by the hormone secretin, and secretin production by S cells in the duodenum is stimulated by low pH. Secretin, which has been released into the bloodstream, also inhibits production of gastrin by G cells of the stomach, decreases gastric emptying, and stimulates acinar cells of the pancreas to produce and release digestive enzymes. In addition, cholecystokinin (CCK), produced by the duodenal "I cells" stimulates a neuronal circuit, which in turn stimulates acinar cells to synthesize and release pancreatic digestive enzymes (Li and Owyang 1994).

The digestive enzymes are synthesized in the pancreas as inactive pro-enzymes or zymogens, and once they are released into the duodenum become activated by enteropeptidase (*PRSS7*). This external activation, which occurs outside of the pancreas, is important to prevent autoactivation of pancreatic proteases, thereby avoiding harmful tissue damage. The role of enteropeptidase in initiating the activation of the digestive enzyme cascade was first discovered more than a century ago by Ivan Pavlov (Zheng et al. 2009). Enteropeptidase is expressed mostly in the duodenum and, at lower levels, in the proximal segment of jejunum.

Enteropeptidase is a type II transmembrane serine protease, and as such, it is expected to remain at the site of expression, consistent with the association of its activity with the brush border of enterocytes (Hermon-Taylor et al. 1977; Lojda and Gossrau 1983). Enteropeptidase may also be shed from the enterocyte surface and released into the small intestine lumen upon exposure to bile-acids or cholescystokinin-pancreozymin (Hadorn et al. 1971). One of the remarkable features of enteropeptidase is its unique substrate specificity, which recognizes Lys at P_1 and a cluster of four Asp residues at P_2–P_5. Patients with enteropeptidase deficiency suffer from diarrhea, vomiting, edema, anemia, and hypoproteinemia, and fail to gain weight in early infancy (Hadorn et al. 1969). Treatment with pancreatic extracts helps to initiate proteolytic reactions and is usually effective for nutrient absorption and weight gain.

Enteropeptidase in the enterocyte brush border activates pancreatic trypsinogen to the active enzyme, trypsin, which catalyzes the break down of proteins by cleavage at basic amino acids. Trypsin can then function to activate additional trypsinogen and also activate other zymogens released from the pancreas, including chymotrypsinogen, pro-carboxypeptidases, several pro-elastases, and pro-lipases in a classic proteolytic cleavage cascade (Eggermont et al. 1971a, b; Yuan et al. 1998). Chymotrypsinogen breaks down proteins at their aromatic amino acids and carboxypeptidase removes the terminal amino acid group from proteins. Pancreatic elastase digests and degrades various kinds of proteins. During an inflammation of the pancreas, human pancreatic elastase is released into the bloodstream, where its quantification in serum is used for diagnosis of acute pancreatitis.

11.2.5 Jejunum Proteases

The jejunum, which is about 8 feet long in adult humans, lies between the duodenum and the ileum (Silk et al. 1974). The pH in the jejunum is usually between 7 and 9. The jejunum and the ileum are suspended by mesentery thereby allowing the small intestine to move about within the abdomen. The jejunum wall contains smooth muscle which helps to move food along by a process known as peristalsis. The luminal surface of the jejunum is covered with villi, which increase the surface area of the epithelial lining available to absorb nutrients. The epithelial cells that line these villi contain even larger numbers of microvilli. The villi in the jejunum are much longer than in the duodenum or ileum. The jejunum contains very few Brunner's glands (found in the duodenum) or Peyer's patches (found in the ileum). However, there are a few jejunal lymph nodes suspended in its mesentery.

The jejunum is the place where most nutrients are actively absorbed, including amino acids, many vitamins and minerals. Absorption in the jejunum involves active transport in which energy is required. Protein carriers or channels interact with the amino acids and small peptides and transport them through the epithelial barrier of the jejunum into the portal vein, which carries them to the liver.

11.2.6 Ileum Proteases

The ileum is the terminal portion of the small intestine and as such has a limited but vital role in absorption (Silk et al. 1974). The ileum is distal to the jejunum and is separated from the cecum by the ileocecal valve (ICV). In humans, the ileum is about 9 feet long, and the pH is usually between 7 and 8. The ileum has abundant Peyer's patches, which are unencapsulated lymphoid nodules. Vitamin B_{12} is selectively absorbed in the ileum, and bile salts are reabsorbed along with the remaining nutrient products of digestion that were not absorbed by the jejunum. The ileum lining has many tiny villi on its surface. In turn, the epithelial cells that line these villi possess even larger numbers of microvilli. The diffuse neuroendocrine system (DNES) cells of the ileum secrete various hormones (gastrin, secretin, cholecystokinin) into the blood. Cells in the lining of the ileum secrete proteases responsible for the final stages of protein digestion in the lumen of the intestine. The villi contain large numbers of capillaries that take the amino acids produced by digestion to the hepatic portal vein and the liver. Layers of circular and longitudinal smooth muscle enable the digested food to be pushed along the ileum by waves of muscle contractions called peristalsis. The undigested food (waste and water) is sent to the colon along with large numbers of lymphocytes and other cells of the immune system.

Meprin metalloproteases are abundant in the ileum, with both meprin alpha and beta isoforms expressed in humans (Sterchi et al. 1988) and rats (Johnson and Hersh 1992). In mice, meprin beta is the predominant subunit expressed in intestine, however expression of meprin alpha is strain-specific; mice with high kidney meprin alpha have higher levels of mRNA expression in the ileum and colon than those with low kidney meprin alpha (Bankus and Bond 1996). For both humans and mice there is a gradient increase in meprin activity from the proximal duodenum to the distal ileum. In addition, meprin alpha is the only meprin subumit expressed in the colon. Notably, the MEP1A allele for meprin A has been implicated as a susceptibility gene for inflammatory bowel disease (IBD) in both humans and mice (Banerjee et al. 2009). Mice deficient in meprin alpha (meprin alpha knockout mice) develop a more severe form of experimental inflammatory disease than wild-type mice; humans with IBD have lower levels of meprin alpha mRNA than healthy humans. Interestingly, mice lacking both meprin alpha and beta develop less severe inflammation than mice lacking only meprin alpha, albeit more severe than wild-type mice (Banerjee et al. 2011). Thus, the balance of meprins A and B affects the progression of IBD. Expression of the meprin alpha and beta subunits in the ileum of wild-type mice increases markedly after birth, however, expression of meprin alpha plummets at the time of weanlng. Although casein is a substrate of meprin, meprin alpha does not seem to be a critical digestive caseinase because meprin alpha-deficient mice thrive during the suckling period (Kumar and Bond 2001). Gastrin and cholecystokinin are good substrates for meprin beta, perhaps serving to regulate concentrations of bioactive peptides in the ileal lumen (Sterchi et al. 2008).

11.2.7 Large Intestine (Colon) Proteases

A critical role of the large intestine is to conserve sodium and water that was not absorbed in the small intestine (Billich and Levitan 1969). About 1 L of fluid per day enters the large intestine, which is about 5 feet long, including its final segments, the colon and the rectum. The foodstuff remaining at this point is primarily fiber. The contents of the large intestine are held for 6–72 h before elimination; in contrast to the other segments of the alimentary tract. On average, food travels through the stomach in 1/2 to 2 h, continues through the small intestine over the next 2–6 h. The large intestine contains an ecosystem of bacteria that can ferment much of the residual fiber, producing a series of short-chain fatty acids, including proprionate, acetate, and butyrate. The myriads of aerobic and anaerobic species of microorganisms in the colon are important for human health and disease.

Some fiber is not fermented, but is important because it provides bulk for stool excretion, which can also bind and remove toxins and waste products. The colon is the main reservoir for feces (mainly in rectum) before its defecation. It is also where the liquid stool becomes solid, by losing its water, and electrolytes. The Epithelial Na^+Channel (ENaC) is critical for sodium aborption, and its function is regulated by proteases including furin, prostasin, CAP2, matriptase, kallikrein, elastase and the beta subunit of meprins, suggesting these proteases may function in salt homeostasis (Kleyman et al. 2009; Garcia-Caballero et al. 2011). The colon also actively secretes bicarbonate and potassium, which is the reason that severe diarrhea can cause metabolic acidosis as well as hypokalemia.

The colon is a repository of a number of microbial cells that exceeds the number of host cells of humans and mice (Tlaskalová-Hogenová et al. 2004). The intestinal microflora contribute by ill-defined processes in the development and progression of intestinal inflammation. Because meprin A has been implicated as a susceptibility gene for IBD, expression of meprin was assessed in samples from germfree mice, re-colonized mice and conventional mice. No differences in expression of meprin alpha or beta were detected among these samples (Bond et al. 2005). Differences between the flora of aerobic and microaerophilic bacteria of wild-type and meprin beta-deficient mice were observed, but these could not be consistently correlated with a microbial species or genus. More bacteria are found in the ileum of mice administered sodium dodecyl sulfate (DSS) in the drinking water than in the ileum of control mice. Soluble meprin A, which is produced normally in meprin beta deficient mice, is the predominate meprin in the colon. Paradoxically, antibiotics such as neomycin or minocyclin increase the severity of IBD when administered prior to DSS but attenuate severity when administered concurrently (Bond et al. 2006). The adverse effect of antibiotic pre-treatment is greater in the meprin beta deficient mouse than the wild type. Substance P, which regulates gastrointestional motility (El-Salhy and Spangeus 1998), is a good substrate of meprin alpha, which is the primary meprin in the colon.

Antimicrobial peptides of the defensin family are part of the innate immune system, and play an important role in the intestine to protect against microbial invasion at mucosal epithelia surfaces (Cunliffe 2003). Defensins are small cationic antimicrobial peptides capable of killing bacteria. In the mouse small intestine about 20 mRNAs have been found for alpha-defensins, also called cryptdins. Beta-defensins are also found in the colon. The activity of defensins is regulated by proteolysis, and one of the best-studied examples is the activation of mouse alpha-defensin by matrilysin (MMP-7) (Wilson et al. 2009). Matrilysin removes alpha-defensin pro-domains resulting in antimicrobial activity. The active defensins are resistant to further proteolysis and can function in the harsh environment of the gastrointestinal tract at infection and inflammation sites, and in diseases such as Crohn's disease and ulcerative colitis.

11.3 Protease Functions in Intestinal Pathobiology

Conditions of the luminal gastrointestinal tract that arise from genetic, environmental and/or lifestyle factors range from malignancies (squamous cell esophageal cancer, Barrett's esophagus and associated esophageal adenocarcinoma, gastric and colorectal cancers), idiopathic inflammatory disorders such as inflammatory bowel diseases, and post-infectious syndromes including post-infectious irritable bowel syndrome, post-infectious dyspepsia and other functional GI disorders (Hall and Crowe 2011).

11.3.1 Regulation of Proteases Under Physiological Condition

Due to the enormity of proteolytic potential, the activities of proteases must be tightly regulated to prevent inappropriate and potentially destructive proteolysis (reviewed in Antalis et al. 2007). Several mechanisms exist in the gastrointestinal tract to regulate proteases under physiological conditions. First, many enterocyte proteases are synthesized as inactive, latent precursors (zymogens) that must undergo proteolytic cleavage and/or other modifications to become active enzymes. In this way, proteolytic potential can be specifically and irreversibly amplified each time a downstream zymogen is activated, unleashing a burst of proteolytic activity. Second, access of enterocyte proteases to their substrates is regulated by spatial and temporal compartmentalization. Apical and basolateral surfaces of intestinal epithelial cells are exposed repeatedly to a variety of proteases that can, through one or more mechanisms, induce rapid changes in mucosal barrier functions. Pericellular proteases are critical for angiogenesis, inflammation, and cell migration in the GI tract. Membrane type 1 matrix metalloprotease (MT1-MMP), meprins, matriptase, and DPP 4, are tethered directly to plasma membranes through transmembrane-spanning domains, thus anchoring proteolysis to the cell surface. Other proteases, like the secreted plasminogen activators, are sequestered within the pericellular environment through binding to specific cell surface receptors, thus increasing the efficiency of plasminogen activation and subsequent plasmin-dependent proteolysis (Plow and Miles 1990). Finally, the termination of proteolytic activities is mediated by protease inhibitors, comprising Kunitz, Kazal, serpin, and TIMP inhibitor families (Medina and Radomski 2006; Roberts et al. 1995).

Some proteases can serve as signal molecules controlling cell functions through G-protein-coupled cell surface receptors, known as Protease Activated Receptors (PARs) (Kawabata 2003; MacNaughton 2005; Steinhoff et al. 2005; Vergnolle 2000). Proteolytic cleavage of PAR amino-terminal domains exposes a new amino-terminus that acts as a tethered ligand that binds to and activates the receptor. Serine proteases present in the intestinal lumen and elaborated by several cell types of the intestinal mucosa are capable of activating PARs (Adams et al. 2011). PAR_1, and PAR_4 are activated by thrombin and PAR_3 is a co-factor for the activation of PAR_4 by thrombin. PAR_2 is activated by several trypsin-like serine proteases including trypsin and mast cell tryptase. Several types of signaling responses may be induced depending on the specific PAR, the receptor location (apical *vs* basolateral side of the intestinal epithelium), the interactions with other surface receptors, the activating protease, the direction and duration of protease exposure, and the presence of regulatory and inhibitory factors. Activation of PAR_2 can evoke both pro- and antiinflammatory responses in the intestine, including vasodilation, smooth muscle relaxation, cytokine upregulation, Immune cell recruitment, and increased nociception and hyperalgesia (Hoogerwerf et al. 2001; Hyun et al. 2008; Roka et al. 2007). PAR1 activation can stimulate epithelial apoptosis, alter intestinal permeability and has recently been shown to promote microbiota-induced intestinal vascular remodeling (Chin et al. 2003; Reinhardt et al. 2012).

11.3.2 Dipeptidyl Peptidase IV (DDP4) and Diabetes, Celiac Disease

DPP4 (also known as CD26) is a key factor in the modulation of immune responses in the gut and other tissues. It is a multifunctional protein, acting as receptor, binding molecule and serine protease (Misumi and Ikehara 2004). It is identical to the T cell activation antigen CD26 and to the adenosine deaminase binding protein (Kameoka et al. 1993). It is widely distributed in the body and found as a glycoprotein on the membranes of most cell types. As a protease DPP-4 cleaves dipeptides from the N-terminus of a wide range of substrates, including growth factors, chemokines, neuropeptides and vasoactive peptides (Bušek et al. 2004). DPP4 preferentially cleaves Xaa–Pro dipeptides from oligopeptides with typical length of 30 amino acids, leading to biological activation or inactivation of regulatory molecules implicated in diabetes mellitus, obesity, tumor growth, and HIV infection (Engel et al. 2003). DPP4 degrades atrial natriuretic peptide but not brain natriuretic peptide (Brandt et al. 2006), oxyntomodulin and pituitary adenylate cyclase-activating polypeptide-(1–38) (Zhu et al. 2003), vasoactive intestinal peptide, gastrin-releasing peptide and neuropeptide Y (Lambeir et al. 2001) and proteins such as gelatin and collagen (Green et al. 2006). DPP-IV degrades glucose-dependent insulinotropic peptide (GIP) and glucagon-like peptide-1 (GLP-1), which is secreted by ileal L cells when nutrients are present in the lumen of the small intestine. GLP-1 has a half-life of less than 2 min as a result of degradation by DPP4. GLP-1 is a potent antihyperglycemic hormone, inducing glucose-dependent stimulation of insulin secretion while suppressing glucagon secretion. Preventing degradation of GLP-1 and GIP by DPP4 inhibition has become an attractive therapeutic strategy for development of a large number of synthetic DPP4 inhibitors (Green et al. 2006). In addition, DPP IV is considered critical for the elimination of polypeptides released by the digestion of ingested gluten in humans who have a genetic propensity to develop Celiac disease (Shan et al. 2002). Gluten is a mixture of glutamine- and proline-rich glutenin and prolamine molecules, which are cleaved by stomach pepsin and pancreatic trypsin. DPP4 and a number of bacterial, fungal and human recombinant "glutenases" have been used to degrade the immunogenic gliadin products of gluten digestion (Lindfors et al. 2012). These enzyme cocktails are used in several ways, including pretreatment of food, administration as an enzyme nutraceutic, or direct infusion into the intestinal tract.

11.3.3 Tissue Remodeling and Wound Repair

The mechanisms governing tissue remodeling, repair and regeneration of gastrointestinal tissue are complex and involve numerous proteases, growth factors and cytokines that stimulate the proliferation and migration of intestinal epithelial cells. Continual remodeling of multiple components of extracellular matrices including fibrin, collagen, gelatin, elastin, fibronectin, proteoglycans and vitronectin occurs during wound repair. Plasminogen activators (tPA and uPA) convert plasminogen to plasmin, a serine protease that degrades fibrin and multiple other ECM components. Plasmin may further activate MMPs to degrade collagen and other ECM proteins, thereby decreasing cell-substratum adhesion and promoting cell migration. Many proteases including collagenase-1 (MMP-1), stromelysin-2 (MMP-10), and MMP-7, are up-regulated by cytokines released during wound repair and contribute to intestinal re-epithelialization *in vivo* (Salmela et al. 2004).

11.3.4 *Inflammatory Intestinal Diseases (e.g., IBD) and Proteases*

Throughout the gastrointestinal tract, a lining of epithelial cells separates the inside of the body from the external environment. This epithelial lining plays a critical role in both absorptive and barrier functions of the intestine. At the junctions of the epithelial cells that form the epithelial lining are paracellular apical junctional complexes (AJC), which consist of tight junctions (TJ) and adherens junctions (AJ) as well as desmosomes. The AJCs are linked structurally and functionally to the intestinal epithelial cell actin cytoskeleton and regulate the passage of ions and macromolecules, while preventing translocation of potentially harmful bacteria, toxins and their products (Bruewer et al. 2003; Laukoetter et al. 2006). Mucosal permeability, in part, depends on the regulation of the dynamic and complex intercellular tight junction. Increased intestinal permeability and AJC dysfunction occur in a variety of clinical conditions, including food allergies, infections of the GI tract, autoimmune diseases, and in inflammatory bowel diseases (IBD), specifically Crohn's disease (CD) and ulcerative colitis (UC) (Marchiando et al. 2010). The increased permeability causes the transport of microbial and other antigenic material from the intestinal lumen into the submucosa, causing mucosal inflammation and inducing epithelial injury, thereby accelerating disease pathogenesis.

Matriptase (also known as MT-SP1, TADG-15, epithin, and SNC19 (List et al. 2006) plays a critical role in promoting formation of the intestinal epithelial barrier and regulating epithelial barrier function (List et al. 2009; Buzza et al. 2010). Matriptase is highly expressed all along the gastrointestinal epithelium. Increasing matriptase expression is seen along the crypt-villus axis with the highest expression occurring in the villus tips and the superficial epithelium of the colon (Oberst et al. 2003; Tsuzuki et al. 2005). Matriptase is a multidomain type II transmembrane serine protease (TTSP) (Hooper et al. 2001). It is synthesized with an amino-terminal cytoplasmic domain and a carboxy-terminal serine protease domain that is exposed to the extracellular environment. Matriptase is produced as an inactive zymogen but is able to autoactivate (Hahner et al. 2005; List et al. 2006), and therefore has been suggested to be an initiator of proteolytic cascades in intestinal epithelium. Matriptase is known to cleave protein substrates at basic amino acid residues, however it is highly selective and its targets are cell and context specific (Antalis et al. 2010). The matriptase protein targets in the gut are not defined. Mice with tissue-specific genetic deletion of matriptase from the murine gastrointestinal tract (Villin-Cre *p*romoter system), demonstrate loss of the intestinal epithelial barrier (List 2009). The phenotype is associated with persistent diarrhea, inflammation, edema and gross disruption of colonic architecture, that leads to rapid tissue degeneration and death within a few days, although at birth, the microarchitecture of the gastrointestinal tract is structurally and morphologically intact (List 2009). When matriptase is inducibly ablated from the GI tract of adult animals, mice develop dramatic IBD-like symptoms, including loss of weight, increased permeability in microbe-colonized colonic epithelium, rapid intestinal tissue degeneration and death (List et al. 2009). Matriptase hypomorphic mice, which express very low levels of matriptase, are able to survive (List et al. 2007) but are highly susceptible to experimental colitis and the development of chronic persistent disease (Kosa et al. 2012; Netzel-Arnett et al. 2012; Szabo et al. 2003).

The meprin alpha gene, on human chromosome 6p (mouse chromosome 17) has been identified as a susceptibility gene for IBD, particularly ulcerative colitis (Banerjee et al. 2009). Polymorphisms in the coding region and 3′-untranslated region of human MEP1A gene were found to be associated with colitis, and meprin alpha mRNA was decreased at inflammation sites of IBD patients. Mouse studies showed that meprin-alpha knockout mice exhibited a more severe form of inflammation in an experimental model of IBD compared to wild-type counterparts. These studies indicate that decreased expression of meprin alpha is associated with intestinal inflammation in mice and humans.

11.3.5 Colon Cancer and Proteases

Dysregulated expression of proteases is a hallmark of cancer. Colonic tumors are believed to develop through the acquisition of genetic alterations in critical genes, such as oncogenes and tumor suppressor genes that provide a prosurvival or growth advantage to mitotic cells. The genetically altered neoplastic cells frequently co-opt physiological host response mechanisms during tumor development that can favor and accelerate their growth. Proteases provide a significant advantage to developing colonic neoplasms, through their abilities to modulate extracellular matrix metabolism and their abilities to modulate the bioavailability of cytokines, chemokines, growth and angiogenic factors, and processing of cell-cell and cell matrix interaction adhesive molecules (Affara et al. 2009). Indeed the establishment and maintenance of intestinal epithelial cell polarity is crucial for intestinal epithelial cell physiology and intestinal homeostasis. Loss of cell polarity and tissue disorganization can promote excessive tumor growth. Colon carcinoma cells express meprin alpha which results in a secreted form of meprins (homomeric meprin A). In normal colon cells, meprin A is secreted from the apical membrane into the lumen of the colon. However, in carcinoma cells homomeric meprin A is secreted from both the apical and basolateral membranes, and this can result in degradation of extracellular matrix contributing to tumor progression, migration and metastasis (Lottaz et al. 1999).

Human colonic tumors display increased levels of many proteases, of which cathepsins, matrix metalloproteinases and plasminogen activators are particularly important for colon tumor invasion and metastases (Liotta et al. 1991). Plasmin generated by plasminogen activation can in turn proteolytically cleave and activate a wide range of proteins, including the inactive pro-forms of meprin A (Rosmann et al. 2002).

It is increasingly clear that proteases can play complex tissue-specific roles in tumor development and progression. Several different activities have been uncovered for the type II transmembrane serine proteases in colonic tumorigenesis. TMPRSS4 is upregulated in prostate, colon and gastric cancers, and was recently shown to promote tumor growth, invasion, metastasis and the epithelial-mesenchymal transition (EMT) (Choi et al. 2009). Deregulated matriptase proteolysis contributes to tumor formation and metastasis (Oberst et al. 2003; Vogel et al. 2006). Matriptase has also attracted significant attention as a potential oncogene, and as a potential diagnostic and prognostic biomarker in breast and prostate tumors (List 2009; Miller and List 2013). Matriptase possesses a strong oncogenic potential since even a modest overexpression in the skin of transgenic mice caused 100 % of the mice to develop tumors, most of which progressed into carcinomas (List et al. 2005). Increased expression of HAI-1 was found to completely negate the oncogenic effects of matriptase overexpression (List et al. 2005). These data indicate that a dysregulated ratio of matriptase to HAI-1 causes malignant transformation. Conversely, matriptase is down-regulated in tumors derived from intestinal epithelium (Zhang et al. 1998) and acts as a suppressor of tumorigenesis in the colonic epithelium. Ablation of matriptase gene from the intestinal epithelium in mice results in the formation of colonic adenocarcinomas with very early onset and high penetrance (Kosa et al. 2012). It has been hypothesized that a decrease in the intestinal paracellular permeability barrier caused by loss of matriptase promotes accelerated neoplastic progression. Thus certain proteases possess the dual ability to promote carcinogenesis in some contexts, while suppressing carcinogenesis in others.

11.4 Conclusions

The proteases of the mammalian digestive system consist of (1) food processing hydrolases, (2) enzymes that regulate the activities of digestive processes by activating proenzymes (zymogens) or hormones; (3) enzymes that remove debris and bacteria in or on the epithelial lining of the alimentary tract, and (4) proteases integrated into metabolic activities unrelated to digestion, such as immune responses, growth factors and signaling pathways. These proteases interact in a lattice of effector molecules rather than as members of linear metabolic pathways. These proteases often act in specific locales, but if spatially or temporally displaced, there are adverse consequences. This chapter described briefly some examples of these complex and intricate interactions.

References

Adams MN, Ramachandran R, Yau MK, Suen JY, Fairlie DP, Hollenberg MD, Hooper JD (2011) Structure, function and pathophysiology of protease activated receptors. Pharmacol Ther 130:248–282

Affara NI, Andreu P, Coussens LM (2009) Delineating protease functions during cancer development. Methods Mol Biol 539:1–32

Andrén A (1992) Production of prochymosin, pepsinogen and progastricsin, and their cellular and intracellular localization in bovine abomasal mucosa. Scand J Clin Lab Invest Suppl 210:59–64

Antalis TM, Shea-Donohue T, Vogel SN, Sears C, Fasano A (2007) Mechanisms of disease: protease functions in intestinal mucosal pathobiology. Nat Clin Pract Gastroenterol Hepatol 4:393–402

Antalis TM, Buzza MS, Hodge KM, Hooper JD, Netzel-Arnett S (2010) The cutting edge: membrane-anchored serine protease activities in the pericellular microenvironment. Biochem J 428:325–346

Banerjee S, Oneda B, Yap LM, Jewell DP, Matters GL, Fitzpatrick LR, Seibold F, Sterchi EE, Ahmad T, Lottaz D, Bond JS (2009) *MEP1A* for meprin A metalloprotease subunit is a susceptibility gene for inflammatory bowel disease. Mucosal Immunol 2:220–231

Banerjee S, Jin G, Bradley SG, Matters GL, Gailey RD, Crisman JM, Bond JS (2011) Balance of meprin A and B in mice affects the progression of experimental inflammatory bowel disease. Am J Physiol Gastrointest Liver Physiol 300:G273–G282

Bankus JM, Bond JS (1996) Expression and distribution of meprin protease subunits in mouse intestine. Arch Biochem Biophys 331:87–94

Barrett AJ, Woessner F, Rawlings N (eds) (2004) Handbook of proteolytic enzymes, 2nd edn. Academic Press, London

Billich CO, Levitan R (1969) Effects of sodium concentration and osmolality on water and electrolyte absorption from the intact human colon. J Clin Invest 48:1336–1347

Bond JS, Matters GL, Banerjee S, Dusheck RE (2005) Meprin metalloprotease expression and regulation in kidney, intestine, urinary tract infections and cancer. FEBS Lett 579:3317–3322

Bond JS, Gailey R, Bradley SG (2006) Meprin metalloproteases in intestinal disease. In: Naim HY, Zimmer K-P (eds) The brush border membrane – from molecular cell biology to clinical pathology. Verlagsgesellschaft mbH, Heilboronn, pp 122–135

Brandt I, Lambeir A-M, Ketelslegers J-M, Vanderheyden M, Scharpé S, De Meester I (2006) Dipeptidyl-peptidase IV converts intact B-Type natriuretic peptide into Its des-SerPro form. Clin Chem 52:82–87

Bruewer M, Luegering A, Kucharzik T, Parkos CA, Madara JL, Hopkins AM, Nusrat A (2003) Proinflammatory cytokines disrupt epithelial barrier function by apoptosis-independent mechanisms. J Immunol 171:6164–6172

Busek P, Malik R, Sedo A (2004) Dipeptidyl peptidase IV activity and/or structure homologues (DASH) and their substrates in cancer. Int J Biochem Cell Biol 36:408–421

Buzza MS, Netzel-Arnett S, Shea-Donohue T, Zhao A, Lin CY, List K, Szabo R, Fasano A, Bugge TH, Antalis TM (2010) Membrane-anchored serine protease matriptase regulates epithelial barrier formation and permeability in the intestine. Proc Natl Acad Sci U S A 107:4200–4205

Chin AC, Vergnolle N, MacNaughton WK, Wallace JL, Hollenberg MD, Buret AG (2003) Proteinase-activated receptor 1 activation induces epithelial apoptosis and increases intestinal permeability. Proc Natl Acad Sci U S A 100:11104–11109

Choi SY, Bertram S, Glowacka I, Park YW, Pohlmann S (2009) Type II transmembrane serine proteases in cancer and viral infections. Trends Mol Med 15:303–312

Craig SS, Mader C, Bond JS (1991) Immunohistochemical localization of the metalloproteinase meprin in salivary glands of male and female mice. J Histochem Cytochem 39:123–129

Cunliffe RN (2003) Alpha-defensins in the gastrointestinal tract. Mol Immunol 40:463–467

Eggermont E, Molla AM, Rutgeerts L, Tytgat G (1971a) The source of human enterokinase. Lancet 2:369
Eggermont E, Molla AM, Tytgat G, Rutgeerts L (1971b) Distribution of enterokinase activity in the human intestine. Acta Gastroenterol Belg 34:655–662
El-Salhy M, Spangeus A (1998) Substance P in the gastrointestinal tract of non-obese diabetic mice. Scand J Gastroenterol 33:394–400
Engel M, Hoffmann T, Wagner L, Wermann M, Heiser U, Kiefersauer R, Huber R, Bode W, Demuth H-U, Brandstetter H (2003) The crystal structure of dipeptidyl peptidase IV (CD26) reveals its functional regulation and enzymatic mechanism. Proc Natl Acad Sci USA 100:5063–5068
Foster C, Aktar A, Kopf D, Zhang P, Guttentag S (2004) Pepsinogen C: a type 2 cell-specific protease. Am J Physiol Lung Cell Mol Physiol 286:L382–L387
Garcia-Caballero A, Ishmael SS, Dang Y, Gillie D, Bond JS, Milgram SL, Stutts MJ (2011) Activation of the epithelial sodium channel by the metalloprotease meprin b subunit. Channels 5:14–22
Goscinski MA, Suo Z, Nesland JM, Chen W-T, Zakrzewska M, Wang J, Zhang S, Flørenes VA, Giercksky K-E (2008a) Seprase, dipeptidyl peptidase IV and urokinase-type plasminogen activator expression in dysplasia and invasive squamous cell carcinoma of the Esophagus A Study of 229 cases from Anyang Tumor Hospital, Henan Province, China. Oncology 75:49–59
Goscinski MA, Suo Z, Nesland JM, Flørenes VA, Giercksky K-E (2008b) Dipeptidyl peptidase IV expression in cancer and stromal cells of human esophageal squamous cell carcinomas, adenocarcinomas and squamous cell carcinoma cell fines. Acta Pathol Microbiol Immunol Scand 116:823–831
Green BD, Flatt PR, Bailey CJ (2006) Dipeptiudyl peptidase IV (DPP IV) inhibitors: a newly emerging class for the treatment of type 2 diabetes. Diab Vasc Dis Res 3:159–165
Gusman H, Travis J, Helmerhorst EJ, Potempa J, Troxler RF, Oppenheim FG (2001) Salivary histatin 5 is an inhibitor of both host and bacterial enzymes implicated in periodontal disease. Infect Immun 69:1402–1408
Hadorn B, Tarlow MJ, Lloyd JK, Wolff OH (1969) Intestinal enterokinase deficiency. Lancet 1:812–813
Hadorn B, Steiner N, Sumida C, Peters TJ (1971) Intestinal enterokinase. Mechanisms of tts "secretion" into the lumen of the small intestine. Lancet 1:165–166
Hahner S, Fassnacht M, Hammer F, Schammann M, Weismann D, Hansen IA, Allolio B (2005) Evidence against a role of human airway trypsin-like protease–the human analogue of the growth-promoting rat adrenal secretory protease–in adrenal tumourigenesis. Eur J Endocrinol 152:143–153
Hall EH, Crowe SE (2011) Environmental and lifestyle influences on disorders of the large and small intestine: implications for treatment. Dig Dis 29:249–254
Hassan MI, Toor A, Ahmad F (2010) Progastriscin: structure, function, and its role in tumor progression. J Mol Cell Biol 2:118–127
Hasty DL, Simpson WA (1987) Effects of fibronectin and other salivary macromolecules on the adherence of *Escherichia coli* to buccal epithelial cells. Infect Immun 55:2103–2109
Helmerhorst EJ, Sun X, Salih E, Oppenheim FG (2008) Identification of Lys-Pro-Gln as a novel cleavage site specificity of saliva-associated proteases. J Biol Chem 283:19957–19966
Hermon-Taylor J, Perrin J, Grant DA, Appleyard A, Bubel M, Magee AI (1977) Immunofluorescent localization of enterokinase in human small intestine. Gut 18:259–265
Hoogerwerf WA, Zou L, Shenoy M, Sun D, Micci MA, Lee-Hellmich H, Xiao SY, Winston JH, Pasricha PJ (2001) The proteinase-activated receptor 2 is involved in nociception. J Neurosci 21:9036–9042
Hooper JD, Clements JA, Quigley JP, Antalis TM (2001) Type II transmembrane serine proteases. Insights into an emerging class of cell surface proteolytic enzymes. J Biol Chem 276:857–860

Hopfer U (2011) Digestion and absorption of basic nutrition constituents. In: Devlin TM (ed) Textbook of biochemistry with clinical correlations, 7th edn. Wiley, Hoboken, NJ, pp 1029–1062
Hyun E, Andrade-Gordon P, Steinhoff M, Vergnolle N (2008) Protease-activated receptor-2-activation: a major actor in intestinal inflammation. Gut 57:1222–1229
Imamura T (2003) The role of gingipains in the pathogenesis of periodontal disease. Periodontol 74:111–118
Inci K, Edebo A, Olbe L, Casselbrant A (2009) Expression of protease-activated-receptor 2 (PAR-2) in human esophageal mucosa. Scand J Gastroenterol 44:664–671
Ingman T, Sorsa T, Konttinen YT, Liede K, Saari H, Lindy O, Suomalainen K (1993) Salivary collagenase, elastase- and trypsin-like proteases as biochemical markers of periodontal tissue destruction in adult and localized juvenile periodontitis. Oral Microbiol Immunol 8:298–305
Johnson GD, Hersh LB (1992) Cloning a rat meprin cDNA reveals the enzyme is a heterodimer. J Biol Chem 267:13505–13512
Kameoka J, Tanaka T, Nojima Y, Schlossman SF, Morimoto C (1993) Direct association of adenosine deaminase with a T cell activation antigen, CD26. Science 261:466–469
Kawabata A (2003) Gastrointestinal functions of proteinase-activated receptors. Life Sci 74:247–254
Kennedy S, Davis C, Abrams WR, Billings PC, Nagashunmugam T, Friedman H, Malamudl D (1998) Submandibular salivary proteases: lack of a role in anti-HIV activity. J Dent Res 77 (7):1515–1519
Kleyman TR, Carattino MD, Hughey RP (2009) ENaC at the cutting edge: regulation of epithelial sodium channels by proteases. J Biol Chem 284:20447–20451
Korot'ko GF, Gotovtseva LP, Bulgakova VA (2002) Postprandial transformations of enzymatic and hormonal properties of saliva and blood. Ross Fiziol Zh Im I M Sechenova 88:396–405 (Article in Russian)
Kosa P, Szabo R, Molinolo AA, Bugge TH (2012) Suppression of Tumorigenicity-14, encoding matriptase, is a critical suppressor of colitis and colitis-associated colon carcinogenesis. Oncogene 31:3679–3695
Kumar JM, Bond JS (2001) Developmental expression of meprin metalloprotease subunits in ICR and C3H/He mouse kidney and intestine in the embryo, postnatally and after weaning. Biochim Biophys Acta 1518:106–114
Lambeir AM, Durinx C, Proost P, Van Damme J, Scharpé S, De Meester I (2001) Kinetic study of the processing by dipeptidyl-peptidase IV/CD26 of neuropeptides involved in pancreatic insulin secretion. FEBS Lett 507:327–330
Laukoetter MG, Bruewer M, Nusrat A (2006) Regulation of the intestinal epithelial barrier by the apical junctional complex. Curr Opin Gastroenterol 22:85–89
Li Y, Owyang C (1994) Endogenous cholecystokinin stimulates pancreatic enzyme secretion via vagal afferent pathway in rats. Gastroenterology 107:525–531
Lindfors K, Rauhavirta T, Stenman S, Mäki M, Kaukinen K (2012) *In vitro* models for gluten toxicity: relevance for celiac disease pathogenesis and development of novel treatment options. Exp Biol Med 237:119–125
Liotta LA, Stetler-Stevenson WG, Steeg PS (1991) Cancer invasion and metastasis: positive and negative regulatory elements. Cancer Invest 9:543–551
List K (2009) Matriptase: a culprit in cancer? Future Oncol 5:97–104
List K, Szabo R, Molinolo A, Sriuranpong V, Redeye V, Murdock T, Burke B, Nielsen BS, Gutkind JS, Bugge TH (2005) Deregulated matriptase causes ras-independent multistage carcinogenesis and promotes ras-mediated malignant transformation. Genes Dev 19:1934–1950
List K, Bugge TH, Szabo R (2006) Matriptase: potent proteolysis on the cell surface. Mol Med 12:1–7

List K, Currie B, Scharschmidt TC, Szabo R, Shireman J, Molinolo A, Cravatt BF, Segre J, Bugge TH (2007) Autosomal ichthyosis with hypotrichosis syndrome displays low matriptase proteolytic activity and is phenocopied in ST14 hypomorphic mice. J Biol Chem 282:36714–36723
List K, Kosa P, Szabo R, Bey AL, Wang CB, Molinolo A, Bugge TH (2009) Epithelial integrity is maintained by a matriptase-dependent proteolytic pathway. Am J Pathol 175:1453–1463
Lojda Z, Gossrau R (1983) Histochemical demonstration of enteropeptidase activity. New method with a synthetic substrate and its comparison with the trypsinogen procedure. Histochemistry 78:251–270
Lottaz D, Maurer CA, Hahn D, Buchler MW, Sterchi EE (1999) Nonpolarized secretion of human meprin alpha in colorectal cancer generates an increased proteolytic potential in the stroma. Cancer Res 59:1127–1133
MacNaughton WK (2005) Epithelial effects of proteinase-activated receptors in the gastrointestinal tract. Mem Inst Oswaldo Cruz 100(Suppl 1):211–215
Marchiando AM, Graham WV, Turner JR (2010) Epithelial barriers in homeostasis and disease. Annu Rev Pathol 5:119–144
Medina C, Radomski MW (2006) Role of matrix metalloproteinases in intestinal inflammation. J Pharmacol Exp Ther 318:933–938
Miller GS, List K (2013) The matriptase-prostasin proteolytic cascade in epithelial development and pathology. Cell Tissue Res 351(2):245–253
Misumi Y, Ikehara Y (2004) Dipeptidyl-peptidase IV. In: Barrett AJ, Rawlings ND, Woessner JF (eds) Cysteine, serine and threonine peptidases, vol 2, Handbook of proteolytic enzymes. Elsevier, San Diego, CA, pp 1905–1909
Moriyama A, Kageyama T, Takahashi K (1983) Identification of monkey lung procathepsin D-II as a pepsinogen-C-like acid protease zymogen. Eur J Biochem 132:687–692
Mukherjee S, Roth MJ, Dawsey SM, Yan W, Rodriguez-Canales J, Erickson HS, Hu N, Goldstein AM, Taylor PR, Richardson AM, Tangrea MA, Chuaqui RF, Emmert-Buck MR (2010) Increased matrix metalloproteinase activation in esophageal squamous cell carcinoma. J Transl Med 8:91, http://www.translational-medicine.com/content/8/1/91
Netzel-Arnett S, Buzza MS, Shea-Donohue T, Desilets A, Leduc R, Fasano A, Bugge TH, Antalis TM (2012) Matriptase protects against experimental colitis and promotes intestinal barrier recovery. Inflamm Bowel Dis 18:1303–1314
Niemi LD, Johansson I (2004) Salivary statherin peptide-binding epitopes of commensal and potentially infectious *Actinomyces* spp. delineated by a hybrid peptide construct. Infect Immun 72:782–787
Nieminen A, Nordlund L, Uitto VJ (1993) The effect of treatment on the activity of salivary proteases and glycosidases in adults with advanced periodontitis. J Periodontol 64:297–301
Oberst MD, Singh B, Ozdemirli M, Dickson RB, Johnson MD, Lin CY (2003) Characterization of matriptase expression in normal human tissues. J Histochem Cytochem 51:1017–1025
Plow EF, Miles LA (1990) Plasminogen receptors in the mediation of pericellular proteolysis. Cell Differ Dev 32:293–298
Rao GJ, Nadler HL (1972) Deficiency of trypsin-like activity in saliva of patients with cystic fibrosis. J Pediatr 4:573–576
Rawlings ND, Salvessen G (eds) (2013) Handbook of proteolytic enzymes, 3rd edn. Academic Press, London
Reid WA, Vongsorasak L, Svasti J, Valler MJ, Kay J (1984) Identification of the acid proteinase in human seminal fluid as a gastricsin originating in the prostate. Cell Tissue Res 236:597–600
Reinhardt C, Bergentall M, Greiner TU, Schaffner F, Ostergren-Lunden G, Petersen LC, Ruf W, Backhed F (2012) Tissue factor and PAR1 promote microbiota-induced intestinal vascular remodelling. Nature 483:627–631
Richter C, Tanaka T, Yada RY (1998) Mechanism of activation of the gastric aspartic proteinases: pepsinogen, progastricsin and prochymosin. Biochem J 335:481–490
Roberts RM, Mathialagan N, Duffy JY, Smith GW (1995) Regulation and regulatory role of proteinase inhibitors. Crit Rev Eukaryot Gene Expr 5:385–436

Roff AN, Panganiban RP, Bond JS, Ishmael FT (2013) Post-transcriptional regulation of meprin a by the RNA binding proteins HuR and TTP. J Biol Chem 288(7):4733–4743

Roka R, Ait-Belgnaoui A, Salvador-Cartier C, Garcia-Villar R, Fioramonti J, Eutamene H, Bueno L (2007) Dexamethasone prevents visceral hyperalgesia but not colonic permeability increase induced by luminal protease-activated receptor-2 agonist in rats. Gut 56:1072–1078

Rosmann S, Hahn D, Lottaz D, Kruse MN, Stocker W, Sterchi EE (2002) Activation of human meprin-alpha in a cell culture model of colorectal cancer is triggered by the plasminogen-activating system. J Biol Chem 277:40650–40658

Salmela MT, Pender SL, Karjalainen-Lindsberg ML, Puolakkainen P, Macdonald TT, Saarialho-Kere U (2004) Collagenase-1 (MMP-1), matrilysin-1 (MMP-7), and stromelysin-2 (MMP-10) are expressed by migrating enterocytes during intestinal wound healing. Scand J Gastroenterol 39:1095–1104

Shan L, Molberg Ø, Parrot I, Hausch F, Filiz F, Gray GM, Sollid LM, Khosla C (2002) Structural basis for gluten intolerance in celiac sprue. Science 297:2275–2279

Silk DBA, Webb JPW, Lane AE, Clark ML, Dawson AM (1974) Functional differentiation of human jejunum and ileum: a comparison of the handling of glucose, peptides and amino acids. Gut 15:444–449

Steinhoff M, Buddenkotte J, Shpacovitch V, Rattenholl A, Moormann C, Vergnolle N, Luger TA, Hollenberg MD (2005) Proteinase-activated receptors: transducers of proteinase-mediated signaling in inflammation and immune response. Endocr Rev 26:1–43

Sterchi EE, Naim HY, Lentze MJ, Hauri HP, Fransen JAM (1988) N-benzoyl-L-tyrosyl-p-aminobenzoic acid hydrolase: a metalloendopeptidase of the human intestinal microvillus membrane which degrades biologically active peptides. Arch Biochem Biophys 265:105–118

Sterchi EE, Stöcker W, Bond JS (2008) Meprins, membrane-bound and secreted astacin metalloproteases. Mol Aspects Med 29:309–328

Sun X, Salih E, Oppenheim FG, Helmerhorst EJ (2009) Kinetics of histatin proteolysis in whole saliva and the effect on bioactive domains with metal-binding, antifungal, and wound-healing properties. FASEB J 23:2691–2701

Szabo R, Hudecz F, Reig F (2003) Interfacial interactions between poly[L-lysine]-based branched polypeptides and phospholipid model membranes. J Colloid Interface Sci 267:18–24

Tlaskalová-Hogenová H, Stepánková R, Hudcovic T, Tucková L, Cukrowska B, Lodinová-Zádníková R, Kozáková H, Rossmann P, Bártová J, Sokol D, Funda DP, Borovská D, Reháková Z, Sinkora J, Hofman J, Drastich P, Kokesová A (2004) Commensal bacteria (normal microflora), mucosal immunity and chronic inflammatory and autoimmune diseases. Immunol Lett 93:97–108

Tsuzuki S, Murai N, Miyake Y, Inouye K, Hirayasu H, Iwanaga T, Fushiki T (2005) Evidence for the occurrence of membrane-type serine protease 1/matriptase on the basolateral sides of enterocytes. Biochem J 388:679–687

Vallejo JA, Ageitos JM, Poza M, Villa TG (2012) Short communication: a comparative analysis of recombinant chymosins. J Dairy Sci 95:609–613

Vergnolle N (2000) Review article: proteinase-activated receptors – novel signals for gastrointestinal pathophysiology. Aliment Pharmacol Ther 14:257–266

Vogel LK, Saebo M, Skjelbred CF, Abell K, Pedersen ED, Vogel U, Kure EH (2006) The ratio of Matriptase/HAI-1 mRNA is higher in colorectal cancer adenomas and carcinomas than corresponding tissue from control individuals. BMC Cancer 6:176

Westerveld BD, Pals G, Bosma A, Defize J, Pronk JC, Frants RR, Eriksson AW, Meuwissen SG (1987) Gastric proteases in Barrett's esophagus. Gastroenterology 93:774–778

Wilson CL, Schmidt AP, Pirila E, Valore EV, Ferri N, Sorsa T, Ganz T, Parks WC (2009) Differential processing of alpha- and beta-defensin precursors by matrix metalloproteinase-7 (MMP-7). J Biol Chem 284:8301–8311

Woods DE, Straus DC, Johanson WG Jr, Bass JA (1981a) Role of salivary protease activity in adherence of Gram-negative bacilli to mammalian buccal epithelial cells in vivo. J Clin Invest 68:1435–1440

Woods DE, Straus DC, Johanson WG Jr, Bass JA (1981b) Role of fibronectin in the prevention of adherence of *Pseudomonas aeruginosa* to buccal cells. J Infect Dis 143:784–790

Yuan X, Zheng X, Lu D, Rubin DC, Pung CY, Sadler JE (1998) Structure of murine enterokinase (enteropeptidase) and expression in small intestine during development. Am J Physiol 274: G342–G349

Zhang Y, Cai X, Schlegelberger B, Zheng S (1998) Assignment1 of human putative tumor suppressor genes ST13 (alias SNC6) and ST14 (alias SNC19) to human chromosome bands 22q13 and 11q24–>q25 by in situ hybridization. Cytogenet Cell Genet 83:56–57

Zheng XL, Kitamoto Y, Sadler JE (2009) Enteropeptidase, a type II transmembrane serine protease. Front Biosci (Elite Ed) 1:242–249

Zhu L, Tamvakopoulos C, Xie D, Dragovic J, Shen X, Fenyk-Melody JE, Schmidt K, Bagchi A, Griffin PR, Thornberry NA, Sinha Roy R (2003) The role of dipeptidyl peptidase IV in the cleavage of glucagon family peptides: in vivo metabolism of pituitary adenylate cyclase activating polypeptide-(1-38). J Biol Chem 278:22418–22423

Chapter 12
Calpains in Health and Disease

John Anagli, Kevin K.W. Wang, Yasuko Ono, and Hiroyuki Sorimachi

12.1 Introduction

Since the first report in 1964 (Guroff 1964) of a cytosolic Ca^{2+}-dependent neutral proteolytic activity, now known as calpain-1 (also called μ-calpain), other members of the calpain family of proteinases have been discovered (Croall and DeMartino 1991; Molinari and Carafoli 1997; Goll et al. 2003; Bertipaglia and Carafoli 2007; Croall and Ersfeld 2007; Sorimachi et al. 2010, 2011a, b). The calpains are defined as having amino acid (aa) sequences significantly similar to that of the protease domain of human calpain-1. Calpains also contain a variety of protein structural motifs including C2, calpain-type β-sandwich (CBSW; once called C2-like), Zn-finger and penta-EF-hand (PEF) domains. This important class of intracellular cysteine proteases is found in almost all eukaryotes and some bacteria. To date, 15 calpain genes have been identified in humans (see Fig. 12.1). However, mammalian calpain-1 and calpain-2 (also called m-calpain, calpain-II or mCANP), sometimes referred to as the "ubiquitous," "conventional," or "classical" calpains, are the best-characterized members of the calpain superfamily (Clan CA, family C02, EC 3.4.22.17). The conventional calpains have a very specific proteinaceous inhibitor *in vivo* called calpastatin. The ubiquitous calpain-1 and -2 and their natural inhibitor calpastatin are the three members of the calpain system that have been

J. Anagli (✉)
Banyan Laboratories, Banyan Biomarkers, Inc., 12085 Research Drive, Alachua, FL, USA
e-mail: johnanagli@gmail.com

K.K.W. Wang (✉)
Center for Neuroproteomics & Biomarkers Research, Departments of Psychiatry, University of Florida, 1149 South Newell Drive, Gainesville, FL 32611, USA
e-mail: kwang@ufl.edu

Y. Ono • H. Sorimachi
Department of Advanced Science for Biomolecules, Tokyo Metropolitan Institute of Medical Science (IGAKUKEN), 2-1-6 Kamikitazawa, Setagaya-ku, Tokyo 156-8506, Japan
e-mail: ono-ys@igakuken.or.jp; sorimachi-hr@igakuken.or.jp

K. Brix and W. Stöcker (eds.), *Proteases: Structure and Function*,
DOI 10.1007/978-3-7091-0885-7_12, © Springer-Verlag Wien 2013

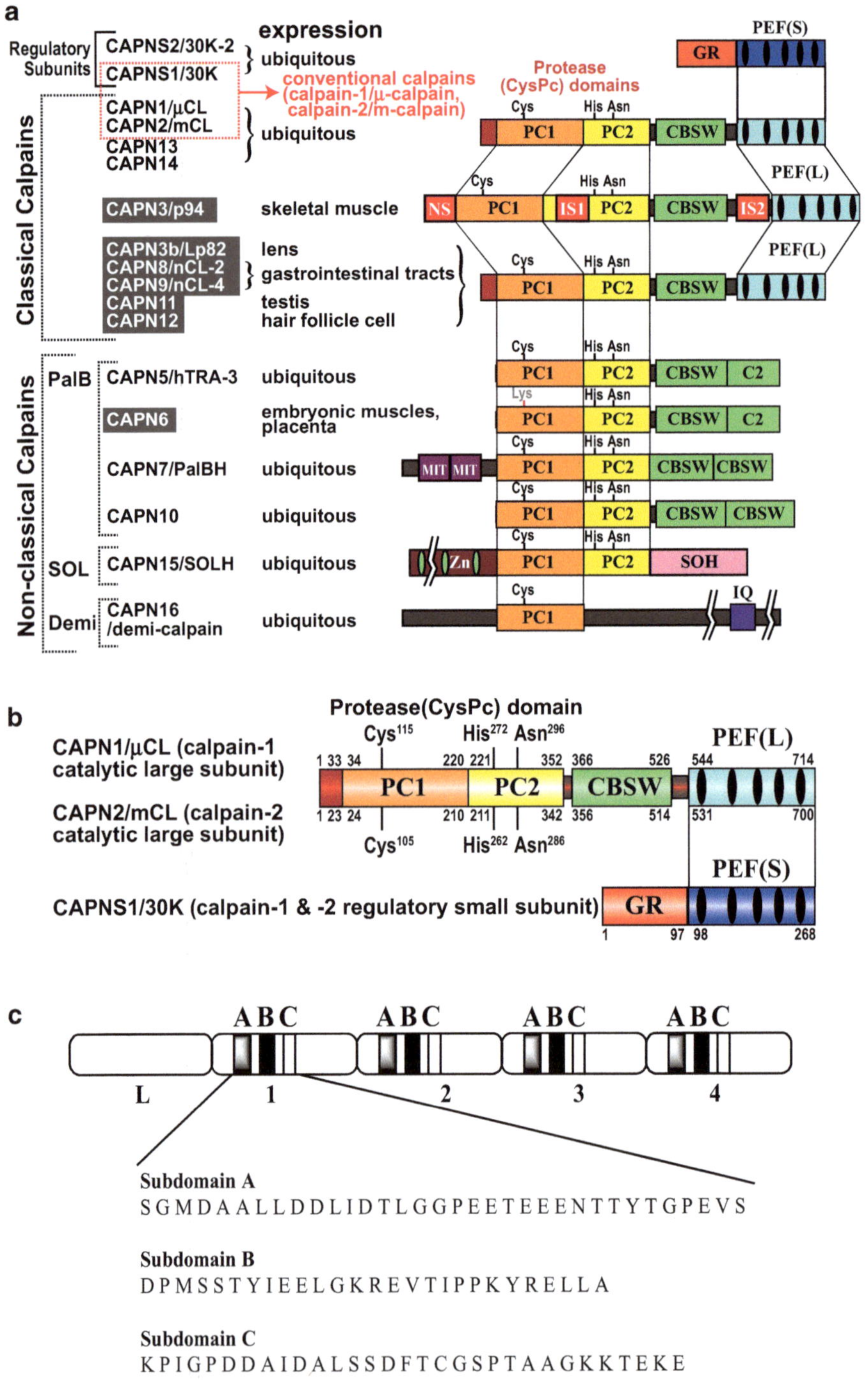

Fig. 12.1 Schematic structures of human calpains and the related molecules. (**a**) Schematic structures of 15 human calpains. Names for calpains are by the general nomenclature of gene product (*e.g.*, *CAPN1* → CAPN1) used in this review, and their previous names, if any, are shown

studied most extensively at the structure-functional level. Under normal physiological conditions, calpains exist at very low activity in cells and are proposed to participate in important cellular activities, including signal transduction, cell motility and apoptosis. However, excessive calpain activation following ischemic and traumatic cell injury or aberrant expression of components of the calpain system results in cell death or impaired cellular function. This review summarizes current knowledge of the biological significance of regulated calpain activity and the pathological consequences of uncontrolled or excessive calpain activation, with special emphasis on the pathophysiological roles of the calpains in ischemic and traumatic brain injury and some tissue-specific diseases.

12.2 Conventional Calpains

12.2.1 Calpain-1/μ-Calpain and Calpain-2/m-Calpain

12.2.1.1 Discovery and Nomenclature

Ca^{2+}-activated neutral proteinase (CANP) activity was first discovered in the soluble fraction of rat brain in 1964 (Guroff 1964). Similar observations were subsequently made in rabbit, human and chicken skeletal muscle as well as in rat liver (Meyer et al. 1964; Huston and Krebs 1968; Busch et al. 1972; Reddy et al. 1975; Kar and Pearson 1976; Takai et al. 1977; Ishuira et al. 1978). Further characterization of the biochemical properties of the Ca^{2+}-dependent proteolytic activity led to the isolation of distinct molecules of the calpain system. The cDNA for calpain was finally cloned by Ohno et al. (1984). Hundreds of calpain-related genes, including tissue-specific and non-classical forms, have been cloned and sequenced since the original cloning of the ubiquitously expressed calpain, but the gene products of a majority of them remain to be characterized. The conventional calpains are represented by two isoforms of calpains, calpain-1 and calpain-2, that are activated at neutral pH by micro- and milli-molar concentrations of Ca^{2+} *in vitro*, respectively, and are ubiquitously expressed in all eukaryotic cells examined so far (Croall and DeMartino 1991; Goll et al. 2003; Croall and Ersfeld 2007;

Fig. 12.1 (continued) after the general names. *Black* and *reversed letters* indicate ubiquitous and tissue/organ-specific calpains, respectively. Symbols: PC1 and PC2, protease core domains 1 and 2 in the calpain protease (CysPc) domain; CBSW calpain-type β-sandwich domain; PEF(L/S), penta-EF-hand domains in the large(L)/small(S) subunit; GR, glycine-rich hydrophobic domain; MIT, microtubule interacting and transport motif; C2, C2 domain; Zn, Zn-finger motif; SOH, SOL-homology domain; IQ, a motif interactive with calmodulin; NS/IS1/IS2, CAPN3/p94-characteristic sequences. (**b**) Domain architecture of human calpain-1 and calpain-2. Calpain-1 and -2 are heterodimers composed of a common regulatory small subunit (CAPNS1/30K) and each catalytic large subunit (CAPN1/μCL and CAPN2/mCL, respectively). See (**a**) for abbreviations of domain names. (**c**) Domain architecture of calpastatin

Sorimachi et al. 2010). They are also called μ-calpain and m-calpain, calpain-I and calpain II, or μCANP and mCANP, respectively. An endogenous specific inhibitor of calpain-1 and -2, called calpastatin, was later discovered in the late 1970s (Dayton et al. 1976; Suzuki and Murachi 1977; Nishiura et al. 1978). To date, 15 human calpain genes have been discovered and numbered *CAPN1–3* and *5–16*, as shown in Fig. 12.1a. μCL and mCL correspond to the gene products of *CAPN1* and *CAPN2*, and the systematic names, calpain-1 and calpain-2, have gained popularity. However, since "calpain" originally was a term for an enzyme consisting of two distinct subunits, this nomenclature is confusing and misleading. Therefore, here, the formal gene product nomenclature, *i.e.*, CAPN1 and CAPN2, is used. Accordingly, calpain-1 (= μ-calpain) is a heterodimer of CAPNS1 and CAPN1. In this review, to include the older name, CAPN1/μCL and CAPN2/mCL, is used for clarity.

12.2.1.2 Structural Features

Both calpain-1 and -2 are heterodimers, composed of a common 28 kDa regulatory subunit (CAPNS1/30K) and two distinct 80 kDa catalytic subunits (CAPN1/μCL and CAPN2/mCL, respectively). The catalytic and regulatory subunits of the conventional calpains are divided into four and two region/domains, respectively (see Fig. 12.1b). The protease (CysPc) domain is the catalytic site containing the cysteine, histidine, asparagine triad (Saido et al. 1994). In the absence of Ca^{2+}, CysPc domain is divided into protease core domains PC1 and PC2, which are folded into one domain upon Ca^{2+} binding (Hanna et al. 2008; Hosfield et al. 1999; Strobl et al. 2000). Domains PC1, PC2, CBSW, PEF(L) and PEF(S) bind at least one Ca^{2+} with varying affinities (Moldoveanu et al. 2002, 2003; Tompa et al. 2001). An extensive knowledge of the structures of the inactive Ca^{2+}-free and the active (Ca^{2+} and calpastatin bound) forms of calpain-2 is available, and has provided useful insights into the mechanism for the Ca^{2+}-dependent activation of the enzyme but the mechanism by which the protease is regulated *in vivo* is still not clear, in spite of many suggestions (Hanna et al. 2008).

The classical calpains normally exist as inactive proteases because their catalytic triad residues are not properly aligned in the Ca^{2+}-free form of the protein. It has been deduced from the crystal structures that upon binding of Ca^{2+} to specific penta-EF-hands in PEF domains and other non-EF-hand sites within PC1, PC2, and CBSW domains, multiple conformational changes occur within the protein to align the catalytic machinery of the active site (Hosfield et al. 1999, 2004; Strobl et al. 2000; Moldoveanu et al. 2002). Furthermore, site-directed mutagenesis studies have allowed the identification of multiple intramolecular sites in the NH_2-terminal peptide (residues 1–20) and the domain CBSW-PEF(L) linker peptide ("transducer") that regulate the response of calpain to Ca^{2+} binding to CysPc, CBSW, PEF(L) and PEF(S) domains (Moldoveanu et al. 2002; Hosfield et al. 2004; Bozoky et al. 2005). The exact sequence of the steps involved in the relay of the Ca^{2+} message is however unknown. On activation, calpain autolyzes by truncating the

N-terminal portion of the large subunit and most of Gly-rich (GR) domain of the small subunit. The activated form appears to have greater calcium sensitivity (Hathaway et al. 1982; Imajoh et al. 1986; Suzuki et al. 1981).

An endogenous proteinaceous inhibitor of calpain-1 and -2, called calpastatin, was discovered in the late 1970s (Dayton et al. 1976; Suzuki and Murachi 1977; Nishiura et al. 1978; Takahashi-Nakamura et al. 1981). Among the other calpain homologues, CAPN8/nCL-2 and CAPN9/nCL-4, but not CAPN3/p94, are also inhibited by calpastatin. The primary structure of calpastatin consists of a non-inhibitory L-domain and four repeating (1–4) inhibitory domains, each having an independent inhibitory activity against calpain (Emori et al. 1988) (see Fig. 12.1c). The repeating inhibitory domains have highly conserved internal subdomains A, B, and C, with subdomain B representing the calpain inhibitory moiety. A 27-mer synthetic peptide corresponding to subdomain 1B of human calpastatin has been found to exhibit potent and specific inhibition toward calpain-1 and -2 *in vitro* (Maki et al. 1989; Betts et al. 2003; Betts and Anagli 2004; Fiorino et al. 2007) and *in vivo* in a rat model of ischemic stroke (Anagli et al. 2009).

12.2.1.3 Biological Significance

Because the conventional calpains are ubiquitously expressed in vertebrate cells, their function is considered to be fundamental and essential. The physiological role of calpain-1 and -2 is not fully understood, but it has been linked to Ca^{2+}-regulated signaling pathways. Their involvement in the regulation of physiological processes such as apoptosis (Benetti et al. 2001; Danial and Korsmeyer 2004; Gomez-Vicente et al. 2005; Norberg et al. 2008; Sanges et al. 2006), signal transduction (Kishimoto et al. 1983; Noguchi et al. 1997; Pontremoli and Melloni 1988), membrane repair (Mellgren and Huang 2007; Mellgren et al. 2007), modulation of cell motility (Satish et al. 2005; Wells et al. 2005) and synaptic function and memory formation (see review by Liu et al. 2008a) has been described. Calpains are somewhat selective with regard to substrate selection; only 5 % of cellular proteins are degraded or fragmented by calpain-1 and -2 (Wang et al. 1989). A number of substrates have been reported for calpains, but few of them have been linked to physiological functions. These include, focal adhesion kinase (FAK) and ezrin, which are involved in fibroblast cell attachment and mobility (Potter et al. 1998; Huang et al. 2004), and phosphotyrosine phosphatase 1 B (PTP1B), which is involved in platelet aggregation (Kuchay et al. 2007). Possible substrates in the central nervous system (CNS) that might be critical to synaptic function and memory formation include αII- and βII-spectrin, the postsynaptic density-95 (PSD95) scaffolding protein, calcium- and calmodulin-dependent protein kinase II (CaMKII), and the *N*-methyl-D-aspartate (NMDA)-type and other glutamate receptors (Bi et al. 1998; Wang 2000; Liu et al. 2008b).

12.2.1.4 Pathologic Role of Calpain

Evidence accumulated over the past 15 years incriminates calpain hyper-activation in neural cell death following severe cellular challenge or damage in cerebral ischemia (Hong et al. 1994; Roberts-Lewis et al. 1994; Markgraf et al. 1998; Yokota et al. 1999; Liebetrau et al. 1999; Kambe et al. 2005; Kawamura et al. 2005; Anagli et al. 2009), traumatic brain injury (TBI) (Taft et al. 1992; Nath et al. 1996; Postmantur et al. 1996; Buki et al. 1999, 2003; Newcomb et al. 1999; Kampfl et al. 1997; Pike et al. 1998; Pike et al. 2000; Ringger et al. 2004), spinal cord injury (SCI) (Banik et al. 1997; Ray et al. 1999, 2001a, b, 2011; Schumacher et al. 2000; Ray et al. 2003; Sribnick et al. 2007; Yu et al. 2008; Zhang et al. 2003), sub-arachnoid hemorrhage (SAH) (Germano et al. 2002; Zhang et al. 2000; Wang and Yuen 1994; Lee et al. 1997; Yamaura et al. 1993) cerebral vasospasm (Minami et al. 1992), multiple sclerosis (Shields and Banik 1999; Schaecher et al. 2001, 2002; Benjamins et al. 2003; Nakanishi 2003; Shields et al. 1999), Parkinson's Disease (Crocker et al. 2003; Ray et al. 2000; Chera et al. 2002; Kim et al. 2003; Samantaray et al. 2008, 2011), Alzheimer's Disease (Veeranna et al. 2004; Chen and Fernandez 2005; Higuchi et al. 2005; Fifre et al. 2006; Rao et al. 2008; Liu et al. 2011), Huntington's Disease (Gafni and Ellerby 2002; Gafni et al. 2004; Bizat et al. 2003, 2005), and amyotrophic lateral sclerosis (ALS) (Das et al. 2008; Tradewell and Durham 2010).

Investigation of the neuronal pathobiology following cerebral ischemia (Hong et al. 1994; Chen et al. 1998; Kristian et al. 1998) and TBI (Povlishock et al. 1983, 1992; Povlishock 1992; Maxwell et al. 1997; Meaney et al. 2001) indicates that the majority of injured neurons are not damaged at the time of impact but show progressive degenerative changes over the ensuing hours or days. The initial injurious event triggers a destructive cascade of biochemical events that results in cell death within the core of the injured tissue and which spreads out into surrounding tissue over time. As the initial step, ischemic (Szatkowski and Attwell 1994; Rothman and Olney 1986) or traumatic (Hayes et al. 1992; McIntosh et al. 1998) insult induces massive release of glutamate from damaged synapses which leads to activation of glutamate receptor-associated and voltage-dependent calcium channels. Such activation induces influx of calcium ions into the neuron (Fineman et al. 1993; Pettus et al. 1994; Pettus and Povlishock 1996) and release of calcium ions from intracellular stores (Berridge 1993) (see Fig. 12.2). Loss of intracellular calcium homeostasis contributes to cell death by (1) activating various enzymes including proteases, kinases, phosphatases, and phospholipases (Siesjo et al. 1989; Wieloch et al. 1991; Morioka et al. 1992; Verity 1992; Nishida et al. 1994; Wang and Yuen 1994; Dash et al. 1995), (2) induction of free radical release (Kontos 1989; Bading et al. 1993; Hall 1998), and (3) mediating detrimental changes in gene expression (Bading et al. 1993; Rink et al. 1995). Integral to the mechanism of calcium-mediated neuronal degeneration in both ischemic (Yamashima 2000; Rami 2003) and traumatic (Kampfl et al. 1997; Newcomb et al. 1997; Posmantur et al. 1997; Saatman et al. 1996a; Wang 2000) brain injury

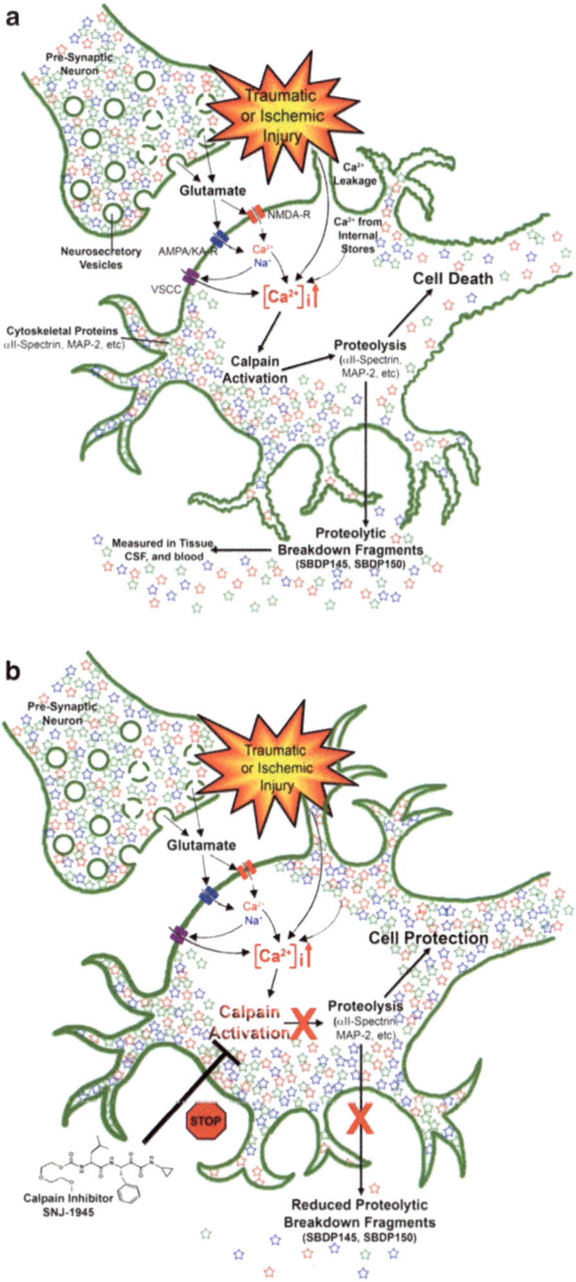

Fig. 12.2 Conventional calpains in disease. (**a**) Schematic of the destructive Ca^{2+}-mediated cascade that results in progressive cell damage following traumatic or ischemic brain injury. Initial injury induces massive glutamate release from damaged synapses leading to hyper-activation of NMDA and AMP/kainite receptors (NMDA-R and AMPA/KA-R) leading to influx of calcium and sodium ions. Ca^{2+} ions can also enter the cell through sodium-activated voltage-sensitive calcium channels (VSCC), via leakage through damaged cell membranes, and they can be released from internal stores. The resulting increase in intracellular Ca^{2+} levels activates a number of Ca^{2+}-sensitive proteases, including calpain. Once activated, calpain proteolyses critical cellular proteins, including the cytoskeletal protein αII-spectrin, resulting in compromise of the cell membrane and cell death. (**b**) Calpain as a therapeutic target for ameliorating brain injury. Calpain inhibitors, such as SNJ-1945 or B27-HYD, can block calpain hyper-activation following injury, thereby shutting down the destructive cascade and preserving cell integrity. Thus, calpain inhibitors are neuroprotective

is the pathological activation of calpain-1 (CAPN1/μCL + CAPNS1/30K) and calpain-2 (CAPN2/mCL + CAPNS1/30K), which results in proteolytic destruction of many cellular proteins including receptor proteins, calmodulin binding proteins, signal transduction enzymes, transcription factors, and cytoskeletal proteins (Wang and Yuen 1997, 1999). Such uncontrolled proteolysis can result in disruption of axonal transport and structural collapse culminating in oncotic and/or apoptotic cell death (Buki et al. 2000; Medana and Esiri 2003) (see Fig. 12.2).

12.2.1.5 Conventional Calpain-Generated Biomarkers of CNS Injury

During brain injury, neural proteins or their breakdown products generated by calpains (calpain-1 and -2) are released into the extracellular environment and eventually reach the cerebrospinal fluid (CSF) in relatively high concentration (Wang et al. 2005). In due time the proteins reach the blood stream either via the compromised blood brain barrier (BBB) or *via* filtration of the CSF (see Fig. 12.3). Clearance and half life of the biomarkers contribute to the final concentration that can be measured in the blood. The CSF volume of an adult human (CSF 125–150 mL) is about 30- to 40-fold less than the blood volume (4.5–5 L) which explains why the brain biomarker concentration is significantly higher in the CSF samples versus blood samples and makes the former valuable for drug development. Enabled by recent technological advances in proteomics, novel brain injury biomarkers that have elevated levels in biofluids such as cerebrospinal fluid or blood after traumatic brain injury have been discovered (Vitzthum et al. 2005; Kobeissy et al. 2006; Liu et al. 2006; Mondello et al. 2011). The most well-studied calpain target is the cytoskeletal structural protein αII-spectrin which is cleaved by calpain into signature breakdown products of 150 kDa (SBDP150) and 145 kDa (SBDP145) (Saido et al. 1993; Nath et al. 1996; Wang et al. 1998; Pike et al. 1998; Pike et al. 2000; Zhao et al. 1999; Newcomb-Fernandez et al. 2001). Calpain-generated neural protein breakdown products (BDPs) such as SBDP150, SBDP145, c-Tau, PARP, as well as MAP2, GFAP and CRMP-2 BDPs can be used to monitor acute necrotic/oncotic neural cell death and diagnose brain injury severity and progression (Siman and Noszek 1988; Siman et al. 1989, 2004; Pike et al. 2001, 2003; Ringger et al. 2004; Liu et al. 2006, 2008a, b, 2011; Anagli et al. 2009; Zhang et al. 2009). In addition, sensitive and selective sandwich ELISAs for such biomarkers have allowed quantification of the extent of brain injury in animal studies and clinical studies of brain injury alike (Liu et al. 2010; McGinn et al. 2009; Mondello et al. 2010; Siman et al. 2009).

12.2.1.6 Calpain-Target-Based Theranostics for CNS Injury

Although brain (ischemic and traumatic) and spinal cord injury is a significant medical crisis, there are no FDA-approved therapies that have been demonstrated to improve functional outcomes. This is the case despite a large number of clinical trials with promising animal model efficacy data. Major pharmaceutical companies and biotech companies have been trying for years to tackle acute brain injury without success. New therapeutic development traditionally has an extremely

Fig. 12.3 Generation and detection of brain injury biomarkers

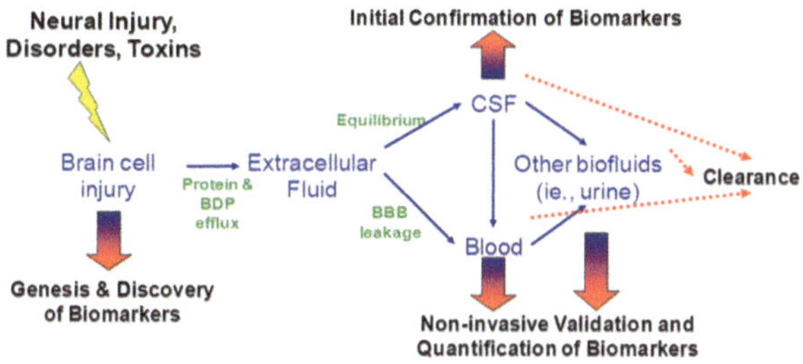

high triage rate because more than 90 % of drugs that advance to Phase I clinical trials fail. Some argue such extreme loss can be overcome by guiding all new therapeutic development and clinical trials with a disease-relevant diagnostic test. Discovery of translational biomarkers (from animal studies to clinical trials) might help to finally deliver the long sought after clinical trial success.

Since the discovery of calpain in the mid-1960s (Guroff 1964; Huston and Krebs 1968), scientists have sought methods to inhibit its activity. Early attempts used calcium chelators, but as knowledge of calpain substrate specificity grew, and inhibitors from natural products screening programs became available, more selective inhibitors were designed and synthesized (Aoyagi and Umezawa 1975; Parkes et al. 1985; Crawford 1987; Crawford et al. 1988; Anagli et al. 1991; Angliker et al. 1992; Wikstrom et al. 1992; Fukiage et al. 1997; Schroder et al. 1993; Wang and Yuen 1997; Shirasaki et al. 2005; Pietsch et al. 2010; Donkor 2011). Although synthetic calpain inhibitors have proved efficacious in *in vivo* models of TBI, SCI and cerebral ischemia they have limited therapeutic value because (1) they inhibit other cysteine and serine proteases, (2) they have poor membrane permeability, and (3) they are readily degraded *in vivo* (Wang and Yuen 1997; Yuen and Wang 1998). The endogenous calpain inhibitor, calpastatin, is the most potent and selective inhibitor identified to date (Murachi 1989; Maki et al. 1991). Other proteins, including the heavy chains of L- and H-kininogens and α_2-macroglobulin, also inhibit calpain (Crawford 1987; Salvesen et al. 1986). However, these high molecular weight inhibitors have limited therapeutic utility due to their poor cell permeability and lack of specificity. Calpain inhibition has the potential to shut down the destructive calcium-mediated cascade, thereby reducing proteolysis of essential cell proteins, resulting in cell preservation and protection (see Fig. 12.4). Previous therapeutic strategies for ameliorating the post-injury calcium-mediated destructive cascade have focused on antagonizing glutamate release, glutamate receptors, or voltage-sensitive calcium channels (Hossman 1988; Siesjo and Bengtsson 1989; Bullock and Fujisawa 1992; McIntosh 1994). These strategies have produced limited success because (1) the glutamate response occurs very rapidly after injury leaving only a narrow window for treatment efficacy, and (2) it is difficult to distinguish which of the various receptor and ion channel subclasses are primarily involved in perpetuating the cytotoxic cascade (Hossman 1988; Saatman et al. 1996b). Treatment aimed at a downstream neuropathological event could provide a longer window of opportunity for effective intervention, and therefore be

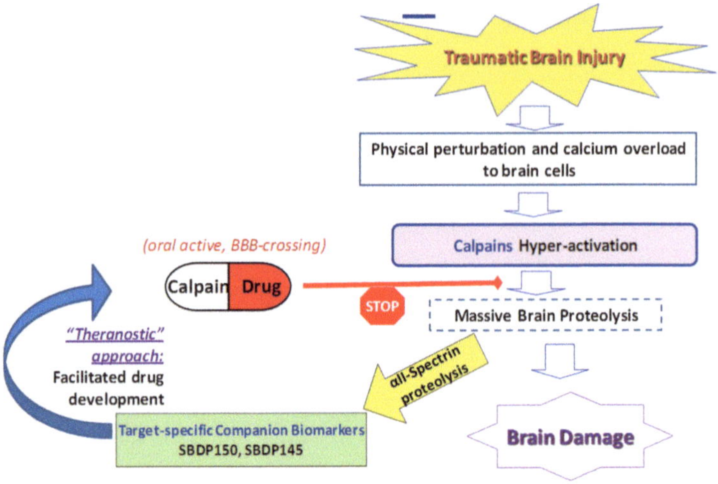

Fig. 12.4 "Theranostic Approach" to treat traumatic brain injury with a calpain inhibitor drug

Table 12.1 Unique characteristics of calpain as a therapeutic target for CNS injury

Unique characteristics of calpain as a therapeutic target for CNS injury	Translation into therapeutic advantage
Calpain has very low basal activity and is only hyperactivated by calcium overload under pathological conditions	Calpain inhibitor should have minimal adverse effects
Calpain activation is downstream from glutamate receptors	Calpain inhibition should provide an extended therapeutic window
Calpain is dually involved in both necrotic and apoptotic cell death	Calpain inhibition can suppress both acutely occurring necrosis and delayed apoptosis
Proteases as a class are well-established therapeutic targets (e.g. angiotensin converting enzyme, HIV proteases)	Calpain inhibitors have high likelihood of being formulated into molecules that can be used as human drugs

valuable for a greater number of patients. Calpain antagonists may have a further advantage over glutamate and calcium channel receptor antagonists in that calpain exists predominantly in its inactive proenzyme form under normal physiological conditions, and only becomes significantly activated under pathological conditions. Therefore, it would be reasonable to assume that calpain inhibition would not lead to any untoward adverse events. On the other hand, glutamate and calcium channel receptors play a critical neurotransmitter role in and outside of the CNS, and their inhibition could be expected to have profound side effects. Indeed, although glutamate receptors have proved to provide neuroprotection in animal models of cerebral ischemia, they have also been shown to have significant psychotomimetic effects (Olney et al. 1991). Therefore, calpain has unique characteristics as a therapeutic target for CNS injury (Liu et al. 2008a, b) (see Table 12.1). Taken together, these data suggest that brain injury is in need of a paradigm shift with regards to therapeutic development. Calpain-target-based therapeutics and companion diagnostic biomarkers might be the missing component that helps guide successful clinical trials for brain injury.

"Theranostics" represents the convergence between Therapeutics and diagnostics. "Theranostics is the term used to describe the proposed process of diagnostic therapy for individual patients—to test them for possible reaction to taking a new medication and to tailor a treatment for them based on the test results" (Warner 2004). Theranostics encompasses the possible utilization of a wide range of procedures including: predictive medicine, personalized medicine, integrated medicine, pharmaco-diagnostics and Dx/Rx partnering. It has been viewed as the parallel use of new therapy and diagnostic tests for a human disease or disorder so as to facilitate drug development and clinical trials and to achieve optimal clinical outcomes in a population of patients. Importantly, in recognizing the emerging role of theranostic approach, FDA has recently drafted a "*Drug-Diagnostic Co-Development Concept Paper*" with the goal of setting guidelines for prospective co-development of a drug or biological therapy (drugs) and a device test in a scientifically robust and efficient way. Lastly, another advantage of the theranostic approach is its built-in ability to achieve a post-marketing personalized medicine paradigm (i.e., drug treatment could be tailored according to a patient's diagnostic biomarker profile over time. Most importantly, as proposed in this review, a novel theranostic approach that combines calpain-generated acute brain injury-tracking biomarkers with potent and selective calpain inhibitor drug candidates could fast-track and improve the chances of successful drug development for CNS injury.

12.3 Unconventional Calpains

12.3.1 Skeletal-Muscle-Specific Calpain

12.3.1.1 Discovery and Nomenclature

In 1989, in the course of the cDNA cloning of human calpain-1 and -2 catalytic large subunits (CAPN1/μCL and CAPN2/mCL, respectively), a cDNA encoding a novel molecule that was similar to but distinct from both was discovered (Sorimachi et al. 1989). In contrast to the ubiquitous expression of CAPN1/μCL and CAPN2/mCL, the identified mRNA was predominantly expressed in skeletal muscle, *i.e.*, skeletal muscle-specific calpain was the first tissue-specific calpain to be discovered. The novel calpain catalytic subunit was about 50 % identical in aa sequence to those of CAPN1/μCL and CAPN2/mCL. It had the same basic domain structure as the known calpain catalytic subunits, consisting of four region/domains: the N-terminal specific region (NS), CysPc, CBSW, and PEF domains (see Fig. 12.1a) (Ono et al. 1999; Sorimachi et al. 2010, 2011a, b; Sorimachi and Suzuki 2001).

Initially, this molecule was called p94 for its putative molecular mass (Sorimachi et al. 1989), because there was no information about its tertiary structure. Since the overall aa sequence of p94 was highly similar to those of CAPN1/μCL and CAPN2/mCL, it was thought likely that an as-yet-unknown regulatory small subunit, similar or identical to CAPNS1/30K, would associate with p94 to make the active enzyme, which would be called a novel calpain with a new name such as n-calpain. Four years later, another novel calpain homolog was identified, clarifying the need for a method to systematize the names of the calpain-related molecules (Sorimachi et al. 1993a). For this purpose, it was proposed that novel calpains, including as-yet-undiscovered ones, would be numbered as n-calpain-1, n-calpain-2, *etc.*, where 'n' stands for novel, and p94 would be the catalytic large subunit of n-calpain-1 (for short, nCL-1). In 1990, the human genes for CAPN1/μCL, CAPN2/mCL, p94/nCL-1, and CAPNS1/30K were identified and named *CAPN1*, *CAPN2*, *CAPN3*, and *CAPN4*, respectively (Ohno et al. 1990). In 1995, a defect of *CAPN3* was shown responsible for limb-girdle muscular dystrophy (LGMD) type 2A (LGMD2A, also called calpainopathy), and p94/nCL-1 was termed calpain 3 after the gene name *CAPN3* (Richard et al. 1995). As discussed in the conventional calpain nomenclature, the skeletal-muscle-specific calpain should now be called CAPN3. In this review, CAPN3/p94 is used.

12.3.1.2 Structural Features

The basic domain structure of CAPN3/p94 is identical to those of CAPN1/μCL and CAPN2/mCL, as shown in Fig. 12.1a, b. However, CAPN3/p94 contains three unique regions, NS, IS1, and IS2 (insertion sequence 1 and 2, respectively), which are neither found in the other calpains nor related to other peptide sequences in the databases.

The NS is a Pro-rich structure of unknown function that is cut off upon the autolysis of CAPN3/p94 (Hayashi et al. 2008; Ono et al. 2004). CysPc domain, as in other calpain homologs, is the most highly conserved. Site-directed mutagenesis identified that Cys129, His334, Asn358 are essential for CAPN3/p94's activity. IS1 is inserted into the PC2 domain.

IS2 locates between the CBSW and PEF domains and contains a nuclear-localization signal-like sequence, PXKKKKXKP. Deletion of either IS1 or IS2 suppresses autolysis, and thus stabilizes CAPN3/p94 expression. In addition, the IS2 region is necessary and sufficient for CAPN3/p94's binding to the N2A region of connectin/titin (Sorimachi et al. 1995).

The CBSW and PEF domains of CAPN3/p94 are highly homologous to those of the conventional calpains, strongly suggesting that these domains bind Ca^{2+} regardless of the Ca^{2+}-independence of CAPN3/p94's protease activity in the presence of enough Na^+ (see below). Many of the pathogenic mutations found in LGMD2A/calpainopathy patients reside in these domains as well as the CysPc domain (Richard et al. 1999).

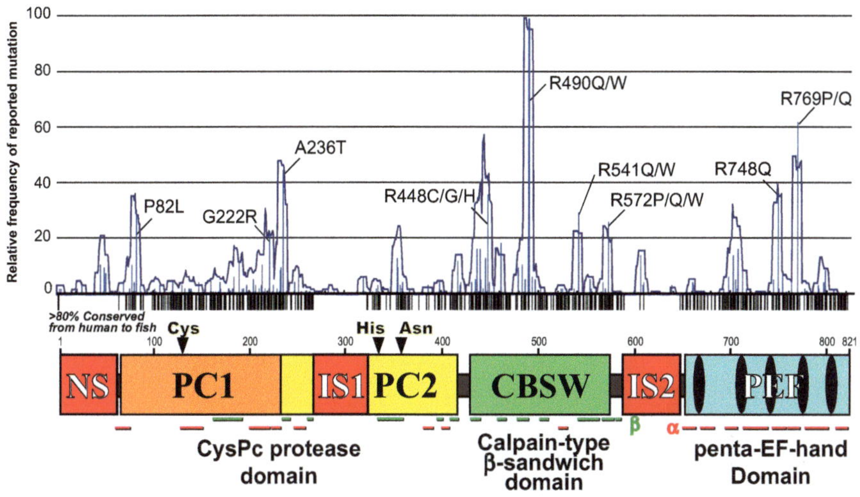

Fig. 12.5 Pathogenic *CAPN3* missense mutations and their frequencies in LGMD2A/calpainopathy patients. Frequencies are relative to the maximum value. *Black vertical lines* indicate aar that are conserved more than 80 % of vertebrates. α (*red*) and β (*green*) indicate α-helix and β-strand secondary structures of the corresponding calpain-2 3D structure. See Fig. 12.1 legend for other symbols

A comparison of the CAPN3/p94s of different vertebrates showed that more than half aa residues (aar) are more than 80 % conserved between species (Sorimachi et al. 2011a). These conserved aar include more than 70 % of the missense loci found in LGMD2A/calpainopathy pathogenic mutations (see below and Fig. 12.5). Conservation within the CAPN3/p94-characterizing regions (NS, IS1, and IS2) is low; however, it should be noted that some missense mutations, such as the R49C/H, P319L, and S606L mutations within the NS, IS1, and IS2 regions, respectively, occur relatively frequently (Sorimachi et al. 2011a) (Fig. 12.5). This observation suggests that strict sequence conservation within the CAPN3/p94-characterizing regions is not necessarily required, but a few important aar, usually near the borders of these regions, must be conserved to maintain proper function.

12.3.1.3 Proteolytic Activity

The mRNA for CAPN3/p94 is expressed predominantly in skeletal muscle, about ten times more abundant than the those for CAPN1/μCL and CAPN2/mCL. For a long time, however, neither the CAPN3/p94 activity nor the protein itself could be detected in skeletal muscle extract by the isolation procedures used successfully for calpain-1 and -2. Then, expression experiments using mutated CAPN3/p94 began to reveal its unique properties (Baghdiguian et al. 1999; Benayoun et al. 2008;

Fukiage et al. 2002; Hayashi et al. 2008; Kramerova et al. 2006, 2008, 2009; Ono et al. 1998, 2004, 2010; Richard et al. 1995; Sorimachi et al. 1993b, 1995; Taveau et al. 2003; Ueda et al. 2001). These studies showed that the CAPN3/p94 protein undergoes extremely rapid autolysis (its half-life *in vitro* is less than 10 min), indicating that CAPN3/p94 has strong protease activity, which, on the other hand, makes studying it at the protein level very difficult. Moreover, this autodegradation is Na^+-dependent in the absence of Ca^{2+}, establishing CAPN3/p94 as the first example of an intracellular Na^+-dependent enzyme. Specific inhibitors for calpain-1 and -2, including calpastatin, E-64, and leupeptin, have little effect on CAPN3/p94's autolytic activity. The *in vitro* substrates of CAPN3/p94 include CAPN3/p94 itself, myotonin protein kinase, fodrin, heat-shock protein (HSP) 60, calpastatin, muscle-specific ankyrin-repeat proteins (MARPs), and connectin/titin, the gigantic filamentous muscle protein.

Although the precise *in vivo* mechanism of CAPN3/p94's regulation remains unclear, it is very likely that the activity is suppressed by the interaction of the enzyme with connectin/titin (Hayashi et al. 2008). The physiological relevance of the Na^+-dependency of CAPN3/p94 is still unclear, but its substrate specificity differs depending on whether it is activated by Ca^{2+} or Na^+ (Ono et al. 2010). One of the splicing variants of *CAPN3* encodes CAPN3:ex1–5|7–14|17–24 (called p94Δ), which lacks both IS1 and IS2, and has markedly reduced autolytic activity. Therefore, recombinant p94Δ could be purified relatively stable, and was used for enzyme characterization studies. The results showed that p94Δ has characteristics very similar to those of conventional calpains, except that it proteolyzes, and is not inhibited by, calpastatin (Ono et al. 2004).

The lens-specific splicing variant, CAPN3:ex1B|2–5|7–14|17–24 (called Lp82), also shows characteristics very similar to those of conventional calpains (Fukiage et al. 2002). Lp82 shows Ca^{2+}-dependent protease activity against βA3 and αB crystallins, and autolyzes at the N-terminus and between the CBSW and PEF domains. Lp82's activity is inhibited by E-64 and iodoacetamide (Ueda et al. 2001). Like p94Δ, Lp82 is not inhibited by calpastatin, and proteolyzes it. Some other splicing variants with structures similar to Lp82 are expressed in embryonic skeletal muscles (Fougerousse et al. 2000). These variants show a number of unique features compared with CAPN3/p94, although their physiological functions are not clear (Herasse et al. 1999; Ojima et al. 2007).

12.3.1.4 Biological Significance

In 1995, human *CAPN3* was identified as a gene responsible for LGMD2A/calpainopathy (Richard et al. 1995). The mutations, including missense point mutations, nonsense mutations, frame-shift mutations, and splice-site mutations, are widely distributed in *CAPN3*. The absence of any 'hot-spot' in *CAPN3* makes the diagnosis of LGMD2A/calpainopathy difficult (Sorimachi et al. 2011a). So far, more than 450 unique mutations have been reported in *CAPN3*. More than half of the *CAPN3* pathogenic mutations are point mutations, and the majority of these are missense mutations, which are distributed throughout the protein at more than 175 independent loci among the 821 aar (Sorimachi et al. 2011a) (Fig. 12.5). Then, knock-out (*Capn3*$^{-/-}$) mice were shown to emulate a human LGMD2A/calpainopathy-like phenotype (although less severe), indicating that defects in *CAPN3*/*Capn3* and LGMD2A/calpainopathy have a cause-effect relationship (Kramerova et al. 2004; Richard et al. 2000). This is the first and only example so far of a clear cause-effect relationship between human disease and calpain gene mutations.

The following evidence showed that LGMD2A/calpainopathy is primarily caused by compromised CAPN3/p94 protease activity, rather than by damaged structural properties. First, it was shown that 10 pathogenic missense mutants of CAPN3/p94 all showed a deficient ability to proteolyze fodrin (Ono et al. 1998). Second, transgenic mice expressing a structurally intact but inactive CAPN3/p94: C129S protein showed an accumulation of the inactive protein and myopathic phenotypes similar to LGMD2A/calpainopathy (Tagawa et al. 2000). Finally, CAPN3/p94 knock-in (*Capn3*$^{CS/CS}$) mice, which express CAPN3/p94:C129S in place of wild-type CAPN3, showed a muscular dystrophy phenotype (Ojima et al. 2010). Intriguingly, however, the *Capn3*$^{CS/CS}$ mice showed a less severe phenotype than *Capn3*$^{-/-}$ mice did, indicating that the proteolytically inactive CAPN3/p94 retains some function. One of possible scenarios is that CAPN3/p94 is a novel structural components of the SR, considering its association with ryanodine receptors (Ojima et al. 2010, 2011). On the other hand, transgenic mice expressing a CAPN3/p94 splicing variant lacking exon 6 (CAPN3:ex1–5|7–24) showed muscle phenotypes that were more severe than those of the knock-out mice (Spencer et al. 2002). These observations indicate that autolytic unstability of CAPN3/p94 is related to a certain aspect of its regulatory mechanism.

Studies using *Capn3*$^{CS/CS}$ mice have provided some insights into the molecular mechanisms underlying the pathogenesis of LGMD2A/calpainopathy. First, CAPN3/p94 shows a stretch-dependent distribution. The amount of CAPN3/p94 at the M-line relative to that in the N2A region of myofibrils decreases as the sarcomere lengthens. This change in localization is delayed when CAPN3/p94 is inactive, as in *Capn3*$^{CS/CS}$ mouse muscle, and is likely to be one of the molecular mechanisms by which LGMD2A/calpainopathy develops. Second, *Capn3*$^{CS/CS}$ mice show an impaired ability to adapt to physical stress, accompanied by a compromised exercise-induced upregulation of MARP2 and HSPs. These findings suggest that the stretch-induced dynamic redistribution of p94, which is dependent on its protease activity, functions in the surveillance of myofibrillar conditions, and that this system is essential for protecting muscle tissue from degeneration, particularly under physical stress conditions (Ojima et al. 2011).

12.3.2 Gastrointestinal-Tract-Specific Calpains

12.3.2.1 Discovery and Nomenclature

In 1993, in the course of cDNA screening for tissue-specific calpains, a cDNA was obtained from rat stomach that encoded a novel molecule similar to but distinct from existing calpain catalytic large subunits, CAPN1/μCL, CAPN2/mCL, and CAPN3/p94 (Sorimachi et al. 1993a). The novel catalytic subunit had the same basic domain structure as these (Sorimachi et al. 2010, 2011a) (see below, Fig. 12.1a). As mentioned above, next to the first tissue-specific calpain, CAPN3/p94, which was called nCL-1 at that time, this stomach-specific calpain was designated nCL-2, and is now called CAPN8 (Sorimachi et al. 1993a). In 1997, Dear and his colleagues discovered the third novel calpain, CAPN5/hTRA-3, and called it nCL-3 in their database submission. However, nCL-3 was not used in the published report, and they named it Capn5 (Dear et al. 1997). Almost simultaneously, the fourth one was found, which was named nCL-4; it is now called CAPN9 (Lee et al. 1998).

Comparative examinations revealed that CAPN8/nCL-2 and CAPN9/nCL-4 show predominant expression in the surface mucus-secreting cells (pit cells) of the stomach, and that smaller amounts are expressed by the goblet cells in the intestines (Hata et al. 2006; Lee et al. 1998). Thus, these calpains are categorized as gastrointestinal-tract-specific calpains (Hata et al. 2010). Their physiological functions and mode of action were unclear until 2010, when they were shown to form a complex with each other in the stomach and to function in gastric mucosal protection against stress-induced ulcer (Hata et al. 2010). For CAPN8/nCL-2 and CAPN9/nCL-4 to be proteolytically active, they must form a complex; this complex is called G-calpain (G stands for gastric). After calpain-1 and -2, G-calpain was the third mammalian calpain enzyme complex shown to function *in vivo*, and the first to be composed of two different calpain catalytic subunits.

12.3.2.2 Structural Features

The basic domain structures of both CAPN8/nCL-2 and CAPN9/nCL-4 are identical to those of CAPN1/μCL and CAPN2/mCL, as shown in Fig. 12.1a. In cultured cells, CAPN9/nCL-4 is active only in the presence of CAPNS1/30K, and, therefore, is probably similar to, if not exactly the same as, those of the conventional calpains (CAPN1/μCL or CAPN2/mCL + CAPNS1/30K). CAPN8/nCL-2, however, is active without CAPNS1/30K *in vitro*, and shows homo-oligomerization *via* the CBSW domain (see below). Some C2-domain-like (β-sandwich) structures are known to form homo-oligomers (Jones et al. 1989; Lu et al. 2010), and the CAPN8/nCL-2 homo-oligomer may have a structure similar to one of these.

In contrast, these calpains show rather different characteristics *in vivo*. Endogenous CAPN8/nCL-2 and CAPN9/nCL-4 in mouse stomach form a hybrid complex,

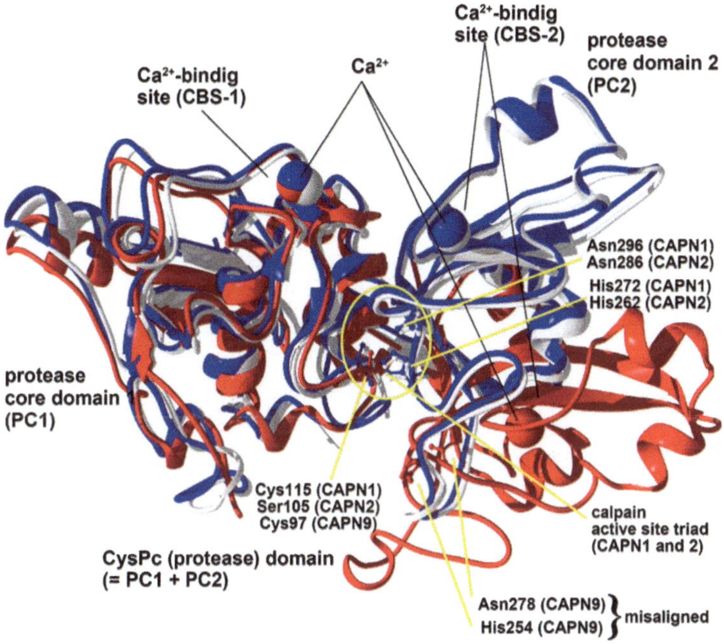

Fig. 12.6 Superimposed schematic 3D structures of CysPc domains of the active CAPN1/μCL, CAPN2/mCL, and CAPN9/nCL-4. Schematic 3D ribbon structures of the active (Ca^{2+}-bound) forms of CysPc domains of CAPN1/μCL (*blue*), CAPN2/mCL (*white*), and CAPN9/nCL-4 (*red*); their PC1 domains are superimposed by Deli server (PDB data: 2ARY, 1MDW, and 1ZIV) (Davis et al. 2007; Moldoveanu et al. 2003). The active sites are *circled* in *yellow*. PC1 subdomains of the three CysPc domains fit very well (Root-Mean-Square Deviation (RMSD) = 1~4 Å). CAPN1/μCL and CAPN2/mCL show overall fitting with very low RMSD (1.0 Å) including Ca^{2+}, but with one significant difference at Trp116 (CAPN1/μCL) and Trp106 (CAPN2/mCL). On the other hand, the PC2 domains of CAPN1/μCL and CAPN9/nCL-4 are markedly displaced. Note that, in the CAPN9/nCL-4 structure, His254 and Asn278 are too far away from Cys97 to form an active triad

G-calpain, and neither CAPN8/nCL-2 nor CAPN9/nCL-4 forms a stable complex with CAPNS1/30K (Hata et al. 2010). Gel filtration analysis of the endogenous mouse stomach G-calpain showed a molecular mass of about 180,000, indicating that G-calpain may contain other small subunit(s) of up to 20,000 (Hata et al. 2010).

The 3D structural analysis of the active (Ca^{2+}-bound) protease domains of CAPN2/mCL and CAPN9/nCL-4 as well as CAPN1/μCL revealed locally distinct structures. Compared with the structure of CAPN1/μCL, that of CAPN2/mCL has a reversible intrinsic silencing mechanism by Trp106 (Moldoveanu et al. 2002, 2003) (Fig. 12.6). In contrast, the structure of CAPN9/nCL-4 shows that its activity can be auto-inhibited by misalignment of the catalytic triad, mediated by large intra-domain movements (Davis et al. 2007) (Fig. 12.6). These findings indicate that, in addition to the common activation mechanism, each calpain family member evolved molecule-specific one, even though they share high levels of sequence conservation.

12.3.2.3 Activity and Substrates

Of the human calpains, CAPN8/nCL-2 shows the highest similarity (~60 % identity in aa sequence) to CAPN2/mCL. However, unlike calpain-2 (CAPN2/mCL + CAPNS1/30K), recombinant CAPN8/nCL-2 expressed in *Escherichia coli* exhibits Ca^{2+}-dependent activity without CAPNS1/30K, and forms a homodimer~oligomer *via* its CBSW domain *in vitro* (Hata et al. 2007). Recombinant mouse CAPN8/nCL-2 shows half-maximal activity at approximately 0.3 mM, which is similar to calpain-2 (Hata et al. 2007). Its autolytic activity in the presence of Ca^{2+} is very rapid, with more than 80 % of the protein being cleaved within 30 s (Hata et al. 2007). Several autolytic sites have been identified in the N-terminal anchor helix region, and in the CBSW and PEF domains (Hata et al. 2007). Although most of these features are unique to CAPN8/nCL-2, their physiological relevance remains unclear. On the other hand, recombinant human CAPN9/nCL-4 requires CAPNS1/30K for its activity *in vitro*, and the activity of this complex is Ca^{2+}-dependent, with half-maximal activity at approximately 0.1 mM Ca^{2+} under the optimal condition (Lee et al. 1999). The activity of both CAPN9/nCL-4 and CAPN8/nCL-2 is inhibited by calpastatin and other cysteine protease inhibitors, similar to the conventional calpains (Hata et al. 2007; Lee et al. 1999).

The β-subunit (β-COP) of the coatomer complex, which is involved in retrograde membrane trafficking from the Golgi to the endoplasmic reticulum, is expressed in the stomach pit cells, and was shown to be proteolyzed by CAPN8/nCL-2 in vitro (Hata et al. 2006). Furthermore, β-COP and CAPN8/nCL-2 co-expressed in COS7 cells co-localize to the Golgi, and Ca^{2+}-ionophore stimulation causes limited proteolysis of β-COP near the linker region after Ser528 (as in NP_057535). This cleavage probably results in the dissociation of β-COP from the Golgi, strongly suggesting that CAPN8/nCL-2 is involved in the membrane trafficking of mucus cells via interactions with coat protein.

12.3.2.4 Biological Significance

Xenopus laevis has a CAPN8/nCL-2 orthologue named xCL-2, which causes severe developmental defects when disrupted (Cao et al. 2001). CAPN9/nCL-4 was reported to be down-regulated in gastric cancer (Yoshikawa et al. 2000), and is involved in the formation of a lumen by breast epithelial cells induced by carcinoembryonic antigen-related cell adhesion molecule 1 (Chen et al. 2010).

In mouse stomach *in vivo*, the disruption of either mouse *Capn8* or *Capn9* causes a down-regulation of the other gene product, similar to the effect of *Capns1*$^{-/-}$ on the levels of CAPN1/μCL and CAPN2/mCL. This suggests a co-dependency in terms of the stability and functionality of both gene products. This point is supported by the finding that the residual CAPN9/nCL-4 in *Capn8*$^{-/-}$ mice does not display autolytic activity, and *vice versa* (Hata et al. 2010). However, *Capn8*$^{-/-}$ and *Capn9*$^{-/-}$ mice appear healthy under normal conditions. Nevertheless, they are significantly more susceptible to ethanol-induced gastric ulcers (Hata et al. 2010). CAPN8/nCL-2 knock-in (*Capn8*$^{CS/CS}$) mice expressing a protease-inactive CAPN8/nCL-2:C105S mutant also show stress-induced gastropathy, indicating that CAPN8/nCL-2 and CAPN9/nCL-4 (in the form of G-calpain) mediate key components of gastric mucosal defense (Hata et al. 2010).

Gastric mucosal defense is a complex process involving mucus secretion and the migration of the pit cells that differentiate from stem-cell progenitors. Considering that β-COP is a substrate for G-calpain, it might be involved in the regulation of mucus secretion. On the other hand, the expression of CAPN8/nCL-2 in cranial neural crest cells, where it is involved in cell motility, is regulated by ADAM13, a transmembrane metalloprotease containing a disintegrin domain (Cousin et al. 2011), suggesting that G-calpain is involved in pit cell migration.

In another study on the physiological functions of G-calpain, a single nucleotide polymorphism (SNP) database search showed that human *CAPN8* and *CAPN9* contain several SNPs that result in aa substitutions (Hata et al. 2010). *In vitro* expression experiments showed that the G-calpain variants resulting from these SNPs have compromised proteolytic activity, suggesting that individuals expressing certain *CAPN8* and *CAPN9* variants may suffer from gastrointestinal dysfunction.

12.3.3 Other Calpains

12.3.3.1 PalB Homologs

PalB was first identified in *Emericella* (*Aspergillus*) *nidulans* as a product of the gene responsible for the fungus's adaptation to alkaline conditions (Denison et al. 1995). Subsequently, its orthologs were identified in *S. cerevisiae* (Rim13/Cpl1) and humans (CAPN7/PalBH) (Futai et al. 1999; Sorimachi et al. 2011a). *E. nidulans*, like many other microorganisms, grows over a wide pH range. The *palB* gene locus was identified as one of the *pal* genes that show defective alkaline adaptation when disrupted. PalB is involved in the proteolytic activation of the PacC transcription factor, a key regulator of pH-dependent gene expression. Here, we refer to this pH adaptation system as the Pal-PacC pathway.

The Pal-PacC pathway contains *palA*, *palB*, *palC*, *palF*, *palH*, *palI*, and *pacC*. PalF is a distant homolog of mammalian arrestin and binds to the large cytosolic domain of PalH. The phosphorylation and ubiquitination of PalF under alkaline conditions depend on PalH and PalI, suggesting that pH-sensing is mediated *via* the interaction between PalF and PalH (similar to that observed for the arrestin receptor, which regulates various processes in mammals, such as photo-sensing). PacC usually adopts an inactive conformation during intramolecular interactions, and is activated by proteolytic removal of its C-terminus. PalB is primarily responsible for the processing of PacC in response to increased pH, which is followed by further proteolysis by the proteasome to yield the active form.

The Pal-PacC pathway is well conserved in yeasts. The yeast calpain, Rim13/Cpl1, was identified as a product of the *RIM* genes. Like PacC and PalB, Rim101 is proteolytically regulated by Rim13/Cpl1 (Futai et al. 1999), although in this case the proteasome is not involved. Under normal conditions, the environment for *Saccharomycetes* is rather acidic and this adaptive system (the Rim pathway) functions in near-neutral pH conditions. One example of a biological event in which the Rim pathway plays a role is the infection and invasion of mammalian skin by pathogenic and saprophytic yeasts such as *Candida albicans* and *Yarrowia lipolytica* at neutral pH. Disrupting the Rim pathway compromises the pathogenicity of these organisms (Mitchell et al. 2007).

The Pal-PacC and Rim pathways relate to membrane trafficking and involve the ESCRT (endosomal sorting complex required for transport) and Vps (vacuolar protein sorting) proteins. The Pal-PacC and Rim pathways are the first examples of genetic studies that are being used to thoroughly elucidate the molecular components and mechanisms underlying calpain-mediated systems (Hayashi et al. 2005; Rodriguez-Galan et al. 2009). These pathways also exemplify the fact that calpains function as modulator proteases, affecting the function of their substrates through limited proteolysis. Theoretical extension of the knowledge of these pathways is anticipated to provide important keys to understanding other calpain systems, especially those that involve CAPN7/PalBH (Osako et al. 2010).

12.3.3.2 The TRA-3 Homologs

TRA-3/CLP-5 was first identified as the product of one of the genes involved in the sex determination cascade of *C. elegans*. The Ca^{2+}-dependent protease activity of TRA-3/CLP-5 is necessary for the processing of TRA-2A, which is required for female development in XX hermaphrodites. On another front, both TRA-3/CLP-5 and CLP-1 are components of a neuronal necrotic death cascade, which acts upstream of the aspartic proteases, ASP-3 and ASP-4, orthologs for mammalian cathepsins D and E (Syntichaki et al. 2002). In addition, an SNP in *tra-3* is reported to be involved in nematode body-size determination (Kammenga et al. 2007).

Mammals have two TRA-3 orthologs, CAPN5/hTRA-3 and CAPN6 (Sorimachi et al. 2011a). CAPN5/hTRA-3 has Ca^{2+}-dependent autolytic activity and is sensitive to several calpain inhibitors (Waghray et al. 2004). *CAPN5* is expressed at varying levels in almost all tissues. Analysis of *Capn5*$^{-/-}$ mice shows that CAPN5/hTRA-3 is expressed by a subset of T cells, but is not required for development. SNPs within CAPN5/hTRA-3 are associated with polycystic ovary syndrome, diastolic blood pressure, and cholesterol levels (Saez et al. 2007), and expression of both *Capn5* and *Capn2* is upregulated in caerulein-induced acute pancreatitis in mice (Nakada et al. 2010).

CAPN6 proteins expressed in eutherians (placental mammals) and schistosomes have a naturally occurring aa substitution at the most important residue in the active site triad (Cys->Lys in humans); strongly suggesting that these CAPN6 proteins have no proteolytic activity. Interestingly, CAPN6 proteins expressed in marsupialia and birds retain all the active-site residues. Moreover, frogs and fish express three TRA-3 homologs, all of which retain the active-site residues (Sorimachi et al. 2011a). Mammalian CAPN6 is predominantly expressed in embryonic muscles, placenta, and in several cultured cell lines. CAPN6 is involved in regulating microtubule dynamics (Tonami et al. 2007) and motility (Tonami et al. 2011) *in cellulo*, although the *in vivo* physiological functions of CAPN6 are still unclear (Tonami et al. 2013).

12.3.3.3 The CAPN10 Homologs

A large-scale genetic association study identified an SNP within intron 3 of *CAPN10* that is linked to susceptibility to non-insulin-dependent diabetes mellitus (NIDDM, or type 2 diabetes) (Horikawa et al. 2000), although there is no clear molecular explanation for the reason. *Capn10* is also a candidate gene responsible for the NIDDM phenotype in the Otsuka Long-Evans Tokushima Fatty (OLETF) rat. Quantitative trait locus (QTL) analyses using *Capn10*$^{-/-}$ mice and two other strains with low and high obesity phenotypes (LG/J and SM/J) show that *Capn10* is a component of the obesity QTL, *Adip1*. This indicates that CAPN10 is involved in obesity in mice (Cheverud et al. 2010).

Studies using Capn10−/− or CAPN10-overexpressing mice suggest that CAPN10 is involved in type 2 ryanodine receptor-mediated apoptosis (Johnson et al. 2004). Although phenotype of Capn10−/− mice was not described, they are not embryonic lethal. CAPN10 is ubiquitously distributed and its cellular localization is dynamic. When localized in the mitochondria, it mediates mitochondrial dysfunction by cleaving Complex I subunits and promotes mitochondrial permeability transition (Arrington et al. 2006). CAPN10 is also involved in GLUT4 vesicle translocation during insulin-stimulated glucose uptake in adipocytes (Paul et al. 2003).

12.3.3.4 The SOL Homologs

The first genetic calpain study identified the *Drosophila* gene, *small optic lobes* (*sol*), in 1991 (Delaney et al. 1991). A mutation in *sol* results in the absence of certain classes of columnar neurons from the optic lobes, the differentiation and/or survival of which is dependent on the protease activity of SOL. SOL comprises an N-terminal six Zn-finger motifs and C-terminal CysPc domain (Delaney et al. 1991; Sorimachi et al. 2011a). Unfortunately, there are no other reports analyzing the molecular mechanisms involved in this process. Mammals express a single SOL ortholog, CAPN15/SOLH, but its physiological role remains unclear. In parallel with the PalB subfamily, the SOL subfamily (including their mammalian homologs) is a group of evolutionarily interesting molecules that warrant further study.

12.3.3.5 Phytocalpains

A plant calpain, phytocalpain, was first identified from the sugarcane expressed sequence tag database (Correa et al. 2001). Subsequently, many other phytocalpains were identified from dicotyledons, monocotyledons, and gymnospermae. A genetic study identified phytocalpain as the *defective kernel 1* (*dek1*) gene product, DEK1, which is required for aleurone cell development in the maize endosperm (Lid et al. 2002).

DEK1 contains a potential signal peptide sequence followed by possible three 7-trans-membrane (TM) regions, and CysPc and CBSW domains (Lid et al. 2002; Sorimachi et al. 2011a). Although the CysPc domain of DEK1 is divergent from mammalian calpains, the C-terminal CBSW domain is significantly similar to that of classical calpains such as CAPN1/μCL, suggesting that DEK1 consists of evolutionarily-different modules. Recombinant maize DEK1 CysPc-CBSW fragment show Ca^{2+}-activated caseinolytic activity, depending on the predicted active site Cys residue. DEK1 is the only calpain homolog in the *Arabidopsis* genome, and it is important for regulating growth in this plant. Although the full-length DEK1 protein localizes to membranes, intramolecular autolytic cleavage releases the CysPc-CBSW domains into the cytoplasm. These domains are sufficient to fully complement *dek1* mutants (Tian et al. 2007).

12.3.3.6 Other Calpain Members

Leishmania and *Trypanosoma* each expresses around 20 calpain homologs, which likely contribute to cell morphogenesis, drug resistance, and stress-response mechanisms (Olego-Fernandez et al. 2009; Sangenito et al. 2009). Some trypanosome calpains have N-terminal domains with weak similarity to calpastatin. As described above, some calpain homologs show substitutions in one or more of the well-conserved active-site triad residues. This non-proteolytic family of calpain homologs includes eutherian and schistosome CAPN6, several of the schistosome and nematode calpains, insect CALPC, and all the *Trypanosoma* homologs (Croall and Ersfeld 2007; Sorimachi et al. 2011a). The evolutionary background for the occurrence of these calpain species is quite interesting, and elucidating their physiological functions will expand our knowledge regarding the functions of the calpain superfamily, *e.g.*, possible non-proteolytic functions (Ojima et al. 2011).

12.4 Concluding Remarks

Here we reviewed pathological and biological situations where involvement of calpains has gained great interest from multiple angles. For example, conventional calpains (calpain-1 and -2) are involved in neurodegeneration, as well as producing several well documented proteolytic biomarkers that can help track brain injury progress, severity and as a theranostic tool for drug development. In addition, other calpains such as tissue-specific calpains (CAPN3 in skeletal-muscle, and CAPN8 and CAPN9 in gastrointestinal tract) and calpain homologs originally identified in non-mammals represent various physiological, cellular and biochemical pathways.

Although many studies suggest that activation of calpain, especially the conventional ones, is involved in pathophysiological processes as an exacerbating factor, it is absolutely false to recognize calpain as an existence for making diseases worse. Genetic manipulations causing constitutive and conditional disruption of *Capn2* in mice result in embryonic lethality and cardiac hypertrophy, respectively. In other words, calpain could be a two-edged sword in that its activity is indispensable for certain aspects of life, yet, its regulation gets somehow fragile within the cells undergoing pathogenic chaos.

Therefore, in addition to investing our efforts in developing novel theranostic approaches aimed at the conventional calpains as useful promising targets for biomarkers and inhibitory drugs, further application fields should be simultaneously pursued with the view that some pathological contexts may be related to insufficient conventional calpain activity. In these cases, appropriate calpain activators, rather than inhibitors, will be candidate drug targets to be developed. In fact, genetic inactivation of tissue-specific calpains causes disease states, indicating that activators/stabilizers for these calpains, if developed, may be a drug for these tissue-specific diseases such as muscular dystrophy and gastropathy.

To date, only the conventional calpains' over-activation has been associated with diseases. It is, however, equally possible that an inappropriate activation of other calpains including tissue-specific ones is laterally or centrally linked to disease aggravation. In any case, calpain protease research is a rich area for both basic research as well as implicational research in biomedical science. The open question is how to extract and profile its activity with a resolution that is consistently competent for both normal and disease conditions.

References

Anagli J, Hagmann J, Shaw E (1991) Investigation of the role of calpain as a stimulus–response mediator in human platelets using new synthetic inhibitors. Biochem J 274(Pt 2):497–502

Anagli J, Han Y, Stewart L, Yang D, Movsisyan A, Abounit K, Seyfried D (2009) A novel calpastatin-based inhibitor improves postischemic neurological recovery. Biochem Biophys Res Commun 385(1):94–99

Angliker H, Anagli J, Shaw E (1992) Inactivation of calpain by peptidyl fluoromethylketones. J Med Chem 35(2):216–220

Aoyagi T, Umezawa H (1975) In: Reich E, Rifkin DB, Shaw E (eds) Proteases and biological control. Cold Spring Harbor Laboratory, Cold Spring Harbor, pp 429–454

Arrington DD, Van Vleet TR, Schnellmann RG (2006) Calpain 10: a mitochondrial calpain and its role in calcium-induced mitochondrial dysfunction. Am J Physiol Cell Physiol 291: C1159–C1171

Bading H, Ginty DD, Greenberg ME (1993) Regulation of gene expression in hippocampal neurons by distinct calcium signaling pathways. Science 260:181–186

Baghdiguian S, Martin M, Richard I, Pons F, Astier C, Bourg N, Hay RT, Chemaly R, Halaby G, Loiselet J, Anderson LV, Lopez de Munain A, Fardeau M, Mangeat P, Beckmann JS, Lefranc G (1999) Calpain 3 deficiency is associated with myonuclear apoptosis and profound perturbation of the IkappaB alpha/NF-kappaB pathway in limb-girdle muscular dystrophy type 2A. Nat Med 5:503–511

Banik NL, Matzelle DC, Gantt-Wilford G, Osborne A, Hogan EL (1997) Increased calpain content and progressive degradation of neurofilament protein in spinal cord injury. Brain Res 752:301–306

Benayoun B, Baghdiguian S, Lajmanovich A, Bartoli M, Daniele N, Gicquel E, Bourg N, Raynaud F, Pasquier MA, Suel L, Lochmuller H, Lefranc G, Richard I (2008) NF-kappaB-dependent expression of the antiapoptotic factor c-FLIP is regulated by calpain 3, the protein involved in limb-girdle muscular dystrophy type 2A. FASEB J 22:1521–1529

Benetti R, Del Sal G, Monte M, Paroni G, Brancolini C, Schneider C (2001) The death substrate Gas2 binds m-calpain and increases susceptibility to p53-dependent apoptosis. EMBO J 20:2702–2714

Benjamins JA, Nedelkoska L, George EB (2003) Protection of mature oligodendrocytes by inhibitors of caspases and calpains. Neurochem Res 28:143–152

Berridge MJ (1993) Inositol trisphosphate and calcium signalling. Nature 361:315–325

Bertipaglia I, Carafoli E (2007) Calpains and human disease. Subcell Biochem 45:29–53

Betts R, Anagli J (2004) The β- and γ-CH_2 of B27-WT's Leu^{11} and Ile^{18} side chains play a direct role in calpain inhibition. Biochemistry 43(9):2596–2604

Betts R, Weinsheimer S, Blouse GE, Anagli J (2003) Structural determinants of the calpain inhibitory activity of calpastatin peptide B27-WT. J Biol Chem 278(10):7800–7809

Bi R, Bi XN, Baudry M (1998) Phosphorylation regulates calpain-regulated truncation of glutamate ionotropic receptors. Brain Res 797:154–158

Bizat N, Hermel JM, Boyer F, Jacquard C, Creminon C, Ouary S, Escartin C, Hantraye P, Kajewski S, Brouillet E (2003) Calpain is a major cell death effector in selective striatal degeneration induced in vivo by 3-nitropropionate: implications for Huntington's disease. J Neurosci 23:5020–5030

Bizat N, Galas MC, Jacquard C, Boyer F, Hermel JM, Schiffmann SN, Hantraye P, Blum D, Brouillet E (2005) Neuroprotective effect of zVAD against the neurotoxin 3-nitropropionic acid involves inhibition of calpain. Neuropharmacology 49:695–702

Bozoky Z, Alexa A, Tompa P, Friedrich P (2005) Multiple interactions of the 'transducer' govern its function in calpain activation by Ca^{2+}. Biochem J 388(Pt 3):741–744

Buki A, Siman R, Trojanowski JQ, Povlishock JT (1999) The role of calpain-mediated spectrin proteolysis in traumatically induced axonal injury. J Neuropathol Exp Neurol 58(4):365–375

Buki A, Okonkwo DO, Wang KK, Povlishock JT (2000) Cytochrome c release and caspase activation in traumatic axonal injury. J Neurosci 20:2825–2834

Buki A, Farkas O, Doczi T, Povlishock JT (2003) Preinjury administration of the calpain inhibitor MDL-28170 attenuates traumatically induced axonal injury. J Neurotrauma 20(3):261–268

Bullock R, Fujisawa H (1992) The role of glutamate antagonists for the treatment of CNS injury. J Neurotrauma 9(Suppl 2):S443–S462

Busch WA, Stromer MH, Goll DE, Suzuki A (1972) Ca^{2+}-specific removal of Z lines from rabbit skeletal muscle. J Cell Biol 52:367–381

Cao Y, Zhao H, Grunz H (2001) xCL-2 is a novel m-type calpain and disrupts morphogenetic movements during embryogenesis in *Xenopus laevis*. Dev Growth Differ 43:563–571

Chen M, Fernandez HL (2005) Mu-calpain is functionally required for alpha-processing of Alzheimer's beta-amyloid precursor protein. Biochem Biophys Res Commun 330:714–721

Chen J, Nagayama T, Jin K, Stetler RA, Zhu RL, Graham SH, Simon RP (1998) Induction of caspase-3-like protease may mediate delayed neuronal death in the hippocampus after transient cerebral ischemia. J Neurosci 18:4914–4928

Chen CJ, Nguyen T, Shively JE (2010) Role of calpain-9 and PKC-delta in the apoptotic mechanism of lumen formation in CEACAM1 transfected breast epithelial cells. Exp Cell Res 316:638–648

Chera B, Schaecher KE, Rocchini A, Imam SZ, Ray SK, Ali SF, Banik NL (2002) Calpain upregulation and neuron death in spinal cord of MPTP-induced parkinsonism in mice. Ann NY Acad Sci 965:274–280

Cheverud JM, Fawcett GL, Jarvis JP, Norgard EA, Pavlicev M, Pletscher LS, Polonsky KS, Ye H, Bell GI, Semenkovich CF (2010) Calpain-10 is a component of the obesity-related quantitative trait locus Adip1. J Lipid Res 51:907–913

Correa GC, Margis-Pinheiro M, Margis R (2001) Identification, classification and expression pattern analysis of sugarcane cysteine proteinases. Genet Mol Biol 24:275–283

Cousin H, Abbruzzese G, Kerdavid E, Gaultier A, Alfandari D (2011) Translocation of the cytoplasmic domain of ADAM13 to the nucleus is essential for Calpain8-a expression and cranial neural crest cell migration. Dev Cell 20:256–263

Crawford C (1987) Inhibition of chicken calpain II by proteins of the cystatin superfamily and alpha 2-macroglobulin. Biochem J 248(5):89–594

Crawford C, Mason RW, Wikstrom P, Shaw E (1988) The design of peptidyldiazomethane inhibitors to distinguish between the cysteine proteinases calpain II, cathepsin L and cathepsin B. Biochem J 253(3):751–758

Croall DE, DeMartino GN (1991) Calcium-activated neutral protease (calpain) system: structure, function, and regulation. Physiol Rev 71(3):813–847

Croall DE, Ersfeld K (2007) The calpains: modular designs and functional diversity. Genome Biol 8:218

Crocker SJ, Smith PD, Jackson-Lewis V, Lamba WR, Hayley SP, Grimm E, Callaghan SM, Slack RS, Melloni E, Przedborski S, Robertson GS, Anisman H, Merali Z, Park DS (2003) Inhibition of calpains prevents neuronal and behavioral deficits in an MPTP mouse model of Parkinson's disease. J Neurosci 23(10):4081–4091

Danial NN, Korsmeyer SJ (2004) Cell death: critical control points. Cell 116:205–219

Das A, Guyton MK, Butler JT, Ray SK, Banik NL (2008) Activation of calpain and caspase pathways in demyelination and neurodegeneration in animal model of multiple sclerosis. CNS Neurol Disord Drug Targets 7(3):313–320, Review

Dash PK, Moore AN, Dixon CE (1995) Spatial memory deficits, increased phosphorylation of the transcription factor CREB, and induction of the AP-1 complex following experimental brain injury. J Neurosci 15:2030–2039

Davis TL, Walker JR, Finerty PJ Jr, Mackenzie F, Newman EM, Dhe-Paganon S (2007) The crystal structures of human calpains 1 and 9 imply diverse mechanisms of action and auto-inhibition. J Mol Biol 366:216–229

Dayton WR, Goll DE, Zeece MG, Robson RM, Reville WJ (1976) A Ca^{2+}-activated protease possibly involved in myofribrillar protein turnover. Purification from porcine muscle. Biochemistry 15:2150–2158

Dear N, Matena K, Vingron M, Boehm T (1997) A new subfamily of vertebrate calpains lacking a calmodulin-like domain: implications for calpain regulation and evolution. Genomics 45:175–184

Delaney SJ, Hayward DC, Barleben F, Fischbach KF, Miklos GL (1991) Molecular cloning and analysis of small optic lobes, a structural brain gene of *Drosophila melanogaster*. Proc Natl Acad Sci U S A 88:7214–7218

Denison SH, Orejas M, Arst HN Jr (1995) Signaling of ambient pH in Aspergillus involves a cysteine protease. J Biol Chem 270:28519–28522

Donkor IO (2011) Calpain inhibitors: a survey of compounds reported in the patent and scientific literature. Expert Opin Ther Pat 21(5):601–636, Review

Emori Y, Kawasaki H, Imajoh S, Minami Y, Suzuki K (1988) All four repeating domains of the endogenous inhibitor of calcium-dependent protease independently retain inhibitory activity. Expression of the cDNA fragments in Escherichia coli. J Biol Chem 263(5):2364–2370

Fifre A, Spoone I, Koziel V, Kriem B, Yen-Potin FT, Bihain BB, Oliver JL, Oster T, Pillot T (2006) Microtubule-associated proteins MAP1A, MAP1B and MAP2 proteolysis during soluble amyloid-beta peptide induced neuronal apoptosis: Synergistic involvement of calpain and caspase-3. J Biol Chem 281(1):229–240

Fineman I, Hovda DA, Smith M, Yoshino A, Becker DP (1993) Concussive brain injury is associated with a prolonged accumulation of calcium: a 45Ca autoradiographic study. Brain Res 624:94–102

Fiorino F, Gil-Parrado S, Assfalg-Machleidt I, Machleidt W, Moroder L (2007) A new cell-permeable calpain inhibitor. J Pept Sci 13(1):70–73

Fougerousse F, Anderson LV, Delezoide AL, Suel L, Durand M, Beckmann JS (2000) Calpain3 expression during human cardiogenesis. Neuromuscul Disord 10:251–256

Fukiage C, Azuma M, Nakamura Y, Tamada Y, Nakamura M, Shearer TR (1997) SJA6017, a newly synthesized peptide aldehyde inhibitor of calpain: amelioration of cataract in cultured rat lenses. Biochim Biophys Acta 1361:304–312

Fukiage C, Nakajima E, Ma H, Azuma M, Shearer TR (2002) Characterization and regulation of lens-specific calpain Lp82. J Biol Chem 277:20678–20685

Futai E, Maeda T, Sorimachi H, Kitamoto K, Ishiura S, Suzuki K (1999) The protease activity of a calpain-like cysteine protease in Saccharomyces cerevisiae is required for alkaline adaptation and sporulation. Mol Gen Genet 260:559–568

Gafni J, Ellerby LM (2002) Calpain activation in Huntington's disease. J Neurosci 22:4842–4849

Gafni J, Hermel E, Young JE, Wellington CL, Hayden MR, Ellerby LM (2004) Inhibition of calpain cleavage of huntingtin reduces toxicity: accumulation of calpain/caspase fragments in the nucleus. J Biol Chem 279:20211–20220

Germano A, Costa C, DeFord SM, Angileri FF, Arcadi F, Pike BR, Bramanti P, Bausano B, Zhao X, Day AL, Anderson DK, Hayes RL (2002) Systemic administration of a calpain inhibitor reduces behavioral deficits and blood–brain barrier permeability changes after experimental subarachnoid hemorrhage in the rat. J Neurotrauma 19:887–896

Goll DE, Thompson VF, Li H, Wei W, Cong J (2003) The calpain system. Physiol Rev 83 (3):731–801

Gomez-Vicente V, Donovan M, Cotter TG (2005) Multiple death pathways in retina-derived 661W cells following growth factor deprivation: cross-talk between caspases and calpains. Cell Death Differ 12:796–804

Guroff G (1964) A neutral calcium-activated proteinase from the soluble fraction of rat brain. J Biol Chem 239:149–155

Hall ED (1998) Antioxidant pharmacotherapy. In: Ginsberg MD, Bogousslavsky J (eds) Cerebrovascular disease: pathophysiology, diagnosis, and management. Blackwell Science, Malden, MA, pp 710–720

Hanna RA, Campbell RL, Davies PL (2008) Calcium-bound structure of calpain and its mechanism of inhibition by calpastatin. Nature 456:409–412

Hata S, Koyama S, Kawahara H, Doi N, Maeda T, Toyama-Sorimachi N, Abe K, Suzuki K, Sorimachi H (2006) Stomach-specific calpain, nCL-2, localizes in mucus cells and proteolyzes the β-subunit of coatomer complex, β-COP. J Biol Chem 281:11214–11224

Hata S, Doi N, Kitamura F, Sorimachi H (2007) Stomach-specific calpain, nCL-2/calpain 8, is active without calpain regulatory subunit and oligomerizes through C2-like domains. J Biol Chem 282:27847–27856

Hata S, Abe M, Suzuki H, Kitamura F, Toyama-Sorimachi N, Abe K, Sakimura K, Sorimachi H (2010) Calpain 8/nCL-2 and calpain 9/nCL-4 constitute an active protease complex, G-calpain, involved in gastric mucosal defense. PLoS Genet 6:e1001040

Hathaway DR, Werth DR, Haeberle JR (1982) Limited proteolysis reduces the Ca^{2+} requirement of a smooth muscle Ca^{2+}-activated protease. J Biol Chem 257:9072–9077

Hayashi M, Fukuzawa T, Sorimachi H, Maeda T (2005) Constitutive activation of the pH-responsive Rim101 pathway in yeast mutants defective in late steps of the MVB/ESCRT pathway. Mol Cell Biol 25:9478–9490

Hayashi C, Ono Y, Doi N, Kitamura F, Tagami M, Mineki R, Arai T, Taguchi H, Yanagida M, Hirner S, Labeit D, Labeit S, Sorimachi H (2008) Multiple molecular interactions implicate the connectin/titin N2A region as a modulating scaffold for p94/calpain 3 activity in skeletal muscle. J Biol Chem 283:14801–14814

Hayes RL, Jenkins LW, Lyeth BG (1992) Neurotransmitter-mediated mechanisms of traumatic brain injury: acetylcholine and excitatory amino acids. J Neurotrauma 9(Suppl 1):S173–S187

Herasse M, Ono Y, Fougerousse F, Kimura E, Stockholm D, Beley C, Montarras D, Pinset C, Sorimachi H, Suzuki K, Beckmann JS, Richard I (1999) Expression and functional characteristics of calpain 3 isoforms generated through tissue-specific transcriptional and posttranscriptional events. Mol Cell Biol 19:4047–4055

Higuchi M, Iwata N, Saido TC (2005) Understanding molecular mechanisms of proteolysis in Alzheimer's disease: progress toward therapeutic interventions. Biochim Biophys Acta 1751:60–67

Hong SC, Lanzino G, Goto Y, Kang SK, Schottler F, Kassell NF, Lee KS (1994) Calcium-activated proteolysis in rat neocortex induced by transient focal ischemia. Brain Res 661:43–50

Horikawa Y, Oda N, Cox NJ, Li X, Orho-Melander M, Hara M, Hinokio Y, Lindner TH, Mashima H, Schwarz PE, del Bosque-Plata L, Horikawa Y, Oda Y, Yoshiuchi I, Colilla S, Polonsky KS, Wei S, Concannon P, Iwasaki N, Schulze J, Baier LJ, Bogardus C, Groop L, Boerwinkle E, Hanis CL, Bell GI (2000) Genetic variation in the gene encoding calpain-10 is associated with type 2 diabetes mellitus. Nat Genet 26:163–175

Hosfield CM, Elce JS, Davies PL, Jia Z (1999) Crystal structure of calpain reveals the structural basis for Ca^{2+}-dependent protease activity and a novel mode of enzyme activation. EMBO J 18:6880–6889

Hosfield CM, Elce JS, Jia Z (2004) Activation of calpain by Ca^{2+}: roles of the large subunit N-terminal and domain III-IV linker peptides. J Mol Biol 343(4):1049–1053

Hossman K (1988) In: Kreiglstein J (ed) Pharmacology of cerebral ischemia. CRC, Boco Raton, FL, pp 53–63

Huang C, Jacobson K, Schaller MD (2004) MAP kinases and cell migration. J Cell Sci 117:4619–4628

Huston RB, Krebs EG (1968) Activation of skeletal muscle phosphorylase kinase by Ca^{2+}. II. Identification of the kinase activating factor as a proteolytic enzyme. Biochemistry 7:2116–2122

Imajoh S, Kawasaki H, Suzuki K (1986) Limited autolysis of calcium-activated neutral protease (CANP): reduction of the Ca^{2+}-requirement is due to the NH_2-terminal processing of the large subunit. J Biochem 100:633–642

Ishuira S, Murofushi H, Suzuki K, Imahori K (1978) Studies of a calcium-activated neutral protease from chicken skeletal muscle. I. Purification and characterization. J Biochem 84:225–230

Johnson JD, Han Z, Otani K, Ye H, Zhang Y, Wu H, Horikawa Y, Misler S, Bell GI, Polonsky KS (2004) RyR2 and calpain-10 delineate a novel apoptosis pathway in pancreatic islets. J Biol Chem 279:24794–24802

Jones EY, Stuart DI, Walker NP (1989) Structure of tumour necrosis factor. Nature 338:225–228

Kambe A, Yokota M, Saido TC, Satokata I, Fujikawa H, Tabuchi S, Kamitani H, Watanabe T (2005) Spatial resolution of calpain-catalyzed proteolysis in focal cerebral ischemia Brain Res 1040(1–2):36–43

Kammenga JE, Doroszuk A, Riksen JA, Hazendonk E, Spiridon L, Petrescu AJ, Tijsterman M, Plasterk RH, Bakker J (2007) A *Caenorhabditis elegans* wild type defies the temperature-size rule owing to a single nucleotide polymorphism in *tra-3*. PLoS Genet 3:e34

Kampfl A, Postmantur RM, Zhao X, Schmutzhard E, Clifton GL, Hayes RL (1997) Mechanisms of calpain proteolysis following traumatic brain injury: implications for pathology and therapy: a review and update. J Neurotrauma 14(3):121–134

Kar NC, Pearson CM (1976) A calcium-activated neutral protease in normal and dystrophic human muscle. Clin Chim Acta 73:293–297

Kawamura M, Nakajima W, Ishida A, Ohmura A, Miura S, Takada G (2005) Calpain inhibitor MDL 28170 protects hypoxic-ischemic brain injury in neonatal rats by inhibition of both apoptosis and necrosis. Brain Res 1037:59–69

Kim SJ, Sung JY, Um JW, Hattori N, Mizuno Y, Tanaka K, Paik SR, Kim J, Chung KC (2003) Parkin cleaves intracellular alpha-synuclein inclusions via the activation of calpain. J Biol Chem 278(43):41890–41899

Kishimoto A, Kajikawa N, Shiota M, Nishizuka Y (1983) Proteolytic activation of calcium-activated, phospholipid-dependent protein kinase by calcium-dependent neutral protease. J Biol Chem 258:1156–1164

Kobeissy FH, Ottens AK, Zhang ZQ, Dave JR, Tortella FC, Hayes RL, Wang KKW (2006) Differential proteomic analysis of traumatic brain injury biomarker study using CAX-PAGE/RPLC-MSMS method. Mol Cell Proteomics 5:1887–1898

Kontos HA (1989) Oxygen radicals in CNS damage. Chem Biol Interact 72:229–255

Kramerova I, Kudryashova E, Tidball JG, Spencer MJ (2004) Null mutation of calpain 3 (p94) in mice causes abnormal sarcomere formation in vivo and in vitro. Hum Mol Genet 13:1373–1388

Kramerova I, Kudryashova E, Wu B, Spencer MJ (2006) Regulation of the M-cadherin-beta-catenin complex by calpain 3 during terminal stages of myogenic differentiation. Mol Cell Biol 26:8437–8447

Kramerova I, Kudryashova E, Wu B, Ottenheijm C, Granzier H, Spencer MJ (2008) Novel role of calpain-3 in the triad-associated protein complex regulating calcium release in skeletal muscle. Hum Mol Genet 17:3271–3280

Kramerova I, Kudryashova E, Wu B, Germain S, Vandenborne K, Romain N, Haller RG, Verity MA, Spencer MJ (2009) Mitochondrial abnormalities, energy deficit and oxidative stress are features of calpain 3 deficiency in skeletal muscle. Hum Mol Genet 18:3194–3205

Kristian T, Gido G, Kuroda S, Schutz A, Siesjo BK (1998) Calcium metabolism of focal and penumbral tissues in rats subjected to transient middle cerebral artery occlusion. Exp Brain Res 120:503–509

Kuchay SM, Kim N, Grunz EA, Fay WP, Chrishti AH (2007) Double knockouts reveal that protein tyrosine phosphatase 1B is a physiological target of calpain-1 in platelets. Mol Cell Biol 27:6038–6052

Lee KS, Yanamoto H, Fergus A, Hong SC, Kang SD, Cappelletto B, Toyoda T, Kassell NF, Bavbek M, Kwan AL (1997) Calcium-activated proteolysis as a therapeutic target in cerebrovascular disease. Ann NY Acad Sci 825:95–103

Lee HJ, Sorimachi H, Jeong SY, Ishiura S, Suzuki K (1998) Molecular cloning and characterization of a novel tissue-specific calpain predominantly expressed in the digestive tract. Biol Chem 379:175–183

Lee HJ, Tomioka S, Kinbara K, Masumoto H, Jeong SY, Sorimachi H, Ishiura S, Suzuki K (1999) Characterization of a human digestive tract-specific calpain, nCL-4, expressed in the baculovirus system. Arch Biochem Biophys 362:22–31

Lid SE, Gruis D, Jung R, Lorentzen JA, Ananiev E, Chamberlin M, Niu X, Meeley R, Nichols S, Olsen OA (2002) The defective kernel 1 (dek1) gene required for aleurone cell development in the endosperm of maize grains encodes a membrane protein of the calpain gene superfamily. Proc Natl Acad Sci U S A 99:5460–5465

Liebetrau M, Stauffer B, Auerswald EA, Gabrijelcic-Geiger D, Fritz H, Zimmermann C, Pfefferkom T, Hamann GF (1999) Increased intracellular calpain detection in experimental focal cerebral ischemia. Neuroreport 10:529–534

Liu MC, Akle V, Zheng WR, Dave JR, Tortella FC, Hayes RL, Wang KKW (2006) Comparing calpain- and caspase-3-degradation patterns in traumatic brain injury by differential proteome analysis. Biochem J 394:715–725

Liu J, Liu MC, Wang KKW (2008a) Physiological and pathological actions of calpains in glutamatergic neurons. Sci Signal 1(23):tr3

Liu J, Liu MC, Wang KKW (2008b) Calpain in the CNS: from synaptic function to neurotoxicity. Sci Signal 1(14):re1
Liu MC, Zheng ZQ, Akinyi L, Oli MW, Zheng WR, Larner SF, Kobeissy F, Papa L, Lu XC, Dave JR, Tortella FC, Hayes RL, Wang KKW (2010) Ubiquitin-C-terminal hydrolase as a novel biomarker for stroke and traumatic brain injury in rats. Eur J Neurosci 31:722–732
Liu MC, Kobeissy F, Zheng W, Zhang Z, Hayes RL, Wang KK (2011) Dual vulnerability of tau to calpains and caspase-3 proteolysis under neurotoxic and neurodegenerative conditions. ASN Neuro 3(1):e00051
Lu J, Zheng J, Liu H, Li J, Chen H, Chen K (2010) Protein profiling analysis of skeletal muscle of a pufferfish, Takifugu rubripes. Mol Biol Rep 37:2141–2147
Maki M, Bagci H, Hamaguchi K, Ueda M, Murachi T, Hatanaka M (1989) Inhibition of calpain by a synthetic oligopeptide corresponding to an exon of the human calpastatin gene. J Biol Chem 264(32):18866–18869
Maki M, Ma H, Takano E, Adachi Y, Lee WJ, Hatanaka M, Murachi T (1991) Calpastatins: biochemical and molecular biological studies. Biomed Biochim Acta 50:509–516
Markgraf CG, Velayo NL, Johnson MP, McCarty DR, Medhi S, Koehl JR, Chmielewski PA, Linnik MD (1998) Six-hour window of opportunity for calpain inhibition in focal cerebral ischemia in rats. Stroke 29:152–158
Maxwell WL, Povlishock JT, Graham DL (1997) A mechanistic analysis of nondisruptive axonal injury: a review. J Neurotrauma 14:419–440
McGinn MJ, Kelley BJ, Akinyi L, Oli MW, Liu MC, Hayes RL, Wang KKW, Povlishock JT (2009) Biochemical, structural and biomarker evidence for calpain-mediated cytoskeletal change following diffuse brain injury uncomplicated by contusion. J Neuropathol Exp Neurol 68(3):241–249
McIntosh TK (1994) Neurochemical sequelae of traumatic brain injury: therapeutic implications. Cerebrovasc Brain Metab Rev 6:109–162
McIntosh TK, Saatman KE, Raghupathi R, Graham DI, Smith DH, Lee VM, Trojanowski JQ (1998) The Dorothy Russell Memorial Lecture. The molecular and cellular sequelae of experimental traumatic brain injury: pathogenetic mechanisms. Neuropathol Appl Neurobiol 24:251–267
Meaney DF, Margulies SS, Smith DH (2001) Diffuse axonal injury. J Neurosurg 95:1108–1110
Medana IM, Esiri MM (2003) Axonal damage: a key predictor of outcome in human CNS diseases. Brain 126:515–530
Mellgren RL, Huang X (2007) Fetuin A stabilizes m-calpain and facilitates plasma membrane repair. J Biol Chem 282:35868–35877
Mellgren RL, Zhang W, Miyake K, McNeil PL (2007) Calpain is required for the rapid, calcium-dependent repair of wounded plasma membrane. J Biol Chem 282:2567–2575
Meyer WL, Fischer EH, Krebs EG (1964) Activation of skeletal muscle phosphorylase kinase by Ca^{2+}. Biochemistry 3:1033–1039
Minami N, Tani E, Maeda Y, Yamaura I, Fukami M (1992) Effects of inhibitors of protein kinase C and calpain in experimental delayed cerebral vasospasm. J Neurosurg 76:111–118
Mitchell BM, Wu TG, Jackson BE, Wilhelmus KR (2007) Candida albicans strain-dependent virulence and Rim13p-mediated filamentation in experimental keratomycosis. Invest Ophthalmol Vis Sci 48:774–780
Moldoveanu T, Hosfield CM, Lim D, Elce JS, Jia Z, Davies PL (2002) A Ca^{2+} switch aligns the active site of calpain. Cell 108:649–660
Moldoveanu T, Hosfield CM, Lim D, Jia Z, Davies PL (2003) Calpain silencing by a reversible intrinsic mechanism. Nat Struct Biol 10:371–378
Molinari M, Carafoli E (1997) Calpain: a cytosolic proteinase active at the membranes. J Membr Biol 156(1):1–8
Mondello S, Robicsek SA, Gabrielli A, Brophy GM, Papa L, Tepas J, Robertson C, Buki A, Scharf D, Jixiang M, Akinyi L, Muller U, Wang KK, Hayes RL (2010) αII-spectrin breakdown

products (SBDPs): diagnosis and outcome in severe traumatic brain injury patients. J Neurotrauma 27(7):1203–1213

Mondello S, Jeromin A, Streeter J, Hayes RL, Wang KKW (2011) Blood-based diagnostics of traumatic brain injuries. Expert Rev Mol Diagn 11(1):65–78

Morioka M, Fukunaga K, Yasugawa S, Nagahiro S, Ushio Y, Miyamoto E (1992) Regional and temporal alterations in Ca2+/calmodulin-dependent protein kinase II and calcineurin in the hippocampus of rat brain after transient forebrain ischemia. J Neurochem 58:1798–1809

Murachi T (1989) Intracellular regulatory system involving calpain and calpastatin. Biochem Int 18:263–294

Nakada S, Tsuneyama K, Kato I, Tabuchi Y, Takasaki I, Furusawa Y, Kawaguchi H, Fujimoto M, Goto H, Hikiami H, Kondo T, Takano Y, Shimada Y (2010) Identification of candidate genes involved in endogenous protection mechanisms against acute pancreatitis in mice. Biochem Biophys Res Commun 391:1342–1347

Nakanishi H (2003) Microglial functions and proteases. Mol Neurobiol 27:163–176

Nath R, Raser KJ, Stafford D, Hajimohammadreza I, Posner A, Allen H, Talanian RV, Yuen P, Gilbertsen RB, Wang KK (1996) Non-erythroid alpha-spectrin breakdown by calpain and interleukin 1 beta-converting-enzyme-like protease(s) in apoptotic cells: contributory roles of both protease families in neuronal apoptosis. Biochem J 319(Pt 3):683–690

Nath R, Probert A Jr, McGinnis KM, Wang KK (1998) Evidence of activation of caspase-3 like protease in excitotoxin- and hypoxia/hypoglycemia-injured neurons. J Neurochem 71(1):186–195

Newcomb JK, Kampfl A, Posmantur RM, Zhao X, Pike BR, Liu SJ, Clifton GL, Hayes RL (1997) Immunohistochemical study of calpain-mediated breakdown products to alpha-spectrin following controlled cortical impact injury in the rat. J Neurotrauma 14:369–383

Newcomb JK, Pike BR, Zhao X, Banik NL, Hayes RL (1999) Altered calpastatin protein levels following traumatic brain injury. J Neurotrauma 16(1):1–11

Newcomb-Fernandez JK, Zhao X, Pike BR, Wang KK, Kampfl A, Beer R, DeFord SM, Hayes RL (2001) Concurrent assessment of calpain and caspase-3 activation after oxygen-glucose deprivation in primary septo-hippocampal cultures. J Cereb Blood Flow Metab 21(11):1281–1294

Nishida A, Emoto K, Shimizu M, Uozumi T, Yamawaki S (1994) Brain ischemia decreases phosphatidylcholine-phospholipase D but not phosphatidylinositol-phospholipase C in rats. Stroke 25:1247–1251

Nishiura I, Tanaka K, Yamamoto S, Murachi T (1978) The occurence of an inhibitor of Ca^{2+}-dependent neutral protease in rat liver. J Biochem 84(6):1657–1659

Noguchi M, Sarin A, Aman MJ, Najajima H, Shores EW, Henkart PA, Leonard WJ (1997) Functional cleavage of the common cytokine receptor gamma chain (γ_c) by calpain. Proc Natl Acad Sci U S A 94:11534–11539

Norberg E, Gogvadze V, Ott M, Horn M, Uhlen P, Orrenius S, Zhivotovsky B (2008) An increase in intracellular Ca^{2+} is required for the activation of mitochondrial calpain to release AIF during cell death. Cell Death Differ 15(12):1857–1864

Ohno S, Emori Y, Imajoh S, Kawasaki H, Kisaragi M, Suzuki K (1984) Evolutionary origin of a calcium-dependent protease by fusion of genes for a thiol protease and a calcium-binding protein? Nature 312(5994):566–570

Ohno S, Minoshima S, Kudoh J, Fukuyama R, Shimizu Y, Ohmi-Imajoh S, Shimizu N, Suzuki K (1990) Four genes for the calpain family locate on four distinct human chromosomes. Cytogenet Cell Genet 53:225–229

Ojima K, Ono Y, Doi N, Yoshioka K, Kawabata Y, Labeit S, Sorimachi H (2007) Myogenic stage, sarcomere length, and protease activity modulate localization of muscle-specific calpain. J Biol Chem 282:14493–14504

Ojima K, Kawabata Y, Nakao H, Nakao K, Doi N, Kitamura F, Ono Y, Hata S, Suzuki H, Kawahara H, Bogomolovas J, Witt C, Ottenheijm C, Labeit S, Granzier H, Toyama-Sorimachi-N, Sorimachi M, Suzuki K, Maeda T, Abe K, Aiba A, Sorimachi H (2010) Dynamic distribution of muscle-specific calpain in mice has a key role in physical-stress adaptation and is impaired in muscular dystrophy. J Clin Invest 120:2672–2683

Ojima K, Ono Y, Ottenheijm C, Hata S, Suzuki H, Granzier H, Sorimachi H (2011) Non-proteolytic functions of calpain-3 in sarcoplasmic reticulum in skeletal muscles. J Mol Biol 407:439–449

Olego-Fernandez S, Vaughan S, Shaw MK, Gull K, Ginger ML (2009) Cell morphogenesis of Trypanosoma brucei requires the paralogous, differentially expressed calpain-related proteins CAP5.5 and CAP5.5V. Protist 160:576–590

Olney JW, Labruyere J, Wang G, Wozniak DF, Price MT, Sesma MA (1991) NMDA antagonist neurotoxicity: mechanism and prevention. Science 254:1515–1518

Ono Y, Shimada H, Sorimachi H, Richard I, Saido TC, Beckmann JS, Ishiura S, Suzuki K (1998) Functional defects of a muscle-specific calpain, p94, caused by mutations associated with limb-girdle muscular dystrophy type 2A. J Biol Chem 273:17073–17078

Ono Y, Sorimachi H, Suzuki K (1999) New aspect of the research on limb-girdle muscular dystrophy 2A: a molecular biologic and biochemical approach to pathology. Trends Cardiovasc Med 9:114–118

Ono Y, Kakinuma K, Torii F, Irie A, Nakagawa K, Labeit S, Abe K, Suzuki K, Sorimachi H (2004) Possible regulation of the conventional calpain system by skeletal muscle-specific calpain, p94/calpain 3. J Biol Chem 279:2761–2771

Ono Y, Ojima K, Torii F, Takaya E, Doi N, Nakagawa K, Hata S, Abe K, Sorimachi H (2010) Skeletal muscle-specific calpain is an intracellular Na^+-dependent protease. J Biol Chem 285:22986–22998

Osako Y, Maemoto Y, Tanaka R, Suzuki H, Shibata H, Maki M (2010) Autolytic activity of human calpain 7 is enhanced by ESCRT-III-related protein IST1 through MIT-MIM interaction. FEBS J 277:4412–4426

Parkes C, Kembhavi AA, Barrett AJ (1985) Calpain inhibition by peptide epoxides. Biochem J 230:509–516

Paul DS, Harmon AW, Winston CP, Patel YM (2003) Calpain facilitates GLUT4 vesicle translocation during insulin-stimulated glucose uptake in adipocytes. Biochem J 376:625–632

Pettus EH, Povlishock JT (1996) Characterization of a distinct set of intra-axonal ultrastructural changes associated with traumatically induced alteration in axolemmal permeability. Brain Res 722:1–11

Pettus EH, Christman CW, Giebel ML, Povlishock JT (1994) Traumatically induced altered membrane permeability: its relationship to traumatically induced reactive axonal change. J Neurotrauma 11:507–522

Pietsch M, Chua KC, Abell AD (2010) Calpains: attractive targets for the development of synthetic inhibitors. Curr Top Med Chem 10:270–293

Pike BR, Zhao X, Newcomb JK, Postmantur RM, Wang KK, Hayes RL (1998) Regional calpain and caspase-3 proteolysis of alpha-spectrin after traumatic brain injury. Neuroreport 9 (11):2437–2442

Pike BR, Zhao X, Newcomb JK, Glenn CC, Anderson DK, Hayes RL (2000) Stretch injury causes calpain and caspase 3 activation and necrotic and apoptotic cell death in septo hippocampal cell cultures. J Neurotrauma 17(4):283–298

Pike BR, Flint J, Johnson E, Glenn CC, Dutta S, Wang KKW, Hayes RL (2001) Accumulation of calpain-cleaved non-erythroid αII-spectrin in cerebrospinal fluid after traumatic brain injury in rats. J Neurochem 78:1297–1306

Pike BR, Flint J, Dave JR, Lu XC, Wang KKW, Tortella FC, Hayes RL (2003) Accumulation of calpain and caspase-3 proteolytic fragments of brain-derived αII-spectrin in CSF after middle cerebral artery occlusion in rats. J Cereb Blood Flow Metab 24(1):98–106

Pontremoli S, Melloni E (1988) The role of calpain and protein kinase C in activation of human neutrophils. Prog Clin Biol Res 282:195–208

Posmantur R, Kampfl A, Siman R, Liu J, Zhao X, Clifton GL, Hayes RL (1997) A calpain inhibitor attenuates cortical cytoskeletal protein loss after experimental traumatic brain injury in the rat. Neuroscience 77:875–888

Postmantur RM, Kampfl A, Liu SJ, Heck K, Taft WC, Clifton GL, Hayes RL (1996) Cytoskeletal derangements of cortical neuronal processes three hours after traumatic brain injury in rats: an immunofluorescence study. J Neuropathol Exp Neurol 55(1):68–80

Potter DA, Tirnauer JS, Janssen R, Croall DE, Hughes CN, Fiacco KA, Mier JW, Maki M, Herman IM (1998) Calpain regulates actin remodeling during cell spreading. J Cell Biol 141:647–662

Povlishock JT (1992) Traumatically induced axonal injury: pathogenesis and pathobiological implications. Brain Pathol 2:1–12

Povlishock JT, Becker DP, Cheng CL, Vaughan GW (1983) Axonal change in minor head injury. J Neuropathol Exp Neurol 42:225–242

Povlishock JT, Erb DE, Astruc J (1992) Axonal response to traumatic brain injury: reactive axonal change, deafferentation, and neuroplasticity. J Neurotrauma 9(Suppl 1):S189–S200

Rami A (2003) Ischemic neuronal death in the rat hippocampus: the calpain-calpastatin-caspase hypothesis. Neurobiol Dis 13:75–88

Rao MV, Mohan PS, Peterhoff CM, Yang DS, Schmidt SD, Stavrides PH, Campbell J, Chen Y, Jiang Y, Paskevich PA, Cataldo AM, Haroutunian V, Nixon RA (2008) Marked calpastatin (CAST) depletion in Alzheimer's disease accelerates cytoskeleton disruption and neurodegeneraton: neuroprotection by CAST overexpression. J Neurosci 28(47):12241–12254

Ray SK, Shields DC, Saido TC, Matzelle DC, Wilford GG, Hogan EL, Banik NL (1999) Calpain activity and translational expression increased in spinal cord injury. Brain Res 816:375–380

Ray SK, Wilford GG, Ali SF, Banik NL (2000) Calpain upregulation in spinal cords of mice with 1-methyl-4-phenyl-1,2,3,6-tetrahydropyridine (MPTP)-induced Parkinson's disease. Ann NY Acad Sci 914:275–283

Ray SK, Matzelle DD, Wilford GG, Hogan EL, Banik NL (2001a) Cell death in spinal cord injury (SCI) requires de novo protein synthesis. Calpain inhibitor E-64-d provides neuroprotection in SCI lesion and penumbra. Ann NY Acad Sci 939:436–449

Ray SK, Matzelle DD, Wilford GG, Hogan EL, Banik NL (2001b) Inhibition of calpain-mediated apoptosis by E-64 d-reduced immediate early gene (IEG) expression and reactive astrogliosis in the lesion and penumbra following spinal cord injury in rats. Brain Res 916:115–126

Ray SK, Hogan EL, Banik NL (2003) Calpain in the pathophysiology of spinal cord injury: neuroprotection with calpain inhibitors. Brain Res Brain Res Rev 42(2):169–185

Ray SK, Samantaray S, Smith JA, Matzelle DD, Das A, Banik NL (2011) Inhibition of cysteine proteases in acute and chronic spinal cord injury. Neurotherapeutics 8(2):180–186

Reddy MK, Etlinger JD, Rabinowitz M, Fischman DA, Zak R (1975) Removal of Z-lines and alpha-actinin from isolated myofibrils by a calcium-activated neutral protease. J Biol Chem 250:4278–4284

Richard I, Broux O, Allamand V, Fougerousse F, Chiannilkulchai N, Bourg N, Brenguier L, Devaud C, Pasturaud P, Roudaut C, Hillaire D, Passos-Bueno M-R, Zats M, Tischfield JA, Fardeau M, Jackson CE, Cohen D, Beckmann JS (1995) Mutations in the proteolytic enzyme calpain 3 cause limb-girdle muscular dystrophy type 2A. Cell 81:27–40

Richard I, Roudaut C, Saenz A, Pogue R, Grimbergen JE, Anderson LV, Beley C, Cobo AM, de Diego C, Eymard B, Gallano P, Ginjaar HB, Lasa A, Pollitt C, Topaloglu H, Urtizberea JA, de Visser M, van der Kooi A, Bushby K, Bakker E, Lopez de Munain A, Fardeau M, Beckmann JS (1999) Calpainopathy-a survey of mutations and polymorphisms. Am J Hum Genet 64:1524–1540

Richard I, Roudaut C, Marchand S, Baghdiguian S, Herasse M, Stockholm D, Ono Y, Suel L, Bourg N, Sorimachi H, Lefranc G, Fardeau M, Sebille A, Beckmann JS (2000) Loss of calpain 3 proteolytic activity leads to muscular dystrophy and to apoptosis-associated IkappaBalpha/nuclear factor kappaB pathway perturbation in mice. J Cell Biol 151:1583–1590

Ringger NC, O'Steen BE, Brabham JG, Silver X, Pineda JA, Wang KKW, Hayes RL (2004) A novel marker for traumatic brain injury: CSF aII-spectrin breakdown product levels. J Neurotrauma 21:1443–1456
Rink A, Fung KM, Trojanowski JQ, Lee VM, Neugebauer E, McIntosh TK (1995) Evidence of apoptotic cell death after experimental traumatic brain injury in the rat. Am J Pathol 147:1575–1583
Roberts-Lewis JM, Savage MJ, Marcy VR, Pinsker LR, Siman R (1994) Immunolocalization of calpain I-mediated spectrin degradation to vulnerable neurons in the ischemic gerbil brain. J Neurosci 14:3934–3944
Rodriguez-Galan O, Galindo A, Hervas-Aguilar A, Arst HN Jr, Penalva MA (2009) Physiological involvement in pH signaling of Vps24-mediated recruitment of Aspergillus PalB cysteine protease to ESCRT-III. J Biol Chem 284:4404–4412
Rothman SM, Olney JW (1986) Glutamate and the pathophysiology of hypoxic–ischemic brain damage. Ann Neurol 19:105–111
Saatman KE, Murai H, Bartus RT, Smith DH, Hayward NJ, Perri BR, McIntosh TK (1996a) Calpain inhibitor AK295 attenuates motor and cognitive deficits following experimental brain injury in the rat. Proc Natl Acad Sci U S A 93:3428–3433
Saatman KE, Bozyczko-Coyne D, Marcy V, Siman R, McIntosh TK (1996b) Prolonged calpain-mediated spectrin breakdown occurs regionally following experimental brain injury in the rat. J Neuropathol Exp Neurol 55:850–860
Saez ME, Martinez-Larrad MT, Ramirez-Lorca R, Gonzalez-Sanchez JL, Zabena C, Martinez-Calatrava MJ, Gonzalez A, Moron FJ, Ruiz A, Serrano-Rios M (2007) Calpain-5 gene variants are associated with diastolic blood pressure and cholesterol levels. BMC Med Genet 8:1
Saido TC, Yokota M, Nagao S, Yamaura I, Tani E, Tsuchiya T, Suzuki K, Kawashima S (1993) Spatial resolution of fodrin proteolysis in postischemic brain. J Biol Chem 268:25239–25243
Saido TC, Sorimachi H, Suzuki K (1994) Calpain: new perspectives in molecular diversity and physiological-pathological involvement. FASEB J 8:814–822
Salvesen G, Parkes C, Abrahamson M, Grubb A, Barrett AJ (1986) Human low-Mr kininogen contains three copies of a cystatin sequence that are divergent in structure and in inhibitory activity for cysteine proteinases. Biochem J 234:429–434
Samantaray S, Ray SK, Banik NL (2008) Calpain as a potential therapeutic target in Parkinson's disease. CNS Neurol Disord Drug Targets 7(3):305–312
Samantaray S, Knaryan VH, Gal L, Ray SK, Banik NL (2011) Calpain inhibition protected spinal cord motoneurons against 1-methyl-4-phenylpyridinium ion and rotenone. Neuroscience 129:263–274
Sangenito LS, Ennes-Vidal V, Marinho FA, Da Mota FF, Santos AL, D'Avila-Levy CM, Branquinha MH (2009) Arrested growth of Trypanosoma cruzi by the calpain inhibitor MDL28170 and detection of calpain homologues in epimastigote forms. Parasitology 136:433–441
Sanges D, Comitato A, Tammaro R, Marigo V (2006) Apoptosis in retinal degeneration involves cross-talk between apoptosis-inducing factor (AIF) and caspase-12 and blocked by calpain inhibitors. Proc Natl Acad Sci U S A 103:17366–17371
Satish L, Blair HC, Glading A, Wells A (2005) Interferon-inducible protein 9 (CXCL11)-induced cell motility in keratinocytes requires calcium flu-dependent activation of mu-calpain. Mol Cell Biol 25(5):1922–1941
Schaecher KE, Shields DC, Banik NL (2001) Mechanism of myelin breakdown in experimental demyelination: a putative role for calpain. Neurochem Res 26:731–737
Schaecher K, Rocchini A, Dinkins J, Matzelle DD, Banik NL (2002) Calpain expression and infiltration of activated T cells in experimental allergic encephalomyelitis over time: increased calpain activity begins with onset of disease. J Neuroimmunol 129:1–9
Schroder E, Phillips C, Garman E, Harlos K, Crawford C (1993) X-ray crystallographic structure of a papain-leupeptin complex. FEBS Lett 315:38–42

Schumacher PA, Siman RG, Fehlings MG (2000) Pretreatment with calpain inhibitor CEP-4143 inhibits calpain I activation and cytoskeletal degradation, improves neurological function, and enhances axonal survival after traumatic spinal cord injury. J Neurochem 74:1646–1655
Shields DC, Banik NL (1999) Pathological role of calpain in experimental demyelination. J Neurosci Res 55(5):533–541
Shields DC, Schaecher KE, Saido TC, Banik NL (1999) A putative mechanism of demyelination in multiple sclerosis by a proteolytic enzyme, calpain. Proc Natl Acad Sci U S A 96:11486–11491
Shirasaki Y, Miyashita H, Yamaguchi M, Inoue J, Nakamura M (2005) Exploration of orally available calpain inhibitors: peptidyl alpha-ketoamides containing an amphiphile at P3 site. Bioorg Med Chem 13:4473–4484
Siesjo BK, Bengtsson F (1989) Calcium fluxes, calcium antagonists, and calcium-related pathology in brain ischemia, hypoglycemia, and spreading depression: a unifying hypothesis. J Cereb Blood Flow Metab 9:127–140
Siesjo BK, Bengtsson F, Grampp W, Theander S (1989) Calcium, excitotoxins, and neuronal death in the brain. Ann NY Acad Sci 568:234–251
Siman R, Noszek JC (1988) Excitatory amino acids activate calpain I and induce structural protein breakdown in vivo. Neuron 1:279–287
Siman R, Noszek JC, Kegerise C (1989) Calpain I activation is specifically related to excitatory amino acid induction of hippocampal damage. J Neurosci 9:1579–1590
Siman R, McIntosh TK, Soltesz KM, Chen Z, Neumar RW, Roberts VL (2004) Proteins released from degenerating neurons are surrogate markers for acute brain damage. Neurobiol Dis 16:311–320
Siman R, Toraskar N, Dang A, McNeil E, McGarvey M, Plaum J, Maloney E, Grady MS (2009) A panel of neuron-enriched proteins as markers for traumatic brain injury in humans. J Neurotrauma 26(11):1867–1877
Sorimachi H, Suzuki K (2001) The structure of calpain. J Biochem 129:653–664
Sorimachi H, Imajoh-Ohmi S, Emori Y, Kawasaki H, Ohno S, Minami Y, Suzuki K (1989) Molecular cloning of a novel mammalian calcium-dependent protease distinct from both m- and mu-types. Specific expression of the mRNA in skeletal muscle. J Biol Chem 264:20106–20111
Sorimachi H, Ishiura S, Suzuki K (1993a) A novel tissue-specific calpain species expressed predominantly in the stomach comprises two alternative splicing products with and without Ca^{2+}-binding domain. J Biol Chem 268:19476–19482
Sorimachi H, Toyama-Sorimachi N, Saido TC, Kawasaki H, Sugita H, Miyasaka M, Arahata K, Ishiura S, Suzuki K (1993b) Muscle-specific calpain, p94, is degraded by autolysis immediately after translation, resulting in disappearance from muscle. J Biol Chem 268:10593–10605
Sorimachi H, Kinbara K, Kimura S, Takahashi M, Ishiura S, Sasagawa N, Sorimachi N, Shimada H, Tagawa K, Maruyama K, Suzuki K (1995) Muscle-specific calpain, p94, responsible for limb girdle muscular dystrophy type 2A, associates with connectin through IS2, a p94-specific sequence. J Biol Chem 270:31158–31162
Sorimachi H, Hata S, Ono Y (2010) Expanding members and roles of the calpain superfamily and their genetically modified animals. Exp Anim 59:549–566
Sorimachi H, Hata S, Ono Y (2011a) Calpain chronicle – an enzyme family under multidisciplinary characterization. Proc Jpn Acad Ser B Phys Biol Sci 87(6):287–327
Sorimachi H, Hata S, Ono Y (2011b) Impact of genetic insights into calpain biology. J Biochem 150:23–37
Spencer MJ, Guyon JR, Sorimachi H, Potts A, Richard I, Herasse M, Chamberlain J, Dalkilic I, Kunkel LM, Beckmann JS (2002) Stable expression of calpain 3 from a muscle transgene in vivo: immature muscle in transgenic mice suggests a role for calpain 3 in muscle maturation. Proc Natl Acad Sci U S A 99:8874–8879

Sribnick EA, Matzelle DD, Banik NL, Ray SK (2007) Direct evidence for calpain involvement in apoptotic death of neurons in spinal cord injury in rats and neuroprotection with calpain inhibitor. Neurochem Res 32(12):2210–2216

Strobl S, Fernandez-Catalan C, Braun M, Huber R, Masumoto H, Nakagawa K, Irie A, Sorimachi H, Bourenkow G, Bartunik H, Suzuki K, Bode W (2000) The crystal structure of calcium-free human m-calpain suggests an electrostatic switch mechanism for activation by calcium. Proc Natl Acad Sci U S A 97:588–592

Suzuki Y, Murachi T (1977) The occurence of a neutral protease and its inhibitor in rat peritoneal macrophages. J Biochem 82(1):215–220

Suzuki K, Tsuji S, Kubota S, Kimura Y, Imahori K (1981) Limited proteolysisof Ca^{2+}-activated neutral protease (CANP) changes its sensitivity to Ca^{2+} ions. J Biochem 90:275–278

Syntichaki P, Xu K, Driscoll M, Tavernarakis N (2002) Specific aspartyl and calpain proteases are required for neurodegeneration in *C. elegans*. Nature 419:939–944

Szatkowski M, Attwell D (1994) Triggering and execution of neuronal death in brain ischaemia: two phases of glutamate release by different mechanisms. Trends Neurosci 17:359–365

Taft WC, Yang K, Dixon CE, Hayes RL (1992) Microtubule-associated protein 2 levels decrease in hippocampus following traumatic brain injury. J Neurotrauma 9(3):281–290

Tagawa K, Taya C, Hayashi Y, Nakagawa M, Ono Y, Fukuda R, Karasuyama H, Toyama-Sorimachi N, Katsui Y, Hata S, Ishiura S, Nonaka I, Seyama Y, Arahata K, Yonekawa H, Sorimachi H, Suzuki K (2000) Myopathy phenotype of transgenic mice expressing active site-mutated inactive p94 skeletal muscle-specific calpain, the gene product responsible for limb girdle muscular dystrophy type 2A. Hum Mol Genet 9:1393–1402

Takahashi-Nakamura M, Tsuji S, Suzuki K, Imahori K (1981) Purification and characterization of an inhibitor of calcium-activated neutral protease from rabbit skeletal muscle. J Biochem 90 (6):1583–1589

Takai Y, Yamamoto M, Inoue M, Kishimoto A, Nishizuka Y (1977) A proenzyme of cyclic nucleotide-independent protein kinase and its activation by calcium-dependent neutral protease from rat liver. Biochem Biophys Res Commun 77:542–550

Taveau M, Bourg N, Sillon G, Roudaut C, Bartoli M, Richard I (2003) Calpain 3 is activated through autolysis within the active site and lyses sarcomeric and sarcolemmal components. Mol Cell Biol 23:9127–9135

Tian Q, Olsen L, Sun B, Lid SE, Brown RC, Lemmon BE, Fosnes K, Gruis DF, Opsahl-Sorteberg HG, Otegui MS, Olsen OA (2007) Subcellular localization and functional domain studies of DEFECTIVE KERNEL1 in maize and Arabidopsis suggest a model for aleurone cell fate specification involving CRINKLY4 and SUPERNUMERARY ALEURONE LAYER1. Plant Cell 19:3127–3145

Tompa P, Emori Y, Sorimachi H, Suzuki K, Friedrich P (2001) Domain III of calpain is a Ca^{2+}-regulated phospholipid-binding domain. Biochem. Biophys Res Commun 280:1333–1339

Tonami K, Kurihara Y, Aburatani H, Uchijima Y, Asano T, Kurihara H (2007) Calpain 6 is involved in microtubule stabilization and cytoskeletal organization. Mol Cell Biol 27:2548–2561

Tonami K, Hata S, Ojima K, Ono Y, Kurihara Y, Amano T, Sato T, Kawamura Y, Kurihara H, Sorimachi H (2013) Calpain-6 Deficiency Promotes Skeletal Muscle Development and Regeneration. PLoS Genet 9:e1003668

Tonami K, Kurihara Y, Arima S, Nishiyama K, Uchijima Y, Asano T, Sorimachi H, Kurihara H (2011) Calpain 6, a microtubule-stabilizing protein, regulates Rac1 activity and cell motility through interaction with GEF-H1. J Cell Sci 124:1214–1223

Tradewell ML, Durham HD (2010) Calpastatin reduces toxicity of SOD1G93A in a culture model of amyotrophic lateral sclerosis. Neuroreport 21(15):976–979

Ueda Y, McCormack AL, Shearer TR, David LL (2001) Purification and characterization of lens specific calpain (Lp82) from bovine lens. Exp Eye Res 73:625–637

Veeranna KT, Boland B, Odrljin T, Mohan P, Basavarajappa BS, Peterhoff C, Cataldo A, Rudnicki A, Amin N, Li BS, Pant HC, Hungund BL, Arancio O, Nixon RA (2004) Calpain mediates calcium-induced activation of the erk1,2 MAPK pathway and cytoskeletal phosphorylation in neurons: relevance to Alzheimer's disease. Am J Pathol 165:795–805

Verity MA (1992) Ca(2+)-dependent processes as mediators of neurotoxicity. Neurotoxicology 13:139–147

Vitzthum F, Behrens F, Anderson NL, Shaw JH (2005) Proteomics: from basic research to diagnostic application. A review of requirements and needs. J Proteome Res 4(4):1086–1097

Waghray A, Wang DS, McKinsey D, Hayes RL, Wang KK (2004) Molecular cloning and characterization of rat and human calpain-5. Biochem Biophys Res Commun 324:46–51

Wang KKW (2000) Calpain and caspase: can you tell the difference? Trends Neurosci 23:20–26

Wang KK, Yuen PW (1994) Calpain inhibition: an overview of its therapeutic potential. Trends Pharmacol Sci 15:412–419

Wang KK, Yuen PW (1997) Development and therapeutic potential of calpain inhibitors. Adv Pharmacol 37:117–152

Wang KK, Posmantur R, Nath R, McGinnis K, Whitton M, Talanian RV, Glantz SB, Morrow JS (1998) Simultaneous degradation of alphaII- and betaII-spectrin by caspase 3 (CPP32) in apoptotic cells. J Biol Chem 273:22490–22497

Wang KKW, Yuen PW (1999) Calpain substrates, assay methods, regulation and its inhibitory agents. In: Wang KKW, Yuen PW (eds) Calpain: pharmacology and toxicology of a calcium-dependent cellular protease. Taylor and Francis, Philadelphia, PA, pp 77–102

Wang KKW, Villalobo A, Roufogalis BD (1989) Calmodulin-binding proteins as calpain substrates. Biochem J 262:693–706

Wang KKW, Ottens A, Liu MC, Lewis SB, Meegan C, Oli MW, Tortella FC, Hayes RL (2005) Proteomic identification of biomarkers of traumatic brain injury. Expert Rev Proteomics 2 (4):603–614

Warner S (2004) *The Scientist* 18(16):38

Wells A, Huttenlocher A, Lauffenburger DA (2005) Calpain proteases in cell adhesion and motility. Int Rev Cytol 245:1–16

Wieloch T, Cardell M, Bingren H, Zivin J, Saitoh T (1991) Changes in the activity of protein kinase C and the differential subcellular redistribution of its isozymes in the rat striatum during and following transient forebrain ischemia. J Neurochem 56:1227–1235

Wikstrom P, Anagli J, Angliker H, Shaw E (1992) Additional peptidyl diazomethyl ketones, including biotinyl derivatives, which affinity-label calpain and related cysteinyl proteinases. J Enzyme Inhib 6(4):259–269

Yamashima T (2000) Implication of cysteine proteases calpain, cathepsin and caspase in ischemic neuronal death of primates. Prog Neurobiol 62:273–295

Yamaura I, Tani E, Saido TC, Suzuki K, Minami N, Maeda Y (1993) Calpain-calpastatin system of canine basilar artery in vasospasm. J Neurosurg 79:537–543

Yokota M, Tani E, Tsubuki S, Yamaura I, Nakagaki I, Hori S, Saido TC (1999) Calpain inhibitor entrapped in liposome rescues ischemic neuronal damage. Brain Res 819:8–14

Yoshikawa Y, Mukai H, Hino F, Asada K, Kato I (2000) Isolation of two novel genes, down-regulated in gastric cancer. Jpn J Cancer Res 91:459–463

Yu CG, Joshi A, Geddes JW (2008) Intraspinal MDL28170 microinjection improves functional and pathological outcome following spinal cord injury. J Neurotrauma 25(7):833–840

Yuen PW, Wang KKW (1998) Calpain inhibitors: Novel neuroprotectants and potential anticataract agents. Drugs Future 23:741–749

Zhang SX, Underwood M, Landfield A, Huang FF, Gison S, Geddes JW (2000) Cytoskeletal disruption following contusion injury to the rat spinal cord. J Neuropathol Exp Neurol 59:287–296

Zhang SX, Bondada V, Geddes JW (2003) Evaluation of conditions for calpain inhibition in the rat spinal cord: effective postinjury inhibition with intraspinal MDL28170 microinjection. J Neurotrauma 20:59–67

Zhang Z, Larner S, Liu MC, Zheng W, Hayes RL, Wang KKW (2009) Multiple αII-spectrin breakdown products distinguish calpain and caspase dominated necrotic and apoptotic cell death pathways. Apoptosis 14(11):1289–1298

Zhao X, Pike BR, Newcomb JK, Wang KK, Posmantur RM, Hayes RL (1999) Maitotoxin induces calpain but not caspase-3 activation and necrotic cell death in primary septo-hippocampal cultures. Neurochem Res 24:371–382

Chapter 13
Metalloproteinases in Cartilage Matrix Breakdown: The Roles in Rheumatoid Arthritis and Osteoarthritis

Hideaki Nagase and Gillian Murphy

13.1 Introduction

The cartilage of articular joints consists of a relatively small number of cells and an abundant extracellular matrix (ECM). The major components of its ECM are collagen fibrils and aggregated proteogylcans, aggrecans, which account for approximately 90 % of the total ECM of the cartilage with each roughly in a similar amount. Collagen fibrils, mainly consisting of type II collagen, together with minor types IX and XI collagens, form a meshwork and provide tensile strength to the tissue. Aggrecans, present as large aggregated complexes interacting with hyaluronan and link proteins, form a hydrated gel within the collagen meshwork and give cartilage its ability to withstand mechanical compression. Cartilage also contains less abundant ECM macromolecules such as fibromodulin, decorin, biglycan, cartilage oligomeric matrix protein, types VI collagen, tenascin C, matrilins and cell surface proteoglycans (Heinegård and Saxne 2011). In normal cartilage the turnover of these molecules is at equilibrium, but in arthritic cartilage the loss of ECM components exceeds their synthesis and cause impairment of joint function.

Rheumatoid arthritis (RA) and osteoarthritis (OA) are the two most common joint disorders, which cause pain and disability. The etiology of the two joint diseases is different, but a shared feature is the destruction of articular cartilage. Many possible causes of cartilage destruction are recognized (e.g., injury, inflammation, mechanical load, oxidative stress, aging, apoptosis and extracellular matrix

H. Nagase (✉)
Nuffield Department of Orthopaedics, Rheumatology and Musculoskeletal Sciences, Kennedy Institute of Rheumatology, University of Oxford, Roosevelt Drive, Headington, Oxford OX3 7FY, UK
e-mail: hideaki.nagase@kennedy.ox.ac.uk

G. Murphy
Department of Oncology, Cancer Research UK, Cambridge Institute, Li Ka Shing Centre, University of Cambridge, Cambridge CB2 0RE, UK

K. Brix and W. Stöcker (eds.), *Proteases: Structure and Function*,
DOI 10.1007/978-3-7091-0885-7_13, © Springer-Verlag Wien 2013

disassembly), but the primary cause is elevated levels of active proteinases that degrade the ECM of cartilage. RA is an inflammation driven autoimmune disease associated with hyperplasia of synovial cells and rheumatoid synovia contain infiltrated T and B lymphocytes, macrophages, mast cells, endothelial cells and fibroblastic synovial lining cells. Neutrophils are often found in the cartilage-pannus junction in early erosive lesions and multinucleated osteoclasts at the site of subchondral bone erosion. The main cellular source of ECM-degrading enzyme are macrophages, synovial lining cells and their production is regulated by inflammatory mediators, cytokines and growth factors generated by activated lymphocytes, macrophages and mast cells (Fig. 13.1, left). In OA, on the other hand, destructive proteinases are derived primarily from the cartilage and inductive stimuli include inflammatory cytokines, mechanical load, tissue injury, reactive oxygen species, degraded matrix and aging (Fig 13.1, right). There are a number of proteases, including serine and cysteine proteases that are found in the joint tissues, but the major proteinases that degrade cartilage matrices are the matrix metalloproteinases (MMPs) and the metalloproteinases with disintegrin and thrombospondin domains (ADAMTSs). Aggrecans are degraded by both MMPs and the so-called "aggrecanases" which belong to the ADAMTS family, whereas collagen fibrils are cleaved by collagenases which are members of the MMP family. In addition, the metalloproeineases with disintegrin domain (ADAMs) that shed cell surface cytokines, growth a factors and their receptors also play key roles in inflammation and tissue damage.

Other proteinases such as plasminogen activator-plasminogen system, matriptase, high temperature requirement A (HtrA), activated protein C, cathespins, calpains and caspases are also involved directly and indirectly in cartilage degradation (Troeberg and Nagase 2012). This chapter focuses on the role of metalloproteinases in RA and OA and their structures.

13.2 MMPs and Metzincins

There are 24 MMP genes in the human genome, but there are 23 expressed proteinases because of the duplication of the *MMP23* gene. The MMPs are also called "matrixins" and they are classified as matrixin subfamily of metalloproteinase family M10 in the MEROPS database (http://www.merops.sanger.ac.uk). All matrixins have a signal peptide, a propeptide domain and a catalytic metalloproteinase (M) domain and most have a C-terminal hemopexin (Hpx) domain, with the exception of MMP-7, MMP-23 and MMP-26. MMP-2 and MMP-9 called gelatinase A and gelatinase B, respectively, have three repeats of fibronectin type II (FN) motifs inserted in the catalytic domain. These fibronectin domains interact with collagen and other ECM molecules and influence substrate specificity. The catalytic M domain of MMPs has the zinc-binding motif HEXXHXXGXXH where three histidines bind to the catalytic Zn^{2+} ion. The Hpx domain in MMPs is important for some members to express biological activities,

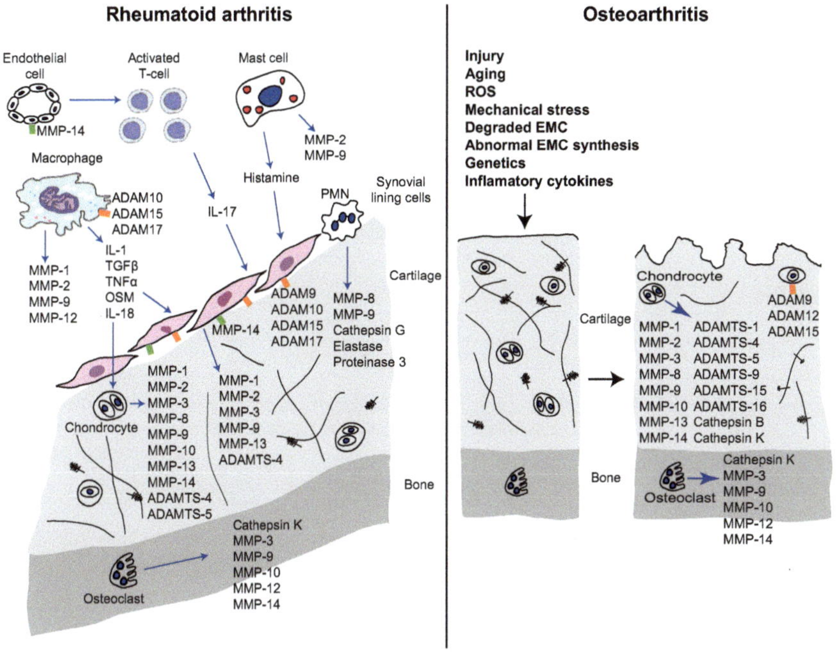

Fig. 13.1 Sources of cartilage matrix-degrading metalloproteinases in RA and OA. In RA (*left*), inflammation or tissue damage drives the recruitment of lymphocytes, macrophages and mast cells and polymorphonuclear leukocytes (PMN). Those cells produce metalloproteinases and/or cytokines that stimulate fibroblastic synovial lining cells and chondrocytes in cartilage to produce metalloproteinases. MMP-14 is expressed on the surface of the synovial lining cells when cells contact to collagen of the cartilage. In OA (*right*), the primary cellular source of matrix-degrading enzymes are the chondrocytes, which respond to tissue injury, aging, mechanical stimuli, inflammatory cytokines, reactive oxygen species and other factors. Bone erosion is due to osteoclasts that migrate from blood vessels into the bony region

e.g, for collagenases to cleave triple helical collagen (Visse and Nagase 2003), and for membrane-type 1 MMP (MT1-MMP or MMP-14) to assume a dimeric form on the cell surface which is crucial for its biological activities, including the activation of proMMP-2, cell migration and collagenolysis (Tochowicz et al. 2011). Most MMPs are secreted from the cell as inactive pro-enzymes, but six are plasma membrane-anchored membrane-type MMPs (MT-MMPs), of which four are type I transmembrane-type harbouring a cytosolic domain and two are glycosylphosphatidyl inositol-anchored. Pro-enzymes secreted from the cell secreted into the ECM are activated by tissue or plasma proteinases or by reactive oxygen species (Hadler-Olsen et al. 2011). Six pro-MT-MMPs and proMMP-11, -21, -23 and -28 are, on the other hand, most likely to be activated intracellularly by furin or a related pro-proteinases (Nagase et al. 2006).

MMPs have been characterized for their abilities to cleave ECM components, and the domain arrangements of MMPs that are found in cartilage and symonvium are shown in Fig. 13.2. They cleave aggrecan core protein at several sites, but the

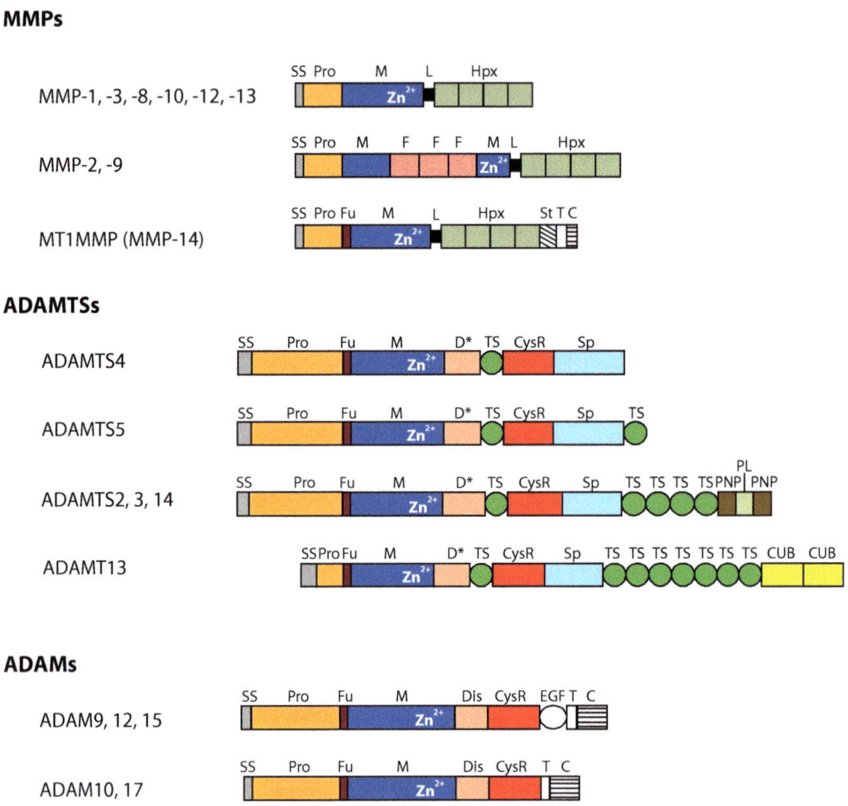

C	Cytoplasmic domain	F	Fibronectin type II domain	Pro	Pro-domain
CUB	Complement C1r/C1s - Urchin epidermal growth factor - Bone morphogenetic protein1 domain	Fu	Furin recognition motif	Sp	Spacer domain
CysR	Cysteine-rich domain	Hpx	Hemopexin-like domain	SS	Signal sequence
D*	Disintegrin domain with a Ch fold	L	Linker region	St	Stalk region
Dis	Disintegrin domain	M	Metalloproteinase domain	T	Transmembrane domain
EGF	Epidermal growth factor domain	Mu	Mucin-like	TS	Thrombospondin type 1 repeat
		PL	Proteinase and lacunin domain		
		PNP	Procollagen N-proteinase specific		

Fig. 13.2 Domain arrangement of MMPs, ADAMTSs and ADAMs that are involved in cartilage matrix degradation. ADAMTS2, 3 and 14 are procollagen N-proteinases. ADAMTS13 is shown as the structure of its non-catalytic domains was used in model ADAMTS4

cleavage at the Asn^{341}–Phe^{342} bond located in the interglobular domain between the N-terminal globular domain (G1) and the second globular domain (G2) has been characterized as a signature of the MMP activity (see Fig. 13.3). ADAMTSs, on the other hand, cleave the the Glu^{373}–Ala^{374} bond, which is the signature of "aggrecanase" activity. Collagenolytic MMPs are so-called collagenases (MMP-1, MMP-8 and MMP-13) and MMP-2 (gelatinase A) and MMP-14 (membrane type-1 MMP). It should also be noted that many MMPs cleave non-ECM

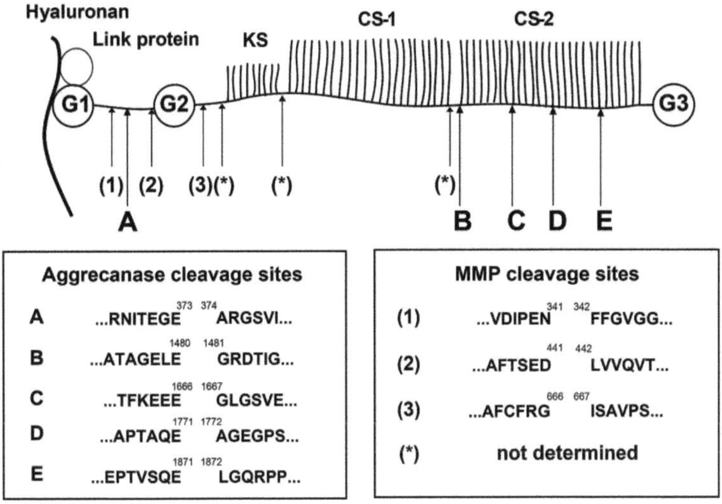

Fig. 13.3 Sites cleaved in aggrecan core protein by aggrecanase and MMPs

molecules such as cytokines, chemokines, growth factors, growth factor binding proteins, proteinase inhibitors, cell surface adhesion molecules and receptors to regulate their biological activities (Overall 2002). MMP activities are inhibited by endogenous inhibitors called TIMPs (tissue inhibitors of metalloproteinases), but only limited members of the ADAMTSs and the ADAMs are inhibited by TIMPs (see below).

Crystal structures of MMPs, a crayfish metalloproteinase astacin, snake venom metalloproteinase adamalysin, and a bacterial metalloproteinase serralysin, all revealed similar polypeptide folds, although their amino acid sequences are not related each other, except the catalytic zinc ion-binding motif HEXXHXXGXXH and the 'Met-turn' located after the zinc-binding motif. This methionine forms a base to create the active site for these metalloproteinases. Those are collectively called 'metzincins' (Bode et al. 1993; Gomis-Rüth 2009). The astacin family includes bone morphogenetic protein 1, tolloid-like metalloproteinases, meprins and ovastacin; the adamalysins include the ADAMTSs and the ADAMs; and the pappalysins including human pregnancy-associated plasma protein A (Boldt et al. 2004) and archeae metalloproteinases ulilysin (Tallant et al. 2006). The serralysins are related bacterial metalloproteinase. Fragilysin 3 from *Bacteroides fragilis* was recently classified as a metzincin metalloproteinase, which shows structural similarity to mammalian ADAMs with little sequence identity (Goulas et al. 2011). Since similar proteins have not been found in species other than *B. fragilis*, it is considered to result from horizontal gene transfer from a eukaryotic cell to a bacterial cell (Goulas et al. 2011).

13.2.1 Three-Dimensional (3D) Structures of MMPs

Numerous 3D structures of the catalytic domains of MMPs, including those of all collagenolytic MMPs (MMP-1, MMP-2, MMP-8, MMP-13 and MMP-14) have been solved by X-ray crystallography and some by NMR spectroscopy [see Maskos (2005) for review]. Full-length proMMP-1 (Jozic et al. 2005), proMMP-2 (Morgunova et al. 1999) and proMMP-2-TIMP-2 complex (Morgunova et al. 2002), activated from of full-length MMP-1 (Iyer et al. 2006; Li et al. 1995) and the full-length MMP-1-triple helical collagen peptide complex (Manka et al. 2012) were also solved. Those studies have helped us understand the mechanisms of peptide hydrolysis by MMPs and design their inhibitors. The overall polypeptide folds of prodomains, catalytic domains and Hpx domains are very similar among MMPs. Figure 13.4 shows the crystal structure of the proMMP-2-TIMP-2 complex to indicate examples of individual domains found in MMPs.

13.2.1.1 Pro-domains

Pro-domains of the secreted proMMPs consist of about 80 amino acids, and they are folded into three α-helices and connecting loops (Becker et al. 1995; Elkins et al. 2002; Jozic et al. 2005; Morgunova et al. 1999). The pro-domain functions as a chaperone as well as keeping the enzyme in an inactive zymogen form. The sequence PRCGXPD called "cysteine switch" is conserved in proMMPs and it lies in the substrate binding groove of the catalytic domain. It forms hydrogen bonds in a similar manner as in those of a peptide substrate deduced from the interactions of peptide inhibitors (Grams et al. 1995). However, the direction of the cysteine switch sequence is opposite from that of a peptide substrate. The SH group of the cysteine interacts with the catalytic Zn^{2+}, which keep the enzyme inactive. Upon activation of a proMMP, the Cys-Zn^{2+} interaction is disrupted, which enables Zn^{2+} to interact with H_2O necessary for peptide hydrolysis. Disruption of the Cys-Zn^{2+} interaction can be triggered by proteolytic processing of the pro-peptide or by structural perturbation of the pro-domain by chemicals such as chaotropic agents, sodium dodecyl sulphate, SH reactive agents, or oxidation of the SH group (Nagase 1997). The activation of proMMPs by reactive oxygen species such as hypochlorous acid generated by myeloperoxidase, or by NO from phagocytes may be important mechanism of proMMP activation (Fu et al. 2004; Gu et al. 2002; Peppin and Weiss 1986). However, a longer reaction may inactivate the enzyme (Fu et al. 2004).

13.2.1.2 M Domain

The catalytic domain consists of a five-stranded β-sheet and three α helices, two Zn^{2+} ions and one to three Ca^{2+} ions. Calcium ions and one Zn^{2+} ion hold the structure

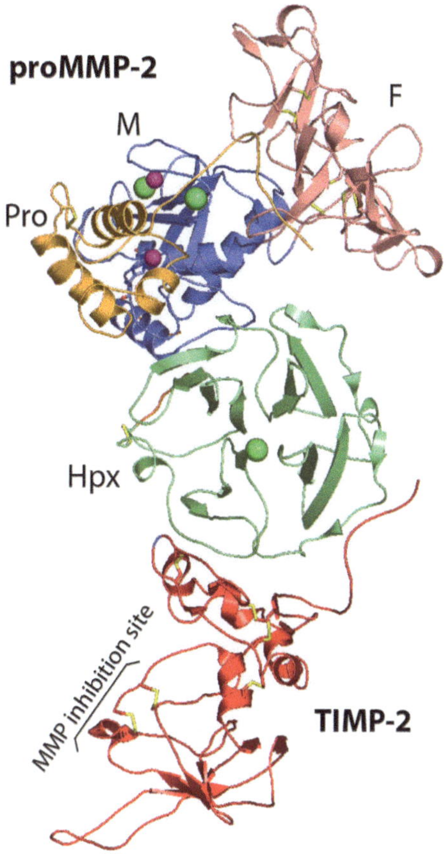

Fig. 13.4 Ribbon structure of the proMMP-2-TIMP-2 complex. The pro-domain is shown in *gold*, catalytic domain in *blue*, fibronectin type II motifs in *salmon*, hemopexin domain in *green*, TIMP-2 in *red*, zinc ions in *purple sphere*, calcium ions in *green sphere*, and disulfide bonds in *yellow*. The image was prepared by Robert Visse based on PDB ID: 1GXY

and one Zn^{2+} in the active site is essential for peptide hydrolysis. The catalytic Zn^{2+} is coordinated with three histidines in the HEXGHXXGXXH sequence conserved in all MMPs. The substrate binding S1′ pocket located adjacent to the catalytic Zn^{2+} is hydrophobic in nature and variable in depth and shape among MMPs. It is therefore one of the determinant factors for substrate specificity of MMPs. However, for MMPs to recognize large extended ECM molecules, extended substrate recognition sites (exosites) are found in both the catalytic and non-catalytic domain (Bertini et al. 2012; Fulcher and Van Doren 2011; Manka et al. 2012; Palmier et al. 2010).

13.2.1.3 Fibronectin Type II Domain

MMP-2 and MMP-9 have three repeats of fibronectin type II (Fn II) domain and three domains participate in their specificity for gelatin, collagen IV and laminin. They are inserted into the loop between the fifth β strand and the second helix of the catalytic M domain. The Fn II domains in MMP-2 and MMP-9 have similar conformations. Each domain consists of two antiparallel β-sheets connected with a short α-helix which are stabilized by two disulfide bonds. However, the placement of the three Fn II domains in the two gelatinases are different in that domain 2 of proMMP-2 has an area that interacts with the M domain, but the corresponding domain of proMMP-9 is rotated and twisted away from the M domain (Elkins et al. 2002). Domain 3 in the two progelatinases makes contact with the propeptide and with the M domain (Elkins et al. 2002; Morgunova et al. 1999).

13.2.1.4 Linker Region

The linker region connects the M domain and the Hpx domain and it is variable in length. As shown for MMP-1, MMP-9 and MMP-12, linker regions are considered to be flexible (Bertini et al. 2008, 2009; Jozic et al. 2005; Rosenblum et al. 2007), but they may have some structural constraint as they contain a number of proline residues and therefore are suited for specific functions. Mutational studies of this region in MMP-1 (Tsukada and Pourmotabbed 2002) and MMP-8 (Knäuper et al. 1997) significantly decreased the collagenolytic activity, supporting the idea that correct movement around the M and the Hpx domain is important for the expression of collagenase activity. The linker region of MMP-9 and MMP-14 are N- and O-glycosylated and these carbohydrates have effects on the cellular biochemistry of these MMPs (Van den Steen et al. 2006; Wu et al. 2004).

13.2.1.5 Hpx Domain

The Hpx domain is found in most MMPs, with the exception of MMP-7, -23 and -26. It is a ellipsoidal disk shape, four-bladed β propeller structure with a single disulphide bond between the first and fourth blades (Gomis-Rüth et al. 1996; Li et al. 1995; Libson et al. 1995). Each blade is made up of four antiparallel β-strands and the four β-sheets have similar scaffolds and are arranged almost symmetrically around a central core axis. The centre of the propeller usually contains one calcium ion and a chloride. The Hpx domains are essential for collagenases to degrade triple helical collagens and MMP-14 to dimerize on the cell surface (Tochowicz et al. 2011).

13.2.2 Mechanisms of Collagenolysis by MMPs

The fibril-forming insterstitial collagens consist of three α chains of approximately 1,000 amino acid residues with repeating Gly-Xaa-Yaa triplets, where Xaa and Yaa are often Pro and Hyp respectively. Due to a high imino acid content and repeated

Gly in every third residue, the α chain adopts a left-handed poly-proline II-like helix and three left-handed chains intertwine to form a right-handed superhelix (Kramer et al. 2001; Ramachandran and Kartha 1955). The triple helical structures make interstitial collagen I, II and III resistant to most proteinases, and in vertebrates, only collagenolytic MMPs (also referred to as "collagenases") and cathepsin K can cleave the helical region of these collagens. While MMP collagenases cleave collagen near neutral pH, cathepsin K, mainly produced by osteoclasts, functions in acidic environments, e.g. in resorbing bones (Garnero et al. 1998). Collagenolytic MMPs are MMP-1 (collagenase 1), MMP-8 (collagenase 2/neutrophil collagenase) and MMP-13 (collagenase 3), MMP-2 (gelatinase A) and MMP-14 (MT1-MMP). They cleave native collagen I at a single site ¾ of the way from the N-terminus and generate characteristic ¾-long N-terminal and ¼-long C-terminal fragments.

However, interstitial collagens in the tissue form insoluble fibrils and fibrils are arranged as suprafibrillar architectures with the non-helical telopeptide regions cross-linked intra- and inter-molecularly. Cartilage collagen fibrils are comprised mainly of collagen II, but contain collagens IX and XI as minor components (Fig. 13.5a). Collagen IX maintains the structural integrity of cartilage by aligning along the surface of collagen II fibril and other ECM molecules through the basic non-collagenous 4 (NC4) domain which projects out from the surface of fibrils (Heinegård and Saxne 2011). Collagen fibrils are much more resistant to mammalian collagenases than acid-extracted monomeric collagens in solution (Welgus et al. 1980). This is explained by the accessibility of the collagenase cleavage site which is largely blocked by the C-telopeptide of neighbouring collagen molecules (Perumal et al. 2008). Thus, proteinases that cleave non-collagenous telopeptide regions increase the accessibility of collagenolytic MMPs. These enzymes are MMP-3, neutrophil elastase, cathepsin G and lysosomal cysteine proteinases (Fig. 13.5a). Once collagen fibrils are cleaved or depolymerized, the helical structures unfold at the body temperature and become susceptible to gelatinases (MMP-2 and MMP-9) and other tissue proteinases.

Mammalian collagenases have the unique ability to cleave triple helical interstitial collagens, which requires both the catalytic M domain and the Hpx domain. However, when the 3D structure of MMP-1 was reported it became apparent that the active site of the enzyme cannot accommodate the triple helical collagens in the active site for the enzyme, because the entrance of the active site is only 5 Å wide and the diameter of the triple helix is 15 Å (Bode 1995). Simple docking of triple helical peptide indicated that the closest peptide bond is ~7 Å away from the catalytic Zn^{2+}. It has therefore been postulated that either the active site of collagenase undergoes large conformational changes to accommodate triple helix or the collagen needs to be unwound to present a single α chain to the active site to initiate collagenolysis.

The evidence that collagenase unwinds triple helical collagen was presented by Chung et al. (2004). They demonstrated that proteinases that do not cleave native collagen I (e.g., MMP-3, neutrophil elastase) can cleave collagen I in the presence of MMP-1(E200A) mutant which lacks the ability to hydrolyze peptide. Thus, it

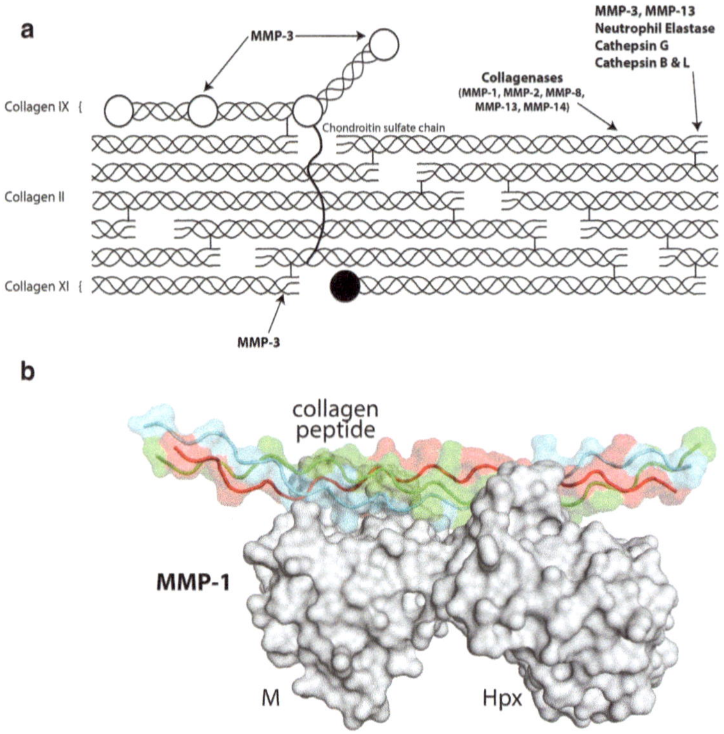

Fig. 13.5 The sites of proteinase action on cartilage collagen fibrils. (**a**) Schematic presentation of cartilage collagen fibrils and the sites of proteinase action. (**b**) MMP-1 and collagen peptide interaction. MMP-1 is represented as a *grey* surface and the three collagen chains are colored in *cyan*, *green* and *red*. Both catalytic M domain and the C-terminal Hpx domain interact with collagen and the collagen peptide is slightly bent through this interaction. The image was prepared by Robert Visse based on PDB ID: 4AUO

was concluded this MMP-1 mutant unwinds collagen without cleaving the peptide bonds. This proposal was challenged by Stultz and colleagues who proposed that collagen around the collagenase-cleavage site is less stable than the rest of the collagen molecule and it tends to be partially unfolded, based on their theoretical molecular simulation (Nerenberg et al. 2008). Such an unfolded "vulnerable" state was considered to be sufficient for collagenases to recognize and hydrolyze triple helical collagens. This model, however, does not explain why MMP-3 or neutrophil elastase alone does not cleave collagens (Chung et al. 2004). Recent NMR studies of MMP-1(E200A) mutant and triple helical collagen peptide demonstrated unfolding of the triple helix upon interacting with MMP-1(E200A) (Bertini et al. 2012). The importance of the Hpx domain for collagenolysis was shown by that the M domain and the Hpx domain cooperatively bind to collagen and it is temperature dependent, suggesting a subtle structural changes in triple helix enhance collagenase-collagen interaction and this subtle changes are prone to be the C-terminal side of collagenase cleavage site in collagen (Manka et al. 2012).

The crystal structure of the MMP-1(E200A)-collagen peptide complex indicated that MMP-1 has an extended collagen binding sites in both the M and the Hpx domains and all three chains of collagen interact with the enzyme (Manka et al. 2012) (Fig. 13.5b). It is suggested a flexing of the M and the Hpx domains help unwind collagen (Bertini et al. 2012; Manka et al. 2012).

13.2.3 MMPs in Arthritis

MMP activities are not readily detected in the normal steady state tissues, but the production of many MMPs is transcriptionally regulated by cytokines, growth factors, hormones, cell-cell and cell-matrix interaction, physical and chemical stimuli (Sternlicht and Werb 2001). MMP activities are also functionally regulated through their activation from their precursors, inhibition by endogenous inhibitors and cellular uptake by endocytic pathways. They are functional according to physiological demands and pathological situations.

The main cellular sources of MMPs in RA are synovial lining cells and macrophages, and their production is greatly influenced by inflammatory mediators, cytokines and growth factors (e.g. IL-1, TNF, IL-17, IL-18, oncostatin M, transforming growth factor-β, histamine) derived primarily from activated lymphocytes, macrophages and mast cells (see Fig. 13.1).The production of some of these MMPs is suppressed by IL-4, IL-10 and IL-13. These MMPs are considered to act mainly on ECM components, but they clearly have a number of other functions that orchestrate cell migration and infiltration of the tissue, and the release of growth factors (Overall 2002). For example, MMP-8-null mice have exacerbated adjuvant-induced arthritis due to suppression of the transcripts of the apoptosis initiator caspase 11 (the murine homologue of human caspases 4 and 5), resulting in delayed neutrophil apoposis, although the mechanism of the reduced transcripts is not known (Cox et al. 2010). MT1-MMP (MMP-14) plays a key role in migration of many cell types. Little MMP-14 is expressed in normal synovial cells, but it is highly expressed in the pannus-cartilage junction in RA joints, where synovial cells are in contact with cartilage collagen, and allows the cells to invade into cartilage (Miller et al. 2009).

In OA the ECM-degrading proteinases are derived primarily from the cartilage. Although inflammation is less prevalent, synovial membrane inflammation may be observed particularly at an early stage (Benito et al. 2005). Thus, inflammatory cytokines may participate in enhancing the production of MMPs. Cartilage explant studies suggest that cytokines such as IL-1, IL-17, TNFα and oncostatin M (Milner and Cawston 2005), proteolytic fragments of fibronectin (Homandberg 1999), endothelin 1 (Roy-Beaudry et al. 2003), mechanical compression (Blain et al. 2001) and mechanical injury (Lee et al. 2005a) are potential stimulatory factors. Instability of the joint, ageing and oxidative stress also contribute to cartilage matrix catabolism (Aigner et al. 2007).

MMPs are able to cleave a number of ECM components of cartilage. MMPs cleave aggrecan at the Glu^{341}–Phe^{342} bond in the interglobular domain of the core protein and MMP-generated G1-VDIPEN^{341} fragments are found in RA and OA cartilage (Lark et al. 1997; Struglics et al. 2006). However, in vitro studies indicate that aggrecanolytic activities of MMPs are much weaker than that of ADAMTS-5 (Durigova et al. 2011). Collagen II fibrils are, on the other hand, degraded primarily by MMPs (Dodge and Poole 1989). Potential candidate enzymes are MMP-1 and MMP-13 in interterritorial collagenolysis and MMP-2 and MMP-14 in pericellular collagenolysis. Collagen IX is cleaved by MMP-3 at NC2 and NC domains. This cleavage of collagen IX may affect the susceptibility of collagen fibrils by collagenolytic MMPs. Ablation of MMP-13 in mice protects the cartilage from degradation upon induction of OA, but not chondrocyte hypertrophy nor osteophyte formation (Little et al. 2009). Prevention of cartilage and bone destruction in animal models of RA and OA by synthetic MMP inhibitors (Conway et al. 1995; Ishikawa et al. 2005; Sabatini et al. 2005) also indicates that MMPs play a key role in joint destruction. However, it is still unclear which MMPs plays key roles in cartilage destruction in humans.

13.3 ADAMTSs and Aggrecanases

The human gemone encodes 19 ADAMTSs. The first ADAMTS (ADMTS1) was discovered by Kuno et al. (1997) in mouse colon adenocarcinoma cells as a cDNA which encodes a signal peptide, a pro-domain and metalloproteinase (M) domain, a disintegrin (Dis) domain, a thrombospondin type I (TS) domain, a cysteine-rich (CysR) domain, a spacer (Sp) domain and two additional TS domains at the C-terminus (Fig. 13.1). The M domain and the Dis domain are similar in amino acid sequence to those of ADAMs, but TS domains are unique. Thus, the terms, “ADAMTS” was coined. The proteolytic activity of ADAMTS1 was first demonstrated as its ability to cleave α2-macroglobulin (Kuno et al. 1999). The second enzyme cloned was ADAMTS2, which process the N-propetide of procollagens I, II, and III (Colige et al. 1997). This enzyme has a unique domain specific to procollagen N-proteinase and a proteinase and lacunin (PLAC) domain at the C-teminus. ADAMTS3 (Fernandes et al. 2001) and ADAMTS14 (Bolz et al. 2001; Colige et al. 2002) are also characterized as procollagen N-proteinase with similar domain arrangements.

ADAMTS4 was the first enzyme characterized as an aggrecanase that cleaves the Glu^{373}–Ala^{374} bond in the interglobular domain of aggrecan core protein and called aggrecanase 1 (Tortorella et al. 1999). Aggrecanase 2 was also reported soon after and originally assigned to ADAMTS11 (Abbaszade et al. 1999), but it was found to be identical to ADAMTS5 (Hurskainen et al. 1999) and therefore ADAMTS11 is no longer used. However, the detection of this unique aggrecan-cleaving activity was reported in the early 1990s in normal and interleukin 1 (IL-1)-treated cartilage (Ilic et al. 1992; Loulakis et al. 1992; Sandy et al. 1991).

Several other fragments of aggrecan resulting from cleavage of Glu^{1666}–Gly^{1667}, Glu^{1771}–Ala^{1771}, and Glu^{1871}–Leu^{1871} were also found (Ilic et al. 1992; Loulakis et al. 1992). Since these fragments were found in synovial fluid and serum of age-matched animals (Ilic et al. 1992), it was suggested that the enzyme responsible for these cleavages was important in both physiological and pathological catabolism of aggrecan. Aggrecan fragments resulting from the cleavage of the Glu^{373}–Ala^{374} bond accumulate in synovial fluids of patients with OA and inflammatory joint injury (Lohmander et al. 1993; Sandy et al. 1992).

Another notable member of the family is ADAMTS13 that selectively cleaves ultra-large von Willebrand factor under sheer flow stress. ADAMTS13 is unique in both structure and enzymatic activity. It is the only ADAMTS that has two C-terminal CUB (complement C1r/C1s—urchin epidermal growth factor-bone morphogenetic protein 1) domains along with seven TS domains after the Sp domain. The control of the size of multimeric von Willebrand factor by ADAMTS13 is important to maintain normal hemostasis, and larger multimers are hemostatically more active (Sadler 1998). Reduced ADAMTS13 activity in plasma causes thrombotic thrombocytopenic purpura (TTP) which characterized by thrombocytopenia, microangiopathic haemolytic anaemia, renal failure, neurologic dysfunction and fever resulting from blood clots formed in micro-capillaries. The deficiency of its activity is caused either by genetic mutation of the gene (Levy et al. 2001; Lotta et al. 2010) or by autoantibodies (Luken et al. 2006) that inhibit the ADAMT13 activity.

13.3.1 3D Structures of ADAMTSs

ADAMTSs are synthesized as pre-proenzymes, and they are intracellularly processed and secreted from the cell. The pro-domain of these enzymes may assist folding of the proteinase (Koo et al. 2007; Longpre and Leduc 2004) and they are most likely processed by a pro-protein convertase intracellularly as they have furin-like proteinase processing sites (Longpre and Leduc 2004; Wang et al. 2004), although proADAMTS5 (Longpre et al. 2009) and proADAMTS9 (Koo et al. 2006) may be activated on the cell surface. The first 3D structure reported in this family was the M domain with the Dis domain of ADAMTS1 in 2007 (Gerhardt et al. 2007). In the following year, the structures of the M-Dis domains of ADAMTS4 and ADAMTS5 (Mosyak et al. 2008) and the M domain of ADAMTS-5 (Shieh et al. 2008) were also reported. Full-length ADATMTS protein structures have not been solved yet, but Akiyama et al. (2009) reported the crystal structure of the Dis-TS-CysR-Sp domains of ADAMTS13. Figure 13.6 shows a composite structure of the M-Dis domain of ADAMTS4 (Mosyak et al. 2008) and the TS-CysR-Sp of ADAMTS13 (Akiyama et al. 2009), illustrating a possible topological arrangement of the domains in this family of enzymes. Since the residues located at the domain-domain interface are concerved among ADAMTSs, the general topological arrangements of Dis-TS-CysR-Sp in ADAMTSs are considered to be similar to that of ADAMTS13 (Akiyama et al. 2009).

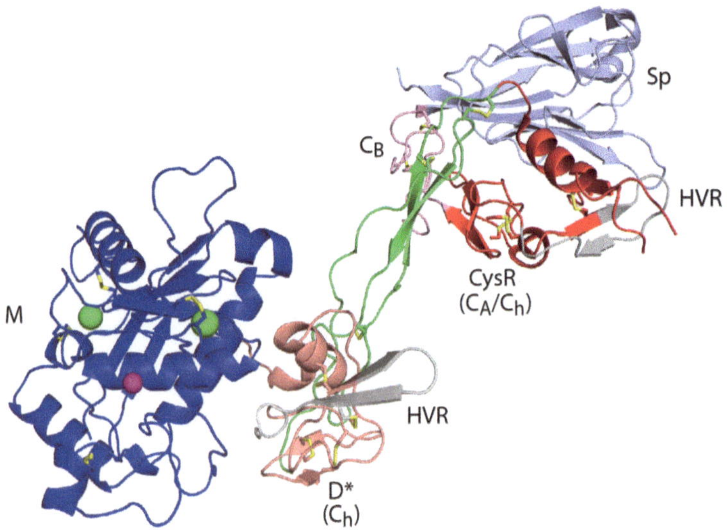

Fig. 13.6 A 3D structural model of ADAMTS. The structure was modelled based on the M and D* domain of ADMTS-5 (PDB ID: 2RJQ) and the D*, TS, CysR and Sp domains of ADMTS13 (PDB ID: 3GHN). Catalytic zinc ion is in *purple sphere*; calcium ions are in *green sphere*; and disulfide bonds in *yellow*. The image was prepared by Robert Visse

13.3.1.1 M Domain

The M domain of ADAMTS1 consists of a twisted central β-sheet of five strands, five α helices and connecting loops. The overall polypeptide fold of the M domain is a typical metzincin metalloproteinase fold (Gerhardt et al. 2007), similar to those of MMPs and ADAMs. The catalytic zinc ion interacts with three His residues in the HEXXHXXGXXH motif, but there is no structural zinc. There are, however, four disulfide bridges and the fourth bridge Cys^{379}-Cys^{462} connects the loop between strands β4 and β5 and a 22-residue connector loop after helix α5 which wraps around the back of the M domain and connects with the Dis domain. This arrangement places the Dis domain on the sites, so-called "primed sites", which interact with substrate residues located at C-terminal side of the scissile bond. An interesting feature noted for the M domains of ADAMTS4 and ADAMTS5 (Mosyak et al. 2008) is that, in the presence of an active site-directed inhibitor, the unique S2′-loop defined by a short disulfide-containing loop with the sequence CGXXXCDTL (322–330 in ADAMTS4) adopts a compact β-turn structure which is stabilized by the disulfide bond and a Ca^{2+}. Without the inhibitor, the S2′-loop is in a markedly different conformation and loses a Ca^{2+}. These changes bring Asp^{328} and Thr^{329} closer to the catalytic Zn^{2+}, and the carboxylate group of Asp^{328} chelates the Zn^{2+} and Thr^{329} fills the space at the mouth of the S1′ pocket. The S2′-loop therefore behaves like an auto-inhibitor of the metalloproteinases. This interaction is reminiscent of the "cysteine switch" found in the pro-domains of MMPs that interacts with the catalytic Zn^{2+} and maintains their inactive zymogen form.

Non-catalytic domains of ADAMTSs play important roles in substrate recognition and catalysis. It may be speculated that the binding of a natural substrate through the non-catalytic ancillary domains of the enzyme may induce such changes of the S2′-loop open and activate the enzyme. This may also explain why only certain substrates that interact with the non-catalytic domain are substrates of this group of enzymes.

13.3.1.2 Dis Domain

The 3D structure of the Dis domain of ADAMTS1, 4 and 5 (Gerhardt et al. 2007; Mosyak et al. 2008) did not show structural homology in 3D to the Dis domain of snake venom metalloproteinase VAP1 (Takeda et al. 2006), or that of ADAM10 (Janes et al. 2005). It is similar to the structure of the "hand" (C_h) segment of the CysR domain of VAP1. The CysR domain of VAP1 is structurally subdivided into "wrist" (C_w) and "hand" (C_h) subdomains (Takeda 2009; Takeda et al. 2006). The C_h subdomain has a core of α/β-fold structure consisting of two antiparallel β strands packed against two of the three α helices and five disulfide bonds (Takeda et al. 2006). The C_h fold of VAP1 is unique, but it is similar to the fold of the Dis domain and an N-terminal part of the CysR domain called "C_A" subdomain of ADAMTS13 (illustrated in red in Fig. 13.6), even though their amino acid sequence identity is low (Akiyama et al. 2009). The Dis domain of ADAMTSs is designated as "D*" by Takeda et al. (2012) to distinguish its structure from the canonical Dis domains found in snake venom metalloproteinases and ADAMs, and this distinction is adopted here. The D* domain in ADAMTS13 contains the loop region encompassing the central core of the D* domain. The sequences corresponding to this region in ADAMs and ADAMTSs are the most divergent and variable in length (11–55 amino acids) and it is referred as the hyper-variable (HVR) region. The HVR is located in close proximity to the active site and forms a part of the S3′ pocket and therefore participates in substrate interaction (Fig. 13.6). This may be the reason why the M domain alone has little proteolytic activity (Gendron et al. 2007; Kashiwagi et al. 2004).

13.3.1.3 TS Domain

The TS domain has an anti-parallel 3-stranded fold with three disulfide bonds, (Akiyama et al. 2009), a very similar structure of that of the prototypical thrombospondin type I repeat in thrombospondin-1 (Tan et al. 2002).

13.3.1.4 CysR Domain

The CysR domain of ADAMTS13 consists of two subdomains, the N-terminal C_A subdomain and the C-terminal C_B subdomain (Fig. 13.6) (Akiyama et al. 2009).

The structure of the C_A domain is similar to that of D* (or C_h), although they have only 17 % identity in amino acid sequence. The C_B subdomain does not have an apparent secondary structure (shown in pink in Fig. 13.6), but it has a series of turns stabilized by a pair of disulfide bonds at its N- and C-termini and forms a rod shape. The C_A subdomain contains an RGD (498–500) sequence and an HVR, but the side-chain of Arg^{498} of RGD is buried and unavailable for protein-protein interaction, but Asp^{500} is exposed to the solvent.

13.3.1.5 Sp Domain

The Sp domain is a long segment without cysteines and the primary structure shows no apparent homology to known structural motifs, but it folds into a single globular domain with 10 β-strands in a jelly-roll topology, forming two anti-parallel β-sheets that lie almost parallel to each other (Akiyama et al. 2009). Many key residues of the Sp domains are conserved among ADAMTS proteins and it is predicted that ADAMTSs share a similar domain architecture. In addition, the N- and C-termini of the Sp domain are in close proximity and therefore the TS2 domain after the Sp domain should be close to the C_A/Sp domain junction.

13.3.2 The Role of Non-catalytic Domain of Aggrecanases (ADAMTS4 and ADAMTS5) in Aggrecanolysis

There are eight ADAMTSs (ADAMTS 1, 4, 5, 8, 9, 15, 17 and 18) that have been reported to cleave aggrecan core protein at the aggrecanase sites (Stanton et al. 2011). Among them ADAMTS4 and ADAMTS5 have been extensively characterized biochemically as they are much more active than other ADAMTSs on aggrecan cleavage. The mature full-length ADAMTS4 and ADAMTS5 cleave aggrecan core protein effectively, but the M-D* domains alone show little aggrecananse activity (Gendron et al. 2007; Kashiwagi et al. 2004), suggesting that non-catalytic ancillary domains regulate their aggrecanolytic activities.

Domain deletion studies of ADAMTS4 by Kashiwagi et al. (2004) showed that the full-length ADAMTS4 has the greatest aggrecan-degrading activity, and the deletion of the Sp domain reduces the activity by 20 %. Further deletion of the CysR reduces it by about 60 %. The subsequent deletion of the TS domain yields a form with only 3 % of the full activity. However, it was reported that full-length ADAMTS4 showed only a very weak ability to cleave the Glu^{373}–Ala^{374} bond in the interglobular domain of aggrecan (Gao et al. 2004; Kashiwagi et al. 2004). Upon removal of the Sp domain by proteolysis (Gao et al. 2004) or mutagenesis (Kashiwagi et al. 2004), ADAMTS4 gained the ability to cleave the Glu^{373}–Ala^{374} bond and activity towards S-carboxymethylated transferrin, fibromodulin and decorin (Kashiwagi et al. 2004). Gao et al. (2004) found that MT4-MMP

(MMP-17) on the cell surface processes the C-terminal Sp domain and activates ADAMTS4 to cleave the Glu^{373}–Ala^{374} bond. Those data were interpreted to mean that the Sp domain inhibits ADAMTS4 activity towards specific cleavage sites. Later, however, Fushimi et al. (Fushimi et al. 2008) found that the suppressed cleavage of the Glu^{373}–Ala^{374} bond by full-length ADAMTS4 was due to the presence of heparin that was used to prevent the full-length enzyme from binding to ECM. Once heparin was chromatographically removed, ADAMTS4 revealed about 20-times higher activity on the Glu^{373}–Ala^{374} bond than the Sp domain-deleted form. It was found that heparin preferentially masks the activity of full-length ADAMTS4 towards the Glu^{373}–Ala^{374} bond, but it is less effective on its activity for the chondroitin sulfate-rich region. This suggests that full-length ADAMTS4 bound to heparan sulfate proteoglycan on the cell surface or the ECM has greatly reduced activity towards the Glu^{373}–Ala^{374} bond in the interglobular damain. When the Sp domain is removed, the enzyme can be released from the cell surface or ECM, and it gains broader proteolytic activity and digest decorin, biglycan, fibromodulin. However, the truncation of the Sp domain from ADAMTS4 results in about 95 % loss of aggrecanase activity including the Glu^{373}–Ala^{374} bond (Fushimi et al. 2008). Thus, removal of the Sp domain from ADAMTS4 shifts the preference of its substrates as well as the site of its action in the tissue.

ADAMTS5 also binds to the cell surface and the ECM. The main ECM binding site of ADAMTS5 is located in the CysR domain (Gendron et al. 2007). Full-length ADAMTS5 is about 30-times more active on aggrecan than full-length ADAMTS4. Domain deletion studies have indicated that the C-terminal TS domain has little effect on aggrecanase activity, but the deletion of the Sp domain lowers the activity by about 99 % for the chondroitin sulfate region, but only by about 50 % for the Glu^{373}–Ala^{374} bond. Further deletion of the CysR domain reduced the activity on the Glu^{373}–Ala^{374} bond by further 25 % (Gendron et al. 2007). Thus, non-catalytic domains are important controllers of the aggrecanolytic activity.

13.3.3 ADAMTSs in Arthritis

The major role of ADAMTSs in arthritis is considered to be their ability to cleave aggrecan and both ADAMTS 4 and 5 are much more effective in this activity than MMPs (Durigova et al. 2011). They also cleave other ECM molecules such as fibromodulin, biglycan and decorin in cartilage in vitro. Again, MMPs are not effective enzymes to cleave decorin (Geng et al. 2006). The treatment of human cartilage with IL-1 or TNFα increases aggrecanase activity, but it has no effect on mRNA levels for ADAMTS1, 4 and 5 (Flannery et al. 1999). mRNA levels for ADAMTS4 and ADAMTS5 in human OA cartilage are not significantly elevated compared to that in normal cartilage (Kevorkian et al. 2004), but ADAMTS5 mRNA levels are higher than ADAMTS4 mRNA (Bau et al. 2002). Treatment of human OA chondrocytes in culture with IL-1 increased ADAMTS4 mRNA levels, but did not alter the levels of ADAMTS5 mRNA (Bau et al. 2002). While some studies confirmed the unaltered ADAMTS5 mRNA level in human chondrocytes upon IL-1 treatment, other studies reported its increase [see Fosang et al. (2008) for review]. Inconsistency among these reports may be due to the age of the tissue, culture conditions of isolated chondrocytes, the time of the transcriptional activity measurement and stability of the mRNA. Nonetheless, an increase in the aggrecan fragments generated by aggrecanases were found in RA and OA cartilage (Lark et al. 1997) and in synovial fluids (Lohmander et al. 1993; Sandy et al. 1992; Struglics et al. 2006), suggesting that ADAMTSs are important enzymes in aggrecanolysis. ADAMTS4-null mice and ADAMTS5-null mice show no obvious abnormality, but when challenged either via surgically induced OA (Glasson et al. 2004, 2005) or antigen-induced inflammatory arthritis (Stanton et al. 2005), the degradation of aggrecan in the cartilage of ADAMTS5-null mice was protected, but not that of ADAMTS4-null mice, indicating that ADAMTS5 is a major aggrecan-degrading enzyme in cartilage, at least in mice. Whether ADAMTS5 is a key aggrecanase in the development of human OA will only be determined by further investigation, as the levels of ADAMTS4 expression in both mRNA and protein correlate with OA progression (Naito et al. 2007).

While the importance of aggrecanases in aggrecanolysis is well recognised, using neoepitope antibodies that detect either MMP cleaved fragments or ADAMTS cleaved fragments, Lark et al. (Lark et al. 1997) showed that both RA and OA cartilages contain aggrecan fragments generated by MMPs (G1-NITEGE^{373} fragment) and ADAMTSs (G1-VDIPEN^{360} fragment). Struglics et al. (2006) confirmed these observations in OA cartilage and suggested that MMP-mediated aggrecanolysis is mostly pericellular while ADAMTSs are both pericellular and in the intraterritorial regions. Based on the fact that MT1-MMP-null mice and MMP-9-null mice cause destruction of articular cartilage, impairment of endochondral-ossification and fracture repair, it is suggested that some MMPs may be important in cartilage matrix homeostasis (Sandy 2006). Kevorkian et al. (2004) showed that ADAMTS16 is elevated in late human OA cartilage, but its function is not known. More recently, Yamamoto et al. (2013) reported that ADAMTS5 activity is regulated by endocytosis mediated by low density-lipoprotein receptor related protein 1 (LRP1) in normal cartilage, but this endocytic pathway is impaired in OA cartilage due to increased shedding of LRP1. This is considered one of the mechanisms that increase aggrecanolytic activity in OA cartilage without changing mRNA levels of ADAMTS5.

13.4 ADAMs

The ADAMs are type I transmembrane proteins with a characteristic domain structure comprised of a pro-peptide, a metalloproteinase (M) domain, a disintegrin (Dis) domain, a cysteine-rich (CysR) domain, an EGF-like (or other) domain, a transmembrane domain and a cytoplasmic domain (Fig. 13.1) (White 2003).

Homologous M, Dis and CysR domains are also found in snake venom metalloproteinases (SVMPs) (Gomis-Rüth 2003; Takeda et al. 2012). Thirteen out of a total of 21 ADAMs in the human genome have a potentially functional catalytic domain with the characteristic metzincin zinc-binding motif, but other ADAMs do not have proteolytic activity as they lack a complete zinc-binding motif. The pro-peptide acts as a molecular chaperone during synthesis (Leonard et al. 2005) and interacts with the catalytic cleft of the M domain, forming the inactive precursor (Gonzales et al. 2004). Proteolytic cleavage of the pro-peptide by a furin-like proteinase in the secretory pathway generates a mature active form of the ADAM in many cases. The isolated pro-domain of ADAM17 inhibits the proteolyic activity of the M domain, but the cysteine residue found in the putative cysteine switch in not important for the inhibition of the enzyme activity (Gonzales et al. 2004). Many of the mammalian ADAM metalloproteinases have been shown to cleave cell surface proteins, such as cytokines, growth factors, growth factor receptors and binding proteins and cell adhesion molecules, a process referred to as "shedding" (Becherer and Blobel 2003; Edwards et al. 2008; Weber and Saftig 2012; White 2003). In rheumatoid arthritis, ADAM17, also called tumor necrosis factor α converting enzyme (TACE), releases the inflammatory mediator TNFα from the plasma membrane, which in turn stimulates synovial cells to produce ECM-degrading enzymes. ADAM9, ADAM10 and ADAM15 are also expressed in rheumatoid synovial cells (Komiya et al. 2005) and ADAM9, ADAM 12 and 15 in OA cartilage (Okada et al. 2008).

13.4.1 3D Structure of the ADAMs

The first 3D structure of this family member solved was adamalysin II from eastern diamondback rattlesnake venom (Gomis-Rüth et al. 1993). Since then numerous structures of snake venom metalloproteinases have been solved including those with the M, Dis and CysR domains, which belong to P-II and P-III SVMPs (Takeda et al. 2012). Among the mammalian ADAMs, structures of the M domains of ADAM17 (Maskos et al. 1998) and ADAM33 (Orth et al. 2004), the Dis-CysR domains of ADAM10 (Janes et al. 2005), and the full ectodomain of ADAM22 (Liu et al. 2009) have been reported. ADAM22 lacks proteolytic activity as its zinc-binding motif of the M domain lacks two out of the three essential histidines. However, the M domain has a typical metalloproteinase fold even without a zinc ion. ADAM22 functions as a postsynaptic receptor for the secreted neurotransmission modulators LGI1 and LGI4 (Fukata et al. 2006; Sagane et al. 2008). It is predominantly expressed in the neuronal tissues and the mice deficient in ADAM22 exhibit ataxia, seizure and hypomyelination in the peripheral nerves (Sagane et al. 2005). Figure 13.7 shows the structure of ADAM22, illustrating that the M, Dis, CysR, EGF (E) domains which are typically found in ADAMs are arranged like a four-leaf clover, each leaf representing one of the four domains. The Dis, CysR and E domains have a continuous hydrophobic core and they appear as an

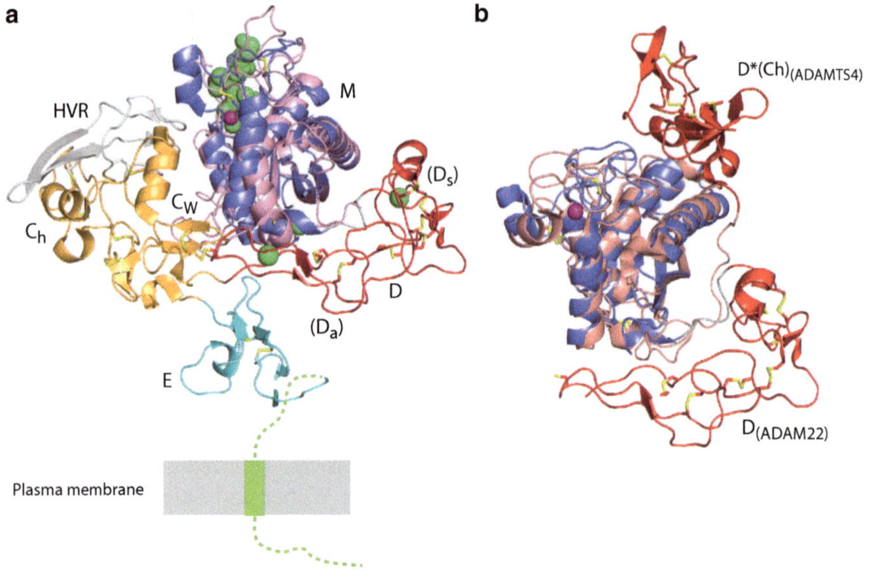

Fig. 13.7 3D structures of ADAMs. (**a**) Superimposition of the M domains of ADAM17 (PDB ID: 1BKC) that of the ectodomain of ADAM22 (PDB ID: 3G5C). The catalytic domain of ADAM17 is in *pink* and that of ADAM22 in *blue*. A hydroxamate inhibitor boud to the active site of ADAM17 is shown as a *green sphere*. The disintegrin (D), cysteine-rich domain consisting of C_w C_h, and hypervariable region (HVR) subdomains, and EGF (E) domain in the ectodomains are shown in *red*, *gold* and *grey*, and *cyan*, respectively. The catalytic zinc ion is in *purple sphere* and calcium ions in *green sphere*. The linker, transmembrane domain and the cytoplasmic tail are schematically shown in *green*. (**b**) Superimposition of the catalytic domains of ADAM22 shown in *blue* and that of ADAMTS4 (PDF ID: 2RJP) in *salmon* shows that the location of the disintegrin domains (D and D*) are located at the opposite sides of the catalytic domains. The images were prepared by Robert Visse

integral module mediated by extensive interface between the domains. The M domain is held in a concave face of the Dis-CysR-E module, and there is a long stretch of interactions between the M domain and the Dis domain and a small patch of interaction between the M domain and the CysR domain. The location of the M, Dis and CysR domains are resemble that in SVMPs, but vascular apoptosis-inducing protein-1 (VAP1) (Takeda et al. 2006) and VAP2B (Igarashi et al. 2007) have an open C-shaped conformation. However, the location of the D* domain of ADAMTSs is essentially opposite from that of the Dis domain of ADAMs (Fig. 13.7b).

13.4.1.1 M Domain

The M domain of ADAM22 and that of ADAM17 which has proteolytic activity are essentially identical, indicating that the catalytic zinc ion is not critical for folding of the M domain. Besides the catalytic Zn^{2+} the M domain has one Ca^{2+} ion.

Topological arrangement of Dis-CysR suggests that ADAM proteinases have the putative substrate binding site at the top of the molecule away from the plasma membrane. However, while ADAM17 has an extended substrate binding grove, one end of the groove of ADAM22 is blocked.

13.4.1.2 Dis Domain

The term "disintegrin" was initially used for a group of cysteine-rich RGD-containing small proteins (49–84 amino acids) from viper venom that inhibit platelet aggregation and integrin-mediated cell adhesion (Gould et al. 1990). The Dis structures of SVMPs and ADAM22 revealed similar structure as the RGD-containing disintegrin trimestatin (Fujii et al. 2003), but the disintegrin loop of ADAMs is not exposed for protein-protein interaction (Takeda 2009). The Dis domain of ADAMs and SVMPs is different from that of the D* domain of ADAMTSs as discussed above, and it is divided into two structurally distinct sub-segments, the 'shoulder' (D_s) and the 'arm' (D_a) subdomains. Both segments consist largely of a series of turns and two short regions of antiparallel β-sheet and constitute a continuous C-shaped 'arm' structure together with the N-terminal region of the CysR domain, designated as the 'wrist' (C_w) subdomain. Both D_s and D_a subdomains have three disulfide bonds and one structural calcium-binding site in each subdomain, which are essential for the structural rigidity of the C-shaped 'arm' structure. The D_a subdomain is connected to the C_w subdomain by an additional disulfide bond. ADAM10 and ADAM17 are atypical members of the ADAM family and they lack calcium-binding sites in the M domain and in D_a subdomain. Instead they have an additional disufide bond in the M domain. They also lack the E domain.

13.4.1.3 CysR Domain

The CysR domain is comprised of the "wrist" C_w subdomain and the "hand" (C_h) subdomain. The C_w subdomain consists of a pair of antiparallel β-sheets and loops. It is tightly associated with the D_a subdomain of the Dis domain on one side and with the C-terminal region of the C_h segment on the other side. The C_h subdomain has a unique fold with no structural homology to currently known proteins, except the corresponding segments of ADAMs and ADAMTSs. It has a core of αβ-fold structure consisting of two antiparallel β-stands packed against two of the three α-helices, five disulfide bonds which stabilize the core, and peripheral loops including an HVR which has a relatively small number of direct contacts with the core, suggesting that they are flexible in solution. Because of its location, opposed to the catalytic site, it is suggested that this region participates in substrate recognition (Takeda 2009).

13.4.1.4 EGF Domain

The E domain located after the C domain is divided into an N-subdomain and a C-subdomain. The N-subdomain consists of a loop region, a 3_{10} helix and a pair of antiparallel β-strands. The C-subdomain consists of two short antiparallel β-strands and the C-terminal flexible region which connect to a linker region and a transmembrane domain of the protein. There are three disulfide bonds in the E domain that stabilize its secondary structure. The alignment of E domains of ADAMs indicates that the loop regions are most diverged, but the overall structures of E domains are similar.

13.4.2 ADAMs in Arthritis

An imbalance in pro- and anti-inflammatory cytokines is thought to be a major feature of RA and may have some role in OA. The major pro-inflammatory cytokines TNFα and IL-1 are produced by synovial macrophages and ADAM17 is thought to be responsible for the solubilisation of the membrane-bound pro form of TNFα, hence it is also called TNFα converting enzyme or TACE. An endogenous inhibitor TIMP-3 inhibits ADAM17 (Amour et al. 1998) and TIMP-3-null mice have shown the importance of ADAM17 in the regulation of TNFα in systemic inflammation (Smookler et al. 2006). When antigen induced arthritis was studied in TIMP-3-null mice, compared to wild-type animals they showed a drastic increase in the initial inflammatory response to intra-articular antigen injection, and serum TNF-α levels were greatly elevated, although these differences in clinical features disappeared by days 7–14 (Mahmoodi et al. 2005). More recently the importance of ADAM17 in inflammatory arthritis has also been shown by the analysis of mice lacking iRHOM2, a catalytically inactive member of the rhomboid family, which controls maturation of ADAM17 in myeloid cells (Adrain et al. 2012; McIlwain et al. 2012). Mice deficient in iRHOM2 were protected from the K/BxN inflammatory arthritis model to a similar level as mice lacking ADAM17 in myeloid cells, or those deficient in TNFα (Issuree et al. 2013).

ADAM17 is activated by a variety of stimuli, including phorbol esters (Reddy et al. 2000), TNFα (Le Gall et al. 2010), IL-1β (Xu and Derynck 2010), LPS (Rousseau et al. 2008), activation of G protein-coupled receptors (Zhang et al. 2006) or FGF receptor 2 (Maretzky et al. 2011), and disulfide isomerization (Willems et al. 2010). ADAM17 also releases a number of other key effectors including TRANCE, fractalkine and various EGF receptor ligands, which may play roles in the development of arthritic disease. In addition, ADAM17 is potentially responsible for the proteolysis of receptors such as TNFR-I and TNFR-II, c-Met, IL1 receptor-II, trkA, IL-6 receptor, etc. The generation of soluble forms of receptors, e.g., IL-6 receptor, may be critical for the down-regulation of ligand function (Scheller et al. 2006). The shedding of cell adhesion molecules such as

L-selectin, VCAM-1, ICAM-1, CD30 and CD40 by ADAM17 may also have implications for cell behaviour within inflammatory and immune pathways.

While the role of ADAM17 is well recognized in RA, potential roles of other ADAMs in arthritic joint tissues have been reported. Examining the mRNA expression levels of 10 different ADAMs in synovial tissues of patients with RA or OA, Komiya et al. (Komiya et al. 2005) reported that ADAM15 mRNA was more elevated in RA samples compared with those in OA, but that negligible mRNA for ADAM8 and ADAM28 could be found whilst ADAM9, 10, and 17 were constitutively expressed in both RA and OA synovial tissues. Immunohistochemistry and *in situ* hybridisation have demonstrated a high level of expression of ADAM15 in RA synovial tissue, compared with normal or OA synovial tissue. It is found in macrophage-like and fibroblast-like synoviocytes, as well as in plasma cells (Böhm et al. 2001). Its expression was shown to have a direct correlated with vascular density and upregulated by VEGF165 (Komiya et al. 2005). The study of aging ADAM15-null mice exhibited accelerated development of OA lesions compared with wild-type mice (Böhm et al. 2005). The results were interpreted as showing that ADAM-15 had a protective role in the maintenance of joint integrity. More recent studies have shown that ADAM15 contributes anti-apoptotic pathway in OA chondrocytes upon serum- and matrix-withdrawal or under genotoxic stress-induced by camptothecin exposure (Böhm et al. 2010). This is through the cytoplasmic domain-mediated activation of survival signal-transducing kinase FAK and concurrent activation of Src (Fried et al. 2012).

In OA cartilage, elevated mRNA levels were reported for ADAM9 and membrane bound ADAM12m, as well as ADAM15, compared to those in normal human articular cartilage, but mRNAs for ADAM8, ADAM17 and ADAM28 were not detected, and ADAM19 was constitutive (Okada et al. 2008). Okada et al. (2008) also found that ADAM12m cleaves insulin-like growth factor-binding protein 5 and promote chondrocyte proliferation in OA cartilage. The production of ADAM12m was increased by TGFβ, which explains an anabolic effect of TGFβ which enhances matrix production and chondrocyte proliferation (Grimaud et al. 2002).

13.5 Endogenous Inhibitors of MMPs, ADAMTSs and ADAMs

Well-characterized endogenous inhibitors of metalloproteinases are α_2-macroglobulin and the TIMPs.

α_2M is a major plasma proteinase inhibitor of 725 kDa, consisting of four identical subunits. It inhibits most endopeptidases from different catalytic classes of proteinase regardless of their substrate specificities (Barrett 1981). The interaction of a proteinases with α_2M is initiated by proteolysis of the so-called "bait" region located in the middle of the subunits. This triggers a conformational change in α_2M that, in turn, entraps the enzyme without blocking the active site (Barrett and

Starkey 1973). The resulting α_2M-proteinase complexes are rapidly endocytosed via a scavenger receptor, LRP1 and degraded (Lillis et al. 2008). The activities of MMPs (Nagase et al. 1994), ADAMTS-1 (Kuno et al. 1999), ADAMTS-4 (Tortorella et al. 2004), and soluble form of ADAM12 (Loechel et al. 1998) have been shown to be regulated by α_2M. Indeed most, if not all, MMPs and ADAMTSs and soluble form of ADAMs are likely to be inhibited by α_2M. The site of action of α_2M is thought to be primarily in the fluid phase. The levels of α_2M in synovial fluids of RA patients are 0.7–1.0 mg/ml, approximately one-third of the normal plasma level (Flory et al. 1982).

In the tissue, TIMPs are considered to be key inhibitors of metalloproteinases. They form 1:1 enzyme-inhibitor complexes. Four TIMPs (TIMPs-1–4) are found in the human genome and they are 21–29 kDa proteins consisting of the N-terminal metalloproteinase inhibitory domain (about 125 amino acids) and the C-terminal domain (about 65 amino acids). The C-terminal domain mediates specific interaction with some MMP zymogens through their C-terminal Hpx domains. The complex formation between proMMP-2 and TIMP-2 through their C-terminal domain is important for proMMP-2 activation by MMP-14 on the cell surface. Four TIMPs have been characterized as MMP inhibitors, but some TIMPs inhibit ADAMTSs and ADAMs (Brew and Nagase 2010): TIMP-3 has the broadest activity and inhibits ADAMTS1, 4, 5, ADAM 10, 12, 28 and 33, and weakly ADAMTS2; TIMP-1 inhibits ADAM10; TIMP-2 inhibits ADAM12; and TIMP-4 inhibits ADAM28 and weakly ADAM33.

Early work by Dean et al. (1989) showed that both MMP levels and TIMP levels were elevated in an early stage of OA cartilage compared with unaffected cartilage, but the total amount of MMPs was slightly higher than that of TIMPs, whereas this balance was reversed in the unaffected cartilage. This subtle difference in the ratio of MMPs and TIMPs is considered to be a cause of the gradual degradation of the cartilage matrix.

13.5.1 Inhibition Mechanisms of TIMPs

The first 3D structure of TIMP reported was the NMR solution structure of the N-terminal inhibitory domain of TIMP-2 (N-TIMP-2), which revealed a five-stranded OB-fold β-barrel and two α-helices (Williamson et al. 1994). Elucidation of the inhibition mechanism of TIMPs had to wait for the crystal structures of the TIMP-1-MMP-3 catalytic domain complex (Gomis-Rüth et al. 1997) (see Fig. 13.8). The structure showed that TIMP is a "wedge-shaped" molecule and where Cys^1 and Cys^{70} and Cys^3-Cys^{99} are disulfide bonded and form a rigid structure, slots into the active site of MMP. The N-terminal α-amino and carboxyl groups of Cys^1 bidentately coordinate the catalytic Zn^{2+} of the enzyme and the N-terminal segment binds to the active site cleft subsite S1–S4′ of the enzyme like the substrate, and Ser^{68} and Val^{69} are orientated towards the S_2 and S_3 subsites, respectively, nearly opposite to the direction of a substrate. The side chain of Thr^2

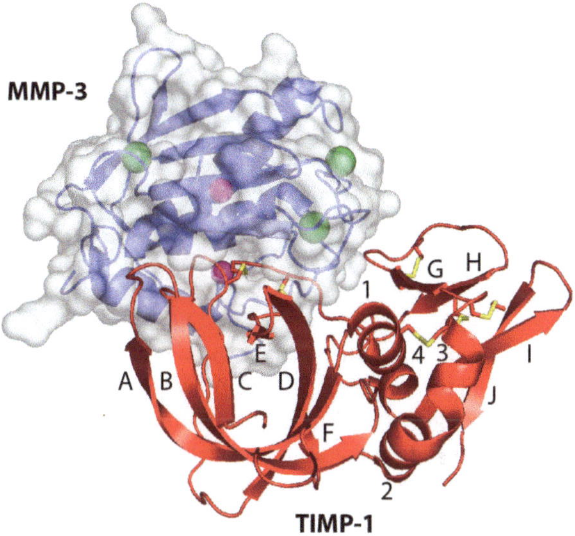

Fig. 13.8 3D structure of the TIMP-1-MMP-3 catalytic domain. TIMP-1 is shown in *red* and disulfide bonds in *yellow*; α-helices and β-strand are indicated. MMP-3 is shown in *blue* ribbon and *grey* surface. Zinc ions and calcium ions in MMP-3 are shown as *purple* and *green spheres*. The image were prepared by Robert Visse based on PDB ID: 1UEA

extends into the large S1′ specificity pocket of the MMP. The structure of the N-TIMP-3-ADAM17 catalytic domain (Wisniewska et al. 2008) has indicated that ADAMs and ADAMTSs may interact with TIMPs in a similar binding mode.

Because the N-terminal α-amino group is essential to chelate the catalytic Zn^{2+} and it plays an important role in inhibition of MMPs, any extension or blocking of the N-terminal drastically reduces the inhibitory activity for MMPs (Higashi and Miyazaki 1999; Troeberg et al. 2002; Wingfield et al. 1999). Similarly the mutation of Thr^2 to Gly leads to greatly reduced inhibitory activity for MMPs (Meng et al. 1999). However, this mutation has a relatively small effect on the affinity of N-TIMP-1 for MMP-9 (Hamze et al. 2007), suggesting that interaction outside of the S1′ pocket are more important with MMP-9 than with other MMPs. Mutation of Thr^2 to Gly of N-TIMP-3 also reduced inhibitory activity for most MMPs, but this mutant retains good inhibitory activity for ADAM17 (Wei et al. 2005) and ADAMTS-4 and -5 (Lim et al. 2010). Addition of one or two extra Ala to the N-terminus ([-1A]N-TIMP-3 or [-2A]N-TIMP-3) also retained inhibitory activity for ADAM17 and ADAMTS-4 and -5 (Lim et al. 2010; Wei et al. 2005). Those mutants are useful to discriminate activities derived from MMPs and those from ADAM17 or ADAMTS-4 and -5. TIMP-1, TIMP-2 and TIMP-4 do not inhibit ADAM17. A series of studies by Lee et al. mutated the TIMPs to inhibit ADAM17, further delineating the segments that can dictate specificity of each TIMP molecule (Lee et al. 2004a, b, 2005b).

13.5.2 Pentosan Polysulfate Is an Exosite Inhibitor of ADAMTS4 and ADAMTS5

Pentosan polysulfate (PPS) is a chemically sulfated xylosan extracted from beechwood with a molecular mass of 4–6 kDa. It has been shown to be an effective anti-arthritic agent in animal models (Ghosh 1999). PPS inhibits ADAMTS4 and ADAMTS5 activity on native aggrecan, but not non-aggrecan substrates, by binding to the Sp domain of ADAMTS4 and the Sp and CysR domains of ADAMTS5, but not to the M domain of the enzymes (Troeberg et al. 2008). PPS is therefore an exosite inhibitor of aggrecanases. Other interesting activities of PPS include its ability to increase the affinity between aggrecanases and TIMP-3 more than 100-fold (Troeberg et al. 2008). A single chain of PPS binds to the Sp domain of ADAMTS5 and to TIMP-3 by forming a trimolecular complex. This interaction depends on the size and 11 saccharide length is sufficient for the trimolecular complex formation, which is driven electrostatically (Troeberg et al. 2012). PPT also bocks the LRP1-mediated endocytosis of TIMP-3 by chrondrocytes or other cell types (Troeberg et al. 2008). The treatment of the cartilage with IL-1 induces aggrecanase activity and aggrecan degradation. This activity is blocked by PPS, which is associated with an increased amount of TIMP-3 in the cartilage without changing its mRNA levels. This is due to blocking TIMP-3 endocytosis and increasing the affinity of TIMP-3 and aggrecanases. PPS is ineffective in blocking of the IL-1-induced aggrecan degradation of the cartilage from TIMP-3-null mice, indicating that chondroprotective activity of PPS is dependent on the presence of TIMP-3 (Troeberg et al. 2008).

13.6 Future Prospects

A number of MMPs, ADAMTSs and ADAMs have been implicated in the progression of RA and OA and they are involved in degradation of cartilage ECM, altering cellular environments and trafficking cells. Therefore, the agents that regulate these metalloproteinase activities are potential therapeutics for RA and OA. The structural studies have assisted to design numerous potent inhibitors and some have been advanced to clinical studies for arthritis, and also for cancer and cardiovascular disease, but it has turned out to be challenging. Whilst all compounds showed good efficacy and tolerability in animal models of arthritis, all have been discontinued citing either lack of efficacy or appearance of musculoskeletal syndrome (MSS) in humans. This is due to poor pharmacokinetics in man and/or poor selectivity. Good advancement has been made in the case of MMP-13 inhibitors that do not have a chelating moiety, but enter into the deep S1′ pocket of the active site (Engel et al. 2005; Johnson et al. 2007). This greatly increases the selectivity of the inhibitor. Another important consideration is the exosites of these enzymes. MMPs, ADAMs and ADAMTSs are multi-domain proteinases and in many cases their activities for the natural ECM substrates are regulated by non-catalytic ancillary domains. Deletion of such domains critically impairs their activity, suggesting molecules that modify their functions (allosteric inhibitors) may specifically affect the enzymatic activity on natural substrates. An example may be seen in autoantibodies found in patients with thrombotic thrombocytopenic purpura that inhibit the activity of ADAMTS-13 to cleave ultra-large von Willebrand factor. Those antibodies are directed to the non-catalytic domains of the proteinase, not toward the catalytic domain (Soejima et al. 2003). Specific antibodies and allosteric inhibition are ideal for selective inhibition of a target proteinase, and such antibodies have been engineered for ADAM17 (Tape et al. 2011; Richards et al. 2012) and MMP-14 (Basu et al. 2012). Therefore, this direction of research will be further developed.

The activities of these metalloproteinases are regulated at the level of transcription, activation from precursors or inactive forms, inhibition by endogenous inhibitors. More recently, it has been shown that the synthesis of ADAM17 is regulated by iRHOMs, and that some of MMPs, ADAMTSs and TIMPs are endocytosed via a scavenger receptor LRP1. Furthermore, under inflammatory conditions shedding of LRP1 increased and the levels of soluble LRP1 is elevated in plasma of patients of rheumatoid arthritis, systemic lupus erythematous (Gorovoy et al. 2010) and liver disease (Quinn et al. 1997). LRP1 protein is largely lost in OA cartilage (Yamamoto et al. 2013). These findings indicate further complexity of regulation of these metalloproteinases, but they also suggest numerous possibilities to develop therapeutic agents for RA and OA, and for other diseases associated with aberrant ECM turnover.

Acknowledgements We thank Ngee Han Lim, Alan Lyons and Rob Visse for preparation of the figures. This work is supported by grant from Arthritis Research UK, European FP7 LIVIMODE programme, NIH/NIAMS grant AR40994 (to HN), and Cancer Research UK and Medical Research Council (to GM).

References

Abbaszade I, Liu RQ, Yang F, Rosenfeld SA, Ross OH, Link JR, Ellis DM, Tortorella MD, Pratta MA, Hollis JM, Wynn R, Duke JL, George HJ, Hillman MC Jr, Murphy K, Wiswall BH, Copeland RA, Decicco CP, Bruckner R, Nagase H, Itoh Y, Newton RC, Magolda RL, Trzaskos JM, Burn TC (1999) Cloning and characterization of ADAMTS11, an aggrecanase from the ADAMTS family. J Biol Chem 274:23443–23450

Adrain C, Zettl M, Christova Y, Taylor N, Freeman M (2012) Tumor necrosis factor signaling requires iRhom2 to promote trafficking and activation of TACE. Science 335:225–228

Aigner T, Soder S, Gebhard PM, McAlinden A, Haag J (2007) Mechanisms of disease: role of chondrocytes in the pathogenesis of osteoarthritis–structure, chaos and senescence. Nat Clin Pract Rheumatol 3:391–399

Akiyama M, Takeda S, Kokame K, Takagi J, Miyata T (2009) Crystal structures of the noncatalytic domains of ADAMTS13 reveal multiple discontinuous exosites for von Willebrand factor. Proc Natl Acad Sci U S A 106:19274–19279

Amour A, Slocombe PM, Webster A, Butler M, Knight CG, Smith BJ, Stephens PE, Shelley C, Hutton M, Knäuper V, Docherty AJP, Murphy G (1998) TNF-Alpha converting enzyme (TACE) is inhibited by TIMP-3. FEBS Lett 435:39–44

Barrett AJ (1981) α2-Macroglobulin. Methods Enzymol 80:737–754

Barrett AJ, Starkey PM (1973) The interaction of α2-macroglobulin with proteinases. Characteristics and specificity of the reaction, and a hypothesis concerning its molecular mechanism. Biochem J 133:709–724

Basu B, Correa de Sampaio P, Mohammed H, Fogarasi M, Corrie P, Watkins NA, Smethurst PA, English WR, Ouwehand WH, Murphy G (2012) Inhibition of MT1-MMP activity using functional antibody fragments selected against its hemopexin domain. Int J Biochem Cell Biol 44:393–403

Bau B, Gebhard PM, Haag J, Knorr T, Bartnik E, Aigner T (2002) Relative messenger RNA expression profiling of collagenases and aggrecanases in human articular chondrocytes in vivo and in vitro. Arthritis Rheum 46:2648–2657

Becherer JD, Blobel CP (2003) Biochemical properties and functions of membrane-anchored metalloprotease-disintegrin proteins (ADAMs). Curr Top Dev Biol 54:101–123

Becker JW, Marcy AI, Rokosz LL, Axel MG, Burbaum JJ, Fitzgerald PM, Cameron PM, Esser CK, Hagmann WK, Hermes JD et al (1995) Stromelysin-1: three-dimensional structure of the inhibited catalytic domain and of the C-truncated proenzyme. Protein Sci 4:1966–1976

Benito MJ, Veale DJ, FitzGerald O, van den Berg WB, Bresnihan B (2005) Synovial tissue inflammation in early and late osteoarthritis. Ann Rheum Dis 64:1263–1267

Bertini I, Calderone V, Fragai M, Jaiswal R, Luchinat C, Melikian M, Mylonas E, Svergun DI (2008) Evidence of reciprocal reorientation of the catalytic and hemopexin-like domains of full-length MMP-12. J Am Chem Soc 130:7011–7021

Bertini I, Fragai M, Luchinat C, Melikian M, Mylonas E, Sarti N, Svergun DI (2009) Interdomain flexibility in full-length matrix metalloproteinase-1 (MMP-1). J Biol Chem 284:12821–12828

Bertini I, Fragai M, Luchinat C, Melikian M, Toccafondi M, Lauer JL, Fields GB (2012) Structural basis for matrix metalloproteinase 1-catalyzed collagenolysis. J Am Chem Soc 134:2100–2110

Blain EJ, Gilbert SJ, Wardale RJ, Capper SJ, Mason DJ, Duance VC (2001) Up-regulation of matrix metalloproteinase expression and activation following cyclical compressive loading of articular cartilage in vitro. Arch Biochem Biophys 396:49–55

Bode W, Gomis-Rüth FX, Stöcker W (1993) Astacins, serralysins, snake venom and matrix metalloproteinases exhibit identical zinc-binding environments (HEXXHXXGXXH and Met-turn) and topologies and should be grouped into a common family, the 'metzincins'. FEBS Lett 331:134–140

Bode W (1995) A helping hand for collagenases: the haemopexin-like domain. Structure 3:527–530

Böhm BB, Aigner T, Blobel CP, Kalden JR, Burkhardt H (2001) Highly enhanced expression of the disintegrin metalloproteinase MDC15 (metargidin) in rheumatoid synovial tissue. Arthritis Rheum 44:2046–2054

Böhm BB, Aigner T, Roy B, Brodie TA, Blobel CP, Burkhardt H (2005) Homeostatic effects of the metalloproteinase disintegrin ADAM15 in degenerative cartilage remodeling. Arthritis Rheum 52:1100–1109

Böhm B, Hess S, Krause K, Schirner A, Ewald W, Aigner T, Burkhardt H (2010) ADAM15 exerts an antiapoptotic effect on osteoarthritic chondrocytes via up-regulation of the X-linked inhibitor of apoptosis. Arthritis Rheum 62:1372–1382

Boldt HB, Kjaer-Sorensen K, Overgaard MT, Weyer K, Poulsen CB, Sottrup-Jensen L, Conover CA, Giudice LC, Oxvig C (2004) The Lin12-notch repeats of pregnancy-associated plasma protein-A bind calcium and determine its proteolytic specificity. J Biol Chem 279:38525–38531

Bolz H, Ramirez A, von Brederlow B, Kubisch C (2001) Characterization of ADAMTS14, a novel member of the ADAMTS metalloproteinase family. Biochim Biophys Acta 1522:221–225

Brew K, Nagase H (2010) The tissue inhibitors of metalloproteinases (TIMPs): an ancient family with structural and functional diversity. Biochim Biophys Acta 1803:55–71

Chung L, Dinakarpandian D, Yoshida N, Lauer-Fields JL, Fields GB, Visse R, Nagase H (2004) Collagenase unwinds triple-helical collagen prior to peptide bond hydrolysis. EMBO J 23:3020–3030

Colige A, Li SW, Sieron AL, Nusgens BV, Prockop DJ, Lapière CM (1997) cDNA cloning and expression of bovine procollagen I N-proteinase: a new member of the superfamily of zinc-metalloproteinases with binding sites for cells and other matrix components. Proc Natl Acad Sci U S A 94:2374–2379

Colige A, Vandenberghe I, Thiry M, Lambert CA, Van Beeumen J, Li SW, Prockop DJ, Lapière CM, Nusgens BV (2002) Cloning and characterization of ADAMTS-14, a novel ADAMTS displaying high homology with ADAMTS-2 and ADAMTS-3. J Biol Chem 277:5756–5766

Conway JG, Wakefield JA, Brown RH, Marron BE, Sekut L, Stimpson SA, McElroy A, Menius JA, Jeffreys JJ, Clark RL et al (1995) Inhibition of cartilage and bone destruction in adjuvant arthritis in the rat by a matrix metalloproteinase inhibitor. J Exp Med 182:449–457

Cox JH, Starr AE, Kappelhoff R, Yan R, Roberts CR, Overall CM (2010) Matrix metalloproteinase 8 deficiency in mice exacerbates inflammatory arthritis through delayed neutrophil apoptosis and reduced caspase 11 expression. Arthritis Rheum 62:3645–3655

Dean DD, Martel-Pelletier J, Pelletier JP, Howell DS, Woessner JF Jr (1989) Evidence for metalloproteinase and metalloproteinase inhibitor imbalance in human osteoarthritic cartilage. J Clin Invest 84:678–685

Dodge GR, Poole AR (1989) Immunohistochemical detection and immunochemical analysis of type II collagen degradation in human normal, rheumatoid, and osteoarthritic articular cartilages and in explants of bovine articular cartilage cultured with interleukin 1. J Clin Invest 83:647–661

Durigova M, Nagase H, Mort JS, Roughley PJ (2011) MMPs are less efficient than ADAMTS5 in cleaving aggrecan core protein. Matrix Biol 30:145–153

Edwards DR, Handsley MM, Pennington CJ (2008) The ADAM metalloproteinases. Mol Aspects Med 29:258–289

Elkins PA, Ho YS, Smith WW, Janson CA, D'Alessio KJ, McQueney MS, Cummings MD, Romanic AM (2002) Structure of the C-terminally truncated human ProMMP9, a gelatin-binding matrix metalloproteinase. Acta Crystallogr D Biol Crystallogr 58:1182–1192

Engel CK, Pirard B, Schimanski S, Kirsch R, Habermann J, Klingler O, Schlotte V, Weithmann KU, Wendt KU (2005) Structural basis for the highly selective inhibition of MMP-13. Chem Biol 12:181–189

Fernandes RJ, Hirohata S, Engle JM, Colige A, Cohn DH, Eyre DR, Apte SS (2001) Procollagen II amino propeptide processing by ADAMTS-3. Insights on dermatosparaxis. J Biol Chem 276:31502–31509

Flannery CR, Little CB, Hughes CE, Caterson B (1999) Expression of ADAMTS homologues in articular cartilage. Biochem Biophys Res Commun 260:318–322

Flory ED, Clarris BJ, Muirden KD (1982) Deposits of α2M in the rheumatoid synovial membrane. Ann Rheum Dis 41:520–526

Fosang AJ, Rogerson FM, East CJ, Stanton H (2008) ADAMTS-5: the story so far. Eur Cell Mater 15:11–26

Fried D, Bohm BB, Krause K, Burkhardt H (2012) ADAM15 protein amplifies focal adhesion kinase phosphorylation under genotoxic stress conditions. J Biol Chem 287:21214–21223

Fu X, Kao JL, Bergt C, Kassim SY, Huq NP, d'Avignon A, Parks WC, Mecham RP, Heinecke JW (2004) Oxidative cross-linking of tryptophan to glycine restrains matrix metalloproteinase activity: specific structural motifs control protein oxidation. J Biol Chem 279:6209–6212

Fujii Y, Okuda D, Fujimoto Z, Horii K, Morita T, Mizuno H (2003) Crystal structure of trimestatin, a disintegrin containing a cell adhesion recognition motif RGD. J Mol Biol 332:1115–1122

Fukata Y, Adesnik H, Iwanaga T, Bredt DS, Nicoll RA, Fukata M (2006) Epilepsy-related ligand/receptor complex LGI1 and ADAM22 regulate synaptic transmission. Science 313:1792–1795
Fulcher YG, Van Doren SR (2011) Remote exosites of the catalytic domain of matrix metalloproteinase-12 enhance elastin degradation. Biochemistry 50:9488–9499
Fushimi K, Troeberg L, Nakamura H, Lim NH, Nagase H (2008) Functional differences of the catalytic and non-catalytic domains in human ADAMTS-4 and ADAMTS-5 in aggrecanolytic activity. J Biol Chem 283:6706–6716
Gao G, Plaas A, Thompson VP, Jin S, Zuo F, Sandy JD (2004) ADAMTS4 (aggrecanase-1) activation on the cell surface involves C-terminal cleavage by glycosylphosphatidyl inositol-anchored membrane type 4-matrix metalloproteinase and binding of the activated proteinase to chondroitin sulfate and heparan sulfate on syndecan-1. J Biol Chem 279:10042–10051
Garnero P, Borel O, Byrjalsen I, Ferreras M, Drake FH, McQueney MS, Foged NT, Delmas PD, Delaisse JM (1998) The collagenolytic activity of cathepsin K is unique among mammalian proteinases. J Biol Chem 273:32347–32352
Gendron C, Kashiwagi M, Lim NH, Enghild JJ, Thogersen IB, Hughes C, Caterson B, Nagase H (2007) Proteolytic activities of human ADAMTS-5: comparative studies with ADAMTS-4. J Biol Chem 282:18294–18306
Geng Y, McQuillan D, Roughley PJ (2006) SLRP interaction can protect collagen fibrils from cleavage by collagenases. Matrix Biol 25:484–491
Gerhardt S, Hassall G, Hawtin P, McCall E, Flavell L, Minshull C, Hargreaves D, Ting A, Pauptit RA, Parker AE, Abbott WM (2007) Crystal structures of human ADAMTS-1 reveal a conserved catalytic domain and a disintegrin-like domain with a fold homologous to cysteine-rich domains. J Mol Biol 373:891–902
Ghosh P (1999) The pathobiology of osteoarthritis and the rationale for the use of pentosan polysulfate for its treatment. Semin Arthritis Rheum 28:211–267
Glasson SS, Askew R, Sheppard B, Carito BA, Blanchet T, Ma HL, Flannery CR, Kanki K, Wang E, Peluso D, Yang Z, Majumdar MK, Morris EA (2004) Characterization of and osteoarthritis susceptibility in ADAMTS-4-knockout mice. Arthritis Rheum 50:2547–2558
Glasson SS, Askew R, Sheppard B, Carito B, Blanchet T, Ma HL, Flannery CR, Peluso D, Kanki K, Yang Z, Majumdar MK, Morris EA (2005) Deletion of active ADAMTS5 prevents cartilage degradation in a murine model of osteoarthritis. Nature 434:644–648
Gomis-Rüth FX (2003) Structural aspects of the metzincin clan of metalloendopeptidases. Mol Biotechnol 24:157–202
Gomis-Rüth FX (2009) Catalytic domain architecture of metzincin metalloproteases. J Biol Chem 284:15353–15357
Gomis-Rüth FX, Kress LF, Bode W (1993) First structure of a snake venom metalloproteinase: a prototype for matrix metalloproteinases/collagenases. EMBO J 12:4151–4157
Gomis-Rüth FX, Gohlke U, Betz M, Knauper V, Murphy G, López-Otìn C, Bode W (1996) The helping hand of collagenase-3 (MMP-13): 2.7 Å crystal structure of its C-terminal haemopexin-like domain. J Mol Biol 264:556–566
Gomis-Rüth FX, Maskos K, Betz M, Bergner A, Huber R, Suzuki K, Yoshida N, Nagase H, Brew K, Bourenkov GP, Bartunik H, Bode W (1997) Mechanism of inhibition of the human matrix metalloproteinase stromelysin-1 by TIMP-1. Nature 389:77–81
Gonzales PE, Solomon A, Miller AB, Leesnitzer MA, Sagi I, Milla ME (2004) Inhibition of the tumor necrosis factor-alpha-converting enzyme by its pro domain. J Biol Chem 279:31638–31645
Gorovoy M, Gaultier A, Campana WM, Firestein GS, Gonias SL (2010) Inflammatory mediators promote production of shed LRP1/CD91, which regulates cell signaling and cytokine expression by macrophages. J Leukoc Biol 88:769–778
Goulas T, Arolas JL, Gomis-Rüth FX (2011) Structure, function and latency regulation of a bacterial enterotoxin potentially derived from a mammalian adamalysin/ADAM xenolog. Proc Natl Acad Sci U S A 108:1856–1861

Gould RJ, Polokoff MA, Friedman PA, Huang TF, Holt JC, Cook JJ, Niewiarowski S (1990) Disintegrins: a family of integrin inhibitory proteins from viper venoms. Proc Soc Exp Biol Med 195:168–171

Grams F, Reinemer P, Powers JC, Kleine T, Pieper M, Tschesche H, Huber R, Bode W (1995) X-ray structures of human neutrophil collagenase complexed with peptide hydroxamate and peptide thiol inhibitors. Implications for substrate binding and rational drug design. Eur J Biochem 228:830–841

Grimaud E, Heymann D, Redini F (2002) Recent advances in TGF-β effects on chondrocyte metabolism. Potential therapeutic roles of TGF-beta in cartilage disorders. Cytokine Growth Factor Rev 13:241–257

Gu Z, Kaul M, Yan B, Kridel SJ, Cui J, Strongin A, Smith JW, Liddington RC, Lipton SA (2002) S-nitrosylation of matrix metalloproteinases: signaling pathway to neuronal cell death. Science 297:1186–1190

Hadler-Olsen E, Fadnes B, Sylte I, Uhlin-Hansen L, Winberg JO (2011) Regulation of matrix metalloproteinase activity in health and disease. FEBS J 278:28–45

Hamze AB, Wei S, Bahudhanapati H, Kota S, Acharya KR, Brew K (2007) Constraining specificity in the N-domain of tissue inhibitor of metalloproteinases-1; gelatinase-selective inhibitors. Protein Sci 16:1905–1913

Heinegård D, Saxne T (2011) The role of the cartilage matrix in osteoarthritis. Nat Rev Rheumatol 7:50–56

Higashi S, Miyazaki K (1999) Reactive site-modified tissue inhibitor of metalloproteinases-2- inhibits the cell-mediated activation of progelatinase A. J Biol Chem 274:10497–10504

Homandberg GA (1999) Potential regulation of cartilage metabolism in osteoarthritis by fibronectin fragments. Front Biosci 4:D713–D730 [Review] [151 refs]

Hurskainen TL, Hirohata S, Seldin MF, Apte SS (1999) ADAM-TS5, ADAM-TS6, and ADAM-TS7, novel members of a new family of zinc metalloproteases. General features and genomic distribution of the ADAM-TS family. J Biol Chem 274:25555–25563

Igarashi T, Araki S, Mori H, Takeda S (2007) Crystal structures of catrocollastatin/VAP2B reveal a dynamic, modular architecture of ADAM/adamalysin/reprolysin family proteins. FEBS Lett 581:2416–2422

Ilic MZ, Handley CJ, Robinson HC, Mok MT (1992) Mechanism of catabolism of aggrecan by articular cartilage. Arch Biochem Biophys 294:115–122

Ishikawa T, Nishigaki F, Miyata S, Hirayama Y, Minoura K, Imanishi J, Neya M, Mizutani T, Imamura Y, Naritomi Y, Murai H, Ohkubo Y, Kagayama A, Mutoh S (2005) Prevention of progressive joint destruction in collagen-induced arthritis in rats by a novel matrix metalloproteinase inhibitor, FR255031. Br J Pharmacol 144:133–143

Issuree PD, Maretzky T, McIlwain DR, Monette S, Qing X, Lang PA, Swendeman SL, Park-Min KH, Binder N, Kalliolias GD, Yarilina A, Horiuchi K, Ivashkiv LB, Mak TW, Salmon JE, Blobel CP (2013) iRHOM2 is a critical pathogenic mediator of inflammatory arthritis. J Clin Invest 123:928–932

Iyer S, Visse R, Nagase H, Acharya KR (2006) Crystal structure of an active form of human MMP-1. J Mol Biol 362(1):78–88

Janes PW, Saha N, Barton WA, Kolev MV, Wimmer-Kleikamp SH, Nievergall E, Blobel CP, Himanen JP, Lackmann M, Nikolov DB (2005) Adam meets Eph: an ADAM substrate recognition module acts as a molecular switch for ephrin cleavage in trans. Cell 123:291–304

Johnson AR, Pavlovsky AG, Ortwine DF, Prior F, Man CF, Bornemeier DA, Banotai CA, Mueller WT, McConnell P, Yan C, Baragi V, Lesch C, Roark WH, Wilson M, Datta K, Guzman R, Han HK, Dyer RD (2007) Discovery and characterization of a novel inhibitor of matrix metalloprotease-13 (MMP13) that reduces cartilage damage in vivo without joint fibroplasia side effects. J Biol Chem 282(38):27781–27791

Jozic D, Bourenkov G, Lim NH, Visse R, Nagase H, Bode W, Maskos K (2005) X-ray structure of human proMMP-1: new insights into procollagenase activation and collagen binding. J Biol Chem 280:9578–9585

Kashiwagi M, Enghild JJ, Gendron C, Hughes C, Caterson B, Itoh Y, Nagase H (2004) Altered proteolytic activities of ADAMTS-4 expressed by C-terminal processing. J Biol Chem 279:10109–10119

Kevorkian L, Young DA, Darrah C, Donell ST, Shepstone L, Porter S, Brockbank SM, Edwards DR, Parker AE, Clark IM (2004) Expression profiling of metalloproteinases and their inhibitors in cartilage. Arthritis Rheum 50:131–141

Knäuper V, Docherty AJP, Smith B, Tschesche H, Murphy G (1997) Analysis of the contribution of the hinge region of human neutrophil collagenase (HNC, MMP-8) to stability and collagenolytic activity by alanine scanning mutagenesis. FEBS Lett 405:60–64

Komiya K, Enomoto H, Inoki I, Okazaki S, Fujita Y, Ikeda E, Ohuchi E, Matsumoto H, Toyama Y, Okada Y (2005) Expression of ADAM15 in rheumatoid synovium: up-regulation by vascular endothelial growth factor and possible implications for angiogenesis. Arthritis Res Ther 7: R1158–R1173

Koo BH, Longpre JM, Somerville RP, Alexander JP, Leduc R, Apte SS (2006) Cell-surface processing of pro-adams9 by furin. J Biol Chem 281(18):12485–12494

Koo BH, Longpre JM, Somerville RP, Alexander JP, Leduc R, Apte SS (2007) Regulation of ADAMTS9 secretion and enzymatic activity by its propeptide. J Biol Chem 282:16146–16154

Kramer RZ, Bella J, Brodsky B, Berman HM (2001) The crystal and molecular structure of a collagen-like peptide with a biologically relevant sequence. J Mol Biol 311:131–147

Kuno K, Kanada N, Nakashima E, Fujiki F, Ichimura F, Matsushima K (1997) Molecular cloning of a gene encoding a new type of metalloproteinase-disintegrin family protein with thrombospondin motifs as an inflammation associated gene. J Biol Chem 272:556–562

Kuno K, Terashima Y, Matsushima K (1999) ADAMTS-1 is an active metalloproteinase associated with the extracellular matrix. J Biol Chem 274:18821–18826

Lark MW, Bayne EK, Flanagan J, Harper CF, Hoerrner LA, Hutchinson NI, Singer II, Donatelli SA, Weidner JR, Williams HR, Mumford RA, Lohmander LS (1997) Aggrecan degradation in human cartilage – evidence for both matrix metalloproteinase and aggrecanase activity in normal, osteoarthritic, and rheumatoid joints. J Clin Invest 100:93–106

Le Gall SM, Maretzky T, Issuree PD, Niu XD, Reiss K, Saftig P, Khokha R, Lundell D, Blobel CP (2010) ADAM17 is regulated by a rapid and reversible mechanism that controls access to its catalytic site. J Cell Sci 123:3913–3922

Lee MH, Rapti M, Knäuper V, Murphy G (2004a) Threonine 98, the pivotal residue of tissue inhibitor of metalloproteinases (TIMP)-1 in metalloproteinase recognition. J Biol Chem 279:17562–17569

Lee MH, Rapti M, Murphy G (2004b) Delineating the molecular basis of the inactivity of tissue inhibitor of metalloproteinase-2 against tumor necrosis factor-alpha-converting enzyme. J Biol Chem 279:45121–45129

Lee JH, Fitzgerald JB, DiMicco MA, Grodzinsky AJ (2005a) Mechanical injury of cartilage explants causes specific time-dependent changes in chondrocyte gene expression. Arthritis Rheum 52:2386–2395

Lee MH, Rapti M, Murphy G (2005b) Total conversion of tissue inhibitor of metalloproteinase (TIMP) for specific metalloproteinase targeting: fine-tuning TIMP-4 for optimal inhibition of tumor necrosis factor-{alpha}-converting enzyme. J Biol Chem 280:15967–15975

Leonard JD, Lin F, Milla ME (2005) Chaperone-like properties of the prodomain of TNFalpha-converting enzyme (TACE) and the functional role of its cysteine switch. Biochem J 387:797–805

Levy GG, Nichols WC, Lian EC, Foroud T, McClintick JN, McGee BM, Yang AY, Siemieniak DR, Stark KR, Gruppo R, Sarode R, Shurin SB, Chandrasekaran V, Stabler SP, Sabio H, Bouhassira EE, Upshaw JD Jr, Ginsburg D, Tsai HM (2001) Mutations in a member of the ADAMTS gene family cause thrombotic thrombocytopenic purpura. Nature 413:488–494

Li J, Brick P, O'Hare MC, Skarzynski T, Lloyd LF, Curry VA, Clark IM, Bigg HF, Hazleman BL, Cawston TE, Blow DM (1995) Structure of full-length porcine synovial collagenase reveals a

C-terminal domain containing a calcium-linked, four-bladed beta-propeller. Structure 3:541–549
Libson AM, Gittis AG, Collier IE, Marmer BL, Goldberg GI, Lattman EE (1995) Crystal structure of the haemopexin-like C-terminal domain of gelatinase A [letter]. Nat Struct Biol 2:938–942
Lillis AP, Van Duyn LB, Murphy-Ullrich JE, Strickland DK (2008) LDL receptor-related protein 1: unique tissue-specific functions revealed by selective gene knockout studies. Physiol Rev 88:887–918
Lim NH, Kashiwagi M, Visse R, Jones J, Enghild JJ, Brew K, Nagase H (2010) Reactive-site mutants of N-TIMP-3 that selectively inhibit ADAMTS-4 and ADAMTS-5: biological and structural implications. Biochem J 431:113–122
Little CB, Barai A, Burkhardt D, Smith SM, Fosang AJ, Werb Z, Shah M, Thompson EW (2009) Matrix metalloproteinase 13-deficient mice are resistant to osteoarthritic cartilage erosion but not chondrocyte hypertrophy or osteophyte development. Arthritis Rheum 60:3723–3733
Liu H, Shim AH, He X (2009) Structural characterization of the ectodomain of a disintegrin and metalloproteinase-22 (ADAM22), a neural adhesion receptor instead of metalloproteinase: insights on ADAM function. J Biol Chem 284:29077–29086
Loechel F, Gilpin BJ, Engvall E, Albrechtsen R, Wewer UM (1998) Human ADAM 12 (meltrin alpha) is an active metalloprotease. J Biol Chem 273:16993–16997
Lohmander LS, Neame PJ, Sandy JD (1993) The structure of aggrecan fragments in human synovial fluid. Evidence that aggrecanase mediates cartilage degradation in inflammatory joint disease, joint injury, and osteoarthritis. Arthritis Rheum 36:1214–1222
Longpre JM, Leduc R (2004) Identification of prodomain determinants involved in ADAMTS-1 biosynthesis. J Biol Chem 279:33237–33245
Longpre JM, McCulloch DR, Koo BH, Alexander JP, Apte SS, Leduc R (2009) Characterization of proADAMTS5 processing by proprotein convertases. Int J Biochem Cell Biol 41:1116–1126
Lotta LA, Garagiola I, Palla R, Cairo A, Peyvandi F (2010) ADAMTS13 mutations and polymorphisms in congenital thrombotic thrombocytopenic purpura. Hum Mutat 31:11–19
Loulakis P, Shrikhande A, Davis G, Maniglia CA (1992) N-terminal sequence of proteoglycan fragments isolated from medium of interleukin-1-treated articular-cartilage cultures. Putative site(s) of enzymic cleavage. Biochem J 284:589–593
Luken BM, Turenhout EA, Kaijen PH, Greuter MJ, Pos W, van Mourik JA, Fijnheer R, Voorberg J (2006) Amino acid regions 572-579 and 657-666 of the spacer domain of ADAMTS13 provide a common antigenic core required for binding of antibodies in patients with acquired TTP. Thromb Haemost 96:295–301
Mahmoodi M, Sahebjam S, Smookler D, Khokha R, Mort JS (2005) Lack of tissue inhibitor of metalloproteinases-3 results in an enhanced inflammatory response in antigen-induced arthritis. Am J Pathol 166:1733–1740
Manka SW, Carafoli F, Visse R, Bihan D, Raynal N, Farndale RW, Murphy G, Enghild JJ, Hohenester E, Nagase H (2012) Structural insights into triple-helical collagen cleavage by matrix metalloproteinase 1. Proc Natl Acad Sci U S A 109:12461–12466
Maretzky T, Evers A, Zhou W, Swendeman SL, Wong PM, Rafii S, Reiss K, Blobel CP (2011) Migration of growth factor-stimulated epithelial and endothelial cells depends on EGFR transactivation by ADAM17. Nat Commun 2:229
Maskos K (2005) Crystal structures of MMPs in complex with physiological and pharmacological inhibitors. Biochimie 87:249–263
Maskos K, Fernandez-Catalan C, Huber R, Bourenkov GP, Bartunik H, Ellestad GA, Reddy P, Wolfson MF, Rauch CT, Castner BJ, Davis R, Clarke HR, Petersen M, Fitzner JN, Cerretti DP, March CJ, Paxton RJ, Black RA, Bode W (1998) Crystal structure of the catalytic domain of human tumor necrosis factor-alpha-converting enzyme. Proc Natl Acad Sci U S A 95:3408–3412
McIlwain DR, Lang PA, Maretzky T, Hamada K, Ohishi K, Maney SK, Berger T, Murthy A, Duncan G, Xu HC, Lang KS, Haussinger D, Wakeham A, Itie-Youten A, Khokha R, Ohashi

PS, Blobel CP, Mak TW (2012) iRhom2 regulation of TACE controls TNF-mediated protection against Listeria and responses to LPS. Science 335:229–232
Meng Q, Malinovskii V, Huang W, Hu YJ, Chung L, Nagase H, Bode W, Maskos K, Brew K (1999) Residue 2 of TIMP-1 is a major determinant of affinity and specificity for matrix metalloproteinases but effects of substitutions do not correlate with those of the corresponding P1' residue of substrate. J Biol Chem 274:10184–10189
Miller M-C, Manning HB, Jain A, Troeberg L, Dudhia J, Essex D, Sandison A, Seiki M, Nanchahal J, Nagase H, Itoh Y (2009) Membrane type 1 matrix metalloproteinase is a crucial promoter of synovial invasion in human rheumatoid arthritis. Arthritis Rheum 60:686–697
Milner JM, Cawston TE (2005) Matrix metalloproteinase knockout studies and the potential use of matrix metalloproteinase inhibitors in the rheumatic diseases. Curr Drug Targets Inflamm Allergy 4:363–375
Morgunova E, Tuuttila A, Bergmann U, Isupov M, Lindqvist Y, Schneider G, Tryggvason K (1999) Structure of human pro-matrix metalloproteinase-2: activation mechanism revealed. Science 284:1667–1670
Morgunova E, Tuuttila A, Bergmann U, Tryggvason K (2002) Structural insight into the complex formation of latent matrix metalloproteinase 2 with tissue inhibitor of metalloproteinase 2. Proc Natl Acad Sci U S A 99:7414–7419
Mosyak L, Georgiadis K, Shane T, Svenson K, Hebert T, McDonagh T, Mackie S, Olland S, Lin L, Zhong X, Kriz R, Reifenberg EL, Collins-Racie LA, Corcoran C, Freeman B, Zollner R, Marvell T, Vera M, Sum PE, LaVallie ER, Stahl M, Somers W (2008) Crystal structures of the two major aggrecan degrading enzymes, ADAMTS4 and ADAMTS5. Protein Sci 17:16–21
Nagase H (1997) Activation mechanisms of matrix metalloproteinases. Biol Chem 378:151–160
Nagase H, Itoh Y, Binner S (1994) Interaction of α 2-macroglobulin with matrix metalloproteinases and its use for identification of their active forms. Ann NY Acad Sci 732:294–302
Nagase H, Visse R, Murphy G (2006) Structure and function of matrix metalloproteinases and TIMPs. Cardiovasc Res 69:562–573
Naito S, Shiomi T, Okada A, Kimura T, Chijiiwa M, Fujita Y, Yatabe T, Komiya K, Enomoto H, Fujikawa K, Okada Y (2007) Expression of ADAMTS4 (aggrecanase-1) in human osteoarthritic cartilage. Pathol Int 57:703–711
Nerenberg PS, Salsas-Escat R, Stultz CM (2008) Do collagenases unwind triple-helical collagen before peptide bond hydrolysis? Reinterpreting experimental observations with mathematical models. Proteins 70:1154–1161
Okada A, Mochizuki S, Yatabe T, Kimura T, Shiomi T, Fujita Y, Matsumoto H, Sehara-Fujisawa-A, Iwamoto Y, Okada Y (2008) ADAM-12 (meltrin alpha) is involved in chondrocyte proliferation via cleavage of insulin-like growth factor binding protein 5 in osteoarthritic cartilage. Arthritis Rheum 58:778–789
Orth P, Reichert P, Wang W, Prosise WW, Yarosh-Tomaine T, Hammond G, Ingram RN, Xiao L, Mirza UA, Zou J, Strickland C, Taremi SS, Le HV, Madison V (2004) Crystal structure of the catalytic domain of human ADAM33. J Mol Biol 335:129–137
Overall CM (2002) Molecular determinants of metalloproteinase substrate specificity: matrix metalloproteinase substrate binding domains, modules, and exosites. MolBiotechnol 22:51–86
Palmier MO, Fulcher YG, Bhaskaran R, Duong VQ, Fields GB, Van Doren SR (2010) NMR and bioinformatics discovery of exosites that tune metalloelastase specificity for solubilized elastin and collagen triple helices. J Biol Chem 285:30918–30930
Peppin GJ, Weiss SJ (1986) Activation of the endogenous metalloproteinase, gelatinase, by triggered human neutrophils. Proc Natl Acad Sci U S A 83:4322–4326
Perumal S, Antipova O, Orgel JP (2008) Collagen fibril architecture, domain organization, and triple-helical conformation govern its proteolysis. Proc Natl Acad Sci U S A 105:2824–2829
Quinn KA, Grimsley PG, Dai YP, Tapner M, Chesterman CN, Owensby DA (1997) Soluble low density lipoprotein receptor-related protein (LRP) circulates in human plasma. J Biol Chem 272:23946–23951
Ramachandran GN, Kartha G (1955) Structure of collagen. Nature 176:593–595

Reddy P, Slack JL, Davis R, Cerretti DP, Kozlosky CJ, Blanton RA, Shows D, Peschon JJ, Black RA (2000) Functional analysis of the domain structure of tumor necrosis factor-alpha converting enzyme. J Biol Chem 275:14608–14614
Richards FM, Tape CJ, Jodrell DI, Murphy G (2012) Anti-tumour effects of a specific anti-ADAM17 antibody in an ovarian cancer model in vivo. PLoS One 7:e40597
Rosenblum G, Van den Steen PE, Cohen SR, Grossmann JG, Frenkel J, Sertchook R, Slack N, Strange RW, Opdenakker G, Sagi I (2007) Insights into the structure and domain flexibility of full-length pro-matrix metalloproteinase-9/gelatinase B. Structure 15:1227–1236
Rousseau S, Papoutsopoulou M, Symons A, Cook D, Lucocq JM, Prescott AR, O'Garra A, Ley SC, Cohen P (2008) TPL2-mediated activation of ERK1 and ERK2 regulates the processing of pre-TNF alpha in LPS-stimulated macrophages. J Cell Sci 121:149–154
Roy-Beaudry M, Martel-Pelletier J, Pelletier JP, M'Barek KN, Christgau S, Shipkolye F, Moldovan F (2003) Endothelin 1 promotes osteoarthritic cartilage degradation via matrix metalloprotease 1 and matrix metalloprotease 13 induction. Arthritis Rheum 48:2855–2864
Sabatini M, Lesur C, Thomas M, Chomel A, Anract P, de Nanteuil G, Pastoureau P (2005) Effect of inhibition of matrix metalloproteinases on cartilage loss in vitro and in a guinea pig model of osteoarthritis. Arthritis Rheum 52:171–180
Sadler JE (1998) Biochemistry and genetics of von Willebrand factor. Annu Rev Biochem 67:395–424
Sagane K, Hayakawa K, Kai J, Hirohashi T, Takahashi E, Miyamoto N, Ino M, Oki T, Yamazaki K, Nagasu T (2005) Ataxia and peripheral nerve hypomyelination in ADAM22-deficient mice. BMC Neurosci 6:33
Sagane K, Ishihama Y, Sugimoto H (2008) LGI1 and LGI4 bind to ADAM22, ADAM23 and ADAM11. Int J Biol Sci 4:387–396
Sandy JD (2006) A contentious issue finds some clarity: on the independent and complementary roles of aggrecanase activity and MMP activity in human joint aggrecanolysis. Osteoarthritis Cartilage 14:95–100
Sandy JD, Neame PJ, Boynton RE, Flannery CR (1991) Catabolism of aggrecan in cartilage explants. Identification of a major cleavage site within the interglobular domain. J Biol Chem 266:8683–8685
Sandy JD, Flannery CR, Neame PJ, Lohmander LS (1992) The structure of aggrecan fragments in human synovial fluid. Evidence for the involvement in osteoarthritis of a novel proteinase which cleaves the Glu373-Ala374 bond of the interglobular domain. J Clin Invest 89:1512–1516
Scheller J, Ohnesorge N, Rose-John S (2006) Interleukin-6 trans-signalling in chronic inflammation and cancer. Scand J Immunol 63:321–329
Shieh HS, Mathis KJ, Williams JM, Hills RL, Wiese JF, Benson TE, Kiefer JR, Marino MH, Carroll JN, Leone JW, Malfait AM, Arner EC, Tortorella MD, Tomasselli A (2008) High resolution crystal structure of the catalytic domain of ADAMTS-5 (aggrecanase-2). J Biol Chem 283:1501–1507
Smookler DS, Mohammed FF, Kassiri Z, Duncan GS, Mak TW, Khokha R (2006) Tissue inhibitor of metalloproteinase 3 regulates TNF-dependent systemic inflammation. J Immunol 176:721–725
Soejima K, Matsumoto M, Kokame K, Yagi H, Ishizashi H, Maeda H, Nozaki C, Miyata T, Fujimura Y, Nakagaki T (2003) ADAMTS-13 cysteine-rich/spacer domains are functionally essential for von Willebrand factor cleavage. Blood 102:3232–3237
Stanton H, Rogerson FM, East CJ, Golub SB, Lawlor KE, Meeker CT, Little CB, Last K, Farmer PJ, Campbell IK, Fourie AM, Fosang AJ (2005) ADAMTS5 is the major aggrecanase in mouse cartilage in vivo and in vitro. Nature 434:648–652
Stanton H, Melrose J, Little CB, Fosang AJ (2011) Proteoglycan degradation by the ADAMTS family of proteinases. Biochim Biophys Acta 1812:1616–1629
Sternlicht MD, Werb Z (2001) How matrix metalloproteinases regulate cell behavior. Annu Rev Cell Dev Biol 17:463–516

Struglics A, Larsson S, Pratta MA, Kumar S, Lark MW, Lohmander LS (2006) Human osteoarthritis synovial fluid and joint cartilage contain both aggrecanase- and matrix metalloproteinase-generated aggrecan fragments. Osteoarthritis Cartilage 14:101–113

Takeda S (2009) Three-dimensional domain architecture of the ADAM family proteinases. Semin Cell Dev Biol 20:146–152

Takeda S, Igarashi T, Mori H, Araki S (2006) Crystal structures of VAP1 reveal ADAMs' MDC domain architecture and its unique C-shaped scaffold. EMBO J 25:2388–2396

Takeda S, Takeya H, Iwanaga S (2012) Snake venom metalloproteinases: structure, function and relevance to the mammalian ADAM/ADAMTS family proteins. Biochim Biophys Acta 1824:164–176

Tallant C, Garcia-Castellanos R, Seco J, Baumann U, Gomis-Rüth FX (2006) Molecular analysis of ulilysin, the structural prototype of a new family of metzincin metalloproteases. J Biol Chem 281:17920–17928

Tan K, Duquette M, Liu JH, Dong Y, Zhang R, Joachimiak A, Lawler J, Wang JH (2002) Crystal structure of the TSP-1 type 1 repeats: a novel layered fold and its biological implication. J Cell Biol 159:373–382

Tape CJ, Willems SH, Dombernowsky SL, Stanley PL, Fogarasi M, Ouwehand W, McCafferty J, Murphy G (2011) Cross-domain inhibition of TACE ectodomain. Proc Natl Acad Sci USA 108:5578–5583

Tochowicz A, Goettig P, Evans R, Visse R, Shitomi Y, Palmisano R, Ito N, Richter K, Maskos K, Franke D, Svergun D, Nagase H, Bode W, Itoh Y (2011) The dimer interface of the membrane type 1 matrix metalloproteinase hemopexin domain: crystal structure and biological functions. J Biol Chem 286:7587–7600

Tortorella MD, Burn TC, Pratta MA, Abbaszade I, Hollis JM, Liu R, Rosenfeld SA, Copeland RA, Decicco CP, Wynn R, Rockwell A, Yang F, Duke JL, Solomon K, George H, Bruckner R, Nagase H, Itoh Y, Ellis DM, Ross H, Wiswall BH, Murphy K, Hillman MC Jr, Hollis GF, Arner EC (1999) Purification and cloning of aggrecanase-1: a member of the ADAMTS family of proteins. Science 284:1664–1666

Tortorella MD, Arner EC, Hills R, Easton A, Korte-Sarfaty J, Fok K, Wittwer A, Liu RQ, Malfait AM (2004) alpha 2-macroglobulin is a novel substrate for adamts-4 and represents an endogenous inhibitor of these enzymes. J Biol Chem 279(17):17554–17561

Troeberg L, Nagase H (2012) Proteases involved in cartilage matrix degradation in osteoarthritis. Biochim Biophys Acta 1824:133–145

Troeberg L, Tanaka M, Wait R, Shi YE, Brew K, Nagase H (2002) E. coli expression of TIMP-4 and comparative kinetic studies with TIMP-1 and TIMP-2: insights into the interactions of TIMPs and matrix metalloproteinase 2 (gelatinase A). Biochemistry 41:15025–15035

Troeberg L, Fushimi K, Khokha R, Emonard H, Ghosh P, Nagase H (2008) Calcium pentosan polysulfate is a multifaceted exosite inhibitor of aggrecanases. FASEB J 22:3515–3524

Troeberg L, Mulloy B, Ghosh P, Lee MH, Murphy G, Nagase H (2012) Pentosan polysulfate increases affinity between ADAMTS-5 and TIMP-3 through formation of an electrostatically driven trimolecular complex. Biochem J 443(1):307–315

Tsukada H, Pourmotabbed T (2002) Unexpected crucial role of residue 272 in substrate specificity of fibroblast collagenase. J Biol Chem 277:27378–27384

Van den Steen PE, Van AI, Hvidberg V, Piccard H, Fiten P, Jacobsen C, Moestrup SK, Fry S, Royle L, Wormald MR, Wallis R, Rudd PM, Dwek RA, Opdenakker G (2006) The hemopexin and O-glycosylated domains tune gelatinase B/MMP-9 bioavailability via inhibition and binding to cargo receptors. J Biol Chem 281:18626–18637

Visse R, Nagase H (2003) Matrix metalloproteinases and tissue inhibitors of metalloproteinases: structure, function, and biochemistry. Circ Res 92:827–839

Wang P, Tortorella M, England K, Malfait AM, Thomas G, Arner EC, Pei D (2004) Proprotein convertase furin interacts with and cleaves pro-ADAMTS4 (Aggrecanase-1) in the trans-Golgi network. J Biol Chem 279:15434–15440

Weber S, Saftig P (2012) Ectodomain shedding and ADAMs in development. Development 139:3693–3709

Wei S, Kashiwagi M, Kota S, Xie Z, Nagase H, Brew K (2005) Reactive site mutations in tissue inhibitor of metalloproteinase-3 disrupt inhibition of matrix metalloproteinases but not tumor necrosis factor-alpha-converting enzyme. J Biol Chem 280:32877–32882

Welgus HG, Jeffrey JJ, Stricklin GP, Roswit WT, Eisen AZ (1980) Characteristics of the action of human skin fibroblast collagenase on fibrillar collagen. J Biol Chem 255:6806–6813

White JM (2003) ADAMs: modulators of cell-cell and cell-matrix interactions. Curr Opin Cell Biol 15:598–606

Willems SH, Tape CJ, Stanley PL, Taylor NA, Mills IG, Neal DE, McCafferty J, Murphy G (2010) Thiol isomerases negatively regulate the cellular shedding activity of ADAM17. Biochem J 428:439–450

Williamson RA, Martorell G, Carr MD, Murphy G, Docherty AJP, Freedman RB, Feeney J (1994) Solution structure of the active domain of tissue inhibitor of metalloproteinases-2. A new member of the OB fold protein family. Biochemistry 33:11745–11759

Wingfield PT, Sax JK, Stahl SJ, Kaufman J, Palmer I, Chung V, Corcoran ML, Kleiner DE, Stetler-Stevenson WG (1999) Biophysical and functional characterization of full-length, recombinant human tissue inhibitor of metalloproteinases-2 (TIMP-2) produced in Escherichia coli – Comparison of wild type and amino-terminal alanine appended variant with implications for the mechanism of TIMP functions. J Biol Chem 274:21362–21368

Wisniewska M, Goettig P, Maskos K, Belouski E, Winters D, Hecht R, Black R, Bode W (2008) Structural determinants of the ADAM inhibition by TIMP-3: crystal structure of the TACE-N-TIMP-3 complex. J Mol Biol 381:1307–1319

Wu YI, Munshi HG, Sen R, Snipas SJ, Salvesen GS, Fridman R, Stack MS (2004) Glycosylation broadens the substrate profile of membrane type 1 matrix metalloproteinase. J Biol Chem 279:8278–8289

Xu P, Derynck R (2010) Direct activation of TACE-mediated ectodomain shedding by p38 MAP kinase regulates EGF receptor-dependent cell proliferation. Mol Cell 37:551–566

Yamamoto K, Troeberg L, Scilabra SD, Pelosi M, Murphy CL, Strickland DK, Nagase H (2013) LRP-1-mediated endocytosis regulates extracellular activity of ADAMTS-5 in articular cartilage. FASEB J 27:511–521

Zhang Q, Thomas SM, Lui VW, Xi S, Siegfried JM, Fan H, Smithgall TE, Mills GB, Grandis JR (2006) Phosphorylation of TNF-alpha converting enzyme by gastrin-releasing peptide induces amphiregulin release and EGF receptor activation. Proc Natl Acad Sci U S A 103:6901–6906

Chapter 14
MMP-Mediated Collagen Remodeling and Vessel Functions

Agnès Noel and Nor Eddine Sounni

14.1 Introduction

Blood vessels are highly organized structures in which the extracellular matrix (ECM) of the vessel wall and the perivascular interstitium contribute substantially to their diverse functions. Synthesis and degradation of ECM components in vascular wall or perivascular stroma are tighly controlled mechanisms. Matrix metabolism perturbation is sufficient to significantly affect vascular system physiology leading to several vascular disorders including diabetic retinopathy (Turley 2001), hemorrhagic telangiectasia (Arteaga-Solis et al. 2000), hypertensive heart diseases (Morwood and Nicholson 2006), atherosclerosis and fibrosis (Diez 2007; Radisky et al. 2007). Vascular abnormalities are also associated with cancer progression and metastatic dissemination (Berk et al. 2007). The ECM in the vascular wall contains a variety of molecules including collagens, elastic fibers, glycoproteins and proteoglycans that provide structural and mechanical support to cells. Vascular cells are connected to these structural matrix components by cell surface receptors of integrins and non-integrins types (Davis and Senger 2008). Matrix receptor interactions influence vascular cell shape, behavior and response to cytokines and growth factors (Boudreau and Jones 1999). Alterations of vessel matrix have thus profound impact on blood vessel integrity affecting thereby the extravasation of fluids and plasma proteins into the interstitium (Wiig et al. 2008). ECM molecules play a major role of in maintaining mature tubular structures in vascular and lymphatic systems (Bergers and Benjamin 2003; Oliver and Detmar

A. Noel • N.E. Sounni (✉)
Laboratory of Tumor and Developmental Biology, Groupe Interdisciplinaire de Génoprotéomique Appliquée-Cancer (GIGA-Cancer), B23, University of Liège, 4000 Liège, Belgium
e-mail: nesounni@ulg.ac.be

K. Brix and W. Stöcker (eds.), *Proteases: Structure and Function*,
DOI 10.1007/978-3-7091-0885-7_14, © Springer-Verlag Wien 2013

2002). For instance, genetic studies show that mutations in *COL3A1* gene lead to Ehlers-Dahlos syndrome type IV, a connective tissue disorder characterized predominantly by arterial dilation and rupture (Arteaga-Solis et al. 2000). On other hand, a mutation in the *Fibrillin-1* (*FBN1*) gene in Marfan syndrome, results in deficiency of elastic fibers that regulate cell attachment in vessel walls to the neighboring ECM (Neptune et al. 2003). Thus proper metabolism, assembly in vascular tissues regulates maintenance of tissue homeostasis. Functional and structural vascular remodeling involve several enzymes among which MMPs produced by endothelial, inflammatory, or malignant cells are key players. The present review will describe the contribution of MMPs in the remodeling of collagens that are essential vessel components. Understanding the complex role of MMPs in vessel wall assembly and functions in homeostasis or diseases is mandatory to provide new therapy to maintain vessel stability and proper function.

14.2 Collagenolytic Activities of MMPs

MMPs are a family of zinc-dependent endopeptidases composed of 24 currently known human enzymes that share several functional domains. MMPs are often referred to soluble (MMPs) or membrane type-MMPs (MT-MMPs) that are anchored to the cell surface through transmembrane domain or glycosylphosphatidylinositol (GPI) linker. For a description of the structure, function and regulation of MMPs and MT-MMPs, the reader is referred to previous reviews (Kessenbrock et al. 2010; Page-McCaw et al. 2007; Sohail et al. 2008; Sounni and Noel 2005; Strongin 2010; Zucker et al. 2003). Interstitial collagenases are the only known mammalian enzymes able of degrading triple-helical fibrillar collagen into distinctive tropocollagen A (TCA) (3/4) and (TCB) (1/4) fragments as a consequence of a specific proteolysis of all three α-chains at a single locus three-quarters from the N-terminus. Collagenolytic MMPs include soluble MMPs (MMP1, MMP8, MMP13) and the membrane-associated MMP14/MT1-MMP. More recently, MMP2 has been identified as an interstitial collagenase that can cleave native type I collagen in a distinctive way from other collagenases without generating the classical TCA fragment (Egeblad et al. 2007). MMP-2 and MMP-9 (initially referred to as gelatinase A and gelatinase B, respectively) cleave type IV collagen present in basement membrane and the degradation products generated by the action of interstitial collagenases. These MMPs produced by endothelial cells, fibroblasts, smooth muscle cells (VSMCs) and inflammatory cells play pivotal roles in multiple physiological and pathological processes involving extensive and aberrant collagenolysis (Fig. 14.1). Latest technological progresses have obviously advanced our consideration of MMPs as key regulators that operate in complex cellular and molecular networks. In addition to the mentioned collagenolytic activities, MMPs contribute to the proteolytic cleavage of an increasing repertoire of substrates that includes at least, almost all ECM components, chemokines/cytokines, growth factors and cell surface receptors (Butler and

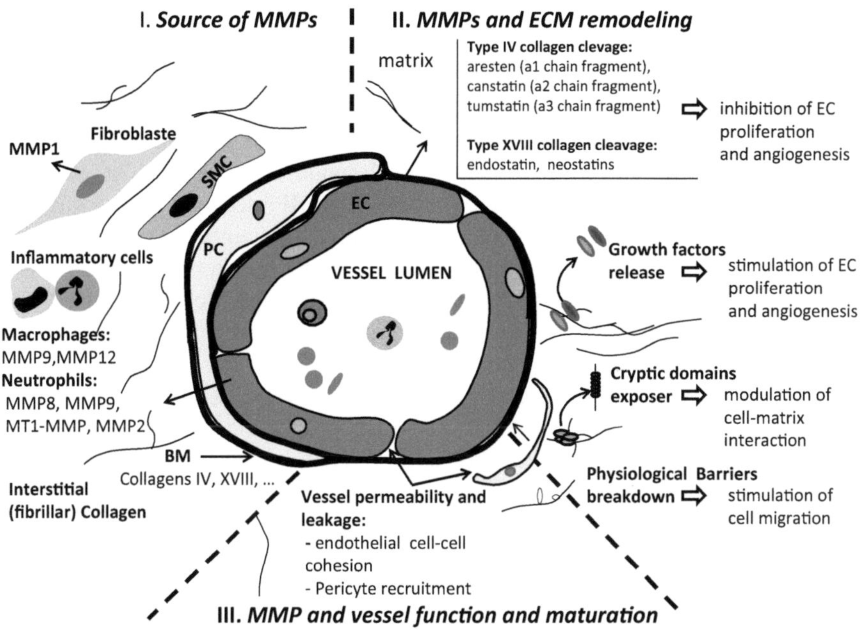

Fig. 14.1 MMP effects on vessel function and maturation. MMPs are produced in perivascular stroma by endothelial (EC), pericyte (PC) inflammatory cells, and fibroblasts (I). They regulate vessel function and maturation through different mechanisms which relay on their proteolytic activity. MMPs increase growth factor availability in perivascular stroma, unmask cryptic sites within extracellular matrix molecules and generate angiogenesis inhibitor form basement membrane (BM) and collagens (II). MMPs activity controls the vascular homeostasis and regulates vessel permeability and leakage in vascular diseases (III)

Overall 2009; Hu et al. 2007). Under normal conditions, MMP activities are controlled through transcriptional regulation, activation of pro-MMP precursor zymogens and inhibition by physiological tissue inhibitor of matrix metalloproteinases (TIMPs) that bind MMP in a 1:1 stoichiometry (Kessenbrock et al. 2010; Lopez-Otin and Overall 2002). Four TIMPs have been identified in vertebrates and their expression is regulated during development and tissue remodeling (Nagase et al. 2006). More recently some MMPs and TIMPs were described as targets of epigenetic regulation via CpG methylation and histones modifications (Chernov et al. 2009; Liu et al. 2007; Pulukuri et al. 2007).

14.3 MMP-Mediated Collagen Remodeling for Vessel Structure and Functions

MMPs contribute to perivascular matrix remodeling during normal angiogenesis (Ucuzian and Greisler 2007). In pathological conditions, the vascular wall and perivascular interstitium are exposed to intensive MMP-mediated remodeling that can create a less restrictive microenvironment. A sustained MMP activity is associated with several vascular pathologies including aneurysm, atherosclerosis, hypertension and restinosis (Mott and Werb 2004; Page-McCaw et al. 2007). MMP-mediated collagen remodeling can regulate tissue architecture through different mechanisms. They can generate extracellular space for cell migration and unmask cryptic sites within ECM molecules modifying thereby cell to matrix adhesion. By promoting the release of matrix-associated growth factors or cytokines, they modulate the activity or bioavailability of signaling molecules during vascular response to physiological or pathological stimuli (Kessenbrock et al. 2010; Page-McCaw et al. 2007). MMP activities result also in the generation of matrix fragments displaying novel biological activity (Kessenbrock et al. 2010; Page-McCaw et al. 2007; Rundhaug 2005). It is now well recognized that collagen proteolysis may release a number of endogenous angiogenesis inhibitors, including type IV (arresten, canstatin, tumstatin), type V (restin) and type XVIII (endostatin, neostatins) collagen fragments among other fragments of ECM proteins that may display anti-angiogenic activity (Egeblad and Werb 2002; Kojima et al. 2008; Nyberg et al. 2005). These angio-inhibitory fragments regulate primarily endothelial cell proliferation and apoptosis by interfering with integrins. The contribution of such endogenous inhibitors in human pathologies is supported by the role played by endostatin in choroidal neovascularization (CNV) associated to age-related macular degeneration (AMD), the leading cause of blindness in elderly patients (Noel et al. 2007). MMPs release and activate endostatin from collagen XVIII present in vascular basement membranes and Bruch membrane (Ferreras et al. 2000; Zatterstrom et al. 2000). Endostatin levels are decreased in human eyes with AMD, and its deficiency was suggested to predispose to CNV formation (Bhutto et al. 2004). Accordingly, both external delivery of endostatin and endogenous endostatin inhibits experimental CNV (Marneros et al. 2007).

14.3.1 Type IV Collagen Remodeling

Type IV collagen is the main component of vascular basement membrane that forms a mesh-like structure with other molecules such as laminin, heparan sulfate proteoglycans, fibronectin and entactin. In addition to the release of endogenous angiogenic inhibitor (Fig. 14.1), specific cryptic type IV collagen epitopes can be exposed upon proteolysis. For instance, a cryptic site hidden within the three-dimensional structure of type IV collagen is exposed by MMP2 and promotes *in vivo* angiogenesis (Xu et al. 2001). The recognition of a cryptic epitope (HU177) present exclusively in type IV collagen by a specific mouse antibody (HUIV26) inhibits endothelial cell proliferation and differentiation into tubule-like structures (Cretu et al. 2007). Interference with this epitope results in the expression of cyclin-dependent kinase (CDK) inhibitor p27kip1 in endothelial cells (Cretu et al. 2007). Cryptic activities embedded within intact type IV collagen molecules are also unmasked as a consequence of MT1-MMP, MT2-MMP and MT3-MMP action during morphogenesis (Rebustini et al. 2009). In eyes, exposure of cryptic collagen type IV epitopes is associated with increased CNV incidence. In line with this finding, the humanized antibody H8 directed against a cryptic collagen type IV epitope inhibits CNV progression (Gocheva et al. 2006).

14.3.2 Type I Collagen Remodeling

Type I collagen fibrils, the most abundant extracellular matrix (ECM) proteins in perivascular stroma (Di Lullo et al. 2002) is a heterotrimer molecule composed of two α1(I) and one α2(I) chains encoded by two separate genes: *Col1a1* and *Col1a2* (Shoulders and Raines 2009). Type I collagen remodeling in perivascular stroma represents an important step in cell migration and endothelial cell reorganization into tubular structure during normal and pathological angiogenesis. We have previously investigated the role of type I collagen proteolysis in perivascular stroma in transgenic mice carrying a mutation in *col1a1* gene (*Col1a1r/r* mice) that renders type I collagen resistant to collagenase-mediated cleavage into the TCA and TCB fragments (Liu et al. 1995). These transgenic mice (*Col1a1r/r*) showed a dramatic reduction in steady state vascular leakage measured by plasma extravasation after quantification of Evans blue dye in interstitial space of skin tissue. Vessel structure analysis showed that these mice have reduced number of vascular openings within vessel wall and failed to develop normal vascular response after chemical stimulation of skin due to the lack of collagen remodelling in the perivascular stroma (Sounni et al. 2010a).

These data suggest that type I collagen remodelling is an important step in vascular response to tissue injuries and damage. Type I collagen is classically cleaved into characteristic TCA and TCB fragments by MMP1, MMP8, MMP13 and MT1-MMP (also called MMP14) (Ohuchi et al. 1997). Perturbations in collagen type I assembly, remodeling and synthesis is associated with increased tissue alterations such as atherosclerosis, fibrosis, and tumor. Type I collagen cleavage by MT1-MMP at endothelial cell surface stimulates migration, guidance and organization of endothelial cells into tubular structures (Collen et al. 2003). In the tumor microenvironment, type I collagen remodeling by MT1-MMP enables cancer cells to escape the mechanical barriers that confine them to collagen matrix and stimulates tumor growth *in vivo* (Hotary et al. 2003). The generation of MT1-MMP knockout mice delineated the role of MT1-MMP in collagen remodeling during development. Indeed, *Mt1-mmp*$^{-/-}$ mice exhibit skeletal defects with craniofacial abnormalities, osteopenia and angiogenesis (Holmbeck et al. 1999; Zhou et al. 2000). MT1-MMP not only acts as an interstitial collagenase, but is also the main activator of pro-MMP2 at the cell surface (Sato et al. 1994; Sounni et al. 2010b). Although MMP2 exhibits gelatinolytic activity, it does not cleave type I collagen in a similar manner than do the classical interstitial collagenases (Detry et al. 2012). Nevertheless, MMP2 contribution in interstitial collagen remodeling has been evidenced *in vivo*, by intercrossing *Col1a1r/r* mice with *Mmp2*$^{-/-}$ mice and generation of double transgenic *Col1a1r/r/Mmp2*$^{-/-}$ mice in FVBn strain. Intriguingly, these *Col1a1r/r/Mmp2*$^{-/-}$ mice display severe developmental defects resembling the skeletal defect syndromes found in *MMP2*-null patients with inactivating mutations in *Mmp2* gene (Egeblad et al. 2007; Martignetti et al. 2001). While *Mmp2*$^{-/-}$ mice have only mild aspects of these abnormalities, collagen metabolism perturbation in double transgenic *Col1a1r/r/Mmp2*$^{-/-}$ mice increases alteration phenotype of *Mmp2*$^{-/-}$ mice, suggesting that MMP2 is important in type I collagen remodelling through a mechanism different than that used by classical interstitial collagenases (MMP1, MMP8, MMP13 and MT1-MMP).

14.4 MMP-Mediated Proteolysis of Endothelial Cell-Cell Contact Molecules

The functional barrier properties of blood vessels are not only dependent on endothelial cell interactions with underlying basement membrane, but also on the resistance and cohesive features of endothelial cells. Cell-cell cohesion involves the adherent complex vascular endothelial-cadherin (VE-cadherin), the inter-endothelial cell tight junction proteins (TJPs) and junctional adhesion molecules (JAMs). TJPs consist of three major families of transmembrane proteins, zonula occludens-1 (ZO-1), claudins and occludins (Matter and Balda 2003). MMP contribution to endothelial cell-cell contact regulation is well documented in several experimental model of cerebral ischemia (Hawkins et al. 2007; Yang et al. 2007b). For instances, MMP activity has been correlated with the breakdown of tight junctions leading to increased blood-brain barrier (BBB) permeability, ischemia, diabetic retinopathy, vericosis and atherosclerosis (Fujimoto et al. 2008; Hu et al. 2007; Navaratna et al. 2007; Yang et al. 2007a). There is direct evidence that MMP2, MMP9 and MT1-MMP activities induce changes in TJPs (claudin-5 and occludin) leading to BBB disruption of rat brain (Yang et al. 2007b). Moreover, MMP-9 gene deletion reduced ZO-1 degradation associated with attenuation of BBB leakage (Asahi et al. 2001). Accordingly, TIMP-1 gene deletion accelerates ZO-1 degradation through increased MMP-9 activity in cerebral ischemia (Fujimoto et al. 2008). Another report supports also the role of MMP2, MMP3 and MMP9 in TNF alpha-induced alteration of the blood cerebrospinal fluid barrier *in vitro* (Zeni et al. 2007).

The role of MMPs in the maintenance of systemic vessel integrity and remodeling in diabetic retinopathy has been linked to the breakdown of occludin (Giebel et al. 2005) and VE-cadherin (Navaratna et al. 2007). Beside MMPs, a number of recent findings point out to the role of ADAMs (a disintegrin and metalloproteinase with thrombospondin motif) in endothelial adherent junction. For instance, proteolytic activity of ADAM10 induces removal and shedding of VE-cadherin from endothelial cell surface, which in turn regulates vascular permeability and inflammation process associated with atherosclerosis (Schulz et al. 2008). ADAM-10 effect on endothelial cell permeability and T cell transmigration involve VEcadherin ectodomain cleavage generating a soluble fragment, while the remaining carboxyterminal membrane bound portion is further cleaved by γ-secretase. ADAM-10-mediated VEcadherin cleavage is induced by thrombin activation of endothelial cells, Ca^{2+} influx, as well as induction of apoptosis by staurosporine treatment. Inhibition of ADAM-10 by GI254023X decreased endothelial cell permeability and transmigration of T cells.

Tight junction proteins such as JAM-A, JAM-B, and JAM-C regulate leukocyte-endothelial cell interaction through their ability to undergo heterophilic binding with leukocyte integrins LFA-1, VLA-4, and Mac1, respectively (Ebnet et al. 2004; Keiper et al. 2005). In addition, their junctional localization in endothelial cells regulates endothelial barrier permeability through the control of actomyosin-dependent contractility and VE-catherin-mediated cell-cell contact in a Rap1-dependent manner (Orlova et al. 2006). A recent study suggests that MMPs could be involved in the generation of a soluble JMA-C form (sJMA-C) appearing as a potent proangiogenic mediator of pathological angiogenesis (Rabquer et al. 2010). However, the individual MMP or ADAM involved in this process remains to be identified and it is not clear whether this process could affect vascular leakage and permeability in tumors.

In addition to these effects on vascular permeability, MMP contribution to vascular contraction is well documented. For instance, MMPs affect membrane Ca^{2+} and/or K^+ channel activity, likely through an interaction with αvβ3 integrin (Miyazaki et al. 2011). MMP inhibitors block Ca^{2+} entry and vascular contraction (Chew et al. 2004). During varicose vein formation, MMP2 induces venous dilation *via* hyperpolarization and activation of K^+ channels (Raffetto et al. 2007). MMP2 also regulates Ca^{2+} entry into smooth muscle (VSM) which in turn causes hyperpolarization and relaxation of venous segments (Raffetto et al. 2010).

14.5 MMP-Mediated Vessel Maturation Through Pericyte Recruitment

During the process of angiogenesis, vascular mural cells (VSMCs or pericytes) are recruited to the newly formed blood vessels where they contribute to vessel maturation. Pericytes, the mural cells of microvessels, extend long cytoplasmic processes on the abluminal surface of the endothelial cells, making tight contacts that are important for blood vessel stabilization. They are important in the formation and/or remodeling of perivascular ECM including the vascular basement membrane, and they regulate vessel function (Armulik et al. 2005; Diaz-Flores et al. 2009; von Tell et al. 2006). Pericytes control vessel permeability through their contractile properties and regulate blood flow in vessels, and consequently they could influence delivery of drug to tumors (Feron 2004; Raza et al. 2010). Pericytes are recruited in newly formed vessels through several molecular axis including PDGFB/PDGFR-β, S1P/endothelial differentiation gene-1 (EDG-1), Ang1/Tie2 and TGF-β/activine-like kinase receptor (ALK5) (Gerhardt and Semb 2008; Jain 2003; Park et al. 2009). For instance, double deletion of PDGFB and PDGFR-β in mice lead to lethal microhaemorages due to lack of vessel coverage by pericytes (Lindahl et al. 1997). Increasing evidences demonstrate that MMPs affect vessel coverage through different mechanisms. MMPs induce pericytes migration through ECM degradation, but also regulate their differentiation and recruitment from bone marrow through the release of angiogenic factors sequestered in the ECM (Chantrain et al. 2006; Sounni and Noel 2005). MMP inhibition in different experimental tumor models of melanoma and neuroblastoma reduced pericyte recruitment and decreased tumor vessel perfusion (Chantrain et al. 2004; Noel et al. 2008; Spurbeck et al. 2002). Indeed, MMP3 and MMP7 promote pericyte recruitment and vessel maturation through shedding of membrane bound HB-EGF and signaling *in vivo* (Suzuki et al. 1997). If inflammatory cell-derived MMP9 promotes tumor angiogenesis by releasing ECMbound VEGF (Bergers et al. 2000; Huang et al. 2002), pericyte coverage along tumor microvessels is directly affected in MMP9 deficient mice xenotransplanted with human neuroblastoma. Interestingly, bone marrow transfer from wild type mice to MMP9$^{-/-}$ mice completely restores tumor vessels maturation (Chantrain et al. 2004). These data suggest that MMP9 is an important regulator of pericyte recruitment from bone marrow.

A number of studies demonstrated that TGFβ promotes investment and differentiation of peri-vascular cells leading to vessel stabilization. TGFβ has been implicated as a regulator of vascular integrity (Gleizes and Rifkin 1999; Pepper 1997; Tuxhorn et al. 2002), vasculogenic and angiogenic processes (Dickson et al. 1995; Pepper 1997), endothelial and mural cell proliferation and/or differentiation (Sato 1995; Vinals and Pouyssegur 2001; Yan and Sage 1998). Moreover, vascular abnormalities have been described in TGFβ null mice (Dickson et al. 1995). In VSMCs, TGFβ can bind to ALK5 and activate SMADs 2 and 3 to promote cell differentiation, increase contractility and ECM synthesis (Chan et al. 2010; Kano et al. 2007). Alternatively, TGFβ may also promote vascular stability by increasing the expression of factors such as Ang-1 that stabilizes vasculature by acting on adjacent endothelial cells (Thurston et al. 1999). Moreover TGFβ-induced fibroblast or smooth muscle cell (SMC) contractility could increase tension on ECM leading to increased interstitial fluid pressure to further restrict capillary outflow (Heldin et al. 2004). The control of TGFβ bioavailability is a post-translational pathway subject to regulation by MMP activity. MMP2, -3, -7 induce *in vitro* the release of TGFβ1 from decorin, a proteoglycan that sequester TGFβ in the ECM (Imai et al. 1997). Furthermore, studies on MT1-MMP revealed its important role in the activation of proTGFβ *in vivo* and control of vascular homeostasis (Sounni et al. 2010a; Tatti et al. 2008).

14.6 MT1-MMP: A Key Regulator of Vessel Function and Physiology

MT1-MMP is likely the most important MMP regulating endothelial cell functions. Among all the MMPs knockout mice generated, *Mt1-mmp* mice only showed impaired angiogenesis leading to delayed ossification and consequently a severe skeletal defect (Sounni and Noel 2005; Zhou et al. 2000). Beside its role in ECM remodeling, MT1-MMP promotes endothelial cell migration, lumen formation and vascular guidance tunnels in collagen matrices (Stratman et al. 2009). It stimulates angiogenesis in fibrin gel more efficiently than other proteases (Collen et al. 2003; Hiraoka et al. 1998; Hotary et al. 2002). MT1-MMP proangiogenic capacities in both physiological and pathological conditions are related to several mechanisms including: (1) ECM remodeling (Hotary et al. 2003), (2) transcriptional and post-translational control of VEGF expression and biodisponibility (Deryugina et al. 2002; Eisenach et al. 2010; Sounni et al. 2002, 2004), (3) interaction with cell surface molecules, such as CD44 (Kajita et al. 2001) and sphingosine 1-phosphate (S1P) (Langlois et al. 2004), (4) hematopoietic progenitor cells mobilization (Vagima et al. 2009), or (5) degradation of anti-angiogenic factors such as decorin in cornea (Mimura et al. 2009). Furthermore, a number of recent reports have shed light of the role of MT1-MMP in TGFβ signaling during angiogenesis and vessel maturation (Hawinkels et al. 2010; Sounni et al. 2011; Tatti et al. 2008).

Spatial and temporal MT1-MMP expression during angiogenesis has been recently followed in the transgenic *Mt1-mmp+/LacZ mice* (Yana et al. 2007). MT1-MMP expression at leading edges of newly developed vessels is a result of coordinated crosstalk between endothelial cells and VSMCs during vascular maturation step of angiogenesis. Works by Lehti and colleagues (Lehti et al. 2005) showed that MT1-MMP is important for PDGFR-β processing and propagating signaling in VSMCs and promoting cell migration. The vascular network of *Mt1-mmp*$^{-/-}$ mice brain have a sever reduction in mural cell density and abnormal vessel morphology, due to impaired PDGFB/PDGFR-β signaling in VSMCs. Moreover, MT1-MMP induces VSMC dedifferentiation and acquisition of migratory and invasive phenotype during vascular injury through low density lipoprotein (LDL) receptor-related protein (LRP) proteolysis that promotes signaling through PDGFB/PDGFR-β axis (Lehti et al. 2009). In addition, MT1-MMP regulates signaling in vascular cells through several mechanisms. MT1-MMP cooperates with platelet-derived S1P to induce endothelial cell migration and morphogenic differentiation (Langlois et al. 2004). It regulates signaling of the advanced glycation end products (AGE)/a receptor for AGE (RAGE) axis in vascular disorders associated with diabetic (Kamioka et al. 2011). Furthermore, a recent report shows that MT1-MMP activity at endothelial cell surface mediates Tie-2 shedding and regulates angiopoietin1 (Ang1)/Tie-2-associated signaling pathways (Onimaru et al. 2010). It is generally accepted that Ang1-mediated activation of Tie-2 promotes vascular stabilization and quiescence, whereas Ang2 acts in opposition to Ang1 to facilitate VEGF mediated angiogenesis (Gale and Yancopoulos 1999). MT1-MMP interference with Ang1-Tie2 reinforces its role in vessel activation a well known process widely attributed to VEGF (Findley et al. 2007).

Recently, we provided evidence for MT1-MMP activity in perivascular stroma *in vivo* and regulation of vascular stability and permeability (Sounni et al. 2010a). By assessing the effect of MMP deletion on vessel permeability and leakage applied to the skin of *Mmp2*$^{-/-}$, *Mmp9*$^{-/-}$, *Mmp8*$^{-/-}$, *Mmp13*$^{-/-}$ and Mt1-*mmp*$^{-/-}$ mice, we found that except Mt1-*mmp*$^{-/-}$ mice, both induced and steady state leakage were not affected in these mice, whereas and Mt1-*mmp*$^{-/-}$ mice have a higher steady-state vascular leakage. Moreover, treatment of wild-type mice with broad spectrum MMP inhibitor GM6001 (Ilomasta, Galardin) significantly increases vascular leakage *in vivo*. These data indicate that MMP inhibition renders vessels more susceptible to induced acute leakage and implies a link between basal MMP activity and vessel function. A link between MT1-MMP and TGFβ pathway in vascular homeostasis maintenance has been demonstrated by our finding on MT1-MMP-mediated control of TGFβ bioavailability and signaling through the TGF-beta receptor type-1 ALK5 in vascular wall *in vivo* (Sounni et al. 2010a). Interestingly, ALK5 inhibitor increased *in vivo* vascular leakage and enhanced macromolecule delivery and biodistribution in two syngenic model of skin carcinoma (K14-HPV16 transgenic mice) and mammary adenocarcinoma (MMTV-PyMT mice). These data shed light on MT1-MMP contribution to the TGFβ-controlled vascular homeostasis and remodeling. They further indicate that TGFβ and/or MT1-MMP-selective antagonists may enhance vascular leakage and therapeutic delivery to tissues where hemodynamic limits efficient drug delivery. Recently, MT1-MMP-dependent shedding of endoglin from endothelial cell surface has been documented and associated to angiogenesis inhibition *in vitro*. This study correlates a decrease of circulating endoglin with colorectal cancer and suggests that the release of soluble endoglin by MT1-MMP could likely act as decoy for TGFβ, blocks downstream signaling in endothelial cells and inhibits tumor angiogenesis (Hawinkels et al. 2010). However, MT1-MMP levels in these samples were not correlated with the level of soluble endoglin and further studies are required to support the controversial anti-angiogenic effect of MT1-MMP *in vivo*. This effect of MT1-MMP interaction with TGFβ signaling on angiogenesis *in vivo*, appears therefore as a complex regulatory mechanism that depends on its spatial distribution either in perivascular stroma or endothelial cell compartment. A direct shedding of endoglin from endothelial cell surface inhibits angiogenesis (Hawinkels et al. 2010), whereas the cleavage of LTBP-1 (Dallas et al. 2002; Tatti et al. 2008) or LAP-TGFβ from the stroma in fibrotic tissue increases TGFβ availability which in turn activates TGFβ signaling in vascular cells (Sato 1995; Sounni et al. 2010a). An additional level of complexity comes from the regulation of MT1-MMP/TGFβ signaling axis by type I collagen. Type I collagen induces MT1-MMP expression which in turn regulates TGFβ signaling in stromal and cancer cells (Gilles et al. 1997; Noel et al. 2001; Ottaviano et al. 2006; Shields et al. 2011). Altogether these observations highlight the multifunctional feature of MT1-MMP. MT1-MMP emerged recently as a key angiomodulator enzyme that controls vessel maturation and/or regression both in normal and pathological conditions.

14.7 Concluding Remarks and Therapeutic Potentials

The ECM in vessel wall is no longer seen as a simple scaffold for vascular cells, but is now viewed as a dynamic actor in vascular cell signaling and physiology control. Deregulation of the balance between ECM synthesis and degradation contributes to vascular pathologies and tumor angiogenesis. Recent progresses have highlighted the role of MMPs in vascular biology and underlined the multiple functions of these proteases in the regulation of vessel functions. MMPs are now recognized as key modulators of angiogenesis in different diseases such as cancer (Lafleur et al. 2003; Lopez-Otin and Matrisian 2007), AMD (Noel et al. 2007), psoriasis (Suomela et al. 2001), diabetic retinopathy (Giebel et al. 2005) and pulmonary oedema (Davey et al. 2011). ECM disruption on the arterial wall is likely a main factor in the pathogenesis of vascular system (Raffetto and Khalil 2008), as a decrease in structural matrix proteins has been demonstrated in atherosclerotic lesions, abdominal and intracranial aneurysms, and vascular dilatation (Ruigrok et al. 2006; Wagsater et al. 2011). It is well accepted that increased macrophage and SMC-derived MMP activities have deleterious effects on atherosclerotic plaque stability leading finally to plaque rupture (Johnson 2007). MMP12 overexpressed in ruptured plaque tissues (Jormsjo et al. 2000) appears as a key target that led to the generation of selective MMP inhibitors with the aim to inhibit plaque rupture of advanced atherosclerosis lesion (Fic et al. 2011).

MT-MMPs effects on vessel wall function and integrity is of great interest for the delivery of therapeutics *in vivo*. MT-MMPs are able to modify the tumor physiology and to affect vessel permeability *in vivo*. At this end, understanding the cellular and molecular mechanisms of MTMMPs and their interaction with ECM and non ECM molecules, permeability and interstitial fluid pressure of tumor (Sounni et al. 2011) will open new targeting opportunities to enhance vascular leakage and delivery of therapeutics to tissues where hemodynamics limit efficient drug delivery. However, therapeutic benefits derived from ECM degrading enzyme inhibition may lead to vessel stabilization in cancer and in hyper active tissues. Particularly, this approach may be beneficial if combined with other target pathway inhibitors or cytoxic drugs for cancer treatment. In addition, increased understanding of regulation and function of cryptic fragments within ECM and their role in the control of vascular cell physiology will likely lead to more effective strategies for therapy. The complexity of targeting proteolytic enzymes due to their pro- and antiangiogenic function in cancer or vascular diseases (Lopez-Otin and Matrisian 2007), renders targeting cryptic collagen sites a highly selective approach for regulating angiogenesis. Further investigations are required to understand how basement membrane molecules regulate vessel permeability and leakage in cancer and vascular diseases processes.

Acknowledgments This work was supported by grants from the Action de Recherche Concertée (ARC) de l'Université de Liège, Belgium, the FP7-HEALTH-2007-A Proposal No. 201279 "MICROENVIMET", the Fonds de la Recherche Scientifique-FNRS (F.R.S.-FNRS, Belgium), the Foundation against Cancer (foundation of public interest, Belgium), the Centre Anticancéreux près l'Université de Liège, the Fonds Léon Fredericq (University of Liège), the Direction Générale Opérationnelle de l'Economie, de l'Emploi et de la Recherche from the S.P.W. (Région Wallonne, Belgium), the Interuniversity Attraction Poles Programme—Belgian Science Policy (Brussels, Belgium).

References

Armulik A, Abramsson A, Betsholtz C (2005) Endothelial/pericyte interactions. Circ Res 97:512–523

Arteaga-Solis E, Gayraud B, Ramirez F (2000) Elastic and collagenous networks in vascular diseases. Cell Struct Funct 25:69–72

Asahi M, Wang X, Mori T, Sumii T, Jung JC, Moskowitz MA, Fini ME, Lo EH (2001) Effects of matrix metalloproteinase-9 gene knock-out on the proteolysis of bloodbrain barrier and white matter components after cerebral ischemia. J Neurosci 21:7724–7732

Bergers G, Benjamin LE (2003) Angiogenesis: tumorigenesis and the angiogenic switch. Nat Rev Cancer 3:401–410

Bergers G, Brekken R, McMahon G, Vu TH, Itoh T, Tamaki K, Tanzawa K, Thorpe P, Itohara S, Werb Z, Hanahan D (2000) Matrix metalloproteinase-9 triggers the angiogenic switch during carcinogenesis. Nat Cell Biol 2:737–744

Berk BC, Fujiwara K, Lehoux S (2007) ECM remodeling in hypertensive heart disease. J Clin Invest 117:568–575

Bhutto IA, Kim SY, McLeod DS, Merges C, Fukai N, Olsen BR, Lutty GA (2004) Localization of collagen XVIII and the endostatin portion of collagen XVIII in aged human control eyes and eyes with age-related macular degeneration. Invest Ophthalmol Vis Sci 45:1544–1552

Boudreau NJ, Jones PL (1999) Extracellular matrix and integrin signalling: the shape of things to come. Biochem J 339:481–488

Butler GS, Overall CM (2009) Proteomic identification of multitasking proteins in unexpected locations complicates drug targeting. Nat Rev Drug Discov 8:935–948

Chan MC, Hilyard AC, Wu C, Davis BN, Hill NS, Lal A, Lieberman J, Lagna G, Hata A (2010) Molecular basis for antagonism between PDGF and the TGFbeta family of signalling pathways by control of miR-24 expression. EMBO J 29:559–573

Chantrain C, Shimada H, Jodele S, Groshen S, Ye W, Shalinsky D, Werb Z, Coussens LM, DeClerck YA (2004) Stromal matrix metalloproteinase-9 regulates the vascular architecture in neuroblastoma by promoting pericyte recruitment. Cancer Res 64:1675–1686

Chantrain CF, Henriet P, Jodele S, Emonard H, Feron O, Courtoy PJ, DeClerck YA, Marbaix E (2006) Mechanisms of pericyte recruitment in tumour angiogenesis: a new role for metalloproteinases. Eur J Cancer 42:310–318

Chernov AV, Sounni NE, Remacle AG, Strongin AY (2009) Epigenetic control of the invasion-promoting MT1-MMP/MMP-2/TIMP-2 axis in cancer cells. J Biol Chem 284:12727–12734

Chew DK, Conte MS, Khalil RA (2004) Matrix metalloproteinase-specific inhibition of Ca2+ entry mechanisms of vascular contraction. J Vasc Surg 40:1001–1010

Collen A, Hanemaaijer R, Lupu F, Quax PH, Van Lent N, Grimbergen J, Peters E, Koolwijk P, Van Hinsbergh VW (2003) Membrane-type matrix metalloproteinasemediated angiogenesis in a fibrin-collagen matrix. Blood 101:1810–1817

Cretu A, Roth JM, Caunt M, Akalu A, Policarpio D, Formenti S, Gagne P, Liebes L, Brooks PC (2007) Disruption of endothelial cell interactions with the novel HU177 cryptic collagen epitope inhibits angiogenesis. Clin Cancer Res 13:3068–3078

Dallas SL, Rosser JL, Mundy GR, Bonewald LF (2002) Proteolysis of latent transforming growth factor-beta (TGF-beta)- binding protein-1 by osteoclasts. A cellular mechanism for release of TGF-beta from bone matrix. J Biol Chem 277:21352–21360

Davey A, McAuley DF, O'Kane CM (2011) Matrix metalloproteinases in acute lung injury: mediators of injury and drivers of repair. Eur Respir J 38(4):959–970

Davis GE, Senger DR (2008) Extracellular matrix mediates a molecular balance between vascular morphogenesis and regression. Curr Opin Hematol 15:197–203

Deryugina EI, Soroceanu L, Strongin AY (2002) Up-regulation of vascular endothelial growth factor by membrane-type 1 matrix metalloproteinase stimulates human glioma xenograft growth and angiogenesis. Cancer Res 62:580–588

Detry B, Erpicum C, Paupert J, Blacher S, Maillard C, Bruyere F, Pendeville H, Remacle T, Lambert V, Balsat C, Ormenese S, Lamaye F, Janssens E, Moons L, Cataldo D, Kridelka F, Carmeliet P, Thiry M, Foidart JM, Struman I, Noel A (2012) Matrix metalloproteinase-2 governs lymphatic vessel formation as an interstitial collagenase. Blood 119:5048–5056

Di Lullo G, Sweeney S, Korkko J, Ala-Kokkor L, San Antonio J (2002) Mapping the ligandbinding sites and disease-associated mutations on the most abundant protein in the human-type I collagen. J Biol Chem 277:4223–4231

Diaz-Flores L, Gutierrez R, Madrid JF, Varela H, Valladares F, Acosta E, Martin- Vasallo P, Diaz-Flores L Jr (2009) Pericytes. Morphofunction, interactions and pathology in a quiescent and activated mesenchymal cell niche. Histol Histopathol 24:909–969

Dickson MC, Martin JS, Cousins FM, Kulkarni AB, Karlsson S, Akhurst RJ (1995) Defective haematopoiesis and vasculogenesis in transforming growth factor-beta 1 knock out mice. Development 121:1845–1854

Diez J (2007) Arterial stiffness and extracellular matrix. Adv Cardiol 44:76–95

Ebnet K, Suzuki A, Ohno S, Vestweber D (2004) Junctional adhesion molecules (JAMs): more molecules with dual functions? J Cell Sci 117:19–29

Egeblad M, Werb Z (2002) New functions for the matrix metalloproteinases in cancer progression. Nat Rev Cancer 2:161–174

Egeblad M, Shen HC, Behonick DJ, Wilmes L, Eichten A, Korets LV, Kheradmand F, Werb Z, Coussens LM (2007) Type I collagen is a genetic modifier of matrix metalloproteinase 2 in murine skeletal development. Dev Dyn 236:1683–1693

Eisenach PA, Roghi C, Fogarasi M, Murphy G, English WR (2010) MT1-MMP regulates VEGF-A expression through a complex with VEGFR-2 and Src. J Cell Sci 123:4182–4193

Feron O (2004) Targeting the tumor vascular compartment to improve conventional cancer therapy. Trends Pharmacol Sci 25:536–542

Ferreras M, Felbor U, Lenhard T, Olsen BR, Delaisse J (2000) Generation and degradation of human endostatin proteins by various proteinases. FEBS Lett 486:247–251

Fic P, Zakrocka I, Kurzepa J, Stepulak A (2011) Matrix metalloproteinases and atherosclerosis. Postepy Hig Med Dosw (Online) 65:16–27

Findley CM, Cudmore MJ, Ahmed A, Kontos CD (2007) VEGF induces Tie2 shedding via a phosphoinositide 3-kinase/Akt dependent pathway to modulate Tie2 signaling. Arterioscler Thromb Vasc Biol 27:2619–2626

Fujimoto M, Takagi Y, Aoki T, Hayase M, Marumo T, Gomi M, Nishimura M, Kataoka H, Hashimoto N, Nozaki K (2008) Tissue inhibitor of metalloproteinases protect bloodbrain barrier disruption in focal cerebral ischemia. J Cereb Blood Flow Metab 28:1674–1685

Gale NW, Yancopoulos GD (1999) Growth factors acting via endothelial cell-specific receptor tyrosine kinases: VEGFs, angiopoietins, and ephrins in vascular development. Genes Dev 13:1055–1066

Gerhardt H, Semb H (2008) Pericytes: gatekeepers in tumour cell metastasis? J Mol Med 86:135–144

Giebel SJ, Menicucci G, McGuire PG, Das A (2005) Matrix metalloproteinases in early diabetic retinopathy and their role in alteration of the blood-retinal barrier. Lab Invest 85:597–607

Gilles C, Polette M, Seiki M, Birembaut P, Thompson EW (1997) Implication of collagen type I-induced membrane-type 1-matrix metalloproteinase expression and matrix metalloproteinase-2 activation in the metastatic progression of breast carcinoma. Lab Invest 76:651–660

Gleizes PE, Rifkin DB (1999) Activation of latent TGF-beta. A required mechanism for vascular integrity. Pathol Biol (Paris) 47:322–329

Gocheva V, Zeng W, Ke D, Klimstra D, Reinheckel T, Peters C, Hanahan D, Joyce JA (2006) Distinct roles for cysteine cathepsin genes in multistage tumorigenesis. Genes Dev 20:543–556

Hawinkels LJ, Kuiper P, Wiercinska E, Verspaget HW, Liu Z, Pardali E, Sier CF, ten Dijke P (2010) Matrix metalloproteinase-14 (MT1-MMP)-mediated endoglin shedding inhibits tumor angiogenesis. Cancer Res 70:4141–4150

Hawkins BT, Lundeen TF, Norwood KM, Brooks HL, Egleton RD (2007) Increased blood-brain barrier permeability and altered tight junctions in experimental diabetes in the rat: contribution of hyperglycaemia and matrix metalloproteinases. Diabetologia 50:202–211

Heldin CH, Rubin K, Pietras K, Ostman A (2004) High interstitial fluid pressure – an obstacle in cancer therapy. Nat Rev Cancer 4:806–813

Hiraoka N, Allen E, Apel IJ, Gyetko MR, Weiss SJ (1998) Matrix metalloproteinases regulate neovascularization by acting as pericellular fibrinolysins. Cell 95:365–377

Holmbeck K, Bianco P, Caterina J, Yamada S, Kromer M, Kuznetsov SA, Mankani M, Robey PG, Poole AR, Pidoux I, Ward JM, Birkedal-Hansen H (1999) MT1- MMP-deficient mice develop dwarfism, osteopenia, arthritis, and connective tissue disease due to inadequate collagen turnover. Cell 99:81–92

Hotary KB, Yana I, Sabeh F, Li XY, Holmbeck K, Birkedal-Hansen H, Allen ED, Hiraoka N, Weiss SJ (2002) Matrix metalloproteinases (MMPs) regulate fibrininvasive activity via MT1-MMP-dependent and -independent processes. J Exp Med 195:295–308

Hotary KB, Allen ED, Brooks PC, Datta NS, Long MW, Weiss SJ (2003) Membrane type I matrix metalloproteinase usurps tumor growth control imposed by the threedimensional extracellular matrix. Cell 114:33–45

Hu J, Van den Steen PE, Sang QX, Opdenakker G (2007) Matrix metalloproteinase inhibitors as therapy for inflammatory and vascular diseases. Nat Rev Drug Discov 6:480–498

Huang S, Van Arsdall M, Tedjarati S, McCarty M, Wu W, Langley R, Fidler IJ (2002) Contributions of stromal metalloproteinase-9 to angiogenesis and growth of human ovarian carcinoma in mice. J Natl Cancer Inst 94:1134–1142

Imai K, Hiramatsu A, Fukushima D, Pierschbacher MD, Okada Y (1997) Degradation of decorin by matrix metalloproteinases: identification of the cleavage sites, kinetic analyses and transforming growth factor-beta1 release. Biochem J 322:809–814

Jain RK (2003) Molecular regulation of vessel maturation. Nat Med 9:685–693

Johnson JL (2007) Matrix metalloproteinases: influence on smooth muscle cells and atherosclerotic plaque stability. Expert Rev Cardiovasc Ther 5:265–282

Jormsjo S, Ye S, Moritz J, Walter DH, Dimmeler S, Zeiher AM, Henney A, Hamsten A, Eriksson P (2000) Allele-specific regulation of matrix metalloproteinase-12 gene activity is associated with coronary artery luminal dimensions in diabetic patients with manifest coronary artery disease. Circ Res 86:998–1003

Kajita M, Itoh Y, Chiba T, Mori H, Okada A, Kinoh H, Seiki M (2001) Membrane-type 1 matrix metalloproteinase cleaves CD44 and promotes cell migration. J Cell Biol 153:893–904

Kamioka M, Ishibashi T, Ohkawara H, Eguchi R, Sugimoto K, Uekita H, Matsui T, Yamagishi SI, Ando K, Sakamoto T, Sakamoto N, Takuwa Y, Wada I, Shiomi M, Maruyama Y, Takeishi Y (2011) Involvement of membrane type 1-matrix metalloproteinase (MT1-MMP) in rage activation signaling pathways. J Cell Physiol 226(6):1554–1563

Kano MR, Bae Y, Iwata C, Morishita Y, Yashiro M, Oka M, Fujii T, Komuro A, Kiyono K, Kaminishi M, Hirakawa K, Ouchi Y, Nishiyama N, Kataoka K, Miyazono K (2007) Improvement of cancer-targeting therapy, using nanocarriers for intractable solid tumors by inhibition of TGF-beta signaling. Proc Natl Acad Sci U S A 104:3460–3465

Keiper T, Santoso S, Nawroth PP, Orlova V, Chavakis T (2005) The role of junctional adhesion molecules in cell-cell interactions. Histol Histopathol 20:197–203

Kessenbrock K, Plaks V, Werb Z (2010) Matrix metalloproteinases: regulators of the tumor microenvironment. Cell 141:52–67

Kojima T, Azar DT, Chang JH (2008) Neostatin-7 regulates bFGF-induced corneal lymphangiogenesis. FEBS Lett 582:2515–2520

Lafleur MA, Handsley MM, Edwards DR (2003) Metalloproteinases and their inhibitors in angiogenesis. Expert Rev Mol Med 5:1–39

Langlois S, Gingras D, Beliveau R (2004) Membrane type 1-matrix metalloproteinase (MT1-MMP) cooperates with sphingosine 1-phosphate to induce endothelial cell migration and morphogenic differentiation. Blood 103:3020–3028

Lehti K, Allen E, Birkedal-Hansen H, Holmbeck K, Miyake Y, Chun TH, Weiss SJ (2005) An MT1-MMP-PDGF receptor-beta axis regulates mural cell investment of the microvasculature. Genes Dev 19:979–991

Lehti K, Rose NF, Valavaara S, Weiss SJ, Keski-Oja J (2009) MT1-MMP promotes vascular smooth muscle dedifferentiation through LRP1 processing. J Cell Sci 122:126–135

Lindahl P, Johansson BR, Leveen P, Betsholtz C (1997) Pericyte loss and microaneurysm formation in PDGF-B-deficient mice. Science 277:242–245

Liu X, Wu H, Byrne M, Jeffrey J, Krane S, Jaenisch R (1995) A targeted mutation at the known collagenase cleavage site in mouse type I collagen impairs tissue remodeling. J Cell Biol 130:227–237

Liu MC, Choong DY, Hooi CS, Williams LH, Campbell IG (2007) Genetic and epigenetic analysis of the TIMP-3 gene in ovarian cancer. Cancer Lett 247:91–97

Lopez-Otin C, Matrisian LM (2007) Emerging roles of proteases in tumour suppression. Nat Rev Cancer 7:800–808

Lopez-Otin C, Overall CM (2002) Protease degradomics: a new challenge for proteomics. Nat Rev Mol Cell Biol 3:509–519

Marneros AG, She H, Zambarakji H, Hashizume H, Connolly EJ, Kim I, Gragoudas ES, Miller JW, Olsen BR (2007) Endogenous endostatin inhibits choroidal neovascularization. FASEB J 21:3809–3818

Martignetti JA, Aqeel AA, Sewairi WA, Boumah CE, Kambouris M, Mayouf SA, Sheth KV, Eid WA, Dowling O, Harris J, Glucksman MJ, Bahabri S, Meyer BF, Desnick RJ (2001) Mutation of the matrix metalloproteinase 2 gene (MMP2) causes a multicentric osteolysis and arthritis syndrome. Nat Genet 28:261–265

Matter K, Balda MS (2003) Signalling to and from tight junctions. Nat Rev Mol Cell Biol 4:225–236

Mimura T, Han KY, Onguchi T, Chang JH, Kim TI, Kojima T, Zhou Z, Azar DT (2009) MT1-MMP-mediated cleavage of decorin in corneal angiogenesis. J Vasc Res 46:541–550

Miyazaki D, Nakamura A, Fukushima K, Yoshida K, Takeda S, Ikeda S (2011) Matrix metalloproteinase-2 ablation in dystrophin-deficient mdx muscles reduces angiogenesis resulting in impaired growth of regenerated muscle fibers. Hum Mol Genet 20:1787–1799

Morwood SR, Nicholson LB (2006) Modulation of the immune response by extracellular matrix proteins. Arch Immunol Ther Exp (Warsz) 54:367–374

Mott JD, Werb Z (2004) Regulation of matrix biology by matrix metalloproteinases. Curr Opin Cell Biol 16:558–564

Nagase H, Visse R, Murphy G (2006) Structure and function of matrix metalloproteinases and TIMPs. Cardiovasc Res 69:562–573

Navaratna D, McGuire PG, Menicucci G, Das A (2007) Proteolytic degradation of VEcadherin alters the blood-retinal barrier in diabetes. Diabetes 56:2380–2387

Neptune ER, Frischmeyer PA, Arking DE, Myers L, Bunton TE, Gayraud B, Ramirez F, Sakai LY, Dietz HC (2003) Dysregulation of TGF-beta activation contributes to pathogenesis in Marfan syndrome. Nat Genet 33:407–411

Noel A, Albert V, Bajou K, Bisson C, Devy L, Frankenne F, Maquoi E, Masson V, Sounni NE, Foidart JM (2001) New functions of stromal proteases and their inhibitors in tumor progression. Surg Oncol Clin N Am 10:417–432, x–xi

Noel A, Jost M, Lambert V, Lecomte J, Rakic JM (2007) Anti-angiogenic therapy of exudative age-related macular degeneration: current progress and emerging concepts. Trends Mol Med 13:345–352

Noel A, Jost M, Maquoi E (2008) Matrix metalloproteinases at cancer tumor-host interface. Semin Cell Dev Biol 19:52–60

Nyberg P, Xie L, Kalluri R (2005) Endogenous inhibitors of angiogenesis. Cancer Res 65:3967–3979

Ohuchi E, Imai K, Fujii Y, Sato H, Seiki M, Okada Y (1997) Membrane type 1 matrix metalloproteinase digests interstitial collagens and other extracellular matrix macromolecules. J Biol Chem 272:2446–2451

Oliver G, Detmar M (2002) The rediscovery of the lymphatic system: old and new insights into the development and biological function of the lymphatic vasculature. Genes Dev 16:773–783

Onimaru M, Yonemitsu Y, Suzuki H, Fujii T, Sueishi K (2010) An autocrine linkage between matrix metalloproteinase-14 and Tie-2 via ectodomain shedding modulates angiopoietin-1-dependent function in endothelial cells. Arterioscler Thromb Vasc Biol 30:818–826

Orlova VV, Economopoulou M, Lupu F, Santoso S, Chavakis T (2006) Junctional adhesion molecule-C regulates vascular endothelial permeability by modulating VEcadherin- mediated cell-cell contacts. J Exp Med 203:2703–2714

Ottaviano AJ, Sun L, Ananthanarayanan V, Munshi HG (2006) Extracellular matrixmediated membrane-type 1 matrix metalloproteinase expression in pancreatic ductal cells is regulated by transforming growth factor-beta1. Cancer Res 66:7032–7040

Page-McCaw A, Ewald AJ, Werb Z (2007) Matrix metalloproteinases and the regulation of tissue remodelling. Nat Rev Mol Cell Biol 8:221–233

Park JS, Seo J, Kim YO, Lee HS, Jo I (2009) Coordinated regulation of angiopoietin-1 and vascular endothelial growth factor by arsenite in human brain microvascular pericytes: implications of arsenite-induced vascular dysfunction. Toxicology 264:26–31

Pepper MS (1997) Transforming growth factor-beta: vasculogenesis, angiogenesis, and vessel wall integrity. Cytokine Growth Factor Rev 8:21–43

Pulukuri SM, Patibandla S, Patel J, Estes N, Rao JS (2007) Epigenetic inactivation of the tissue inhibitor of metalloproteinase-2 (TIMP-2) gene in human prostate tumors. Oncogene 26:5229–5237

Rabquer BJ, Amin MA, Teegala N, Shaheen MK, Tsou PS, Ruth JH, Lesch CA, Imhof BA, Koch AE (2010) Junctional adhesion molecule-C is a soluble mediator of angiogenesis. J Immunol 185:1777–1785

Radisky DC, Kenny PA, Bissell MJ (2007) Fibrosis and cancer: do myofibroblasts come also from epithelial cells via EMT? J Cell Biochem 101:830–839

Raffetto JD, Khalil RA (2008) Matrix metalloproteinases and their inhibitors in vascular remodeling and vascular disease. Biochem Pharmacol 75:346–359

Raffetto JD, Ross RL, Khalil RA (2007) Matrix metalloproteinase 2-induced venous dilation via hyperpolarization and activation of K+ channels: relevance to varicose vein formation. J Vasc Surg 45:373–380

Raffetto JD, Barros YV, Wells AK, Khalil RA (2010) MMP-2 induced vein relaxation via inhibition of [Ca2+]e-dependent mechanisms of venous smooth muscle contraction. Role of RGD peptides. J Surg Res 159:755–764

Raza A, Franklin MJ, Dudek AZ (2010) Pericytes and vessel maturation during tumor angiogenesis and metastasis. Am J Hematol 85:593–598

Rebustini IT, Myers C, Lassiter KS, Surmak A, Szabova L, Holmbeck K, Pedchenko V, Hudson BG, Hoffman MP (2009) MT2-MMP-dependent release of collagen IV NC1 domains regulates submandibular gland branching morphogenesis. Dev Cell 17:482–493

Ruigrok YM, Rinkel GJ, van't Slot R, Wolfs M, Tang S, Wijmenga C (2006) Evidence in favor of the contribution of genes involved in the maintenance of the extracellular matrix of the arterial wall to the development of intracranial aneurysms. Hum Mol Genet 15:3361–3368

Rundhaug JE (2005) Matrix metalloproteinases and angiogenesis. J Cell Mol Med 9:267–285

Sato Y (1995) Activation of latent TGF-beta at the vascular wall–roles of endothelial cells and mural pericytes or smooth muscle cells. J Atheroscler Thromb 2:24–29

Sato H, Takino T, Okada Y, Cao J, Shinagawa A, Yamamoto E, Seiki M (1994) A matrix metalloproteinase expressed on the surface of invasive tumour cells. Nature 370:61–65

Schulz B, Pruessmeyer J, Maretzky T, Ludwig A, Blobel CP, Saftig P, Reiss K (2008) ADAM10 regulates endothelial permeability and T-Cell transmigration by proteolysis of vascular endothelial cadherin. Circ Res 102:1192–1201

Shields MA, Dangi-Garimella S, Krantz SB, Bentrem DJ, Munshi HG (2011) Pancreatic cancer cells respond to Type I collagen by inducing snail expression to promote membrane Type 1 matrix metalloproteinase-dependent collagen invasion. J Biol Chem 286:10495–10504

Shoulders MD, Raines RT (2009) Collagen structure and stability. Annu Rev Biochem 78:929–958

Sohail A, Sun Q, Zhao H, Bernardo MM, Cho JA, Fridman R (2008) MT4-(MMP17) and MT6-MMP (MMP25), a unique set of membrane-anchored matrix metalloproteinases: properties and expression in cancer. Cancer Metastasis Rev 27:289–302

Sounni NE, Noel A (2005) Membrane type-matrix metalloproteinases and tumor progression. Biochimie 87:329–342

Sounni NE, Devy L, Hajitou A, Frankenne F, Munaut C, Gilles C, Deroanne C, Thompson EW, Foidart JM, Noel A (2002) MT1-MMP expression promotes tumor growth and angiogenesis through an up-regulation of vascular endothelial growth factor expression. FASEB J 16:555–564

Sounni NE, Roghi C, Chabottaux V, Janssen M, Munaut C, Maquoi E, Galvez BG, Gilles C, Frankenne F, Murphy G, Foidart JM, Noel A (2004) Up-regulation of vascular endothelial growth factor-A by active membrane-type 1 matrix metalloproteinase through activation of Src-tyrosine kinases. J Biol Chem 279:13564–13574

Sounni NE, Dehne K, van Kempen L, Egeblad M, Affara NI, Cuevas I, Wiesen J, Junankar S, Korets L, Lee J, Shen J, Morrison CJ, Overall CM, Krane SM, Werb Z, Boudreau N, Coussens LM (2010a) Stromal regulation of vessel stability by MMP14 and TGFbeta. Dis Model Mech 3:317–332

Sounni NE, Rozanov DV, Remacle AG, Golubkov VS, Noel A, Strongin AY (2010b) Timp-2 binding with cellular MT1-MMP stimulates invasion-promoting MEK/ERK signaling in cancer cells. Int J Cancer 126:1067–1078

Sounni NE, Paye A, Host L, Noël A (2011) MT-MMPs as regulators of vessel stability associated with angiogenesis. Front Pharmacol 2:111

Spurbeck WW, Ng CY, Strom TS, Vanin EF, Davidoff AM (2002) Enforced expression of tissue inhibitor of matrix metalloproteinase-3 affects functional capillary morphogenesis and inhibits tumor growth in a murine tumor model. Blood 100:3361–3368

Stratman AN, Saunders WB, Sacharidou A, Koh W, Fisher KE, Zawieja DC, Davis MJ, Davis GE (2009) Endothelial cell lumen and vascular guidance tunnel formation requires MT1-MMP-dependent proteolysis in 3-dimensional collagen matrices. Blood 114:237–247

Strongin AY (2010) Proteolytic and non-proteolytic roles of membrane type-1 matrix metalloproteinase in malignancy. Biochim Biophys Acta 1803:133–141

Suomela S, Kariniemi AL, Snellman E, Saarialho-Kere U (2001) Metalloelastase (MMP- 12) and 92-kDa gelatinase (MMP-9) as well as their inhibitors, TIMP-1 and -3, are expressed in psoriatic lesions. Exp Dermatol 10:175–183

Suzuki M, Raab G, Moses MA, Fernandez CA, Klagsbrun M (1997) Matrix metalloproteinase-3 releases active heparin-binding EGF-like growth factor by cleavage at a specific juxtamembrane site. J Biol Chem 272:31730–31737

Tatti O, Vehvilainen P, Lehti K, Keski-Oja J (2008) MT1-MMP releases latent TGF-beta1 from endothelial cell extracellular matrix via proteolytic processing of LTBP-1. Exp Cell Res 314:2501–2514

Thurston G, Suri C, Smith K, McClain J, Sato TN, Yancopoulos GD, McDonald DM (1999) Leakage-resistant blood vessels in mice transgenically overexpressing angiopoietin-1. Science 286:2511–2514

Turley EA (2001) Extracellular matrix remodeling: multiple paradigms in vascular disease. Circ Res 88:2–4

Tuxhorn JA, McAlhany SJ, Yang F, Dang TD, Rowley DR (2002) Inhibition of transforming growth factor-beta activity decreases angiogenesis in a human prostate cancer-reactive stroma xenograft model. Cancer Res 62:6021–6025

Ucuzian AA, Greisler HP (2007) In vitro models of angiogenesis. World J Surg 31:654–663

Vagima Y, Avigdor A, Goichberg P, Shivtiel S, Tesio M, Kalinkovich A, Golan K, Dar A, Kollet O, Petit I, Perl O, Rosenthal E, Resnick I, Hardan I, Gellman YN, Naor D, Nagler A, Lapidot T (2009) MT1-MMP and RECK are involved in human CD34+ progenitor cell retention, egress, and mobilization. J Clin Invest 119:492–503

Vinals F, Pouyssegur J (2001) Transforming growth factor beta1 (TGF-beta1) promotes endothelial cell survival during in vitro angiogenesis via an autocrine mechanism implicating TGF-alpha signaling. Mol Cell Biol 21:7218–7230

von Tell D, Armulik A, Betsholtz C (2006) Pericytes and vascular stability. Exp Cell Res 312:623–629

Wagsater D, Zhu C, Bjorkegren J, Skogsberg J, Eriksson P (2011) MMP-2 and MMP-9 are prominent matrix metalloproteinases during atherosclerosis development in the Ldlr-/-Apob100/100 mouse. Int J Mol Med 28:247–253

Wiig H, Gyenge C, Iversen PO, Gullberg D, Tenstad O (2008) The role of the extracellular matrix in tissue distribution of macromolecules in normal and pathological tissues: potential therapeutic consequences. Microcirculation 15:283–296

Xu J, Rodriguez D, Petitclerc E, Kim JJ, Hangai M, Moon YS, Davis GE, Brooks PC, Yuen SM (2001) Proteolytic exposure of a cryptic site within collagen type IV is required for angiogenesis and tumor growth in vivo. J Cell Biol 154:1069–1079

Yan Q, Sage EH (1998) Transforming growth factor-beta1 induces apoptotic cell death in cultured retinal endothelial cells but not pericytes: association with decreased expression of p21waf1/cip1. J Cell Biochem 70:70–83

Yana I, Sagara H, Takaki S, Takatsu K, Nakamura K, Nakao K, Katsuki M, Taniguchi S, Aoki T, Sato H, Weiss SJ, Seiki M (2007) Crosstalk between neovessels and mural cells directs the site-specific expression of MT1-MMP to endothelial tip cells. J Cell Sci 120:1607–1614

Yang R, Liu H, Williams I, Chaqour B (2007a) Matrix metalloproteinase-2 expression and apoptogenic activity in retinal pericytes: implications in diabetic retinopathy. Ann NY Acad Sci 1103:196–201

Yang Y, Estrada EY, Thompson JF, Liu W, Rosenberg GA (2007b) Matrix metalloproteinase-mediated disruption of tight junction proteins in cerebral vessels is reversed by synthetic matrix metalloproteinase inhibitor in focal ischemia in rat. J Cereb Blood Flow Metab 27:697–709

Zatterstrom UK, Felbor U, Fukai N, Olsen BR (2000) Collagen XVIII/endostatin structure and functional role in angiogenesis. Cell Struct Funct 25:97–101

Zeni P, Doepker E, Schulze-Topphoff U, Huewel S, Tenenbaum T, Galla HJ (2007) MMPs contribute to TNF-alpha-induced alteration of the blood-cerebrospinal fluid barrier in vitro. Am J Physiol Cell Physiol 293:C855–C864

Zhou Z, Apte SS, Soininen R, Cao R, Baaklini GY, Rauser RW, Wang J, Cao Y, Tryggvason K (2000) Impaired endochondral ossification and angiogenesis in mice deficient in membrane-type matrix metalloproteinase I. Proc Natl Acad Sci U S A 97:4052–4057

Zucker S, Pei D, Cao J, Lopez-Otin C (2003) Membrane type-matrix metalloproteinases (MT-MMP). Curr Top Dev Biol 54:1–74

Chapter 15
Proteases in Cancer: Significance for Invasion and Metastasis

Bonnie F. Sloane, Karin List, Barbara Fingleton, and Lynn Matrisian

15.1 Introduction

There is an extensive body of literature documenting the association of proteases with cancer. Indeed, a search of PubMed for the phrase "proteases in cancer" brings up a list of ~73,000 papers, including >7,200 reviews. Nonetheless, the protease community still has not identified and validated all of the proteases and proteolytic pathways that play causal roles in neoplastic progression, nor determined which proteases would be appropriate therapeutic targets in pre-malignant lesions as compared to end-stage cancers or in any one type of cancer. Furthermore, more than one catalytic type of protease has been implicated in the progression of human tumors, as have interactions among proteases of more than one catalytic type. How vast a repertoire of proteases has been implicated in cancer is evident from a

B.F. Sloane (✉)
Department of Pharmacology, School of Medicine, Wayne State University, 540 E. Canfield, Detroit, MI 48201, USA
e-mail: bsloane@med.wayne.edu

K. List
Barbara Ann Karmanos Cancer Institute, Wayne State University School of Medicine, 540 E. Canfield, Detroit, MI 48201, USA
e-mail: klist@med.wayne.edu

B. Fingleton
Vanderbilt University Medical Center, 771 Preston Research Building, Nashville, TN 37232-6840, USA
e-mail: barbara.fingleton@vanderbilt.edu

L. Matrisian
Vanderbilt University Medical Center, 23rd and Pierce Avenues, 771 Preston Research Building, Nashville, TN 37232-6840, USA

Pancreatic Research Action Network, 1500 Rosecrans Ave, Manhattan Beach, CA 90266
e-mail: lmatrisian@pancan.org

K. Brix and W. Stöcker (eds.), *Proteases: Structure and Function*,
DOI 10.1007/978-3-7091-0885-7_15, © Springer-Verlag Wien 2013

comprehensive tome on the many proteases that comprise the cancer degradome (Edwards et al. 2008b).

The initial working hypothesis for those studying proteases in cancer was that invasive processes (local and during metastatic spread) require degradation of extracellular matrices by proteases. The roles of proteases in cancer are now known to be much broader, as will be discussed here. Furthermore, the proteases themselves derive not only from tumor cells, but also from other cells that make up the tumor microenvironment, e.g., fibroblasts, macrophages, mast cells, neutrophils and endothelial cells. A critical factor to remember when considering whether a protease plays a causal role in malignant progression is that a purified protease capable of cleaving a protein substrate *in vitro* may not be the protease or the only protease responsible for degradation of that substrate *in vivo*. Transgenic mice deficient in specific proteases have helped elucidate the *in vivo* functions of proteases, but have also confirmed that there is redundancy and compensation. This along with the large number of proteases in the human genome, the interplay of proteases with one another as well as with their endogenous inhibitors and activators and the complexity of their biological roles [for review, see (Doucet and Overall 2008; Ordonez et al. 2009; Puente et al. 2003; Rawlings et al. 2010)] suggest that we should employ multiple technologies to identify proteases and proteolytic pathways that are associated with the processes of tumor initiation and progression (Fig. 15.1) and that may be targets for therapeutic intervention.

15.2 Metalloproteases

Metalloproteases are a heterogeneous group of proteolytic enzymes that use a metal ion to polarize water and hydrolyze proteins. The MA clan, the largest clan of metalloproteases, binds zinc using a HExxH motif. The ADAM and ADAMTS proteases are included in the adamalysin subfamily of metzincins within the MA clan, and the MMPs are in the matrixin subfamily. These three metalloprotease families have been implicated in a number of human pathologies; their various roles in cancer development and progression are summarized below.

15.2.1 Matrix Metalloproteases

15.2.1.1 The MMP Family

The 23 MMPs expressed in humans all contain the minimal domains of a signal sequence that directs protein secretion, a pro-domain to maintain latency, and a Zn-containing catalytic domain. Two of the MMP family members (MMP-7,26) contain only the minimal domains and the rest possess additional domains that facilitate substrate or inhibitor binding such as the hemopexin and fibronectin-like domains, and/or domains that anchor the enzyme to the cell surface such as the

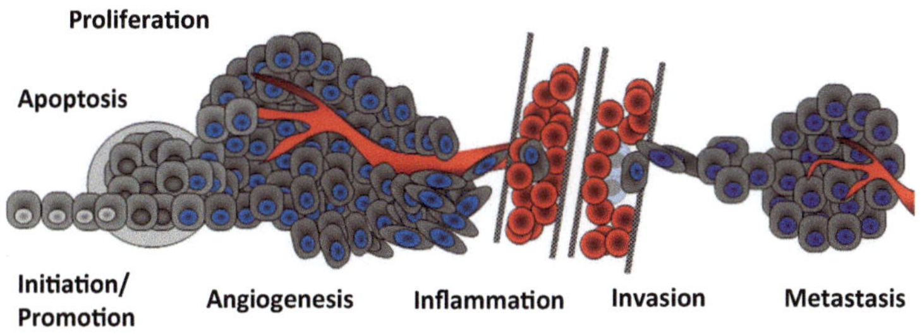

Fig. 15.1 Steps in tumor development, from initiation to metastasis, in which proteases have been shown to participate

transmembrane or GPI anchor domains [see (Brinckerhoff and Matrisian 2002) for review].

"Matrix" metalloproteinases are so named because the prototypic family member, MMP-1/collagenase-1, was initially purified as the enzyme responsible for degradation of fibrillar collagen in the tadpole tail [see (Brinckerhoff and Matrisian 2002) for review]. Several MMP family members possess the rare ability to cleave collagen in its native triple helical structure and are collectively referred to as collagenases (MMP-1,8,13, and the membrane-type MMP-14). The stromelysins (MMP-3,7,10) are MMP family members with greater activity against proteoglycans and glycoproteins such as laminin and fibronectin; gelatinases (MMP-2,9) cleave basement membrane and denatured collagens efficiently; and MMP-12/metalloelastase is particularly potent against elastin. As a result of their combined ability to degrade virtually all components of the extracellular matrix, MMPs have been implicated in diseases characterized by extensive matrix destruction, which includes basement membrane dissolution as benign tumors progress to malignant, invasive cancers.

15.2.1.2 MMPs in Cancer

The enzymatic activities described by Liotta in 1980 as being associated with tumor metastasis (Liotta et al. 1980) were eventually ascribed to MMP-2/gelatinase A and MMP-9/gelatinase B. The ease of identifying these two MMPs, in particular by gelatin zymography, resulted in an abundant literature documenting their expression in cancerous tissues and their association with cancer invasion and metastasis. Our view of the role of MMPs in cancer has, however, greatly expanded over the past decade and is no longer restricted to invasive activity as a result of gelatinase-mediated matrix degradation. Virtually all MMP family members have been detected in tumor tissue, and a causal association for at least half of them has been determined by genetic manipulation of MMP levels. In Fig. 15.2 we illustrate the association of some MMPs with various stages of tumor development. MMP substrates have expanded to include growth factors, cytokines, cell surface receptors and adhesion molecules, and the biological activity of these factors can

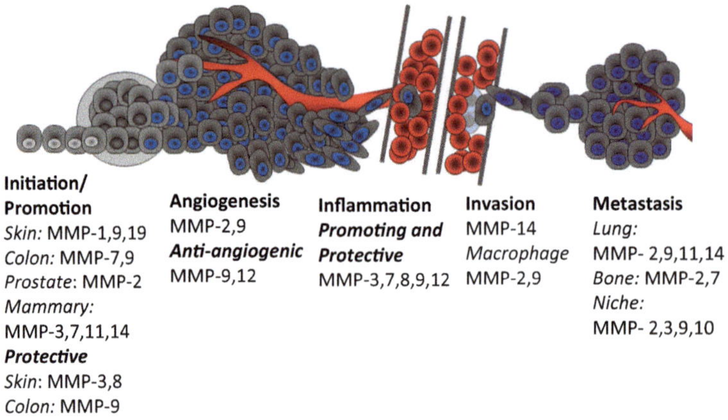

Fig. 15.2 Matrix metalloproteinases in cancer: representative examples of family members demonstrated by *in vivo* experimentation to play roles in the indicated stages of tumor progression

be linked to all stages of cancer progression. MMP cleavage can either activate or inactivate these substrates, influencing fundamental activities of proliferation, apoptosis, angiogenesis and inflammation in addition to invasion and metastasis. The role of MMPs in cancer is discussed in detail in several excellent recent reviews (Fingleton 2006; Kessenbrock et al. 2010), and there is a comprehensive review of biologically relevant MMP substrates (Cauwe et al. 2007).

MMPs in tumor initiation and/or promotion: The earliest steps of tumorigenesis are classically defined as tumor initiation, a mutational event, which is followed by tumor promotion, a process that selects and expands the population of mutated cells. A number of experiments using genetic manipulation of specific MMPs in mouse models of cancer have linked MMPs to these early steps of carcinogenesis, although in general the two processes of initiation and promotion have not been dissected.

The development of benign papillomas and malignant squamous cell carcinomas in the skin of mice treated with a single dose of the carcinogen DMBA (7,12-dimethylbenzanthracene, a tumor initiator) followed by multiple treatments with the phorbol ester tumor promoter TPA (12-0-tetradecanoylphorbal-13-acetate) is a classic model of multistage carcinogenesis. The overexpression of MMP-1 (collagenase-1) in the skin of transgenic mice produces hyperplastic lesions and increases the number of DMBA/TPA-induced tumors (D'Armiento et al. 1995). Using the viral oncogenes of the human papilloma virus HPV16 as initiator and promoter, the ablation of the gelatinase MMP-9 decreases tumor incidence in the skin, but interestingly increases the progression of the resulting tumors to metastasis (Coussens et al. 2000). Mice null for MMP-19, a soluble MMP with potent activity again basement membrane and cartilage, show decreased sensitivity to methylcholanthrene-induced skin fibrosarcomas (Pendas et al. 2004).

In contrast to the documented pro-tumorigenic effects of MMPs-1, 9, and 19, several MMPs have been shown to have protective effects in skin carcinogenesis models. Genetic ablation of MMP-8 (collagenase-2) increases the number of tumors in male mice undergoing a DMBA/TPA skin carcinogenesis regimen (Balbin et al. 2003), a result that contrasts with the promoting effect of MMP-1 and indicates that, although MMP-1 and MMP-8 are both collagenases, they have opposing effects on early stages of skin tumorigenesis. Genetic ablation of MMP-3 (stromelysin-1) also revealed a protective effect of this MMP on skin carcinogenesis with an increase in tumor growth rate in both the DMBA/TPA model and enhanced growth and progression in mice treated repeatedly with the complete carcinogen MNNG (1-methyl-3-nitrosoguanidine) (McCawley et al. 2004). The protective effect of MMP-3 in skin carcinogenesis was confirmed in transgenic mice overexpressing MMP-3 in skin keratinocytes; these mice demonstrate increased keratinocyte differentiation and decreased carcinogen-induced tumor multiplicity (McCawley et al. 2008). Collectively, these observations reinforce the notion that MMPs are much more sophisticated than simple "bulldozers" that destroy matrix and can act as signaling molecules with specific downstream effects that either enhance or retard individual steps of tumor progression.

Early stages of colorectal cancer are commonly initiated by the loss of function of the APC gene, an event that is recapitulated in the Multiple Intestinal Neoplasia (Min) mouse model of intestinal tumorigenesis. Using genetically deficient mice, a tumor promoting role for MMP-7 (matrilysin) and MMP-9, but not MMP-2, 12, or 19, was identified in the development of Min adenomas (Sinnamon et al. 2008; Wilson et al. 1997). Treatment with the carcinogen azoxymethane followed by the inflammatory agent dextran sodium sulfate results in colon cancer that is associated with colitis. In contrast to its role in Min-mediated colon cancer, MMP-9 plays a suppressive role in colitis-associated colon cancer; MMP9-null mice demonstrate a significant increase in tumor number and size (Garg et al. 2010). This result is surprising in that there are well known pro-inflammatory effects of MMP-9 activity (see below). The effect was traced to MMP-9 cleavage of Notch-1 in the epithelial cells of the intestine and the modulation of cellular apoptosis.

The contributions of MMPs to prostate tumor progression were examined in transgenic mice expressing SV40 large T antigen targeted to prostatic neuroendocrine cells (Littlepage et al. 2010). The results from experiments using mice deficient in the gelatinase MMP-2 suggest an effect of this MMP on prostate cancer cell growth, angiogenesis, and metastasis to the lung. In contrast, mice deficient for MMP-7 and MMP-9 do not exhibit effects on primary tumor growth, although a reduction in angiogenesis is observed in both cases, and MMP-9-null mice also demonstrate reduced invasive foci and perivascular invasion.

The role of MMPs in mammary carcinogenesis has been particularly well studied. The overexpression of both MMP-3 and the membrane-type collagenase MMP-14 in the mammary gland is sufficient to induce mammary adenocarcinoma, suggesting both initiating and promoting roles for these MMPs (Ha et al. 2001; Sternlicht et al. 1999). MMP-7 may play a similar role, although in this study the hyperplastic lesions that develop do not progress to frank adenocarcinoma (Rudolph-Owen et al. 1998). The activity of MMP-3 that results in tumor initiation was traced to the generation of cellular reactive oxygen species that induce DNA damage and genomic instability (Radisky et al. 2005). In MMTV-Wnt1-driven mammary carcinogenesis, inhibition of MMP activity by the overexpression of a natural MMP inhibitor, TIMP-2, influences both initiation and promotion as measured by the number of tumors and their growth rate; this effect can be attributed to MMPs-2, 3, 9, 13, and/or 14 (Blavier et al. 2006). In MMTV-*neu*-induced carcinogenesis, the overexpression of MMP-7 increases the number and growth rate of mammary tumors (Rudolph-Owen et al. 1998). Despite the effects of MMP overexpression on mammary tumor progression, genetic ablation of MMP-3 or MMP-7 in MMTV-polyoma middle T-induced mammary tumors has no effect on primary tumor development, suggesting that the effect of these MMPs on initiation and promotion are relatively modest and their expression is not required for the activity of a strong oncogene (Martin et al. 2008). Similarly, genetic ablation of MMP-9 and 13 has no effect on the development of MMTV-polyoma middle T-induced primary mammary tumors (Martin et al. 2008; Nielsen et al. 2008). The effect of MMPs on mammary tumors that develop following initiation with the carcinogen DMBA has also been examined. Ablation of MMP-11 (stromelysin-3) from stromal fibroblasts decreases DMBA-induced mammary tumor formation (Masson et al. 1998). The overexpression of MMP-3 in the mammary gland has been reported to decrease tumor incidence following DMBA treatment, a result that challenges the tumor promoting activity of MMP-3, yet may be explained by the elimination of initiated cells through elevated cellular turnover (Witty et al. 1995).

There is abundant evidence that MMPs can alter cellular proliferation and apoptosis, both key components of the process of tumor promotion. The mechanism underlying this activity is often, but not exclusively, associated with the "sheddase" function of MMPs and the release of growth- and apoptosis-factors from the cell surface. The role of MMPs in cell proliferation and apoptosis and the relevant substrates has been extensively reviewed (Fingleton 2006; Kessenbrock et al. 2010). The anti-promoting effect of MMPs can also be related to cell autonomous-effects, but has been particularly associated with modulation of inflammation as a result of either proteolytic processing or cleavage-mediated inactivation of inflammatory cytokines. These functions of MMPs are discussed in more detail below.

15.2.1.3 MMPs and Angiogenesis

The ability of tumor cells to stimulate the production of new blood vessels from existing local vessels or bone marrow progenitor cells is an adaptive mechanism that facilitates tumor growth and provides an avenue for tumor cell dissemination. Angiogenesis is influenced by MMPs through a wide variety of mechanisms. Perhaps the most obvious is facilitating endothelial cell invasion and migration, an activity that has been ascribed to MMP-2 and requires its association with the integrin alphaVbeta3 since inhibitors of this interaction disrupt angiogenesis in the chick chorioallantoic membrane (Brooks et al. 1998). MMP-2-null mice show reduced angiogenesis and reduced lung and melanoma tumor growth (Itoh et al. 1998). An important mechanism is modulating the availability and signaling of angiogenic factors. MMP-9, but not MMP-2, null mice are impaired in their ability to vascularize tumors in the pancreatic islets induced by the SV40 T-antigen (Bergers et al. 2000). In this model, MMP-9 increases the bioavailability of Vascular Endothelial Cell Growth Factor (VEGF), presumably by releasing it from matrix stores (Bergers et al. 2000), although MMP-9 can also directly cleave VEGF to remove domains responsible for matrix binding (Lee et al. 2005). The basic fibroblast growth factor 2 pathway is activated by neutrophil MMP-9 and induces angiogenesis in a collagen implant assay (Ardi et al. 2009). MMP-9 is also essential for vasculogenesis—the generation of tumor vessels from bone marrow precursors. MMP-9-null mice lack bone marrow-derived endothelial cells in neuroblastoma-associated blood vessels (Jodele et al. 2005), and MMP-9-positive myeloid cells can restore vasculogenesis in tumors transplanted into irradiated tissues that are incapable of angiogenesis (Ahn and Brown 2008). Bone marrow-derived Myeloid-Derived Suppressor Cells (MDSCs) lose their tumor-promoting function and fail to differentiate into tumor-associated endothelial cells in the absence of MMP-9 (Yang et al. 2004). In addition, the recruitment of pericytes and subsequent maturation of tumor blood vessels in a neuroblastoma model is dependent on MMP-9 (Chantrain et al. 2004). Thus, MMP effects on angiogenesis include increasing the concentration of pro-angiogenic factors, delivering signals to critical bone marrow cells, facilitating endothelial cell invasion, and regulating pericyte function to influence vascular stability and permeability.

There are also anti-angiogenic properties of MMPs, in particular related to their ability to generate small inhibitory peptides from higher molecular weight precursors. Angiostatin, a potent suppressor of angiogenesis, is generated from plasminogen most efficiently by MMP-12 (Cornelius et al. 1998), an activity that explains the protective effect of MMP-12 in mouse models of lung cancer (Acuff et al. 2006; Houghton et al. 2006). MMP-9-mediated generation of angiostatin also influences lung tumor growth (Chen et al. 2005). MMP-9-deficient mice can also demonstrate a surprising acceleration in tumor growth, a result that can be attributed to the cleavage of the alpha3 chain of IV collagen to generate tumstatin (Hamano et al. 2003). Endostatin is generated by cleavage of type XVIII collagen by MMPs-3, 7, 9, 13, and 20 (Cauwe et al. 2007; Heljasvaara et al. 2005). Thus, the angiogenic activity of MMPs is balanced by the release of anti-angiogenic factors, an effect that is likely to be part of a normal homeostatic mechanism.

15.2.1.4 MMPs and Inflammation

The multifaceted role of the innate immune response and inflammation in cancer initiation, promotion, and progression has been increasingly apparent (Grivennikov et al. 2010). There are now many examples of MMP effects in *in vivo* models that are mediated by modulation of the inflammatory process [reviewed in (Fingleton 2006; Kessenbrock et al. 2010; Parks et al. 2004)]. The complexity of MMP actions on the inflammatory response is layered upon the complexity of the effect of inflammation on cancer, so that both pro- and anti-inflammatory activities of MMPs have been documented. For example, in a transgenic mouse study MMP-12 expression directed to lung epithelial cells results in increased inflammatory cell infiltration, emphysema, and the development of bronchioalveolar adenocarcinoma (Qu et al. 2009). MMP-12 is also upregulated in human chronic obstructive pulmonary disease and lung cancer, and is speculated to play a critical role in the transition from emphysema to lung cancer. In contrast, in Helicobacter pylori-induced gastric cancer, the expression of MMP-7 in the gastric epithelial cells has a protective effect on H. pylori-induced inflammation. MMP-7-null mice demonstrate an increase in Th1- and Th17-mediated inflammation and accelerated epithelial cell turnover mediated by increases in both proliferation and apoptosis (Ogden et al. 2010). Similarly, and as discussed previously, MMP-3 and MMP-8 have protective effects in skin carcinogenesis, and in both cases this effect correlates with a reduction in inflammation (Balbin et al. 2003; McCawley et al. 2004). MMP-8 has been implicated in the resolution of acute inflammation, and the loss of MMP-8 results in chronic inflammation that may contribute to its tumor suppressive effects (Gutierrez-Fernandez et al. 2007). Interestingly, loss-of-function mutations in MMP-8 have been reported in melanoma, suggesting that the inability to resolve inflammation may also contribute to melanoma development (Palavalli et al. 2009).

The mechanisms underlying the effects of MMPs on inflammation and innate immunity are diverse, but one that stands out is the influence of MMPs on chemokine gradients that attract inflammatory cells. For example, the CCL monocyte-attracting chemokines, CCL2, -7, -8, and -13, are cleaved by MMPs-1 and -3, which converts them into receptor antagonists and reduces monocyte chemotaxis and the inflammatory phenotype (McQuibban et al. 2002). MMPs-8, -9, and -12 similarly cleave CXCL11, a T lymphocyte chemotactic factor. This first inactivates CXCL11, then reverses the inactivation, but removes the domain responsible for matrix binding and alters its ability to generate a haptotactic gradient (Cox et al. 2008). MMP-7 modulates the neutrophil chemoattractant CXCL1 indirectly through the cleavage of syndecan-1 from cell surfaces, releasing chemotactic CXCL1/syndecan-1 complexes (Li et al. 2002). Many other inflammatory cytokines and chemokines, including Tumor Necrosis Factor-alpha and CXCL12/SDF-1, are processed by MMPs, resulting in a wide range of effects on inflammatory cell infiltration and biological activities [see (Cauwe et al. 2007; Parks et al. 2004) for review].

15.2.1.5 MMPs in Cancer Invasion

Invasion, the migration of tumor cells from a central cluster into surrounding tissue and into lymphatic or blood vessels can occur either at a single cell or multicellular level. The latter, referred to as collective invasion, is dependent on the destruction of fibrillar collagen fibers (Friedl and Wolf 2008). Thus, MMPs with the specialized ability to cleave fibrillar collagens, making distinct ¾ and ¼ fragments in collagen chains and unraveling the fibrillar structure, are likely to contribute to this process. These include MMP-1, 8, 13, and 14, but of these MMP-14 is implicated as the rate-limiting step in invasion through collagen-containing ECM (Sabeh et al. 2004). In contrast to the other three collagenases, MMP-14 is membrane anchored and (along with the transmembrane serine proteinase seprase) is a critical component of invadopodia, actin-rich protrusions that focalize matrix-degrading activity to cell: substratum contact points [(Weaver 2006) for review]. An additional well-known function of MMP-14 is to activate the secreted gelatinase MMP-2. MMP-2 and MMP-9 are also found associated with invadopodia, perhaps functioning to facilitate further degradation of the denatured collagen/gelatin fragments.

The role of macrophages in assisting tumor cells in invading into surrounding tissue has been elegantly demonstrated by Pollard, Condeelis, and colleagues [(Qian and Pollard 2010) for review]. Macrophages produce a number of proteases that have been associated with their ability to invade tissues, and thus are candidates for facilitating the invasive activity of macrophage-mediated tumor cell invasion. Of the MMPs, MMP-2 and MMP-9 produced by immature myeloid cells are required for the collective invasion of sheets of tumor cells in a mouse model of colon cancer (Kitamura et al. 2007). These same enzymes have been implicated in macrophage infiltration in a model of autoimmune disease (Agrawal et al. 2006) and may thus represent a general mechanism for inflammatory cell infiltration.

15.2.1.6 MMPs in Metastasis

Based on the original concept of MMP activity in cancer, there is extensive documentation in experimental metastasis assays of a role for MMP activity following genetic manipulation or pharmacological inhibition. Since many of these studies were performed in immunodeficient mice, there is reason to focus on authochthonous or syngeneic mouse models of cancer metastasis so the pleiotropic effects of MMPs on the immune system are considered in this multistep process. The MMTV-polyoma middle T model of mammary tumorigenesis has revealed a role for MMP-9 and -14 in spontaneous metastasis to the lung, but no effect of MMPs-3,7, or 13 (Martin et al. 2008; Nielsen et al. 2008; Szabova et al. 2008). MMP-14 deficiency does not eliminate the development of hyperplastic lesions in bigenic mice and in fact seems to enhance their growth following transplantation into cleared mammary fat pads of recipient animals. MMP-14-null glands, however, display a markedly reduced ability to metastasize to the lungs (Szabova et al. 2008). MMP-9 deficiency has no effect on the development of primary tumors, but markedly inhibits lung tumor burden. This effect is only observed in C57Bl/6 mice and not in FVB/N mice, a dependency on genetic background that was confirmed with a small molecule MMP-9 inhibitor (Martin et al. 2008). In the SV40-large T antigen model of prostate cancer, MMP-2, but not MMP-7 or 9, deficiency reduces metastasis to the lung (Littlepage et al. 2010).

The effect of ablating the stromal production of several MMP family members on metastasis has been examined in syngeneic models of experimental metastasis, which examine primarily the ability of the cells to establish lesions in distant organs. The injection of syngeneic melanoma or lung cells into MMP-2- (Itoh et al. 1998) and MMP-9- (Itoh et al. 1999) null mice results in reduced metastatic foci in the lung following intravenous injection. Syngeneic colon cancer cells injected into MMP11-deficient mice reduce the number of lung foci, but paradoxically favor their growth (Brasse et al. 2010). The injection of syngeneic breast cancer cells into the tibia of MMP-7 mice greatly reduces their ability to establish lesions in the bone microenvironment (Thiolloy et al. 2009). MMP-2 ablation has a similar effect on reducing the establishment and growth of bone lesions, although stromal MMP-9 ablation has no effect on this process (Thiolloy et al. 2009; Thiolloy et al. 2012).

Tumor cells can stimulate bone marrow cells to populate distant organs and prepare a pre-metastatic niche that fosters the establishment and growth of tumor cells in that organ. The gelatinase MMP-9 plays an important role in the formation of this niche. MMP-9 ablation destroys the pre-metastatic niche (Kaplan et al. 2005), an effect likely be attributed to MMP-9 releasing soluble Kit-ligand (Heissig et al. 2002) and/or VEGF (Bergers et al. 2000). There is evidence that it is the disruption of vessels in the metastatic organ that leads to the enhanced intravasation of both bone marrow-derived cells and circulating tumor cells. The stromelysins MMP-3 and MMP-10 play an important role in disrupting blood vessel integrity, and targeted downregulation of these MMPs reduces vessel permeability, reduces infiltration of myeloid cells, and reduces the ability of human breast cancer cells implanted in the mammary fat pad to spontaneously metastasize to the lung (Huang et al. 2009). The production of MMP-2 by myeloid cells that are recruited to the pre-metastatic niche results in collagen cleavage that, in combination with the effect of the collagen crosslinker lysyl oxidase, enhances the invasion and recruitment of additional bone marrow derived cells and metastasizing tumor cells (Erler et al. 2009). Thus, MMPs do contribute to tumor metastasis, but the mechanisms behind this effect are much more diverse than originally envisioned.

15.2.2 ADAM and ADAM-TS Proteases

The ADAM and ADAM-TS families of metalloproteinases are so-named for their multiple domains: A Disintegrin And Metalloprotease (ADAM) with Thrombospondin motif (ADAM-TS). Despite similar structure, proteolytic activity is absent in some family members. These family members are thought to be especially important in cell adhesion events such as sperm-egg fusion (Klein and Bischoff 2011). Altogether in man, there are 22 ADAMs, 12 of which have protease activity, and 19 ADAM-TS proteins. Proteolytically active members of both families have critical physiological roles in diverse processes such as shedding/solubilization of membrane proteins ('sheddase' activity), matrix protein processing and degradation and control of coagulation factor activation. Unlike the closely related MMPs, genetic deficiency of several individual ADAMs or

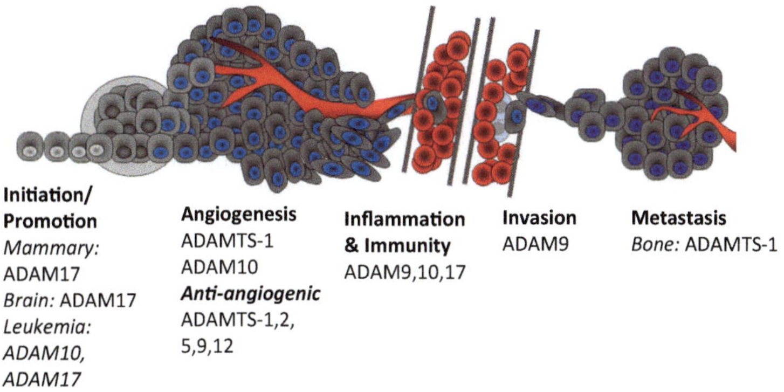

Fig. 15.3 ADAMs and ADAM-TSs in cancer: representative examples of family members demonstrated by *in vitro* and *in vivo* experimentation to play roles in the indicated stages of tumor development and progression

ADAM-TSs in animals or humans has resulted in severe phenotypes including embryonic lethality, underlying their essential roles in normal homeostasis. The reader is referred to several recent articles for an overview of these families of proteases (Apte 2009; Edwards et al. 2008a; Klein and Bischoff 2011). Here, we discuss specific activities that relate to cancer development, progression, treatment or detection. Representative examples are illustrated in Fig. 15.3. More extensive discussion relating to some of these activities can be found in several comprehensive reviews (Edwards et al. 2008a; Klein and Bischoff 2011; Murphy 2008; Turner et al. 2009; Wagstaff et al. 2011).

ADAMs and ADAM-TSs in initiation, promotion and growth of tumors: ADAM-17, also known as TACE or tumor necrosis factor alpha converting enzyme, is a widely expressed sheddase responsible for the regulated cleavage of several important signaling molecules such as TNF-a, L-selectin and various epidermal growth factor (EGF) ligands (Blobel 2005; Edwards et al. 2008a). Studies from the laboratory of Mina Bissell have indicated that increased ADAM17 activity is sufficient to provide an autocrine oncogenic stimulus in a 3D breast cancer model, without mutation of any proto-oncogene (Kenny and Bissell 2007). In these cells, the primary substrate for ADAM17 responsible for tumorigenesis is transforming growth factor alpha, an EGF receptor ligand. A cortical astrocyte cell line, originally developed from normal brain, also can be converted to a malignant tumor line by overexpression of ADAM17 (Katakowski et al. 2009). Again, generation of active EGF receptor ligand is the identified mechanism.

Notch signaling is a critical pathway for development of multiple cell lineages. Interaction of the Notch receptor with ligand (members of the Jagged or Delta families) leads to a confirmation change and subsequent exposure of proteolytic cleavage sites. ADAMs-10 and -17 are both Notch1 cleaving enzymes; however, ligand-induced N1 signaling requires ADAM-10, whereas ligand independent N1 signaling requires ADAM-17 (Bozkulak and Weinmaster 2009). In several tumor types, particularly leukemias, mutations in Notch are associated with disease. Many such mutations result in constitutive exposure of the S2 ADAM cleavage site thereby removing the requirement of ligand binding for Notch activity (van Tetering et al. 2009) and both ADAM-10 and -17 participate in signaling in leukemias (Bozkulak and Weinmaster 2009). Overexpression of ADAM10 in B cells has a similar result of excessive Notch signaling and a block in B cell differentiation (Gibb et al. 2011).

15.2.2.1 ADAMs and ADAM-TSs in Angiogenesis

Several ADAM-TS proteases have been identified as regulators of tumor angiogenesis. Much of this regulation is negative, likely due to the presence of the thrombospondin motifs, since thrombospondins are known angiogenesis inhibitors (Ribatti 2009). Although originally identified as an angiogenesis inhibitor (Iruela-Arispe et al. 2003), the activity of ADAM-TS1 is more complex as it appears to enhance tumor vasculature development in a fibrosarcoma model. Rather than through classical angiogenesis, this occurs by promotion of tumor cell plasticity toward an endothelial phenotype resulting in a vascular mimicry phenomenon (Casal et al. 2010). ADAM-TS1 has also been shown to lead to accelerated tumor growth *in vivo* through induction of a profound stromal response (Rocks et al. 2008). Recruitment of activated fibroblasts can then enhance classical angiogenesis. ADAM-TS12-null mice, which appear normal and viable, allow the accelerated growth of implanted tumors as compared to wildtype counterparts (El Hour et al. 2010). This is related to enhanced vascularity of the tumors. In complementary *in vitro* assays, addition of ADAM-TS12 to aortic rings suppresses angiogenic outgrowth. As might be expected, this anti-angiogenic function of ADAMTS-12 is independent of its proteolytic activity. Similarly, ADAM-TS2 shows anti-angiogenic activity that is independent of proteolysis, and possibly mediated through binding to the cell surface receptor nucleolin (Dubail et al. 2010). In ADAM-TS5, the anti-angiogenic property has been mapped specifically to only the first of the thrombospondin-like domains (Sharghi-Namini et al. 2008). ADAM-TS9 has been implicated as an anti-angiogenic molecule in head and neck cancers (Lo et al. 2010). Here, the antiangiogenic activity appears related to reduced generation of the pro-angiogenic factors VEGF and MMP9.

ADAM10, in addition to its major role in Notch signaling, is also a sheddase for several other cell surface proteins. One of these, osteoactivin (also known as GPNMB), is highly expressed in triple negative breast cancers, a subtype of breast cancer that is associated with shorter median times to relapse and death. In an effort to determine whether osteoactivin is related to the aggressiveness of this tumor type, Rose et al. (2010) showed that vessel density within tumors is significantly associated with higher osteoactivin levels, that ADAM10 is also expressed and can cleave osteoactivin and finally, that the shed protein can promote endothelial migration.

15.2.2.2 ADAMs and ADAM-TSs in Invasion and Metastasis

ADAM9 expression has been linked to increased invasive and metastatic activity in many cancers. This is a protein with splice variants that appear to have opposite effects in some tumor types. ADAM9-S is a secreted version that can promote migration of breast cancer cells in a manner dependent on its proteolytic activity (Fry and Toker 2010). On the other hand ADAM9-L is a transmembrane protein that suppresses migration of breast cancer cells independent of its proteolytic activity (Fry and Toker 2010). ADAM9 appears to be regulated by oxidative stress, with increased activity resulting in increased invasiveness under conditions of higher stress (Mongaret et al. 2011). One way in which ADAM9 can promote invasion is through increasing the levels of other proteases, e.g., MMPs. Such a mechanism has been reported in melanoma, where ADAM9 mediates binding between tumor cells and adjacent stromal fibroblasts with resulting up-regulation of MMPs-1 and 2. Blocking ADAM9 is sufficient to inhibit invasion in this system (Zigrino et al. 2011).

ADAM-TS1 is one of the molecules that appears to regulate osteolytic bone-metastasis in breast cancer (Lu et al. 2009). The mechanism is related to release of epidermal growth factor receptor (EGFR) ligands such as amphiregulin and HB-EGF from tumor cells that then bind to receptors on osteoblasts, activating a signaling pathway that downregulates osteoprotegerin production and subsequent enhanced osteoclast differentiation.

15.2.2.3 ADAMs and ADAM-TSs in Inflammation and Immunity

The sheddase activity of ADAM proteases is an important regulator of immune system function. As discussed previously, overexpression of ADAM10 in hematopoietic cells can lead to elevated Notch signaling and decreased B-cell development. A further consequence is a switch of lineage from lymphoid to myeloid with a particular increase in the myeloid derived suppressor cell (MDSC) group (Gibb et al. 2011). As MDSCs are considered important drivers of the tumor-associated immunosuppressive phenotype, this consequence of increased ADAM10 could be significantly beneficial for tumor development. Further amplifying this effect is expression of another ADAM, ADAM17, by MDSCs, which enhances tumor immune evasion through shedding of L-selectin from T cells (Hanson et al. 2009). Without cell surface L-selectin, neither helper (CD4+) nor cytotoxic (CD8+) T cells home to lymph nodes for activation or to tumors. Natural killer (NK) cells represent an arm of the immune system that has profound anti-tumor effects, but which is frequently suppressed in cancer. They respond to one of a limited number of surface antigens frequently found on tumor cells, one example of which is MHC class I-related chain A (MICA). In two different studies, Kohga et al. have found that both ADAM9 and ADAM10 can shed MICA from hepato-cellular carcinoma cells, thus preventing NK-mediated clearance (Kohga et al. 2009, 2010). These authors found that two different anti-cancer drugs, epirubicin and sorafenib, inhibit ADAM10 and ADAM9-mediated MICA shedding, respectively. They thus suggest that a secondary effect of these agents is the boosting of NK cell anti-tumor activity.

In contrast to the aforementioned roles in promoting immune evasion, ADAM17 may also be beneficial for stimulating anti-tumor immunity. In breast, ovarian and prostate cancers, ADAM17 on tumor cells can be processed and presented as an HLA A2 restricted epitope that is not present on normal cells (Sinnathamby et al. 2011). This offers the possibility of ADAM17 as a suitable antigen for a cancer vaccine.

15.2.2.4 ADAMs and ADAM-TSs as Markers or Therapeutic Targets in Cancer

As with many of the other proteases discussed in this chapter, there is a large body of literature that describes correlations between levels of particular ADAM or ADAM-TS proteases and cancer progression. Overviews of these studies can be found in recent excellent review articles (Edwards et al. 2008a; Murphy 2008; Turner et al. 2009; Wagstaff et al. 2011). One particularly exciting study from the laboratory of Marsha Moses uses urinary levels of ADAM12 as a predictive marker of breast cancer risk (Pories et al. 2008). As a truly non-invasive test that can be used to detect early stage disease, this has the potential to really impact treatment decisions for patients.

In addition to potential use as biomarkers for assessing cancer presence, progression or response to therapy, ADAMs and ADAM-TS proteases also represent potential targets for therapeutic development. As is evident from the previous paragraphs, these families of proteases have multiple functions, both pro- and anti-tumorigenic. Therefore, as is the case for inhibitors of MMPs, ADAM or ADAM-TS protease inhibitors will require careful testing to ensure no or minimal blocking of beneficial activities. ADAM17 is considered a particularly appealing target in cancer, mainly because of its roles in EGFR ligand processing (Kenny 2007). ADAM10 is also considered a therapeutic target in cancer, largely due to its Notch cleavage ability as well as EGFR processing (Crawford et al. 2009). Several small molecule inhibitors have been developed for these targets (Kenny and Bissell 2007; Ludwig et al. 2005; Murumkar et al. 2010; Zhou et al. 2006). While such drugs have shown some efficacy in pre-clinical models, particularly when combined with EGFR inhibitors (Kenny and Bissell 2007; Witters et al. 2008; Zhou et al. 2006), they suffer from specificity problems due to the similarities in the protease domain across this family of enzymes. Thus, as is happening with MMP inhibitors, a new approach that is being explored is the targeting of other parts of the ADAM or ADAM-TS molecules using antibodies (Tape et al. 2011).

In some cases, protease activity is of benefit for therapeutic efficacy through processing targets of other drugs. A recently reported example is the monoclonal antibody DN30, which targets c-Met, the hepatocyte growth factor (HGF) receptor.

In order to be efficacious, the antibody must bind to the receptor and induce its shedding from the tumor cell, a process that is dependent on ADAM10 activity (Schelter et al. 2010).

15.3 Serine Proteases

There are 175 predicted serine proteases in humans with the vast majority being secreted (Puente et al. 2005). Secreted proteases play essential roles in a multitude of physiological processes, which include digestion, blood clotting, and immunity. Many studies have defined pivotal roles of secreted serine proteases in pathological conditions including inflammation, cancer progression and metastasis. A subgroup of serine proteases is directly anchored to plasma membranes, a finding that substantially expanded the known repertoire of enzymes that execute proteolytic cleavage reactions within the pericellular and extracellular environments. Although the role of the majority of both secreted and membrane serine proteases in tumorigenesis is still unexplored, important information has been garnered on several proteolytic systems and individual members. The association of selected serine proteases with various stages of tumor development is depicted in Fig. 15.4. In the sections below we will focus on particular serine proteases, which have been reported to be critical players in cancer progression and metastasis.

15.3.1 Secreted Serine Proteases

15.3.1.1 The Plasminogen Activation System

The plasminogen activation (PA) system has been extensively studied in cancer and it has been reported that several of the components play an important role, mainly in metastasis. The proteolytic effector enzyme is plasmin, which has broad substrate specificity and is capable of degrading various extracellular matrix proteins including fibrin, fibronectin and laminin. Plasmin can also indirectly promote matrix degradation through the activation of certain pro-MMPs. In addition, plasmin may be involved in functions unrelated to matrix degradation, e.g., in the activation of growth factor precursors such as transforming growth factor beta (TGF-β). Plasmin is produced in the liver and is present at high concentrations in the plasma (~2 μM). Two mammalian plasminogen activators are known, urokinase type-plasminogen activator (uPA) and tissue-type plasminogen activator (tPA). Studies on tPA have mainly focused on its role in thrombolysis whereas the function of uPA has been thoroughly studied in both cancer progression and fibrinolysis. uPA is inhibited by two specific plasminogen activator inhibitors, PAI-1 and PAI-2 belonging to the serpin family of serine protease inhibitors. uPA binds to a cell

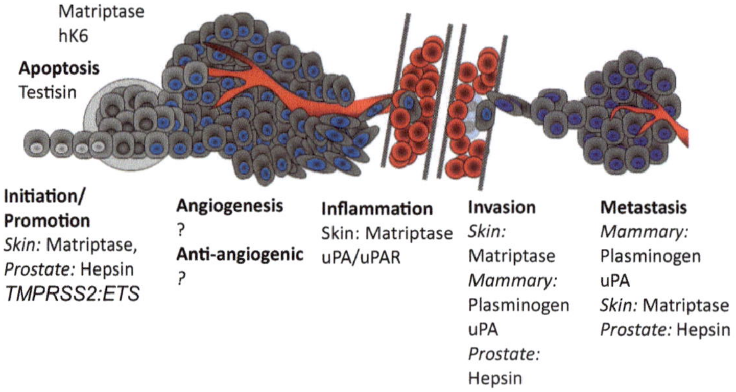

Fig. 15.4 Serine proteases in cancer: representative examples of family members demonstrated by *in vitro* and *in vivo* experimentation to play roles in the indicated stages of tumor progression

surface-bound receptor, uPAR, and the bound form induces plasminogen activation much more rapidly than does fluid-phase uPA (Ellis et al. 1989).

uPA, uPAR and PAI-1 are consistently expressed in the invasive areas of a broad range of carcinomas. As a result, these proteins have been studied in detail for their potential as prognostic markers. uPA was the first protease shown to be a prognostic marker in human malignancy. Duffy et al. showed that patients with breast tumors containing high levels of uPA enzyme activity have a significantly shorter disease-free interval than do patients with low levels (Duffy et al. 1988). Later, high uPA antigen levels were found to correlate with a shortened overall survival in this disease (Duffy 1990). Since then, high levels of uPA, uPAR and PAI-1 have been demonstrated to be predictive of poor patient outcome in many malignancies including cancers of the lung, esophagus, breast, stomach, and colorectum [reviews in (Dass et al. 2008; Duffy 2004; Gandolfo et al. 1996; Han et al. 2005; Schmitt et al. 1997)]. In contrast, high levels of PAI-2 have mainly been associated with decreased tumor growth and metastasis [reviewed in (Croucher et al. 2008)]. Breast cancer is the most frequently studied cancer type in which the prognostic value of PAI-2 expression has been assessed and the majority of studies demonstrate a significant association between PAI-2 expression and prognosis. Specifically, relatively high tumor-associated PAI-2 expression is linked with prolonged survival, decreased metastasis, or decreased tumor size (Croucher et al. 2008). The reason for an apparently paradoxical action of two similar serpins is not fully understood, but it has been proposed that key structural differences control interactions with components of the extracellular matrix and endocytosis–signaling co-receptors.

Carcinomas contain a variety of stromal cells, including fibroblasts, endothelial cells and macrophages, in addition to the malignant epithelial compartment. PA system components are often expressed by stromal cells rather than by epithelial cancer cells, in patterns that are characteristic for each type of cancer (Almholt and

Johnsen 2003; Hewitt and Dano 1996; Johnsen et al. 1998). In ductal mammary carcinoma, uPA is primarily expressed by myofibroblasts, yet is also found in some macrophages and endothelial cells, as well as, in rare instances, in a small subpopulation of cancer cells. PAI-1 is also chiefly expressed by myofibroblasts, whereas uPAR is mainly seen in macrophages (Almholt et al. 2003).

Functional studies using genetic mouse loss-of-function models have provided valuable insights into defining the role of the PA system in cancer progression *in vivo*. Studies of the consequence of uPA deficiency on tumor progression in the MMTV-PymT transgenic breast cancer model revealed that tumor incidence, latency, growth rate and final primary tumor burden are not significantly affected by uPA deficiency. In contrast, metastasis to both lungs and brachial lymph nodes is significantly reduced in uPA null mice as compared to wild-type controls (Almholt et al. 2003). Similar results were observed in plasminogen deficient mice (Bugge et al. 1998) indicating that both uPA and plasminogen are primarily involved in invasion and metastatic dissimination and not in primary tumor growth. In contrast, neither uPAR nor PAI-1 deficiency significantly affects primary tumor growth and lung metastatic burden in the MMTV-PymT model; vascular density is also unaffected by PAI-1 status (Almholt et al. 2003, 2007). Other transgenic cancer models that have been employed to study the function of PA system proteases and inhibitors include the RIP1-Tag2 pancreatic cancer model and the HPV16 skin cancer model. In the RIP1-Tag2 model, induction of angiogenesis and the number of pancreatic islet tumors are not affected by uPA deficiency (Bergers et al. 2000). Recently, it was reported that skin carcinogenesis in K14-transgenic mice deficient for PAI-1 is unimpeded (Masset et al. 2011). In another study focusing on skin, transgenic mice overexpressing either uPA or uPAR in basal epidermis cells and hair follicles have no detectable cutaneous alterations (Zhou et al. 2000). In contrast, bi-transgenic mice overexpressing both uPA and uPAR develop extensive alopecia induced by involution of hair follicles, epidermal thickening and sub-epidermal blisters. The phenotype is due to uPA catalytic activity since combined overexpression of uPAR and uPA that binds to uPAR but is catalytically inactive in the same tissue is not detrimental in another bi-transgenic line (Zhou et al. 2000). These data indicate that combined overexpression of uPA and uPAR acts in synergy to promote pathogenic extracellular proteolysis. It is important to note, however, that no malignant lesions were observed in the bi-transgenice mice.

In sum, there is clear evidence for direct involvement of the PA system, particularly uPA and plasminogen, in cancer invasion and metastasis.

15.3.1.2 Tissue Kallikreins

Human tissue kallikrein (KLK) genes represent the largest contiguous group of proteases within the human genome comprising 15 members. These genes, and their encoded proteins, share a high degree of homology and are expressed in a variety of different tissues. In relation to cancer, the most prominent member of the family is human kallikrein (hK) 3—also known as prostate-specific antigen (PSA).

PSA is the best known cancer biomarker in clinical medicine, being widely used for the early detection as well as management of prostate cancer [review by (Diamandis 1998)]. Several other kallikreins, including hK2 and hK11, are emerging as complementary prostate cancer biomarkers. Along with these kallikreins, several others have been implicated in other types of cancer. For example, hK5, 6, 7, 10, 11, and 14 are emerging biomarkers for ovarian cancer [reviewed in (Paliouras et al. 2007)].

Despite their substantial use in clinical applications, the involvement of the hK family, in the pathophysiology of cancer is just beginning to be understood. Accumulating evidence indicates that the *hK* family is dysregulated in cancer. Numerous studies report aberrant amounts of kallikrein transcripts and/or proteins in cancer cells, particularly adenocarcinomas derived from steroid-hormone-regulated tissues, as compared with their normal, benign and/or pre-malignant tissue counterparts. The most striking example is the concurrent upregulation of 12 *KLK* genes in ovarian carcinoma [reviewed in (Borgono and Diamandis 2004). Recent *in vitro* studies implicate kallikreins in cancer-related processes, including cell-growth regulation, angiogenesis, invasion and metastasis. They have been shown to promote or inhibit neoplastic processes, acting individually on growth factors or growth factor receptors and/or in cascades with other kallikreins and members of other protease families *in vitro* (Borgono and Diamandis 2004; Borgono et al. 2004). For example, several reports indicate that hK2 and hK3 might represent important regulators of the insulin-like growth factor (IGF) axis in prostate carcinogenesis. IGF1 and IGF2 are important mitogenic peptides involved in regulating normal and malignant cellular proliferation, differentiation, apoptosis and transformation. hK4 can activate pro-hK3/PSA and pro-uPA (Takayama et al. 2001) and hK2 can activate pro-uPA and inactivate PAI-1 (Mikolajczyk et al. 1999). Tumor-suppressive actions have been assigned to certain kallikreins. hK3/PSA is able to inhibit the growth of the estrogen-receptor-positive MCF-7 breast cancer cell line by stimulating the conversion of the potent estrogen estradiol to the less potent estrone (Lai et al. 1996). In contrast in prostate cancer, hK3/PSA protein does not affect growth of prostate cancer cells *in vitro* or prostate cancer xenografts *in vivo* (Denmeade et al. 2003).

Only a few studies to date have used genetic animal models to analyze the function of kallikreins in cancer. In a mouse model, both hK3/PSA and hK2 were overexpressed in the prostate by transgenesis. Those double transgenic mice produce more active PSA than single transgenic mice indicating that hK2 is activating the pro-form of hK3/PSA *in vivo*. In a longitudinal evaluation over a 2-year period, no morphologic changes (i.e., no PIN or prostate cancer) can be observed due to hK3/PSA or hK3/PSA plus KLK2 double transgene expression when compared to non-transgenic mice (Williams et al. 2010). Transgenic expression of hK6 in skin keratinocytes of mice causes increased keratinocyte proliferation and migration and decreased levels of E-cadherin protein (Klucky et al. 2007). This observation, combined with the finding that hK6 is up-regulated in squamous skin tumors of human patients, may implicate hK6 in the early events of squamous cell carcinogenesis (Klucky et al. 2007).

Thus far, the *in vivo* role of most kallikreins has not been elucidated. Future studies using loss-of-function or gain-of-function animal models may shed light on their potential roles in tumor growth and metastasis.

15.3.2 Membrane-Anchored Serine Proteases

Over the last decade a novel subgroup of membrane serine proteases has emerged as an important component of the human degradome, garnering significant attention for their roles in tissue homeostasis and in cancer progression and metastasis. The membrane anchored serine proteases are tethered to the cell membrane via three different means: (1) a carboxy terminal glycophosphatidylinositol (GPI) anchor, which is added posttranslationally, (2) a carboxy-terminal transmembrane domain (Type I), or (3) an amino terminal proximal transmembrane domain (Type II transmembrane serine proteases—TTSPs).

The catalytic domain of all membrane-anchored serine proteases belongs to the S1 peptidase family, which includes the prototypic chymotrypsin and trypsin (Rawlings and Barrett 1993). The biochemical properties of the catalytic domain have been studied in great detail [see (Hedstrom 2002; Page and Di Cera 2008) for reviews]. Importantly, all membrane-anchored serine proteases show a strong preference for cleavage after arginine or lysine residues. Membrane-anchored serine proteases are synthesized as catalytically inactive zymogens that are irreversibly converted to active proteases by autocatalytic or heterocatalytic cleavage after an arginine or lysine residue located in a conserved activation motif within the catalytic domain. Regulation of proteolytic activity is often attributed to shedding of the protease from the cell surface upon complex formation with membrane associated or secreted serine protease inhibitors or by internalization followed by degradation within lysosomes [reviewed in (Bugge et al. 2009)].

Although the role of the majority of membrane serine proteases in tumorigenesis is still unexplored, important information has, in recent years, been obtained about several of the members.

15.3.2.1 GPI-Anchored Proteases: Testisin

Testisin (also known as PRSS21) was first described as a candidate tumor suppressor in the testis where it is expressed in premeiotic testicular germ cells but is lost in testicular germ cell tumors (Hooper et al. 1999). Subsequently, it was demonstrated that hypermethylation of the 5′ CpG island of the gene encoding testisin promotes its loss in testicular tumorigenesis (Manton et al. 2005). Testisin expression in normal tissues appears to be highly restricted in both mice and humans and has to date only been detected in testis and in human eosinophils (Hooper et al. 1999, 2000). Interestingly, in a study aiming to identify proteases that were overexpressed in ovarian cancer, the testisin transcript was found to be abundant in ovarian carcinoma but not detected in the normal ovary. Furthermore, there is a positive correlation between increased testisin mRNA levels and tumor stage (Shigemasa et al. 2000). In cell culture studies, *si*RNA–mediated knockdown of endogenous testisin mRNA and protein expression in an ovarian tumor cell line (CaOv3) leads to increased apoptosis. Conversely, overexpression of testisin in the ovarian cell line SKOV3 causes an increase in subcutaneous tumor growth in severe combined immunodeficient mice (Tang et al. 2005). These data indicate that testisin may be involved in ovarian tumor progression possibly by facilitating anti-apoptotic signals; however, the molecular mechanism and candidate substrates for testisin in this context are yet to be identified.

15.3.2.2 GPI-Anchored Proteases: Prostasin

Prostasin (or PRSS8) is a serine protease with trypsin-like substrate specificity. Prostasin is also known as channel-activating protease (CAP)-1 and is the first of several membrane serine peptidases found to activate the epithelial sodium channel (ENaC) (Vallet et al. 1997). Prostasin was first identified in human seminal fluid, but also is expressed in epithelial cells and ducts of the prostate gland (Yu et al. 1994). Prostasin is also found at lower levels in normal colon, lung, kidney, pancreas, salivary gland, liver, and bronchi (Yu et al. 1994). Later, it was reported that endogenous prostasin is GPI anchored to the cell surface and is secreted from the cell upon cleavage of the GPI anchor by endogenous GPI-specific phospholipase D1 (Chen et al. 2001).

Expression studies in human prostate cancer cell lines and tumors demonstrated that prostasin is expressed in normal human prostate epithelial cells and a non-invasive human prostate cancer cell line, whereas it is not found in invasive human prostate cancer cell lines (Chen et al. 2001). Others found that prostasin mRNA and protein expression is down-regulated in high-grade prostate cancer or hormone-refractory human prostate cancers (Chen et al. 2001; Takahashi et al. 2003). The regulatory mechanisms for prostasin expression and activity are not completely understood. Prostate cell culture experiments indicate that prostasin is regulated at the transcriptional level by DNA methylation and growth factors and at the posttranslational level by the serpin protease nexin-1 (PN-1) (Chen et al. 2004).

Interestingly, similarly to testisin, prostasin is detected at higher levels in ovarian epithelial tumors than in normal ovarian tissue (Costa et al. 2009; Mok et al. 2001). Furthermore, prostasin can be detected in serum from ovarian cancer patients and postoperative prostasin levels are significantly lower than preoperative levels (Mok et al. 2001). These data suggest that prostasin is shed from the cell surface of ovarian cancer cells and may be a candidate serum marker for ovarian cancer.

In bladder cancer, a phenomenon similar to that in prostate cancer is observed. Prostasin is expressed in the normal human urothelium and in a normal human urothelial cell line, but is significantly down-regulated in high-grade transitional cell carcinoma (TCC) and lost in TCC cell lines. Loss of prostasin expression in the TCC cell lines correlates with loss of or reduced E-cadherin expression, loss of epithelial morphology, and promoter DNA hypermethylation (Chen et al. 2009). In colorectal cancer the mRNA level of prostasin is slightly, but significantly decreased in both mild/moderate dysplasia and severe dysplasia, and in carcinomas as compared to normal tissue from the same individual. The mRNA level of the prostasin inhibitor, PN-1, is more that two-fold elevated in colorectal cancer tissue as compared to healthy individuals and elevated in both mild/moderate dysplasia, severe dysplasia and in colorectal cancer tissue as compared to normal tissue from the same individual (Selzer-Plon et al. 2009). In gastric cancer, prostasin mRNA and protein expression are significantly down-regulated in tumor tissues. In those patients whose tumor expresses prostasin mRNA at reduced levels, a shorter survival is observed (Sakashita et al. 2008). The specific physiologic substrates for prostasin are as yet undefined and the mechanisms whereby the proteolytic activity of prostasin participates in tumor cell biology are unknown. In cell culture based studies, prostasin has, in addition to being involved in ENaC processing, been proposed to mediate proteolytic modification of the epidermal growth factor receptor and regulation of iNOS and cyclin D1 expression by modulating protease-activated receptor-2 signaling (Chen et al. 2008, 2009).

15.3.2.3 Type II Transmembrane Serine Proteases

The type II serine proteases (TTSPs) have garnered significant attention for their roles in epidermal development and homeostasis, iron metabolism and hearing. Defining the role of TTSP's in carcinogenesis and metastasis is still ongoing; however, recent studies have implicated several members of this family as critical players in neoplastic progression. Several TTSPs display aberrant expression in a variety of cancers, often in correlation with poor patient prognosis [reviewed in (Antalis et al. 2011; List et al. 2009; Magee et al. 2001; Netzel-Arnett et al. 2003; Wallrapp et al. 2000; Webb et al. 2011)]. Of particular interest are the proteins hepsin, matriptase and TMPRSS2, which have been linked to epithelial carcinogenesis and metastasis in the prostate and in the skin.

TTSPs: Hepsin

Hepsin/TMPRSS1 is the first TTSP to be cloned and was named based on its original identification in hepatocytes and hepatoma (Leytus et al. 1988; Tsuji et al. 1991). Hepsin has been implicated in oncogenesis due to its overexpression in several human cancers including renal cell carcinoma, hepatocellular carcinoma, breast cancer, endometrial cancer and ovarian cancer. Hepsin has been most extensively studied in prostate cancer where its over-expression is among the most consistent biomarkers for malignant transformation from benign prostate hyperplasia. Additionally, high hepsin expression correlates with a high Gleason score for prostate tumors and poor clinical outcome with relapse following prostatectomy (Saleem et al. 2006; Stamey et al. 2001; Stephan et al. 2004; Wu and Parry 2007).

The role of hepsin in prostate cancer has been studied experimentally using both genetic mouse models and cell culture-based systems. Transgenic overexpression of hepsin in the prostate epithelium of mice causes disorganization of the basement membrane, promotion of primary prostate cancer progression and metastasis to distant organs (Klezovitch et al. 2004). Interestingly, the prostate of transgenic mice does not display increased cell proliferation, indicating that hepsin may affect cancer cell migration, invasion and metastasis more than tumor growth. In support of this hypothesis, antibodies neutralizing the protease activity of hepsin do not impact cell growth, but do inhibit invasion by prostate and ovarian tumor cells in culture (Xuan et al. 2006). Hepsin may be involved in direct degradation of extracellular matrix proteins and/or activation of other proteases and growth factors or growth factor receptors. Several candidate substrates have been identified including laminin-322 and the pro-forms of uPA and HGF. Nonetheless, the physiologically relevant substrates are, as yet, not known (Herter et al. 2005; Moran et al. 2006; Tripathi et al. 2008).

TTSPs: Matriptase

This protease is among the most extensively studied TTSP with more than 100 published papers characterizing various aspects of matriptase and its role in cancer, including expression profiles, gene-regulation, regulation by cognate inhibitors, identification of proteolytic substrates, and *in vivo* studies using tumor transplantation models or genetically engineered mouse models.

Matriptase expression is deregulated in breast, prostate, endometrial, cervical, colorectal, gastric, and pancreatic carcinoma and in tumors of the lung, liver, and kidney among others [reviewed in (Bugge et al. 2007; List et al. 2009; Webb et al. 2011)]. An increase in matriptase expression correlates with the malignancy of tumors in the breast and prostate. *De novo* expression is found in ovarian and cervical carcinomas and expression levels in these tumors also correlate with histopathological grade. It has been hypothesized that a disturbance of the balance between matriptase and its two cognate inhibitors hepatocyte growth factor activator inhibitor 1 and 2 (HAI-1 and HAI-2) may be critical for cancer progression. In support of this hypothesis, the ratio of matriptase:HAI-1 mRNA is increased in both ovarian and colorectal cancer, and the expression of HAI-1 and HAI-2 is reduced in prostate and endometrial carcinoma when compared with normal tissue (Bergum and List 2010; Nakamura et al. 2011; Oberst et al. 2002; Vogel et al. 2006).

The functional role of matriptase is being studied by manipulating levels of expression or activity *in vitro* and *in vivo*. Inhibition of endogenous matriptase synthesis in cultured colon and prostate carcinoma cells by specific siRNAs significantly reduces invasiveness *in vitro* (Forbs et al. 2005). The same effect is observed after treatment of the cells with synthetic matriptase inhibitors. HGF has been proposed to be functionally involved since inhibition of matriptase in these cell lines impairs the conversion of pro-HGF into HGF on the cell surface and inhibits cell scattering upon stimulation with pro-HGF. In a human prostate cancer xenograft model, a selective small molecule matriptase inhibitor reduces growth of established tumors. The inhibitory mechanism of tumor expansion is proposed to be through impairment of invasion rather than inhibition of tumor cell proliferation, since matriptase inhibition has no effect on prostate cell growth rate *in vitro*, whereas invasion is significantly reduced (Galkin et al. 2004).

The role of matriptase *in vivo* has been studied in detail in squamous cell carcinogenesis of the skin. Interestingly, matriptase is expressed in all stages of carcinogenesis: hyperplasia, dysplasia, carcinoma *in situ*, invasive lesions and lymph node metastases. Spatial deregulation of matriptase expression is critical for epidermal carcinogenesis. Thus, in normal murine epidermis, matriptase expression is found solely in differentiated, nonproliferating keratinocytes; however, treatment with carcinogens induces *de novo* expression in the proliferating basal layer of cells, where the target cells for carcinogenesis reside (List et al. 2006). The expression of matriptase in the proliferative layer of the epidermis plays a causal role in tumor initiation and promotion in mice in which there is transgenic expression in the basal layer of the epidermis (keratin-5 matriptase mice). These transgenic mice develop spontaneous squamous cell carcinomas, severe dermal inflammation, and lymph node metastases and display dramatically increased susceptibility to carcinogen-induced tumorigenesis (List et al. 2005). Importantly, the oncogenic properties of matriptase are completely abolished in double transgenic mice that also express the inhibitor HAI-1 in the basal layer, providing evidence that a disturbance in the matriptase:HAI-1 balance in the epidermis is causative for carcinogenesis (List et al. 2005). Recently, a critical molecular target for matriptase in epidermal carcinogenesis has been identified using mouse genetic models (Szabo et al. 2011). When keratin-5 matriptase transgenic mice are crossed to mice with conditional epidermal deletion of c-Met, the membrane receptor for pro-HGF/HGF, the matriptase-mediated premalignant and malignant lesions are strongly abrogated, demonstrating that the oncogenic potential of matriptase in mouse epidermis requires c-Met. Pro-HGF binds to c-Met, but requires proteolytic cleavage to activate c-Met and elicit downstream signaling events. To this end, it was demonstrated that elevated levels of matriptase increase processing of pro-HGF, which results in initiation of the c-Met-mTOR signaling pathway to induce proliferation and migration in primary epithelial cells. From a cancer intervention standpoint, it is encouraging that epidermal carcinogenesis is efficiently inhibited by rapamycin, a pharmacological inhibitor of the mTOR pathway (Szabo et al. 2011). Accordingly, matriptase is being actively pursued as a candidate for pharmacological intervention as well as for a biomarker of disease onset and progression.

TMPRSS2

The major role of TMPRSS2 in cancer results from genetic fusion with members of the erythroblast transformation specific (ETS) transcription factors [reviewed in (Clark and Cooper 2009; Shah and Small 2010)]. TMPRSS2-ETS fusions occur via complex chromosomal translocation events that combine the 5′ regulatory region of TMPRSS2 in frame with either the entire or partial coding sequence of ETS genes, resulting in androgen-induced expression of the transcription factor. Four TMPRSS-ETS somatic fusions have been identified, with fusions to ERG (ETS-related gene) being the most prevalent. Both the prevalence and prognostic significance of the TMPRSS2–ERG fusion have been examined in multiple studies with some discrepancy in results. Although TMPRSS2–ERG fusions have typically been reported as prevalent in 40–50 % of prostate tumors, the range has varied by as much as 25–60 % (Barwick et al. 2010). Normal TMPRSS2 expression is localized to the prostate epithelium, and TMPRSS-ETS fusions result in the over-expression of the ETS member proto-oncogene in the prostate (Tomlins et al. 2006, 2007). Furthermore, TMPRSS2-ERG fusion events appear to be induced in both malignant and non-malignant prostate cells through androgen signaling, which can cause topoisomerase II beta-mediated DNA double strand breaks at the regions of genetic fusion (Bastus et al. 2010). ETS transcription factors are known to regulate cell growth, and TMPRSS2-ERG fusions most likely play a role early in prostate cancer development and/or progression. Targeted expression of ERG in luminal prostate epithelial cells of transgenic mice results in initiation of prostate neoplasia observed as the development of focal precancerous PIN lesions (Klezovitch et al. 2008; Tomlins et al. 2008). TMPRSS2-ETS fusion proteins are currently being assessed for both their potential as therapeutic targets and as biomarkers for early detection of disease (Tomlins et al. 2009).

15.3.3 Serine Protease Summary

A wide spectrum of serine proteases is deregulated in cancer, and may contribute to both tumor formation and metastasis through diverse molecular mechanisms. The loss of precise regulation of serine protease expression and/or catalytic activity may contribute to the etiology of several types of cancers. There is a strong impetus to understand the events that lead to deregulation at the gene and protein level, and how this precipitates various stages of tumorigenesis. Additionally, many serine proteases have yet to be assessed for physiological and pathological functions, and the mechanisms by which many characterized members affect tumorigenesis are largely unknown. Progress in these areas will add to our burgeoning knowledge of how proteases are involved in cancer progression and metastasis.

15.4 Cysteine Proteases

The cysteine cathepsins are lysosomal proteases that are part of clan CA, which contains several families of cysteine proteases that are structurally similar to the papaya cysteine protease papain (Rawlings et al. 2010). Other families in clan CA that have been implicated in cancer include a number of cytosolic proteases: calpains [for review, see (Storr et al. 2011)], autophagins (Marino et al. 2007) and a variety of deubiquitinating enzymes or DUBs (Colland 2010; Fang et al. 2010; Harhaj and Dixit 2011; Hymowitz and Wertz 2010; Sacco et al. 2010; Sun 2010), e.g., A20, BAP1, CYLD, USP28. The DUB family [for review, see (Komander et al. 2009; Reyes-Turcu et al. 2009)], which currently has 95 members, is intriguing as it consists of 80 clan CA cysteine proteases and 15 metalloproteases and in addition provides a link to the threonine proteases of the proteasome. Clan CD cysteine proteases (Rawlings et al. 2010), also have been associated with cancer: legumain (Lewen et al. 2008; Luo et al. 2006), a lysosomal protease originally identified in plants, and the cytosolic caspases of the apoptotic cascade (Demon et al. 2009; Ghavami et al. 2009). Here we will focus on cysteine cathepsins and cancer.

15.4.1 Cysteine Cathepsins

There are 11 human cysteine cathepsins: B, C, F, H, L, K, O, S, V, W and X (Rawlings et al. 2010). Their localization within lysosomes and late endosomes, vesicular compartments often referred to as the digestive tract of the cell, may have influenced research on these enzymes and the conclusions that we have reached about the roles they play in cancer. The 11 human cysteine cathepsins degrade a wide array of substrates within lysosomes and late endosomes and are stable and active at the acidic pH found in these vesicular compartments. For reviews on these enzymes see Vasiljeva and Turk (2008) and Reiser et al. (2010). Seven of the cysteine cathepsins are endopeptidases and thus presumed to participate in complete digestion of proteins within the lysosomes. On the other hand, four of the cysteine cathepsins are exopeptidases, cathepsin C is an aminopeptidase and cathepsin X is a carboxypeptidase, whereas cathepsins H and B have both endopeptidase and exopeptidase (aminopeptidase and carboxydipeptidase, respectively) activities. The two activities of cathepsin B depend on the position of an occluding loop that covers the active site pocket. At acidic pH such as in the lysosome the occluding loop limits access to the active site so that cathepsin B acts as a carboxydipeptidase. At neutral pH the occluding loop is displaced so that cathepsin B acts as an endopeptidase. The observations that cathepsin X (Sevenich et al. 2010b; Vasiljeva et al. 2006), a carboxypeptidase, and cathepsin L (Felbor et al. 2002; Stahl et al. 2007), an endopeptidase, are able to compensate for the absence of cathepsin B in mouse models would be consistent with cathepsin B

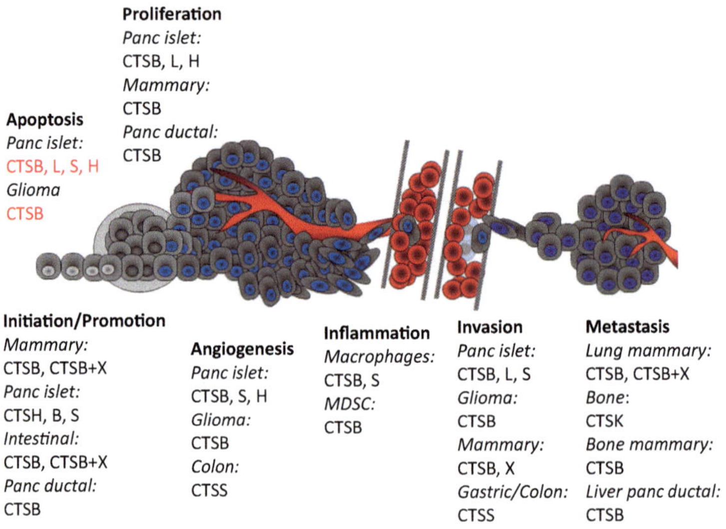

Fig. 15.5 Cysteine cathepsins in cancer: representative examples of family members demonstrated by *in vivo* experimentation to play roles in the indicated stages of tumor progression. An absence of the enzymes listed in *red font* under the heading apoptosis has been shown to induce apoptosis in pancreatic islet tumors and gliomas

exhibiting both exo- and endopeptidase activities *in vivo*. Although cysteine cathepsins normally function intracellularly in lysosomes and late endosomes in a number of developmental and pathological processes, individual cysteine cathepsins have been found to be secreted, associated with the cell membrane, released into the cytoplasm and present in the nucleus [for review, see (Mohamed and Sloane 2006; Reiser et al. 2010; Turk et al. 2012; Vasiljeva and Turk 2008)] (Fig. 15.5). These enzymes have been shown to participate in degradative processes at these extra-lysosomal sites, indicating that their stability and activity can be maintained under conditions not normally found in the lysosome, i.e., at a neutral pH and in a non-reducing environment. Since the roles of cysteine cathepsins in cancer have been previously reviewed (Mohamed and Sloane 2006), we concentrate here on those studies published more recently.

15.4.1.1 Cathepsin B

Among the cysteine cathepsins, cathepsin B has been the one most often linked to cancer. Expression is elevated in a wide variety of human cancers [for review, see (Jedeszko and Sloane 2004; Mohamed and Sloane 2006)]. The mechanisms responsible include amplication of the cathepsin B gene, which is localized on chromosome 8 p22. This region is a novel amplicon that is associated with amplification and overexpression of cathepsin B in Barrett's esophagus, a premalignant lesion that predisposes to esophageal cancer (Hughes et al. 1998). Additional mechanisms for altered expression include elevated transcription, use of alternative promoters and alternative splicing. These molecular changes lead to increased cathepsin B protein levels and in turn redistribution, secretion and increased activity (Yan and Sloane 2003).

Studies by Joyce and colleagues (Gocheva et al. 2006) in which cathepsin B knockout mice were crossed with a transgenic model for pancreatic islet cell carcinogenesis have established roles for cathepsin B at multiple points in the development of these tumors, i.e., initiation, proliferation, angiogenesis, invasion and apoptosis. The absence of cathepsin B reduces initiation, proliferation, angiogenesis and invasion, but increases apoptosis. Peters, Reinheckel and colleagues (Vasiljeva et al. 2006, 2008) have crossed cathepsin B knockout mice with transgenic models for mammary carcinoma. These studies demonstrated a role for cathepsin B in initiation, proliferation, invasion and metastasis in these tumors, including a role for cathepsin B in macrophages in lung metastases. Interestingly, there was compensation for the loss of cathepsin B in the mammary tumors and specifically for the loss of cathepsin B on the tumor cell surface by a redistribution of cathepsin X to the tumor cell surface (Vasiljeva et al. 2006). As cathepsin X is exclusively a carboxypeptidase, this suggests that the function of cathepsin B on the tumor cell surface is through its carboxypeptidase activity. A single deficiency of cathepsin B exerts partial reciprocal compensation for cathepsin X; double deficiencies of cathepsins B and X revealed additive effects, indicative of the two enzymes contributing to tumor initiation, progression, invasion and metastasis. Compensation by cathepsin X has also been observed in the intestinal polyps that form in mice with mutations in the adenomatous polyposis coli gene (Gounaris et al. 2008). In these mice, the active cathepsins could be used to image the polyps and myeloid derived suppressor cells and macrophages within the polyps. Polyposis and infiltration of pro-inflammatory cells could be reduced by ablation of cathepsin B. A compensatory role for cathepsin B has recently been observed in two mouse models of pancreatic neuroendocrine carcinogenesis (Shchors et al. 2013). In these tumors this is due to infiltration of myeloid derived suppressor cells expressing cathepsin B to the invasive front of the tumors, infiltration of which is associated with an increased invasive phenotype. In a transgenic mouse model of pancreatic ductal adenocarcinoma that was crossed with cathepsin B knockout mice, a role for cathepsin B in tumor initiation, proliferation and progression has been identified (Gopinathan et al. 2012). In these tumors cathepsin L was upregulated, once again showing compensation by one protease for deficiency in another.

Downregulation of cathepsin B by antisense, siRNA and shRNA technologies has established a causal role for this enzyme in a variety of tumors. For a review of the studies by Rao and colleagues in glioma, see (Rao 2003; Vink and Boomsma 2002). A recent study in a murine mammary carcinoma model found that shRNA knockdown of cathepsin B reduces degradation of type I collagen in vitro and bone metastasis in vivo (Withana et al. 2012). Deficiencies in both cathepsins B and X are synergistic, significantly delaying mammary tumor development and inhibiting metastasis (Sevenich et al. 2010b). Studies in which cathepsin B is downregulated in concert with other proteases/protease receptors have shown that cathepsin B is part of a proteolytic network that involves the plasminogen activator cascade and MMPs. Rao and colleagues have found that downregulation of cathepsin B and uPAR in glioma induces apoptosis (Malla et al. 2010, 2012) and of cathepsin B and MMP9 in meningioma reduces growth in vitro and in vivo and reduces invasion, angiogenesis and regulation of downstream kinase signaling pathways in vitro (Tummalapalli et al. 2007). They reported similar findings when cathepsin B, MMP9 and uPAR were downregulated in prostate carcinoma (Nalla et al. 2010) and in glioma, a finding linked to downregulation of integrins (Veeravalli et al. 2010). Also in glioma specific inhibition of alpha3beta1 integrin and the tetraspanin CD151 occurs as a result of knockdown of cathepsin B and uPAR (Rao Malla et al. 2012). An interplay between integrins and proteases including cathepsin B has also been shown in breast and prostate tumor cells (Sameni et al. 2008). In those studies downregulation of beta1-integrin reduced secretion of cathepsin B and MMP13, expression of MMP14 and degradation of type IV collagen. In gliomas, downregulation of cathepsin B and uPAR decreases proliferation due to G0/G1 arrest and reduced phosphorylation of PI3K/Akt (Gopinath et al. 2010; To et al. 2010) and also inhibits tumor-induced angiogenesis (Malla et al. 2011). Comparable reductions of angiogenesis in meningiomas could be reversed by overexpression of cathepsin B and uPAR, a finding linked to increased FAK phosphorylation (Gupta et al. 2011). A complex signaling network that increases transcription of cathepsin B in association with the invasive phenotype induced by ErbB2 in breast cancer has been identified (Rafn et al. 2012). These studies provide strong evidence that proteases act in pathways that extend beyond a single family of proteases and that those pathways are activated by and also modulate downstream signaling pathways. Furthermore, the roles played by proteases are not constant. In the studies cited above from the laboratories of Joyce and Rao, downregulation of cathepsin B was shown to induce apoptosis, yet in other studies cathepsin B and other lysosomal cysteine cathepsins have been shown to mediate apoptosis [for a review discussing these opposing functions, see Vasiljeva and Turk (2008)].

15.4.1.2 Cathepsin X

In melanomas and mammary carcinomas, the presence of cathepsin X in tumor cells has been associated with malignant progression (Rumpler et al. 2003; Sevenich et al. 2010b; Vasiljeva et al. 2006). As noted above, cathepsin X, which is an exopeptidase, compensates for a deficiency in cathepsin B in mammary carcinoma cells, including becoming associated with the surface of the carcinoma cells (Vasiljeva et al. 2006). Deficiencies in both cathepsins X and B delay development of early and advanced mammary carcinomas and reduce the number and the size of their lung metastases (Sevenich et al. 2010b). In contrast, studies in lung tumors found cathepsin X in macrophages rather than in tumor cells (Kos et al. 2005). Similarly, cathepsin X functions in gastric epithelial cells in the inflammatory response to *Helicobacter pylori* infection (Bernhardt et al. 2010), which increases the risk for gastric cancer (Leung 2006).

15.4.1.3 Cathepsins L and V

Determining the relationship between cathepsins L and V and cancer is made more difficult by the existence of two similar enzymes in human, i.e., cathepsins L and V (sometimes referred to as L2). Cathepsin V is the human orthologue of mouse cathepsin L (Dennemarker et al. 2010). Reinheckel and colleagues (Sevenich et al. 2010b) postulate that functional differences between the two enzymes are determined by their tissue specific expression patterns, cathepsin V being expressed primarily in thymus, cornea, testis and epidermis (Hagemann et al. 2004). In support of this postulate, either human cathepsin L or human cathepsin V can compensate for a deficiency of mouse cathepsin L in the thymus of cathepsin L knockout mice (Sevenich et al. 2010a). Nonetheless, tissue specific expression patterns do not explain why in a single tissue, i.e., in human cervical carcinomas cathepsin V is upregulated while cathepsin L is downregulated (Vazquez-Ortiz et al. 2005). If both are performing the same function, it seems counterintuitive for expression of one to go up and expression of the other to go down. There is an extensive literature implicating secretion of procathepsin L in malignant transformation, literature that predates the identification of cathepsin V. Indeed the major excreted protein of malignantly transformed murine fibroblasts is procathepsin L (Kane and Gottesman 1990). In agreement with a role for cathepsin L in malignancy, the deletion of cathepsin L in a transgenic model for pancreatic islet cell carcinogenesis reduces proliferation and invasion and increases apoptosis (Gocheva et al. 2006). In contrast in another transgenic model, i.e., a model of multistage squamous carcinogenesis, cathepsin L plays a protective role as the progression of tumors in mice crossed with mice deficient in cathepsin L is enhanced (Dennemarker et al. 2010). A comparable finding was made in regard to a protective role for cathepsin L in chemically-induced skin carcinogenesis (Benavides et al. 2012). These two findings may be indicative of functional differences due to tissue specific expression patterns of mouse cathepsin L and its human orthologue cathepsin V in the epidermis.

Cathepsin L may play roles other than degradation of extracellular matrices such as activation of other enzymes that degrade matrices. For example, cathepsin L has been shown to activate pro-heparanase; heparanase itself degrades heparan sulfate in the extracellular matrix and has pro-metastatic, pro-angiogenic and pro-inflammatory activities (Abboud-Jarrous et al. 2008). There is evidence from several laboratories that a nuclear form of cathepsin L plays a role in cell cycle progression through the processing of CDP/Cux transcription factor (Goulet et al. 2004, 2006, 2007), drug resistance through the cleavage of topoisomerase II alpha (Zheng et al. 2009; Murray et al. 2008), in colon cancer through degradation of histone H1 (Puchi et al. 2010) and in thyroid cancer through the modification of DNA-associated proteins (Tedelind et al. 2010). Nuclear cathepsin L has been observed in human colorectal tumors in both the tumor cells and stromal cells, increasing in parallel with stage in the tumor cells (Sullivan et al. 2009). The nuclear cathepsin in thyroid cancers has been identified as cathepsin V rather than cathepsin L (Tedelind et al. 2010). Upregulation of cathepsin V is also observed in human breast carcinomas where it has been shown to be one of two protease genes that comprise a predictive invasive signature for these tumors (Paik et al. 2004), in particular for node-negative breast tumors. This invasive signature is part of the gene panel used clinically in the Oncotype DX test to make treatment decisions.

15.4.1.4 Cathepsin H

Cathepsin H is upregulated in early stage, node negative human lung tumors and in lung tumors in a transgenic mouse model (Linnerth et al. 2005). In human tumors, cathepsin H is upregulated in immunological cells surrounding basal cell carcinoma nodules (Frohlich et al. 2004), in transitional cell carcinoma of the bladder (Staack et al. 2004), in brain tumors (Hunter and Moreno 2002) and in hepatomas where it is associated with microvascular invasion and kinase activation (Wu et al. 2011). Deletion of cathepsin H in a mouse model of pancreatic islet carcinoma reduces tumor progression, growth, burden and angiogenesis, thus demonstrating a tumor-promoting role for this enzyme (Gocheva et al. 2010).

15.4.1.5 Cathepsin C

Cathepsin C is highly expressed in several tumors: rat esophageal adenocarcinoma (Cheng et al. 2005), murine pancreatic islet cell carcinoma (Gocheva et al. 2006) and human non-small cell lung carcinoma (Cordes et al. 2009) and Hodgkin lymphoma (Ma et al. 2008). Whether cathepsin C plays a causal role in malignant progression depends on the tumor. Downregulation of cathepsin C in the RIP-Tag model of pancreatic islet cell carcinoma had no effect on tumor formation or progression (Gocheva et al. 2006). Similarly in murine mammary carcinomas in which cathepsin C expression is increased as a result of infiltration of macrophages, deletion of cathepsin C had no effect on any parameters of primary or metastatic tumor development. On the other hand in a transgenic model of multistage squamous carcinogenesis a deficiency of cathepsin C reduces tumor progression (Ruffell et al. in press).

15.4.1.6 Cathepsin K

Cathepsin K is highly expressed in osteoclasts where its predominant role is in bone resorption (Saftig et al. 1998) and therefore many cathepsin K inhibitors have been developed for treatment of osteoporosis (Lewiecki 2011). Indeed 50 patent applications were filed between 2004 and 2010 (Wijkmans and Gossen 2011). Cathepsin K is viewed as a therapeutic target for the osteolytic metastases associated with a variety of human tumors, e.g., prostate (Podgorski et al. 2009), breast (Tomita et al. 2008), lung (Rapa et al. 2006; Tomita et al. 2008), multiple myeloma (Hecht et al. 2008). The source of cathepsin K in tumors may be tumor cells as in melanomas (Quintanilla-Dieck et al. 2008), stromal fibroblasts and osteoclasts activated by breast and lung tumors (Tomita et al. 2008) and stromal macrophages and fibroblasts in squamous cell carcinomas (Yan et al. 2011). Secretion of cathepsin K by stromal fibroblasts increases invasiveness of breast tumor cells, a paracrine interaction that can be blocked with cathepsin K inhibitors (Kleer et al. 2008). In cathepsin K knockout mice, the progression of prostate tumor bone metastases is reduced as a result of a deficiency of cathepsin K in bone marrow-derived macrophages (Herroon et al. 2013).

15.4.1.7 Cathepsin S

Cathepsin S is highly expressed in a number of tumors. In astrocytomas, the level of expression is highest in grade IV tumors (Flannery et al. 2003) and in glioblastomas is associated with significantly shorter survival (Flannery et al. 2006). In uveal melanomas, cathepsin S expression and activity is increased, the latter in parallel with decreases in expression of cystatin C, an endogenous inhibitor (Paraoan et al. 2009). The source of cathepsin S may vary from tumor to tumor. In a mouse model of prostate carcinoma, cathepsin S is found in tumor-associated macrophages (Lindahl et al. 2009), whereas in human hepatocellular carcinoma (Xu et al. 2009) and gastric carcinoma (Yang et al. 2010), cathepsin S is associated with the tumor cells. Cathepsin S has been shown to participate in tumor progression. Tumor proliferation and angiogenesis are reduced in a mouse model for pancreatic islet carcinoma crossed with cathepsin S deficient mice (Wang et al. 2006). Another study reported reduced tumor formation and angiogenesis and increased apoptosis in the same model crossed with cathepsin S deficient mice (Gocheva et al. 2006). In yet other studies in which cathepsin S was downregulated a causal role for this enzyme in migration and invasion of gastric carcinomas has been established (Yang et al. 2010).

15.4.1.8 Cathepsins F and W

Cathepsins F and W form an evolutionarily related subgroup of the cysteine cathepsins (Wang et al. 1998; Wex et al. 1999). Cathepsin F has a longer propeptide domain than the other cysteine cathepsins and is present at high levels in only a few cancer cell lines (Santamaria et al. 1999), e.g., cervical carcinoma lines (Vazquez-Ortiz et al. 2005). In contrast, cathepsin F expression is low in highly metastatic osteosarcoma lines (Husmann et al. 2008). Cathepsin F is also found in macrophages (Kaakinen et al. 2007; Shi et al. 2000), but to our knowledge has not been studied in tumor-associated macrophages. Cathepsin W, which is also called lymphopain, is one of a series of genes associated with cytotoxic function that are upregulated in large granular lymphocyte leukemia (Kothapalli et al. 2003).

15.4.1.9 Cathepsin O

Cathepsin O has been identified as a potential marker for metastases of tongue carcinoma.

15.4.1.10 Cysteine Cathepsins as Therapeutic Targets

A broad-spectrum inhibitor of cysteine cathepsins, JPM-OEt, has been shown to be efficacious in a transgenic mouse model for pancreatic islet tumors (Bell-McGuinn et al. 2007). Tumor burden, vascularity and invasion were all reduced by JPM-OEt, findings consistent with cysteine cathepsins being therapeutic targets in this tumor model. A note of caution is needed in regard to how these preclinical findings will translate to clinical use of cysteine cathepsin inhibitors as broad-spectrum inhibitors of MMPs failed in clinical trial despite showing efficacy in preclinical studies in a wide variety of tumor models (see section on MMPs). A potential problem with use of broad-spectrum inhibitors is that some cysteine cathepsins also protect against tumor development just as has been shown for MMPs, e.g., MMP-12 in lung cancer (Acuff et al. 2006; Houghton et al. 2006). Bioavailability of the inhibitors may also impact their effectiveness as has been shown in studies comparing the use of JPM-OEt in mouse models for pancreatic islet tumors and mammary tumors. In the former inhibition of cysteine cathepsins reduced tumor development, yet not in the latter model (Schurigt et al. 2008). Poor bioavailability of the inhibitor in the mammary carcinomas and the lungs, yet not in the liver, kidney or pancreas of the mammary-tumor bearing mice was shown to account for the differences in efficacy in the two preclinical models.

van Noorden and colleagues have reviewed the case for cathepsin L being a therapeutic target and in so doing raise a concern that selective inhibition of cathepsin L will lead to compensatory changes in other cysteine cathepsins (Lankelma et al. 2010). This has been shown in a transgenic mammary carcinoma model in which deficiency in cathepsin B, an enzyme localized on the tumor cell surface (Mai et al. 2000; Sloane et al. 1994), is compensated for not only by increased expression of cathepsin X, but also by an association of cathepsin X with the tumor cell surface (Vasiljeva et al. 2006). van Noorden and colleagues further postulate that combination therapy with a cathepsin L inhibitor and conventional chemotherapy may be a better strategy than just use of a cathepsin L inhibitor (Lankelma et al. 2010). Thus it is of interest that inhibition of cathepsin S enhances the therapeutic effect of chemotherapy in colorectal carcinoma (Burden et al. 2012) and the broad-spectrum inhibitor JPM-OEt used in combination with chemotherapeutic regimens exhibits a greater therapeutic response (tumor regression, decreased invasion and increased survival) than is observed when only cysteine cathepsins are inhibited. Despite these promising findings with JPM-OEt, this is an irreversible inhibitor and therefore will not be used in patients. More relevant in terms of clinical applicability is that a reversible broad-spectrum cysteine cathepsin inhibitor has also been shown to exhibit effectiveness in the pancreatic islet tumor model (Elie et al. 2010).

Among the cysteine cathepsins, the enzyme most widely acknowledged as an important therapeutic target is cathepsin K due to its critical role in bone resorption by osteoclasts (Onishi et al. 2010; Rachner et al. 2012; Sturge et al. 2011). A number of selective inhibitors for this enzyme have been developed and have been shown to decrease bone resorption by breast tumors as well as bone metastases in preclinical studies (Jensen et al. 2010; Le Gall et al. 2007). Selective inhibition of other cysteine cathepsins has also shown some efficacy in preclinical studies. For example, a highly selective cathepsin B inhibitor suppresses metastasis of mammary tumors to bone (Withana et al. 2012). An inhibitory antibody to cathepsin S inhibits invasion and angiogenesis of colon tumors (Burden et al. 2009; Ward et al. 2010). These studies suggest that cathepsins K, B and S may be therapeutic targets; however, whether selective inhibition of any individual cysteine cathepsin will prove to be of value in the clinic will await further study.

Tumor-associated increases in expression and secretion of cysteine cathepsins occur in both tumor cells and inflammatory cells found at the invasive margins. The elevated expression of cysteine cathepsins in tumor-associated macrophages at both primary and metastatic sites means that targeting these enzymes will require targeting both the tumor microenvironment and the tumor. This was the rationale in part for the development of a novel drug delivery system in which magnetic nanoparticles were incorporated into liposomes, termed ferri-liposomes, along with the broad-spectrum cysteine cathepsin inhibitor JPM-OEt (Mikhaylov et al. 2011). The ferri-liposomes could be targeted to tumor xenografts in mice using an external magnet with their delivery to the tumor site being imaged non-invasively by MRI. How such a delivery system would work for human tumors is less clear. Nonetheless, delivery of JPM-OEt to mouse mammary fatpads containing tumor xenografts significantly reduced tumor size in contrast to the lack of efficacy of JPM-OEt not incorporated in liposomes. Cysteine cathepsins and in particular the nuclear form of cathepsin L has been implicated in drug resistance of tumors (Zheng et al. 2009). Resistance to doxorubicin could be abrogated by a cathepsin L inhibitor and combination therapy with doxorubicin and a cathepsin L inhibitor suppressed the proliferation of drug-resistant tumors in a preclinical mouse model. Doxorubicin resistance as well as resistance to a number of other chemotherapeutics has recently been linked to the cysteine cathepsins B and S in tumor-associated macrophages (Shree et al. 2011). By combining the inhibition of macrophage cysteine cathepsins with chemotherapeutics, there was a significant increase in the chemotherapeutic response at both primary and metastatic tumor sites, indicating the importance of macrophages and cysteine cathepsins at both sites as well as the need to develop therapeutic strategies that target tumors in the context of their microenvironment.

15.5 Threonine Proteases

The threonine proteases are part of a specialized group of enzymes in which the N-terminal amino acid acts as the nucleophile that attacks the peptide bonds of substrates. In mammals, the predominant threonine proteases are constituents of the proteasome complex. The mammalian proteasome is a multi-constituent complex that resembles a stack of rings, with each ring having seven subunits. This is an ancient structure with a simpler version, also possessing threonine protease activity, present in archaebacteria (Coux et al. 1996). Initial naming of the proteasome was based on size and appearance so that the designation '26S proteasome' referring to an estimated sedimentation coefficient is now the accepted term for a complex consisting of both the 20S proteolytic and a single 19S regulatory subunit (Dahlmann 2005). The proteolytic part can be separately referred to as the '20S proteasome'. There are multiple other names used in the literature including 'cylindrin', 'high molecular weight protease', '19s ring-type particles' among others. The reader is referred to the chapter 'Eukaryotic 20S proteasome' in the encyclopedic 'Handbook of Proteolytic Enzymes, 2nd Edition' for a more complete description of the history and biochemistry of the proteasome (Seemuller et al. 2004). One point that is critical for understanding the vast array of proteasome substrates is that there are three enzymatic activities within the subunits of the 20S proteasome. These are referred to as caspase-like, trypsin-like and chymotrypsin-like activities (Adams 2004a). Here we focus on the importance of the proteolytic activity of the proteasome in cancer development and progression.

15.5.1 The Proteasome in Tumor Initiation/Promotion

Since the proteasome is the major regulator of protein stability in the cell, it is unsurprising that its activity can impact tumor development at many levels. A number of cancer-associated proteins, either directly oncogenic or constituents of oncogenic signaling pathways, are regulated by proteasomal degradation. Mutations in these molecules, as occurs in tumorigenesis, often lead to proteins that have significantly altered turnover. An example of an oncogene whose half-life is greatly increased by mutation leading to a prolonged oncogenic signal is c-myc (Bahram et al. 2000; Gavine et al. 1999). In contrast, Smads 2 and 4, components of the TGF-β signal transduction pathway that are often mutated in colorectal cancers, have a much shorter half-life when mutated, thus impeding tumor suppressive signaling by TGF-β (Xu and Attisano 2000). The driving mutation in the majority of colon cancers occurs in the adenomatous polyposis coli (APC) gene or components of a complex involving APC (Voutsadakis 2008). A major function of this APC complex is to target the protein beta-catenin for proteasomal degradation (Easwaran et al. 1999). In the absence of proteasomal degradation, beta-catenin accumulates, translocates to the nucleus and, in partner with DNA binding transcription factors such as lymphoid enhancer factor-1, effects transcription of oncogenic proteins including c-myc and cyclin D1.

Gastric cancer, although relatively rare in the United States, is the second leading cause of cancer-related mortality worldwide (Jemal et al. 2010). One of the major causes is chronic infection with the bacterium *Helicobacter pylori* (Piazuelo et al. 2010). There are multiple mechanisms by which chronic *H. pylori* infection could result in gastric cancer. One suggestion is via proteasomal-mediated degradation of the cell cycle inhibitor $p27^{kip1}$ in gastric epithelial cells exposed to *H. pylori* (Eguchi et al. 2003). Interestingly, this appears to be a differently regulated proteasomal degradation pathway than that which occurs in normal physiology. Thus, it seems that bacterial infection can alter proteasome activity to result in continuous cycling of infected cells, which may favor development of neoplasias. The existence of an alternatively regulated proteasomal degradation pathway for $p27^{kip1}$ had previously been demonstrated in an elegant mouse model (Malek et al. 2001). Investigators 'knocked-in' a mutant $p27^{kip1}$ in which the regulatory threonine residue is replaced with an alanine and were surprised that effects of this mutation are relatively subtle. The results point to a novel pathway, but the stimulus for this pathway remains elusive. Another mechanism of cell cycle regulation is also linked to the proteasome. Cyclin degradation by proteasomal activity is important in preventing carcinogenic transformation of bronchial epithelial cells, and in determining responsiveness to chemopreventative agents such as all-trans-retinoic acid (Dragnev et al. 2004). In fact, regulation of cyclins and cyclin-dependent kinase inhibitors is a critical function of the proteasome that can be perturbed by proteasome inhibitors in cancer cells (Adams 2004b).

The principal result of therapeutic proteasome inhibitors in cancer therapy is increased apoptosis. This has been attributed to increased stability of several substrates including NF-κB, p53, death receptors and pro-apoptotic Bcl2 family members (Adams 2004b; Drexler 1997; Drexler et al. 2000). Thus, in the absence of inhibition, proteasome activity leads to increased survival of tumor cells (Hu et al. 2004). Another important mechanism of action is increased endoplasmic reticulum (ER) stress associated with the 'unfolded protein response' or UPR (Zhang and Kaufman 2006). There have been several suggestions that proteasome activity is increased in cancer cells as compared to non-transformed cells (An et al. 1998; Testa 2009) and, with the increase in cell cycling discussed above, in addition to increased survival, outgrowth of a cancer is favored.

15.5.2 The Proteasome in Angiogenesis

The process of angiogenesis is complex involving multiple components such as endothelial cell proliferation, migration and tube formation as well as recruitment of supporting mural cells, generation of appropriate matrix and orchestration of an array of signaling molecules to achieve these results. Proteasomal regulation of almost every aspect has been demonstrated using combinations of *in vivo* and *in vitro* approaches. Many studies use chemical inhibitors of proteasomal activity, e.g., lactacystin or MG-132 as well as newer inhibitors, to indicate a role for the proteasome (Drexler et al. 2000; Matsuo et al. 2010; Oikawa et al. 1998; Sunwoo et al. 2001). Recently, Chou et al. (2010) dissected an elegant angiogenic signaling pathway present in many tumor types, that is ultimately dependent on the proteasome. They showed that stability of the tumor suppressive protein Von-Hippel Lindau can be regulated by oncogenic Src, ultimately leading to increased levels of VEGF and angiogenesis. The stability and signaling of the two major receptors for VEGF, VEGFR1 and VEGFR2, are also controlled through ubiquitination and proteasomal degradation (Bruns et al. 2010), although there is some disagreement on which is the more sensitive target of the proteasome (Meissner et al. 2009). One possible explanation for the discrepancy is the presence of the ligand VEGF, which can simultaneously promote increased stability of VEGFR1 and increased degradation of VEGFR2, overall leading to enhanced endothelial cell survival (Zhang et al. 2010).

Despite the clear role for altered protein stability in promotion of angiogenesis, inhibition of proteasome activity *in vivo* can actually result in enhanced angiogenesis and increased tumor burden in a mouse model of pancreatic adenocarcinoma (Marten et al. 2008). One possibility is that the effect is related to the proteasome inhibitor used, i.e., bortezomib, as a different type of proteasome inhibitor with more limited activity called argyrin A significantly attenuates angiogenesis in a colon cancer model (Nickeleit et al. 2008). The activity of argyrin against the proteasome is actually more broad-spectrum than bortezomib as it inhibits all three of the enzymatic activities whereas bortezomib targets the chymotrypsin-like activity. Nevertheless, argyrin A appears to have fewer off-target effects (Nickeleit et al. 2008).

15.5.3 The Proteasome in Inflammation

One of the best-known functions of the proteasome is in regulation of the transcription factor nuclear factor kappa B (NFκB) (Amit and Ben-Neriah 2003). While it is increasingly clear that NFκB plays multiple roles in different cell types, it is especially associated with promoting inflammation. Indeed pro-inflammatory cytokines, induced by NFkB signaling, are considered tumor promoters in many types of cancer [reviewed in (Ben-Neriah and Karin 2011; Cortes Sempere et al. 2008; Grivennikov et al. 2010)].

In addition to direct effects on tumor cells, altered proteasomal activity can affect other cells of the tumor microenvironment. These effects may be anti-tumorigenic. For example, *in vitro* analysis of macrophages exposed to the endotoxin lipopolysaccharide (LPS) indicates that proteasomal activity is required for the generation of the pro-inflammatory, classically activated macrophage (Cuschieri et al. 2004). When proteasomal activity is inhibited, LPS-stimulated macrophages have a different cytokine profile that is more anti-inflammatory. Since alternatively activated, anti-inflammatory macrophages are thought to promote tumor progression in many types of cancer (Qian and Pollard 2010), this study may indicate that proteasomal inhibitors could have unanticipated side-effects in the tumor stroma. Such thinking has led to attempts to specifically alter NFκB signaling in tumor-associated macrophages *in vivo* (Hagemann et al. 2008).

15.5.4 The Proteasome in Drug Resistance

Resistance of cancer cells to chemotherapeutic drugs is a significant clinical problem in oncology. Many mechanisms have been identified that can contribute to this phenomenon including increased expression of efflux pumps like multidrug resistance (MDR)-1, mutation of drug targets, increased expression of anti-apoptotic molecules or up-regulation of alternative signaling pathways (Fodale et al. 2011; McCormick 2011; Wilson et al. 2009). The proteasome is a significant player in drug resistance, partly through controlling the balance of pro and anti-apoptotic signals, but also in other ways (Adams 2004a). One example is that which occurs with the DNA topoisomerase II poisons, a class of drugs that includes etoposide and doxorubicin, which are widely used against many types of cancer. These drugs are often rendered ineffective through proteasomal-mediated degradation of their target topoisomerase, which occurs when stress conditions such as low glucose and hypoxia are encountered (Tsuruo et al. 2003).

Autophagy is a cellular survival process that is thought to be a major contributor to drug resistance, although this is controversial (Maycotte and Thorburn 2011). Autophagy is normally induced by environmental stresses such as nutrient or oxygen deprivation. The degradation of cellular proteins that occurs in autophagy generally involves proteolysis within the lysosomes and this is considered a complementary mechanism to proteasome-mediated degradation. One potential problem with overcoming drug resistance through inhibiting the proteasome is that autophagic survival mechanisms may be enhanced. For this reason, new combinations of proteasome inhibitors and agents that prevent autophagy have been suggested (Wu et al. 2010). Such combinations would be useful in multiple situations where cancer cells are placed under stress, including stress induced by chemo- or radiation therapy, as well as by naturally occurring nutrient limitation.

15.5.5 Proteasome Inhibitors in Cancer Therapy

As is evident from the foregoing paragraphs, proteosomal activity is increased in tumor versus normal cells and contributes to the tumor phenotype in multiple ways. Therefore, it is not surprising that the proteasome should emerge as a primary 'target' for targeted therapies. Bortezomib (Velcade™, Takeda Millennium Pharmaceuticals) is the first of these targeted therapies to be approved (Adams

2004a), gaining fast-track approval status from the Food and Drug Administration in the USA based on especially encouraging results in a Phase II trial in refractory multiple myeloma patients (Chen et al. 2011). Bortezomib is termed a dipeptidyl boronate inhibitor, which describes the chemical moiety that interacts with the active site threonine (Adams 2004a). It particularly binds to the active site threonine in the chymotryptic-like subunit of the 20S proteolytic core. Recent evidence has suggested that bortezomib is not, however, specific for the proteasome, but has off-target effects (McConkey 2008; Testa 2009). While bortezomib is often used in hematological malignancies, such as multiple myeloma, it has not demonstrated significant efficacy in solid tumors. This has been attributed to a number of factors including development of resistance, dose-limiting side-effects and interaction with natural chemicals that interfere with its activity (Chen et al. 2011; McConkey and Zhu 2008). Since resistance to many chemotherapeutic agents, as outlined above, appears related to proteasomal activity, the combination of a proteasome inhibitor with traditional chemotherapy remains an attractive therapeutic strategy (Adams 2004a). Hence, development of newer generation proteasome inhibitors has received much attention. One particularly promising candidate is marizomib (NPI-0052), an irreversible inhibitor that shows stronger effects at lower doses over a longer time (Chauhan et al. 2005; Potts et al. 2011). Given the inevitable development of resistance seen with bortezomib (McConkey and Zhu 2008), one very attractive feature of marizomib is that it is unaffected by transporter proteins such as members of the MDR family (Obaidat et al. 2011). Other possible proteasomal inhibitory drugs include argyrin A (Nickeleit et al. 2008), as well as a number of natural products (Cecarini et al. 2011). Overall, the clear contribution of threonine protease activity to cancer makes this class of enzyme an important and rapidly evolving drug target.

15.6 Conclusions

We are still far from having a thorough understanding of the functions of proteases in cancer: this includes understanding the roles of the proteases discussed in this chapter, that those roles are dynamic and may change during the course of malignant progression, whether the proteases playing critical roles in malignant progression come from tumor, stromal or inflammatory cells, whether the critical proteases are affected by interactions of the tumor with its microenvironment, etc. In short, we do not yet know which protease(s) is the most appropriate target for anti-protease therapies or when anti-protease therapies might prove most effective. Elevated expression of proteases can be documented at the transcript and protein levels in many tumors, as described here. This elevated expression does not, however, necessarily translate into elevated activity as most proteases are synthesized as inactive proenzymes that require activation. In addition, there may be changes in tumors in the expression of endogenous protease inhibitors that impact protease activity. Thus to assess whether a protease is active and has the potential of playing a functional role, we need techniques that allow us to assess the activity of that protease, both in cell culture-based assays and in whole organisms. We also need such techniques to assess the efficacy of therapies that alter protease activity. One of the difficulties we face in evaluating why the clinical trials of MMP inhibitors did not meet the promise of the preclinical trials is that the clinical trials did not include any surrogate endpoints to assess whether the activity of the MMPs targeted was actually reduced. For discussions of the implications of these clinical trials, including why imaging is important, the reader is referred to the following articles (Coussens et al. 2002; Egeblad and Werb 2002; Marlowe 2005; Scherer et al. 2008a; Zucker et al. 2000; Chau et al. 2003; Li and Anderson 2003; McIntyre and Matrisian 2003; Moin et al. 2012). At the time of the MMP inhibitor trials, there were few methodologies to image protease activity *in vivo* or for that matter *in vitro*. This is an expanding field with a variety of imaging modalities and probes now available for functional imaging of protease activity *in vivo* [for review, see (auf dem Keller and Schilling 2010; Li and Anderson 2003; Moin et al. 2007; Scherer et al. 2008a)].

For *in vivo* imaging of protease activity, Matrisian and her coworkers have developed optical proteolytic beacons (Scherer et al. 2008b) and MRI contrast agents (Lepage et al. 2007) for MMP-7 and optical proteolytic beacons for MMP-9 (Fig. 15.6) or MMPs in general (McIntyre et al. 2010). These reagents are based on known MMP substrates. A recent study also used an MMP substrate, but in this case a triple helical peptide substrate, to develop near-infrared probes that detect activity of MMP-2 and -9 in vivo in mouse tumors (Akers et al. 2012). An alternative strategy is the use of small molecule inhibitors of MMPs, a strategy being developed for in vivo molecular imaging of active MMPs by SPECT and PET (Wagner et al. 2006).

Activity of serine proteases has been imaged in vivo by optical methods using target-activatable self-assembled gold nanoparticle probes (Mu et al. 2010) and fluorescently-labeled blocking antibodies (Darragh et al. 2010). Bogyo and colleagues have pioneered the use of activity-based probes (ABPs), which are based on small molecule cysteine protease inhibitors that bind covalently to their target enzymes, for *in vivo* optical imaging of the activity of several classes of cysteine proteases, including cysteine cathepsins (Blum et al. 2007) and caspases (Edgington et al. 2009). In a side-by-side comparison (Blum et al. 2009) of inhibitor-based probes and substrate-based probes, which were pioneered by Weissleder and colleagues [see for example (Weissleder et al. 1999)], for cysteine cathepsins, Bogyo and colleagues show that both can be used for *in vivo* imaging; however, the rapid uptake and clearance of ABPs reduces background, thus improving image contrast. For further discussion of the use of PET in conjunction with ABPs and the advantages/disadvantages of substrate-based probes versus ABPs, see auf dem Keller and Schilling (2010). Nonetheless, the potential clinical applicability of substrate-based probes has been established as colonic adenocarcinomas in mice could be detected non-invasively with micro-fiberoptic catheters due to the presence in these tumors of the cysteine protease cathepsin B (Alencar et al. 2007). Multimodal activatable optical (fluorescence) and MRI

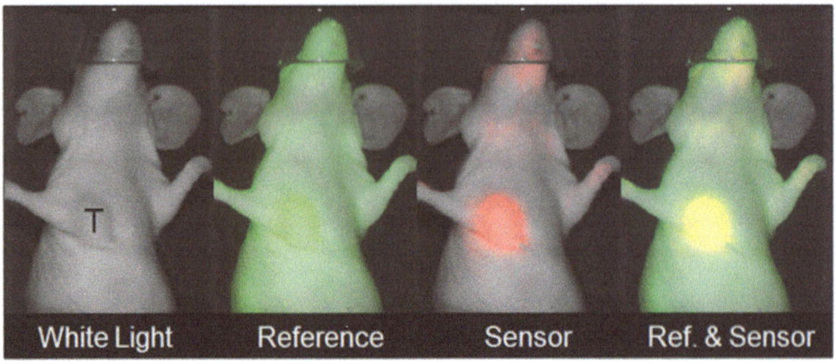

Fig. 15.6 *In vivo* imaging of tumor-associated MMP-9 activity with a *two-color* NIR proteolytic beacon, PB-M9nir. An orthotopic MDA-MB231 tumor, established in a *nu/nu* mouse, was imaged using a Pearl® Impulse imaging system (LI-COR Biosciences: White light, 800 nm Reference and 700 nm Sensor) ~4 h post retro-orbital administration of ~2 nmol PB-M9nir (Samuelson, Matrisian and McIntyre, unpublished studies)

probes for caspases have been developed, but there are not yet reports on their use in vivo (Kikuchi 2010). While the aforementioned imaging reagents/methods will be important for defining the roles of proteases *in vivo* and determining the efficacies of therapies that alter protease activities, they will require further validation before moving to clinical use. On the other hand, there are now reports that imaging protease activity during surgery allows one to distinguish tumor margins and improve excision of the tumors. Tsien and colleagues have shown that protease activatable cell penetrating peptides injected i.v., allow one to visualize the invasive margins of tumors in mice and by so doing during tumor removal improves tumor-free and overall survival (Nguyen et al. 2010). These dendrimeric probes detect activities of MMP-2 and -9 by either fluorescence or MRI (Olson et al. 2010). Recently, topical application of quenched fluorescent activity-based probes has been shown to detect active cysteine cathepsins and shown useful for the visualization and resection of gliomas in mice (Cutter et al. 2012). This latter methodology may be better applicable to the clinic as when these authors compared topical and systemic delivery of the probes they showed that topically applied probes were activated within 5 min and differentiated tumor margins from normal tissue. These examples of the successful use of active proteases/protease activity to monitor and improve surgical removal of tumors are particularly exciting developments as they have the potential for rapid translation to the clinic and for making a positive impact on cancer treatment.

Overall, this chapter summarizes the strong rationale for focusing on proteases in cancer. As described above, recent advances in imaging technology will greatly improve our ability to determine which proteolytic activities should be inhibited, as well as provide potential biomarkers for disease presence and response to therapy, including surgery. Continued efforts to develop efficacious inhibitors will hopefully yield targeted-therapies with minimal side-effects, which can be combined with existing drugs or with each other to treat multiple types of cancer.

References

Abboud-Jarrous G, Atzmon R, Peretz T, Palermo C, Gadea BB, Joyce JA, Vlodavsky I (2008) Cathepsin L is responsible for processing and activation of proheparanase through multiple cleavages of a linker segment. J Biol Chem 283:18167–18176

Acuff HB, Sinnamon M, Fingleton B, Boone B, Levy SE, Chen X, Pozzi A, Carbone DP, Schwartz DR, Moin K, Sloane BF, Matrisian LM (2006) Analysis of host- and tumor-derived proteinases using a custom dual species microarray reveals a protective role for stromal matrix metalloproteinase-12 in non-small cell lung cancer. Cancer Res 66:7968–7975

Adams J (2004a) The development of proteasome inhibitors as anticancer drugs. Cancer Cell 5:417–421

Adams J (2004b) The proteasome: a suitable antineoplastic target. Nat Rev Cancer 4:349–360

Agrawal S, Anderson P, Durbeej M, van Rooijen N, Ivars F, Opdenakker G, Sorokin LM (2006) Dystroglycan is selectively cleaved at the parenchymal basement membrane at sites of leukocyte extravasation in experimental autoimmune encephalomyelitis. J Exp Med 203:1007–1019

Ahn GO, Brown JM (2008) Matrix metalloproteinase-9 is required for tumor vasculogenesis but not for angiogenesis: role of bone marrow-derived myelomonocytic cells. Cancer Cell 13:193–205

Akers WJ, Xu B, Lee H, Sudlow GP, Fields GB, Achilefu S, Edwards WB (2012) Detection of MMP-2 and MMP-9 activity in vivo with a triple-helical peptide optical probe. Bioconjug Chem 23:656–663

Alencar H, Funovics MA, Figueiredo J, Sawaya H, Weissleder R, Mahmood U (2007) Colonic adenocarcinomas: near-infrared microcatheter imaging of smart probes for early detection–study in mice. Radiology 244:232–238

Almholt K, Johnsen M (2003) Stromal cell involvement in cancer. In: Molecular staging of cancer, vol 162. Springer, Berlin, pp 31–42

Almholt K, Nielsen BS, Frandsen TL, Brunner N, Dano K, Johnsen M (2003) Metastasis of transgenic breast cancer in plasminogen activator inhibitor-1 gene-deficient mice. Oncogene 22:4389–4397

Almholt K, Green KA, Juncker-Jensen A, Nielsen BS, Lund LR, Romer J (2007) Extracellular proteolysis in transgenic mouse models of breast cancer. J Mammary Gland Biol Neoplasia 12:83–97

Amit S, Ben-Neriah Y (2003) NF-kappaB activation in cancer: a challenge for ubiquitination- and proteasome-based therapeutic approach. Semin Cancer Biol 13:15–28

An B, Goldfarb RH, Siman R, Dou QP (1998) Novel dipeptidyl proteasome inhibitors overcome Bcl-2 protective function and selectively accumulate the cyclin-dependent kinase inhibitor p27 and induce apoptosis in transformed, but not normal, human fibroblasts. Cell Death Differ 5:1062–1075

Antalis TM, Bugge TH, Wu Q (2011) Membrane-anchored serine proteases in health and disease. Prog Mol Biol Transl Sci 99:1–50

Apte SS (2009) A disintegrin-like and metalloprotease (reprolysin-type) with thrombospondin type 1 motif (ADAMTS) superfamily: functions and mechanisms. J Biol Chem 284:31493–31497

Ardi VC, Van den Steen PE, Opdenakker G, Schweighofer B, Deryugina EI, Quigley JP (2009) Neutrophil MMP-9 proenzyme, unencumbered by TIMP-1, undergoes efficient activation in vivo and catalytically induces angiogenesis via a basic fibroblast growth factor (FGF-2)/FGFR-2 pathway. J Biol Chem 284:25854–25866

auf dem Keller U, Schilling O (2010) Proteomic techniques and activity-based probes for the system-wide study of proteolysis. Biochimie 92:1705–1714

Bahram F, von der Lehr N, Cetinkaya C, Larsson LG (2000) c-Myc hot spot mutations in lymphomas result in inefficient ubiquitination and decreased proteasome-mediated turnover. Blood 95:2104–2110

Balbin M, Fueyo A, Tester AM, Pendas AM, Pitiot AS, Astudillo A, Overall CM, Shapiro SD, Lopez-Otin C (2003) Loss of collagenase-2 confers increased skin tumor susceptibility to male mice. Nat Genet 35:252–257

Barwick BG, Abramovitz M, Kodani M, Moreno CS, Nam R, Tang W, Bouzyk M, Seth A, Leyland-Jones B (2010) Prostate cancer genes associated with TMPRSS2-ERG gene fusion and prognostic of biochemical recurrence in multiple cohorts. Br J Cancer 102:570–576

Bastus NC, Boyd LK, Mao X, Stankiewicz E, Kudahetti SC, Oliver RTD, Berney DM, Lu Y-J (2010) Androgen-induced TMPRSS2:ERG fusion in nonmalignant prostate epithelial cells. Cancer Res 70:9544–9548

Bell-McGuinn KM, Garfall AL, Bogyo M, Hanahan D, Joyce JA (2007) Inhibition of cysteine cathepsin protease activity enhances chemotherapy regimens by decreasing tumor growth and invasiveness in a mouse model of multistage cancer. Cancer Res 67:7378–7385

Benavides F, Perez C, Blando J, Contreras O, Shen J, Coussens LM, Fischer SM, Kusewitt DF, Digiovanni J, Conti CJ (2012) Protective role of cathepsin L in mouse skin carcinogenesis. Mol Carcinog. 51:352–361

Ben-Neriah Y, Karin M (2011) Inflammation meets cancer, with NF-kappaB as the matchmaker. Nat Immunol 12:715–723

Bergers G, Brekken R, McMahon G, Vu TH, Itoh T, Tamaki K, Tanzawa K, Thorpe P, Itohara S, Werb Z, Hanahan D (2000) Matrix metalloproteinase-9 triggers the angiogenic switch during carcinogenesis. Nat Cell Biol 2:737–744

Bergum C, List K (2010) Loss of the Matriptase Inhibitor HAI-2 During Prostate Cancer Progression. Prostate 70:1422–1428

Bernhardt A, Kuester D, Roessner A, Reinheckel T, Krueger S (2010) Cathepsin X-deficient gastric epithelial cells in co-culture with macrophages: characterization of cytokine response and migration capability after Helicobacter pylori infection. J Biol Chem 285:33691–33700

Blavier L, Lazaryev A, Dorey F, Shackleford GM, DeClerck YA (2006) Matrix metalloproteinases play an active role in Wnt1-induced mammary tumorigenesis. Cancer Res 66:2691–2699

Blobel CP (2005) ADAMs: key components in EGFR signalling and development. Nat Rev Mol Cell Biol 6:32–43

Blum G, von Degenfeld G, Merchant MJ, Blau HM, Bogyo M (2007) Noninvasive optical imaging of cysteine protease activity using fluorescently quenched activity-based probes. Nat Chem Biol 3:668–677

Blum G, Weimer RM, Edgington LE, Adams W, Bogyo M (2009) Comparative assessment of substrates and activity based probes as tools for non-invasive optical imaging of cysteine protease activity. PLoS One 4:e6374

Borgono CA, Diamandis EP (2004) The emerging roles of human tissue kallikreins in cancer. Nat Rev Cancer 4:876–890

Borgono CA, Michael IP, Diamandis EP (2004) Human tissue kallikreins: physiologic roles and applications in cancer. Mol Cancer Res 2:257–280

Bozkulak EC, Weinmaster G (2009) Selective use of ADAM10 and ADAM17 in activation of Notch1 signaling. Mol Cell Biol 29:5679–5695

Brasse D, Mathelin C, Leroux K, Chenard MP, Blaise S, Stoll I, Tomasetto C, Rio MC (2010) Matrix metalloproteinase 11/stromelysin-3 exerts both activator and repressor functions during the hematogenous metastatic process in mice. Int J Cancer 127:1347–1355

Brinckerhoff CE, Matrisian LM (2002) Matrix metalloproteinases: a tail of a frog that became a prince. Nat Rev Mol Cell Biol 3:207–214

Brooks PC, Silletti S, von Schalscha TL, Friedlander M, Cheresh DA (1998) Disruption of angiogenesis by PEX, a noncatalytic metalloproteinase fragment with integrin binding activity. Cell 92:391–400

Bruns AF, Herbert SP, Odell AF, Jopling HM, Hooper NM, Zachary IC, Walker JH, Ponnambalam S (2010) Ligand-stimulated VEGFR2 signaling is regulated by co-ordinated trafficking and proteolysis. Traffic 11:161–174

Bugge TH, Lund LR, Kombrinck KK, Nielsen BS, Holmback K, Drew AF, Flick MJ, Witte DP, Dano K, Degen JL (1998) Reduced metastasis of Polyoma virus middle T antigen-induced mammary cancer in plasminogen-deficient mice. Oncogene 16:3097–3104

Bugge TH, List K, Szabo R (2007) Matriptase-dependent cell surface proteolysis in epithelial development and pathogenesis. Front Biosci 12:5060–5070

Bugge TH, Antalis TM, Wu Q (2009) Type II transmembrane serine proteases. J Biol Chem 284:23177–23181

Burden RE, Gormley JA, Jaquin TJ, Small DM, Quinn DJ, Hegarty SM, Ward C, Walker B, Johnston JA, Olwill SA, Scott CJ (2009) Antibody-mediated inhibition of cathepsin S blocks colorectal tumor invasion and angiogenesis. Clin Cancer Res 15:6042–6051

Burden RE, Gormley JA, Kuehn D, Ward C, Kwok HF, Gazdoiu M, McClurg A, Jaquin TJ, Johnston JA, Scott CJ, Olwill SA (2012) Inhibition of Cathepsin S by Fsn0503 enhances the efficacy of chemotherapy in colorectal carcinomas. Biochimie 94:487–493

Casal C, Torres-Collado AX, Plaza-Calonge Mdel C, Martino-Echarri E, Ramon YCS, Rojo F, Griffioen AW, Rodriguez-Manzaneque JC (2010) ADAMTS1 contributes to the acquisition of an endothelial-like phenotype in plastic tumor cells. Cancer Res 70:4676–4686

Cauwe B, Van den Steen PE, Opdenakker G (2007) The biochemical, biological, and pathological kaleidoscope of cell surface substrates processed by matrix metalloproteinases. Crit Rev Biochem Mol Biol 42:113–185

Cecarini V, Cuccioloni M, Mozzicafreddo M, Bonfili L, Angeletti M, Eleuteri AM (2011) Targeting proteasomes with natural occurring compounds in cancer treatment. Curr Cancer Drug Targets 11:307–324

Chantrain CF, Shimada H, Jodele S, Groshen S, Ye W, Shalinsky DR, Werb Z, Coussens LM, DeClerck YA (2004) Stromal matrix metalloproteinase-9 regulates the vascular architecture in neuroblastoma by promoting pericyte recruitment. Cancer Res 64:1675–1686

Chau I, Rigg A, Cunningham D (2003) Matrix metalloproteinase inhibitors–an emphasis on gastrointestinal malignancies. Crit Rev Oncol Hematol 45:151–176

Chauhan D, Catley L, Li G, Podar K, Hideshima T, Velankar M, Mitsiades C, Mitsiades N, Yasui H, Letai A, Ovaa H, Berkers C, Nicholson B, Chao TH, Neuteboom ST, Richardson P, Palladino MA, Anderson KC (2005) A novel orally active proteasome inhibitor induces apoptosis in multiple myeloma cells with mechanisms distinct from Bortezomib. Cancer Cell 8:407–419

Chen LM, Skinner ML, Kauffman SW, Chao J, Chao L, Thaler CD, Chai KX (2001) Prostasin is a glycosylphosphatidylinositol-anchored active serine protease. J Biol Chem 276:21434–21442

Chen LM, Zhang XC, Chai KX (2004) Regulation of prostasin expression and function in the prostate. Prostate 59:1–12

Chen X, Su Y, Fingleton B, Acuff H, Matrisian LM, Zent R, Pozzi A (2005) Increased plasma MMP9 in integrin alpha1-null mice enhances lung metastasis of colon carcinoma cells. Int J Cancer 116:52–61

Chen M, Chen LM, Lin CY, Chai KX (2008) The epidermal growth factor receptor (EGFR) is proteolytically modified by the Matriptase-Prostasin serine protease cascade in cultured epithelial cells. Biochim Biophys Acta Mol Cell Res 1783:896–903

Chen LM, Hatfield ML, Fu YY, Chai KX (2009) Prostasin regulates iNOS and cyclin D1 expression by modulating protease-activated receptor-2 signaling in prostate epithelial cells. Prostate 69:1790–1801

Chen D, Frezza M, Schmitt S, Kanwar J, Dou QP (2011) Bortezomib as the first proteasome inhibitor anticancer drug: current status and future perspectives. Curr Cancer Drug Targets 11:239–253

Cheng P, Gong J, Wang T, Chen J, Liu GS, Zhang R (2005) Gene expression in rats with Barrett's esophagus and esophageal adenocarcinoma induced by gastroduodenoesophageal reflux. World J Gastroenterol 11:5117–5122

Chou MT, Anthony J, Bjorge JD, Fujita DJ (2010) The von Hippel-Lindau tumor suppressor protein is destabilized by src: implications for tumor angiogenesis and progression. Genes Cancer 1:225–238

Clark JP, Cooper CS (2009) ETS gene fusions in prostate cancer. Nat Rev Urol 6:429–439

Colland F (2010) The therapeutic potential of deubiquitinating enzyme inhibitors. Biochem Soc Trans 38:137–143

Cordes C, Bartling B, Simm A, Afar D, Lautenschlager C, Hansen G, Silber RE, Burdach S, Hofmann HS (2009) Simultaneous expression of Cathepsins B and K in pulmonary adenocarcinomas and squamous cell carcinomas predicts poor recurrence-free and overall survival. Lung Cancer 64:79–85

Cornelius LA, Nehring LC, Harding E, Bolanowski M, Welgus HG, Kobayashi DK, Pierce RA, Shapiro SD (1998) Matrix metalloproteinases generate angiostatin: effects on neovascularization. J Immunol 161:6845–6852

Cortes Sempere M, Rodriguez Fanjul V, Sanchez Perez I, Perona R (2008) The role of the NFkappaB signalling pathway in cancer. Clin Transl Oncol 10:143–147

Costa FP, Batista Junior EL, Zelmanowicz A, Svedman C, Devenz G, Alves S, Martins da Silva AS, Garicochea B (2009) Prostasin, a potential tumor marker in ovaria cancer – a pilot study. Clinics 64:641–644

Coussens LM, Tinkle CL, Hanahan D, Werb Z (2000) MMP-9 supplied by bone marrow-derived cells contributes to skin carcinogenesis. Cell 103:481–490

Coussens LM, Fingleton B, Matrisian LM (2002) Matrix metalloproteinase inhibitors and cancer: trials and tribulations. Science 295:2387–2392

Coux O, Tanaka K, Goldberg AL (1996) Structure and functions of the 20S and 26S proteasomes. Annu Rev Biochem 65:801–847

Cox JH, Dean RA, Roberts CR, Overall CM (2008) Matrix metalloproteinase processing of CXCL11/I-TAC results in loss of chemoattractant activity and altered glycosaminoglycan binding. J Biol Chem 283:19389–19399

Crawford HC, Dempsey PJ, Brown G, Adam L, Moss ML (2009) ADAM10 as a therapeutic target for cancer and inflammation. Curr Pharm Des 15:2288–2299

Croucher DR, Saunders DN, Lobov S, Ranson M (2008) Revisiting the biological roles of PAI2 (SERPINB2) in cancer. Nat Rev Cancer 8:535–545

Cuschieri J, Gourlay D, Garcia I, Jelacic S, Maier RV (2004) Implications of proteasome inhibition: an enhanced macrophage phenotype. Cell Immunol 227:140–147

Cutter JL, Cohen NT, Wang J, Sloan AE, Cohen AR, Panneerselvam A, Schluchter M, Blum G, Bogyo M, Basilion JP (2012) Topical application of activity-based probes for visualization of brain tumor tissue. PLoS One 7:e33060

Dahlmann B (2005) Proteasomes. Essays Biochem 41:31–48

D'Armiento J, DiColandrea T, Dalal SS, Okada Y, Huang MT, Conney AH, Chada K (1995) Collagenase expression in transgenic mouse skin causes hyperkeratosis and acanthosis and increases susceptibility to tumorigenesis. Mol Cell Biol 15:5732–5739

Darragh MR, Schneider EL, Lou J, Phojanakong PJ, Farady CJ, Marks JD, Hann BC, Craik CS (2010) Tumor detection by imaging proteolytic activity. Cancer Res 70:1505–1512

Dass K, Ahmad A, Azmi AS, Sarkar SH, Sarkar FH (2008) Evolving role of uPA/uPAR system in human cancers. Cancer Treat Rev 34:122–136

Demon D, Van Damme P, Vanden Berghe T, Vandekerckhove J, Declercq W, Gevaert K, Vandenabeele P (2009) Caspase substrates: easily caught in deep waters? Trends Biotechnol 27:680–688

Denmeade SR, Litvinov I, Sokoll LJ, Lilja H, Isaacs JT (2003) Prostate-specific antigen (PSA) protein does not affect growth of prostate cancer cells InVitro or prostate cancer xenografts in vivo. Prostate 56:45–53

Dennemarker J, Lohmuller T, Mayerle J, Tacke M, Lerch MM, Coussens LM, Peters C, Reinheckel T (2010) Deficiency for the cysteine protease cathepsin L promotes tumor progression in mouse epidermis. Oncogene 29:1611–1621

Diamandis EP (1998) Prostate-specific antigen: its usefulness in clinical medicine. Trends Endocrinol Metab 9:310–316

Doucet A, Overall CM (2008) Protease proteomics: revealing protease in vivo functions using systems biology approaches. Mol Aspects Med 29:339–358

Dragnev KH, Pitha-Rowe I, Ma Y, Petty WJ, Sekula D, Murphy B, Rendi M, Suh N, Desai NB, Sporn MB, Freemantle SJ, Dmitrovsky E (2004) Specific chemopreventive agents trigger proteasomal degradation of G1 cyclins: implications for combination therapy. Clin Cancer Res 10:2570–2577

Drexler HC (1997) Activation of the cell death program by inhibition of proteasome function. Proc Natl Acad Sci U S A 94:855–860

Drexler HC, Risau W, Konerding MA (2000) Inhibition of proteasome function induces programmed cell death in proliferating endothelial cells. FASEB J 14:65–77

Dubail J, Kesteloot F, Deroanne C, Motte P, Lambert V, Rakic JM, Lapiere C, Nusgens B, Colige A (2010) ADAMTS-2 functions as anti-angiogenic and anti-tumoral molecule independently of its catalytic activity. Cell Mol Life Sci 67:4213–4232

Duffy MJ (1990) Plasminogen activators and cancer. Blood Coagul Fibrinol 1:681–687

Duffy MJ (2004) The urokinase plasminogen activator system: Role in malignancy. Curr Pharm Des 10:39–49

Duffy MJ, Ogrady P, Devaney D, Osiorain L, Fennelly JJ, Lijnen HJ (1988) Urokinase-plasminogen activator, a marker for aggressive breast carcinomas – a preliminary report. Cancer 62:531–533

Easwaran V, Song V, Polakis P, Byers S (1999) The ubiquitin-proteasome pathway and serine kinase activity modulate adenomatous polyposis coli protein-mediated regulation of beta-catenin-lymphocyte enhancer-binding factor signaling. J Biol Chem 274:16641–16645

Edgington LE, Berger AB, Blum G, Albrow VE, Paulick MG, Lineberry N, Bogyo M (2009) Noninvasive optical imaging of apoptosis by caspase-targeted activity-based probes. Nat Med 15:967–973

Edwards DR, Handsley MM, Pennington CJ (2008a) The ADAM metalloproteinases. Mol Aspects Med 29:258–289

Edwards DR, Hoyer-Hansen G, Blasi F, Sloane BF (eds) (2008b) The cancer degradome – proteases and cancer biology. Springer, New York

Egeblad M, Werb Z (2002) New functions for the matrix metalloproteinases in cancer progression. Nat Rev Cancer 2:161–174

Eguchi H, Herschenhous N, Kuzushita N, Moss SF (2003) Helicobacter pylori increases proteasome-mediated degradation of p27(kip1) in gastric epithelial cells. Cancer Res 63:4739–4746

El Hour M, Moncada-Pazos A, Blacher S, Masset A, Cal S, Berndt S, Detilleux J, Host L, Obaya AJ, Maillard C, Foidart JM, Ectors F, Noel A, Lopez-Otin C (2010) Higher sensitivity of Adamts12-deficient mice to tumor growth and angiogenesis. Oncogene 29:3025–3032

Elie BT, Gocheva V, Shree T, Dalrymple SA, Holsinger LJ, Joyce JA (2010) Identification and pre-clinical testing of a reversible cathepsin protease inhibitor reveals anti-tumor efficacy in a pancreatic cancer model. Biochimie 92:1618–1624

Ellis V, Scully MF, Kakkar VV (1989) Plasminogen activation initiated by single-chain urokinase-type plasminogen-activator – potentiation by U937 monocytes. J Biol Chem 264:2185–2188

Erler JT, Bennewith KL, Cox TR, Lang G, Bird D, Koong A, Le QT, Giaccia AJ (2009) Hypoxia-induced lysyl oxidase is a critical mediator of bone marrow cell recruitment to form the premetastatic niche. Cancer Cell 15:35–44

Fang Y, Fu D, Shen XZ (2010) The potential role of ubiquitin c-terminal hydrolases in oncogenesis. Biochim Biophys Acta 1806:1–6

Felbor U, Kessler B, Mothes W, Goebel HH, Ploegh HL, Bronson RT, Olsen BR (2002) Neuronal loss and brain atrophy in mice lacking cathepsins B and L. Proc Natl Acad Sci U S A 99:7883–7888

Fingleton B (2006) Matrix metalloproteinases: roles in cancer and metastasis. Front Biosci 11:479–491

Flannery T, Gibson D, Mirakhur M, McQuaid S, Greenan C, Trimble A, Walker B, McCormick D, Johnston PG (2003) The clinical significance of cathepsin S expression in human astrocytomas. Am J Pathol 163:175–182

Flannery T, McQuaid S, McGoohan C, McConnell RS, McGregor G, Mirakhur M, Hamilton P, Diamond J, Cran G, Walker B, Scott C, Martin L, Ellison D, Patel C, Nicholson C, Mendelow D, McCormick D, Johnston PG (2006) Cathepsin S expression: an independent prognostic factor in glioblastoma tumours–a pilot study. Int J Cancer 119:854–860

Fodale V, Pierobon M, Liotta L, Petricoin E (2011) Mechanism of cell adaptation: when and how do cancer cells develop chemoresistance? Cancer J 17:89–95

Forbs D, Thiel S, Stella MC, Sturzebecher A, Schweinitz A, Steinmetzer T, Sturzebecher J, Uhland K (2005) In vitro inhibition of matriptase prevents invasive growth of cell lines of prostate and colon carcinoma. Int J Oncol 27:1061–1070

Friedl P, Wolf K (2008) Tube travel: the role of proteases in individual and collective cancer cell invasion. Cancer Res 68:7247–7249

Frohlich E, Mohrle M, Klessen C (2004) Cathepsins in basal cell carcinomas: activity, immunoreactivity and mRNA staining of cathepsins B, D, H and L. Arch Dermatol Res 295:411–421

Fry JL, Toker A (2010) Secreted and membrane-bound isoforms of protease ADAM9 have opposing effects on breast cancer cell migration. Cancer Res 70:8187–8198

Galkin AV, Mullen L, Fox WD, Brown J, Duncan D, Moreno O, Madison EL, Agus DB (2004) CVS-3983, a selective matriptase inhibitor, suppresses the growth of androgen independent prostate tumor xenografts. Prostate 61:228–235

Gandolfo GM, Conti L, Vercillo M (1996) Fibrinolysis components as prognostic markers in breast cancer and colorectal carcinoma. Anticancer Res 16:2155–2159

Garg P, Sarma D, Jeppsson S, Patel NR, Gewirtz AT, Merlin D, Sitaraman SV (2010) Matrix metalloproteinase-9 functions as a tumor suppressor in colitis-associated cancer. Cancer Res 70:792–801

Gavine PR, Neil JC, Crouch DH (1999) Protein stabilization: a common consequence of mutations in independently derived v-Myc alleles. Oncogene 18:7552–7558

Ghavami S, Hashemi M, Ande SR, Yeganeh B, Xiao W, Eshraghi M, Bus CJ, Kadkhoda K, Wiechec E, Halayko AJ, Los M (2009) Apoptosis and cancer: mutations within caspase genes. J Med Genet 46:497–510

Gibb DR, Saleem SJ, Kang DJ, Subler MA, Conrad DH (2011) ADAM10 overexpression shifts lympho- and myelopoiesis by dysregulating site 2/site 3 cleavage products of Notch. J Immunol 186:4244–4252

Gocheva V, Zeng W, Ke D, Klimstra D, Reinheckel T, Peters C, Hanahan D, Joyce JA (2006) Distinct roles for cysteine cathepsin genes in multistage tumorigenesis. Genes Dev 20:543–556

Gocheva V, Chen X, Peters C, Reinheckel T, Joyce JA (2010) Deletion of cathepsin H perturbs angiogenic switching, vascularization and growth of tumors in a mouse model of pancreatic islet cell cancer. Biol Chem 391:937–945

Gopinathan A, Denicola GM, Frese KK, Cook N, Karreth FA, Mayerle J, Lerch MM, Reinheckel T, Tuveson DA (2012) Cathepsin B promotes the progression of pancreatic ductal adenocarcinoma in mice. Gut 61(6):877–884

Gopinath S, Malia RR, Gondi CS, Alapati K, Fassett D, Klopfenstein JD, Dinh DH, Gujrati M, Rao JS (2010) Co-depletion of cathepsin B and uPAR induces G0/G1 arrest in glioma via FOXO3a mediated p27 upregulation. PLoS One 5:e11668

Goulet B, Baruch A, Moon NS, Poirier M, Sansregret LL, Erickson A, Bogyo M, Nepveu A (2004) A cathepsin L isoform that is devoid of a signal peptide localizes to the nucleus in S phase and processes the CDP/Cux transcription factor. Mol Cell 14:207–219

Goulet B, Truscott M, Nepveu A (2006) A novel proteolytically processed CDP/Cux isoform of 90 kDa is generated by cathepsin L. Biol Chem 387:1285–1293

Goulet B, Sansregret L, Leduy L, Bogyo M, Weber E, Chauhan SS, Nepveu A (2007) Increased expression and activity of nuclear cathepsin L in cancer cells suggests a novel mechanism of cell transformation. Mol Cancer Res 5:899–907

Gounaris E, Tung CH, Restaino C, Maehr R, Kohler R, Joyce JA, Ploegh HL, Barrett TA, Weissleder R, Khazaie K (2008) Live imaging of cysteine-cathepsin activity reveals dynamics of focal inflammation, angiogenesis, and polyp growth. PLoS One 3:e2916

Grivennikov SI, Greten FR, Karin M (2010) Immunity, inflammation, and cancer. Cell 140:883–899

Gupta R, Nalla AK, Gogineni VR, Chetty C, Bhoopathi P, Klopfenstein JD, Tsung AJ, Mohanam S, Rao JS (2011) uPAR/cathepsin B overexpression reverse angiogenesis by rescuing FAK phosphorylation in uPAR/cathepsin B down regulated meningioma. PLoS One 6: e17123

Gutierrez-Fernandez A, Inada M, Balbin M, Fueyo A, Pitiot AS, Astudillo A, Hirose K, Hirata M, Shapiro SD, Noel A, Werb Z, Krane SM, Lopez-Otin C, Puente XS (2007) Increased inflammation delays wound healing in mice deficient in collagenase-2 (MMP-8). FASEB J 21:2580–2591

Ha HY, Moon HB, Nam MS, Lee JW, Ryoo ZY, Lee TH, Lee KK, So BJ, Sato H, Seiki M, Yu DY (2001) Overexpression of membrane-type matrix metalloproteinase-1 gene induces mammary gland abnormalities and adenocarcinoma in transgenic mice. Cancer Res 61:984–990

Hagemann S, Gunther T, Dennemarker J, Lohmuller T, Bromme D, Schule R, Peters C, Reinheckel T (2004) The human cysteine protease cathepsin V can compensate for murine cathepsin L in mouse epidermis and hair follicles. Eur J Cell Biol 83:775–780

Hagemann T, Lawrence T, McNeish I, Charles KA, Kulbe H, Thompson RG, Robinson SC, Balkwill FR (2008) "Re-educating" tumor-associated macrophages by targeting NF-kappaB. J Exp Med 205:1261–1268

Hamano Y, Zeisberg M, Sugimoto H, Lively JC, Maeshima Y, Yang C, Hynes RO, Werb Z, Sudhakar A, Kalluri R (2003) Physiological levels of tumstatin, a fragment of collagen IV alpha3 chain, are generated by MMP-9 proteolysis and suppress angiogenesis via alphaV beta3 integrin. Cancer Cell 3:589–601

Han B, Nakamura M, Mori I, Nakamura Y, Kakudo K (2005) Urokinase-type plasminogen activator system and breast cancer (Review). Oncol Rep 14:105–112

Hanson EM, Clements VK, Sinha P, Ilkovitch D, Ostrand-Rosenberg S (2009) Myeloid-derived suppressor cells down-regulate L-selectin expression on CD4+ and CD8+ T cells. J Immunol 183:937–944

Harhaj EW, Dixit VM (2011) Deubiquitinases in the regulation of NF-kappaB signaling. Cell Res 21:22–39

Hecht M, von Metzler I, Sack K, Kaiser M, Sezer O (2008) Interactions of myeloma cells with osteoclasts promote tumour expansion and bone degradation through activation of a complex signalling network and upregulation of cathepsin K, matrix metalloproteinases (MMPs) and urokinase plasminogen activator (uPA). Exp Cell Res 314:1082–1093

Hedstrom L (2002) Serine protease mechanism and specificity. Chem Rev 102:4501–4524

Heissig B, Hattori K, Dias S, Friedrich M, Ferris B, Hackett NR, Crystal RG, Besmer P, Lyden D, Moore MA, Werb Z, Rafii S (2002) Recruitment of stem and progenitor cells from the bone marrow niche requires MMP-9 mediated release of kit-ligand. Cell 109:625–637

Heljasvaara R, Nyberg P, Luostarinen J, Parikka M, Heikkila P, Rehn M, Sorsa T, Salo T, Pihlajaniemi T (2005) Generation of biologically active endostatin fragments from human collagen XVIII by distinct matrix metalloproteases. Exp Cell Res 307:292–304

Herroon MK, Rajagurubandara E, Rudy D, Chalasani A, Hardaway AL, Podgorski I (2013) Macrophage cathepsin K promotes tumor progression in bone. Oncogene 32(12):1580–1593

Herter S, Piper DE, Aaron W, Gabriele T, Cutler G, Cao P, Bhatt AS, Choe Y, Craik CS, Walker N, Meininger D, Hoey T, Austin RJ (2005) Hepatocyte growth factor is a preferred

in vitro substrate for human hepsin, a membrane-anchored serine protease implicated in prostate and ovarian cancers. Biochem J 390:125–136
Hewitt R, Dano K (1996) Stromal cell expression of components of matrix-degrading protease systems in human cancer. Enzyme Protein 49:163–173
Hooper JD, Nicol DL, Dickinson JL, Eyre HJ, Scarman AL, Normyle JF, Stuttgen MA, Douglas ML, Loveland KAL, Sutherland GR, Antalis TM (1999) Testisin, a new human serine proteinase expressed by premeiotic testicular germ cells and lost in testicular germ cell tumors. Cancer Res 59:3199–3205
Hooper JD, Bowen N, Marshall H, Cullen LM, Sood R, Daniels R, Stuttgen MA, Normyle JF, Higgs DR, Kastner DL, Ogbourne SM, Pera MF, Jazwinska EC, Antalis TM (2000) Localization, expression and genomic structure of the gene encoding the human serine protease testisin. Biochim Biophys Acta Gene Struct Expr 1492:63–71
Houghton AM, Grisolano JL, Baumann ML, Kobayashi DK, Hautamaki RD, Nehring LC, Cornelius LA, Shapiro SD (2006) Macrophage elastase (matrix metalloproteinase-12) suppresses growth of lung metastases. Cancer Res 66:6149–6155
Hu MC, Lee DF, Xia W, Golfman LS, Ou-Yang F, Yang JY, Zou Y, Bao S, Hanada N, Saso H, Kobayashi R, Hung MC (2004) IkappaB kinase promotes tumorigenesis through inhibition of forkhead FOXO3a. Cell 117:225–237
Huang Y, Song N, Ding Y, Yuan S, Li X, Cai H, Shi H, Luo Y (2009) Pulmonary vascular destabilization in the premetastatic phase facilitates lung metastasis. Cancer Res 69:7529–7537
Hughes SJ, Glover TW, Zhu XX, Kuick R, Thoraval D, Orringer MB, Beer DG, Hanash S (1998) A novel amplicon at 8p22-23 results in overexpression of cathepsin B in esophageal adenocarcinoma. Proc Natl Acad Sci U S A 95:12410–12415
Hunter SB, Moreno CS (2002) Expression microarray analysis of brain tumors: what have we learned so far. Front Biosci 7:c74–c82
Husmann K, Muff R, Bolander ME, Sarkar G, Born W, Fuchs B (2008) Cathepsins and osteosarcoma: expression analysis identifies cathepsin K as an indicator of metastasis. Mol Carcinog 47:66–73
Hymowitz SG, Wertz IE (2010) A20: from ubiquitin editing to tumour suppression. Nat Rev Cancer 10:332–341
Iruela-Arispe ML, Carpizo D, Luque A (2003) ADAMTS1: a matrix metalloprotease with angioinhibitory properties. Ann NY Acad Sci 995:183–190
Itoh T, Tanioka M, Yoshida H, Yoshioka T, Nishimoto H, Itohara S (1998) Reduced angiogenesis and tumor progression in gelatinase A-deficient mice. Cancer Res 58:1048–1051
Itoh T, Tanioka M, Matsuda H, Nishimoto H, Yoshioka T, Suzuki R, Uehira M (1999) Experimental metastasis is suppressed in MMP-9-deficient mice. Clin Exp Metastasis 17:177–181
Jedeszko C, Sloane BF (2004) Cysteine cathepsins in human cancer. Biol Chem 385:1017–1027
Jemal A, Center MM, DeSantis C, Ward EM (2010) Global patterns of cancer incidence and mortality rates and trends. Cancer Epidemiol Biomarkers Prev 19:1893–1907
Jensen AB, Wynne C, Ramirez G, He W, Song Y, Berd Y, Wang H, Mehta A, Lombardi A (2010) The cathepsin K inhibitor odanacatib suppresses bone resorption in women with breast cancer and established bone metastases: results of a 4-week, double-blind, randomized, controlled trial. Clin Breast Cancer 10:452–458
Jodele S, Chantrain CF, Blavier L, Lutzko C, Crooks GM, Shimada H, Coussens LM, Declerck YA (2005) The contribution of bone marrow-derived cells to the tumor vasculature in neuroblastoma is matrix metalloproteinase-9 dependent. Cancer Res 65:3200–3208
Johnsen M, Lund LR, Romer J, Almholt K, Dano K (1998) Cancer invasion and tissue remodeling: common themes in proteolytic matrix degradation. Curr Opin Cell Biol 10:667–671
Kaakinen R, Lindstedt KA, Sneck M, Kovanen PT, Oorni K (2007) Angiotensin II increases expression and secretion of cathepsin F in cultured human monocyte-derived macrophages: an angiotensin II type 2 receptor-mediated effect. Atherosclerosis 192:323–327
Kane SE, Gottesman MM (1990) The role of cathepsin L in malignant transformation. Semin Cancer Biol 1:127–136

Kaplan RN, Riba RD, Zacharoulis S, Bramley AH, Vincent L, Costa C, MacDonald DD, Jin DK, Shido K, Kerns SA, Zhu Z, Hicklin D, Wu Y, Port JL, Altorki N, Port ER, Ruggero D, Shmelkov SV, Jensen KK, Rafii S, Lyden D (2005) VEGFR1-positive haematopoietic bone marrow progenitors initiate the pre-metastatic niche. Nature 438:820–827

Katakowski M, Jiang F, Zheng X, Gutierrez JA, Szalad A, Chopp M (2009) Tumorigenicity of cortical astrocyte cell line induced by the protease ADAM17. Cancer Sci 100:1597–1604

Kenny PA (2007) Tackling EGFR signaling with TACE antagonists: a rational target for metalloprotease inhibitors in cancer. Expert Opin Ther Targets 11:1287–1298

Kenny PA, Bissell MJ (2007) Targeting TACE-dependent EGFR ligand shedding in breast cancer. J Clin Invest 117:337–345

Kessenbrock K, Plaks V, Werb Z (2010) Matrix metalloproteinases: regulators of the tumor microenvironment. Cell 141:52–67

Kikuchi K (2010) Design, synthesis and biological application of chemical probes for bio-imaging. Chem Soc Rev 39:2048–2053

Kitamura T, Kometani K, Hashida H, Matsunaga A, Miyoshi H, Hosogi H, Aoki M, Oshima M, Hattori M, Takabayashi A, Minato N, Taketo MM (2007) SMAD4-deficient intestinal tumors recruit CCR1+ myeloid cells that promote invasion. Nat Genet 39:467–475

Kleer CG, Bloushtain-Qimron N, Chen YH, Carrasco D, Hu M, Yao J, Kraeft SK, Collins LC, Sabel MS, Argani P, Gelman R, Schnitt SJ, Krop IE, Polyak K (2008) Epithelial and stromal cathepsin K and CXCL14 expression in breast tumor progression. Clin Cancer Res 14:5357–5367

Klein T, Bischoff R (2011) Active metalloproteases of the A Disintegrin and Metalloprotease (ADAM) family: biological function and structure. J Proteome Res 10:17–33

Klezovitch O, Chevillet J, Mirosevich J, Roberts RL, Matusik RJ, Vasioukhin V (2004) Hepsin promotes prostate cancer progression and metastasis. Cancer Cell 6:185–195

Klezovitch O, Risk M, Coleman I, Lucas JM, Null M, True LD, Nelson PS, Vasioukhin V (2008) A causal role for ERG in neoplastic transformation of prostate epithelium. Proc Natl Acad Sci U S A 105:2105–2110

Klucky B, Mueller R, Vogt I, Teurich S, Hartenstein B, Breuhahn K, Flechtenmacher C, Angel P, Hess J (2007) Kallikrein 6 induces E-cadherin shedding and promotes cell proliferation, migration, and invasion. Cancer Res 67:8198–8206

Kohga K, Takehara T, Tatsumi T, Miyagi T, Ishida H, Ohkawa K, Kanto T, Hiramatsu N, Hayashi N (2009) Anticancer chemotherapy inhibits MHC class I-related chain a ectodomain shedding by downregulating ADAM10 expression in hepatocellular carcinoma. Cancer Res 69:8050–8057

Kohga K, Takehara T, Tatsumi T, Ishida H, Miyagi T, Hosui A, Hayashi N (2010) Sorafenib inhibits the shedding of major histocompatibility complex class I-related chain A on hepatocellular carcinoma cells by down-regulating a disintegrin and metalloproteinase 9. Hepatology 51:1264–1273

Komander D, Clague MJ, Urbe S (2009) Breaking the chains: structure and function of the deubiquitinases. Nat Rev Mol Cell Biol 10:550–563

Kos J, Sekirnik A, Premzl A, Zavasnik Bergant V, Langerholc T, Turk B, Werle B, Golouh R, Repnik U, Jeras M, Turk V (2005) Carboxypeptidases cathepsins X and B display distinct protein profile in human cells and tissues. Exp Cell Res 306:103–113

Kothapalli R, Bailey RD, Kusmartseva I, Mane S, Epling-Burnette PK, Loughran TP Jr (2003) Constitutive expression of cytotoxic proteases and down-regulation of protease inhibitors in LGL leukemia. Int J Oncol 22:33–39

Lai LC, Erbas H, Lennard TWJ, Peaston RT (1996) Prostate-specific antigen in breast cyst fluid: Possible role of prostate-specific antigen in hormone-dependent breast cancer. Int J Cancer 66:743–746

Lankelma JM, Voorend DM, Barwari T, Koetsveld J, Van der Spek AH, De Porto AP, Van Rooijen G, Van Noorden CJ (2010) Cathepsin L, target in cancer treatment? Life Sci 86:225–233

Le Gall C, Bellahcene A, Bonnelye E, Gasser JA, Castronovo V, Green J, Zimmermann J, Clezardin P (2007) A cathepsin K inhibitor reduces breast cancer induced osteolysis and skeletal tumor burden. Cancer Res 67:9894–9902

Lee S, Jilani SM, Nikolova GV, Carpizo D, Iruela-Arispe ML (2005) Processing of VEGF-A by matrix metalloproteinases regulates bioavailability and vascular patterning in tumors. J Cell Biol 169:681–691

Lepage M, Dow WC, Melchior M, You Y, Fingleton B, Quarles CC, Pepin C, Gore JC, Matrisian LM, McIntyre JO (2007) Noninvasive detection of matrix metalloproteinase activity in vivo using a novel magnetic resonance imaging contrast agent with a solubility switch. Mol Imaging 6:393–403

Leung WK (2006) Helicobacter pylori and gastric neoplasia. Contrib Microbiol 13:66–80

Lewen S, Zhou H, Hu HD, Cheng T, Markowitz D, Reisfeld RA, Xiang R, Luo Y (2008) A Legumain-based minigene vaccine targets the tumor stroma and suppresses breast cancer growth and angiogenesis. Cancer Immunol Immunother 57:507–515

Lewiecki EM (2011) New targets for intervention in the treatment of postmenopausal osteoporosis. Nat Rev Rheumatol 7:631–638

Leytus SP, Loeb KR, Hagen FS, Kurachi K, Davie EW (1988) A novel trypsin-like serine protease (hepsin) with a putative transmembrane domain expressed by human-liver and hepatoma-cells. Biochemistry 27:1067–1074

Li WP, Anderson CJ (2003) Imaging matrix metalloproteinase expression in tumors. Q J Nucl Med 47:201–208

Li Q, Park PW, Wilson CL, Parks WC (2002) Matrilysin shedding of syndecan-1 regulates chemokine mobilization and transepithelial efflux of neutrophils in acute lung injury. Cell 111:635–646

Lindahl C, Simonsson M, Bergh A, Thysell E, Antti H, Sund M, Wikstrom P (2009) Increased levels of macrophage-secreted cathepsin S during prostate cancer progression in TRAMP mice and patients. Cancer Genomics Proteomics 6:149–159

Linnerth NM, Sirbovan K, Moorehead RA (2005) Use of a transgenic mouse model to identify markers of human lung tumors. Int J Cancer 114:977–982

Liotta LA, Tryggvason K, Garbisa S, Hart I, Foltz CM, Shafie S (1980) Metastatic potential correlates with enzymatic degradation of basement membrane collagen. Nature 284:67–68

List K, Szabo R, Molinolo A, Sriuranpong V, Redeye V, Murdock T, Burke B, Nielsen BS, Gutkind JS, Bugge TH (2005) Deregulated matriptase causes ras-independent multistage carcinogenesis and promotes ras-mediated malignant transformation. Genes Dev 19:1934–1950

List K, Szabo R, Molinolo A, Nielsen BS, Bugge TH (2006) Delineation of matriptase protein expression by enzymatic gene trapping suggests diverging roles in barrier function, hair formation, and squamous cell carcinogenesis. Am J Pathol 168:1513–1525

List K, Kosa P, Szabo R, Bey AL, Wang CB, Molinolo A, Bugge TH (2009) Epithelial integrity is maintained by a matriptase-dependent proteolytic pathway. Am J Pathol 175:1453–1463

Littlepage LE, Sternlicht MD, Rougier N, Phillips J, Gallo E, Yu Y, Williams K, Brenot A, Gordon JI, Werb Z (2010) Matrix metalloproteinases contribute distinct roles in neuroendocrine prostate carcinogenesis, metastasis, and angiogenesis progression. Cancer Res 70:2224–2234

Lo PH, Lung HL, Cheung AK, Apte SS, Chan KW, Kwong FM, Ko JM, Cheng Y, Law S, Srivastava G, Zabarovsky ER, Tsao SW, Tang JC, Stanbridge EJ, Lung ML (2010) Extracellular protease ADAMTS9 suppresses esophageal and nasopharyngeal carcinoma tumor formation by inhibiting angiogenesis. Cancer Res 70:5567–5576

Lu X, Wang Q, Hu G, Van Poznak C, Fleisher M, Reiss M, Massague J, Kang Y (2009) ADAMTS1 and MMP1 proteolytically engage EGF-like ligands in an osteolytic signaling cascade for bone metastasis. Genes Dev 23:1882–1894

Ludwig A, Hundhausen C, Lambert MH, Broadway N, Andrews RC, Bickett DM, Leesnitzer MA, Becherer JD (2005) Metalloproteinase inhibitors for the disintegrin-like metalloproteinases

ADAM10 and ADAM17 that differentially block constitutive and phorbol ester-inducible shedding of cell surface molecules. Comb Chem High Throughput Screen 8:161–171
Luo Y, Zhou H, Krueger J, Kaplan C, Lee SH, Dolman C, Markowitz D, Wu W, Liu C, Reisfeld RA, Xiang R (2006) Targeting tumor-associated macrophages as a novel strategy against breast cancer. J Clin Invest 116:2132–2141
Ma Y, Visser L, Roelofsen H, de Vries M, Diepstra A, van Imhoff G, van der Wal T, Luinge M, Alvarez-Llamas G, Vos H, Poppema S, Vonk R, van den Berg A (2008) Proteomics analysis of Hodgkin lymphoma: identification of new players involved in the cross-talk between HRS cells and infiltrating lymphocytes. Blood 111:2339–2346
Magee JA, Araki T, Patil S, Ehrig T, True L, Humphrey PA, Catalona WJ, Watson MA, Milbrandt J (2001) Expression profiling reveals hepsin overexpression in prostate cancer. Cancer Res 61:5692–5696
Mai J, Finley RL Jr, Waisman DM, Sloane BF (2000) Human procathepsin B interacts with the annexin II tetramer on the surface of tumor cells. J Biol Chem 275:12806–12812
Malek NP, Sundberg H, McGrew S, Nakayama K, Kyriakides TR, Roberts JM (2001) A mouse knock-in model exposes sequential proteolytic pathways that regulate p27Kip1 in G1 and S phase. Nature 413:323–327
Malla R, Gopinath S, Alapati K, Gondi CS, Gujrati M, Dinh DH, Mohanam S, Rao JS (2010) Downregulation of uPAR and cathepsin B induces apoptosis via regulation of Bcl-2 and Bax and inhibition of the PI3K/Akt pathway in gliomas. PLoS One 5:e13731
Malla RR, Gopinath S, Gondi CS, Alapati K, Dinh DH, Gujrati M, Rao JS (2011) Cathepsin B and uPAR knockdown inhibits tumor-induced angiogenesis by modulating VEGF expression in glioma. Cancer Gene Ther 18:419–434
Malla RR, Gopinath S, Gondi CS, Alapati K, Dinh DH, Tsung AJ, Rao JS (2012) uPAR and cathepsin B downregulation induces apoptosis by targeting calcineurin A to BAD via Bcl-2 in glioma. J Neurooncol 107:69–80
Manton KJ, Douglas ML, Netzel-Arnett S, Fitzpatrick DR, Nicol DL, Boyd AW, Clements JA, Antalis TM (2005) Hypermethylation of the 5′ CpG island of the gene encoding the serine protease Testisin promotes its loss in testicular tumorigenesis. Br J Cancer 92:760–769
Marino G, Salvador-Montoliu N, Fueyo A, Knecht E, Mizushima N, Lopez-Otin C (2007) Tissue-specific autophagy alterations and increased tumorigenesis in mice deficient in Atg4C/autophagin-3. J Biol Chem 282:18573–18583
Marlowe DE (2005) Nanotechnology and the U.S. Food and Drug Administration. Stand News 33:29–31
Marten A, Zeiss N, Serba S, Mehrle S, von Lilienfeld-Toal M, Schmidt J (2008) Bortezomib is ineffective in an orthotopic mouse model of pancreatic adenocarcinoma. Mol Cancer Ther 7:3624–3631
Martin MD, Carter KJ, Jean-Philippe SR, Chang M, Mobashery S, .Thiolloy S, Lynch CC, Matrisian LM, Fingleton B (2008) Effect of ablation or inhibition of stromal matrix metalloproteinase-9 on lung metastasis in a breast cancer model is dependent on genetic background. Cancer Res 68:6251–6259
Masset A, Maillard C, Sounni NE, Jacobs N, Bruyere F, Delvenne P, Tacke M, Reinheckel T, Foidart J-M, Coussens LM, Noel A (2011) Unimpeded skin carcinogenesis in K14-HPV16 transgenic mice deficient for plasminogen activator inhibitor. Int J Cancer 128:283–293
Masson R, Lefebvre O, Noel A, Fahime ME, Chenard MP, Wendling C, Kebers F, LeMeur M, Dierich A, Foidart JM, Basset P, Rio MC (1998) In vivo evidence that the stromelysin-3 metalloproteinase contributes in a paracrine manner to epithelial cell malignancy. J Cell Biol 140:1535–1541
Matsuo Y, Sawai H, Ochi N, Yasuda A, Sakamoto M, Takahashi H, Funahashi H, Takeyama H, Guha S (2010) Proteasome inhibitor MG132 inhibits angiogenesis in pancreatic cancer by blocking NF-kappaB activity. Dig Dis Sci 55:1167–1176
Maycotte P, Thorburn A (2011) Autophagy and cancer therapy. Cancer Biol Ther 11:127–137

McCawley LJ, Crawford HC, King LE Jr, Mudgett J, Matrisian LM (2004) A protective role for matrix metalloproteinase-3 in squamous cell carcinoma. Cancer Res 64:6965–6972

McCawley LJ, Wright J, LaFleur BJ, Crawford HC, Matrisian LM (2008) Keratinocyte expression of MMP3 enhances differentiation and prevents tumor establishment. Am J Pathol 173:1528–1539

McConkey DJ (2008) A novel role for a familiar protein in apoptosis induced by proteasome inhibition. Cancer Cell 14:1–2

McConkey DJ, Zhu K (2008) Mechanisms of proteasome inhibitor action and resistance in cancer. Drug Resist Updat 11:164–179

McCormick F (2011) Cancer therapy based on oncogene addiction. J Surg Oncol 103:464–467

McIntyre JO, Matrisian LM (2003) Molecular imaging of proteolytic activity in cancer. J Cell Biochem 90:1087–1097

McIntyre JO, Scherer RL, Matrisian LM (2010) Near-infrared optical proteolytic beacons for in vivo imaging of matrix metalloproteinase activity. Methods Mol Biol 622:279–304

McQuibban GA, Gong JH, Wong JP, Wallace JL, Clark-Lewis I, Overall CM (2002) Matrix metalloproteinase processing of monocyte chemoattractant proteins generates CC chemokine receptor antagonists with anti-inflammatory properties in vivo. Blood 100:1160–1167

Meissner M, Reichenbach G, Stein M, Hrgovic I, Kaufmann R, Gille J (2009) Down-regulation of vascular endothelial growth factor receptor 2 is a major molecular determinant of proteasome inhibitor-mediated antiangiogenic action in endothelial cells. Cancer Res 69:1976–1984

Mikhaylov G, Mikac U, Magaeva AA, Itin VI, Naiden EP, Psakhye I, Babes L, Reinheckel T, Peters C, Zeiser R, Bogyo M, Turk V, Psakhye SG, Turk B, Vasiljeva O (2011) Ferri-liposomes as an MRI-visible drug-delivery system for targeting tumours and their microenvironment. Nat Nanotechnol 6:594–602

Mikolajczyk SD, Millar LS, Kumar A, Saedi MS (1999) Prostatic human kallikrein 2 inactivates and complexes with plasminogen activator inhibitor-1. Int J Cancer 81:438–442

Mohamed MM, Sloane BF (2006) Cysteine cathepsins: multifunctional enzymes in cancer. Nat Rev Cancer 6:764–775

Moin K, McIntyre OJ, Matrisian LM, Sloane BF (2007) Fluorescent imaging of tumors. In: Shields A, Price P (eds) In vivo imaging of cancer therapy. Humana, Totawa, pp 281–302

Moin K, Sameni M, Victor BC, Rothberg JM, Mattingly RR, Sloane BF (2012) 3D/4D functional imaging of tumor-associated proteolysis: impact of microenvironment. Methods Enzymol 506:175–194

Mok SC, Chao J, Skates S, Wong KK, Yiu GK, Muto MG, Berkowitz RS, Cramer DW (2001) Prostasin, a potential serum marker for ovarian cancer: identification through microarray technology. J Natl Cancer Inst 93:1458–1464

Mongaret C, Alexandre J, Thomas-Schoemann A, Bermudez E, Chereau C, Nicco C, Goldwasser F, Weill B, Batteux F, Lemare F (2011) Tumor invasion induced by oxidative stress is dependent on membrane ADAM 9 protein and its secreted form. Int J Cancer 129:791–798

Moran P, Li W, Fan B, Vij R, Eigenbrot C, Kirchhofer D (2006) Pro-urokinase-type plasminogen activator is a substrate for hepsin. J Biol Chem 281:30439–30446

Mu CJ, Lavan DA, Langer RS, Zetter BR (2010) Self-assembled gold nanoparticle molecular probes for detecting proteolytic activity in vivo. ACS Nano 4:1511–1520

Murphy G (2008) The ADAMs: signalling scissors in the tumour microenvironment. Nat Rev Cancer 8:929–941

Murray KT, Merriman CS, Adamson C (2008) Use of the HESI Admission Assessment to predict student success. Comput Inform Nurs 26:61S–66S

Murumkar PR, DasGupta S, Chandani SR, Giridhar R, Yadav MR (2010) Novel TACE inhibitors in drug discovery: a review of patented compounds. Expert Opin Ther Pat 20:31–57

Nakamura K, Hongo A, Kodama J, Hiramatsu Y (2011) The role of hepatocyte growth factor activator inhibitor (HAI)-1 and HAI-2 in endometrial cancer. Int J Cancer 128:2613–2624

Nalla AK, Gorantla B, Gondi CS, Lakka SS, Rao JS (2010) Targeting MMP-9, uPAR, and cathepsin B inhibits invasion, migration and activates apoptosis in prostate cancer cells. Cancer Gene Ther 17:599–613

Netzel-Arnett S, Hooper JD, Szabo R, Madison EL, Quigley JP, Bugge TH, Antalis AM (2003) Membrane anchored serine proteases: a rapidly expanding group of cell surface proteolytic enzymes with potential roles in cancer. Cancer Metastasis Rev 22:237–258

Nguyen QT, Olson ES, Aguilera TA, Jiang T, Scadeng M, Ellies LG, Tsien RY (2010) Surgery with molecular fluorescence imaging using activatable cell-penetrating peptides decreases residual cancer and improves survival. Proc Natl Acad Sci U S A 107:4317–4322

Nickeleit I, Zender S, Sasse F, Geffers R, Brandes G, Sorensen I, Steinmetz H, Kubicka S, Carlomagno T, Menche D, Gutgemann I, Buer J, Gossler A, Manns MP, Kalesse M, Frank R, Malek NP (2008) Argyrin a reveals a critical role for the tumor suppressor protein p27(kip1) in mediating antitumor activities in response to proteasome inhibition. Cancer Cell 14:23–35

Nielsen BS, Egeblad M, Rank F, Askautrud HA, Pennington CJ, Pedersen TX, Christensen IJ, Edwards DR, Werb Z, Lund LR (2008) Matrix metalloproteinase 13 is induced in fibroblasts in polyomavirus middle T antigen-driven mammary carcinoma without influencing tumor progression. PLoS One 3:e2959

Obaidat A, Weiss J, Wahlgren B, Manam RR, Macherla VR, McArthur K, Chao TH, Palladino MA, Lloyd GK, Potts BC, Enna SJ, Neuteboom ST, Hagenbuch B (2011) Proteasome regulator marizomib (NPI-0052) exhibits prolonged inhibition, attenuated efflux, and greater cytotoxicity than its reversible analogs. J Pharmacol Exp Ther 337:479–486

Oberst MD, Johnson MD, Dickson RB, Lin CY, Singh B, Stewart M, Williams A, al-Nafussi A, Smyth JF, Gabra H, Sellar GC (2002) Expression of the serine protease matriptase and its inhibitor HAI-1 in epithelial ovarian cancer: Correlation with clinical outcome and tumor clinicopathological parameters. Clin Cancer Res 8:1101–1107

Ogden SR, Noto JM, Allen SS, Patel DA, Romero-Gallo J, Washington MK, Fingleton B, Israel DA, Lewis ND, Wilson KT, Chaturvedi R, Zhao Z, Shyr Y, Peek RM Jr (2010) Matrix metalloproteinase-7 and premalignant host responses in Helicobacter pylori-infected mice. Cancer Res 70:30–35

Oikawa T, Sasaki T, Nakamura M, Shimamura M, Tanahashi N, Omura S, Tanaka K (1998) The proteasome is involved in angiogenesis. Biochem Biophys Res Commun 246:243–248

Olson ES, Jiang T, Aguilera TA, Nguyen QT, Ellies LG, Scadeng M, Tsien RY (2010) Activatable cell penetrating peptides linked to nanoparticles as dual probes for in vivo fluorescence and MR imaging of proteases. Proc Natl Acad Sci U S A 107:4311–4316

Onishi T, Hayashi N, Theriault RL, Hortobagyi GN, Ueno NT (2010) Future directions of bone-targeted therapy for metastatic breast cancer. Nat Rev Clin Oncol 7:641–651

Ordonez GR, Puente XS, Quesada V, Lopez-Otin C (2009) Proteolytic systems: constructing degradomes. Methods Mol Biol 539:33–47

Page MJ, Di Cera E (2008) Serine peptidases: classification, structure and function. Cell Mol Life Sci 65:1220–1236

Paik S, Shak S, Tang G, Kim C, Baker J, Cronin M, Baehner FL, Walker MG, Watson D, Park T, Hiller W, Fisher ER, Wickerham DL, Bryant J, Wolmark N (2004) A multigene assay to predict recurrence of tamoxifen-treated, node-negative breast cancer. N Engl J Med 351:2817–2826

Palavalli LH, Prickett TD, Wunderlich JR, Wei X, Burrell AS, Porter-Gill P, Davis S, Wang C, Cronin JC, Agrawal NS, Lin JC, Westbroek W, Hoogstraten-Miller S, Molinolo AA, Fetsch P, Filie AC, O'Connell MP, Banister CE, Howard JD, Buckhaults P, Weeraratna AT, Brody LC, Rosenberg SA, Samuels Y (2009) Analysis of the matrix metalloproteinase family reveals that MMP8 is often mutated in melanoma. Nat Genet 41:518–520

Paliouras M, Borgono C, Diamandis EP (2007) Human tissue kallikreins: the cancer biomarker family. Cancer Lett 249:61–79

Paraoan L, Gray D, Hiscott P, Garcia-Finana M, Lane B, Damato B, Grierson I (2009) Cathepsin S and its inhibitor cystatin C: imbalance in uveal melanoma. Front Biosci 14:2504–2513
Parks WC, Wilson CL, Lopez-Boado YS (2004) Matrix metalloproteinases as modulators of inflammation and innate immunity. Nat Rev Immunol 4:617–629
Pendas AM, Folgueras AR, Llano E, Caterina J, Frerard F, Rodriguez F, Astudillo A, Noel A, Birkedal-Hansen H, Lopez-Otin C (2004) Diet-induced obesity and reduced skin cancer susceptibility in matrix metalloproteinase 19-deficient mice. Mol Cell Biol 24:5304–5313
Piazuelo MB, Epplein M, Correa P (2010) Gastric cancer: an infectious disease. Infect Dis Clin North Am 24:853–869, vii
Podgorski I, Linebaugh BE, Koblinski JE, Rudy DL, Herroon MK, Olive MB, Sloane BF (2009) Bone marrow-derived cathepsin K cleaves SPARC in bone metastasis. Am J Pathol 175:1255–1269
Pories SE, Zurakowski D, Roy R, Lamb CC, Raza S, Exarhopoulos A, Scheib RG, Schumer S, Lenahan C, Borges V, Louis GW, Anand A, Isakovich N, Hirshfield-Bartek J, Wewer U, Lotz MM, Moses MA (2008) Urinary metalloproteinases: noninvasive biomarkers for breast cancer risk assessment. Cancer Epidemiol Biomarkers Prev 17:1034–1042
Potts BC, Albitar MX, Anderson KC, Baritaki S, Berkers C, Bonavida B, Chandra J, Chauhan D, Cusack JC Jr, Fenical W, Ghobrial IM, Groll M, Jensen PR, Lam KS, Lloyd GK, McBride W, McConkey DJ, Miller C, Neuteboom STC, Oki Y, Ovaa H, Pajonk F, Richardson PG, Roccaro AM, Sloss CM, Spear MA, Valashi E, Younes A, Palladino MA (2011) Marizomib, a proteasome inhibitor for all seasons: preclinical profile and a framework for clinical trials. Curr Cancer Drug Targets 11:254–284
Puchi M, Garcia-Huidobro J, Cordova C, Aguilar R, Dufey E, Imschenetzky M, Bustos P, Morin V (2010) A new nuclear protease with cathepsin L properties is present in HeLa and Caco-2 cells. J Cell Biochem 111:1099–1106
Puente XS, Sanchez LM, Overall CM, Lopez-Otin C (2003) Human and mouse proteases: a comparative genomic approach. Nat Rev Genet 4:544–558
Puente XS, Sanchez LM, Gutierrez-Fernandez A, Velasco G, Lopez-Otin C (2005) A genomic view of the complexity of mammalian proteolytic systems. Biochem Soc Trans 33:331–334
Qian BZ, Pollard JW (2010) Macrophage diversity enhances tumor progression and metastasis. Cell 141:39–51
Qu P, Du H, Wang X, Yan C (2009) Matrix metalloproteinase 12 overexpression in lung epithelial cells plays a key role in emphysema to lung bronchioalveolar adenocarcinoma transition. Cancer Res 69:7252–7261
Quintanilla-Dieck MJ, Codriansky K, Keady M, Bhawan J, Runger TM (2008) Cathepsin K in melanoma invasion. J Invest Dermatol 128:2281–2288
Rachner TD, Hadji P, Hofbauer LC (2012) Novel therapies in benign and malignant bone diseases. Pharmacol Ther 134:338–344
Radisky DC, Levy DD, Littlepage LE, Liu H, Nelson CM, Fata JE, Leake D, Godden EL, Albertson DG, Nieto MA, Werb Z, Bissell MJ (2005) Rac1b and reactive oxygen species mediate MMP-3-induced EMT and genomic instability. Nature 436:123–127
Rafn B, Nielsen CF, Andersen SH, Szyniarowski P, Corcelle-Termeau E, Valo E, Fehrenbacher N, Olsen CJ, Daugaard M, Egebjerg C, Bottzauw T, Kohonen P, Nylandsted J, Hautaniemi S, Moreira J, Jaattela M, Kallunki T (2012) ErbB2-driven breast cancer cell invasion depends on a complex signaling network activating myeloid zinc finger-1-dependent cathepsin B expression. Mol Cell 45:764–776
Rao JS (2003) Molecular mechanisms of glioma invasiveness: the role of proteases. Nat Rev Cancer 3:489–501
Rao Malla R, Gopinath S, Alapati K, Gorantla B, Gondi CS, Rao JS (2012) Knockdown of cathepsin B and uPAR inhibits CD151 and alpha3beta1 integrin-mediated cell adhesion and invasion in glioma. Mol Carcinog. doi:10.1002/mc.21915

Rapa I, Volante M, Cappia S, Rosas R, Scagliotti GV, Papotti M (2006) Cathepsin K is selectively expressed in the stroma of lung adenocarcinoma but not in bronchioloalveolar carcinoma. A useful marker of invasive growth. Am J Clin Pathol 125:847–854

Rawlings ND, Barrett AJ (1993) Evolutionary families of peptidases. Biochem J 290 (Pt 1):205–218

Rawlings ND, Barrett AJ, Bateman A (2010) MEROPS: the peptidase database. Nucleic Acids Res 38:D227–D233

Reiser J, Adair B, Reinheckel T (2010) Specialized roles for cysteine cathepsins in health and disease. J Clin Invest 120:3421–3431

Reyes-Turcu FE, Ventii KH, Wilkinson KD (2009) Regulation and cellular roles of ubiquitin-specific deubiquitinating enzymes. Annu Rev Biochem 78:363–397

Ribatti D (2009) Endogenous inhibitors of angiogenesis: a historical review. Leuk Res 33:638–644

Rocks N, Paulissen G, Quesada-Calvo F, Munaut C, Gonzalez ML, Gueders M, Hacha J, Gilles C, Foidart JM, Noel A, Cataldo DD (2008) ADAMTS-1 metalloproteinase promotes tumor development through the induction of a stromal reaction in vivo. Cancer Res 68:9541–9550

Rose AA, Annis MG, Dong Z, Pepin F, Hallett M, Park M, Siegel PM (2010) ADAM10 releases a soluble form of the GPNMB/Osteoactivin extracellular domain with angiogenic properties. PLoS One 5:e12093

Rudolph-Owen LA, Chan R, Muller WJ, Matrisian LM (1998) The matrix metalloproteinase matrilysin influences early-stage mammary tumorigenesis. Cancer Res 58:5500–5506

Ruffell B, Affara NI, Cottone L, Junankar S, Johansson M, DeNardo DG, Korets L, Reinheckel T, Sloane BF, Bogyo M, Coussens LM (2013) Cathepsin C is a tissue-specific regulator of squamous carcinogenesis. Genes Dev (in press)

Rumpler G, Becker B, Hafner C, McClelland M, Stolz W, Landthaler M, Schmitt R, Bosserhoff A, Vogt T (2003) Identification of differentially expressed genes in models of melanoma progression by cDNA array analysis: SPARC, MIF and a novel cathepsin protease characterize aggressive phenotypes. Exp Dermatol 12:761–771

Sabeh F, Ota I, Holmbeck K, Birkedal-Hansen H, Soloway P, Balbin M, Lopez-Otin C, Shapiro S, Inada M, Krane S, Allen E, Chung D, Weiss SJ (2004) Tumor cell traffic through the extracellular matrix is controlled by the membrane-anchored collagenase MT1-MMP. J Cell Biol 167:769–781

Sacco JJ, Coulson JM, Clague MJ, Urbe S (2010) Emerging roles of deubiquitinases in cancer-associated pathways. IUBMB Life 62:140–157

Saftig P, Hunziker E, Wehmeyer O, Jones S, Boyde A, Rommerskirch W, Moritz JD, Schu P, von Figura K (1998) Impaired osteoclastic bone resorption leads to osteopetrosis in cathepsin-K-deficient mice. Proc Natl Acad Sci U S A 95:13453–13458

Sakashita K, Mimori K, Tanaka F, Tahara K, Inoue H, Sawada T, Ohira M, Hirakawa K, Mori M (2008) Clinical significance of low expression of prostasin mRNA in human gastric cancer. J Surg Oncol 98:559–564

Saleem M, Adhami VM, Zhong WX, Longley BJ, Lin CY, Dickson RB, Reagan-Shaw S, Jarrard DF, Mukhtar H (2006) A novel biomarker for staging human prostate adenocarcinoma: overexpression of matriptase with concomitant loss of its inhibitor, hepatocyte growth factor activator inhibitor-1. Cancer Epidemiol Biomarkers Prev 15:217–227

Sameni M, Dosescu J, Yamada KM, Sloane BF, Cavallo-Medved D (2008) Functional live-cell imaging demonstrates that beta1-integrin promotes type IV collagen degradation by breast and prostate cancer cells. Mol Imaging 7:199–213

Santamaria I, Velasco G, Pendas AM, Paz A, Lopez-Otin C (1999) Molecular cloning and structural and functional characterization of human cathepsin F, a new cysteine proteinase of the papain family with a long propeptide domain. J Biol Chem 274:13800–13809

Schelter F, Kobuch J, Moss ML, Becherer JD, Comoglio PM, Boccaccio C, Kruger A (2010) A disintegrin and metalloproteinase-10 (ADAM-10) mediates DN30 antibody-induced shedding of the met surface receptor. J Biol Chem 285:26335–26340

Scherer RL, McIntyre JO, Matrisian LM (2008a) Imaging matrix metalloproteinases in cancer. Cancer Metastasis Rev 27:679–690

Scherer RL, VanSaun MN, McIntyre JO, Matrisian LM (2008b) Optical imaging of matrix metalloproteinase-7 activity in vivo using a proteolytic nanobeacon. Mol Imaging 7:118–131

Schmitt M, Harbeck N, Thomssen C, Wilhelm O, Magdolen V, Reuning U, Ulm K, Hofler H, Janicke F, Graeff H (1997) Clinical impact of the plasminogen activation system in tumor invasion and metastasis: prognostic relevance and target for therapy. Thromb Haemost 78:285–296

Schurigt U, Sevenich L, Vannier C, Gajda M, Schwinde A, Werner F, Stahl A, von Elverfeldt D, Becker AK, Bogyo M, Peters C, Reinheckel T (2008) Trial of the cysteine cathepsin inhibitor JPM-OEt on early and advanced mammary cancer stages in the MMTV-PyMT-transgenic mouse model. Biol Chem 389:1067–1074

Seemuller E, Dolenc I, Lupas A (2004) Eukaryotic 20S proteasome. In: Barrett AJ, Rawlings ND, Woessner JF (eds) Handbook of proteolytic enzymes, vol 2, 2nd edn. Elsevier, London, pp 2068–2077

Selzer-Plon J, Bornholdt J, Friis S, Bisgaard HC, Lothe IMB, Tveit KM, Kure EH, Vogel U, Vogel LK (2009) Expression of prostasin and its inhibitors during colorectal cancer carcinogenesis. BMC Cancer 9:201

Sevenich L, Hagemann S, Stoeckle C, Tolosa E, Peters C, Reinheckel T (2010a) Expression of human cathepsin L or human cathepsin V in mouse thymus mediates positive selection of T helper cells in cathepsin L knock-out mice. Biochimie 92:1674–1680

Sevenich L, Schurigt U, Sachse K, Gajda M, Werner F, Muller S, Vasiljeva O, Schwinde A, Klemm N, Deussing J, Peters C, Reinheckel T (2010b) Synergistic antitumor effects of combined cathepsin B and cathepsin Z deficiencies on breast cancer progression and metastasis in mice. Proc Natl Acad Sci U S A 107:2497–2502

Shah S, Small E (2010) Emerging biological observations in prostate cancer. Expert Rev Anticancer Ther 10:89–101

Sharghi-Namini S, Fan H, Sulochana KN, Potturi P, Xiang W, Chong YS, Wang Z, Yang H, Ge R (2008) The first but not the second thrombospondin type 1 repeat of ADAMTS5 functions as an angiogenesis inhibitor. Biochem Biophys Res Commun 371:215–219

Shchors K, Nozawa H, Xu J, Rostker F, Swigart-Brown L, Evan G, Hanahan D (2013) Increased invasiveness of MMP-9-deficient tumors in two mouse models of neuroendocrine tumorigenesis. Oncogene 32(4):502–513

Shi GP, Bryant RA, Riese R, Verhelst S, Driessen C, Li Z, Bromme D, Ploegh HL, Chapman HA (2000) Role for cathepsin F in invariant chain processing and major histocompatibility complex class II peptide loading by macrophages. J Exp Med 191:1177–1186

Shigemasa K, Underwood LJ, Beard J, Tanimoto H, Ohama K, Parmley TH, O'Brien TJ (2000) Overexpression of testisin, a serine protease expressed by testicular germ cells, in epithelial ovarian tumor cells. J Soc Gynecol Investig 7:358–362

Shree T, Olson OC, Elie BT, Kester JC, Garfall AL, Simpson K, Bell-McGuinn KM, Zabor EC, Brogi E, Joyce JA (2011) Macrophages and cathepsin proteases blunt chemotherapeutic response in breast cancer. Genes Dev 25:2465–2479

Sinnamon MJ, Carter KJ, Fingleton B, Matrisian LM (2008) Matrix metalloproteinase-9 contributes to intestinal tumourigenesis in the adenomatous polyposis coli multiple intestinal neoplasia mouse. Int J Exp Pathol 89:466–475

Sinnathamby G, Zerfass J, Hafner J, Block P, Nickens Z, Hobeika A, Secord AA, Lyerly HK, Morse MA, Philip R (2011) ADAM metallopeptidase domain 17 (ADAM17) is naturally processed through major histocompatibility complex (MHC) class I molecules and is a potential immunotherapeutic target in breast, ovarian and prostate cancers. Clin Exp Immunol 163:324–332

Sloane BF, Moin K, Sameni M, Tait LR, Rozhin J, Ziegler G (1994) Membrane association of cathepsin B can be induced by transfection of human breast epithelial cells with c-Ha-ras oncogene. J Cell Sci 107(Pt 2):373–384

Staack A, Tolic D, Kristiansen G, Schnorr D, Loening SA, Jung K (2004) Expression of cathepsins B, H, and L and their inhibitors as markers of transitional cell carcinoma of the bladder. Urology 63:1089–1094
Stahl S, Reinders Y, Asan E, Mothes W, Conzelmann E, Sickmann A, Felbor U (2007) Proteomic analysis of cathepsin B- and L-deficient mouse brain lysosomes. Biochim Biophys Acta 1774:1237–1246
Stamey TA, Warrington JA, Caldwell MC, Chen ZX, Fan ZB, Mahadevappa M, McNeal JE, Nolley R, Zhang ZM (2001) Molecular genetic profiling of gleason grade 4/5 prostate cancers compared to benign prostatic hyperplasia. J Urol 166:2171–2177
Stephan C, Yousef GM, Scorilas A, Jung K, Jung M, Kristiansen G, Hauptmann S, Kishi T, Nakamura T, Loening SA, Diamandis EP (2004) Hepsin is highly over expressed in and a new candidate for a prognostic indicator in prostate cancer. J Urol 171:187–191
Sternlicht MD, Lochter A, Sympson CJ, Huey B, Rougier JP, Gray JW, Pinkel D, Bissell MJ, Werb Z (1999) The stromal proteinase MMP3/stromelysin-1 promotes mammary carcinogenesis. Cell 98:137–146
Storr SJ, Carragher NO, Frame MC, Parr T, Martin SG (2011) The calpain system and cancer. Nat Rev Cancer 11:364–374
Sturge J, Caley MP, Waxman J (2011) Bone metastasis in prostate cancer: emerging therapeutic strategies. Nat Rev Clin Oncol 8:357–368
Sullivan S, Tosetto M, Kevans D, Coss A, Wang L, O'Donoghue D, Hyland J, Sheahan K, Mulcahy H, O'Sullivan J (2009) Localization of nuclear cathepsin L and its association with disease progression and poor outcome in colorectal cancer. Int J Cancer 125:54–61
Sun SC (2010) CYLD: a tumor suppressor deubiquitinase regulating NF-kappaB activation and diverse biological processes. Cell Death Differ 17:25–34
Sunwoo JB, Chen Z, Dong G, Yeh N, Crowl Bancroft C, Sausville E, Adams J, Elliott P, Van Waes C (2001) Novel proteasome inhibitor PS-341 inhibits activation of nuclear factor-kappa B, cell survival, tumor growth, and angiogenesis in squamous cell carcinoma. Clin Cancer Res 7:1419–1428
Szabo R, Rasmussen AL, Moyer AB, Kosa P, Schafer JM, Molinolo AA, Gutkind JS, Bugge TH (2011) c-Met-induced epithelial carcinogenesis is initiated by the serine protease matriptase. Oncogene 30:2003–2016
Szabova L, Chrysovergis K, Yamada SS, Holmbeck K (2008) MT1-MMP is required for efficient tumor dissemination in experimental metastatic disease. Oncogene 27:3274–3281
Takahashi S, Suzuki S, Inaguma S, Ikeda Y, Cho YM, Hayashi N, Inoue T, Sugimura Y, Nishiyama N, Fujita T, Chao J, Ushijima T, Shirai T (2003) Down-regulated expression of prostasin in high-grade or hormone-refractory human prostate cancers. Prostate 54:187–193
Takayama TK, McMullen BA, Nelson PS, Matsumura M, Fujikawa K (2001) Characterization of hK4 (prostase), a prostate-specific serine protease: activation of the precursor of prostate specific antigen (pro-PSA) and single-chain urokinase-type plasminogen activator and degradation of prostatic acid phosphatase. Biochemistry 40:15341–15348
Tang T, Kmet M, Corral L, Vartanian S, Tobler A, Papkoff J (2005) Testisin, a glycosylphosphatidylinositol-linked serine protease, promotes malignant transformation in vitro and in vivo. Cancer Res 65:868–878
Tape CJ, Willems SH, Dombernowsky SL, Stanley PL, Fogarasi M, Ouwehand W, McCafferty J, Murphy G (2011) Cross-domain inhibition of TACE ectodomain. Proc Natl Acad Sci U S A 108:5578–5583
Tedelind S, Poliakova K, Valeta A, Hunegnaw R, Yemanaberhan EL, Heldin NE, Kurebayashi J, Weber E, Kopitar-Jerala N, Turk B, Bogyo M, Brix K (2010) Nuclear cysteine cathepsin variants in thyroid carcinoma cells. Biol Chem 391:923–935
Testa U (2009) Proteasome inhibitors in cancer therapy. Curr Drug Targets 10:968–981
Thiolloy S, Edwards JR, Fingleton B, Rifkin DB, Matrisian LM, Lynch CC (2012) An osteoblast-derived proteinase controls cell survival via TGF-beta activation in the bone microenvironment. PLoS One 7:e29862

Thiolloy S, Halpern J, Holt GE, Schwartz HS, Mundy GR, Matrisian LM, Lynch CC (2009) Osteoclast-derived matrix metalloproteinase-7, but not matrix metalloproteinase-9, contributes to tumor-induced osteolysis. Cancer Res 69:6747–6755

To S, Rodda SJ, Rathjen PD, Keough RA (2010) Modulation of CP2 family transcriptional activity by CRTR-1 and sumoylation. PLoS One 5:e11702

Tomita A, Kasaoka T, Inui T, Toyoshima M, Nishiyama H, Saiki H, Iguchi H, Nakajima M (2008) Human breast adenocarcinoma (MDA-231) and human lung squamous cell carcinoma (Hara) do not have the ability to cause bone resorption by themselves during the establishment of bone metastasis. Clin Exp Metastasis 25:437–444

Tomlins SA, Rhodes DR, Perner S, Dhanasekaran SM, Mehra R, Sun XW, Varambally S, Cao XH, Tchinda J, Lee C, Shah RB, Rubin MA, Chinnaiyan AM (2006) Recurrent fusion of TMPRSS2 and ETS transcription factors in prostate cancer. FASEB J 20:A1327

Tomlins SA, Laxman B, Dhanasekaran SM, Helgeson BE, Cao X, Morris DS, Menon A, Jing X, Cao Q, Han B, Yu J, Wang L, Montie JE, Rubin MA, Pienta KJ, Roulston D, Shah RB, Varambally S, Mehra R, Chinnaiyan AM (2007) Distinct classes of chromosomal rearrangements create oncogenic ETS gene fusions in prostate cancer. Nature 448:595–599

Tomlins SA, Laxman B, Varambally S, Cao X, Yu J, Helgeson BE, Cao Q, Prensner JR, Rubin MA, Shah RB, Mehra R, Chinnaiyan AM (2008) Role of the TMPRSS2-ERG gene fusion in prostate cancer. Neoplasia (New York, NY) 10:177–188

Tomlins SA, Bjartell A, Chinnaiyan AM, Jenster G, Nam RK, Rubin MA, Schalken JA (2009) ETS gene fusions in prostate cancer: from discovery to daily clinical practice. Eur Urol 56:275–286

Tripathi M, Nandana S, Yamashita H, Ganesan R, Kirchhofer D, Quaranta V (2008) Laminin-332 is a substrate for hepsin, a protease associated with prostate cancer progression. J Biol Chem 283:30576–30584

Tsuji A, Torresrosado A, Arai T, Lebeau MM, Lemons RS, Chou SH, Kurachi K (1991) Hepsin, a cell membrane-associated protease – characterization, tissue distribution, and gene localization. J Biol Chem 266:16948–16953

Tsuruo T, Naito M, Tomida A, Fujita N, Mashima T, Sakamoto H, Haga N (2003) Molecular targeting therapy of cancer: drug resistance, apoptosis and survival signal. Cancer Sci 94:15–21

Tummalapalli P, Spomar D, Gondi CS, Olivero WC, Gujrati M, Dinh DH, Rao JS (2007) RNAi-mediated abrogation of cathepsin B and MMP-9 gene expression in a malignant meningioma cell line leads to decreased tumor growth, invasion and angiogenesis. Int J Oncol 31:1039–1050

Turk V, Stoka V, Vasiljeva O, Renko M, Sun T, Turk B, Turk D (2012) Cysteine cathepsins: from structure, function and regulation to new frontiers. Biochim Biophys Acta 1824:68–88

Turner SL, Blair-Zajdel ME, Bunning RA (2009) ADAMs and ADAMTSs in cancer. Br J Biomed Sci 66:117–128

Vallet V, Chraibi A, Gaeggeler HP, Horisberger JD, Rossier BC (1997) An epithelial serine protease activates the amiloride-sensitive sodium channel. Nature 389:607–610

van Tetering G, van Diest P, Verlaan I, van der Wall E, Kopan R, Vooijs M (2009) Metalloprotease ADAM10 is required for Notch1 site 2 cleavage. J Biol Chem 284:31018–31027

Vasiljeva O, Turk B (2008) Dual contrasting roles of cysteine cathepsins in cancer progression: apoptosis versus tumour invasion. Biochimie 90:380–386

Vasiljeva O, Papazoglou A, Kruger A, Brodoefel H, Korovin M, Deussing J, Augustin N, Nielsen BS, Almholt K, Bogyo M, Peters C, Reinheckel T (2006) Tumor cell-derived and macrophage-derived cathepsin B promotes progression and lung metastasis of mammary cancer. Cancer Res 66:5242–5250

Vasiljeva O, Korovin M, Gajda M, Brodoefel H, Bojic L, Kruger A, Schurigt U, Sevenich L, Turk B, Peters C, Reinheckel T (2008) Reduced tumour cell proliferation and delayed development of high-grade mammary carcinomas in cathepsin B-deficient mice. Oncogene 27:4191–4199

Vazquez-Ortiz G, Pina-Sanchez P, Vazquez K, Duenas A, Taja L, Mendoza P, Garcia JA, Salcedo M (2005) Overexpression of cathepsin F, matrix metalloproteinases 11 and 12 in cervical cancer. BMC Cancer 5:68

Veeravalli KK, Chetty C, Ponnala S, Gondi CS, Lakka SS, Fassett D, Klopfenstein JD, Dinh DH, Gujrati M, Rao JS (2010) MMP-9, uPAR and cathepsin B silencing downregulate integrins in human glioma xenograft cells in vitro and in vivo in nude mice. PLoS One 5:e11583

Vink JM, Boomsma DI (2002) Gene finding strategies. Biol Psychol 61:53–71

Vogel LK, Saebo M, Skjelbred CF, Abell K, Pedersen EDK, Vogel U, Kure EH (2006) The ratio of Matriptase/HAI-1 mRNA is higher in colorectal cancer adenomas and carcinomas than corresponding tissue from control individuals. BMC Cancer 6:176

Voutsadakis IA (2008) The ubiquitin-proteasome system in colorectal cancer. Biochim Biophys Acta 1782:800–808

Wagner S, Breyholz HJ, Faust A, Holtke C, Levkau B, Schober O, Schafers M, Kopka K (2006) Molecular imaging of matrix metalloproteinases in vivo using small molecule inhibitors for SPECT and PET. Curr Med Chem 13:2819–2838

Wagstaff L, Kelwick R, Decock J, Edwards DR (2011) The roles of ADAMTS metalloproteinases in tumorigenesis and metastasis. Front Biosci 16:1861–1872

Wallrapp C, Hahnel S, Muller-Pillasch F, Burghardt B, Iwamura T, Ruthenburger M, Lerch MM, Adler G, Gress TM (2000) A novel transmembrane serine protease (TMPRSS3) overexpressed in pancreatic cancer. Cancer Res 60:2602–2606

Wang B, Shi GP, Yao PM, Li Z, Chapman HA, Bromme D (1998) Human cathepsin F. Molecular cloning, functional expression, tissue localization, and enzymatic characterization. J Biol Chem 273:32000–32008

Wang B, Sun J, Kitamoto S, Yang M, Grubb A, Chapman HA, Kalluri R, Shi GP (2006) Cathepsin S controls angiogenesis and tumor growth via matrix-derived angiogenic factors. J Biol Chem 281:6020–6029

Ward C, Kuehn D, Burden RE, Gormley JA, Jaquin TJ, Gazdoiu M, Small D, Bicknell R, Johnston JA, Scott CJ, Olwill SA (2010) Antibody targeting of cathepsin S inhibits angiogenesis and synergistically enhances anti-VEGF. PLoS One 5:e12543

Weaver AM (2006) Invadopodia: specialized cell structures for cancer invasion. Clin Exp Metastasis 23:97–105

Webb SL, Sanders AJ, Mason MD, Jiang WG (2011) Type II Transmembrane Serine Protease (TTSP) deregulation in cancer. Front Biosci 16:539–552

Weissleder R, Tung CH, Mahmood U, Bogdanov A Jr (1999) In vivo imaging of tumors with protease-activated near-infrared fluorescent probes. Nat Biotechnol 17:375–378

Wex T, Levy B, Wex H, Bromme D (1999) Human cathepsins F and W: a new subgroup of cathepsins. Biochem Biophys Res Commun 259:401–407

Wijkmans J, Gossen J (2011) Inhibitors of cathepsin K: a patent review (2004–2010). Expert Opin Ther Pat 21:1611–1629

Williams SA, Xu Y, De Marzo AM, Isaacs JT, Denmeade SR (2010) Prostate-specific antigen (PSA) is activated by KLK2 in prostate cancer ex vivo models and in prostate-targeted PSA/KLK2 double transgenic mice. Prostate 70:788–796

Wilson CL, Heppner KJ, Labosky PA, Hogan BL, Matrisian LM (1997) Intestinal tumorigenesis is suppressed in mice lacking the metalloproteinase matrilysin. Proc Natl Acad Sci U S A 94:1402–1407

Wilson TR, Johnston PG, Longley DB (2009) Anti-apoptotic mechanisms of drug resistance in cancer. Curr Cancer Drug Targets 9:307–319

Withana NP, Blum G, Sameni M, Slaney C, Anbalagan A, Olive MB, Bidwell BN, Edgington L, Wang L, Moin K, Sloane BF, Anderson RL, Bogyo MS, Parker BS (2012) Cathepsin B inhibition limits bone metastasis in breast cancer. Cancer Res 72:1199–1209

Witters L, Scherle P, Friedman S, Fridman J, Caulder E, Newton R, Lipton A (2008) Synergistic inhibition with a dual epidermal growth factor receptor/HER-2/neu tyrosine kinase inhibitor and a disintegrin and metalloprotease inhibitor. Cancer Res 68:7083–7089

Witty JP, Lempka T, Coffey RJ Jr, Matrisian LM (1995) Decreased tumor formation in 7,12-dimethylbenzanthracene-treated stromelysin-1 transgenic mice is associated with alterations in mammary epithelial cell apoptosis. Cancer Res 55:1401–1406

Wu QY, Parry G (2007) Hepsin and prostate cancer. Front Biosci 12:5052–5059

Wu SK, Sakamoto KM, Milani M, Aldana-Masankgay G, Fan D, Wu K, Lee CW, Cho CH, Yu J, Sung JJ (2010) Macroautophagy modulates cellular response to proteasome inhibitors in cancer therapy. Drug Resist Updat 13:87–92

Wu SM, Huang YH, Yeh CT, Tsai MM, Liao CH, Cheng WL, Chen WJ, Lin KH (2011) Cathepsin H regulated by the thyroid hormone receptors associate with tumor invasion in human hepatoma cells. Oncogene 30:2057–2069

Xu J, Attisano L (2000) Mutations in the tumor suppressors Smad2 and Smad4 inactivate transforming growth factor beta signaling by targeting Smads to the ubiquitin-proteasome pathway. Proc Natl Acad Sci U S A 97:4820–4825

Xu J, Li D, Ke Z, Liu R, Maubach G, Zhuo L (2009) Cathepsin S is aberrantly overexpressed in human hepatocellular carcinoma. Mol Med Report 2:713–718

Xuan JA, Schneider D, Toy P, Lin R, Newton A, Zhu Y, Finster S, Vogel D, Mintzer B, Dinter H, Light D, Parry R, Polokoff M, Whitlow M, Wu QY, Parry G (2006) Antibodies neutralizing hepsin protease activity do not impact cell growth but inhibit invasion of prostate and ovarian tumor cells in culture. Cancer Res 66:3611–3619

Yan S, Sloane BF (2003) Molecular regulation of human cathepsin B: implication in pathologies. Biol Chem 384:845–854

Yan X, Takahara M, Xie L, Oda Y, Nakahara T, Uchi H, Takeuchi S, Tu Y, Moroi Y, Furue M (2011) Stromal expression of cathepsin K in squamous cell carcinoma. J Eur Acad Dermatol Venereol 25:362–365

Yang L, DeBusk LM, Fukuda K, Fingleton B, Green-Jarvis B, Shyr Y, Matrisian LM, Carbone DP, Lin PC (2004) Expansion of myeloid immune suppressor Gr+CD11b+ cells in tumor-bearing host directly promotes tumor angiogenesis. Cancer Cell 6:409–421

Yang Y, Lim SK, Choong LY, Lee H, Chen Y, Chong PK, Ashktorab H, Wang TT, Salto-Tellez M, Yeoh KG, Lim YP (2010) Cathepsin S mediates gastric cancer cell migration and invasion via a putative network of metastasis-associated proteins. J Proteome Res 9:4767–4778

Yu JX, Chao L, Chao J (1994) Prostasin is a novel human serine proteinase from seminal fluid – purification, tissue distribution, and localization in prostate-gland. J Biol Chem 269:18843–18848

Zhang K, Kaufman RJ (2006) The unfolded protein response: a stress signaling pathway critical for health and disease. Neurology 66:S102–S109

Zhang Z, Neiva KG, Lingen MW, Ellis LM, Nor JE (2010) VEGF-dependent tumor angiogenesis requires inverse and reciprocal regulation of VEGFR1 and VEGFR2. Cell Death Differ 17:499–512

Zheng X, Chu F, Chou PM, Gallati C, Dier U, Mirkin BL, Mousa SA, Rebbaa A (2009) Cathepsin L inhibition suppresses drug resistance in vitro and in vivo: a putative mechanism. Am J Physiol Cell Physiol 296:C65–C74

Zhou HM, Nichols A, Meda P, Vassalli JD (2000) Urokinase-type plasminogen activator and its receptor synergize to promote pathogenic proteolysis. EMBO J 19:4817–4826

Zhou BB, Peyton M, He B, Liu C, Girard L, Caudler E, Lo Y, Baribaud F, Mikami I, Reguart N, Yang G, Li Y, Yao W, Vaddi K, Gazdar AF, Friedman SM, Jablons DM, Newton RC, Fridman JS, Minna JD, Scherle PA (2006) Targeting ADAM-mediated ligand cleavage to inhibit HER3 and EGFR pathways in non-small cell lung cancer. Cancer Cell 10:39–50

Zigrino P, Nischt R, Mauch C (2011) The disintegrin-like and cysteine-rich domains of ADAM-9 mediate interactions between melanoma cells and fibroblasts. J Biol Chem 286:6801–6807

Zucker S, Cao J, Chen WT (2000) Critical appraisal of the use of matrix metalloproteinase inhibitors in cancer treatment. Oncogene 19:6642–6650

Index

K. Brix and W. Stöcker (eds.), *Proteases: Structure and Function*,
DOI 10.1007/978-3-7091-0885-7, © Springer-Verlag Wien 2013

T

U

V

X

Z

MIX
Papier aus verantwortungsvollen Quellen
Paper from responsible sources
FSC® C105338

If you have any concerns about our products,
you can contact us on
ProductSafety@springernature.com

In case Publisher is established outside the EU,
the EU authorized representative is:
Springer Nature Customer Service Center GmbH
Europaplatz 3, 69115 Heidelberg, Germany

Printed by Libri Plureos GmbH
in Hamburg, Germany